ORGANIC CHEMISTRY

ORGANIC CHEMISTRY
an experimental approach

JAMES S. SWINEHART
State University of New York
College at Cortland

PRENTICE-HALL, INC.
Englewood Cliffs, New Jersey

Printed in the United States of America

ISBN: 0-13-640649-1

Library of Congress Catalog Card Number: 70-79176

10 9 8 7 6 5 4 3 2 1

PRENTICE-HALL INTERNATIONAL, INC., *London*
PRENTICE-HALL OF AUSTRALIA, PTY. LTD., *Sydney*
PRENTICE-HALL OF CANADA, LTD., *Toronto*
PRENTICE-HALL OF INDIA PRIVATE LIMITED, *New Delhi*
PRENTICE-HALL OF JAPAN, INC., *Tokyo*

Organic Chemistry: An Experimental Approach
is one of a series of books published in cooperation with E. K. Georg Landsberger.

Preface

In the past two decades organic chemistry has experienced a rapid evolution, or some might say revolution. New reagents give increased yields, reduce the number of operations in multistep synthesis, and in some cases make possible the synthesis of compounds that were not achieved by many attempts with "classical" methods. Study of structure and mechanisms and their relation to properties and reactions have transformed organic chemistry from an empirical science to an exact science.

The biggest advances have come in the application of instrumentation, especially gas chromatography and the many spectroscopic techniques. Almost every organic chemist would agree with Dr. Hammett's recent reference to the development of these analytical techniques as the "most marvelous thing that has happened in chemistry in my lifetime."

The changes in organic chemistry were accompanied by changes in its teaching, quickly in many graduate schools but more slowly in most undergraduate courses. The change made a quantum jump approximately ten years ago with the publication of the first modern organic textbook which could be read and understood by the large majority of those studying organic chemistry for the first time. Other modern texts followed, some using spectroscopy as a routine tool *in lecture.*

Improvement has also been evident in organic laboratory manuals published in recent years. These are more up to date and generally better than the laboratory manuals published a decade ago. There has, however, not been as radical a change in laboratory manuals as there has been in lecture texts. Current manuals do not acquaint the student with the instrumentation he will find in industrial and graduate or medical school laboratories.

This text is an attempt at a new approach to the undergraduate organic laboratory. Novelty is not embraced for its own sake, but new techniques are introduced in the way they will be used by the student in his graduate professional experiences. Useful classical techniques are not discarded. It is very seldom that a compound's melting point gives as much information as its infrared spectrum, but a melting point is easier to determine (even if a Model 700 is available) and it gives different information. Just as the various spectroscopic techniques give complementary information, so do instrumental and classical techniques. One or more physical properties or classification tests usually aids the interpretation of spectra and often is required for identification even if several types of spectra are available. For example, together the infrared spectra and results of the permanganate and Br_2 in CCl_4 tests clearly indicate the unsaturation and structural features of $(CH_3)_2C = CHCH_2CH_3$, $(CH_3)_2C = C(CH_3)_2$, $(C_6H_5)_2C = C(C_6H_5)_2$, $(CH_3)_2C = CHCH = C(CH_3)_2$, $(NO_2C_6H_5)_2C = CHCO_2H$, CH_3CH_2CHO, but each of these methods alone would give misleading results about the unsaturation of several of these compounds. In addition to the pragmatic use of classification tests as supplements to instrumental methods, there is considerable pedagogical value in using these reactions for the identification of organic compounds in an *introductory* course.

There is more than sufficient material here to allow selection of experiments to match the material presented in lecture and the resources of the department. This manual can be used in several approaches to the teaching of organic laboratory courses. The use of instrumentation is a prime example of its flexibility. Chapters 8, 18, 20, 27, 37, 63, and 65 deal primarily with gas chromatography and infrared and ultraviolet absorption spectroscopy; however, none of the other chapters *require* the use of these instruments. The instructor can thus organize the laboratory with the experiments he thinks are essential, whether or not he currently has the instrumentation to utilize the optional parts of most of the later chapters. As funds become available gas chromatography, infrared, or ultraviolet instrumentation can be added without radical changes in the laboratory schedule. For example, gas chromatography would require only Chapter 8 as a new assignment. The gas chromatography options at the end of the chapters used the previous year could then be assigned with these chapters.

A department may wish to emphasize synthesis because other courses in the curriculum emphasize organic qualitative analysis and instrumentation. There are more than sufficient synthesis experiments to allow the entire course to consist of nothing but syntheses and introductory techniques. The large number of experiments allows even for variation in emphasis on synthesis. One is to perform as many syntheses as possible. Another is to do fewer syntheses and study them more thoroughly. For the latter choice instructions are given for identification of by-products, intermediates, and determination of the amounts of products, reactants, and by-products at various stages of the synthesis and purification.

Many of the synthetic experiments are written and arranged in the proper sequence so that they can be done as a single multistep synthesis. For example, 17 and 19 can be combined to synthesize *n*-propylbenzene from benzene, 17 and 54 to synthesize 1-phenyl-1-propanol, 21 and 30 to synthesize methyl butyl ether from 1-butanol, 13, 43, and 48 to synthesize *m*-nitrophenol from benzene, 39 and 40 to synthesize sulfanilamide from aniline, 39, 40, and 41 to synthesize 2,6-dichloroaniline from aniline, 42, 39, and 53 combined to synthesize *m*-bromotoluene from *p*-toluidine, and 42 combined with 50 to synthesize *o*-chlorotoluene or with 51 and 52 to synthesize *o*-toluic acid from *o*-toluidine.

The emphasis may be on organic qualitative analysis because the course that was once classical qualitative analysis is now spectroscopic applications to organic chemistry. There are twelve chapters (14, 23, 28, 32, 35, 41, 44, 46, 49, 56, 61, 64) on reactions and properties of various groups of compounds written to provide a continuous introduction to qualitative analysis with or without instrumentation. For example, the permanganate and bromine in carbon tetrachloride tests are first assigned in Chapter 14 "Reactions of Hydrocarbons." These reactions are discussed further in later chapters when other groups of compounds are studied that react with these reagents as, for example, in the chapters on amines and aldehydes and ketones.

Tables of possible unknowns with physical properties and derivatives are included at the end of each qualitative analysis chapter. The chapters are written so that the number of possible unknowns is cumulative; each chapter on qualitative analysis adds another set of unknowns to the list previously assigned. The subject matter is introduced gradually so as not to overwhelm the beginning student; for example, sodium fusion is not introduced until Chapter 23, derivatives in Chapter 32, the tests for elemental nitrogen and sulfur in Chapter 35. In addition to these chapters there are short tables in the "technique chapters" (for example, Distillation, Melting Points, Crystallization, and Paper Chromatography) that allow the assignment of an unknown which can be quickly solved by application of the technique introduced in that chapter. This procedure not only serves as an early introduction to qualitative analysis but also greatly increases student interest in these experiments. In many of the synthesis experiments identification of by-products is similar to identification of unknowns by classical or instrumental techniques.

This laboratory manual can be used to emphasize instrumentation heavily. Such emphasis might be especially desirable if organic chemistry is the institution's terminal chemistry course for most of the students as, for example, with biology and medical technology majors, premeds, or junior college students. Very thorough instruction is provided for gas chromatography, infrared spectroscopy, ultraviolet-visible spectroscopy, and polarimetry in Chapters 8, 18, 26, 37, and 64. Instrumental quantitative analysis with these methods is introduced in Chapters 8, 9, 20, and 37. Special accessories and techniques are described in Chapters 27, 65, and Appendices III and IV. Most of the syntheses chapters have optional sections with instructions for application of one or more instrumental techniques to analyze the product, intermediates, and by-products. The chapter order is such that gas chromatography can be applied to all unknowns, infrared spectroscopy to all but the first one, and ultraviolet spectroscopy to all but the first two unknowns without inverting chapter order. The function, application, and limitation of each instrumental technique are thoroughly explained and compared with other techniques. Instruments are not black boxes. Their limitations are as real and definite as their applications are useful. The student who does not have at least rudimentary understanding of the instrument he uses will not appreciate or fully understand his results or perhaps even obtain results. The understanding and use of such instruments is not difficult, especially if taken a step at a time. With this manual the instructor can assign or omit what is needed to make instrumentation as thorough or as empirical as he wishes.

Noninstrumental techniques are not neglected; see for example Chapters 3, 4, 5, 6, 11, 12, 25, 26 (first part), and 57. Thin layer, column, and paper chromatography are thoroughly introduced. There are experiments on countercurrent extraction and fractional crystallization; the latter separates a pair of N-arylbenzamides that are *not* easily separated by column adsorption chromatography or gas chromatography.

Recently there has been much discussion about "cookbook style" instruction versus independent study. It is the author's opinion that much of the criticism of the modern laboratory manuals has been unjustifiably harsh and often exaggerated. Without the proper precautions the organic laboratory is dangerous. In addition there must be considerable specific instruction in various techniques. The laboratory manuals, however, have not been as imaginative as the lecture texts. The better lecture texts have demonstrated how well problems can be used to help the students *learn* what they have studied. This manual uses problems wherever possible not only to help the student *learn* techniques but also to help *do* the laboratory work. Many of the experimental instructions are posed as questions to attempt to stimulate the student to devise his own instructions and techniques. The answers to critical questions are given in the back of the book. Many of the experiments are open ended or offer open ended options; see for example Chapters 19, 26, 27, or 62 for several approaches. All qualitative analysis chapters can be regarded as independent study assignments. Perhaps most important for independent study is that emphasis in this text is on understanding not rote memory, on the why and how of techniques not mechanical manipulations, on safety not haphazard steps, on function and limitations of instruments not magic black boxes.

This laboratory manual is intended to be used with a *modern* lecture text. Discussion material found in most modern texts is not duplicated here. Discussion material found in none or only some of the modern texts is included where it is essential to the experiment. For example, general spectra interpretation is not included because this is more than adequately covered in almost all lecture texts likely to be used with this manual. Chapter 18 does contain discussion of the out-of-plane C-H bending vibrations and 2000-1600 cm^{-1} region that are used to identify alkenes and aromatic compounds because this is essential for some of the experiments and this particular subject is not treated as thoroughly as needed in a *few* of the modern textbooks.

As with almost any work of this size there are contributors other than the one whose name appears on the cover. First and foremost appreciation is expressed to the many students at Wagner College and The American University who first performed the experiments. Their enthusiastic response to most of the experiments, criticism of portions that needed revision, and many comments were essential for the development of this manual. This manual is primarily the result of their inspiration, interest, and efforts. So many of my colleagues in teaching and industry made helpful suggestions that a complete listing would be impractical. Special appreciation is due Dr. Robert W. Hannah who read most of the sections on spectroscopy and made many important comments which were incorporated in the final manuscript. Sidney Kelley and the Technical Writing Department at Perkin-Elmer furnished many fine illustrations some of which were used directly and some of which were modified by the author for specific instructions in this manual.

The author's wife, Ann, made the largest individual contribution. This included typing and editing most of the manuscript several times, mimeographing and collating chapters for student trial, drawing many of the illustrations, testing a few of the final revised experiments when the author did not have ready access to beginning organic classes, offering many suggestions that improved clarity, proofreading the galleys and final text, and preparing the index. Without her contributions this manual obviously never would have been published.

J. S. S.

Contents

1

General Laboratory Procedures

SAFETY IN THE LABORATORY

The organic chemistry laboratory has more potential hazards than any other laboratory the undergraduate student is likely to encounter. In spite of these *potential* hazards all experiments in this text (and most other organic laboratory manuals) can be completed without accidents, even minor ones, if all the instructions are carefully followed *and* reasonable care and common sense are used.

Chemicals: Toxic Hazards

Almost all the chemicals encountered in an organic chemistry laboratory are toxic or flammable (some explosively so) or both. For example, if one were to submerge his hand up to his wrist for five minutes in a container of carbon tetrachloride, enough of this solvent would be absorbed through the skin to cause death. Although carbon tetrachloride is more toxic than most compounds encountered, every possible precaution should be taken to avoid swallowing, inhaling, or spilling organic compounds. Many organic compounds are readily absorbed through the skin and can be just as toxic when absorbed as when swallowed.

Particular caution must be observed in the organic chemistry laboratory to avoid exposure to toxic substances because there are so many potential sources. These toxic substances may act synergistically, that is, a mixture may be more toxic than would be expected from the toxicities of each component. For example, carbon tetrachloride is 5 to 20 times as toxic to an individual also exposed to ethanol (which is only slightly toxic by itself) as to one not so exposed.

If any material is spilled on the skin, wash it off immediately with large volumes of water. If the material is a sticky solid or viscous liquid, use soap and water to dislodge it. For first aid treatment after exposure to specific noxious substances see the inside front cover.

Do not taste any chemicals, solutions, or other material unless specifically instructed to do so. If it is necessary to smell an organic substance, hold the container 6 to 10 inches from the nose, wave the vapors toward the nose, and cautiously sniff. If no odor is detected, move the container a little closer and sniff a little more strongly but do not inhale deeply. For substances with weak odors it may be necessary to hold the container very close to the nose to detect anything; however, all substances should be smelled first from a distance.

Every possible precaution should be taken to keep the air free from contaminants. Among other measures the following will help achieve this.

a. Immediately after use replace all tops and stoppers on containers. This not only reduces evaporation but also helps keep the chemicals free from contamination.
b. Unless specifically requested to do so, do not remove chemicals or reagents from the areas designated for their storage and measurement. Weigh or pour the correct amounts of chemicals into a graduated cylinder,

beaker, weighing paper, or other container and take *this* to your desk. This is essential for noxious reagents such as bromine and phosphorus pentachloride that are kept or measured in a hood.
c. Never use more than the quantities indicated. This will help to ensure successful experiments as well as minimize evaporation of hazardous substances.
d. When disposing of any liquids in the sink, pour them directly into the drain and flush them down with very large quantities of water. In some laboratories the pouring of various liquids into the drain is not permitted. Which liquids are not permitted depends upon the condition of the plumbing, local ordinances, and other factors. The instructions of the laboratory assistant on disposal of these liquids should be carefully followed. Do not put water-insoluble *solids* in the sink or drain.
e. When distilling (see Chapters 3, 4, and 5) or heating a mixture under reflux (see Chapter 10), be certain water is flowing through the condenser so that even volatile liquids are condensed. Do not heat volatile liquids so strongly that they are not completely condensed. When liquids with a boiling point less than 40° above room temperature are distilled, the receiving flask should be chilled in an ice-water bath. When heating liquids that are not being distilled, carefully follow the precautions listed on pages 66 and 89 on the use of reflux condensers, heating sources, and hoods.

Chemicals: Fire Hazards

Where the potential for fire is very high, warnings and precautions are usually listed in the text; however, even when such specific warnings are not mentioned, there are still fire hazards present. Flames should never be used within 10 ft of containers used for the storage of volatile liquids. There should be no flames in the laboratory when large quantities (20 ml or more) of very flammable liquids are used by most of the class. If these rules and the precautions listed above to avoid the build-up of toxic substances are followed, there will be no fires. The maximum allowable concentrations of organic vapors determined by their toxicities are usually lower than the concentrations at which they are fire or explosion hazards. Among the exceptions to this are alkanes such as petroleum ether, ligroin, heptane, and cyclohexane. The alkanes are relatively nontoxic, and the volatile ones are very flammable.

Chemicals: Strong Acids and Bases

You will be required to use strongly corrosive substances almost every period. Although you have undoubtedly used some of these materials in general chemistry, extra precaution is required here because of the greater variety, quantity, and extensive use of strong acids and bases in this course. Specific warnings are included in the text when most of these new materials are first used. There are no warnings, however, when more common corrosive materials such as sulfuric acid and sodium hydroxide are used. The following precautions should be observed.

a. If any of these corrosive materials comes in contact with the skin, wash off immediately with large quantities of water. See inside of front cover for counter measures for specific chemicals.
b. Use the quantities called for in the instructions; do not use even a small excess.
c. Wear eye protection when working with corrosive substances.
d. When mixing a strong acid with water, always add the acid slowly to the water while stirring the mixture.

Eye Protection

It should be obvious that the eyes are more easily injured by hazards encountered in the chemistry laboratory than any other part of the anatomy. These hazards include acids and other corrosive substances, solvents, flying objects, and toxic substances particularly harmful to the eyes such as methanol (wood alcohol). While most students begin their study of chemistry with two eyes in reasonable working condition, even the most reckless and irresponsible student will do everything to safeguard both eyes unless he is also a complete and absolute fool.

Most injuries, including those to eyes, occur from the unexpected. The eyes should be protected *at all times* in the laboratory. Safety glasses, usually those with case-hardened[1] lens, are preferable but not absolutely necessary. Ordinary prescription glasses or those with plain window glass are satisfactory and stop most hazardous materials before they reach the eyes. Sun glasses are not satisfactory because they may make the interpretation of color changes difficult. Contact lenses give no or very little protection. When these are used, regular glasses or a face shield must also be worn.

When especially hazardous experiments are being conducted, full eye protection such as is given by full or half face shields over safety glasses should be used. More important than full protection in a few isolated instances is almost full protection *all the time*. Even if you are aware of *all* the potential hazards (which is unlikely for a beginning organic chemistry student), can you predict what your neighbors are going to do? **Always wear eye protection when in the organic chemistry laboratory.**

EFFICIENCY IN THE LABORATORY

The students (and scientists) that are the most efficient in the laboratory not only learn more but also enjoy the work more. A complete and thorough knowledge of what is being done is needed for maximum efficiency. *Before* the laboratory period begins the student should read the assignment. If available, a free period just before the laboratory begins is an ideal time for such study. From this study a rough outline should be written for as many of the steps as possible. Draw up a list of equipment, chemicals, other supplies, and, if necessary, a list of points for your laboratory instructor to clarify.

When you walk into the laboratory, you should know exactly what you are going to do. In many of the experiments what is done in the later steps is often dependent on what happened or was observed earlier, and in such situations it is impossible to plan everything ahead. However, in so far as your knowledge of organic chemistry permits, you should have the different possibilities and alternatives in mind.

In the organic chemistry laboratory there are many operations that require considerable time but little effort and attention. When these operations (heating, refluxing, crystallization, elution, cooling, and some distillation) are required, it is necessary to do other operations simultaneously in order to complete the work in the allotted time. These may consist of cleaning the glassware, rereading the instructions, preparing reagents, planning for the next step, starting the next experiment, or analyzing an unknown. The most efficient use of time in a laboratory is achieved by giving attention and priority to a primary experiment or series of operations and having one or two other operations ready to do as time permits.

PROBLEMS

There are many problems in the instructions for the experiments. An attempt should be made to answer all these problems either before or during the experiment. The answers to these problems should be verified in the back of the book before proceeding further because the answers often include instructions. At the end of many chapters are exercises in a separate section. Unless instructed otherwise do the exercises after completion of the experimental work.

GLASSWARE

General

Figure 1-1 illustrates and names the different pieces of glassware used in the beginning organic chemistry course. Most of the glassware used is of borosilicate glass and is reasonably resistant to thermal shock. Care

[1] For those who are planning lifetime careers in science and wear prescription glasses, it would be wise to have a prescription filled with case-hardened lenses. These have the advantage over ordinary lenses in that they protect against flying objects with fairly high momentum. Their slightly higher cost is compensated for by their greater durability.

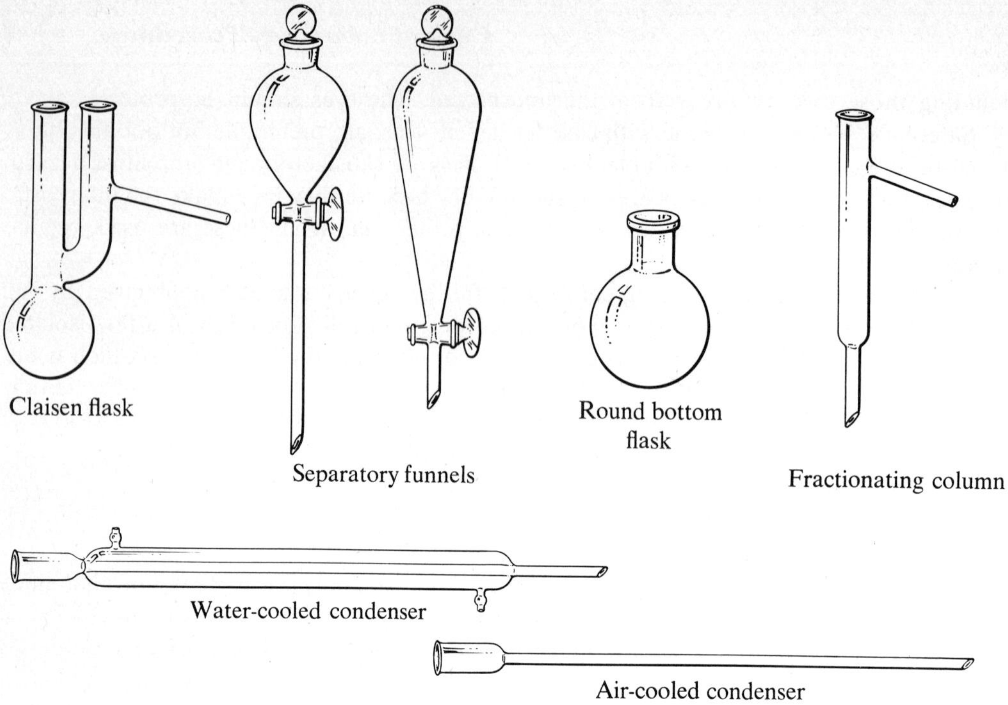

Glassware without ground glass joints

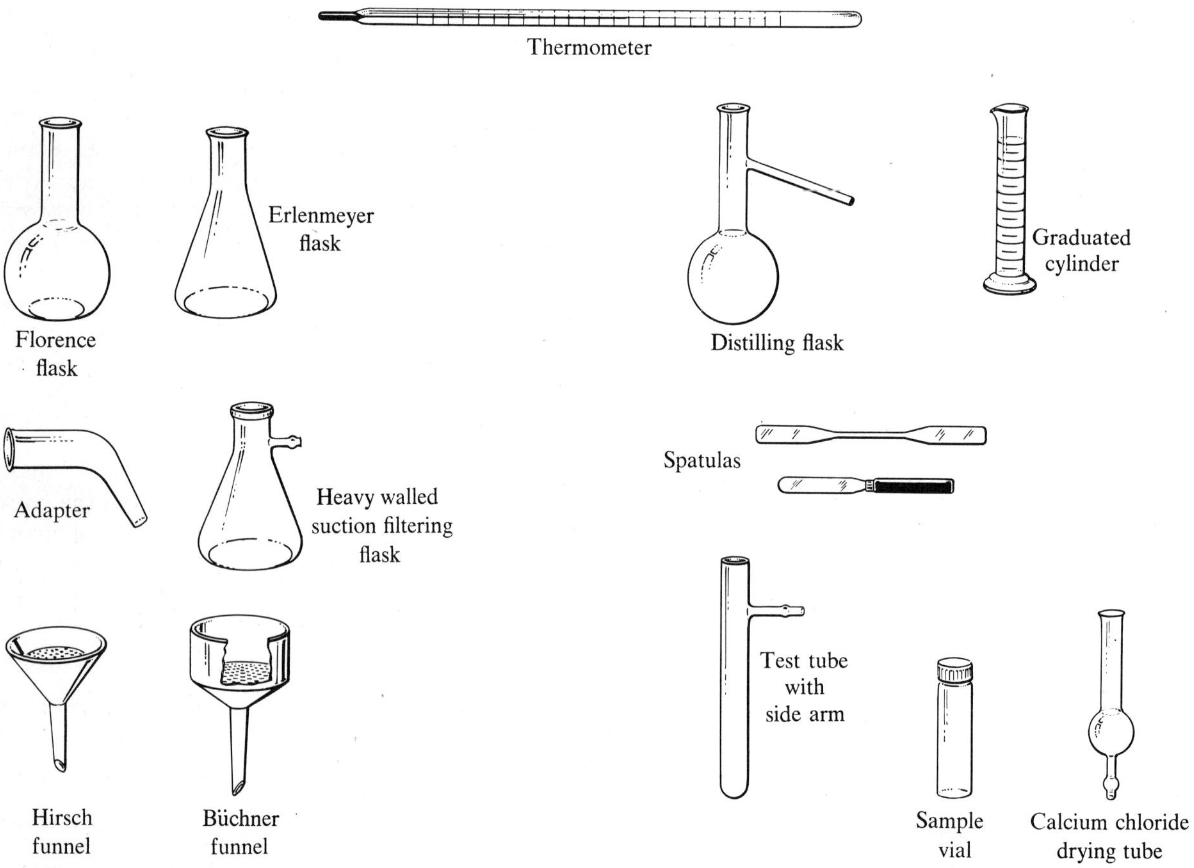

Equipment that may be in lockers with either ground glass or nonground glass equipment

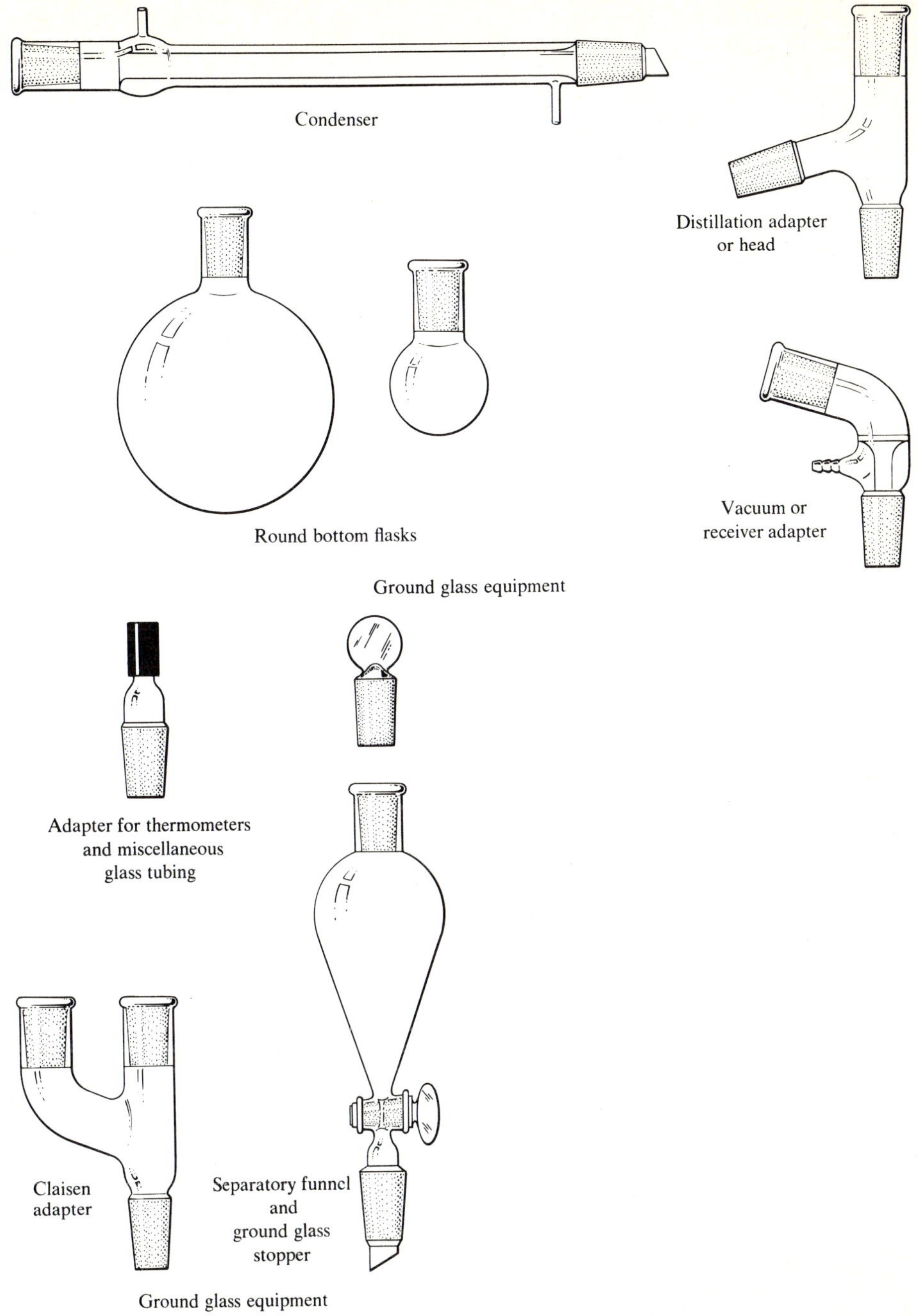

Figure 1-1 Glassware commonly used in the organic chemistry laboratory.

should be taken, however, to avoid extreme and sudden changes in temperature such as pouring cold water on a piece of glass that has been recently heated. Usually wire gauze is used between a burner flame and the glassware. Empty glassware should not be heated with a burner flame except possibly for a few seconds when necessary for quick drying.

Care of Ground Glass Equipment

If equipment with ground glass joints is used, the following section should be studied and carefully followed. If ground glass equipment is not used, this section may be omitted.

The use of equipment with ground glass joints can save considerable time and effort if the joints are properly cared for. It not properly used, the joints may leak, introduce impurities, and stick or "freeze." Considerable effort is often required to separate frozen joints, and they are sometimes broken in the process.

Before use, all joints should be cleaned of dirt, solid compounds, and old, hardened joint lubricant. The joint should then be *lightly*, completely, and evenly lubricated or greased for about 60 to 70% of its length from top (Fig. 1-2). With a cloth wetted in chloroform immediately wipe off any lubricant that is on any part of the glassware other than the joints. Lubricant on glass surfaces inside the apparatus should be especially avoided because it will contaminate the contents. It is usually necessary to lubricate only one of the two joints to be placed together; almost always it is easier to lubricate the male joint. When placing the joints together, turn one piece in the other for two complete (360°) turns to ensure smooth application of the lubricant.

When putting equipment away until the next laboratory period, disassemble all joints and clean them by wiping with a dry paper or cloth towel and then with a cloth dampened with chloroform. In some experiments it may be necessary or convenient to leave a drying tube or a stopper in a flask containing an intermediate or a product or, if locker size permits, to store an assembled apparatus. In such cases the joints need not be cleaned or relubricated before the flask is emptied unless they become dirty or lose most of the lubricant.

Great care must be taken to keep lubricant from the contents of the glassware. When transferring liquid or solid into a flask or separatory funnel, use a funnel for the transfer (Fig. 1-3). If a powder funnel (one with a wide stem) is not available for transfer of solids, one can be made out of paper. When the contents of a flask are to be poured out through a glass joint, all lubricant should first be cleaned from the joint.

If ground glass joints do become stuck or frozen, they should be separated as soon as possible. This can usually be accomplished by heating the joint with steam for a few minutes. Pass the steam through flexible

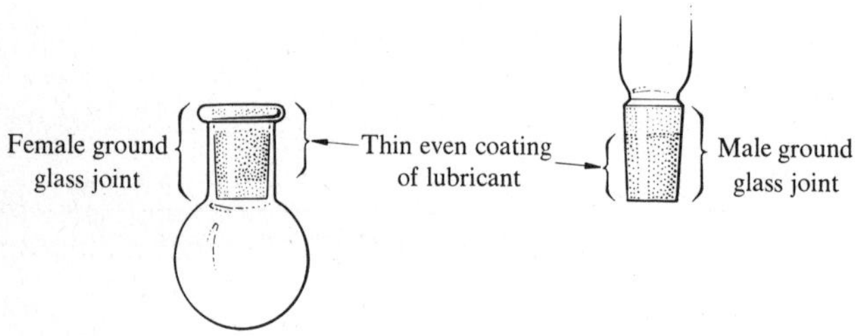

Figure 1-2 Lubrication of joints.

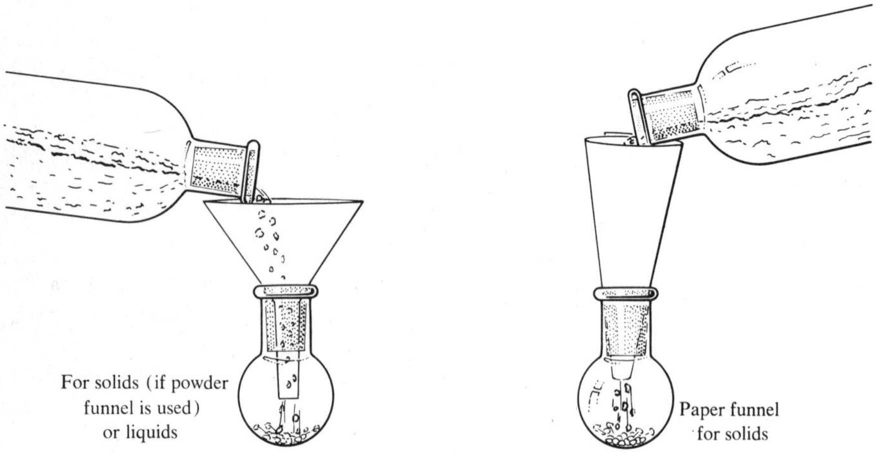

Figure 1-3 Transfer of material through a greased joint.

tubing so that it can be directed over the entire surface of the joint. Gently twist the hot joint. **Caution:** wrap both pieces of glass in a towel so that you are less likely to cut your hand if the glass breaks.

If this is not successful, it will be necessary to heat the joint with a flame. Completely remove any flammable organic material from the flask, rinsing several times with water. All water should be removed that might flow on the joint while it is being heated. Even a few drops contacting the hot joint can cause thermal shock sufficient to crack the joint. Heat the joint with a yellow, oxygen-starved flame. Try to heat briefly the entire joint. Attempt separation immediately.

Occasionally even this procedure fails to separate a frozen joint. Sometimes soaking the joint for several days in a solution of Alconox or other detergent will loosen it sufficiently so that it can be separated. The detergent lowers the surface tension of the solution so that it can seep into the joint.

Care of Stopcocks

The stopcocks to separatory funnels require special care. If they are Teflon, they should not be lubricated. If they are glass, lightly lubricate the plug (Fig. 1-4). Keep grease out of the bore. Place the stopcock in the shell and turn it several times so that all the surfaces are lubricated evenly. Remove any excess lubricant that has accumulated near the bore, in the stem, or in the bottom of the separatory funnel. Place the guard on the stopcock. The stopcock should turn without much resistance. If it does not, it is dirty or the lubricant has been washed away. If either of these happens, the plug should be washed and relubricated.

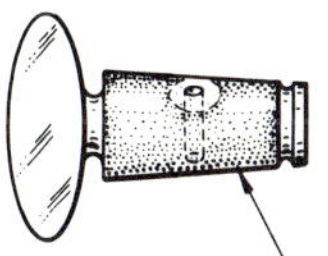

Figure 1-4 Lubrication of ground-glass stopcocks.

If a glass stopcock becomes frozen, it can be freed by the same procedure used to loosen ground glass joints. A frozen Teflon stopcock can be loosened by submerging it (only the stopcock assembly, not the entire separatory funnel) in a beaker of crushed ice and water for 5 min and then turning. Do not heat a frozen Teflon stopcock.

Problem 1-1. What property allows frozen Teflon joints and plugs to be freed in this manner?

CORKS AND STOPPERS

Even if glassware with ground glass joints is available, it is still often necessary to use cork or rubber stoppers The use of these is made easier if a few precautions are followed:

a. Do not use rubber stoppers with liquid hydrocarbons and their halogen derivatives. Use neoprene or cork stoppers with these liquids.
b. Do not use corks for vacuum distillations.
c. Compress and soften corks in a cork roller *before* boring a hole.
d. When inserting tubing or thermometers through holes in stoppers use a *small* amount of lubricant on the glass surface. Lubricants of choice are glycerine (glycerol) for rubber and neoprene, soapy water for cork

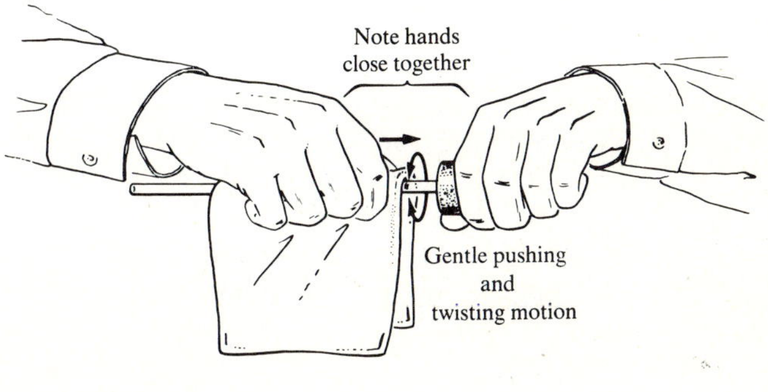

Figure 1-5 Technique or putting glass tubing and thermometers through stoppers.

When insertion is completed, remove the excess lubricant.

e. Grip the tubing or thermometer as shown in Figure 1-5 with a towel to minimize the possibility of cuts. Note how the towel protects *both* hands. The tubing should be gripped as close to the stopper as possible because this reduces the probability of breakage and the possibility of cutting the hand should the tube or thermometer break. Insert the tubing into the stopper with a gentle twisting motion.

f. If it is necessary to remove a stopper from glass tubing, a thermometer, or other equipment and the stopper does not easily slide off, cut it off with a razor blade. Do not force it off as you are likely to break the apparatus and cut your hands by this procedure.

g. Do not use the bottom of a distilling flask or one leg of bent tubing as a lever to force such glassware through a stopper. Hold the tubing with a towel close to the stopper while rotating the tube and applying gentle pressure (Fig. 1-6).

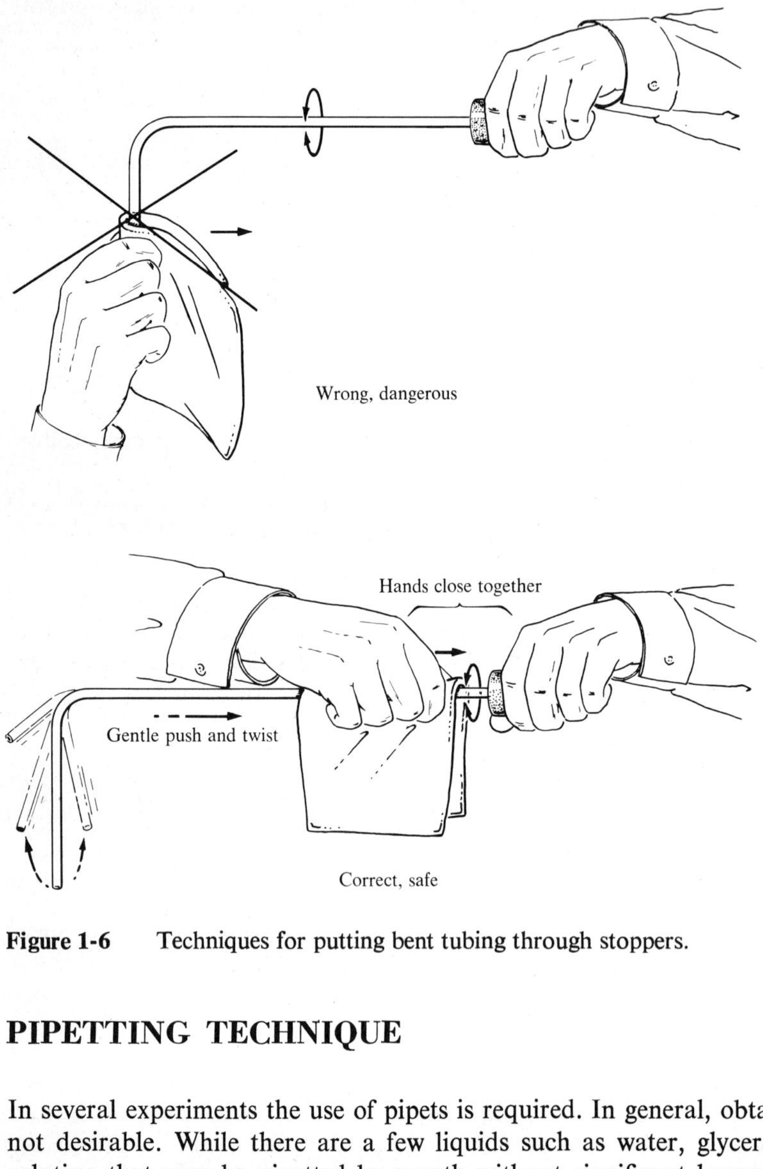

Figure 1-6 Techniques for putting bent tubing through stoppers.

PIPETTING TECHNIQUE

In several experiments the use of pipets is required. In general, obtaining suction from the mouth and lungs is not desirable. While there are a few liquids such as water, glycerol, ethanol, and aqueous sodium chloride solution that may be pipetted by mouth without significant hazard, most liquids encountered in this course are toxic or corrosive or both. Dilute bases, for example, can cause painful mouth burns probably because they remove fats from the tissues. Even if liquid is not accidentally drawn into the mouth, vapors of the liquid could enter the respiratory and digestive systems this way. When suction is supplied by mouth, there is not only danger of contamination of the experimenter by a toxic liquid but also of contamination of the liquid by saliva from the experimenter.

Suction should be supplied by vacuum tubing from an aspirator or a house vacuum line or by a rubber bulb (Fig. 1-7). Care should be taken to avoid getting liquid in the rubber bulb as this will contaminate other samples that are pipetted. Hydrocarbons and their halogen derivatives may also dissolve part of the rubber in the bulb. Special care must be taken with semimicropipets and medicine droppers used as pipets because the length of glass tubing between the top mark and the bulb is so short.

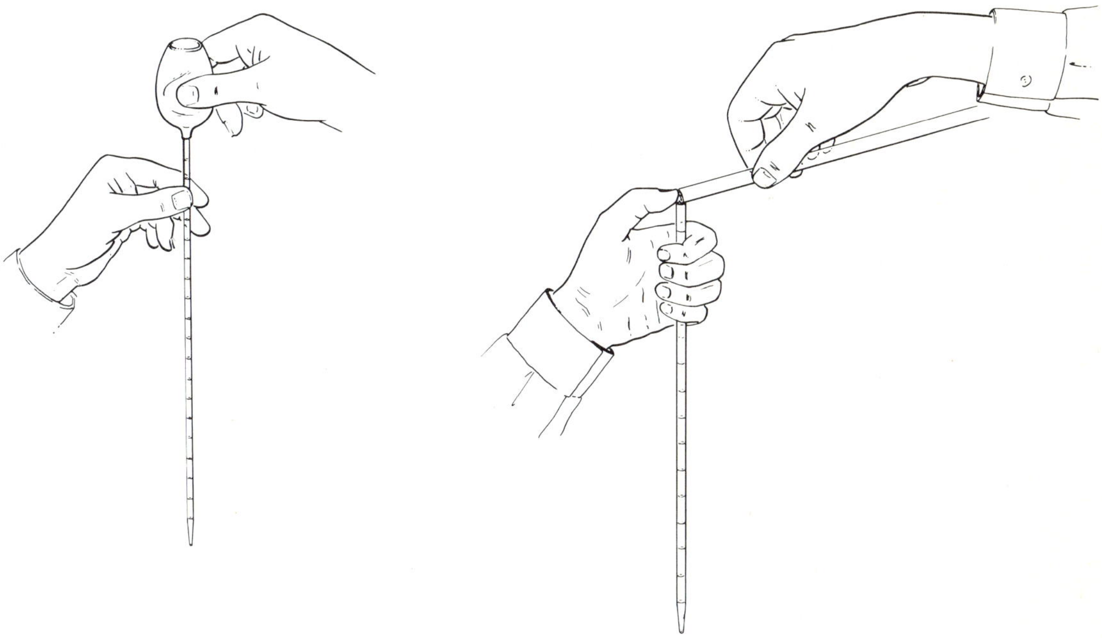

Figure 1-7 When applying suction to a pipet use a bulb or vacuum line. Do not use your mouth.

When measuring the amount of liquid align the *bottom* of the meniscus with the calibration mark (Fig. 1-8). If a vacuum line is used, draw liquid to 1.5 to 2.5 cm above the calibration mark. Place your finger over

Figure 1-8 Proper alignment of calibration mark with the meniscus.

the pipet as you quickly remove the vacuum line. Allow just enough air to leak past your finger so that the bottom of the meniscus falls to the calibration mark. Touch the pipet to the side of the container from which you withdrew the liquid; this removes the hanging drop. Place the pipet slightly into the container that is to receive the liquid. Touch the tip to the inside of the container (this prevents splashing) and allow the liquid to flow into the container. Wait 10 sec, then touch the tip to the side of the flask. Do not blow the remaining liquid out. Follow the same general technique if a rubber bulb is used except that once the pipet is filled, remove it from the liquid and carefully release any pressure on the bulb so that air bubbles do not shoot up and force liquid into the bulb.

Most pipets are calibrated to deliver the volume indicated if liquid is allowed to flow out without any pressure being applied at the top of the pipet. Some of the liquid that remains is removed by touching the pipet to the inside container wall. Occasionally your instructor may tell you it is necessary to remove the liquid in the tip by applying pressure. This may occur with liquids with high viscosity or high adhesion for glass. In general, the small amount of liquid remaining in the tip should not be forced out unless you are given specific instructions to do so.

2

Intermolecular Forces and Physical Properties

INTRODUCTION

A laboratory course in organic chemistry involves the study of the chemical and physical properties of organic compounds. Physical properties are used to detect or illustrate most of the chemical properties and changes. It is therefore appropriate that an introduction to physical properties is the first subject studied in this laboratory text.

This chapter is intended to supplement the material usually available in beginning organic texts on intermolecular forces and their interrelation with structure and physical properties. For more extensive discussions on some of the subjects outlined here the student should consult his lecture text.

STRUCTURE-PROPERTY RELATIONS

In the organic chemistry laboratory much of the time is used observing physical properties and isolating or purifying compounds on the basis of their physical properties. Organic compounds are identified by their melting points, boiling points, solubilities, spectra, and other physical properties. Organic compounds are separated by exploiting differences in their vapor pressures, solubilities, adsorptions, partition coefficients, and other physical properties.

For many compounds most physical properties are known and recorded where they can be easily found. For many other compounds, however, the values of the desired physical properties are either not readily available or are completely unknown. Fortunately, approximate physical constants can often be estimated by applying what is known about intermolecular forces, the structure of liquid and solids, and the structure of the compounds in question. The estimation of physical constants can be reasonably accurate if there are compounds of known physical constants with structures similar to those of the compounds in question. Often from the physical constants valuable information can be obtained about a new or unknown compound's identity.

INTERMOLECULAR ATTRACTIONS

All intermolecular[1] attractions are electrostatic; that is, they result from the attraction of unlike charges. The strength of intermolecular forces is dependent upon three factors: (a) the magnitude of the electrostatic charges involved, (b) the distance between the charges, and (c) the permanence of the charges. The three types of charges—ionic, polar, and induced—are summarized in Table 2-1.

[1] For convenience and brevity the word *intermolecular* in this section will often include interionic, ion-molecular, and atom-molecular.

TABLE 2-1 ELECTROSTATIC CHARGES CAUSING INTERMOLECULAR ATTRACTIONS

Type	Cause	Examples	Strength
IONIC	From groups or atoms that have more or less than the number of electrons necessary for a neutral particle	$Li^{+}\bar{C}H_2(CH_2)_2CH_3$ $Na^{+}\ F^{-}$	Strongest
POLAR	From the unequal distribution of electrons between atoms in a molecule	$\overset{\delta+}{H}\ \overset{\delta-}{Cl}$	Intermediate
Dipole	The vector sum of the polar bond moments is not zero (the centers of positive and negative charges do not coincide)	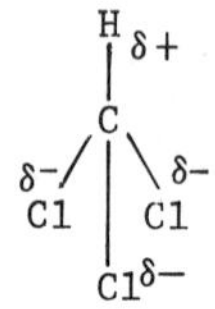	
Quadrupole	The molecule has bonds of significant polarity but the vector sum of bond moments is zero (the centers of positive and negative charge coincide)	$\overset{\delta-}{O}=\overset{\delta+}{C}=\overset{\delta-}{O}$	Usually weaker than dipole forces
INDUCED CHARGES	Temporary charges that are induced by charges in neighboring molecules	(See Fig. 2-5)	Usually weaker than polar

The intermolecular forces caused by these charges may be classified according to the strongest charge involved.

Ionic

Ion-Ion Attractions. These result from attractions of the relatively large opposite charges on anions and cations. These are the strongest forces and occur in all ionic substances. Examples:

$$K^{+}\ Br^{-} \qquad Ca^{++}\ Cl^{-}\ Cl^{-}$$

Ion-Polar Attractions. This type results from the attractions of ions for oppositely charged polar portions of molecules. These are the second strongest of all intermolecular attractions. Examples:

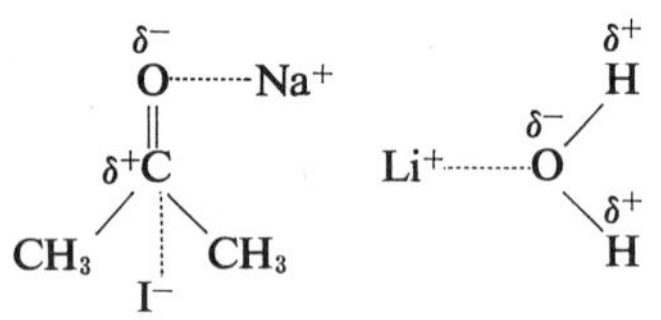

Ion-Induced Dipole Attractions. These result from the attraction of ions for the oppositely charged end of induced dipoles. The dipole is induced by the charge on the neighboring ion. As a positive ion approaches a neutral molecule, it will attract electrons to that part of the molecule nearest the ion. As a negative ion approaches a neutral molecule, it will repel electrons from that part of the molecule nearest the ion. The schematic example in Figure 2-1 illustrates this.

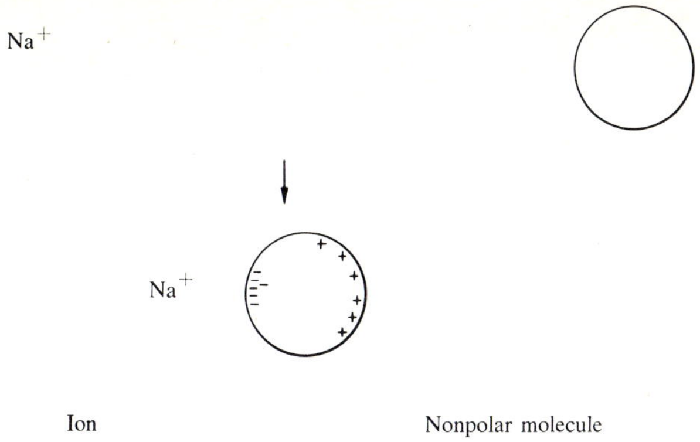

Figure 2-1 Ion-induced dipole attractions.

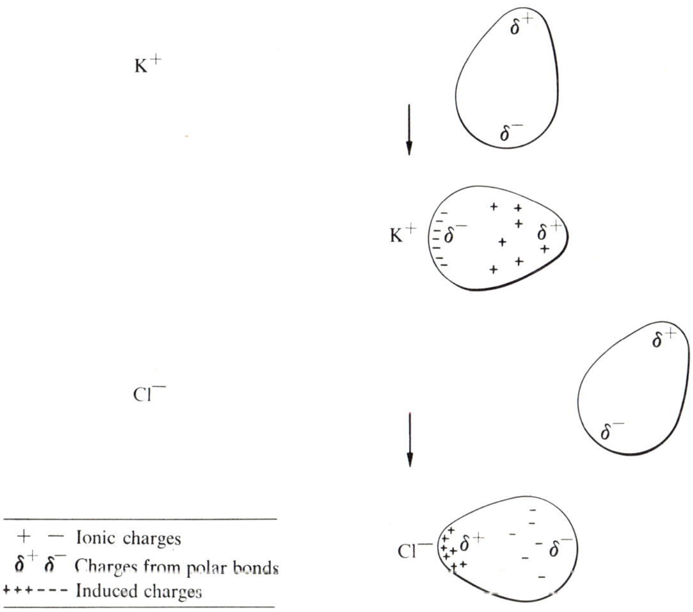

Figure 2-2 Combination ion-polar and ion-induced dipole attractions.

Usually ion-induced dipole attractions add to attractive forces in ion-polar attractions. That is, as a positive ion approaches a negative portion of a polar molecule, electrons are attracted toward the negative portion. This added negative charge causes a greater attractive force than would exist if only the permanent negative pole were there alone. A similar attraction results from the opposite movement of electrons in a negative ion-dipole interaction. Schematic examples of the combination ion-dipole and ion-induced dipole attractions are shown in Figure 2-2.

Polar Attractions

Pole-Pole Attractions.[2] This type results from the attraction of oppositely charged atoms or groups on different molecules. The charges result from polar bonds. Examples are

[2] Often called dipole-dipole attractions but this would seem to exclude other pole-pole attractions such as those among carbon dioxide molecules or between a carbon dioxide molecule and a water molecule. The dipole-dipole attractions are usually stronger and more important than the dipole-quadrupole and quadrupole-quadrupole attractions.

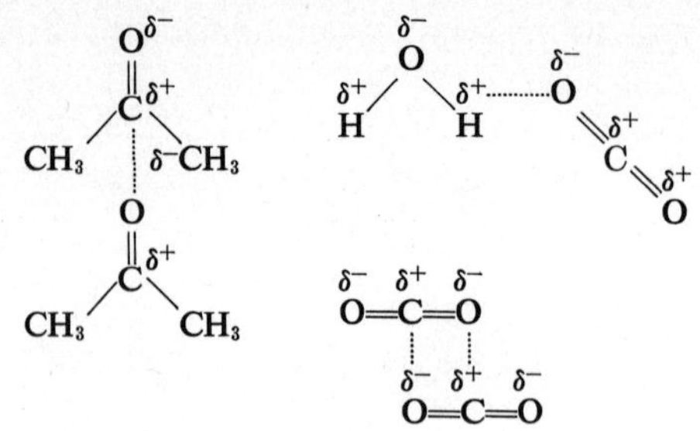

Pole-Induced Dipole Attractions. These are similar to the ion-induced dipole attractions explained above except that the inducing charge is weaker and the resulting attraction is weaker. Schematic examples of this are shown in Fig. 2-3. Pole-induced dipole attraction further increases the attraction that occurs between two molecules with pole-pole attraction. This is illustrated schematically in Figure 2-4.

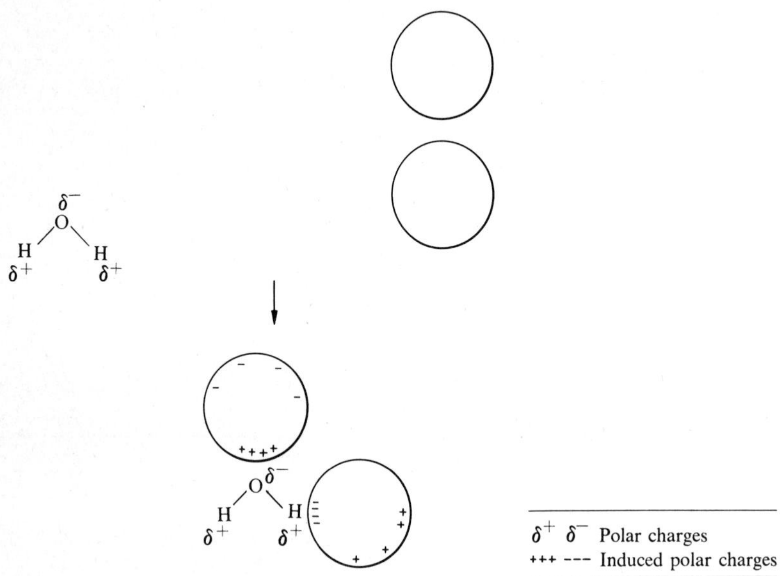

Figure 2-3 Pole-induced dipole attractions.

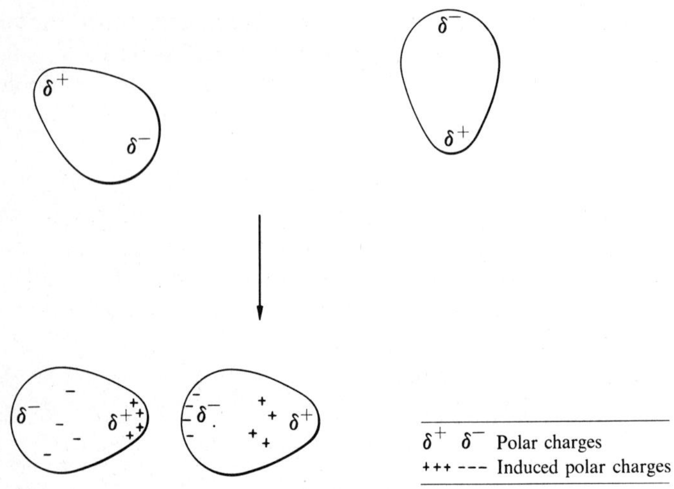

Figure 2-4 Combination pole-pole and pole-induced dipole attractions.

Induced Dipole-Induced Dipole Attractions

These attractions are often called London dispersion forces.[3] Even among molecules with no dipoles or polar bonds there exist significant attractions. Although the *average* charge distribution in these molecules gives no *permanent* centers of positive and negative charge, the movement of electrons causes small, temporary positive and negative charges. A local excess of electrons (negative charge) in a molecule *A* tends to repel electrons in a neighboring molecule *B*. This leaves a deficiency of electrons (positive charge) in *B* near a negative charge in *A*. The attraction between the unlike charges in *A* and *B* is an example of a London dispersion force. In a similar manner a temporary positive charge in that part of molecule *A* near *B* will induce a temporary negative charge in *B*. There will again be attraction between *A* and *B* because of the attractions between unlike charges. A schematic example of London dispersion forces is shown in Figure 2-5.

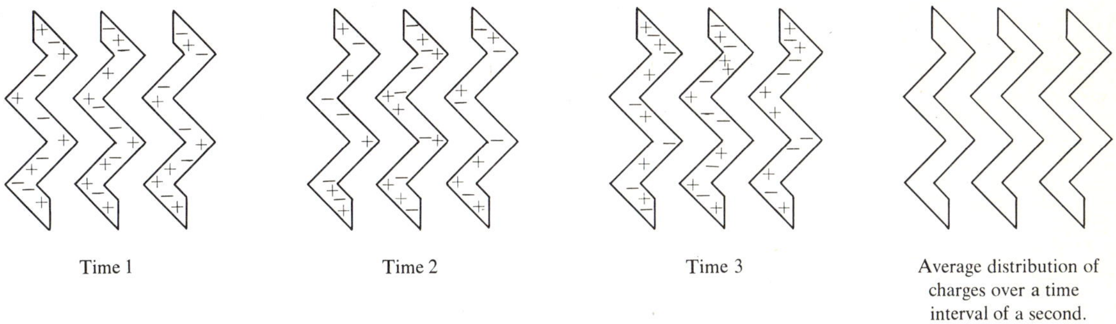

Figure 2-5 London dispersion forces among three hexane molecules at three different times.

PHYSICAL STATES OF MATTER

Solid State

Crystalline solids (almost all substances that we call solids are crystalline solids, glass being one of the exceptions) have their units in a *permanent*, closely packed, *regular*, *ordered* arrangement or lattice. The units that make up the lattice may be atoms, molecules, or ions. The units always maintain the same average position with relation to each other, and the movement about these positions is relatively small (Fig. 2-6). The attractive forces between units are large.

Amorphous solids, such as glasses, are similar to crystalline solids in the closeness of the particles to each other and in their immobility. The amorphous solids, however, are not so highly ordered or organized.

Liquids

In liquids the units are closely packed, but there is no *permanent* order to the arrangement of the units. A molecule or other unit of liquid has no permanent position and is no more likely to return to a position it once occupied than to some other neighboring position. The attractive forces among molecules in a liquid are strong, but usually are considerably weaker than the intracrystalline forces of the same substance. The mean free path[4] is longer than in the corresponding solids but is not much greater than the molecular diameter except at temperatures slightly below the critical temperature. In liquids as in solids the molecules or other units touch or almost touch their neighbors. Liquids, therefore, are almost incompressible.

[3] These are also referred to as London forces, van der Waals forces, or dispersion forces.

[4] The mean free path is the average distance a particle moves in a straight line. In liquids and gases, particles usually travel in straight lines until they collide with another particle. In solids the particles are prevented from traveling far in straight lines not only by interference of neighboring particles but by the lattice forces overcoming the kinetic energy of motion.

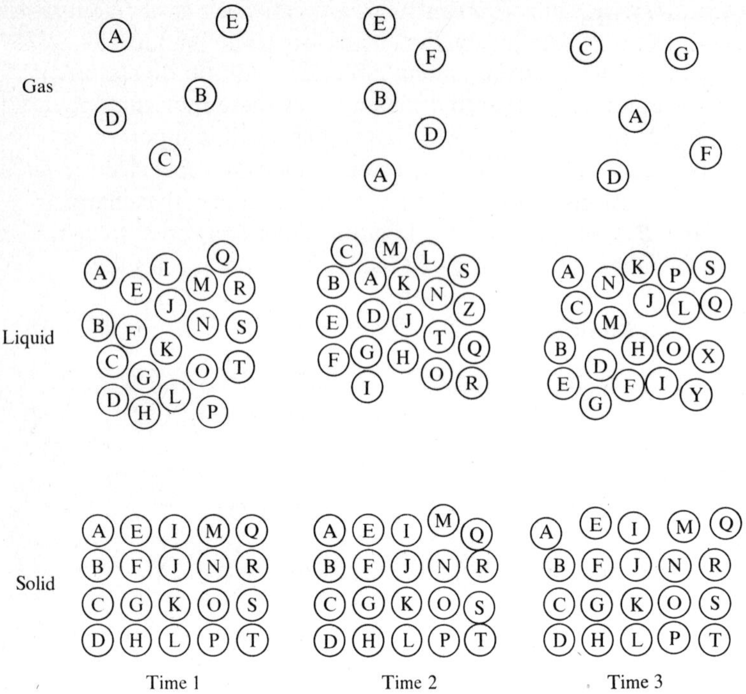

Figure 2-6 The solid, liquid, and gaseous states at three successive time intervals. The molecules are lettered so that relative positions can be followed.

The most important difference between liquids and solids is that liquids have no *permanent* structure, and their arrangement of molecules or other particles is fairly chaotic (Fig. 2-7).

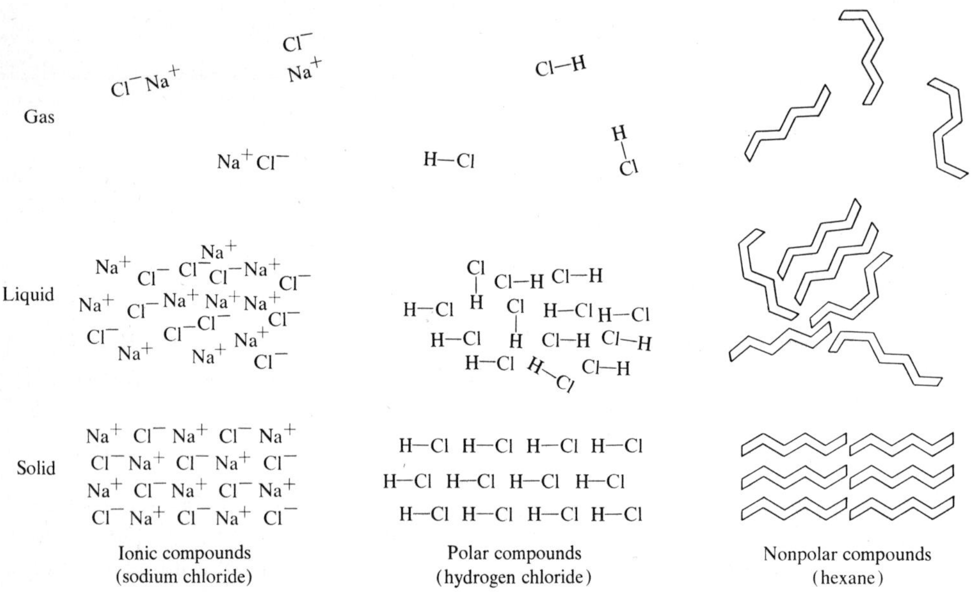

Figure 2-7 Two-dimensional illustration of structure of solid, liquid, and gaseous phases for different types of compounds.

Gases

In gases the attractive forces between molecules are very small (none in perfect gases), there is no order except possibly near the critical point, and there is no permanence of structure even if one uses the smallest possible time interval for observation. The distances between molecules in the gaseous phase are very large and the mean free paths of the molecules are long. Gases are readily compressible; Boyle's law, $V \propto 1/P$, is a mathematical expression of this compressibility. Table 2-2 summarizes the characteristics of solids, liquids, and gases.

TABLE 2-2 SOME CHARACTERISTICS OF SOLIDS, LIQUIDS, AND GASES

CHARACTERISTIC	PHYSICAL STATE			
	Solids-Crystalline	Solids-Noncrystalline	Liquids	Gases
Intermolecular forces	Very strong	Strong to intermediate	Intermediate	very small
Regularity of structure	Very ordered	Usually as disorganized as liquids	Some regular structure	Very little if any.
Permanence of structure	Permanent	Permanent	Temporary	Temporary or none
Ratio of mean free path to molecular diameter	Very small	Very small	About 1	Very large
Path of movement of particle	About a definite point which is its average location		Random	Random
Translational energy*	Zero		Moderate	High
Compressibility	Almost none		Almost none	High $PV=P^1V^1$
Shape	Retains its shape unless finely divided		Takes the shape of its container as long as expansion is not required	Takes the shape and size of any container

*Energy of motion from one position to another.

PHASE CHANGES

Melting Point

The transformation of a solid into a liquid requires that the intracrystalline forces be broken and that the organization of the crystal be considerably scrambled. As a solid is heated, the particles move with greater velocity around their average position; therefore, their kinetic energy (KE $= \frac{1}{2} mv^2$) increases. When the kinetic energy exceeds the intracrystalline or lattice energy, the solid melts.

As a solid is heated, its temperature increases. When the temperature reaches the melting point, heat is supplied to break the intracrystalline forces until all the solid melts. No increase in temperature occurs until all the solid has melted (see Fig. 2-8).

The reverse process happens on solidification; that is, energy (heat) is given off to the environment as the intracrystalline forces are established. There may be some difference in the process because of the organization required in crystals. This difference which is evident as supercooling is explained in Chapter 11.

The stronger the intracrystalline forces (i.e., the higher the lattice energies), the more energy that must be

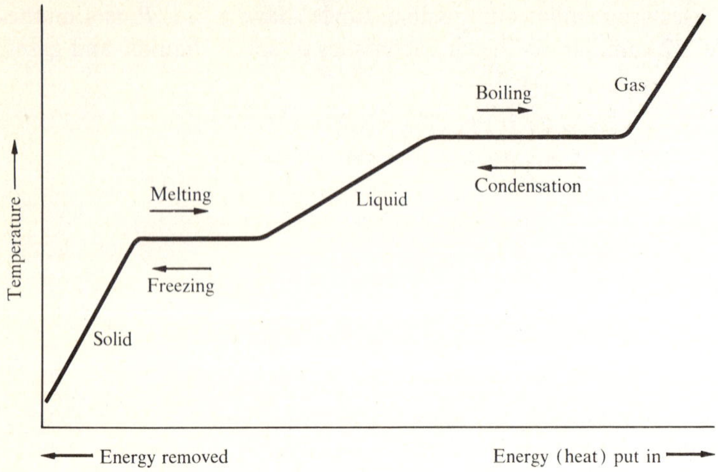

Figure 2-8 Energy and phase changes.

supplied to overcome them. Ionic solids, with strong ion-ion attractions, have very high lattice energies. Considerable energy is required to disrupt the ionic intracrystalline attractive forces and, as a result, ionic solids have very high melting points. If polar attractions are responsible for the principal intracrystalline forces, the forces are moderately strong; hence, polar compounds have intermediate melting points. If the only intracrystalline forces are from the weak London dispersion forces, relatively little energy is required to break these forces. As a result nonpolar compounds have low melting points.

Boiling, Evaporation, and Condensation

In liquids the same forces that are responsible for the intracrystalline forces cause attractions among the ions or molecules. For a given compound the forces holding a liquid together are somewhat weaker than the intracrystalline attractions but are still strong.

As a substance is heated, energy is transferred to it from the environment. As a substance absorbs more heat, the kinetic energy of its ions or molecules increases; the heat energy *is* the kinetic energy (of motion) of the ions or molecules that comprise the solid, liquid, or gas. For liquids (and gases) a good part of this energy is translational energy, that is, energy of movement from one point to another as opposed to rotational and vibrational energy which cause no change in position.

If a molecule near the surface of the liquid has enough kinetic energy to leave the surface of the liquid and overcome the attractive forces of the liquid, it will evaporate or move into the gas phase. Once in the gas phase the molecule may remain there or return to the liquid phase. The more molecules present above the liquid, the greater the possibility of collisions which can return the evaporating molecules to the liquid surface (Fig. 2-9). Therefore, the higher the pressure of the gas above a liquid surface, the slower the evaporation of liquid.

As the temperature of the liquid increases, the kinetic energy of its molecules increases. The more kinetic energy the molecules have, the faster they leave the surface of the liquid or evaporate. The faster they evaporate, the higher their concentration or partial pressure (which equals the vapor pressure of the liquid) in the gas phase. When the vapor pressure of the liquid equals the total pressure of the gases above it (760 mm or 1 atm at sea level in an open container), the evaporation of the liquid becomes very rapid and bubbles of vapor form inside the liquid, that is, the liquid boils.

When a liquid evaporates, ions or molecules leave the surface of the liquid and enter the gas or vapor phase. In the gas phase the molecules are very far apart, and attractive forces among the molecules[5] are very small

[5] Ions in the gas phase usually exist as ion pairs or other small groupings. The attraction between a positive and negative ion in such a pair is very large. Such an ionic grouping is the equivalent of a molecule. The only conditions where there are such molecular equivalents of ionic compounds such as Li^+Br^- and $Ca^{++}2Cl^-$ are in the gas phase and in solution in nonpolar solvents. In the gas phase, ions in ion pairs or other such molecular groupings have no significant attraction with ions outside the grouping. Thus, there are considerably fewer ion-ion attractions in the gas phase than occur in the liquid where an ion may be surrounded by and strongly attracted to six or more oppositely charged ions.

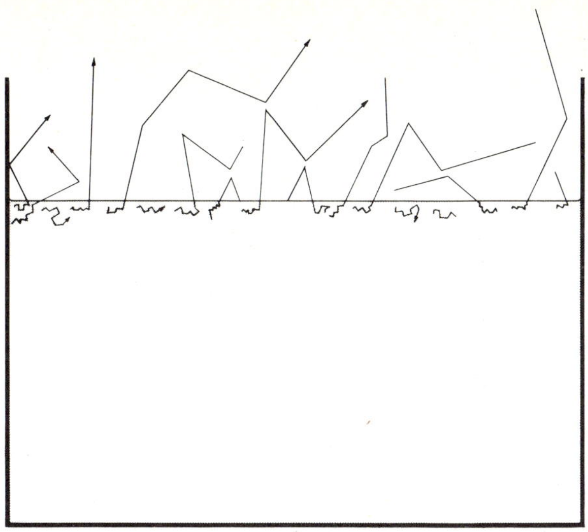

Figure 2-9 Illustration of movements of liquid molecules near the surface and their evaporation to the vapor phase.

and ideally are zero. As molecules or ions enter the gas phase, considerable energy must be supplied to pull the molecules apart and overcome the intermolecular attraction.

More energy is needed to evaporate liquids whose ions or molecules have strong attractions for each other. Compounds with strong intermolecular attractive forces have high boiling points. Similar to the melting points of solids the boiling points of liquids increase in the order nonpolar compounds < polar compounds < ionic compounds.

Effect of Molecular Weight

The discussions concerning the effect of intermolecular attractions on boiling and melting points assume comparisons of compounds with roughly the same molecular weights. The higher the molecular weight of polar and nonpolar compounds the higher the boiling and melting points. There are several reasons for this; among them are: (a) usually the higher the molecular weight, the larger the surface of the molecule and the greater the possibility of intermolecular attractions. This is especially important where London dispersion forces supply the principal attractions. (b) The heavier the molecule the more kinetic energy needed to give it the translation velocity necessary for it to escape from the crystal lattice or liquid surface.

There are many high molecular weight, nonpolar compounds that boil and melt at higher temperatures than low or moderate molecular weight, polar compounds. There are even a few high molecular weight, polar compounds that melt and boil above some ionic compounds. Molecular weight does not have as definite an effect on the melting and boiling points of ionic compounds. There are many high molecular weight, organic ionic compounds that have lower melting and boiling points than most low molecular weight, inorganic ionic compounds.

Other Physical Properties

The principles outlined in this chapter are used in later chapters to explain other physical properties such as solubility and density. In addition, later chapters illustrate and further define how structure may be used to predict and rationalize melting and boiling points.

EXERCISES

2-1. Indicate which of these compounds are nonpolar, polar, and ionic: (a) LiCl, CH_4, NH_3, H_2O; (b) $C_{10}H_{22}$, C_6H_{14}, C_3H_8, $C_4H_{10}O_4$, $C_3H_8O_3$, KBr.

2-2. Arrange the compounds in Exercise 2-1 in order of their boiling points, keeping series a and b separate.

2-3. What are the relative melting points of the compounds in the series a and b of Exercise 2-1?

3

Distillation and Boiling Points

INTRODUCTION

Distillation is a technique used to purify compounds, to separate mixtures (completely or partially), and to determine boiling points.

The boiling point is a constant that is useful for the identification of compounds and is often an indication of the purity of a liquid. A liquid that has a constant boiling point is pure unless it is an azeotropic mixture or a mixture of two or more compounds that have the same or nearly the same boiling points.

A liquid may be said to be constant boiling if its boiling point changes less than 2.5° while a given sample of the liquid is completely distilled. An azeotropic mixture is one that boils *either* above *or* below *all* the boiling points of the components of the mixture.

EXPERIMENTAL PROCEDURE

Macro Method

With a 25-ml flask, distilling head, condenser, adapter, and thermometer assemble the apparatus shown in Figure 3-1. Note carefully the position of the thermometer bulb in relation to the side arm of the distilling head. Placement of the thermometer bulb higher or lower is likely to cause the boiling point to be determined less accurately. Be sure the thermometer bulb is centered as well as possible in the distilling head. If it is placed carelessly, the thermometer bulb or the liquid that collects on it may touch the glass walls. Such contact (and sometimes even a near contact) may cause the observed boiling point to be significantly lower than the actual value.

Place a boiling chip and 12 ml of water in the flask. A boiling chip is a piece of very porous material that is added to ensure smooth boiling and prevent superheating of the liquid. Heat the mixture until it is boiling gently. Observe how the vapor condenses on the wall of the flask and then on the side of the distilling head. A ring of this condensing vapor travels up the distilling head as more heat is applied to the flask. **Warning:** each time heat is removed and the boiling stops for a minute or more add a new boiling chip *before* more heat is applied.

When the ring of vapor reaches the thermometer bulb, the mercury thread rises rapidly. If possible, the heat should be controlled so that the vapor refluxes (i.e., condenses and returns) on the thermometer bulb until the mercury thread stops rising. When the temperature stops rising, heat the flask more strongly until the liquid distills. Continue distilling until only about 2 ml of liquid remains in the flask. What and when were the highest and lowest temperatures during the distillation? If there was any sharp rise in the temperature at the very be-

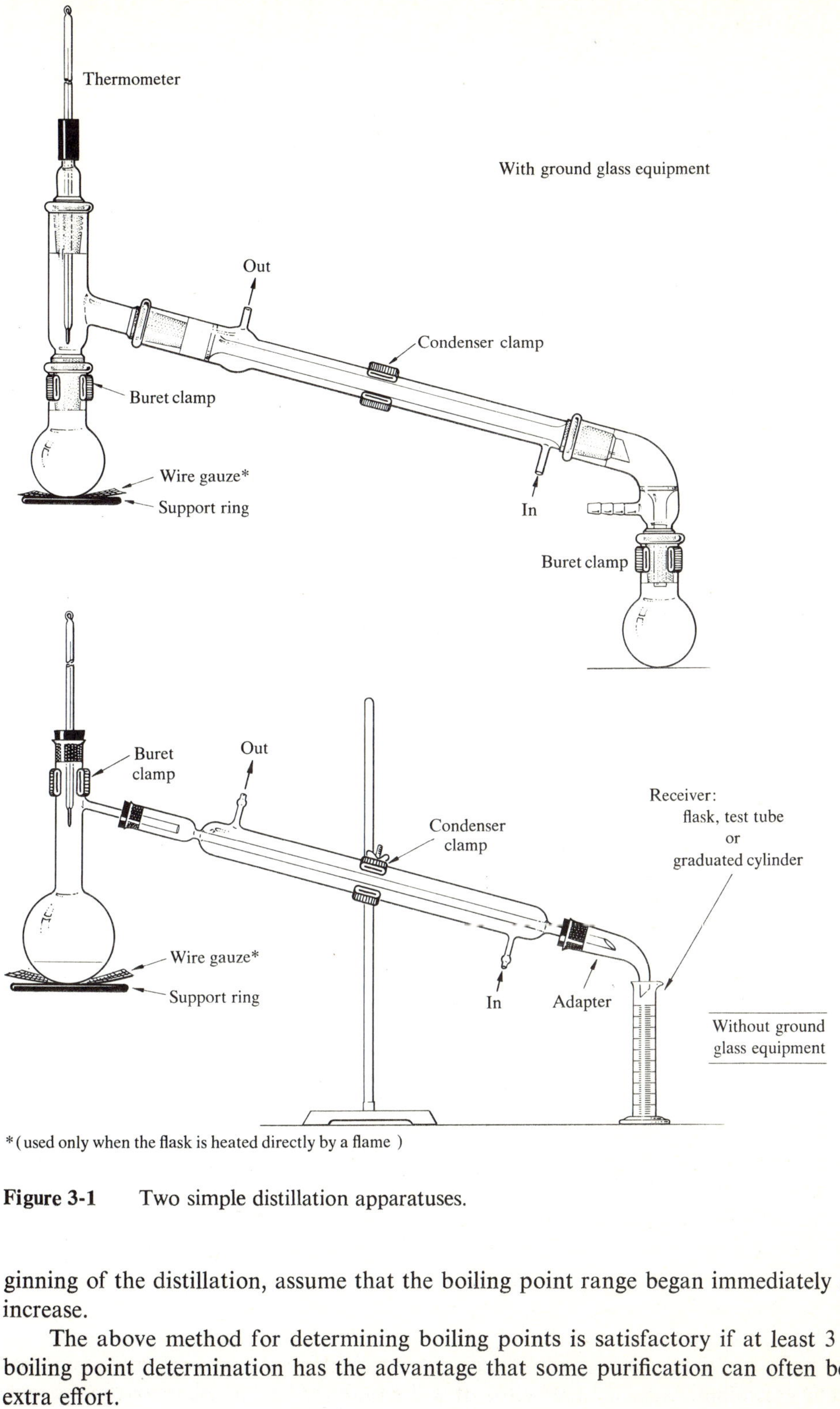

Figure 3-1 Two simple distillation apparatuses.

ginning of the distillation, assume that the boiling point range began immediately after this sharp temperature increase.

The above method for determining boiling points is satisfactory if at least 3 ml is used. This method of boiling point determination has the advantage that some purification can often be accomplished without any extra effort.

Micro Method

Sometimes there is an insufficient amount of liquid to determine its boiling point by the macro method. If less than 3 ml is available, a micro method should be used.

Into a 50 × 5-mm tube sealed at one end put 3 to 6 drops of liquid. Seal a capillary tube about 5 mm from one end. Place the capillary tube with the open end down into the larger tube. Fasten the bottom of the larger tube to the thermometer with a string or label placed above the oil level. The sealed end of the larger tube should be slightly above the bottom of the thermometer bulb (Fig. 3-2). This assembly is placed in an oil

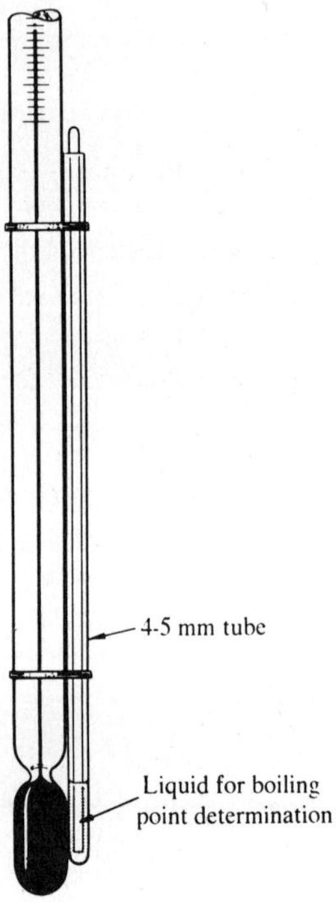

Figure 3-2 Micro boiling point determination.

bath. The bath is heated (with stirring if it is not a Thiele tube type). Bubbles will stream rapidly from the capillary tube when the temperature is a few degrees above the boiling point. Keep the liquid at this temperature for at least a minute. Remove the flame and allow the bath to cool 10 to 12°. What is in the capillary tube?

Heat again at a rate of 3 to 4° per minute for the first 6° and then at about 2° per minute until bubbles start to stream out again. What is the composition of these bubbles? Remove the flame and note the temperature at which the bubbling stops. This is the boiling point of the liquid. What is the pressure in the capillary tube at the boiling point? Why?

By following these instructions the initial boiling point of a boiling point range can be determined; however, the micro method will not help to determine the purity of a liquid mixture—a macro method often can.

Problem 3-1. How could the micro boiling point determination be modified to obtain information about the purity of a liquid?

UNKNOWN

Hand a clean, dry, and labeled test tube in to your instructor. It will be returned with one of the compounds listed in Table 3-1 in it. Determine the unknown's boiling point and report the identity of the unknown.

TABLE 3-1 BOILING POINT UNKNOWNS

Compound	B.p. (°C)	Compound	B.p. (°C)
Pentane	36	2-Methylheptane	118
Cyclopentane	50	Octane	125
2-Methylpentane	60	Ethylbenzene	135
Methyl alcohol	64.7	o-Xylene	142
2-Methylhexane	90	Isopropylbenzene	153
Heptane	98	p-Chlorotoluene	162
Toluene	111	Decane	173

If the barometric pressure of the laboratory is greater than 768 mm or less than 752 mm, the observed boiling point should be corrected to the boiling point at atmospheric pressure (760 mm). Tables used for correcting boiling points for many organic liquids are found in Appendix I. Tables for many other organic compounds are in the various chemical handbooks. If a boiling point correction table for the compound of interest cannot be found, use the boiling point correction table for a compound of similar polarity and boiling point.

Boiling points should be corrected for atmospheric pressure deviations of ± 8 mm or greater whenever the boiling point is used to identify an organic compound or for any other analytical purpose.

EXERCISES

3-1. What is superheating of a liquid? Redraw Figure 2-8 to show superheating of a liquid.

3-2. What is the purpose of a boiling chip? How does a boiling chip produce the desired effect?

3-3. Why should a new boiling chip be added when it is necessary to boil a liquid that has previously been heated with a boiling chip? If a liquid is heated near its boiling point without a boiling chip, why should a boiling chip never be added until the liquid has cooled?

4

Distillation of Mixtures

INTRODUCTION

The behavior of liquid mixtures on distillation can be better understood by considering ideal and nonideal relations among boiling point, vapor pressure, and composition.

DISCUSSION

Ideal Liquid Solutions

An ideal liquid mixture can be defined as one that obeys Raoult's law. This law states that the partial pressure P_x of a component x is proportional to N_x, the mole fraction of x present. Or

$$P_x = P_x^{\circ} N_x , \tag{4-1}$$

where P_x° is the vapor pressure of pure x.

The vapor pressure of two- and three-component ideal mixtures are given by

$$P = P_x^{\circ} N_x + P_y^{\circ} N_y , \tag{4-2a}$$

$$P = P_x^{\circ} N_x + P_y^{\circ} N_y + P_z^{\circ} N_z . \tag{4-2b}$$

Most mixtures do not behave completely as ideal mixtures. Most mixtures, however, that are not azeotropic follow Raoult's law sufficiently closely so that Equations 4-1 and 4-2 can be used.

A graph of the relation of the vapor pressure and composition of a binary mixture, methanol and water, is given in Figure 4-1. The heavy lines show the actual relations; the dotted lines show the relations an ideal binary mixture would have.

The composition of a vapor can be calculated from Dalton's law. If the vapor is in equilibrium with the liquid, then

$$M_x = \frac{P_x}{P_x + P_y + P_z + \cdots} , \tag{4-3}$$

where M_x is the mole fraction of component x in the vapor phase and $P_x + P_y + P_z + \cdots$ is the total pressure.

By combining Raoult's and Dalton's laws, Equations 4-4a and b are obtained for the binary mixture methanol and water at atmospheric pressure:

$$M_{CH_3OH} = \frac{P_{CH_3OH}N_{CH_3OH}}{760}, \quad (4\text{-}4a)$$

$$M_{H_2O} = \frac{P_{H_2O}N_{H_2O}}{760}. \quad (4\text{-}4b)$$

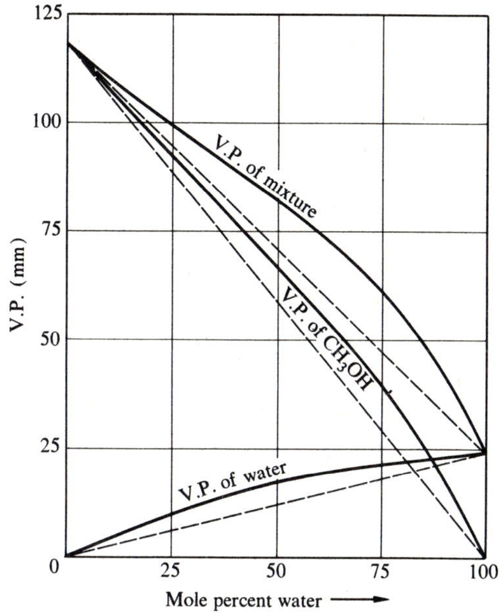

Figure 4-1 Vapor pressure-composition diagram of methanol-water mixture at 25 °C. Solid lines: actual vapor pressure; dotted lines: ideal vapor pressure from Raoult's law.

Problem 4-1. (a) With Equations 4-4a and b and Figure 4-1, calculate the amount (in mole fractions) of methanol and water in the vapor phase above the water-methanol mixtures that have 0.0, 0.1, 0.3, 0.5, 0.7, 0.9, and 1.0 mole fractions of methanol at 760 mm (atmospheric pressure) and 25°. At 400 mm and 25°. (b) Is the ratio of methanol to water higher in the vapor or liquid phase? Could this finding be generalized to all binary liquid mixtures that approximately follow Raoult's law?

TABLE 4-1 VAPOR PRESSURE-TEMPERATURE RELATIONS OF METHANOL, ETHANOL, AND WATER

Temperature (°C)	water	Vapor pressure (mm Hg) for methanol	ethanol
110	1074.5	3562	2361
105	901	2963	1976
100	760	2625	1694
95	634	2128	1434
90	526	1895	1187
85	433.5	1637	986
80	355	1341	812.5
78.32	332	1291	760
75	289	1013	666
70	234	928	542.5
65	187.5	769	449
64.7	185	760	441
61.2	158	656	372
60	149.5	625	353
56.3	125	532	296
55	118	508	281
50	92.5	406	222

As defined in Chapter 2 the boiling point of a substance is the temperature at which its vapor pressure equals the environment's pressure. If the pressure of the environment is 1 atmosphere or 760 mm, a liquid will boil when its vapor pressure reaches 760 mm. Any binary mixture that follows Raoult's law has a vapor pressure between that of the two components; hence, the mixture's boiling point is between that of the two components.

Figure 4-2 shows the relation between boiling point and the composition of the vapor and liquid of carbon tetrachloride-toluene mixtures. This mixture was chosen because it closely approaches ideal behavior as described by Raoult's law.

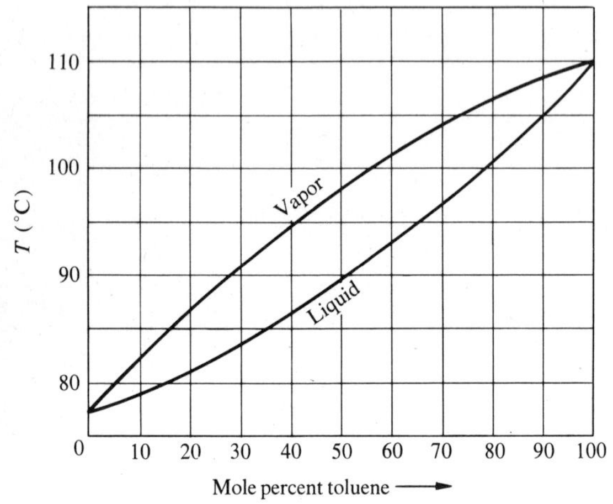

Figure 4-2 Boiling point-composition diagram of carbon tetrachloride-toluene mixtures.

A horizontal line drawn from 95° intercepts the vapor line at 43 mole percent of toluene. This indicates that there are 43 molecules of toluene in the vapor for every 57 molecules of carbon tetrachloride if the boiling point of the mixture is 95°. As the line at 95° is followed to the right-hand side of the chart, it crosses the liquid line at 65 mole % toluene, the amount of toluene in the liquid when the boiling point is 95°.

Problem 4-2. Using Figure 4-2 calculate the amount of toluene and carbon tetrachloride in a boiling mixture of the two in mole fractions and weight percents in the liquid and vapor phases at 80° and 100°. Compare the answer to that of Problem 4-1b.

Problem 4-3. Using Table 4-1 determine the composition of the vapor over boiling methanol-water mixtures when the temperature is 64.7°, 70°, 80°, 85°, 90°, and 100°. Assume 1 atm pressure.

Problem 4-4. Figure 4-3 is the boiling point composition diagram for methanol and water similar to Figure 4-2 for toluene and carbon tetrachloride except that the vapor equilibrium line has been omitted From the answer to Problem 4-3 plot the vapor equilibrium line for methanol-water on Figure 4-3.

Often mixtures of miscible liquids deviate so much from ideal behavior that their composition boiling point diagrams have a minimum or a maximum (see Figs. 4-4 to 4-6). Such a mixture is said to be azeotropic.

When a mixture with a maximum boiling point azeotrope is distilled, the temperature increases until the maximum is reached. At this temperature the compositions of the vapor and the liquid phases are identical, and the remainder of the mixture distills at this temperature. The mixture with this composition is said to be the azeotropic mixture. When a mixture with a minimum boiling point azeotrope is distilled, the initial boiling point is the azeotropic minimum. The compositions of the liquid and vapor phases are not identical unless the liquid has the same composition as the azeotrope. The liquid distills as the azeotrope until one of the components has completely distilled; then the remaining component distills.

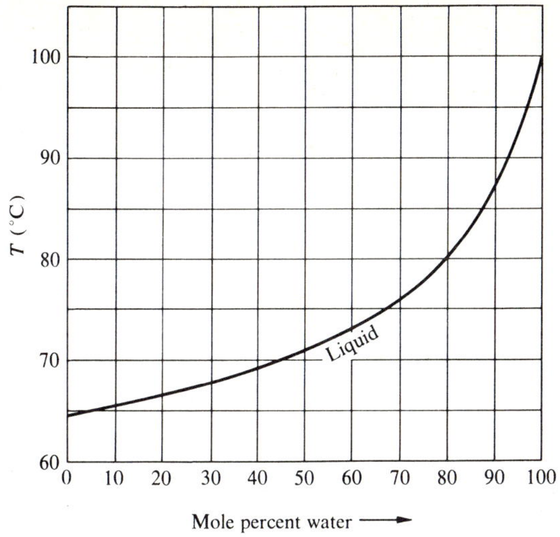

Figure 4-3 Boiling point-composition diagram of methanol-water mixtures.

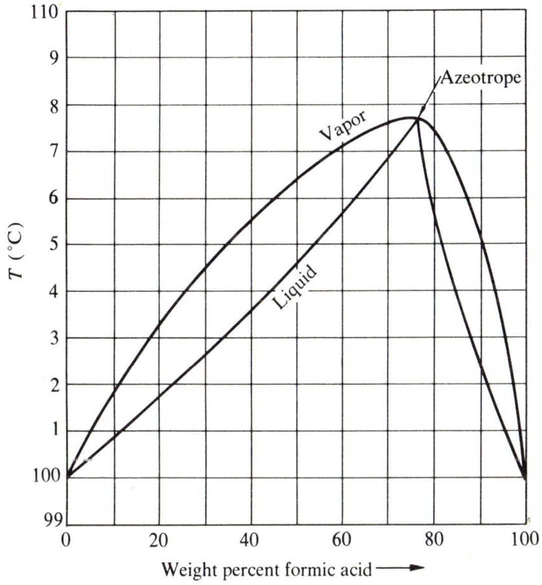

Figure 4-4 Boiling point-composition diagram of formic acid-water mixtures.

For simplicity this discussion assumes that there is a wide difference between the boiling points of the azeotrope and the components. Actually this rarely occurs, and the mixture distills as if it were a mixture of two liquids. One is the liquid in excess of that required for the formation of the azeotrope and the other is a liquid with a boiling point the same as the azeotropic mixture. For example, a mixture of 80 wt % of ethanol and 20 wt % water (Fig. 4-6) behaves as if it were a mixture of 83.77 wt % of a liquid boiling at 78.4° (4.5% water-95.5% ethanol azeotrope) and 16.23 wt % of water. If neither liquid is in excess, the mixture boils as if it were a pure compound.

Problem 4-5. Refer to Figures 4-4 and 4-5. What is the composition of a formic acid-water liquid mixture that boils at 107.6°? What is the composition of the vapor that is in equilibrium with the liquid at 102.3°? What is the boiling point of a 50% solution of formic acid and water? What is the composition of the vapor distilling from this liquid? Describe what happens to the temperature and the composition of the liquid and vapor as such a mixture is distilled.

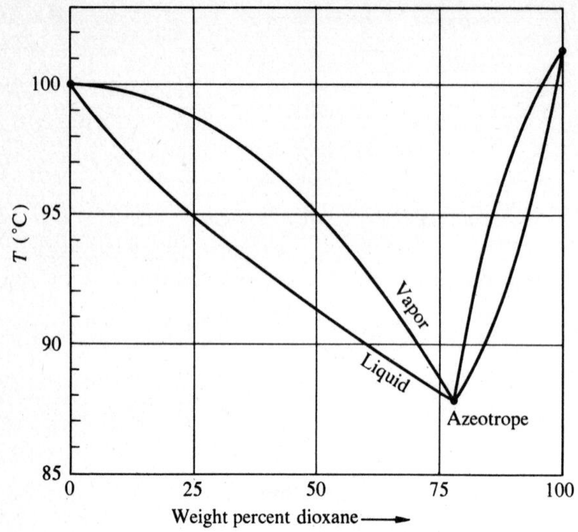

Figure 4-5 Boiling point-composition diagram of dioxane-water mixtures.

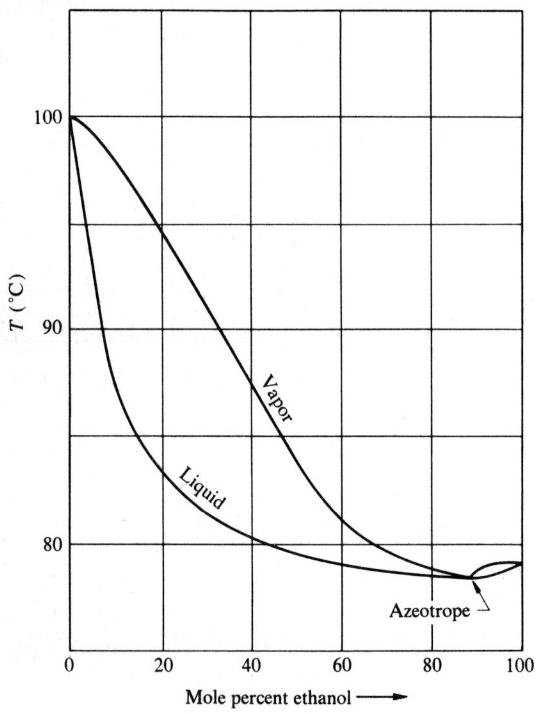

Figure 4-6 Boiling point-composition diagram of ethanol-water mixtures.

Problem 4-6. What is the boiling point of a 5% water in dioxane solution? What is the composition of the vapor distilling from this liquid? Describe what happens to the temperature and the composition of the liquid and vapor as such a mixture is distilled. What is the boiling point of a solution of 40 wt % of water and 60 wt % of dioxane? What is the composition of the vapor distilling from this liquid? Describe what happens to the temperature and composition of the liquid and vapor as such a mixture is distilled.

EXPERIMENTAL PROCEDURE

Put 25 ml of methanol, 25 ml of water, and a boiling chip into a 100-ml round bottom flask. Use this flask for a distillation apparatus similar to that shown in Figure 3-1 except that a 10-ml graduate should be placed under the adapter.

Distil the mixture slowly and collect nine 5-ml fractions. Carefully observe the boiling range for each fraction. Assume the tenth or undistilled fraction has a boiling point range from the highest temperature of the last (ninth) fraction that was distilled to the boiling point of water. Record the boiling point data in a table similar to Table 4-2. From this table plot a graph of boiling point versus milliliters distilled using the boiling point of methanol (64.7°) for the temperature of 0 ml distilled.

Warning: methanol is an inflammable, toxic liquid. All flammable liquids should be treated carefully and should not be allowed to evaporate into the air. See Chapter 1 for more detailed precautions for the handling of dangerous substances.

TABLE 4-2 DISTILLATES FROM METHANOL-WATER MIXTURES

Fraction number	ml collected (x-axis)	B.p. (°C) (y-axis)	Composition mole% MeOH	Mole% H_2O
0	0	64.7*		
1	5			
2	10			
3	15			
4	20			
5	25			
6	30			
7	35			
8	40			
9	45			
10	50	100		

*May vary

EXERCISES

4-1. From the solution to Problem 4-4 estimate the composition of the distillate at each of the eleven temperatures you recorded in Table 4-2. Record your answers in the table you constructed. Note that the temperatures you observed are those of the vapor.

4-2. Under what conditions may a liquid have a narrow boiling point range? Summarize the relationship between boiling point range and purity of a liquid.

5

Fractional Distillation

INTRODUCTION

A liquid mixture can be separated into its components (unless they form azeotropes) by repeated distillation or by use of a fractionating column.

EXPERIMENTAL PROCEDURE

The distillations in only this part of the experiment are to be done with partners. Refer to Figure 5-1 to follow the procedure and to divide the operation between the two partners.

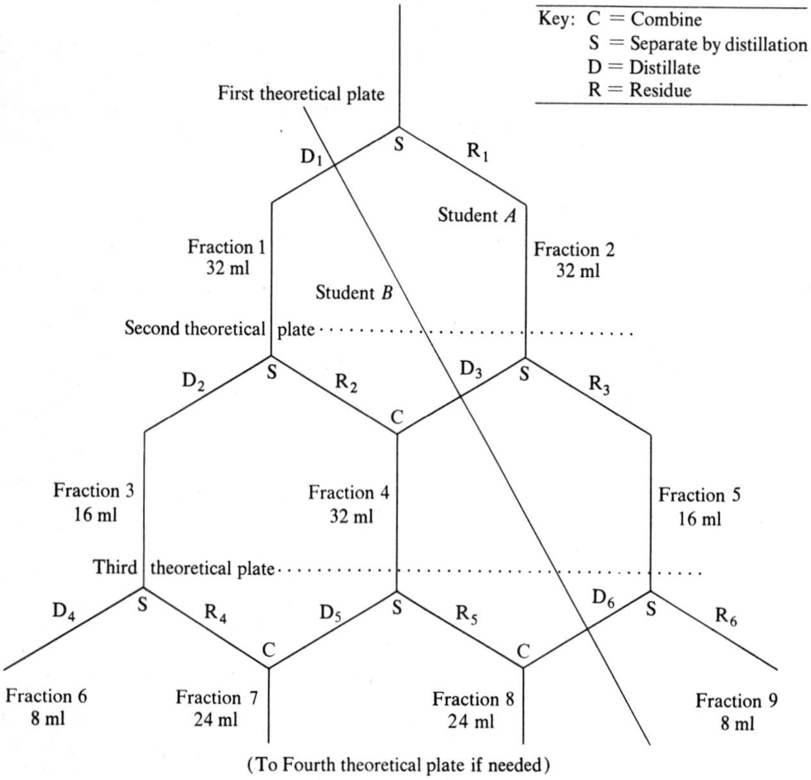

Figure 5-1 Diagram of fractional distillation done with simple distillation apparatus.

Place 64 ml of a 50% solution of methanol in water and a boiling chip into a distillation apparatus (100-ml distilling flask) with a graduated cylinder as a receiver (Fig. 3-1). Heat the mixture until 32 ml of distillate collects. Record carefully the temperature range of each distillation.

The student (A) who did the first distillation takes the residue and collects 16 ml by distillation. His partner (B) takes the distillate and places it in a 50-ml flask and distils 16 ml. Student B takes the residue (R_2) from his first distillation, combines it with the distillate (D_3) from A's second distillation, and distils this combination into two equal fractions (D_5 and R_5).

Refer to Figure 5-1 for completion of the distillation. Each student does a total of three distillations. Record the boiling range for each distillate in your laboratory notebook in a table similar to Table 5-1.

TABLE 5-1 BOILING POINT RANGES OF DISTILLATES

Distillate	B.p. range (°C)	Estimated percent composition
D_1		
D_2		
D_3		
D_4		
D_5		
D_6		

From the appropriate data in Chapter 4 estimate the percentage composition of each of the distillates and enter this in your table.

Problem 5-1. Which distillate has the highest concentration of methanol? Which distillate has the highest concentration of water? Which *fraction* as numbered on the chart has the highest concentration of methanol? Which fraction has the highest concentration of water?

Problem 5-2. Which distillate has a composition most similar to fraction 2? If fraction 7 were distilled into two equal fractions, what would the composition of the distillate be? (*Do not* distil fraction 7 to solve this question.)

Problem 5-3. Describe what operations would have to be done to fractions 7, 8, and 9 to obtain at least 9 ml more of a fraction that has as high a concentration of methanol as fraction 6.

DISCUSSION

The process just described is a fractional distillation. Complex mixtures can be separated by use of this procedure. Each equilibrium of a liquid with its vapor is a *theoretical plate.* The more theoretical plates, the better the separation that is possible. If the components of a mixture have widely different boiling points, only a small number of theoretical plates (as few as one) is required to separate them. If the boiling points of the components are only a degree or so apart, a hundred or more theoretical plates are needed for separation.

Problem 5-4. In the fractional distillation described in the above section, six distillations were required for three theoretical plates. How many distillations are required for four theoretical plates? How many distillations are required for five theoretical plates?

Figure 5-2 is a vapor-liquid diagram for methanol and water. A to B is the first distillation. B to C represents the condensation of D_1. The composition of D_1 is given by points B and C. Line CE represents the second distillation to give D_2, and line EF the condensation of D_2. Note that this applies only to the first few drops of distillate. If any significant volume is distilled, the composition of mixture changes to that with a higher percentage of the least volatile component because the more volatile component is removed. For example, when a 50/50 methanol-water mixture is distilled, the composition of the mixture changes to contain 55% or even some higher fraction of water. Figure 5-2 can be used to determine the *approximate* composition of the distillate and residue if their relative amounts are known. For example, if 50% is distilled A′, the midway point between A and B is used as a reference; if 25% is distilled A″, 25% of the distance from B to A is the reference point. Therefore, to determine the approximate composition of D_1 from distillation of half of the 50/50 mixture of methanol and water, use horizontal line GH through A′. G is the average composition (63% methanol) of the vapor from A, and H is the condensed distillate D_1 and is the starting point (similar to A) for determining the composition of the distillate and residue from distillation of D_1.

Problem 5-5. Copy Figure 5-2 in your notebook. Diagram the distillation that gives D_4 on this figure. What is the composition of D_4 as shown by Figure 5-2?

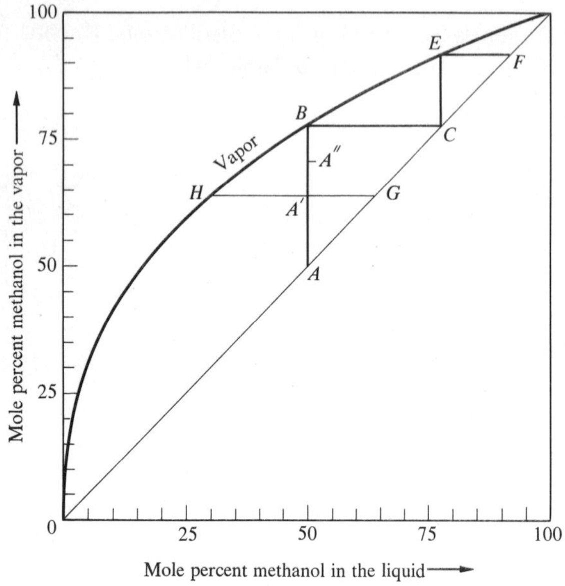

Figure 5-2 Vapor-liquid diagram for methanol and water.

Problem 5-6 Show on Figure 5-2 how the composition of the residue R_1 can be calculated? In the same manner calculate the composition of R_3, D_3, and D_6. Do the compositions obtained this way for R_3, D_3, and D_6 agree with what could be predicted from Figure 5-1? Do they agree with the estimations from the observed boiling points (recorded in Table 5-1)?

Even a small number of theoretical plates requires a considerable effort with this method. Fortunately, fractional distillation can be mechanized by distilling through a *fractionating column.*

There are at least three types of fractionating columns: bubble cap columns which resemble a series of distilling flasks more than other columns do (see Fig. 5-3), spinning band columns, and packed columns. The packed columns are the most common and are the ones to be used in the experiments described in this book. The packing furnishes a large surface area on which vapor and liquid can be in equilibrium.

For maximum effectiveness there should be no heat transfer to or from the column as this disturbs the equilibrium. The column packing is heated either by the refluxing vapor or by external heaters. For complete separation of two components it is necessary (but not sufficient for compounds that boil within a few degrees of each other) that the bottom of the column has the same temperature as the boiling point of the less volatile component and the top of the column has the same temperature as the boiling point of the more volatile component.

As vapor ascends the column, it is continuously enriched with the more volatile component from the descending condensed liquid and loses the less volatile to the descending liquid. Ideally, the two phases are in equilibrium with each other for the entire length of the column. Optimum separation can be obtained only if the equilibrium on the packing surface is not disturbed significantly. To achieve this, considerably more distillate must be condensed and returned down the column (refluxed) than is removed. The ratio of material returned down the column to the amount taken off is called the *reflux ratio.*

Figure 5-4 shows a typical distillation head that is used on top of large fractionating columns. One of the purposes of such heads is to accurately control and adjust the reflux ratio. For good separation reflux ratios as high as 100 to 1 are used. In the experiments described in this text such complicated apparatus is not used, and the reflux ratio can be controlled only by careful heating of the boiling flask.

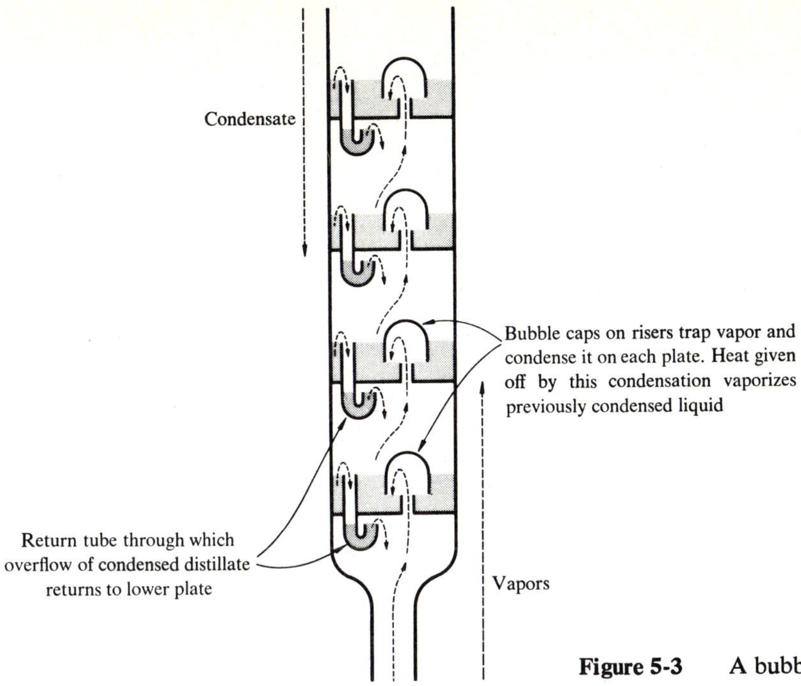

Figure 5-3 A bubble cap column.

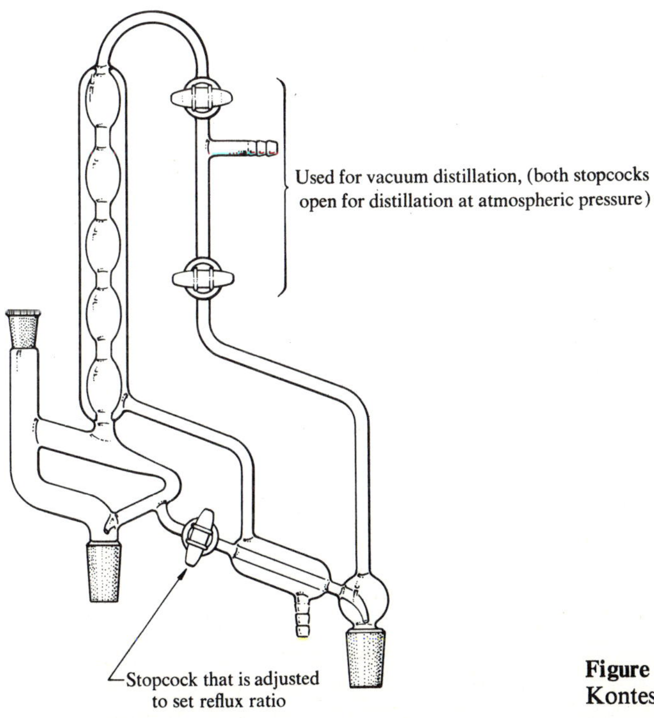

Figure 5-4 A distillation head. Courtesy of Kontes Glass.

EXPERIMENTAL PROCEDURE

Cut glass tubing into approximately half-inch lengths. Pack the fractionating column with the cut tubing. Allow about one-fourth to one-half inch clearance between the bottom of the thermometer bulb and the packing.

Assemble the distillation apparatus shown in Figure 5-5. The top of the thermometer bulb should be level with the bottom of the side arm leading to the condenser. The remarks in Chapter 3 about centering the thermometer bulb apply here.

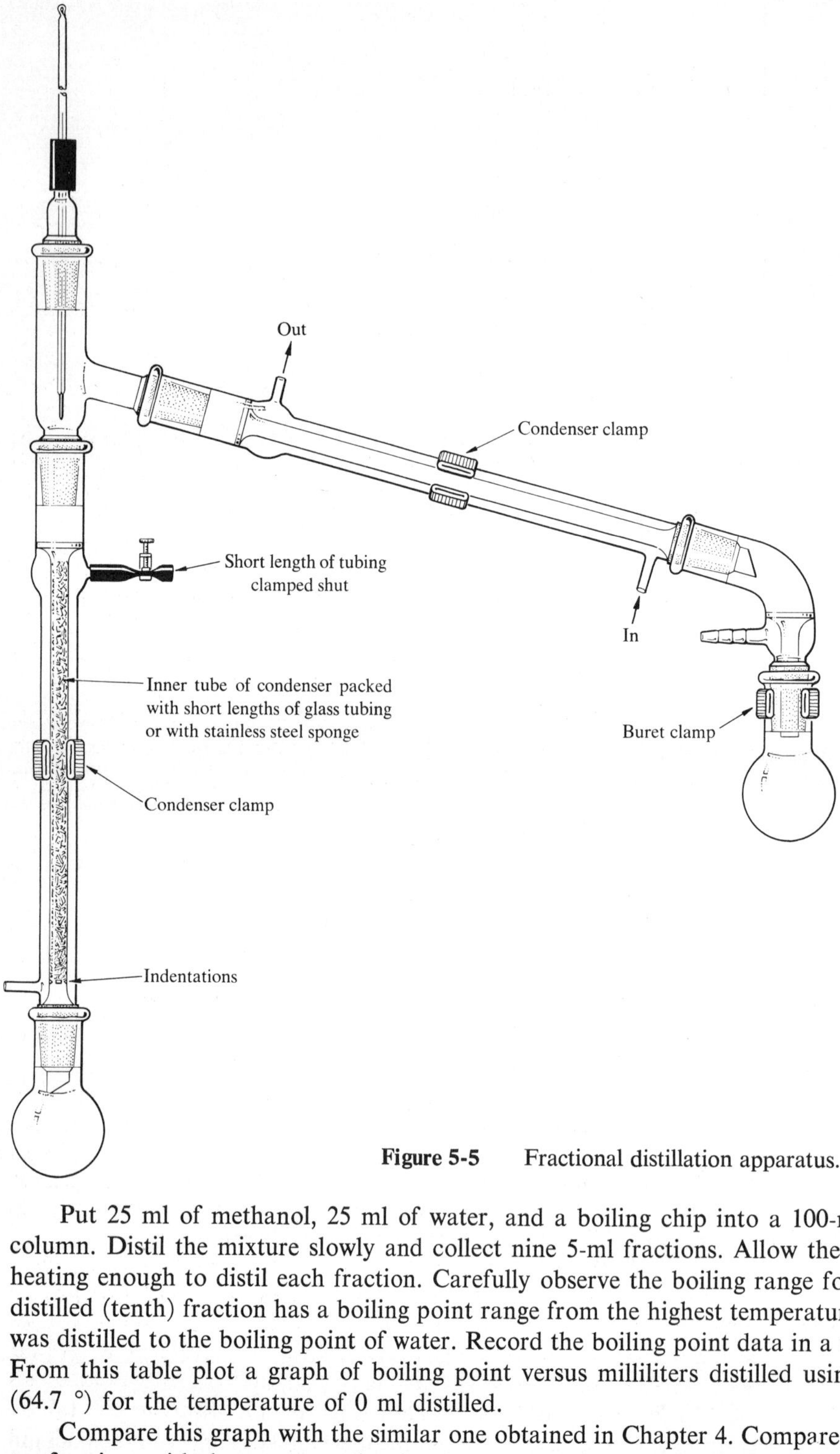

Figure 5-5 Fractional distillation apparatus.

Put 25 ml of methanol, 25 ml of water, and a boiling chip into a 100-ml round bottom flask under the column. Distil the mixture slowly and collect nine 5-ml fractions. Allow the mixture to reflux a little before heating enough to distil each fraction. Carefully observe the boiling range for each fraction. Assume the undistilled (tenth) fraction has a boiling point range from the highest temperature of the last (ninth) fraction that was distilled to the boiling point of water. Record the boiling point data in a table similar to that of Table 4-2. From this table plot a graph of boiling point versus milliliters distilled using the boiling point of methanol (64.7 °) for the temperature of 0 ml distilled.

Compare this graph with the similar one obtained in Chapter 4. Compare the estimated composition of the ten fractions with the corresponding fractions obtained from the experiment in Chapter 4 and with the estimated composition of the various fractions obtained in the fractional distillation (Table 5-1) done as partners.

6

Extraction

INTRODUCTION

Extraction is another widely used technique for the separation, isolation, and purification of organic compounds. Extraction is based upon the higher solubility of one or more of the compounds of a mixture. This experiment demonstrates liquid-liquid extraction which depends upon the higher solubility of a solute in one of an immiscible solvent pair.

The simplest technique usually involves dissolving the organic substance in one of the solvents or obtaining such a solution from a reaction mixture, natural source, etc. This solution is poured into a separatory funnel and the other solvent is added. The mixture is shaken thoroughly and allowed to separate. An equilibrium of the solute or solutes is established between the two phases. The solute (or one or more of the solutes) is removed by the added solvent.

Extraction may be considered as similar to distillation. Distillation involves equilibria between two phases, gas and liquid; extraction involves equilibria between two phases, usually two immiscible liquids. In distillation the more volatile components are removed by evaporation into the gas phase. In extraction components that are more soluble in one liquid phase concentrate in that phase. Two substances can be completely separated by a single distillation only if their boiling points are very different. Similarly, two substances can be completely separated by a single extraction only if their solubilities are very different.

DISCUSSION

Solubility

Liquid-liquid extraction requires two solvents that are relatively insoluble in each other. In addition, the substances that are to be separated must have different solubilities in at least one of the liquid phases.

The fact that some liquids are soluble in certain liquids and insoluble in others is readily established without entering a chemistry laboratory. Gasoline, other petroleum products, and cooking oils such as corn oil are insoluble in water. Alcohol, glycerine, and ethylene glycol (permanent antifreeze) are very soluble in water. Gasoline and cooking oils are soluble in carbon tetrachloride and in ethyl alcohol.

The solubility of one liquid in another can be predicted rather well by considering the intermolecular forces and the structures of compounds involved. The molecules in a liquid touch or almost touch each other and are attracted to each other by fairly strong forces. When one liquid dissolves in another, the molecules of each liquid slip between the other. Figure 6-1 illustrates this for two liquids partially soluble in each other and two liquids completely soluble in each other. If the molecules of liquid *A* have very strong attractions for each other but not for molecules of liquid *B*, then not much *B* will dissolve in *A* because considerable energy would be needed to separate the molecules of *A* to allow molecules of *B* to slip between them.

Figure 6-1 Solubility of liquids in each other.

An example of this insolubility is a mixture of hexane and water. The water molecules have a very strong type of dipole-dipole attraction (called a hydrogen bond) for each other. In order to dissolve in water, molecules of hexane must break the strong dipole attractions and come between the water molecules. The hexane and water molecules attract each other only by the dipole-induced dipole attractions. The formation of these relatively weak attractions cannot supply enough energy to break the dipole attractions among the water molecules. As a result hexane does not dissolve in water.

To dissolve in hexane a water molecule must be separated from the other water molecules. A similar separation occurs when water molecules evaporate. The energy required to dissolve a given quantity of water in hexane is roughly the same as the energy required to evaporate the same quantity of water. The weak dipole-induced dipole attractions between water and hexane do not supply sufficient energy to overcome the attraction of the water molecules for one another. As a result very little water dissolves in hexane.

Problem 6-1. Are hexane and water more or less soluble in each other at higher temperatures?

Molecules of compounds such as methanol, ethanol, ethylene glycol, acetone, and methyl ethyl ketone (Fig. 6-2) form strong dipole attractions with each other and with water because they have polar groups and the nonpolar part of the molecules is not very large. These polar molecules slip between the water molecules forming strong dipole-dipole attractions with them. These attractions replace (at least partially) the strong attractions that the water molecules have for each other. For this reason polar organic compounds as those in Figure 6-2 are soluble in water.

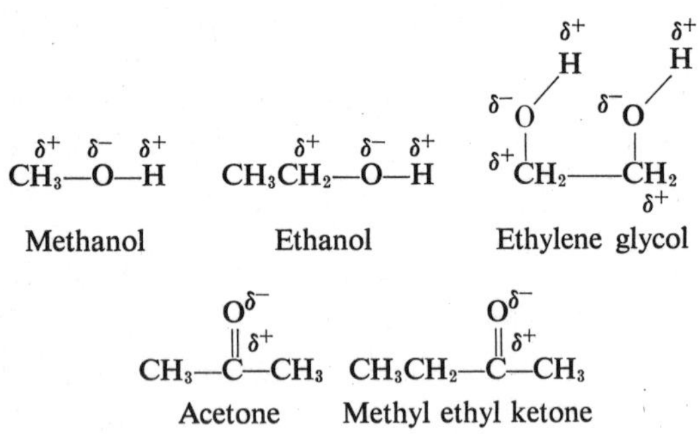

Figure 6-2 Some polar organic molecules.

Similarly, water dissolves in polar organic liquids such as acetone and methanol because the new dipole-dipole attractions supply enough energy to break the attractions between some of the water molecules. Two nonpolar liquids such as hexane and benzene are soluble in each other because the weak London dispersion forces among the hexane molecules are easily replaced with other weak London dispersion forces between hexane and benzene molecules.

A general rule for the solubility of substances in each other is *like dissolves like,* that is, polar compounds dissolve polar compounds and nonpolar compounds dissolve nonpolar compounds. Compounds that have polar and nonpolar moieties such as acetone and ethanol are often soluble in both polar and nonpolar liquids.

Distribution Coefficients

The distribution of a solute (or solutes) between two immiscible[1] solvents (A and O) at equilibrium and at the stated temperature is given by the distribution coefficient K:

$$K = \frac{\text{Conc. of } S \text{ in } O}{\text{Conc. of } S \text{ in } A} = \frac{S_o/O \text{ in ml}}{S_a/A \text{ in ml}},$$

where S_o is the weight in grams of S in O, and S_a is the weight in grams of S in A.

The distribution coefficient usually can be closely approximated by taking the ratio of solubilities of the substance in the two solvents:

$$K = \frac{\text{Sol. of } S \text{ in } O}{\text{Sol. of } S \text{ in } A}.$$

The less the mutual solubility of the two liquids in question, the closer such a calculated distribution coefficient is to the actual one. However, even with such solvents as ether and water which show significant solubility in each other, distribution coefficients calculated from solubility data are usually accurate enough to use for most calculations.

EXPERIMENTAL PROCEDURE

Use of Separatory Funnel

To prevent possible spillages and other accidents the procedure described below should be followed for all extractions. Grease *lightly* the stopcock and ground glass stopper (p. 7) of a 125-ml separatory funnel. Dissolve 9.0 g of butyric acid in 45 ml of ether. (**Danger! Ether is very flammable. There must be no flames in the laboratory while ether is being used. All procedures described in Chapter 1 for handling and disposing of flammable liquids are to be consistently and carefully followed.**) Pour the mixture into a 125-ml separatory funnel. Use a conical funnel to aid in the pouring.[2] Add 45 ml of water and while the separatory funnel is still supported by the ring stand, swirl it gently with its top off as in Figure 6-3.[3] If much gas is evolved, keep swirling the separatory funnel until the gas evolution has decreased significantly.

Remove the separatory funnel from the support and shake the funnel as shown in Figure 6-4(a). Be certain the top is held on and that the stopcock is closed. After *one* shake vent the separatory funnel by opening the stopcock as shown in Figure 6-4(b). If gas is heard escaping, close the stopcock, shake a few more times, and vent the funnel again. Repeat this until no more gas hisses out, and then shake the funnel vigorously.

Return the separatory funnel to its support.[4] Set the top in loosely (Fig. 6-5) so that the separatory funnel is not completely closed. Allow the two layers to separate and drain the bottom, aqueous layer into a 50-ml

[1] Actually very few liquids are *completely* immiscible in each other, but for most purposes nonpolar organic solvents such as benzene and chloroform may be considered immiscible with water.

[2] One of the purposes of the conical funnel is to prevent the washing of stopcock grease from the ground glass surface with resulting contamination of the ether and possible freezing of the stopper.

[3] This step helps to prevent high pressure from building up in the separatory funnel on the first shake. This pressure could become very high if the water were much warmer than the ether, if gas were formed, or if there were considerable heat of reaction or solution when the two phases are brought together. If both liquids have low vapor pressure (b.p. above 95°) and no gas can be formed, this first step can be omitted.

[4] If a separatory funnel with solution in it is to be left unattended, a beaker should be placed beneath it (Figs. 6-3 and 6-5) to catch any liquid that may drip through. The beaker should be wider than the stopcock so that any liquid leaking out of the ends of the stopcock will also be caught by the beaker.

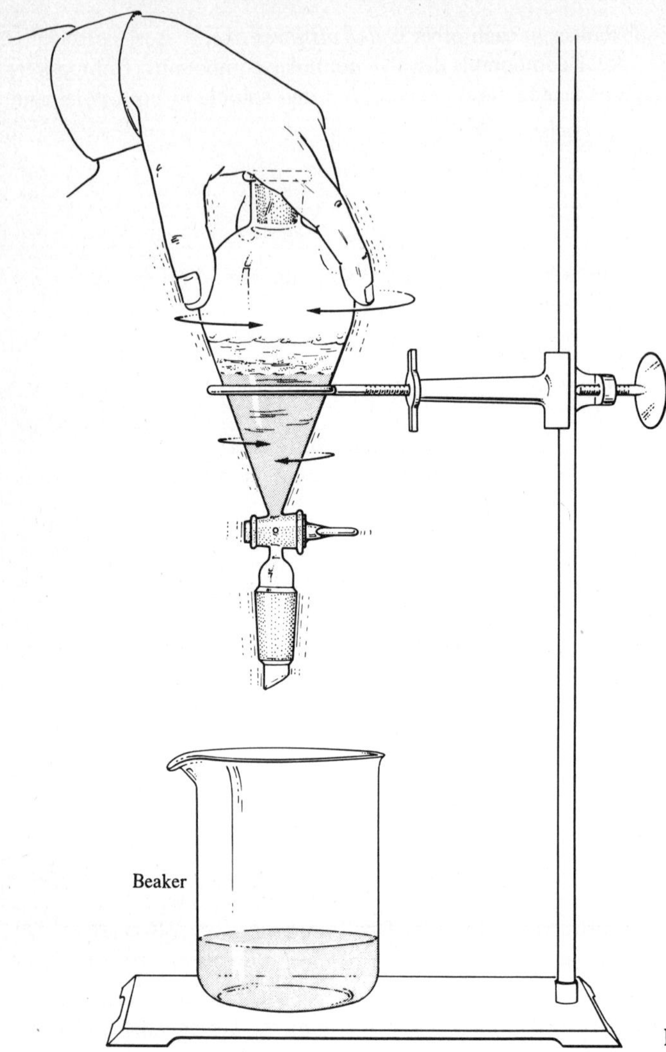

Figure 6-3 Swirling of separatory funnel.

Erlenmeyer flask. When the ether reaches the stopcock, close the stopcock and shake to remove any aqueous globules. Allow the mixture to settle and drain off any water that has collected. Drain all water from the bore of the stopcock, taking care to make a sharp cut between the ether and water layers. Return the top to the separatory funnel, tip the funnel, and carefully open the stopcock so that the ether from the bore runs back into the separatory funnel. Drain a 15-ml aliquot of the ether into a small graduated cylinder.

The two phases in a separatory funnel can also be removed by draining the lower phase as outlined above and pouring the upper phase out the top of the separatory funnel. If the upper phase is nonaqueous, such as benzene or ether, this method is advantageous because the nonpolar phase has less contact with the stopcock and will dissolve less stopcock lubricant. If this method is used, the ground glass surface inside the top of the separatory funnel should be thoroughly cleaned of any lubricant before pouring the liquid out the top. For this separation, however, the first method is preferred because it is easier to measure exactly 15 ml by draining through the stopcock. Cautiously smell the water layer. Is acid present? Verify your answer by testing the water layer with pH paper. Cautiously smell the ether layer and test it with *wet* pH paper. Is butyric acid present? Verify by evaporating a 15-ml aliquot of the ether extract on a steam bath. Use a 25-ml flask that has been tared with a boiling chip in it.

Caution! If 20 ml or less of ether is to be removed and an *efficient* hood can be used, simple evaporation will be sufficient. However, if an efficient hood is not available or large quantities of a volatile solvent are to

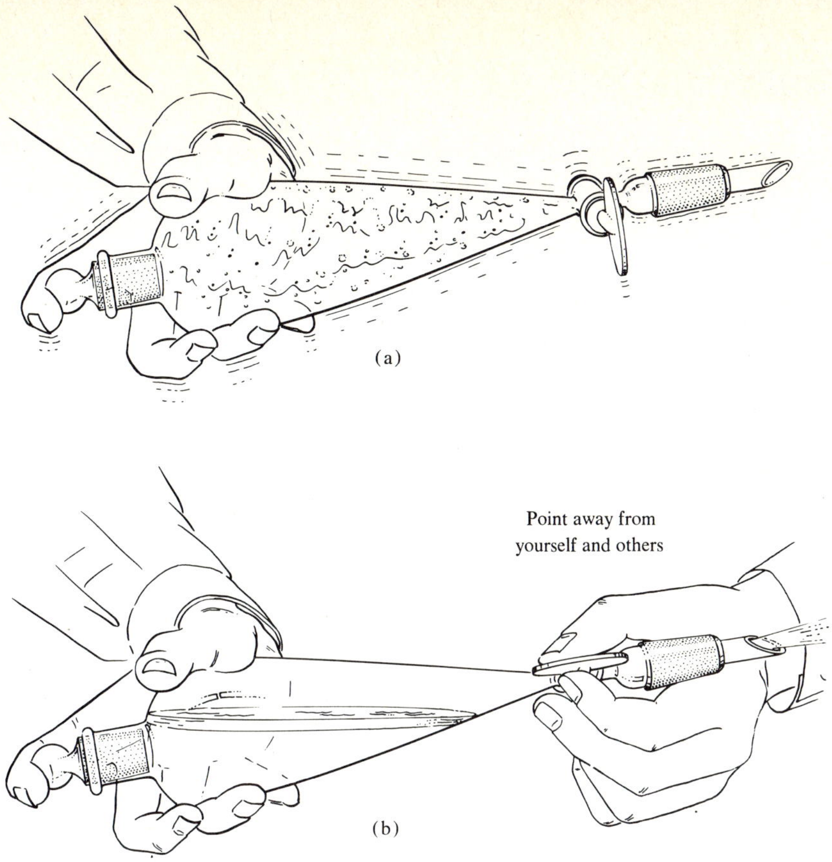

Figure 6-4 Shaking and venting of a separatory funnel.

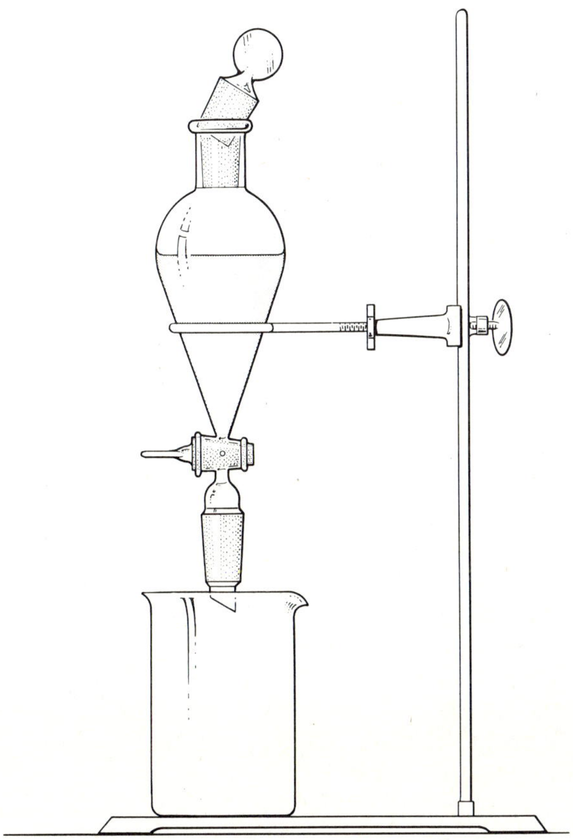

Figure 6-5 Separatory funnel in ring stand.

be evaporated, a distillation apparatus with condenser (Fig. 3-1) must be used for the distillation. As with all distillations or evaporations a boiling chip should be used. When a very volatile liquid such as ether is distilled, an ice bath is used to cool the receiver and the distillate. What is the weight of the residue in the flask? How much butyric acid remained in the ether?

> **Problem 6-2.** Assume all the butyric acid not in the ether was removed by the water extraction and that successive water extractions would remove the same percentage of butyric acid. How many 45-ml water extractions would be necessary to remove 95% of the butyric acid from the ether? How much water would be necessary to remove 95% of the butyric acid if only one water extraction were used?

Carefully following the procedure outlined above, extract the ether that remains in the separatory funnel with 45 ml of 10% sodium carbonate solution. Does the odor of the ether layer give any evidence of butyric acid? Check the ether layer with *wet* pH paper. Distil the ether layer as described above. Is there any residue? How can your observations of the two extractions be reconciled?

Multiple Fractional Extraction

Put 3 drops of 0.01% aqueous methylene blue and 3 drops of 0.5% methyl red into a test tube. Add 1 ml of ether and 1 ml of water, stopper the tube with a cork, shake it vigorously. Allow to settle and remove the ether layer with a small pipet as shown in Figure 6-6 and place it in another test tube. Add 1 ml of distilled

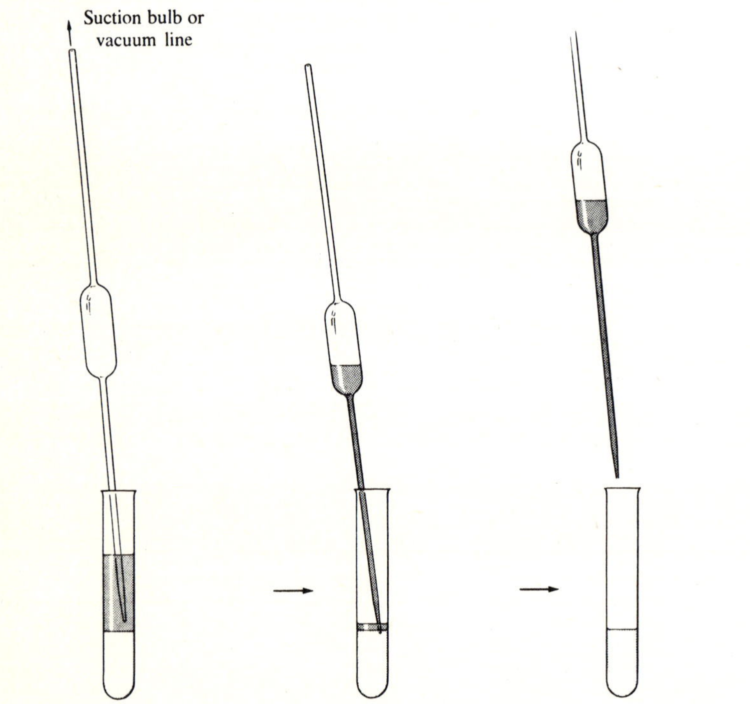

Place tip just above interface of the two phases and remove most of top phase.

Lower pipet to just below interface and remove liquid until air bubbles enter.

Remove pipet to just above test tube. Allow phases to separate in the pipet. Release vacuum enough to allow the small amount of denser phase to return to the test tube.

Figure 6-6 Separation of two phases with a pipet.

water to the ether layer and 1 ml of ether to the distilled water layer. Shake each test tube vigorously and allow to settle. Separate and combine the ether extraction of the *first water* layer with the water extraction of the first ether layer. Add 1 ml of ether to the original aqueous layer and add 1 ml of water to the original ether layer and repeat the extraction process.

The schematic for this procedure is shown in Figure 6-7. Before attempting the separation reread the above paragraph while looking at the schematic.

DISCUSSION

This procedure is an example of multiple fractional extraction. Basically it is similar to the multiple fractional distillation described in Chapter 5. This can be readily verified by the distribution of the dyes among the different fractions. As in fractional distillation each equilibrium between two phases is called a theoretical plate. Each horizontal stratum in Figure 6-7 is a theoretical plate; there are three theoretical plates in Figure 6-7.

As with fractional distillation there are methods of mechanizing multiple fractional extraction. There are large glass counter current distribution apparatuses (Craig machines, Fig. 6-8) with the equivalent of several hundred separatory funnels arranged so that all shaking and transfers are done automatically. Partition of solute between two phases is also the basis of paper chromatography (Chapter 59), some types of column chromatography, and gas-liquid chromatography (Chapter 8).

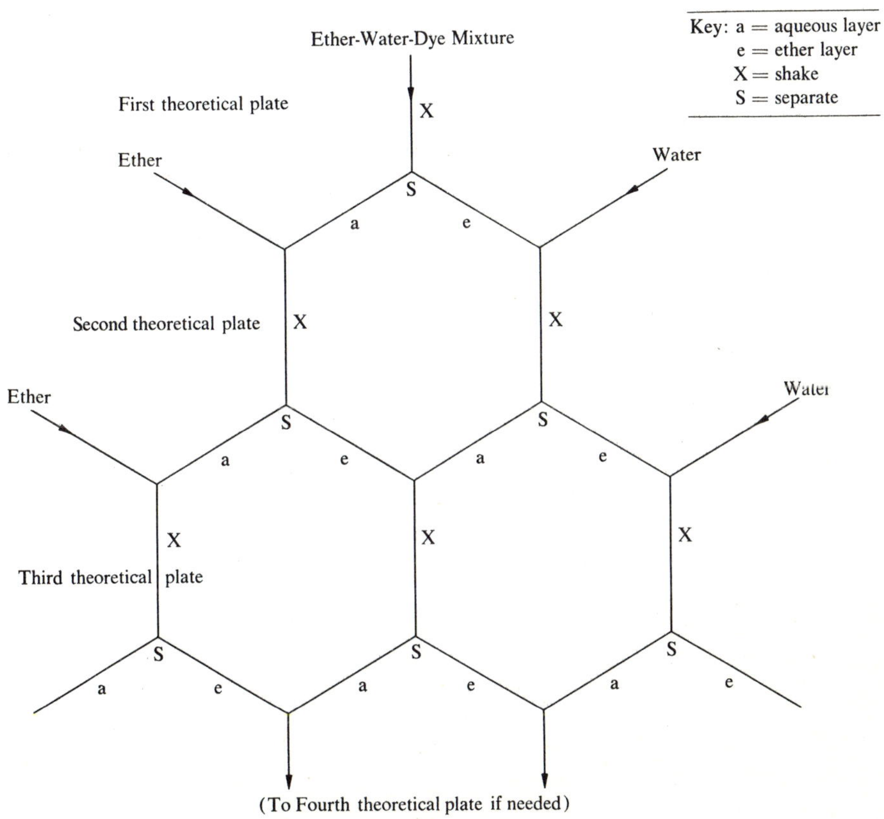

Figure 6-7 Multiple fractional extraction.

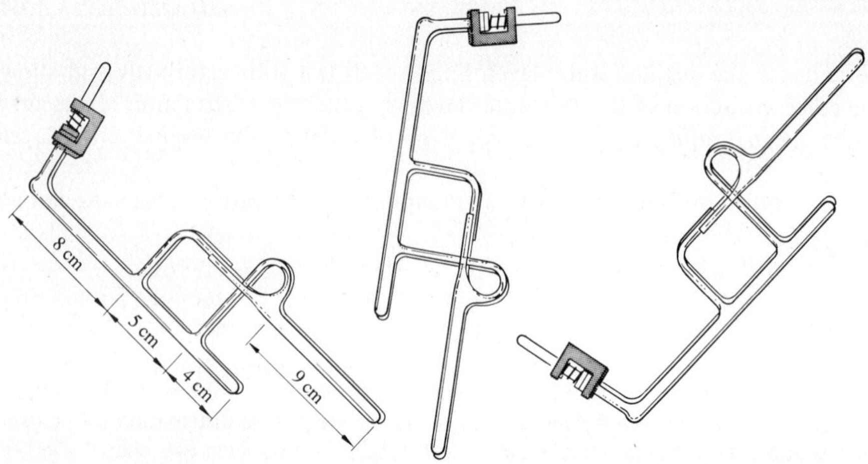

Single tube or theoretical plate and positions during operation of apparatus

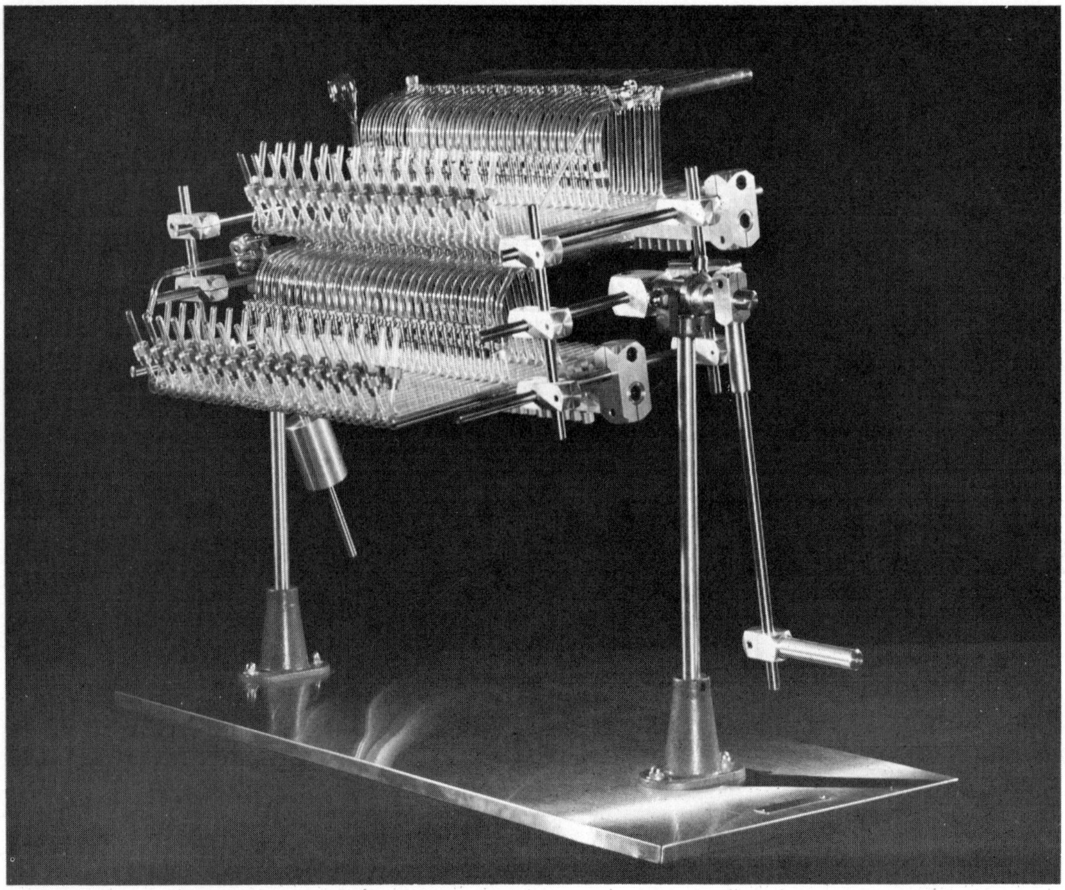

Figure 6-8 Mechanical counter-current distribution apparatus. Courtesy of H. O. Post Scientific Instrument Co. Inc.

EXERCISES

6-1. Of the compounds shown in Figure 6-2 one is much less soluble than the others in nonpolar liquids. Which one is it and why?

6-2. At 25° the solubility of malononitrile, $CH_2(CN)_2$, in ether is 20 g/100 ml and in water is 13.3 g/100 ml. (a) What is the approximate distribution coefficient between ether and water for malononitrile? (b) If a solution of 50 g of malononitrile in 450 ml of water is extracted with 450 ml of ether, how much malononitrile will be in the aqueous phase and how much will be in the ether phase? (c) How much malononitrile will be extracted from 50 g of malononitrile in 450 ml of water by three successive extractions with 150-ml portions of ether? (d) On the basis of these calculations is it better to extract with a large portion of solvent or to extract with several successive portions of solvent assuming the same total volume of solvent is used in the two methods? See also Problem 6-2.

6-3. Determine the distribution coefficient of butyric acid between water and ether from your results in the first part of the experiment. Note that because of the rather crude experimental conditions the value obtained will be only an approximate one.

6-4. Derive the expression

$$F = \frac{C_f}{C_i} = \left(\frac{K\,V_a}{K\,V_a + V_s}\right)^n,$$

where F is the fraction of solute remaining in solvent A of volume V_a after n number of extractions of volume V_s of solvent S, C_f is the final concentration of solute in A, C_i is the initial concentration of solute in solvent A, and K is the distribution coefficient of solute between solvents A and S.

7

Preparation of Ethylene
Preparation of 1,2-Dibromoethane

An *E*2 Elimination and Electrophilic Addition to an Alkene

$$C_2H_5OH \xrightarrow{H^+} C_2H_5\overset{+}{O}H_2 \xrightarrow[(-H_3O^+)]{} \underset{\text{B.p. } -103.7^\circ}{C_2H_4}$$

$$C_2H_4 + \underset{D_4^{20}\ 3.12}{Br_2} \rightarrow \overset{+}{C}_2H_4Br + \bar{Br} \rightarrow \underset{\substack{\text{B.p. } 131.7^\circ \\ D_4^{20}\ 2.180}}{BrCH_2CH_2Br}$$

INTRODUCTION

This experiment illustrates the use of a gas generator and purification train for the synthesis of ethylene by dehydration. The second part illustrates the decolorization of bromine by a gaseous olefin to synthesize ethylene bromide. Extraction and distillation are illustrated in purification of the ethylene bromide.

EXPERIMENTAL PROCEDURE

Assemble a gas generator as shown in Figure 7-1. Place the output into the safety valve and connect this to the gas washing and reaction train (Fig. 7-2). Check the apparatus thoroughly before proceeding further. Are *all* connections gas tight? Is the water in the safety valve from 3 to 4 cm above the glass outlet tube? Are the inlet tubes in the wash bottles and reaction tubes 1 cm above the bottoms of the containers? Is the first wash bottle about one third filled with 20% sodium hydroxide solution and the second wash bottle one third filled with water?

Pour 23 ml of 95% ethanol into the gas generator. Slowly and cautiously add 38 ml of concentrated sulfuric acid to the ethanol. Gently agitate the flask during the addition. While this mixture is cooling, put 4 ml of bromine in the first reaction tube and 2 ml of bromine into the second reaction tube. **Danger!** Bromine is a poisonous and very corrosive liquid. Bromine should always be poured and stored in a hood. Keep all bromine containers stoppered whenever possible. Do not inhale bromine or allow it to come in contact with the skin or

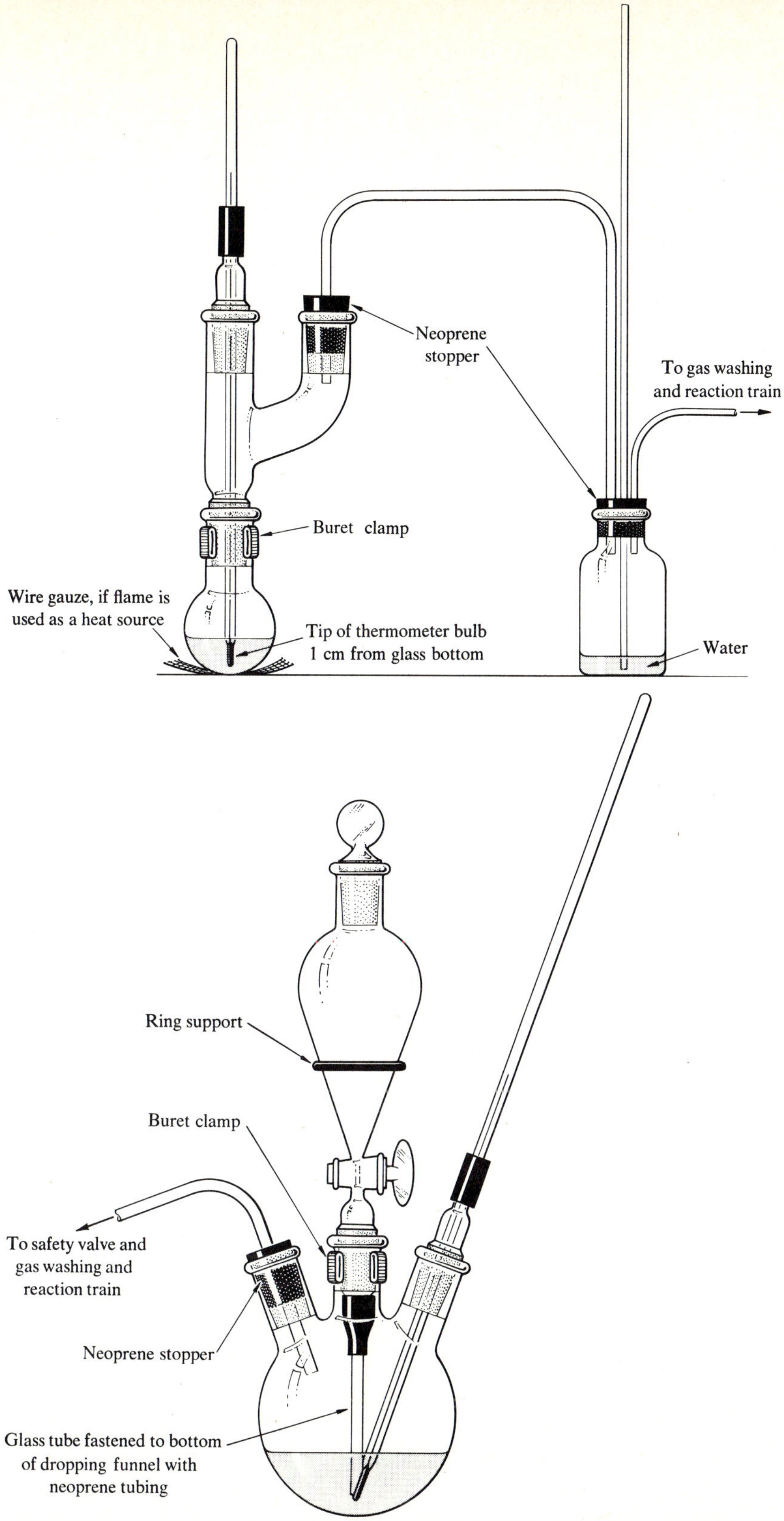

Figure 7-1 (a) Gas generator with safety valve used to synthesize ethylene in this experiment. (b) Gas generator with dropping funnel. Use if several reactions of ethylene are studied.

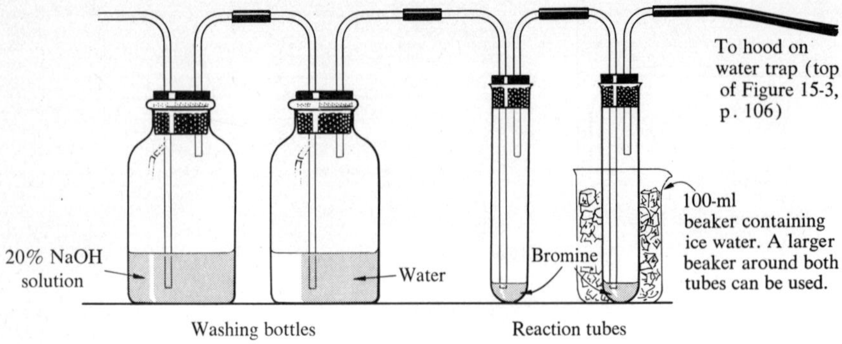

Figure 7-2 Gas washing and reaction train.

clothing. If any spills on the skin, wash off immediately with large quantities of water. Refer to section on antidotes and first aid if necessary (inside front cover).

Add 2 g of powdered pumice[1] to the sulfuric acid–ethanol mixture. Check all connections again; be sure the thermometer in the gas generator is 1.0 to 1.5 cm above the bottom of the flask. Gently heat the mixture until a smooth evolution of gas begins. This usually occurs at 160 to 170°, but it varies somewhat with the specific composition and mesh size of the catalyst. Do not heat above 200° under any condition, and do not heat so rapidly that water is forced out of the tube in the safety valve. If even gentle heating forces water out of the tube, stop the heating and check the system for any obstructions.

Continue heating until the bromine color disappears. Disconnect the gas generator and allow it to cool while proceeding with the experiment. When it has cooled to room temperature, pour the contents of the distilling flask down the drain with a large amount of water.

Transfer the contents of the reaction tubes to a separatory funnel and extract successively with water, 5% sodium carbonate, and finally water again, using 30-ml portions. Be sure not to throw the ethylene bromide away. Is the ethylene bromide the top or bottom layer? Very carefully follow the instructions given in Chapter 6 for extraction.

Dry the ethylene bromide over 2 to 3 g of calcium chloride for about 20 min. Every 5 min swirl the flask

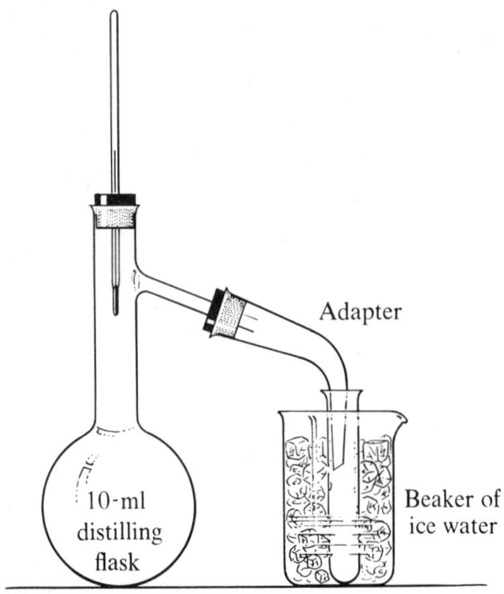

Figure 7-3 Semimicro distillation apparatus.

[1] Acid dehydration of alcohols is catalyzed by finely divided siliceous materials. Fuller's earth, kieselguhr, diatomaceous earth, and sand may also be used. Since the catalytic effect of these substances is due at least in part to their large surface area, sand should be used only as a last choice.

gently. Decant the ethylene bromide into a 25-ml distilling flask. With an adapter and ice-cooled receiver assemble the semimicro distillation apparatus shown in Figure 7-3. Heat the mixture gently and collect the portion that boils at 128 to 135° in a tared yield bottle.

Problem 7-1. A cloudy fraction is often obtained that distills in the 80 to 90° range. What is the composition of this fraction? What could you do to this fraction to improve the yield? If you obtain such a fraction, check your answer in the back of the book before proceeding further.

PERCENT YIELD

The maximum possible yield (*theoretical yield*) of any reaction is determined by the reactant that is least plentiful in terms of equivalents. If 1 mole of butadiene is reacted in the dark with *any* amount of bromine greater than 2 moles, the maximum yield of 1,2,3,4-tetrabromobutadiene is 1 mole.

The actual yield is less than the theoretical yield in almost all organic reactions. A convenient way of expressing the yield is in percent yield (% yield) which is given by the equation

$$\% \text{ yield} = \frac{(\text{actual yield})\ (100)}{(\text{theoretical yield})}$$

In the experiments described in this chapter the % yield would be calculated as follows if there were an excess of bromine:

$$\text{Weight} = (\text{ml})(\% \text{ of EtOH})(\text{density}) = (16)(0.95)(0.807) \qquad x$$

$$C_2H_5OH \rightarrow CH_2BrCH_2Br$$

$$\text{Molecular weight (mol. wt.)} \qquad 46 \qquad 188$$

$$\text{Theoretical yield} = x = \frac{(\text{wt. EtOH})\ (\text{mol. wt. } BrCH_2CH_2Br}{(\text{mol. wt. of EtOH})} = \frac{(12.3)\ (188)}{(46)}$$

$$\text{Theoretical yield} = 50.4 \text{ g.}$$

Where there are two or more successive reactions as in this experiment, the yield of the final product based on the weight of the starting material of the first reaction is often called the over-all yield.

Problem 7-2. What is the theoretical yield of ethylene in grams? In volume at STP?

Problem 7-3. What would be the % yield of the reaction if 10 ml of ethylene bromide were obtained and there were an excess of bromine?

Problem 7-4. What is the minimum volume of bromine that will react with all the ethylene that could possibly be produced?

Problem 7-5. In the synthesis of ethylene bromide does the quantity of ethanol or the quantity of bromine limit the theoretical yield?

Problem 7-6. What are the theoretical yields of the following reactions: (a) chlorine (10 g) and neopentane (10 g) to give 1-chloro-2,2-dimethylpropane, (b) $NaC{\equiv}CNa$ (20 g) and C_2H_5Br (45 g) to give 3-hexyne, (c) 1-octene (30 g) and chlorine (25 g) to give 1,2-dichlorooctane, (d) 1-octene (30 g) and chlorine (25 g) in a light-catalyzed reaction to give 3-chloro-1-octene?

Figure 7-4 shows how to make a label so that yields and other important data are readily apparent. Figure 7-5 is a suggested method for recording the data on synthesis products in your notebook. Weigh the product, calculate the % yield, and hand in the labeled product to the laboratory instructor.

Name of Compound	
Wt.	Percent yield
M.p., b.p., or other physical constants	
	Name of Student

Benzoic acid	
5.3 g	62%
m.p. 118-20°	b.p. 245-8°
	R. Walters

Figure 7-4 Label for organic products.

$$\begin{array}{llll} & 2g & & \text{Theoretical yield} \\ & 2Br_2 & + HC{\equiv}CH \rightarrow & HCBr_2CHBr_2 \\ \text{Mol. wt.} & 160 & 26 & 346 \end{array}$$

Theoretical yield = 2·346/26 = 26.6g
Wt. of filled container = 48.6g
Wt. of empty container = 30.2g
Yield = 18.4g
% yield = 18.4/26.4 (100) = 69.7%

	Observed	Literature value*
M.p.	−3 to −4°	−1°
B.p.	237–39°d	239–42°d
n_D^{20}	1.6372	1.6380
D_4^{20}	2.95	2.9672

* *Chemical Rubber Handbook*, 45th ed., p. C-308.

Figure 7-5 Suggested method for recording data on synthesis products.

SOME OTHER REACTIONS OF ETHYLENE

Replace the two tubes containing bromine with as many of the following as your instructor directs you: 1) one test tube containing 8 ml of 2% aqueous potassium permanganate, b) one test tube containing 10 ml of 2% silver nitrate in ethanol, 3) one test tube containing 8 ml 0.1% iodine in ethanol. Bubble ethylene through these tubes one at a time.

Problem 7-7. (a) Is there any evidence for reactions in 1, 2, or 3 above? What is the evidence and exactly what does it indicate? (b) Write all the chemical equations involved.

EXERCISES

7-1. What by-products are likely to be produced in this reaction? How are they removed?

7-2. What is the purpose of the sodium hydroxide solution wash bottle? What is the purpose of the water wash bottle?

7-3. What by-product would be likely if the gas generator were not heated to a high enough temperature? If too little sulfuric acid were used?

7-4. If by-products boiling higher than the ethylene bromide are found, what are they and how are they formed? What added precautions could be followed to minimize their formation?

8

Gas Chromatography

INTRODUCTION

Gas chromatography is undoubtedly the most versatile, flexible, and universally applicable of all analytical methods. The rapid application (the method was first introduced as a practical analytical tool in 1952) of gas chromatography to all branches of chemistry and to many of the other sciences is an indication of the usefulness of the method.

Chromatography involves separation of substances by repeated distribution between a moving phase and a stationary phase that are in equilibria. The process is called chromatography because the first widely publicized separations using this basic method involved colored compounds. Most separations based on distribution between moving and stationary phases are of colorless compounds; however, they are still called chromatograms (or chromatographic separations) because of the historic precursors. Of the chromatograms described in this text only Chapters 25, 26, 43, 47, and 48 involve the separation of colored compounds.

Problem 8-1. Both fractional distillation (Chapter 5) and multiple fractional extraction or counter-current extraction (Chapter 6) also involve separation of substances by repeated distribution between two phases. How do they differ from chromatographic separation?

DISCUSSION

Classifications of Chromatography

Adsorption and Partition. Chromatographic separations are based on partition and adsorption and occasionally a combination of the two. In adsorption chromatography the stationary phase adsorbs compounds (on its *surface*) from the moving phase. In partition chromatography the compounds dissolve *in* (not on) the stationary phase. Partition chromatography is not a surface phenomenon, whereas adsorption chromatography is. In partition chromatography the stationary phase is usually a liquid; however, some of the new stationary phases for gas chromatography are apparently solids.

In gas chromatography (GC), also called vapor phase chromatography or VPC, the mobile phase is a gas. GC can be either adsorption chromatography (sometimes called gas-solid chromatography or GSC) or partition chromatography (sometimes called gas-liquid chromatography, GLC, or occasionally gas-liquid partition chromatography, GLPC). GLC is more widely used than GSC and is the type discussed in this introduction to gas chromatography.

Analytical and Preparative. Most chromatograms can be classed as either analytical (diagnostic) or preparative. Many chromatograms, however, have both purposes. Analytical chromatograms determine identity and/or concentration. Any isolation of compounds is incidental. Analytical chromatograms usually involve separation of 10^{-3} to 10^{-15} g of material. Preparative chromatograms have as their objective the purification and isolation of relatively large quantities (usually 10^{-3} to 10^{4} g) of compound for *further* experimentation. This may involve isolation of only a milligram or less for infrared and ultraviolet spectra, 0.2 to 10 g for determination of physical properties, or several kilograms as a commercial intermediate or end-product.

Partition chromatography has been used mainly for analysis, and adsorption chromatography has been used, at least until thin layer chromatography was introduced, mainly for preparation. Although preparative gas chromatography is gaining wider usage, the principal application of gas chromatography has been to the separation of 0.001 to 100 *micro*grams of material.

Equipment

Column. In GLC the liquid or stationary phase is spread on an inert solid support.[1] The liquid-solid support mixture is packed in the column, and the column placed in an oven or heating bath where the temperature can be closely controlled.

The mobile phase is a stream of gas (usually helium or nitrogen, sometimes hydrogen or argon, or occasionally other gases) that passes through the column. The sample is injected into the gas stream and the gas stream carries it onto the column. Figure 8-1 is a schematic cross section of a typical column

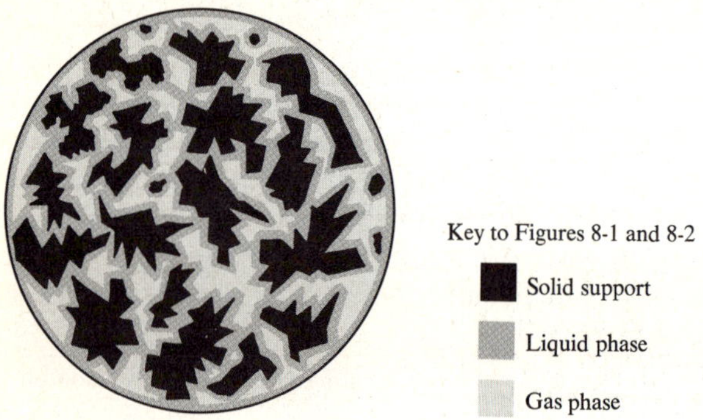

Figure 8-1 Schematic cross section of a typical GLC column.

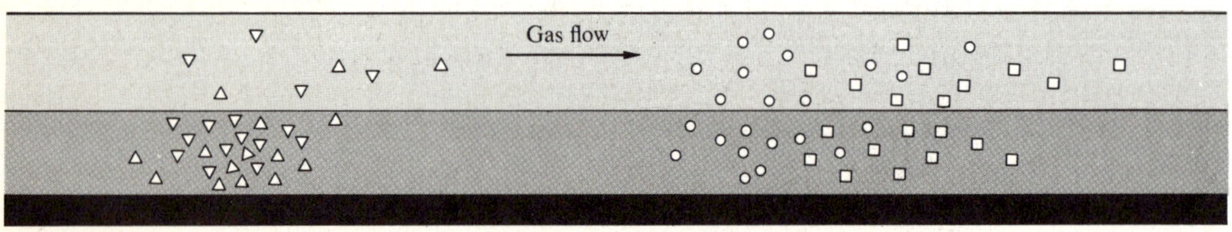

Figure 8-2 Simplified schematic of a chromatographic column showing two partially separated compounds (○ and □) and one completely separated compound (△).

with column wall, solid support, stationary phase (liquid), and mobile phase (gas). Figure 8-2 is a simplified schematic of a chromatographic column showing the mobile and stationary phases, the solid support, two partially separated compounds, and further down the column a compound completely separated from the mixture.

Chromatograph. Figure 8-3 shows a schematic of a gas chromatograph. The components in the schematic are correlated with the controls shown on the front panel of a typical gas chromatograph in

[1] For Golay or capillary columns the only solid support is the column wall.

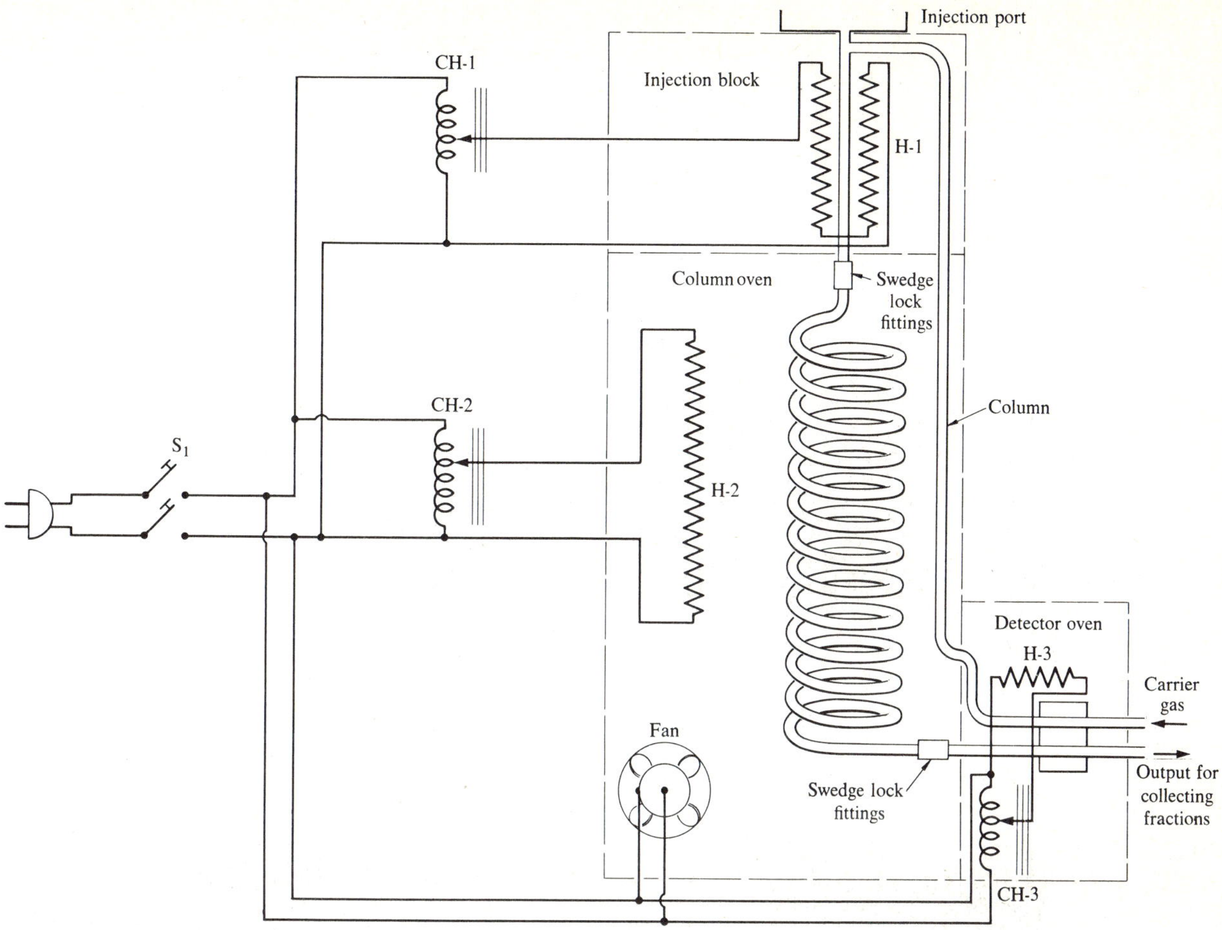

Figure 8-3 Gas chromatograph schematic.

Figure 8-4. All chromatographs have this same basic construction although many are considerably more elaborate. The sample is injected through a septum[2] into an injection port where it is vaporized and mixed with the mobile phase or carrier gas.

Some gas chromatographs do not have injection chambers and require on-column injection. The injection needle is placed directly into the column opening and the contents injected directly onto the column. This technique is satisfactory if the sample is small enough to be vaporized as rapidly as it would in an injection chamber. On-column injection is sometimes required when the hot metal injection port would partially pyrolyze the sample. There can be on-column injection in chromatographs with injection chambers provided a long needle can reach from the septum to the column.

The carrier gas passes through the column, and the compounds to be separated are partitioned between the vapor and liquid phases. Immediately after the column there is a detector which is capable of sensing very small changes in the mobile phase. Some chromatographs have two or more detectors in series or parallel. These detectors are often selectively sensitive to different groups of compounds.

Sometimes various devices are attached after the detector that aid the collection of fractions or serve as interfaces between the gas chromatograph and a mass or infrared spectrometer. A few chromatographs have no separate heater on the detector (H-3, Fig. 8-3). The detector(s) are placed in the column oven compartment, and both detectors and column are served by the same heater (H-2). This is an undesirable arrangement as the detector and column are always at the same temperature. When used, most columns lose or bleed small amounts of liquid phase. This liquid phase and compounds of low vola-

[2] Sometimes gas samples are injected through special gas sampling valves.

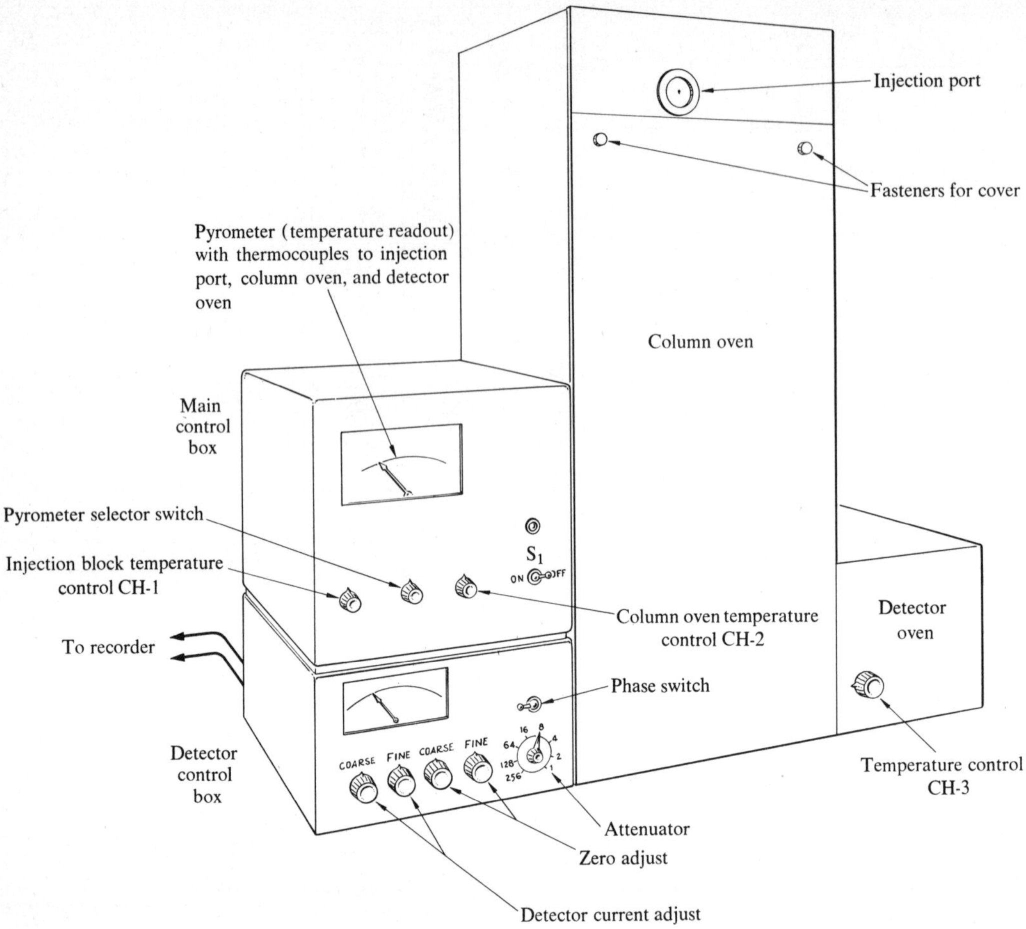

Figure 8-4 A typical gas chromatograph.

tility that are chromatographed are likely to deposit in the detector unless the detector is 20 to 40° hotter than the column. Unless removed these deposits cause the detector to become less sensitive and more noisy. Cleaning the detector may be difficult and time consuming. It is not necessary, however, to have the detector in a separate box as the chromatograph shown in Figure 8-4.

Thermal conductivity and flame ionization detectors are the most widely used on gas chromatographs. Thermal conductivity units are more likely to be used on the chromatographs found in student laboratories. These detectors have a sensing element (wire or thermister) in the path of the effluent (gas stream coming off the column). A voltage is applied to the sensing element and a current passes through it; the current heats the element. A constant temperature is reached when the heat lost by thermal conduction to the gas stream equals the electrical energy heating the sensing element. If the carrier gas is flowing uniformly past the sensing element, there is no variation in the heat lost. If at intervals some other gases or vapors flow past the sensing element, the amount of heat lost by the sensing element changes when the other substances have different thermal conductivities (as is likely). If the heat loss changes, the temperature of the sensing element changes. This temperature change causes a change in resistance; the resistance change causes a change in the current passing through the sensing element. The current change is translated by the bridge circuit of the detector into a voltage change. This is shown in schematic form in Figure 8-5. **Warning. The sensing elements are easily damaged by excess current. Carefully read the instructions in the experimental section before operating the chromatograph.**

The detector is connected to a recorder, often through an amplifier. Typical graphs or chromatograms obtained from a recorder are shown in Figures 8-6 through 8-8. The recorder is often the most

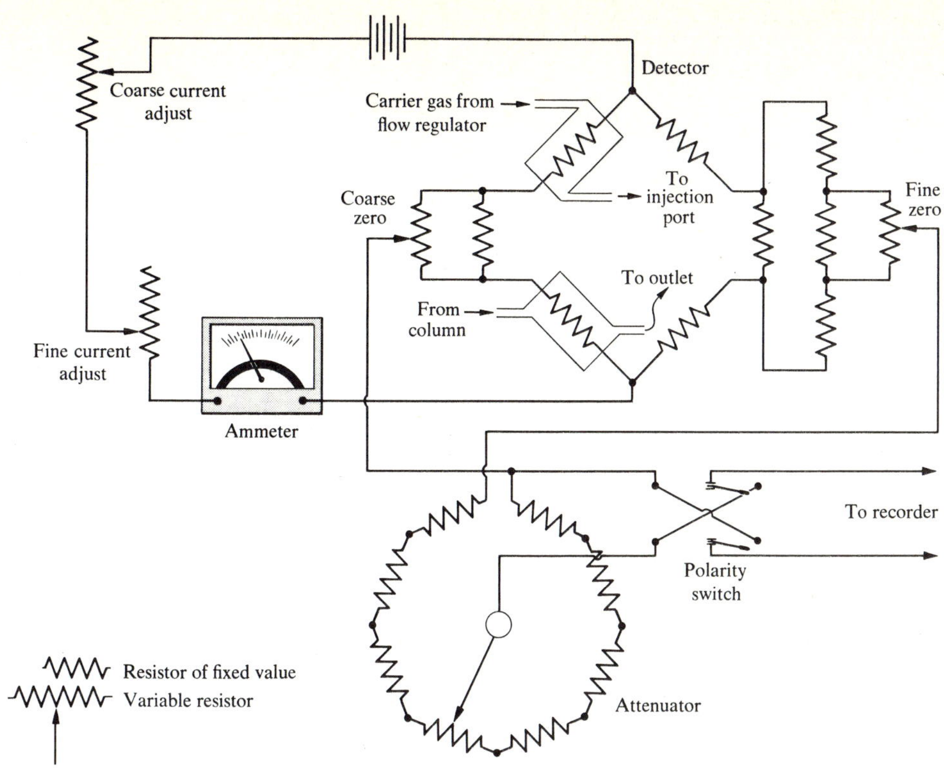

Figure 8-5 Schematic of a thermal conductivity detector and associated circuit. Many variations of this are used.

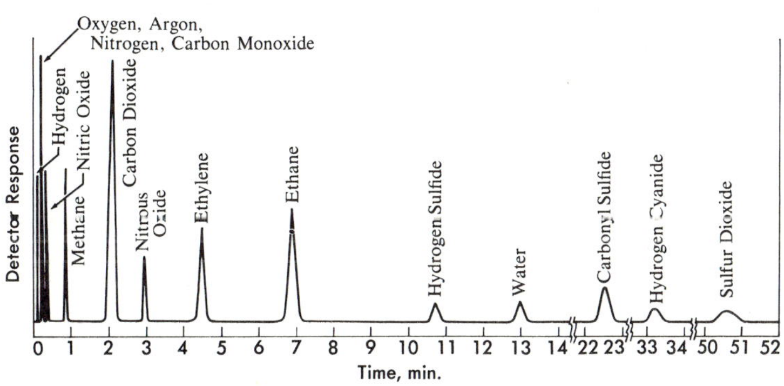

Figure 8-6 Chromatogram of synthetic gas mixture similar to probable Martian atmosphere on a column of 6′ Porapak Q and 2″ Porapak R at 25°. (W. F. Wilhite and O. L. Hollis, *Journal of Gas Chromatography*, Feb. 1968, reprinted by permission of the editors.)

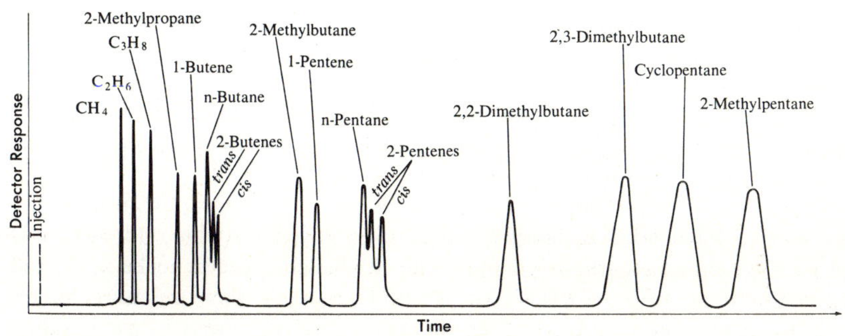

Figure 8-7 Chromatogram of low molecular weight hydrocarbons on 30′ × ⅛″ 5% squalene on Chromosorb G at 10°.

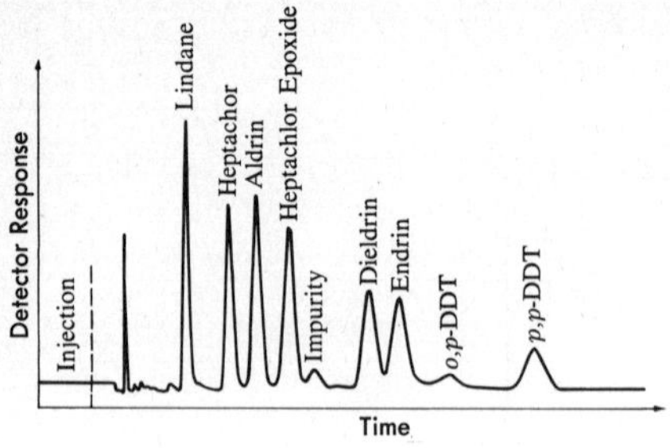

Figure 8-8 Chromatogram of insecticide mixture on 6 × 44 mm of 10% DC 200 on Anakrom SD at 180°. Analabs, Hamden, Conn.

expensive component of a gas chromatograph. On a few gas chromatographic installations there is a digital readout in addition to or instead of the recorder.

Problem 8-2. When helium is used as a carrier gas with a chromatograph having a thermal conductivity detector, all chart deflections (which indicate a voltage from the detector bridge) are in one direction for all compounds. When nitrogen is used a carrier gas, thermal conductivity detectors give deflections in one direction for some compounds and in the other direction for other compounds. Suggest an explanation for this.

Comparisons with Other Analytical and Separation Methods

Fractional Distillation. Gas chromatography has many advantages over other analytical and separation methods. In some respects it is similar to fractional distillation. In distillation the compounds being separated form the liquid and gas phases. The entire column must be filled with the compounds being separated. Therefore, distillation requires a much larger sample, and there are usually considerable overlap fractions between the compounds being separated. The size of the overlap fraction depends on the holdup volume of the column and the efficiency with which the compounds are separated. Overlap fractions in gas chromatography are considerably less likely since the compounds to be separated make up only a very small percentage of the liquid and gas phases.

Compounds can be separated by gas chromatography much quicker than by fractional distillation. The maximum number of theoretical plates with laboratory fractionating columns is several hundred and usually is much less (between 20 to 40 plates). Several thousand theoretical plates are not uncommon with chromatographic columns and may be as high as twenty *thousand or more.* In gas liquid chromatography solubility in the liquid phase as well as vapor pressure determines how long a compound is retained by a chromatography column. Compounds with identical boiling points can be widely separated if an appropriate liquid phase is chosen. If two compounds cannot be separated on one liquid phase, another liquid phase can usually be found that will give wide separation. Examples of this are given below in the section on optimum operation and in Chapter 37. There is nothing even approaching this with distillation techniques. As a result of this and the larger number of theoretical plates available with GLC many compounds can be separated by GLC that cannot be separated by other methods. Many mixtures cannot be separated by fractional distillation no matter how many theoretical plates are available because of azeotrope formation. There is nothing comparable to azeotrope formation in gas chromatography.

Distillation requires less expensive equipment than gas chromatography and until recently has been able to separate much larger volumes than GLC. Recent technological advances in preparative scale GLC indicate that even here distillation may loose its advantage.

Other Chromatographic Methods. Later chapters introduce other chromatographic techniques. Gas chromatography is superior to these other techniques if: (a) the equipment (which admittedly is considerably more expensive for gas chromatography) is available, (b) the compounds to be separated or their readily made derivatives are reasonably volatile to avoid exceptionally long retention times. Reasonable volatility for most organic compounds means a vapor pressure of 10 mm or more below

360° *and* below their decomposition point. Decomposition in a gas chromatographic column usually occurs at a higher temperature than values found in the literature because the compound is in an inert gas such as helium, nitrogen, or argon and because the compound is present in low concentration.

Within the above limitations gas chromatography is superior to other chromatographic techniques because: (a) Gases have low viscosity. This leads to high flow rates so that compounds can be separated rapidly and very long columns with thousands of theoretical plates can be used. (b) Gases diffuse rapidly. This also leads to high flow rates and fast chromatograms. (c) There are many detectors that are sensitive to and respond rapidly to small changes of composition of gases. These detectors are readily adapted to recorders, digital readouts, and various automatic fraction collectors, signals, and switches. Because of the easy adaptation of detector output to recorders and digital readouts or even computers, quantitative analysis by gas chromatography is usually more accurate and almost always easier than by other chromatographic methods.

OPTIMUM OPERATION

Efficiency

Gas chromatography is an incredibly efficient analytical tool. It is so efficient that although most GLC analyses and separations are probably done at less than 20% efficiency, the separations could not be done better by any other method. In spite of this reserve every possible step should be taken to make gas chromatographic separations as efficient as possible. This section is intended to introduce the beginner to some of the variables and their effect on separation.

Theoretical Considerations

Among the equations which aid in the understanding of gas chromatography and selection of optimum operating conditions is

$$\frac{RT_x}{RT_y} = \frac{a_y P_y^o}{a_x P_x^o} = \alpha, \tag{8-1}$$

where RT_x is the retention time for compound X, RT_y is the retention time for compound Y, where the retention time is the time from injection to emergence of the maximum concentration of the compound (Fig. 8-13). P^o is the vapor pressure of compound in question, a is the activity coefficient or everything that affects the partition between moving and stationary phases except vapor pressure, and α is the separation factor.

As P^o and a increase, the retention time decreases. It is principally the solubility of a compound in the stationary phase that controls its a factor. The lower a is the higher the solubility. It is the a factor and the large number of theoretical plates that contribute to the high selectivity and high resolving ability of GLC mentioned on page 54 when GLC was compared with fractional distillation.

Much of the effort in obtaining more efficient chromatographic separation is directed towards optimizing the activity coefficients. Different a values favor the separation of compounds with similar vapor pressures. Often good separations are favored by low a values which favor relatively high concentration in the stationary phase. The general principal of "like dissolves like" is useful in determining if a compound has a low a value in a stationary phase (review Chapter 2 and pp. 36–37). In general, similar compounds have similar a values. The stronger the attractive forces of the liquid phase for the molecules being separated, the lower their a values. The effect of liquid phase and solid support on separation efficiency and retention times will be discussed further in Chapter 37 after more structure-solubility relations are studied.

Problem 8-3. What would the order of retention times of hexane, heptane, octane, and pentane be on paraffin wax?

Effect of Column Temperature

Higher column temperatures result in faster retention times and usually in poorer resolutions. *Very* low column temperatures, while often affording better resolutions, give such long retention times that the chromatographic separations take too long to be practical. Increasing the column temperature by 30° decreases most retention times by approximately 50%; decreasing the column temperature by 30° approximately doubles most retention times.

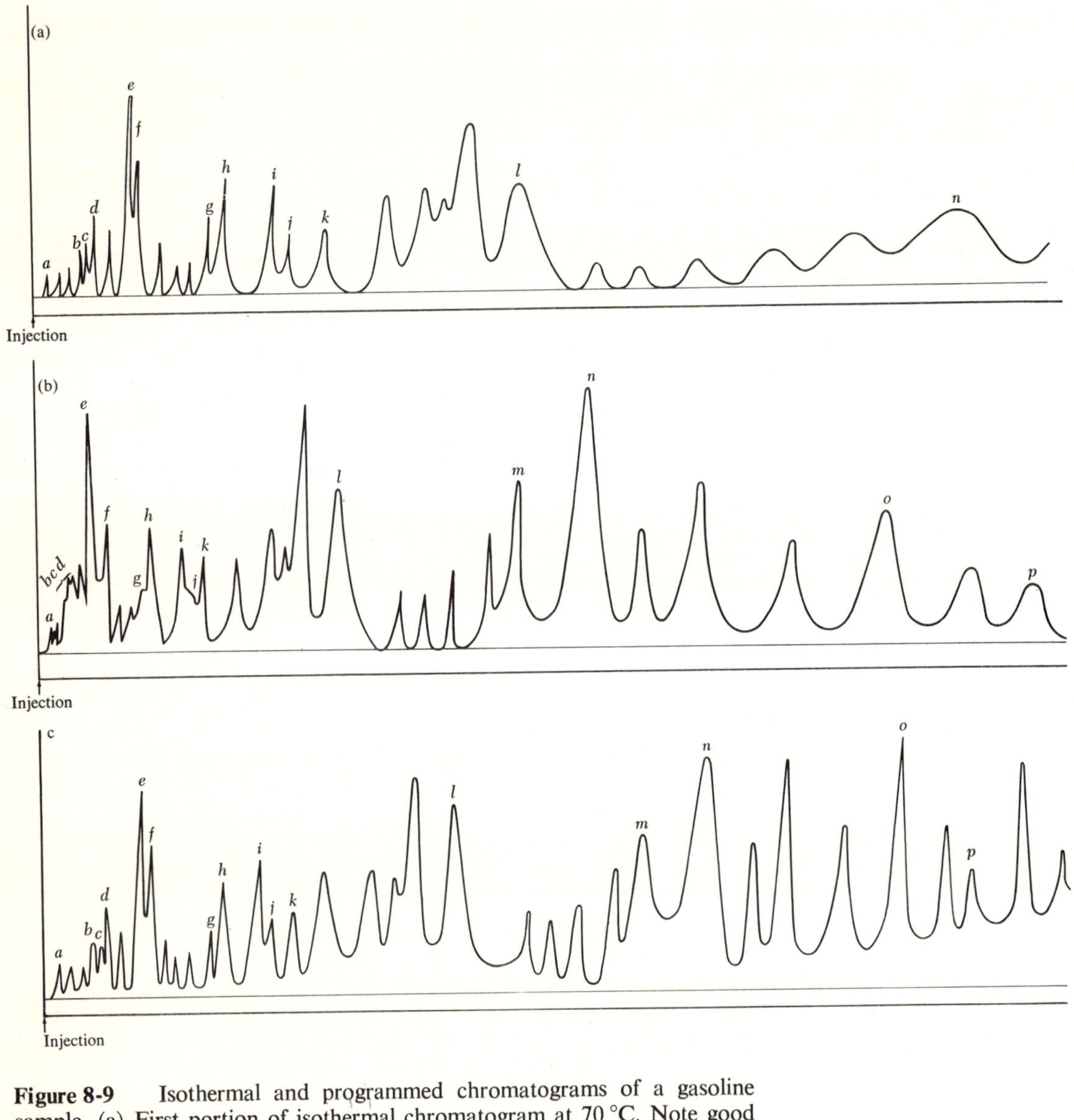

Figure 8-9 Isothermal and programmed chromatograms of a gasoline sample. (a) First portion of isothermal chromatogram at 70 °C. Note good separation of the hydrocarbons that are eluted first and the broad peaks near the end of the portion shown. (b) First portion of isothermal chromatogram at 95 °C. Note poor separation of early peaks; compare peaks *b, c, d, e, f, g, h,* and *i* in the two chromatograms. Note that peaks *m* and *n* are eluted earlier and are sharper. (c) First portion of chromatogram programmed at 5°/min from 70 to 130 °C. Note good separation of early peaks *and* faster elution and sharpness of the later peaks.

Compounds with very long retention times are highly diluted with carrier gas and usually give very broad, low peaks; this causes difficulty in determining the exact retention time and causes minor components to be undetected. If the retention times of compounds to be chromatographed do not have too broad a range, there is usually a temperature that is low enough for adequate separation but high enough to allow the chromatography to be completed in a reasonable length of time. If the differences in retention times are too great, the mixture can be chromatographed at two or more different temperatures or on two or more columns for complete analysis of all the compounds.

Many gas chromatographs are equipped with temperature programmers. These are devices that increase the column temperature at definite, reproducible rates. Compounds with short retention times can be separated at low temperature, and compounds with very long retention times (at low temperature) can be eluted in a convenient amount of time and have peaks that are relatively narrow. Figure 8-9 compares two isothermal chromatograms and a temperature programmed chromatogram of a sample of gasoline. Operation of a temperature programmer is described in Appendix IV.

Effect of Flow Rate

The faster the flow of carrier gas the shorter the retention times. The separating efficiency of a chromatography column is maximum at a rather low flow rate, often 12 to 20 ml/min but occasionally as high as 30 ml/min (Fig. 8-10). The exact flow rate for maximum separation is dependent upon the

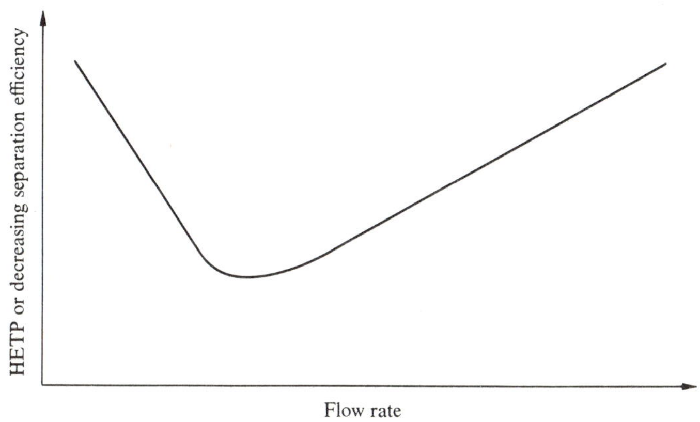

Figure 8-10 Effect of flow rate on number of theoretical plates in a given column distance, called height per equivalent theoretical plate or HETP.

column and temperature used and the substances being separated. For most chromatographic separations, flow rates higher than the flow rate requisite for maximum separation are used. To obtain a good compromise between high separation efficiency and quick chromatographic separations it is usually better to use a high flow rate and a low temperature.

Quantitative Analysis

The response of a detector to a compound is usually proportional to the amount of compound passing through the detector. Under constant conditions that do not allow significant changes in retention time or changes in the peak height to peak width ratio, the height of a peak is proportional to the amount of compound present. If peak height is plotted versus amount of compound, a straight or gently curving line is obtained. This plot can then be used to determine the amounts of compound in a mixture. Examples of this are given in later chapters. The use of peak height for quantitative analysis requires high precision in the reproduction of retention times.

Quantities are also proportional to peak area. Peak areas can be obtained by calculating from the peak dimensions, tedious counting of squares on the chart paper, cutting out the peaks and weighing them, using a planimeter, or by using one of the mechanical or electronic integrators that can be attached to a recorder or to a chromatograph. Although quantities calculated from peak areas are less subject to error from varying retention times than quantities calculated from peak height, they may be more subject to error from adsorption than those from peak height (Chapter 37).

EXPERIMENTAL PROCEDURE

Precautions

On chromatographs using thermal conductivity detectors the detecting elements are easily burned out if too much current flows through them. The amount of current permissible depends on the particular detecting elements being used, the temperature of the detector, and the carrier gas. The higher the thermal conductivity of the gas, the more current that can be safely used. If helium is the carrier gas, a higher current is permitted than when nitrogen is used. The higher the temperature of the detector, the lower the filament current that can be used.

Do not increase the detector current above the value set by the instructor. Turn off the filament current before turning off the carrier gas. Do not turn the filament current on unless the laboratory instructor tells you to do so.

Problem 8-4. What conditions must be met before turning on the detector filament current?

The maximum filament current consistent with long detector life should be used. The higher the filament current, the more sensitive the detector.

Other precautions should be followed even though they are not as critical as the use of the correct filament current. Neither the column nor the detector should be heated for long periods of time without carrier gas flowing through. The column should not be heated above its upper temperature limit as this is likely to ruin the column and foul the detector.

Problem 8-5. Why should the detector heater not be turned off until after the column has cooled at least halfway to room temperature?

Injection into a Gas Chromatograph

a. To ensure complete vaporization of the sample the injection port of the gas chromatograph should be at least 20° hotter than the boiling point of the highest boiling material being chromatographed and 40 to 60° higher if quantities greater than 50 μl are being injected. The injection port should not be so hot that it decomposes the sample.

b. Wash the syringe with sample two or three times.

c. Overfill the syringe taking care not to get any gas bubbles into it. Hold the syringe with needle up (keep it in this position until injection), and eject the excess sample (Fig. 8-11). This procedure removes air bubbles from the needle so the volume in the syringe is not less than indicated. Wipe the needle with a lintless tissue.

d. Just before injection pull back the plunger *slightly* so that air fills the needle. This prevents premature vaporization from the tip of the needle. Push the needle as far as it will go through the center of the septum.

e. The instant the recorder pen crosses one of the heavy lines on the chart paper quickly inject the sample by rapidly pushing the plunger. A slow injection causes the peaks to be spread and diffuse. Mark the injection time on the recorder chart. Be sure to press the plunger straight into the syringe as force applied sideways bends the plunger and ruins the syringe.

f. When high pressures in the injection port are used or when large gas syringes are used, it may be necessary to exert considerable pressure on the plunger to keep it from being pushed back.

g. Withdraw the needle quickly so that no carrier gas escapes and *none* of the sample still in the needle vaporizes.

h. After use wash the syringe with solvent two or three times, remove the plunger, and dry the syringe. It can be conveniently dried by placing the needle through a septum placed over a glass tube going into the trap of a water aspirator (Figs. 8-12 and 12–5) and pulling air through the syringe.

Be careful in handling the syringes; they are precision made and very expensive. Replace the septum after 20 to 30 injections and more often if high inlet pressures or large diameter needles are being used.

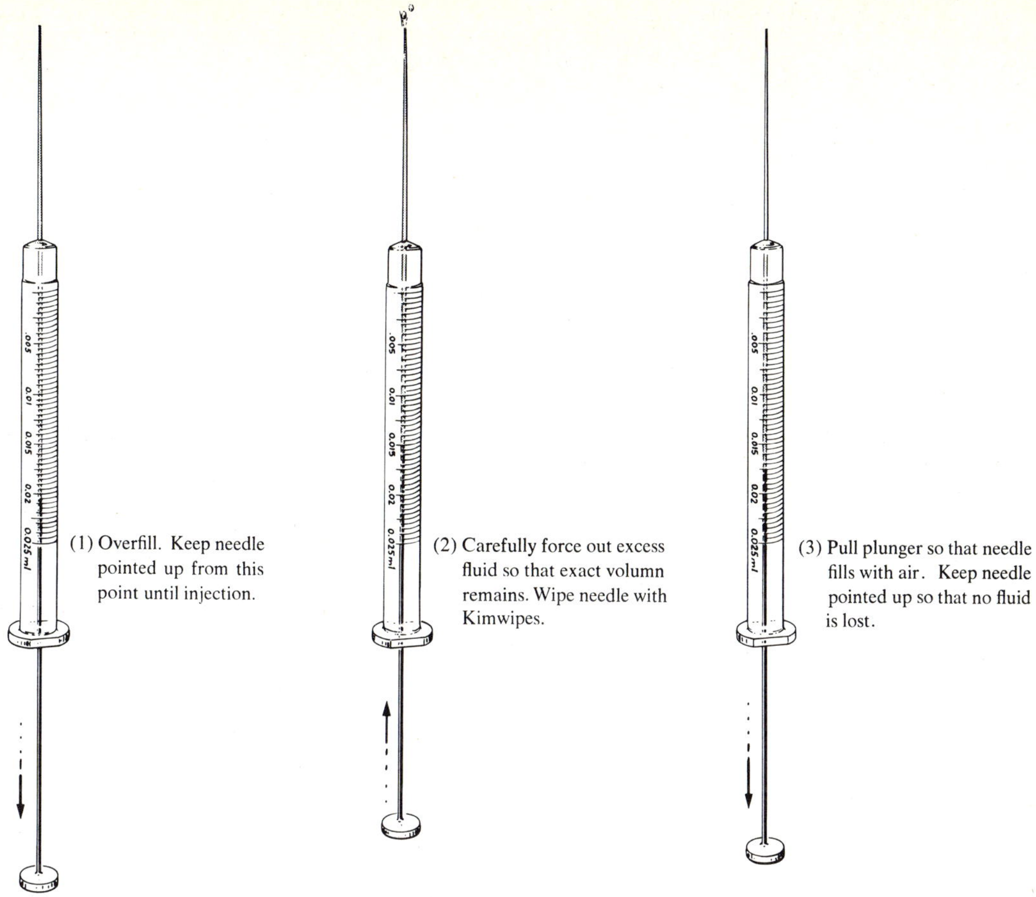

Figure 8-11 Filling a syringe prior to injection.

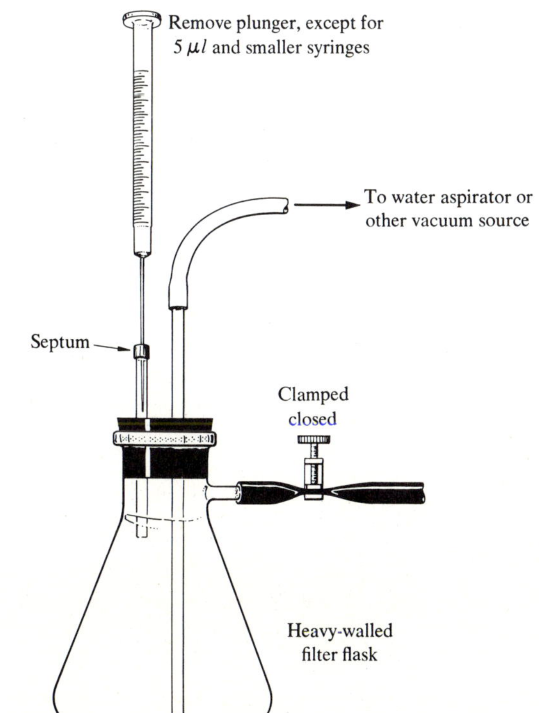

Figure 8-12 Drying a syringe.

Determination of Retention Times and Peak Heights

Peaks of compounds with short retention times are usually sufficiently sharp so that the tip of the peak can be used to locate both the peak height and the retention time. If the chromatography is isothermal and sometimes if it is programmed, the peaks of compounds with medium and long retention times are usually sufficiently rounded and broadened so that determination of exact peak height and retention time is difficult. If a peak is rounded or broadened, the straight part of the peaks should be extended by lines drawn tangent to the long straight segments of the peaks. The intercept of these tangential lines should be used to locate the retention time and peak height (Fig. 8-13).

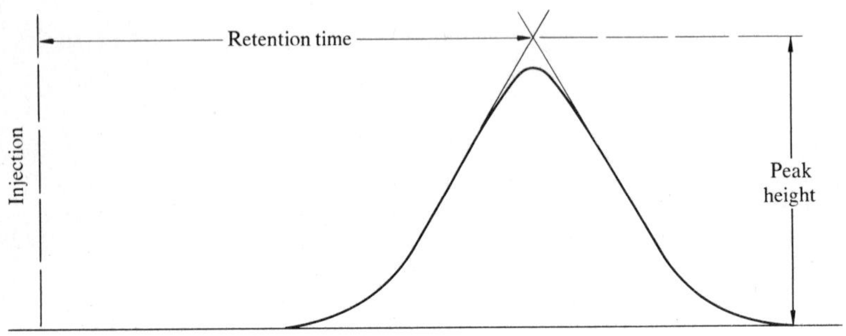

Figure 8-13 Drawing of tangential lines for location of peak height and retention time.

Gas Chromatography of Homologous Series

Inject each compound from *one* (assigned by your instructor) of the following homologous series into a gas chromatograph using as many of the columns in Table 8-1 as are available:

a. *n*-pentane, *n*-hexane, *n*-heptane, *n*-octane
b. benzene, toluene, ethylbenzene, *n*-butylbenzene
c. methanol, ethanol, 1-propanol, 1-butanol

TABLE 8-1 COLUMNS FOR USE IN CHAPTER 8

Liquid phase composition	Percent*	Solid support	Temp. (°C)	Flow rate (ml/min)	Length+ (ft)
Bentone 34 - diisooctyl phthalate (50/50)	12	60-80 mesh, acid washed, silanized Chromosorb G	90	35	15
Apiezon L	10	60-80 mesh, Chromosorb W-AW	70	45	8
Carbowax 400	12	60-80 mesh Chromosorb W-AW-DWCS	70	45	6

* Of total column packing, liquid phase and solid support.

\+ Of 1/4 in. diameter column.

Inject 5 μl of each compound separately; do not inject as a mixture. Determine the retention times (RT) and peak heights for each compound. Prepare a table with headings shown in Table 8-2 (for as many columns as were available) for the series of compounds that you chromatographed.

The instructor will assign students into groups of three; each student of the group chromatographs a different series of the compounds listed on page 60. Compare the retention times and shapes of the peaks of the three series of compounds on the different liquid phases, using the data obtained by the other two members of your group in addition to your own data.

TABLE 8-2 CHROMATOGRAPHY OF HOMOLOGOUS SERIES

	Apiezon			Carbowax 400			Bentone 34 - diisooctyl phthalate		
Compound	RT	Peak ht.	log RT	RT	Peak ht.	log RT	RT	Peak ht.	log RT

Problem 8-6. How can the difference of the retention times on the various liquid phases be rationalized?

Problem 8-7. For the series of compounds you chromatographed plot the log of the retention time (y-axis) versus the number of carbon atoms (x-axis) for each compound in the series. What kind of curve is obtained? Make this plot available to the other two students in your group.

Problem 8-8. Calculate the retention times on at least one of the liquid phases given in Table 8-1 for each of the following compounds: 1-pentanol, *n*-butane, *n*-nonane, *n*-propylbenzene.

Chromatography of Heating Gas

Place the needle of a gas syringe[3] far into the outlet of a gas jet. Turn on the gas and slowly withdraw 5 ml of the gas; turn off the gas. Chromatograph the gas sample on the Apiezon column[4] and on the other two listed in Table 8-1 if they are available. From the retention times of gases supplied by the instructor determine the identity of as many compounds in the gas as possible. Is the gas natural or manufactured?

Repeat the chromatography on the Apiezon column using 0.5 ml of heating gas. If possible, the attenuator should be set so that the response of the detector is ten times more sensitive for this smaller sample than it was for the first chromatogram. See Figure 9-2 for operation of attenuator.

Problem 8-9. Note that with the smaller sample the peaks are sharper than with the larger sample. The peak height times the attenuation for the compounds of the small sample is more than one-tenth as much of the same quantity for the compounds of the large sample even though the ratio of the samples was ten to one. Explain these two facts.

Gas Chromatographic Analysis of Tobacco Smoke (Class Project)

The material to be chromatographed is obtained by having several of the smokers in the class partially blow up balloons by exhaling smoke. As soon as possible after obtaining the smoke, sample by placing the

[3] These are expensive precision syringes with Teflon plungers carefully matched to machined glass barrels. If these are unavailable, ordinary medicinal hypodermic syringes may be used. These all-glass syringes may have gas leaks and should not be used for quantitative analysis of gases.

[4] For better resolution in the chromatographic analysis of the heating gas the Apiezon column should be at 40°. If heat leakage from the injector and detectors prevents this low a temperature from being used, the column oven temperature should be set 4 to 9° *above* the lowest temperature possible.

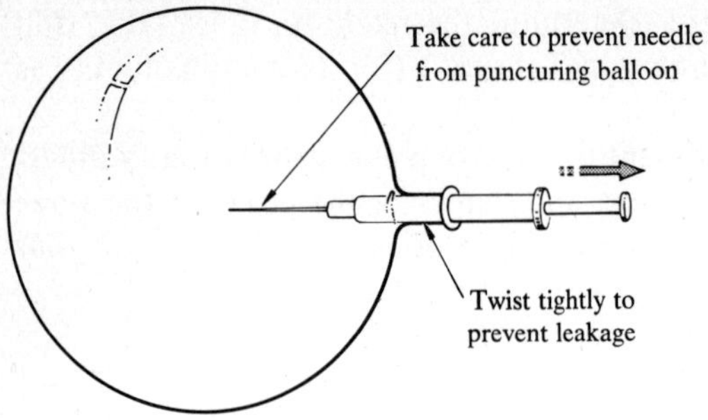

Figure 8-14 Sampling gas from a balloon.

opening of the balloon around the tip of the syringe with the needle well inside (Fig. 8-14) and withdrawing 5 ml. Chromatograph this smoke sample as soon as possible.

Each person in the class should chromatograph one smoke sample on one chromatographic column. The class should chromatograph smoke from at least two different brands of cigarettes and, if possible, samples of cigar and pipe smoke. Each type of smoke should be chromatographed on as many different liquid phases as possible.

From retention times of standard compounds supplied by the instructor identify as many compounds as possible. A compound should not be considered identified until retention times coincident with the retention times of a standard are found on at least two, and preferably three, liquid phases.

EXERCISES

8-1. In addition to the advantages of GLC over fractional distillation given earlier, nothing comparable to azeotrope formation has been found with GLC. Suggest an explanation for this.

8-2. What is the difference between *adsorption* and *absorption?*

9

Preparation of 2-Methyl-2-butene

An *E*1 Reaction

$$\underset{\substack{\text{B.p. } 102.5^\circ \\ D_4^{20}\ 0.812}}{\underset{\text{OH}}{\mathrm{CH_3CH_3\overset{|}{C}(CH_3)_2}}} \xrightarrow{H^+} \underset{\oplus\text{OH}_2}{\mathrm{CH_3CH_2\overset{|}{C}(CH_3)_2}} \xrightarrow[(-H_2O)]{} \mathrm{CH_3CH_2\overset{\oplus}{C}(CH_3)_2}$$

$$\mathrm{CH_3CH_2\overset{\oplus}{C}(CH_3)_2} \xrightarrow[(-H^+)]{} \underset{\substack{\text{B.p. } 38.4^\circ \\ D_4^{20}\ 0.6620}}{\mathrm{CH_3CH{=}C(CH_3)_2}}$$

INTRODUCTION

This experiment illustrates the acid-catalyzed dehydration of a tertiary alcohol to a volatile alkene. A separatory funnel and a fractionating column are used for the preparation and purification of a volatile liquid.

EXPERIMENTAL PROCEDURE

Reaction

Into a 100-ml round bottom flask put 17 ml of ice water. Slowly add 13 ml of concentrated sulfuric acid while swirling the flask in a beaker of ice water. Add 18 ml (14.6 g, 0.33 moles) of *t*-amyl alcohol (2-methyl-2-butanol) to the acid-water mixture. Add two boiling chips and place the flask in the fractional distillation apparatus shown in Figure 5-5.

The water level in the beaker in which the round bottom flask is placed should be about 1 cm above the level of the contents of the flask and should be kept at this level during the heating. The beaker in which the receiving flask is placed should be filled with ice water. Heat the water bath to about 85°; some of the product should begin to collect in the receiver. Maintain the water bath temperature at about 85° for 10 min. Slowly raise the temperature of the water bath until it begins to boil. Maintain the water bath at the boiling point until no more product collects in the receiver.

Separation and Purification

Pour the receiver contents into a separatory funnel and drain any aqueous layer. Wash with two 20-ml portions of water, first using the water washes to rinse any remaining product from the receiving flask into the separatory funnel. Follow the procedure given in Chapter 6 for use of the separatory funnel. Add about 1 g of calcium chloride granules to the washed product in the separatory funnel and shake intermittently for 5 min. Dry the stem of the separatory funnel with a pipe cleaner and draw the product into a 10-ml distilling flask.

Distil through the shortest condenser available into a tared yield bottle which has been placed in an ice bath (Fig. 9-1). Collect the fraction that boils at 37 to 39°. Calculate the yield and % yield. Label the container as instructed in Chapter 7 and hand it in to your instructor.

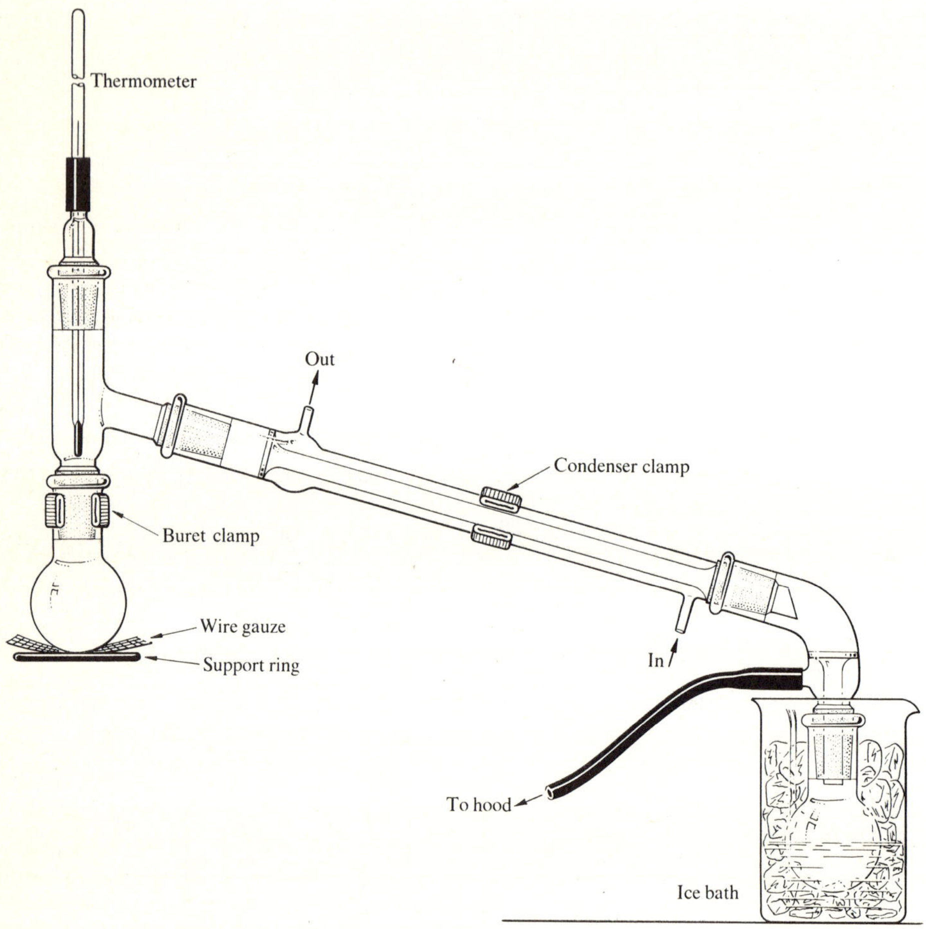

Figure 9-1 Distillation of volatile compounds.

Problem 9-1. What by-products are likely to be produced by this reaction?

Analysis of Product (Optional)

Chromatograph 10 μl of purified product on the Apiezon column using those conditions specified for chromatography of heating gas in Chapter 8. The attenuator should be set so that the largest peak is attenuated 20 to 50 times to ensure that all the by-products can be seen. As soon as a large peak passes, reduce the attenuation to its original value. See Figure 9-2

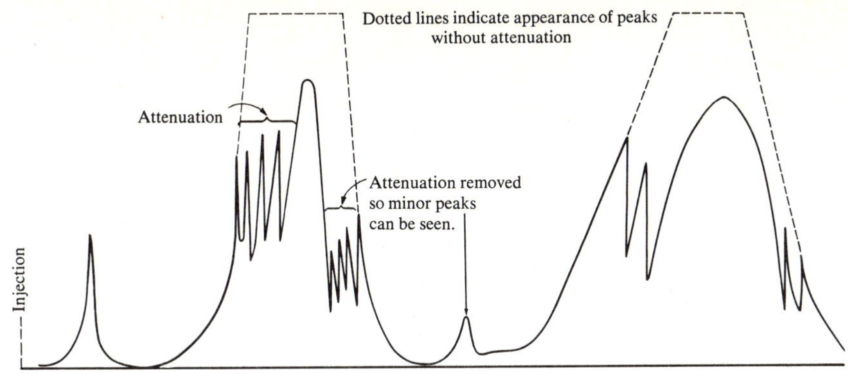

Figure 9-2 Chromatogram of a mixture with two major components and two minor components. Peak 2 from the major component is attenuated twenty times. Peak 4 from the second most prominent component is attenuated five times. This particular chromatograph had attenuation steps of 2, 5, 10, 20, 50, and so forth.

Problem 9-2. What elimination by-products are possible? Verify your answer in the answer section. In what order would you expect these compounds to be eluted from an Apiezon column? Hint: the van der Waals attractions between the molecules of Apiezon wax and the molecules of the several alkenes are probably similar. Differences in retention times among these alkenes are probably due to other attractive forces (pp. 12–15) and factors not related to the solubility of the alkene in the Apiezon wax (see pp. 14 and 36).

Your instructor will supply the retention times and peak height-concentration data for a number of compounds. Identify the compounds in your product and estimate their concentration. Does what you find agree with your answer to Problem 9-1?

You may be asked to save a portion of your product for analysis by infrared spectroscopy when Chapter 18 is studied. If so, be sure the bottle is tightly sealed because the product is very volatile.

EXERCISES

9-1. What is the purpose of the fractionating column?

9-2. What is the purpose of the two water washes?

10

Electrophilic Addition to Alkenes

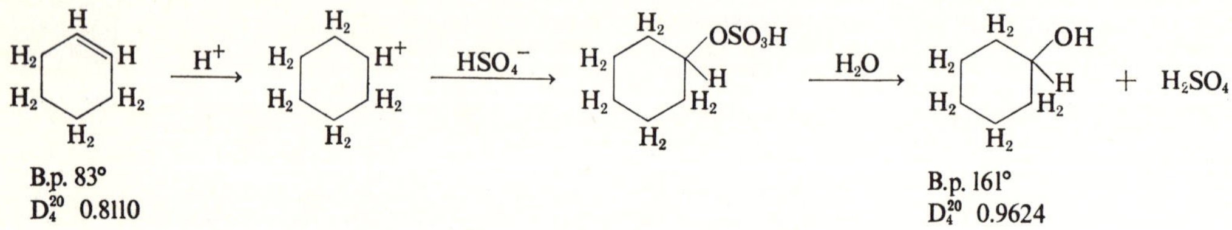

Cyclohexanol

INTRODUCTION

A secondary alcohol is synthesized by addition of sulfuric acid to an alkene and hydrolysis of the resulting ester. The use of a reflux condenser is introduced. Refluxing is boiling a liquid, condensing the vapors, and returning the liquid so obtained to the boiling liquid. It is a technique that allows the heating of a mixture without loss of the components by evaporation.

EXPERIMENTAL PROCEDURE

Clamp a 250 ml round bottom flask to a ring stand so that it can be cooled by an ice bath or refluxed. Keep the flask in this position during the synthesis except when the vigorous agitation is required. Add 7 ml of water and place an ice bath around the flask (Fig. 10-1).

Slowly add 22 ml of concentrated sulfuric acid while stirring the mixture with a glass rod. When the mixture has cooled to room temperature or below, remove the ice bath but keep it close as it may be needed in the next step. Slowly add 29 ml (23.5 g) of cyclohexene to the flask while gently swirling the ring stand. Stopper the flask and shake the mixture vigorously until the separate aqueous and organic phases become homogeneous. This usually requires 12 to 20 min. Stop shaking the flask every minute and open the stopper to release any pressure. If the flask becomes too hot to hold, cool it briefly in the ice bath. After the mixture becomes homogeneous, allow it to stand for 5 min and then slowly add 95 ml of water. Fit the flask with a reflux condenser (Fig. 10-1), add a boiling chip, and reflux the mixture for 25 min, i.e., heat the mixture strongly enough that it boils and the vapors are condensed and returned to the liquid. Allow to cool for a few minutes and then place the ice bath again in position.

Pour the reaction mixture into a separatory funnel and drain the bottom layer. Extract with a 10-ml portion of water, a 5-ml portion of 10% aqueous sodium hydroxide, and two 5-ml portions of water. Follow the procedure outlined in Chapter 6. Pour the cyclohexanol from the *top* of the separatory funnel into a 50-ml Erlenmeyer

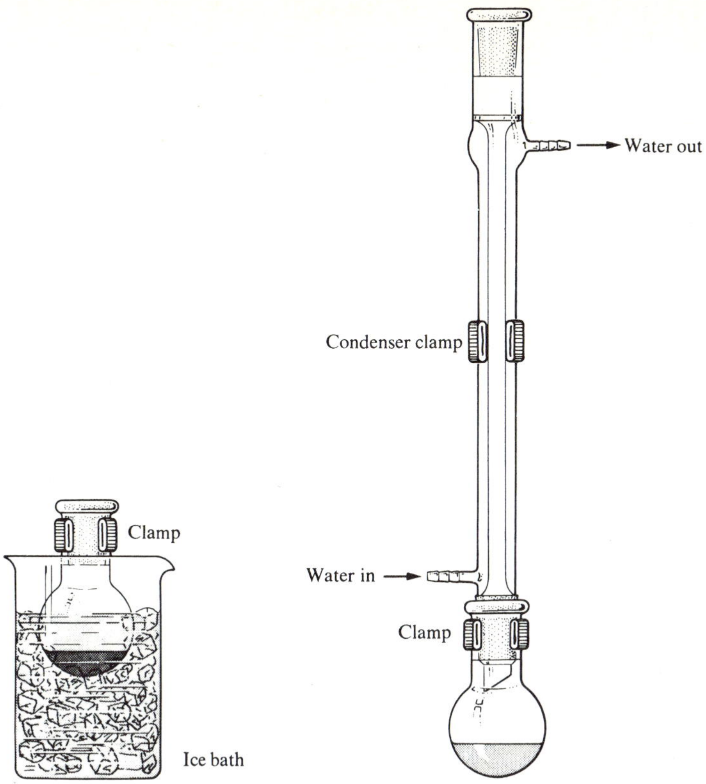

Figure 10-1 Flask with ice bath followed by change to a reflux apparatus.

flask which has approximately 2.5 g of magnesium sulfate in it. Be sure any stopcock grease is wiped from the ground glass surface before the liquid is poured. Swirl the flask intermittently for 10 min and filter the cyclohexanol through a *very small* amount of glass wool in the tip of a small funnel into a 25-ml distilling flask. (Use a 25-ml round bottom flask with a ground glass joint if standard taper equipment is being used.) Press the wool with a glass rod to squeeze out the last few drops of cyclohexanol.

Attach a condenser to form the distillation apparatus shown in Figure 3-1. Distil (as in all distillations be sure there is a fresh boiling chip in the distilling flask); collect the fraction boiling at 155 to 160° in a tared yield bottle. Calculate the theoretical yield and the % yield, label the bottle according to the instructions in Chapter 7, and hand it in to the instructor.

EXERCISES

10-1. Why is the ratio of sulfuric acid to water critical in this experiment?

10-2. What is the purpose of the sodium hydroxide wash? What might happen if it were omitted?

10-3. Why is magnesium sulfate used instead of calcium chloride as the drying agent?

11

Melting Points

INTRODUCTION

Melting points are useful for identifying and estimating the purity of organic compounds. Melting points are readily obtained with great precision on simple apparatus. As a result the melting point is the most widely determined and useful physical constant. The accuracy of melting point determinations are limited mainly by the accuracy of the thermometer used.

EXPERIMENTAL PROCEDURE

Capillary Tubes

Heat a piece of glass tubing about 40 cm long and 4 to 7 mm in diameter in the flame of a Fisher burner or Bunsen burner equipped with a wing tip. Hold the tubing in the flame as shown in Figure 11-1, slowly turning the tube as you heat it. When the tube becomes soft, *take it out* of the flame and *immediately* pull the ends of the tube without twisting them.

A little practice will show how hard the tube should be pulled and if any pause is needed after removing the tube from the flame to obtain the proper diameter for various uses. A capillary the right size for finely divided solids can be obtained by pulling the tube as far as possible. For coarse or tacky compounds wider capillaries are needed. Usually the wider the tube the thicker the walls; thick walls retard heat transfer from the oil bath to the compound and are to be avoided.

Cut the capillaries into 20-cm lengths. Seal both ends by holding them *briefly* in a Bunsen flame. When a capillary is needed, cut one of these in two. The advantage of this procedure is that no dirt can enter the tubes until they are cut.

Filling the Tube

Scrape a small amount of dry, finely divided powder into the open end of the capillary. Tap the closed end of the tube gently on the bench until the solid falls to the bottom of the capillary tube. If the material does not

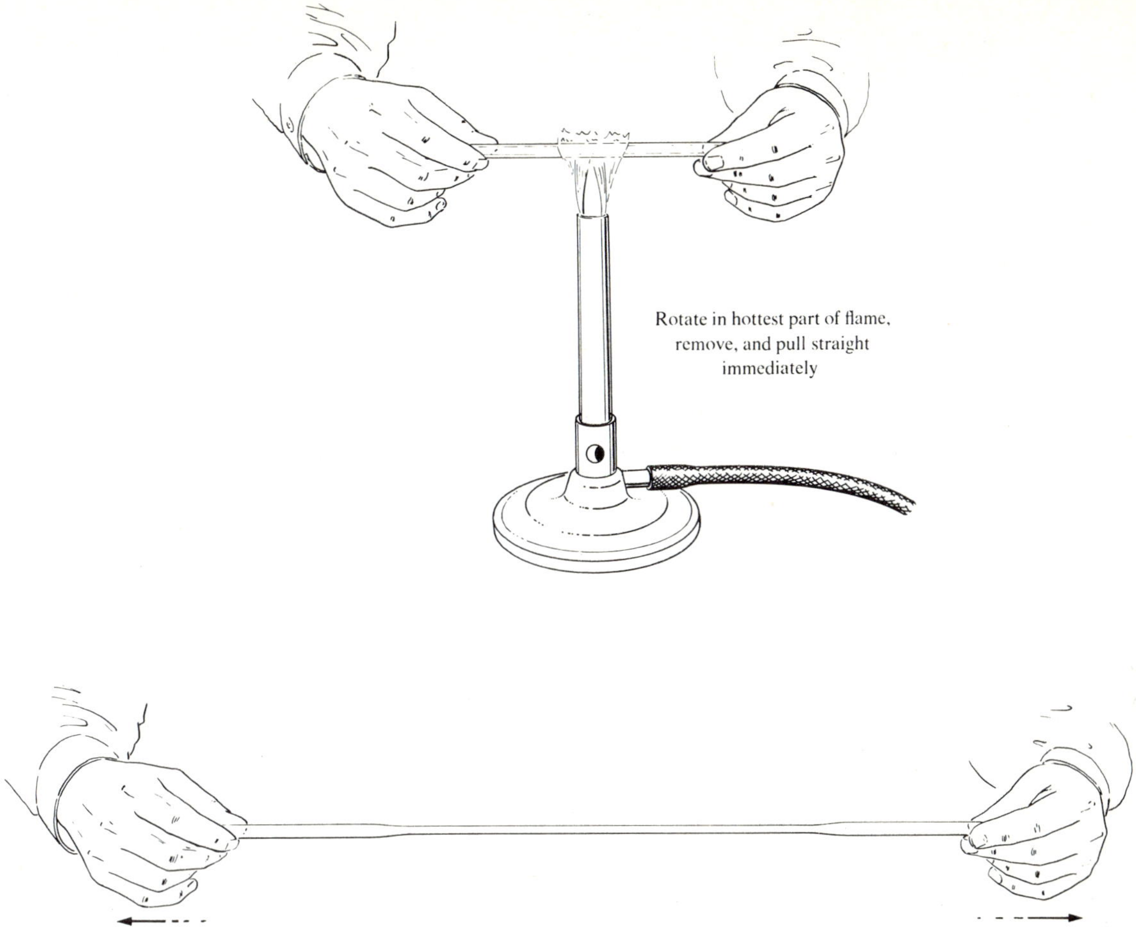

Figure 11-1 Making capillary tubes.

fall to the bottom, vibrate the tube by gently drawing a file across the top of the tube. Tap the tube on the desk again. If the material still sticks in the top of the tube, more forceful procedures are required. Drop the capillary tube sealed end down through a 3-ft piece of glass tubing held on the bench top (Fig. 11-2). If this is not successful, it may be necessary to take a fine piece of glass tubing and ram the compound down the tube. This last method may prove necessary for only a few types of solids such as paraffin hydrocarbons. Usually failure of the solid to fall to the bottom of the capillary when dropped through the glass tube indicates that the solid is impure or not dry, and it should be dried or purified before a melting point is determined.

Apparatus

Figure 11-3 shows some different types of melting point apparatuses. The ones used most often in student laboratories are Thiele tubes or the manually stirred beakers.

Silicone oils are most suitable for bath fluids; however, these are expensive. The best choice when stability, safety, and cost are considered is a thick type of mineral oil or *Nujol*. This material can be used repeatedly at temperatures up to 220° without significant darkening or other decomposition. It can even occasionally be heated as high as 270° for short periods of time without too much darkening. Glycerine is sometimes used, but if heated too much it gives acrolein which is a very potent lachrymator. If the oil bath gets so dark (usually a

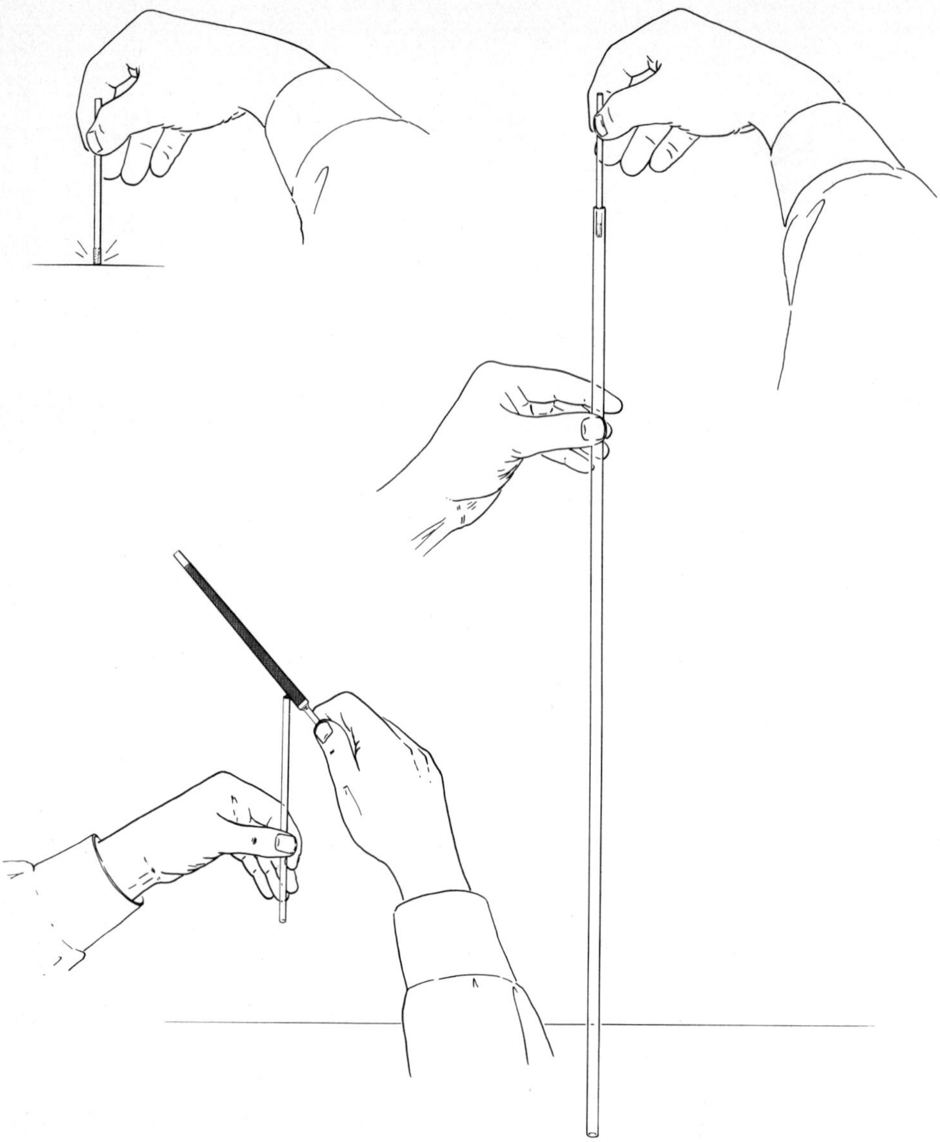

Figure 11-2 Filling capillarytubes.

deep yellow) that the melting point and color changes are not observed easily, it should be thrown out. The melting point baths should not be filled above the dotted lines shown in Figure 11-3 when the oil is at room temperature. Like all liquids the melting point bath fluids expand when heated. Wipe any spilled oil from the side of your melting point bath before heating.

Problem 11-1. If brought into direct contact with a flame, the mineral oil burns; however, mineral oil is not dangerously flammable. Mineral oil is a mixture of aliphatic hydrocarbons. Many other hydrocarbon mixtures such as gasoline and petroleum ether are very dangerously flammable. Why is minera oil so much less flammable?

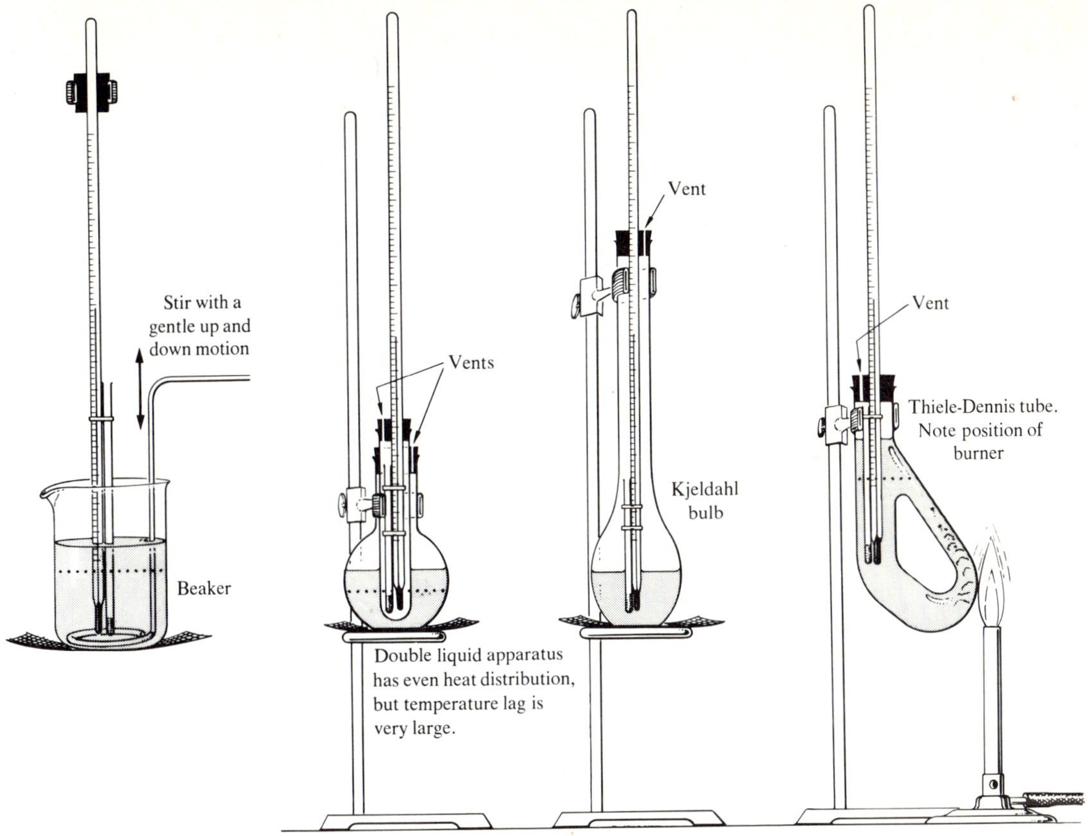

Figure 11-3 Melting point apparatuses.

Observation of Melting Points

The capillary tube is fastened to the thermometer as shown in Figure 11-4. Note that the bottom of the capillary with the compound in it is touching the thermometer bulb. If it were placed a significant distance from the bulb, there could be an error because of a temperature gradient in the liquid.

Note how the *top* of the capillary is fastened to the thermometer. A string, piece of tape, label, or rubber band may be used. Use a fresh rubber band for each laboratory period as the mineral oil vapors attack rubber. If a piece of rubber falls into the hot mineral oil, it will quickly dissolve and foul the oil.

See Figure 11-3 for correct location of thermometer and burner flame for the type of apparatus that you are using. Duplicate the location of the thermometer in the diagrams as closely as possible especially if you are using a Thiele tube.

Heat the melting point bath very rapidly until the temperature is 10 to 12° below the melting point of the solid. At this point heat the bath at about 2° per minute. At 1 to 3° below the melting point of the solid heat the bath approximately 1° per minute until the solid has completely melted. If the melting point is unknown, considerable time usually can be saved by heating the bath very rapidly to obtain the approximate melting point and then following the above procedure.

Determine the melting points of naphthalene, cinammic acid, and sucrose. When the compound has completely melted, remove the flame and allow the bath to cool. Observe carefully what happens as the compound cools through its melting point.

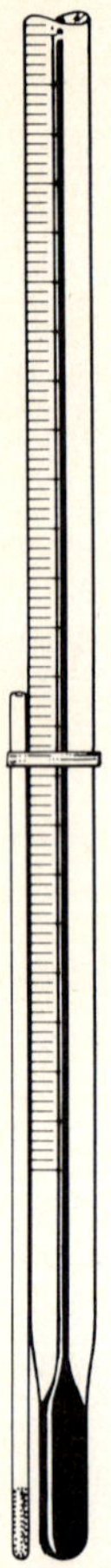

Figure 11-4 Attachment of capillary tube to the thermometer.

Record the data in your notebook for each of the compounds. The melting point should always be reported as a range; the beginning of the range is the temperature at which liquid first appears, and the end of the range is the temperature at which the last solid melts. Occasionally a compound is contaminated with a small amount of an inorganic salt or another very high melting compound. The presence of a *few* grains of such material, which usually has a different texture than the solid that is melting, should be ignored when reporting the melting point range. The solidification or freezing point should be recorded as the temperature at which the last trace of liquid solidifies to the temperature at which the first solid appears.

Many compounds do not melt before they decompose. When this occurs, the decomposition point is recorded as the melting point. To indicate such behavior d is often written after the melting range as 104 to 105.5°d. The decomposition is evident when there is a marked color change, an evolution of gas, or an explosion.

Sometimes the compound may soften, contract, or undergo a phase change. The temperature of such change should be noted as this may help characterize the compound; however, this should not be confused with the beginning of the melting range.

DISCUSSION

Within certain limits the melting point is a very reproducible constant. Among the factors causing error are:

a. Impurities usually cause a sample to melt below the pure compound and have a wide melting range.
b. Too rapid heating causes uneven heat transfer and the observed temperature to be different from that in the sample. Rapid heating may also cause an error of observation because the mercury thread rises so rapidly.
c. Decomposition points are often influenced by the *length* of the time the sample is heated.
d. Thermometer errors are the limiting factors if proper techniques are used in isolating, purifying, and melting the sample. These can be minimized by calibration of the thermometer.

The freezing point is often sharper and more accurately determined than the melting point; however, it has many limitations:

a. It can not be used on compounds that have decomposition rather than melting points.
b. Supercooling (see below).
c. Sometimes only a small amount of the compound decomposes or reacts on melting and the decomposition is not evident. The compound formed by the decomposition contaminates the original and the less pure compound usually freezes at a lower temperature than it melted.

When the freezing range is higher or narrower, but not lower, than the melting range, the freezing range is usually the more accurate of the two and should be reported.

SUPERCOOLING AND ORGANIZATION

Liquids will not always solidify when cooled to or below their freezing point. The temperature of the surroundings may be at or below that required for heat to flow from the liquid as the intracrystalline forces are established, but unless the molecules are correctly *organized,* they cannot form a crystal lattice. The more complicated and flexible the molecules, the more organizational demands there are in forming the crystal lattice. Compare the crystal and liquid phases of hexane in Figure 2-7 and the isomers of C_5H_{12}, C_4H_9OH, and $C_5H_{11}OH$ on pages 217–221.

The converse never happens; compounds are never solids above their freezing points. Disorganization is spontaneous; organization is not.

MELTING POINTS OF SOLIDS BETWEEN −11° AND AMBIENT TEMPERATURES

Experimental Procedure

Place a thin capillary containing about 5 mm of liquid in a beaker melting point apparatus (Fig. 11-3) half full with water. Add a few pieces of ice at a time and stir the mixture until the liquid in the capillary freezes. If it is necessary to go below 0°, pour out some of the water and add calcium chloride and more ice.

Record the freezing range. When the liquid solidifies, remove all the ice except one small piece. Allow the mixture to warm while stirring. If the compound melts above 18°, it may be necessary to warm the back and the sides of the beaker with your hand. Determine the melting points of benzene and indene.

Discussion

Freezing points determined this way are not subject to errors from decomposition. Supercooling, however, is a greater problem with compounds that have low melting points because molecular motion is slower at low temperatures and the molecules orient themselves more slowly to give the organized crystals.

MELTING POINTS OF MIXTURES

Experimental Procedure

Make mixtures of urea and cinnamic acid in ratios (by weight) of 1:3, 1:1, and 3:1. Using a small spatula chop and grind each mixture on a porcelain plate or crucible cover until it is finely divided and homogeneous. Determine the melting point ranges of these mixtures. Use this and the data obtained earlier on the pure compounds to plot a graph. Plot melting points on the ordinate (y-axis) and composition on the abscissa (x-axis). Note that units are weight percent as the measurements were by weight. Start with 100% urea at the origin.

Discussion

Figures 11-5 through 11-9 show possible variations of melting point with composition of binary mixtures. Figures 11-5, 11-6, and 11-7 show the behavior if there is compound or complex formation

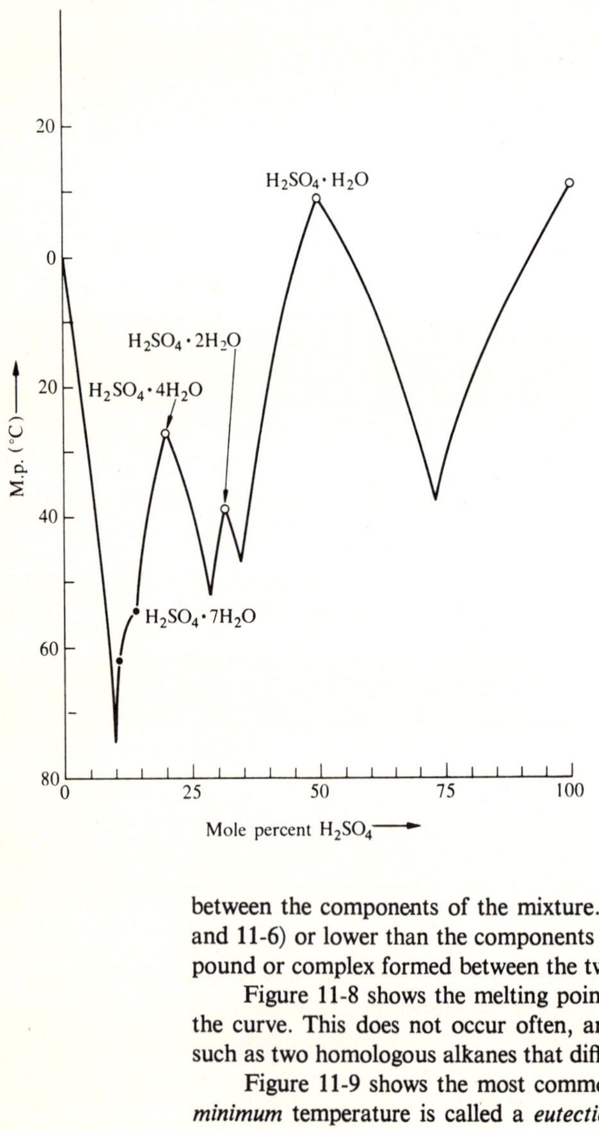

Figure 11-5 Melting point-composition diagram for sulfuric acid and water.

between the components of the mixture. The new compounds or complex may melt higher (Figs. 11-5 and 11-6) or lower than the components (Figs. 11-5 ,11-6, and 11-7). There may be more than one compound or complex formed between the two components (Fig. 11-5).

Figure 11-8 shows the melting point-composition relation if there is no maximum or minimum in the curve. This does not occur often, and when it does, the two compounds are usually very similar, such as two homologous alkanes that differ by only a few methylene groups.

Figure 11-9 shows the most common melting point composition curve for a binary mixture. The *minimum* temperature is called a *eutectic* temperature, and the mixture is called a eutectic mixture or

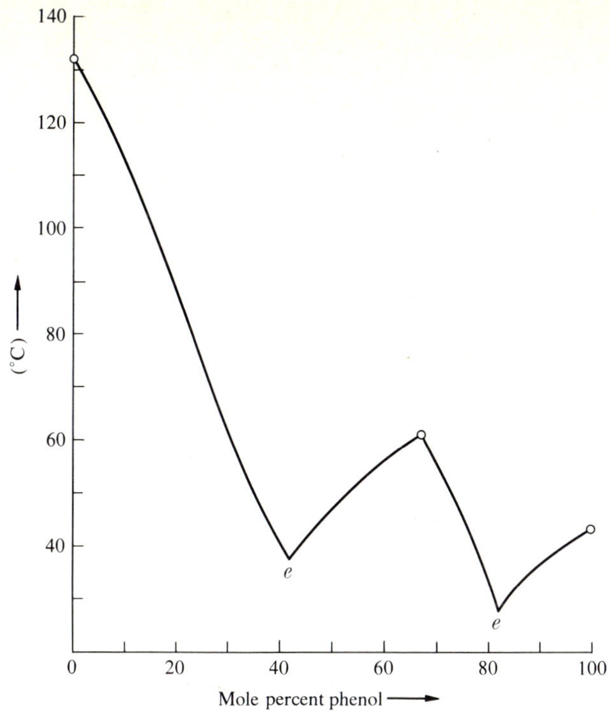

Figure 11-6 Melting point-composition diagram for urea and phenol.

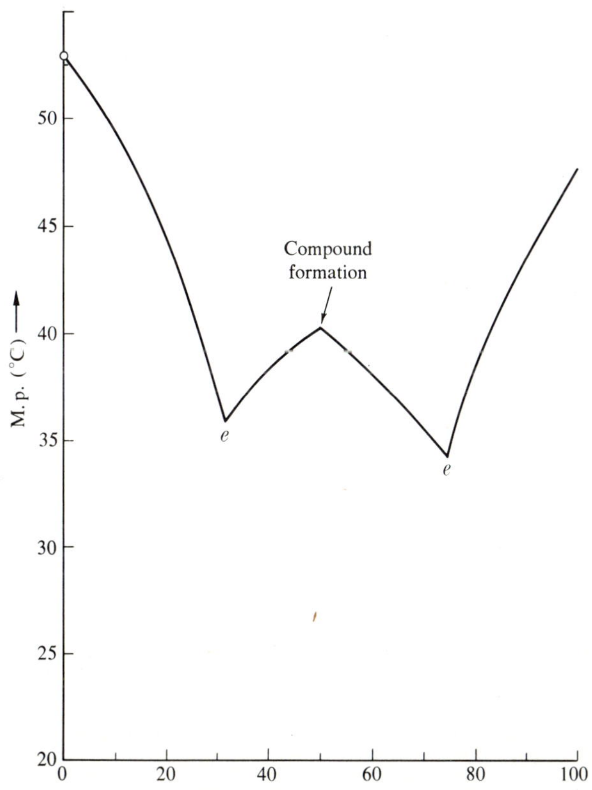

Figure 11-7 Melting point-composition diagram for diphenylamine and benzophenone.

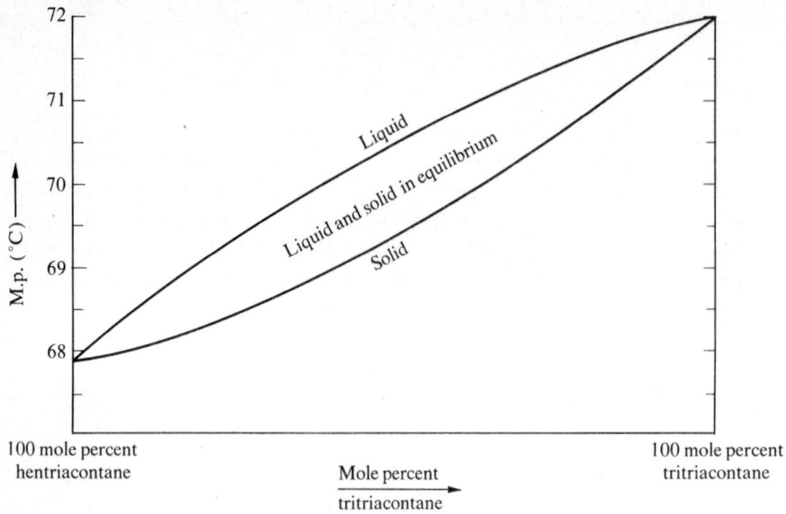

Figure 11-8 Melting point-composition diagram for hentriacontane and tritriacontane.

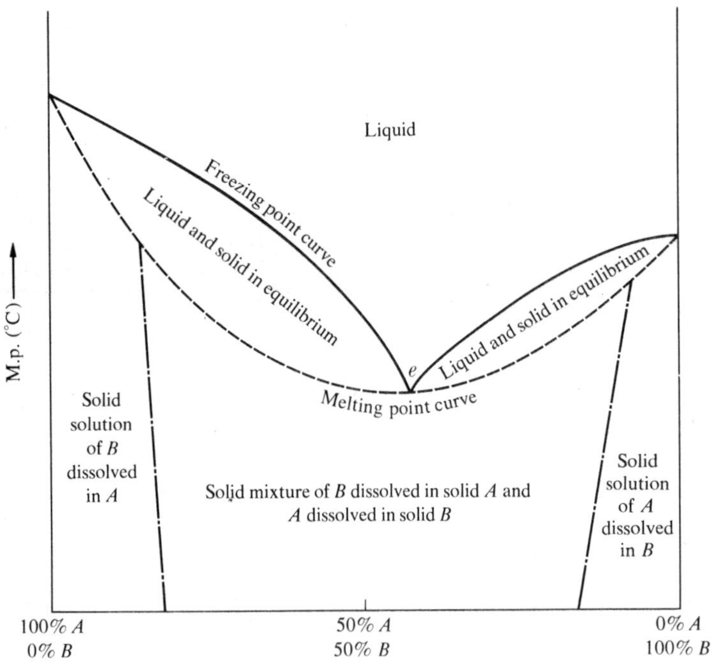

Figure 11-9 Most common type of melting point-composition diagram for a binary mixture.

sometimes a eutectic. If two organic solids randomly selected are mixed, the chances are probably better than 10 to 1 that they form a eutectic. This is basic to the fact mentioned earlier in this chapter that an impure compound usually melts lower and over a wider range[1] than the same compound when pure. The minima found in composition-melting point diagrams that show evidence of compound or complex formation are also called eutectics (point *e* on Figs. 11-6, 11-7, and 11-9).

[1] A wide melting range is an indication but not proof of an impure compound. Wide melting ranges are observed for some pure compounds.

MIXED MELTING POINTS

Discussion

A mixed melting point (sometimes called a mixture melting point) is useful in the identification of a solid. If an unknown compound is mixed thoroughly with a known that melts at approximately the same temperature, the mixture will not have a depressed melting point if the two samples are the same. If they are not the same, the mixture probably will have a depressed melting point. Or conversely, if a mixture of two compounds with similar melting points melts below that of the pure compounds, they are *not* identical; if the mixture melts at the same temperature as the pure compounds, they have a 90% chance of being identical.

Experimental Procedure

Place the unknown and known compounds in separate capillary tubes and label each one. Take a small amount of each, mix and crush them together thoroughly on a porcelain plate or watch class. Place this in capillary and label it.

Attach the three tubes to a thermometer (Fig. 11-10), place them in the melting point apparatus, and observe

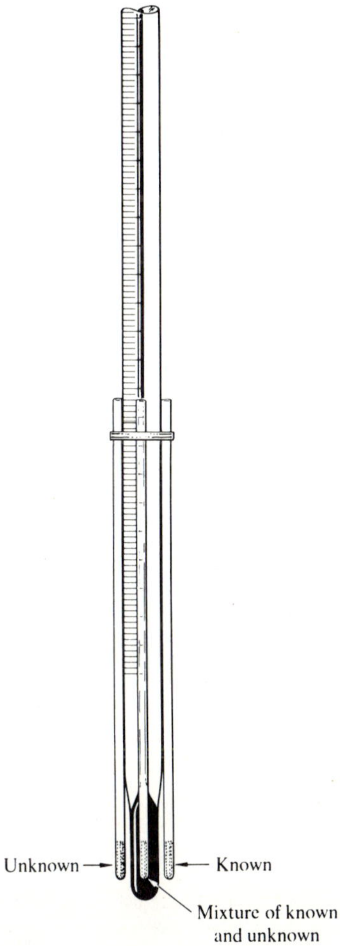

Figure 11-10 Arrangement of melting point tubes for a mixed melting point.

all three melting points. The advantage in observing all three at once is that almost all errors likely in melting point determinations, as those due to temperature gradients, observation, and the thermometer, do not affect the *relative* melting points of the mixture and its two components. Melting point depressions as small as 1° can be detected with ease and certainty by this method.

Hand in a clean, dry stoppered test tube to your instructor. It will be returned with one of the compounds of Table 11-1 in it. Identify the unknown.

TABLE 11-1 MELTING POINT UNKNOWNS

Compound	M.p. (°C)	Compound	M.p. (°C)
n-Eicosane	36.8	Naphthalene	80.5
o-Methylbenzylalcohol	36	4-Hydroxy-3-methoxybenzaldehyde	81
p-Cresol	36	Acrylamide	84
1,6-Diaminohexane	42	o-Hydroxybenzylalcohol	86.5
2,6-Xylenol	42	D-Ribose	87
Pinacol	43	Semicarbazide	96
Lauric acid	43	Acenaphthene	96.7
4-Methaniline	45	Glutaric acid	98
Benzophenone	47	Acetanilide	114
1-Chloro-2,4-dinitrobenzene	51.5	Fluorene	114-6
4-Methoxy-2-hydroxyacetophenone	52.5	Quinone	114-6
2,2-Dimethyl-1-propanol	52.5	Benzoic acid	122.5
1,2-Diphenylethane	53	2-Naphthol	123
N-Phenylurethane	53	Xanthydrol	123d
4-Methoxyphenol	55.5	Urea	133
Benzhydrol	68	trans-Cinnamic acid	133
Biphenyl	69	p-Nitroaniline	174
Stearic acid	70.5	Cholesterol	148
8-Hydroxyquinoline	73	d-Mannitol	166
Glycolic acid	79	p-Bromoacetanilide	166.5

EXERCISES

11-1. Redraw Figure 2-8 to include supercooling of a liquid.

11-2. (a) Propose a reason for the fact that most mixtures of compounds melt below the melting points of any of the components of the mixture. Hint: what is required for crystal formation other than sufficient intracrystalline attractions? (b) Why might you expect homologous alkanes that differ by only a few CH_2 groups not to depress the melting point of each other?

12

Crystallization and Solubility

INTRODUCTION

Crystallization is a very useful technique for the purification of organic solids. Knowledge of a compound's solubility aids in the selection of an optimum solvent for crystallization. An estimation of the solubility of a solid in a solvent can be made from the structures of the solid and the solvent and the intermolecular forces involved.

DISCUSSION

Solubility of Solids in Liquids

Energy Required. When a substance is dissolved, its molecules or ions are dispersed intimately among the solvent molecules. Energy must be supplied to overcome the crystal lattice energy. If a dilute solution is formed, approximately the same energy is required to vaporize a solid as to dissolve it. For example in ionic solids the distances between neighboring ions are of the order of 3 Å. In a one molar solution, ions of unlike charges are approximately an average of 12 Å apart, and in more dilute solutions the distance is greater. Strong electrostatic attraction must be overcome to separate ions of unlike charge. This is very difficult, as is shown by the very high melting points of most ionic substances. If a solid dissolves in a liquid, the new solvent-solute interactions must supply enough energy to overcome the crystal lattice energy.

Solubility of Ionic Compounds. The work required to separate two oppositely charged objects is lowered by the introduction of matter between them by a factor which is called the *dielectric constant* or ϵ of the medium. (See Appendix II for an explanation of dielectric constant.) Water, with a high dielectric constant of 80 which facilitates the separation of ions, dissolves many ionic substances readily. Ether (ϵ = 4.4), benzene (ϵ = 1.9), and similar compounds are poor solvents for salts because of the greater work required to separate charges in a medium with a low dielectric constant.

Water molecules, which are small dipoles, orient themselves between oppositely charged ions (or the charged plates of a condenser) in such a way as to partially neutralize the charges and thus stabilize the system. Or, stated somewhat differently, water is such a good solvent for most ionic substances because of the ion-dipole and ion-induced dipole attractions that exist when ions are dissolved in water. The formation of these ion-dipole attractions supplies the energy necessary to overcome the ion-ion attractions so that the ions can leave the crystal lattice (Fig. 12-1).

Solvating ability and dielectric constant are not exactly parallel since a high dielectric constant is usually necessary but not sufficient for an effective ion solvent. For example, hydrogen cyanide with a dielectric constant of 116 is a poor solvent for most salts. Also ionic substances such as silver bromide,

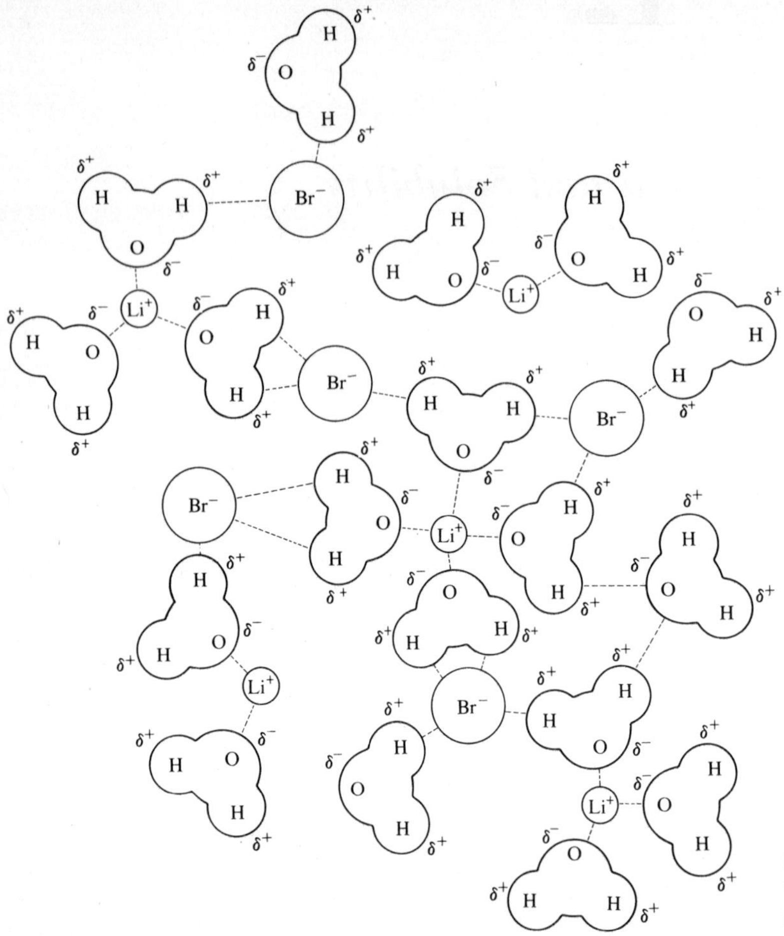

Figure 12-1 Ion-dipole attractions between water and lithium bromide (two-dimensional schematic, not drawn to scale).

cadmium sulfide, and barium sulfate do not dissolve in water, and some substances of only moderate dielectric constant like sulfur dioxide (ϵ = 16) are good solvents for most ionic compounds.

Solubility of Nonionic Compounds. Many organic compounds (all polar) dissolve in water because of polar attractions between the water and the solute molecules (Fig. 12-2).[1] Water is a good solvent for organic compounds that contain 20% or more oxygen or nitrogen such as carbohydrates and low molecular weight acids, alcohols, and amines. The formation of polar attractions between these compounds and water releases enough energy to break the intracrystalline forces. Also, the strong polar attractions between these compounds and the water molecules are able to replace some of the attractive forces between water molecules and allow the water molecules to separate sufficiently to accommodate the solute molecules.

Nonpolar liquids are poor solvents for polar and ionic compounds because the relatively weak ion-induced dipole or dipole-induced dipole attractions between solvent and solute are not strong enough to supply sufficient energy to break the ion-ion or dipole-dipole attractions of the pure compound. This discussion can be generalized into the rule discussed earlier that "like dissolves like." That is, polar solvents generally dissolve only polar and ionic compounds readily, and nonpolar solvents dissolve only nonpolar compounds readily.

Problem 12-1. Sometimes a solid dissolves in a solvent to give a solution that is considerably colder than the components and the surroundings. (Potassium iodide in water is an example of such a system.)

[1] Hydrogen bonding is the principal polar attraction between hydoxylic solvents, such as water and the alcohols, and many compounds containing oxygen and nitrogen.

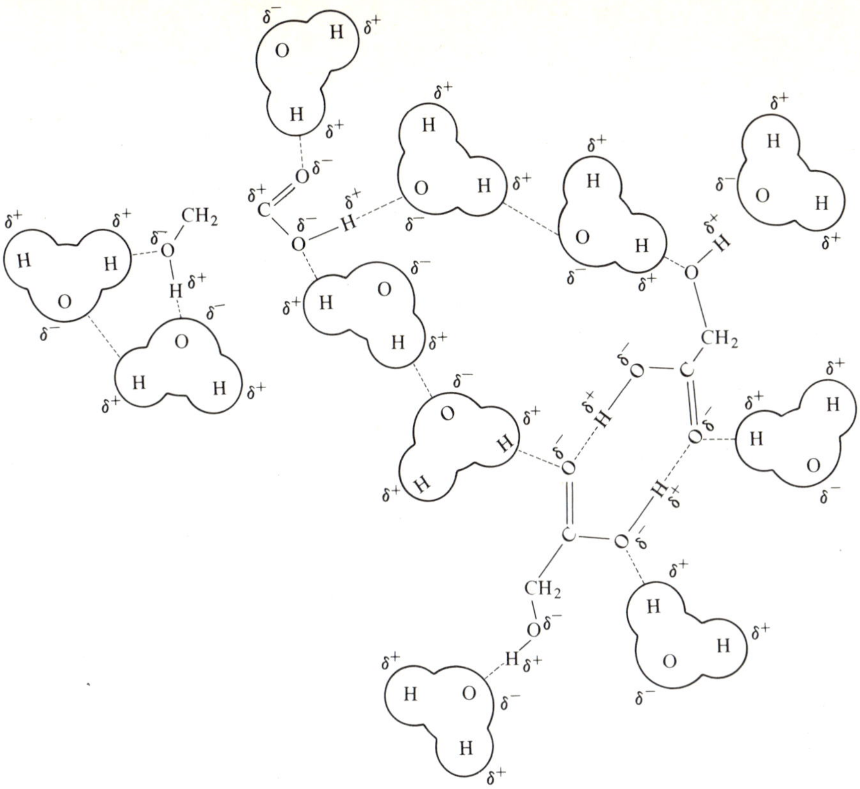

Figure 12-2 Polar attractions between water and glycolic acid.

Why does this occur? What happens when a solution becomes warmer than its components? (Sodium hydroxide in water is an example of this type of system.)

Problem 12-2. Dipole-induced dipole attractions of polar solvents for nonpolar molecules such as naphthalene and *n*-eicosane are stronger than the weak London dispersion forces that supply the only intracrystalline forces for nonpolar compounds. That is, the solvent-solute attractions supply more than enough energy to overcome the lattice energies. Why then do nonpolar compounds usually not dissolve in water or other very polar solvents?

Recrystallization

Description. Most solids are more soluble in most solvents at higher temperatures than at lower temperatures. The usual crystallization process involves dissolving enough solid in a hot solvent to give a saturated or nearly saturated solution and allowing the mixture to cool to room temperature or below. On cooling, the solid crystallizes from solution. The *mother liquor* (the solute that has not crystallized and the solvent) is removed from the crystals by filtering, centrifuging, or occasionally decanting from large crystals or a compact crystal mass. Usually the solvent is heated to or near its boiling point in order to dissolve the maximum amount of solid.

Often before being cooled the solution is filtered to remove any impurity that is insoluble in the hot solvent. Before the hot solution is filtered the mixture is often *clarified* by the addition of carbon black or some other material with a large surface area. *Clarification* is purification by adsorption of colored, resinous, or other easily adsorbed substances from solution.

Principles. One of the reasons for the success of this method is that crystal growth is usually very selective. Unless an impurity has molecules of similar size, shape, and polarity to the molecules of a growing crystal, not very much of the impurity is likely to be incorporated in the growing crystal. Contamination by occlusion or coprecipitation is not likely unless mixed crystals or isomorphism can occur.

To illustrate how crystallization purifies a solid assume there is a mixture of two substances *A* and *B*. The solubilities in ethanol are for *A*, 15 g/100 ml at 25° and 60 g/100 ml at 70°; and for *B*, 8 g/100 ml at 25° and 50 g/100 ml at 70°. If a mixture of 40 g of *B* and 15 g of *A* is dissolved in 100 ml of ethanol at 70° and the mixture allowed to cool to 25° the precipitate would be 32 g of *B* and the mother liquor would contain 15 g of *A* and 8 g of *B*. *B* would theoretically be 100% pure and 80% of it would have been recovered. In actual solutions the relations are not quite so simple as the example given here. The solubilities of a compound in a pure solvent and in a solution of that solvent (with another compound) are usually not the same. The differences, however, are often not large, and the figures given as examples above and in the problems illustrate what usually occurs in crystallizations even if the values given for solubilities would deviate somewhat.

Problem 12-3. Suppose there were 40 g of *A* and 40 g of *B* in a mixture. (a) What would be the composition of the precipitate if the mixture were dissolved in 100 ml of ethanol at 70° and cooled to 25°? (b) With the same temperature limits how much ethanol would be needed to dissolve the mixture and obtain the maximum amount of pure *B*? (c) From your answer to part b what would be the percent recovery of pure *B*? (d) How could pure *A* be recovered from the mother liquor in the procedure described in answer to part b?

Problem 12-4. What is the least optimum solubility ratio of *A* and *B* in a solvent at room temperature (assume this is the lowest temperature used) that is necessary to recover 80% of *pure B* from a mixture of 40 g of *A* and 40 g of *B* with only *one* crystallization? Assume no coprecipitation, adsorption, or mechanical loss.

Selection of Solvent. The above problems and discussion illustrate that proper selection of solvent is very important for optimum purification and separation by crystallization. Selection of the proper solvent is greatly facilitated by an understanding of the relations of structure, intermolecular forces, and physical properties. The discussion of these began in Chapters 2 and 6 and is continued in the first part of this chapter. Further discussions follow in later chapters as a better knowledge of structure is obtained in the lecture part of this course.

Even with a thorough knowledge of structure-solubility relations, it is sometimes necessary to search empirically for a good solvent. A convenient procedure for this is to place a few crystals of the compound into several 7.5-cm test tubes. Add 1 to 2 ml of a different solvent to each and heat to boiling. If the mixture does not dissolve, try crushing it with a glass rod while the mixture is boiling. Allow to cool to room temperature and observe the crystal formation. If no crystals form, chill in an ice bath and scratch the side of the tube. The purest crystals are probably those that are least colored (unless the compound is colored) and show the best growth and most definite structure.

From the results of these tests and what is known about the structure of the compound and the solvents, a suitable solvent or solvent mixture can be selected so that the following criteria are met: (a) The solvent should not react with the compound being crystallized. (b) The compound should be very soluble in the solvent near the boiling point and comparatively insoluble in the solvent at room temperature. (c) The impurities should either be insoluble in the solvent at high temperature or very soluble at low temperature. (d) The solvent should have a boiling point below the melting point of the compound. A solvent suitable in other respects but not in criterion d may still be used if the solution is not heated to a temperature above the melting point of the solid. If a solution is heated much above the melting point of the compound being recrystallized, the compound may separate from the solution as an oil rather than as crystals. (e) The solvent should have a boiling point within the range of 40 to 90°. If safety hot plates are available, the upper limit can be increased. The lower limit results from requirement b and can be made as low as room temperature if crystallization can be induced by cooling the solution below room temperature with ice-salt or dry ice-chloroform mixtures.

It has also been stated that, if possible, the solvent should be nontoxic, nonflammable, and inexpensive. Since the only solvents that meet these requirements are water and some of the Freons,[2] which

[2] Freon is a trade name for a group of polyfluoro and polyfluorochloro derivatives of low molecular weight aliphatic hydrocarbons made by DuPont.

are poor solvents for most organic compounds, some compromise is necessary. If, however, the choice for the best recrystallizing solvents are, say, the very toxic carbon tetrachloride and the flammable but almost nontoxic cyclohexane, there should not be much doubt that cyclohexane would be the better choice. The toxicity and flammability should be noted not only to select the least noxious solvent but also to ensure that proper safety precautions are followed.

Two-Solvent Systems. Very often better recrystallizations can be obtained from a solvent mixture than from a pure solvent. The two solvents must be completely miscible in each other in all proportions. There is usually a significant difference in the polarity of the two solvents. Typical solvent pairs used for recrystallization are: acetone-water, ethanol-water, dioxane-water, and benzene-methanol.

It is, of course, very difficult to reproduce exactly the final solvent compositions used in such a procedure. Exact replication of the solvent mixture, however, is usually unnecessary to give reproducible crystallizations. Solids tend to form oils and reprecipitate (rather than recrystallize) more from two solvent mixtures than a single solvent. If the directions given in the experimental procedure are followed carefully, oil formation and reprecipitation are kept to a minimum.

Supersaturation, Oils, and Colloids. Both supersaturation and oils result from the failure of enough molecules to orient themselves correctly for crystal formation when crystal formation is favored thermodynamically. The molecules usually do not organize in a crystal lattice at or below the temperature where all solute present is not completely soluble. Supersaturation occurs when there is no precipitation as a solution is cooled below the temperature at which the solution is saturated. Crystallization can be started in supersaturated solutions by scratching the inside of the glass tube or flask, by cooling the solution, by adding a small seed crystal, or by a combination of these methods. When a solid is being crystallized, it is always prudent to set aside a few small seed crystals of crude solid.

Oils are caused by the separating of the compound from the solution as a liquid phase rather than as crystals. These oils are supercooled, usually impure, liquids (p. 73), often mixtures of solvent and solutes. This type of precipitation is not so selective as crystal formation, and the oil is usually contaminated with the other solutes and with solvent. Because of this contamination the best procedure is to redissolve the oil and make another attempt at crystallization by adding a few seed crystals at the temperature where the oil first appeared or adding 20 to 30% more solvent so that the compound is precipitating at a lower temperature and from a more dilute solution. If this does not yield crystals, the oil is redissolved by heating the solution, and the solution cooled. As soon as the oil is formed, the mother liquor is decanted. The mother liquor is cooled to room temperature or below and the solution filtered. (If no crystals form, this tack must be abandoned, and the procedure described in the next paragraph followed). If a very small first crop of crystals is obtained, about 20% more solvent should be added to the filtrate. The oil is dissolved by heating it with the augmented filtrate. The mixture is cooled, the oil removed by decanting, the solution cooled further, filtered, and the crystals obtained are added to the first crop.

Sometimes the only method of obtaining crystals from an oil is to grind and mash it in contact with the mother liquor with a glass rod. Seed crystals can be added and the mixture chilled if necessary. Sometimes when all these procedures fail, the oil will crystallize on standing overnight or longer in a refrigerator, especially in contact with a few seed crystals. Crystals obtained in this way should be crystallized at least once more as they usually are not so pure as those grown from a hot solution. Once some purification has been accomplished oil formation is not so likely, and the second crystallization usually occurs without oil formation. Figure 12-3 is a flow chart for methods used to obtain crystals from oils.

Problem 12-5. Sometimes crystals are *more* difficult to obtain from a supersaturated solution or an oil if the mixture is below rather than at room temperature. Why?

Problem 12-6. Consider the cause of supersaturated solutions and oils. What effect would the rate of cooling have on the probability of oil formation or supersaturation?

Colloid formation usually interferes with crystallization by preventing normal crystal growth. The compound may precipitate entirely as a colloid, the precipitate may be contaminated with the colloid, or the colloid may prevent enough molecules from orienting correctly to start crystal growth and a supersaturated solution or an oil may result. Colloids can often be prevented by cooling the solution slowly or by clarification of the hot solution. High molecular species which stabilize colloids are preferentially adsorbed. If colloids can not be prevented, they can often be coagulated by heating, chilling, or if the

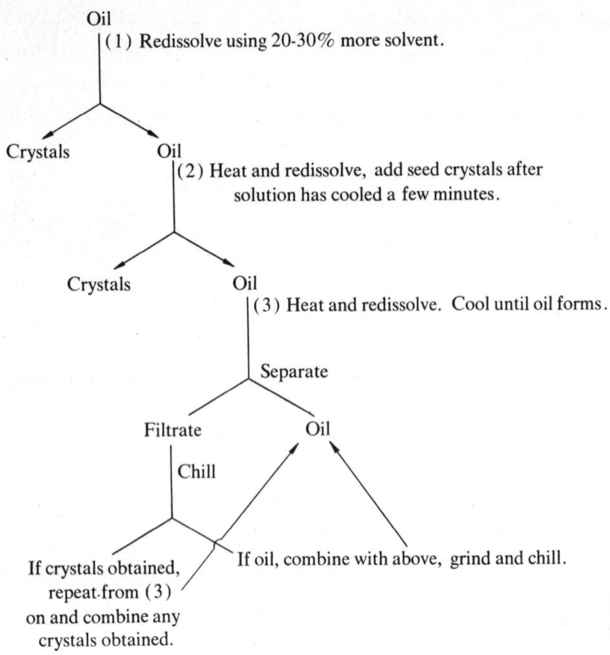

Figure 12-3 Flow chart for obtaining crystals from oils.

solvent is sufficiently polar, by dissolving an electrolyte, such as sodium sulfate or calcium chloride, in the solution.

Efficient Recrystallization. A more efficient crystallization has a higher yield of product and a purer product. The more efficient each recrystallization is, the fewer crystallizations needed and the smaller the loss of compound. By use of the proper techniques, most crystallizations can be made very efficient.

The four sources of impurities in crystals are: (a) occlusion or coprecipitation, (b) adsorption of impurities on the surface of the crystals, (c) precipitation of other species of crystals along with the desired crystals, and (d) postprecipitation. Source a and to a lesser extent sources b and c can be decreased by allowing the solution to cool slowly to room temperature. A towel wrapped around the flask usually insulates it enough to ensure a sufficiently slow rate of cooling. Occlusion and to a limited extent the other three sources of impurities can be decreased by using a more dilute solution, although this also results in a smaller yield of crystals unless the compound is almost insoluble in the solvent at the filtration temperature. Filtering as soon as the solution has cooled to room temperature and crystal growth has stopped minimizes impurities from postprecipitation.

Chilling the solution below room temperature in an ice bath precipitates more crystals, but these crystals usually have more impurities than the ones obtained by cooling to room temperature. Sometimes chilling the crystals to 0° is the only way a satisfactory yield can be obtained. If these crystals are significantly less pure than the first crop obtained by cooling the mixture to room temperature, they should be crystallized again before being combined with the first crop.

Problem 12-7. How would you determine if crystals obtained by cooling to 0° were significantly more impure than the first crop?

Washing the crystals once or twice removes some of the adsorbed and postprecipitated impurities. Washing also prevents further postprecipitation from mother liquor still wetting the filtrate. Chilling the wash a few degrees below the temperature of the mixture when it is filtered helps to prevent any of the product from redissolving in the wash. Redissolving of the product can also be prevented by using a slightly poorer solvent, e.g., a wash of 30% methanol in benzene for a nonpolar solid that has precipitated from 20% methanol in benzene. This could be carried too far and the wash liquid would not dissolve any impurities and even cause precipitation of impurities from the small amount of mother liquor that still wets the crystals.

EXPERIMENTAL PROCEDURE

Two-Solvent Systems

Dissolve the compound in the solvent in which it is more soluble. If needed, clarify the solution at this point. If the second (poorer) solvent has a lower boiling point than the one used to dissolve the compound, cool the solution below the boiling point of the second solvent before it is added. Now add the poorer solvent to the hot, stirred solution. When the solution becomes slightly cloudy, warm the solution with stirring to or near its boiling point. This cloudiness is the beginning of precipitation of the solute. Add more of the second solvent a drop at a time until the cloudiness persists after stirring. Warm the solution until the cloudiness disappears; if necessary, add a drop or two of the first solvent to remove all the cloudiness. If all the cloudiness does not disappear, a colloid or an oil is more likely to form. Allow the mixture to cool and the solute to crystallize.

Purification of Acetanilide

Determine the melting point of impure acetanilide. Heat 6 g of the impure acetanilide, 60 ml of water, and a boiling chip in a 250-ml flask until the mixture boils. Note that some of the acetanilide dissolves in the water and some water dissolves in the acetanilide and forms an oil. Add hot water 3 to 5 ml at a time while heating the mixture to its boiling point and stirring it with a glass rod. When all the acetanilide has dissolved in the boiling water, add 10 ml more.

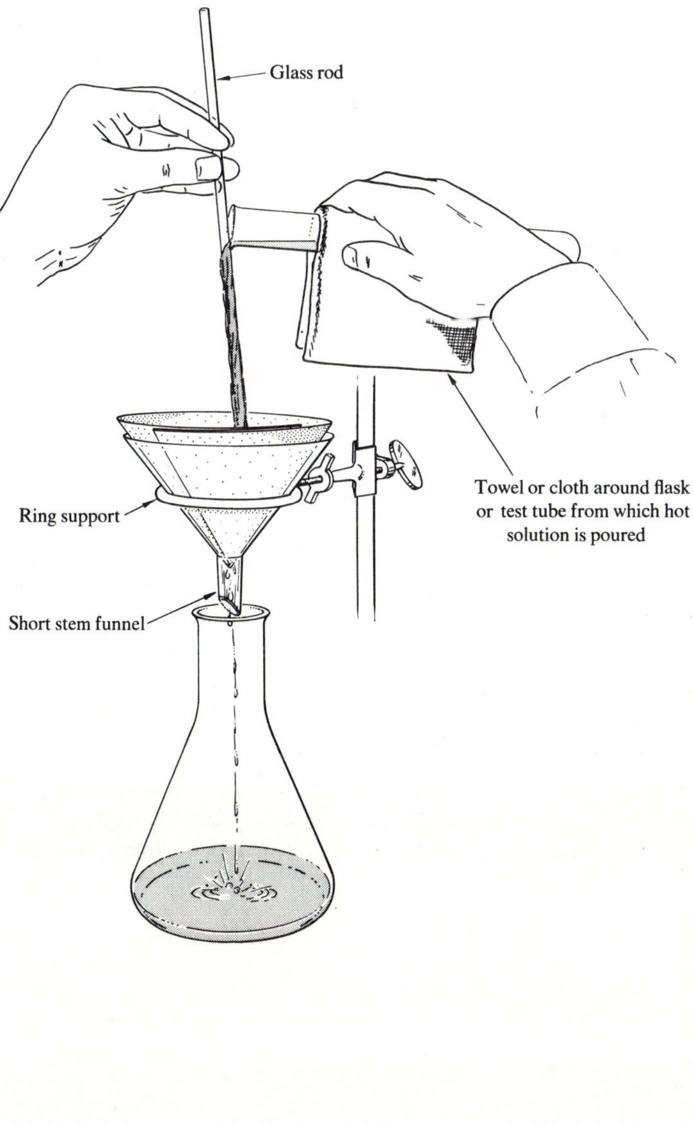

Figure 12-4 Filtration of a hot solution.

Allow the mixture to cool a few minutes, note the color of the solution, and add 0.7 to 1 g of decolorizing carbon (bone black, Nuchar, Norit). Add a very small fraction of this at first and stir. If no boiling or excessive foaming occurs, add the rest of the carbon. Bring the mixture to boiling while stirring, and allow the mixture to stand a few minutes. Bring to boiling with stirring again, and filter immediately through a stemless or very short stemmed funnel (Fig. 12-4) that has been warmed on a steam bath. When the funnel has filled, return the Erlenmeyer flask to be heated by the flame or hot plate until all the liquid has been poured into the funnel. Wash the flask with 10 ml of hot water and pour the hot wash through the filter paper. Note the color of the filtrate and compare with the color of the hot solution before the carbon black was added.

Problem 12-8. If a solution has been heated to or near its boiling point, why should it be cooled a few minutes before the decolorizing carbon is added and the first bit of carbon added cautiously?

Problem 12-9. Why is the funnel preheated before the filtration? Why is the solution kept as hot as possible during filtration?

While the mixture is cooling to room temperature, prepare one of the aspirator water traps shown in Figure 12-5. A trap should be in the vacuum line after the aspirator whenever it is used as a vacuum source. The

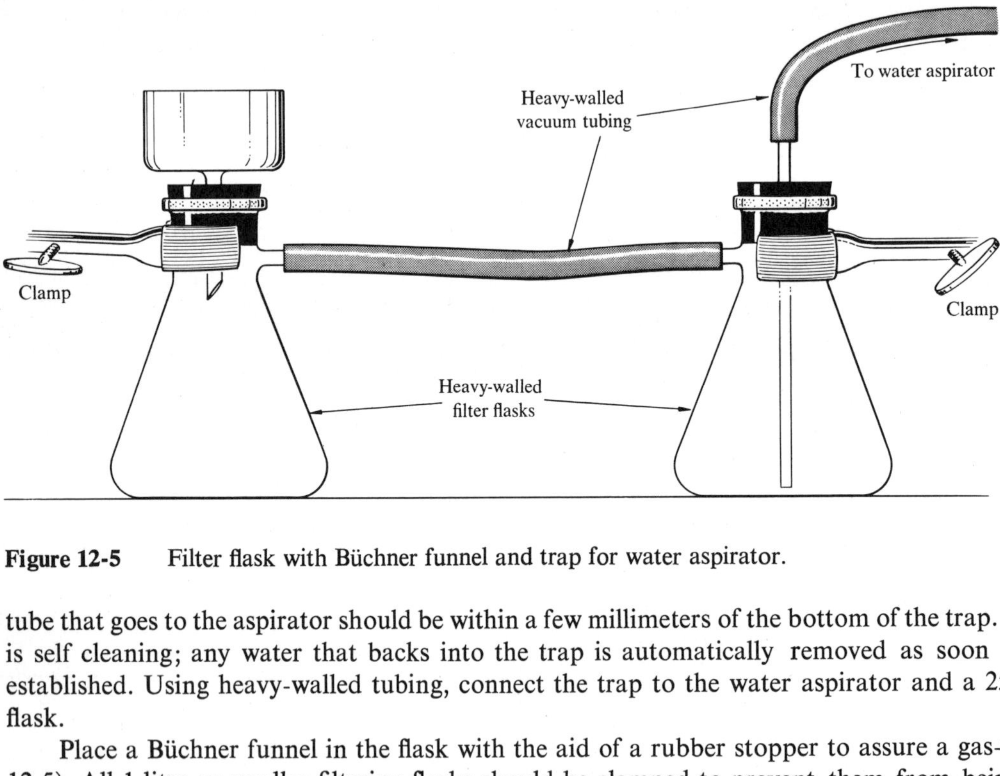

Figure 12-5 Filter flask with Büchner funnel and trap for water aspirator.

tube that goes to the aspirator should be within a few millimeters of the bottom of the trap. A trap so constructed is self cleaning; any water that backs into the trap is automatically removed as soon as the vacuum is re-established. Using heavy-walled tubing, connect the trap to the water aspirator and a 250- or 500-ml filtering flask.

Place a Büchner funnel in the flask with the aid of a rubber stopper to assure a gas-tight connection (Fig. 12-5). All 1-liter or smaller filtering flasks should be clamped to prevent them from being pulled over by the heavy tubing or accidently pushed over. Note that the heavy funnel raises the center of gravity of the assembly.

Be sure the filter paper is the correct size for the funnel, that it is correctly centered so part of it is not up the side of the funnel or folded under (Fig. 12-6), and that it covers all the holes. The vacuum is turned on and the filter paper is usually wetted with a small amount of solvent a second before the filtration is begun so that the paper clings to the funnel and none of the mixture slips around the paper without being filtered. Filter the mixture when it has cooled to room temperature. Turn the vacuum off as soon as almost all the mother liquor is in the filter flask. Air pulled through the crystals evaporates the mother liquor left on them and causes the impurities to be deposited. When the mixture has filtered, use about 20 ml of filtrate to rinse crystals clinging

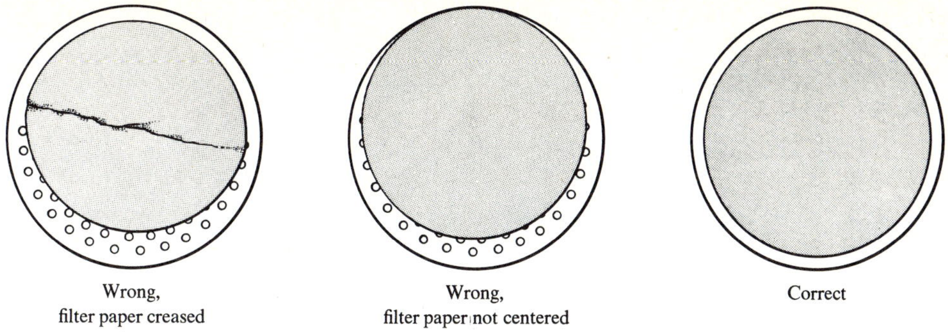

Figure 12-6 Filter paper in a Büchner funnel.

to the side of the flask into the Büchner funnel. Immediately after this wash the crystals with two 5-ml portions of water that have been chilled to 10 to 15°. If necessary, use each wash to rinse any remaining crystals from the Erlenmeyer flask into the Büchner funnel. Mash and stir the crystals with a spatula during each wash, but be careful not to tear or move the filter paper while doing this. It is very important that the crystals do not become dry until *after* the last wash. While the crystals are still wet from the last wash, pack them down with a spatula that has been bent to form an L, a large clean cork, or the bottom of a 50-ml beaker. Keep the vacuum on for 2 to 3 min after the crystals appear dry. Continue packing while the vacuum is on. Remove the boiling chip if it was transferred to the Büchner funnel.

Remove the crystals to a large, fresh, clean filter paper, press the crystals between the filter paper and spread them out on it to dry. If possible, the drying should be completed in a vacuum oven or desiccator. A vacuum desiccator can be constructed from a filter flask and a water bath (Fig. 12-7). If the compound is not

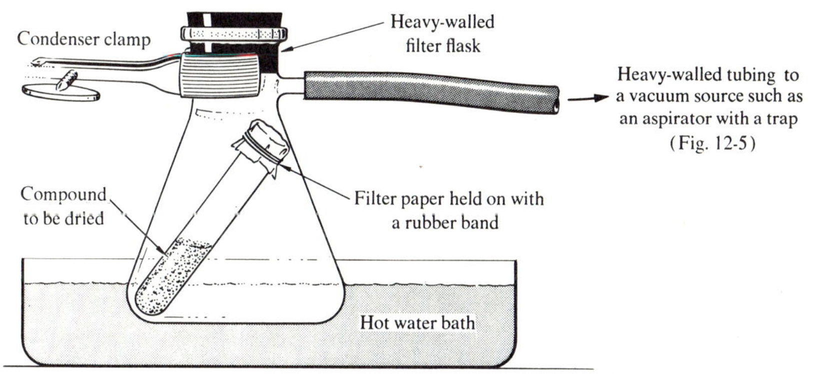

Figure 12-7 Student-assembled vacuum desiccator.

sensitive to air oxidation, it can be dried under an infrared lamp. Compounds may also be dried by placing them on a watch glass over a steam bath (Fig. 12-8). In drying pure solids it is usually advisable not to heat them so high that they melt. Weigh the crystals; determine the percent recovery and melting point of the crystals.

While the first crop of crystals is drying, chill the filtrate in an ice bath for at least 20 min with occasional stirring. Filter and wash, following the instructions above, except that the two washes should be at 0°. Dry and weigh the crystals. If a desiccator or oven is used for drying, the crystals should be placed in a tared yield bottle and the crystals dried in the bottle. Determine their melting point and the percent recovery.

Recrystallize the first crop of crystals (obtain by cooling to room temperature) from the minimum amount of water. Follow the instructions for crystallizing, filtering, and washing described above, except omit the treat-

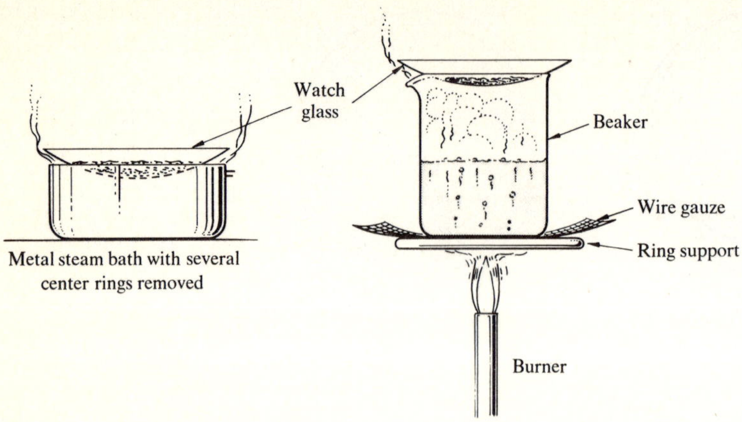

Figure 12-8 Drying compounds with watch glasses on steam baths.

ment with carbon black. Dry (in a tared yield bottle if a desiccator or oven is used), weigh, and determine the melting point. Turn in these crystals and the second crop in separate, labeled containers to your instructor. Record and report the data from this recrystallization in a table similar to Table 12-1.

TABLE 12-1 RESULTS OF RECRYSTALLIZATION OF ACETANILIDE

	Wt. (g)	Percent Recovery	M.p. (°C)
First crop			
Second crop			
Total recovery			
Recrystallized first crop			

DISCUSSION OF RESULTS OF RECRYSTALLIZATION OF ACETANILIDE

Figure 12-9 is a phase diagram of acetanilide-water mixtures. The broken lines outline areas of different phases or different mixtures of phases. In each of the phase areas on Figure 12-9 there is also a vapor phase. The dashed line (*BEFD*) encloses the two liquid phases. This area should be avoided because purification is much less efficient if the liquid phase (or oil) of acetanilide forms (see pp. 83 and 84). Precipitation should not begin in the two-phase (solid-liquid) region (*ABEFGH*) enclosed by the solid line, parts of the dashed lines, and the ordinates of the diagram because not all the acetanilide can dissolve at these concentrations and temperatures.

Problem 12-10. What is the maximum concentration of acetanilide *dissolved in water* from which acetanilide crystallizes without oil formation?

Problem 12-11. How could a mixture of concentration X_1 be crystallized?

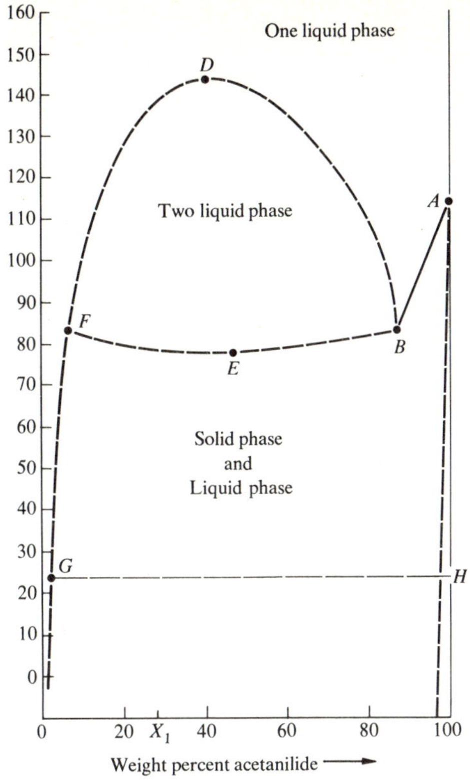

Figure 12-9 Phase diagram of acetanilide-water mixtures.

CRYSTALLIZATION OF NAPHTHALENE

Safety Precautions

Warning! All common solvents except water are either toxic or flammable or both. Every possible precaution should be taken to prevent any solvent vapors from escaping into the laboratory. The solvent should be heated in a hood or under reflux. If no hood is available, a substitute can be constructed with an aspirator, a funnel, and some tubing (Fig. 12-10). Follow carefully the indicated positions of the funnel and the flask. The funnel type of hood should not be used for evaporation of any amount of solvent greater than 10 ml nor should it be used for boiling a volume of liquid greater than 100 ml for short periods nor greater than 50 ml for long periods.

If more than 250 ml of nonaqueous solvent is used, a reflux apparatus is aways needed. A reflux apparatus is also used for recrystallizing from volumes of nonaqueous solvent greater than 100 ml if: (a) only an aspirator hood is available, (b) a flammable solvent with a boiling point greater than 100° is used, or (c) flammable solvents are used and no steam is available. A reflux apparatus is also necessary if no significant solvent loss can be tolerated. For example, a very expensive solvent may be used or the exact quantity of solvent used may be critical.

Solvents of low boiling point should be used only as a last resort not only because there is a greater fire or toxicity hazard with such solvents but also because the difference of most solutes' solubility between the low boiling point of these solvents and room temperature is not very large.

A burner flame should not be used to heat flammable liquids directly. The preferred heat sources for flammable solvents are steam (for solvents boiling below 100°) or heating mantles. Hot plates with enclosed heating coils are a second choice. If none of these heat sources is available, the solvent can be heated on a hot plate made from a Bunsen burner and a steam bath (Fig. 12-11). Note that the top of the burner is above the bottom of the shell formed by the hot plate.

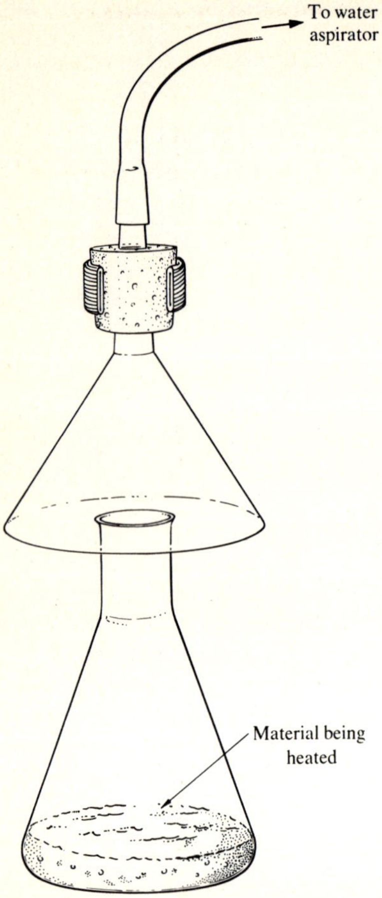

Figure 12-10 Small hood made from funnel.

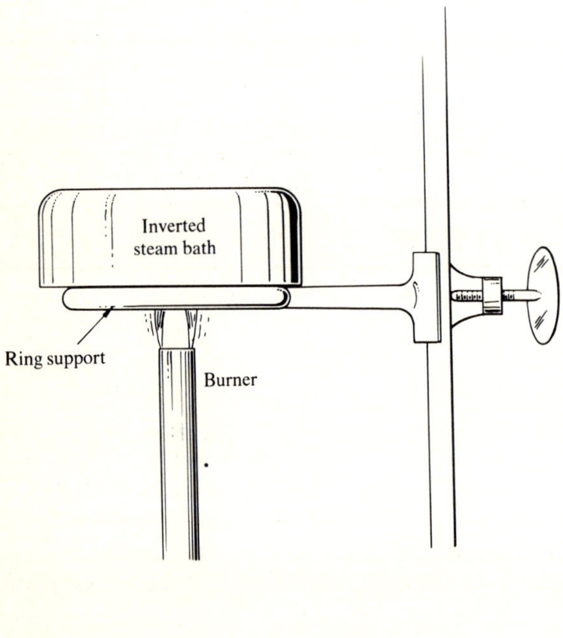

Figure 12-11 Hot plate from burner and steam bath.

Experimental Procedure

Dissolve 5 g of impure naphthalene mixture in the minimum amount of methanol; add 10 ml more methanol. Clarify the mixture with approximately $\frac{1}{2}$ g of carbon black. Add water and follow the procedures given on pages 85, 86, and 87 for recrystallization with a *two-solvent system*, washing, and drying. Determine the melting point and turn in the purified naphthalene to your instructor.

UNKNOWN

Turn in a clean, dry test tube or vial to your instructor. It will be returned with a mixture that has 60 to 80% of a compound from Table 11-1 in it. Identify the major constituent.

EXERCISES

12-1. Arrange the following solvents in order of flammability, placing the most flammable first. Arrange series a and b separately. (a) ethylene glycol, tetrachlorethane, benzene, toluene, pentane, chlorotoluene, butanol. (b) ethyl ether, methanol, ethanol, acetone.

12-2. Indicate whether each of the following solvents is nontoxic, slightly toxic, toxic, or very toxic and non-flammable, flammable, or very flammable: toluene, benzene, carbon tetrachloride, petroleum ether, ethanol, methanol, chloroform, dioxane, cyclohexane.

12-3. Pentane is found to be an ideal solvent for crystallizing a compound. What could be done to make crystallization of a very large quantity of this compound as safe as possible?

13

Nitration of Aromatic Compounds

Synthesis of Nitrobenzene and *m*-Dinitrobenzene

$$HNO_3 + 2H_2SO_4 \rightarrow NO_2^+ + H_3O^+ + 2HSO_4^-$$

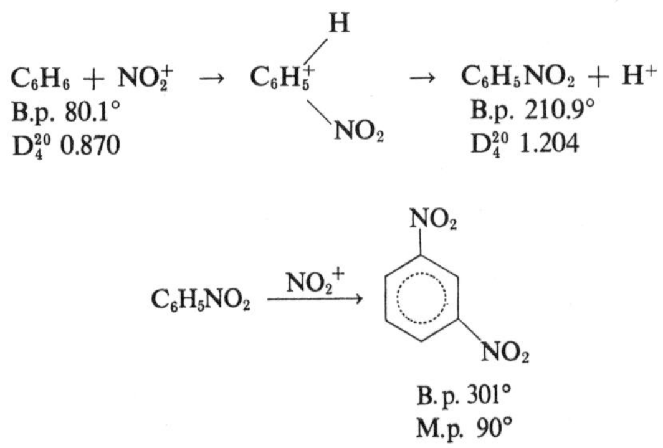

INTRODUCTION

Aromatic hydrocarbons can be nitrated with a mixture of nitric and sulfuric acids. Nitric acid alone will nitrate more reactive aromatic compounds (Chapter 47). Sulfuric acid-fuming nitric acid mixture and longer heating times are required for less reactive aromatic compounds. Although the nitro group is electron-withdrawing and deactivates aromatic systems toward oxidation, the strong acid and oxidizing properties of the nitric-sulfuric acid mixtures can cause extensive degradations of aromatic compounds at high temperatures. Therefore, the conditions for nitration of less active systems are often a compromise between those forcing enough for good nitration and those mild enough to preclude degradation.

EXPERIMENTAL PROCEDURE

Nitrobenzene

Warning! The acids used in this experiment are very corrosive to skin, clothing, metals, etc. Any acid spilled on you should be flushed away with large quantities of water. Acid spilled on the desk top should be neutralized with sodium bicarbonate solution. Slowly add 48 ml of concentrated sulfuric acid to 42 ml of chilled concentrated nitric acid in a 250-ml Erlenmeyer flask in an ice bath. Swirl the flask during the addition.

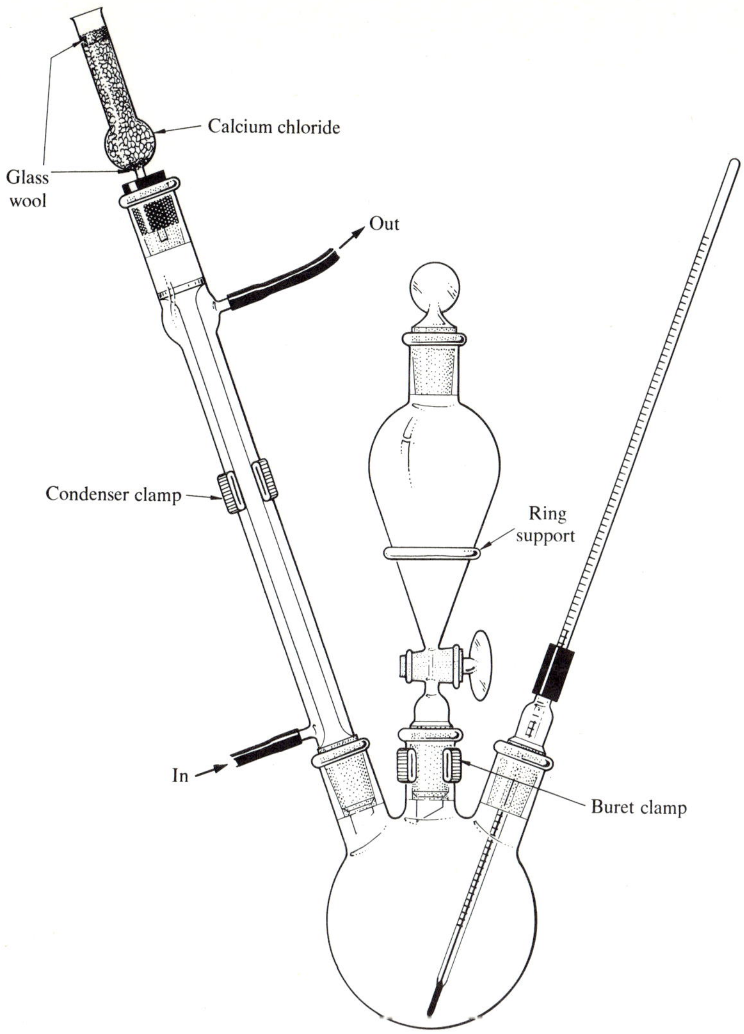

Figure 13-1 Round bottom three-necked flask with thermometer, dropping funnel, reflux condenser, and drying tube.

Place 45 ml of benzene and 1.0 g of urea in a three-necked 500-ml flask equipped with a thermometer, dropping funnel, and reflux condenser (Fig. 13-1). The bottom of the thermometer bulb should be about 1 cm from the bottom of the flask. Slowly add the acid mixture to the benzene while agitating the mixture as vigorously as possible by shaking the ring stand. The agitation is very important to ensure complete reaction because the acid mixture and benzene are very insoluble in each other. If the temperature exceeds 55°, stop the addition and place an ice bath around the flask while continuing the agitation. Remove the ice bath and start the addition again when the temperature falls below 43°. Attempt to keep the temperature between 45° and 55° during the addition.

When all the acid mixture has been added, heat the mixture to 55 to 65° for ½ hr with a hot water bath. If the temperature exceeds 65°, remove the hot water bath. If it exceeds 68°, cool to below 55° with the ice bath. Agitate the mixture every 3 to 5 min during this ½ hr of heating.

Warning! Aromatic nitro compounds are very toxic. They are readily absorbed through the skin. Every precaution should be taken to avoid spilling nitro compounds on the skin and to avoid inhaling their vapors.

Problem 13-1. What possible side reactions may occur? What steps in the synthesis avoid the side reactions?

Problem 13-2. Why is a higher temperature allowed and needed at the end of the reaction?

Cool the mixture to 0 to 5° in an ice bath. Transfer the contents of the flask into a separatory funnel and separate the acid layer as quickly as possible before the mixture warms to room temperature. Wash the nitrobenzene with 50 ml of water, 40 ml of 5% sodium hydroxide, and then three 40-ml portions of water.

Back-extract each of these five aqueous washes with one 20-ml portion of ether. Extract successively the first water wash, then the sodium hydroxide wash, and then the three water washes. Combine the ether with the nitrobenzene.

Problem 13-3. Why should the acid mixture from the reaction not be extracted with ether?

The organic layer may be partly or completely emulsified with water during the sodium hydroxide or last three water extractions. The emulsion can be recognized as a very light grey opaque mixture. Emulsions can be broken in several ways: (a) Heating the emulsion near the boiling point of the lower boiling component. (b) If one of the phases is aqueous, the emulsion can be washed with a saturated sodium chloride solution. (c) The emulsion can be warmed with some anhydrous calcium chloride. This is the equivalent of a combination of a and b. (d) Vacuum filtration of the mixture, through Celite or filter aid if possible, often breaks the emulsion when no other method succeeds. (e) Centrifugation of the emulsion breaks it readily if there is a significant difference in the density of the two liquids. Sometimes high speed centrifugation for long periods of time is the only way to break persistent emulsions.

Problem 13-4. If a centrifuge with a sufficient capacity were available, could it be used advantageously here?

Add approximately 15 g of anhydrous calcium chloride to the nitrobenzene mixture, swirl the stoppered container, and allow to stand for 5 min. Decant the organic solution from the calcium chloride into a 100-ml flask, and add a distilling head and condenser to construct the apparatus shown in Figure 3-1. Distil and collect the fraction that boils at 203 to 212° in a tared yield bottle. **Warning! Do not distil higher than 212° and do not heat the flask excessively. Do not distil to dryness as dinitrobenzene (which is probably present) explodes if heated above 220°.** Calculate the % yield based on the total amount of *pure* nitrobenzene that was collected. If the synthesis of *m*-dinitrobenzene is assigned, do not give the product to the laboratory instructor until this synthesis is begun.

m-Dinitrobenzene

Warning! During all the addition, heating, and stirring described below keep the test tube pointed away from yourself and others. Use a long glass rod so that any splattering from the acid mixture is less likely to hit your fingers. If any acid splashes on your skin, wash it off immediately with large quantities of water and then treat with aqueous sodium bicarbonate.

Cool 5 ml of yellow fuming nitric acid (sp. gr. 1.5) in a 20-cm test tube in an ice bath. Carefully add 5 ml of concentrated sulfuric acid while stirring the mixture with a long glass rod. Remove the mixture from the ice bath and add 3 ml of nitrobenzene taken from the product of the above synthesis. Add the nitrobenzene slowly, a few drops at a time, while agitating the mixture with a long glass rod. After all the nitrobenzene has been added, heat the mixture for 35 min in a water bath maintained at 95°. Stir about every 5 min with the glass rod.

Allow the tube to cool a few minutes and pour the mixture onto 60 g of ice. Collect the crystals by filtration through a Hirsh funnel or the smallest Büchner funnel available. Wash three times with ice water and recrystallize from methanol. Dry and weigh the crystals in a tared yield bottle.

14

Reactions of Hydrocarbons

Organic Qualitative Analysis I

INTRODUCTION

The various classes of hydrocarbons react differently with many reagents. Some of these differences are easily observed, and when this occurs, the reactions can be used as classification tests. Unknown compounds can be identified by a combination of classification tests and physical constant determinations.

CLASSIFICATION TESTS

Concentrated Sulfuric Acid, H_2SO_4

Discussion. Concentrated sulfuric acid is a useful reagent for distinguishing certain compounds, e.g., alkanes and less reactive aromatic hydrocarbons from most other compounds. Hydrocarbons will respond in three different ways to concentrated sulfuric acid.

a. Inert compounds such as alkanes, benzene, monoalkyl- and some dialkylbenzenes undergo no physical or chemical change. For convenience these are sometimes called Class I compounds.
b. More reactive compounds as polyalkylbenzenes, most alkenes, most alkynes, and cyclopropanes dissolve in concentrated sulfuric acid. These compounds dissolve because they react with concentrated sulfuric acid to form sulfonates (from the aromatic hydrocarbons) or sulfates (from the alkenes, alkynes, and cyclopropanes). Sulfates and sulfonates, which are both ionic and strongly polar, are soluble in concentrated sulfuric acid, a very polar solvent. (See Chapters 6 and 12.)

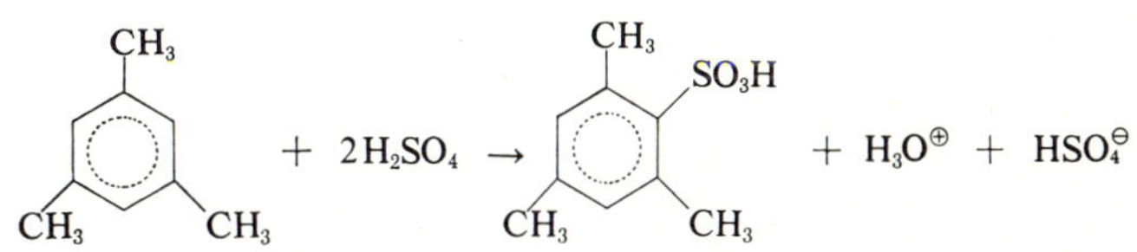

(14–1)

$$RCH{=}CH_2 + H_2SO_4 \rightarrow RCH(OSO_3H){-}CH_3 \quad (14\text{–}2)$$

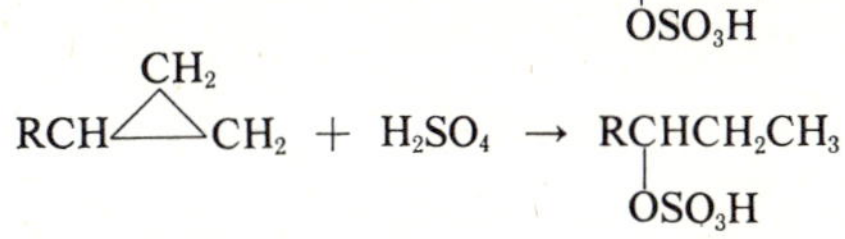

(14–3)

c. Many alkenes and alkynes undergo carbonium ion polymerization when treated with sulfuric acid:

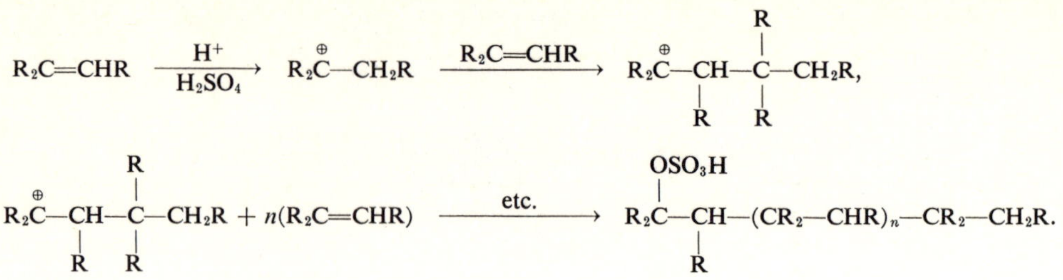

(14-4)

These polymers are insoluble in concentrated sulfuric acid, but their production is accompanied by heat and the solid polymers differ in appearance from the unsaturated hydrocarbons (usually liquids). These compounds with sulfuric acid-insoluble products are often said to be soluble since only their reaction products are insoluble. If a polymer does form, take note because polymerization provides a clue to the identity of the organic compound. Alkyl aromatic hydrocarbons and many alkenes and alkynes do not give insoluble products under the conditions of the test.

Experimental Procedure. Using a 3-ml delivering or 5-ml graduated pipet measure 3 ml of concentrated sulfuric acid into a 7.5-cm test tube. **Carefully** add 0.15 to 0.18 g (weigh on a balance) of solid or 0.2 ml of liquid (about 4 drops) in several portions and agitate with a thin glass rod. Test the following and record the results: toluene, 2-pentene, *p*-xylene, styrene, mesitylene, and naphthalene.

Fuming Sulfuric Acid, $H_2SO_4 \cdot SO_3$

Discussion. Aromatic compounds insoluble in sulfuric acid are sulfonated rapidly enough by fuming sulfuric acid to allow this reagent to be used to distinguish aromatic sulfuric-acid insoluble compounds from aliphatic sulfuric-acid insoluble compounds which do not react with this reagent under the test conditions:

$$C_6H_6 + SO_3 \xrightarrow{H_2SO_4} C_6H_5SO_3H. \qquad (14\text{-}5)$$

The sulfonated aromatic compounds dissolve, and the unreactive alkanes float on top of the fuming sulfuric acid. Many aromatic solids such as *p*-dibromobenzene and naphthalene react so slowly they may appear not to dissolve.

Problem 14-1. Why is it a waste of time to try this test on compounds that are soluble in concentrated sulfuric acid?

Experimental Procedure. Put 2 ml of fuming sulfuric acid in a clean dry test tube and carefully add 0.5 ml (0.3 g if a solid) of the test substance. Agitate vigorously but carefully for about 30 sec; allow to stand for about 3 min. Note whether or not the unknown has dissolved. Try this test on cyclohexane, benzene, naphthalene, and paraffin wax.

Chloroform and Aluminum Chloride, $CHCl_3 + AlCl_3$

Discussion. All compounds suspected of being hydrocarbons or their halogen derivatives should be tested with chloroform and aluminum trichloride. Aromatic compounds that do not have strong deactivating groups, as —NO_2, —CXO, —CN, etc., react with aluminum trichloride and chloroform to give a variety of characteristic colors. These colors are due to triarylcarbonium ions (III) which are formed by a hydride transfer (Eq. 14-6g) from the triarylmethane (II) to the less stable diarylcarbonium ions (I). These triarylmethanes are formed by three successive Friedel-Crafts reactions of the chloroform with the aromatic compound (Eqs. 14-6a–f).

$$CHCl_3 + AlCl_3 \rightarrow CHCl_2^{\oplus} + AlCl_4^{\ominus} \qquad (14\text{-}6a)$$

$$CHCl_2^{\oplus} + ArH \rightarrow \overset{\oplus}{Ar}\left\langle\begin{matrix}CHCl_2\\ H\end{matrix}\right. \rightarrow ArCHCl_2 + H^{\oplus} \qquad (14\text{-}6b)$$

$$ArCHCl_2 + AlCl_3 \rightarrow Ar\overset{\oplus}{C}HCl + AlCl_4^{\ominus} \qquad (14\text{-}6c)$$

$$ArH + Ar\overset{\oplus}{C}HCl \rightarrow \overset{\oplus}{Ar}\left\langle\begin{matrix}CClHAr\\ H\end{matrix}\right. \rightarrow Ar_2CHCl + H^{\oplus} \qquad (14\text{-}6d)$$

$$Ar_2CHCl + AlCl_3 \rightarrow Ar_2CH^{\oplus}\ (I) + AlCl_4^{\ominus} \qquad (14\text{-}6e)$$

$$Ar_2\overset{\oplus}{C}H + ArH \rightarrow \overset{\oplus}{Ar}\left\langle\begin{matrix}CHAr_2\\ H\end{matrix}\right. \rightarrow Ar_3CH\ (II) + H^{\oplus} \qquad (14\text{-}6f)$$

$$Ar_3CH + Ar_2\overset{\oplus}{C}H \rightarrow Ar_2CH_2 + Ar_3\overset{\oplus}{C}\ (III) \qquad (14\text{-}6\ g)$$

The colors characteristic of the different aromatic systems possible are given in Table 14-1. Alkyl and simple halogen derivatives usually give the same colors as the parent ring system. Aliphatic hydrocarbons give either no colors or only a light straw yellow color.

TABLE 14-1 COLOR REACTIONS OF AROMATIC COMPOUNDS WITH $AlCl_3$ AND $CHCl_3$

Aromatic system	Color
Benzene	orange to red
Naphthalene	blue
Biphenyl	purple
Anthracene	green
Phenanthrene	purple

Experimental Procedure. **Danger! Aluminum chloride is a strong corrosive acid and poison. It reacts violently with water.** The instructor will demonstrate the violence of the reaction of aluminum chloride with water.

It is very important that all reagents *and* test tubes be completely dry since traces of water may cause false negative tests. If possible, aluminum chloride should be taken from a bottle that has not been open longer than three days (fewer in humid weather). The aluminum chloride bottle should be tightly stoppered except when reagent actually is being withdrawn.

Try the test on cyclohexane, cyclohexene, toluene, styrene, and anthracene. Add 0.1 ml (0.05 to 0.1 g if a solid) of substance to be tested to 2 ml of $CHCl_3$ in a 7.5-cm test tube. Agitate vigorously. Incline the test tube to wet the walls with the solution. Add 0.5 g of *dry* anhydrous aluminum chloride so that some goes into the solution and some adheres to the wet portion of the test tube wall. Note the color of both the solid on the wall and the solution.

Tests for Unsaturation

It is usually expedient to test *all* organic compounds except those insoluble in concentrated acid for unsaturation. Two reagents are useful for this purpose: 5% bromine in carbon tetrachloride and 2% potassium permanganate in water (Baeyer's reagent).

$$3R_2C{=}CR_2 + 4H_2O + 2KMnO_4 \rightarrow 2KOH + 2MnO_2 + 3R_2\underset{\displaystyle OH}{\underset{|}{C}}{-}\underset{\displaystyle OH}{\underset{|}{C}}R_2 \qquad (14\text{-}7)$$

$$R_2C{=}CR_2 + Br_2 \rightarrow R_2\underset{\displaystyle Br}{\underset{|}{C}}{-}\overset{\displaystyle Br}{\overset{|}{C}}R_2 \qquad (14\text{-}8)$$

When it is desirable to test for double or triple bonds, both these tests should be performed on the unknown because:

a. Both tests are positive for compounds other than alkenes and alkynes.
b. The test with bromine in carbon tetrachloride is often negative for alkenes and alkynes substituted with electron-withdrawing groups.
c. Baeyer's test is occasionally negative for a few complicated high molecular weight compounds with carbon to carbon double bonds.

Bromine in Carbon Tetrachloride. Discussion: carbon tetrachloride is used for a solvent in this reaction because it is a good solvent for bromine and a poor solvent for hydrogen bromide which usually forms when this reagent reacts with compounds other than alkenes and alkynes. Being insoluble, hydrogen bromide can be readily detected if formed. The decolorization of carbon tetrachloride without evolution of hydrogen bromide is a positive test for alkenes and alkynes.

The rate-determining step of this reaction involves attack of a bromium ion or its equivalent on the pi cloud of the carbon to carbon multiple bond.

$$R{-}CH{=}CH_2 \quad Br{-}Br \rightarrow R{-}\underset{\oplus}{C}H{-}CH_2Br + Br^{\ominus} \qquad (14\text{–}9)$$

$$R{-}\underset{\oplus}{C}H{-}CH_2Br + Br^{\ominus} \rightarrow R{-}\underset{\displaystyle Br}{\underset{|}{C}}H{-}CH_2Br \qquad (14\text{–}10)$$

Groups that diminish the availability of the pi electrons and also do not increase significantly the stability of the carbonium ion cause the reaction to proceed slowly. If the electron withdrawal is sufficiently strong, the reaction with bromine in carbon tetrachloride is slow enough so that the test is negative. For this reason compounds such as $ArCH{=}CHAr$ and $Ar_2C{=}CAr_2$ give negative tests.

Problem 14-2. Predict the relative rates of reaction of bromine in carbon tetrachloride with the following series of compounds. Put 1 by the most reactive, 2 by the next most reactive, etc. Treat series a and b separately. (a) $CH_2{=}CH_2$, $CH_2{=}CHCH_2CH_3$, $CH_2{=}C(CH_3)C_2H_5$, $C_2H_5CH{=}CHC_2H_5$, $C_2H_5CH{=}C(CH_3)C_2H_5$, $C_2H_5C(CH_3){=}C(CH_3)C_2H_5$, $C_6H_5CH{=}CHC_6H_5$, $(C_6H_5)_2C{=}C(C_6H_5)_2$, $C_6H_5C(CH_3){=}C(CH_3)C_6H_5$. (b) $C_6H_5CH{=}CH_2$, $C_6H_5CH{=}CHCH_3$, p-$CH_3C_6H_4CH{=}CH_2$, p-$CH_3C_6H_4CH{=}CHCH_3$, p-$CH_3OC_6H_4CH{=}CH_2$, p-$ClC_6H_4CH{=}CH_2$, m-$ClC_6H_4CH{=}CH_2$, p-$NO_2C_6H_4CH{=}CH_2$.

Problem 14-3. Account for the relative rates of reaction of the following compounds with bromine in carbon tetrachloride: $C_6H_5CH{=}CH_2 > C_6H_{13}CH{=}CH_2 > C_6H_5CH{=}CHC_6H_5$.

Problem 14-4. Predict the relative rates of reaction of the following with bromine in carbon tetrachloride: $C_6H_5CH{=}CHC_6H_5$, $(C_6H_5)_2C{=}CHC_6H_5$, $(C_6H_5)_2C{=}C(C_6H_5)_2$.

Experimental procedure: dissolve or disperse about 0.1 g (0.1 to 0.2 ml of liquid) of the compound to be tested in 2 ml of carbon tetrachloride in a 7.5-cm test tube. Add the 5% bromine in carbon tetrachloride a drop at a time, shaking the tube vigorously but briefly after the addition of each drop. The test is positive if two or more drops are decolonized in less than a minute. Try this test on cyclohexene, styrene, and toluene.

Problem 14-5. This test should not be done in direct sunlight or under very strong artificial lighting. Why?

Dilute Aqueous Potassium Permanganate. Discussion: dilute potassium permanganate gives positive tests with more groups of compounds than bromine in carbon tetrachloride does (Chapters 44, 46, 56, and 61), and there are fewer negative tests with unsaturated compounds. If it is being used as a test for alkenes and alkynes, there are fewer false negatives and more false positives. For instance, 1,2-diphenylethylene, cinnamic acid, and 1,2-dibromoethylene give positive tests with this reagent but not with bromine in carbon tetrachloride.

$$3C_6H_5CH{=}CHC_6H_5 + 4H_2O + 2KMnO_4 \rightarrow 3C_6H_5\underset{\mathrm{OH}}{\underset{|}{HC}}{-}{-}\underset{\mathrm{OH}}{\underset{|}{C}}HC_6H_5 + 2KOH + 2MnO_2 \qquad (14\text{-}11)$$

$$3C_6H_5CBr{=}CBrC_6H_5 + 4H_2O + 2KMnO_4 \rightarrow 3C_6H_5\underset{\mathrm{OH}}{\underset{|}{C}}Br{-}\underset{\mathrm{OH}}{\underset{|}{C}}BrC_6H_5 + 2KOH + 2MnO_2 \qquad (14\text{-}12)$$

Experimental procedure: dissolve 0.05 g (0.1 ml if liquid) of the compound to be tested in 2 ml of ethanol in a test tube. Add the 2% aqueous potassium permanganate solution one drop at a time. Shake the tube vigorously after the addition of each drop. If the purple color of the permanganate ion disappears immediately after the addition of two drops of reagent, the presence of an alkene or alkyne (or any compound with easily oxidizable functional groups) is indicated. In addition to the disappearance of the purple color there is usually a brown precipate of MnO_2. Sometimes the MnO_2 will remain suspended, and a reddish-brown color gradually replaces the purple color. Try this test on benzene, cyclohexene, toluene, turpentine, and naphthalene and record the results.

Problem 14-6. The ethanol used must be free from any impurities that reduce the permanganate ion, such as aldehydes. How would one be certain the ethanol was pure enough for this test?

Test for Terminal Triple Bonds

Discussion. There is no simple qualitative method for distinguishing between alkynes and alkenes in as much as alkynes decolorize both bromine in carbon tetrachloride and dilute aqueous permanganate in the same manner as alkenes:

$$R{-}C{\equiv}C{-}R + 2Br_2 \rightarrow R{-}CBr_2{-}CBr_2{-}R \qquad (14\text{-}13)$$

$$R{-}C{\equiv}C{-}R' + 2MnO_4^- \rightarrow RCO_2^- + R'CO_2^- + 2MnO_2 \qquad (14\text{-}14)$$

Compounds with a terminal alkyne group, however, can be distinguished from alkenes and other alkynes because of the acidity of their terminal hydrogen. One such test uses their reaction with sodium (see Chapter 28):

$$2RC{\equiv}CH + 2Na \rightarrow R{-}C{\equiv}\overset{\ominus}{C}{:}\overset{\oplus}{Na} + H_2\uparrow \qquad (14\text{-}15)$$

Another test makes use of the fact that heavy metal ions form insoluble acetylides:

$$R{-}C{\equiv}CH + Cu(NH_3)_2^+ \rightarrow R{-}C{\equiv}C{-}Cu + NH_4^+ + NH_3 \qquad (14\text{-}16)$$

$$R{-}C{\equiv}CH + Ag(NH_3)_2^+ \rightarrow R{-}C{\equiv}C{-}Ag + NH_4^+ + NH_3 \qquad (14\text{-}17)$$

Experimental Procedure. To prepare the reagent, $Cu(NH_3)^{2+}\ Cl^-$, dissolve 0.3 g of sodium hydrogen sulfite in 3 ml of 5% aqueous sodium hydroxide. In a separate test tube, dissolve 1.2 g of $CuSO_4 \cdot 5H_2O$ and 0.3 g of NaCl in 4 ml of warm water. Slowly add the bisulfite solution to the copper sulfate solution with stirring. Cool the mixture to room temperature. Decant the supernatant liquid and wash the precipitate (CuCl) with two 2-ml portions of 5% aqueous NaOH solution. Dissolve the precipitate in a mixture of 3 ml of water and 4 ml of concentrated ammonium hydroxide and pour the solution into a 10-ml Erlenmeyer flask. To prevent air oxidation of the reagent keep the flask tightly stoppered except when actually removing reagent. Vigorously shake 0.05 g (or 0.1 ml) of unknown with 1 ml of ethanol and 1 ml of ammoniacal cuprous chloride reagent. Notice the color of any precipitate. Try this test on 1-hexyne, cyclohexene, phenylacetylene, and styrene.

Danger! Dry cuprous acetylides are very sensitive explosives. Destroy any acetylide residues by heating with dilute nitric acid. Do not leave any acetylides in containers in your desk for they probably will dry out by the next period and explode at the slightest shock.

UNKNOWN

Hand in a clean dry test tube labeled with your name. It will be returned with the hydrocarbon unknown in it. The unknown will be one of the hydrocarbons from Tables 14-2 to 14-5. The first step in the identification of an organic unknown usually is to determine its boiling point and/or melting point. If either of these shows a wide range, the compound should be purified by distillation or crystallization. Figure 7-3 shows an apparatus useful for distilling small amounts of liquids and for determining the boiling points of small quantities of liquids that are too impure to have their boiling points determined by the micro method.

TABLE 14-2 HYDROCARBONS: PARAFFINS AND CYCLOPARAFFINS

Compound	B.p. (°C)	D_4^{20}	M.p. (°C)
Isopentane	31	0.620	
Pentane	36	0.626	
Cyclopentane	50	0.740	
2-Methylpentane	60	0.653	
n-Hexane	68	0.659	
Cyclohexane	81	0.778	
2,3-Dimethylpentane	90	0.695	
2,2,4-Trimethylpentane	99	0.692	
n-Octane	125	0.703	
Ethylcyclohexane	132	0.788	
n-Nonane	150	0.718	
2,7-Dimethyloctane	159	0.724	
n-Decane	173	0.730	
cis-Decalin	194	0.896	
n-Dodecane	215	0.749	
n-Tridecane	230		
n-Tetradecane	245.5		5.5
n-Pentadecane	268		10
n-Hexadecane	287.5		18
n-Octadecane	317		28
n-Eicosane			36.5
n-Hentriacontane			69

TABLE 14-3 HYDROCARBONS: ALKENES AND ALKYNES

Compound	B.p. (°C)	M.p. (°C)	D_4^{20}	Derivatives' M.p. (all acids are produced by oxidation) (°C)
1-Pentene	30		0.6100	
3-Methyl-3-Buten-1-yne	34			
trans-2-Pentene	36.0		0.6486	
1-Pentyne	40		0.6945	Mercuride 118
Cyclopentene	46		0.7736	Glutaric acid 97
1,5-Hexadiene (diallyl)	59		0.690	Succinic acid 188
1-Hexyne	70.5		0.719	Mercuride 99
2,4-Dimethyl-2-Butene	72.5		0.7081	
Cyclohexene	83.5		0.8088	Adipic acid 152
1-Heptene	93		0.6976	
1-Heptyne	100		0.7338	Mercuride 61
4-Methylcyclohexene	103		0.7985	β-Methyladipic acid 93
Cycloheptene	115		0.8828	Pimelic acid 103-105
1-Octene	122.5		0.7195	
Phenylacetylene	140		0.9282	Mercuride 125 Benzoic acid 121
Styrene	145.5		0.9074	Dibromide 73 Benzoic acid 121
Allylbenzene	156			Benzoic acid 121
d-Pinene	156		0.860	Dibromide 164
l-Camphene	160	42		Dibromide 89
Myrecene	166.5		0.7982	
Limonene	177		0.8446	Tetrabromide 104
Indene	180		0.9915	Phthalic acid 206d
Dipentene	181		0.8501	Tetrabromide 124
Dihydronaphthalene	212	15	0.998	Phthalic acid 206d
Stilbene	306	125		Dibromide 237 Benzoic acid 121
1-Octadecene	316	17		Magaric acid 60-61
1,4-Diphenylbutadiene (trans)	350	148		Benzoic acid 121

TABLE 14-4 HYDROCARBONS: AROMATIC (LIQUIDS)

Compound	B.p. (°C)	M.p. (°C)	D_4^{20}	Nitration product position	Nitration product M.p. (°C)	Oxidation of side chains to acid, M.p. (°C)
Benzene	80	5.5	0.8790	1,3	89	
Toluene	110.6	-93	0.8670	2,4	70	121
Ethylbenzene	136	-93	0.8671	2,4,6	37	121
o-Xylene	143-4	-28	0.8802			206d
Isopropylbenzene (Cumene)	152.5		0.8618	2,4,6	109	121
Mesitylene	164	-57	0.8652	2,4	86	379
p-Cymene	176	-73.5	0.8573	2,6	54	300
1,2,3-Trimethylbenzene	176		0.8944			190
n-Butylbenzene	182		0.8601			121
m-Diethylbenzene	182		0.8641			315
1,2,3,5-Termethyl-benzene (Isodurene)	197		0.8899	4,6	157	218-237d
Tetralin	206		0.9711	5,7	95	206d
1,3,5-Triethylbenzene	218		0.860	2,4,6	108	379
Cyclohexylbenzene	236		0.9502	4	58	121
α-Methylnaphthalene	241		1.020	4	71	
1,1-Diphenylethane	269		1.003			

TABLE 14-5 HYDROCARBONS: AROMATIC (SOLIDS)

Compound	M.p. (°C)	B.p. (°C)	Nitration product position	M.p. (°C)
Diphenylmethane	26	263	2,4,2',4'	172
β-Methylnaphthalene	33	240	1	81
Pentamethylbenzene	51	231	6	154
Dibenzyl	52.50	284	4,4'	180
			2,2',4,4'	169
Biphenyl	70	255	4,4'	237
			2,2',4,4'	165
Durene	79	195	3,6	205
Naphthalene	80	218	1	61
Triphenylmethane	92	360	4,4',4"	206
Acenaphthene	95.5	288	5	101
Phenanthrene	100.5	340		
Fluorene	115.5	294	2,7	199
Hexaethylbenzene	129	305		
Hexamethylbenzene	164	264		
p-Terphenyl	171			
s-Tetraphenylethane	211	382	4,4,4,4,	143
Anthracene	216	354		

The next step will be to determine whether or not the unknown reacts with or is soluble in concentrated sulfuric acid. The results of this test will indicate which other tests should be done. For instance, it should be obvious that if the unknown is insoluble in concentrated sulfuric acid, there is no reason to test the unknown with bromine in carbon tetrachloride. On the other hand, if the unknown is soluble in sulfuric acid, it would be a waste of time to test it with fuming sulfuric acid. Organic unknowns can be identified easily only by thorough understanding of the chemistry of each classification test. Otherwise, unnecessary work will be done, many results will be misinterpreted, and compounds may be identified incorrectly. This type of analytical work can not be done cookbook style!

EXERCISES

14-1. A compound insoluble in sulfuric acid boils at 80 to 82°. What possible hydrocarbons could it be? How would you distinguish among the different possibilities? Explain what you would do and observe.

14-2. An unknown had a boiling range of 151 to 152° and was insoluble in sulfuric acid. What structures are possible? How would you distinguish among the different possibilities? Explain what you would do and observe.

14-3. An unknown had a boiling range of 144 to 146° and reacted with concentrated sulfuric acid to give a polymer. What is the unknown? How would you verify this? If the unknown had reacted with concentrated sulfuric acid to give a homogeneous solution, what could the unknown be? Is there more than one possibility? If so, how would you distinguish among them? If the unknown appeared not to react with concentrated sulfuric acid, what are the possible structures?

15

Friedel-Crafts Alkylation

Synthesis of *t*-Butylbenzene and *p*-Di-*t*-butylbenzene

$$(CH_3)_3CCl + AlCl_3 \rightarrow (CH_3)_3C^{\oplus} + AlCl_4^{\ominus}$$

B.p. 50–51°
D_4^{20} 0.846

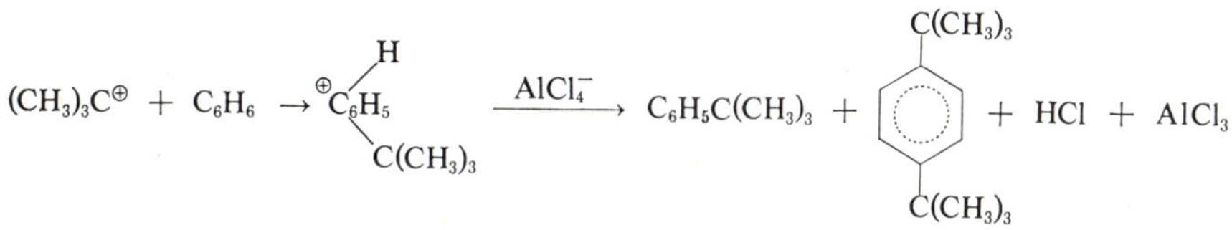

B.p. 80.7°
M.p. 5.6°
D_4^{20} 0.88

B.p. 168–169.5°
D_4^{20} 0.867

M.p. 78°
B.p. 234–7°

INTRODUCTION

Friedel-Crafts alkylation is used to synthesize *t*-butylbenzene. Codistillation is used to dry one of the reactants and to isolate and purify the products.

DISCUSSION

When two immiscible liquids are in a container, each has the same vapor pressure it would have if the other liquid were not present. The total vapor pressure of the liquid mixture is, therefore, always greater than the vapor pressure of either liquid alone. The boiling point of such a mixture is below that of either component.

Problem 15-1. Refer to Table 4-1 for the vapor pressure of water and to the tables in Appendix I for the vapor pressures of organic liquids. What are the vapor pressures of the following mixtures at 50°? The composition ratios are mole percent. (a) Benzene and water, 50:50 and 10:90; (b) toluene and water, 50:50; (c) benzene and toluene, 10:90 and 20:80.

Problem 15-2. What are the boiling points of the following 10:90 and 50:50 mole percent mixtures? (a) Hexane and water, (b) toluene and water, (c) hexane and toluene.

The codistillation of any liquid with water is called *steam distillation.* Since the boiling point of a mixture of two immiscible liquids is always below the boiling point of either component, such a mixture is often called azeotropic. There are some chemists, however, that prefer to restrict the term *azeotropic* to those *single phase* mixtures that have minimum or maximum boiling points. See Chapter 4.

When a mixture codistils, the molar concentration of the components in the distillate is proportional to their vapor pressure:

$$\frac{M_A}{M_B} = \frac{P_A}{P_B}, \tag{15-1}$$

where P_A is the vapor pressure of A at the boiling point of the mixture, and M_A is the molar concentration of A in the distillate. P_B is the vapor pressure of B at the boiling point of the mixture, and M_B is the molar concentration of B in the distillate. Where the vapor pressure of the components is known, the composition of the distillate from the mixture boiling at atmospheric pressure is

$$\frac{M_A}{M_B} = \frac{P_A}{760 - P_A}. \tag{15-2}$$

Expressed in weight W_A and W_B,

$$\frac{W_A \text{ (mol. wt. } B)}{W_B \text{ (mol. wt. } A)} = \frac{P_A}{760 - P_A}. \tag{15-3}$$

Problem 15-3. How much benzene and water in weight, volume, and moles is in 100 g of distillate from a benzene-water mixture?

Problem 15-4. How much hexane will codistil with 36 ml of water at atmospheric pressure? At 380 mm pressure?

Problem 15-5. How much water is required to steam distil 1 liter of toluene at atmospheric pressure?

EXPERIMENTAL PROCEDURE

Synthesis

The Friedel-Crafts reaction is adversely affected by even traces of water. Be sure that all equipment and reagents are thoroughly dried, and take every possible precaution to keep moisture from entering the reaction.

Assemble the distillation apparatus from a three-necked flask, separatory funnel (stopcock closed), and condenser as shown in Figure 15-1. Add 50 ml of benzene and a boiling chip to the flask. Distil 10 ml of benzene. Allow the mixture to cool a few minutes and remove the the condenser and distilling head, covering the exposed neck with a drying tube (Fig. 15-2) while rinsing the condenser with 5 ml of acetone followed by two 5-ml washes of benzene.

Problem 15-6. (a) What is the purpose of this distillation? (b) What does a cloudy distillate indicate? (c) What does a clear distillate indicate?

Problem 15-7. What is the purpose of the acetone wash? What is the purpose of two benzene washes?

Using a *dry* pipet transfer 5 ml of benzene from the three-necked flask to the separatory funnel and restopper the funnel immediately. (Do not use your mouth to supply the suction. Use a vacuum line, a hose to a well trapped aspirator, or a suction bulb.) Replace the drying tube with the condenser set for reflux. Put the drying tube in the top of the condenser (Fig. 13-1). Attach a piece of tubing to the drying tube and lead to a good desk hood. If a good hood is not available, lead the tubing to a combination air and water trap (Fig. 15-3). If the simpler trap made from an Erlenmeyer flask is used, do not connect it until the first 15 drops of *t*-butyl chloride has been

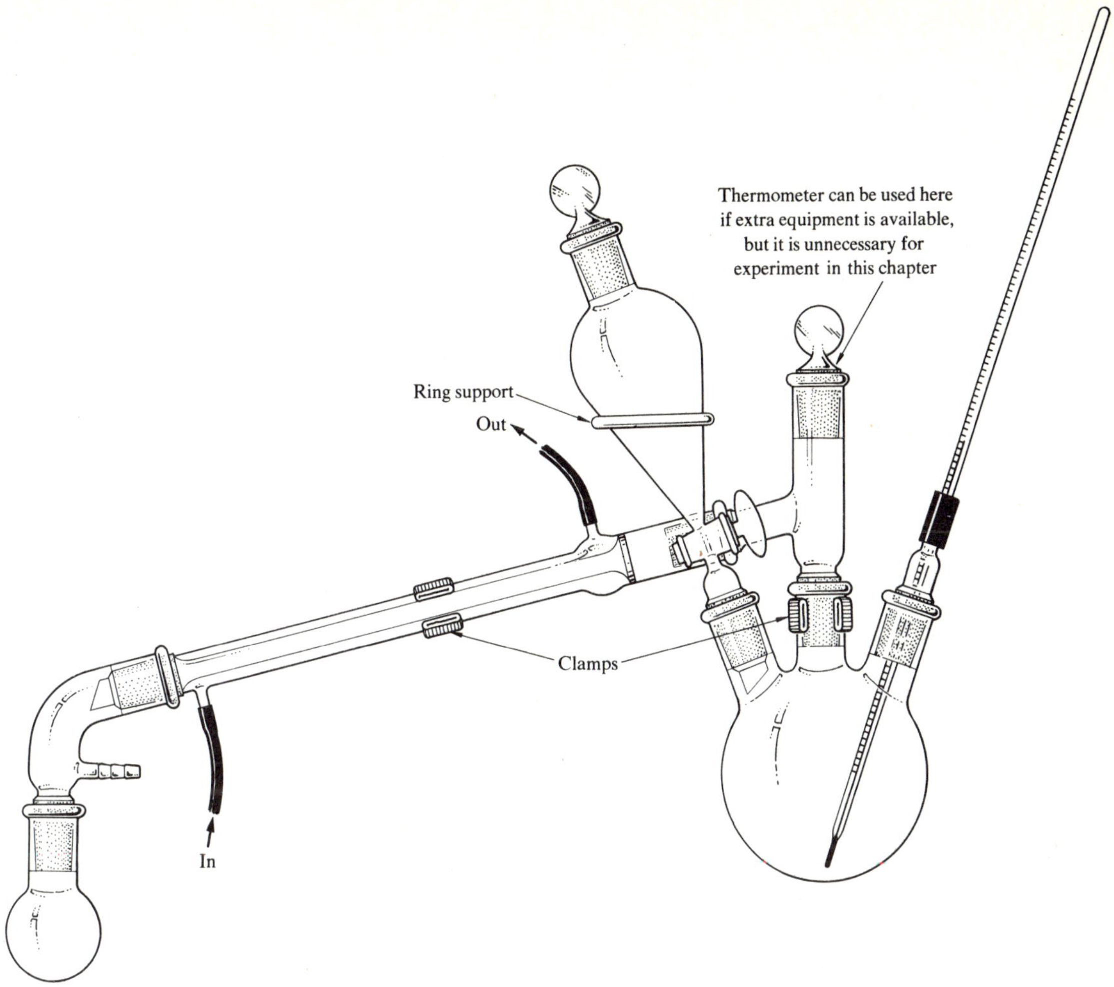

Figure 15-1 Distillation apparatus with dropping funnel.

added. If either trap is used, disconnect the tube from the condenser just before the 25 g of ice is added to the three-necked reaction flask.

Warning! Aluminum chloride is a very strong acid and should be handled very carefully. See Chapter 1.

Add 3 to 4 g of *anhydrous* aluminum chloride to the benzene with the aid of a powder funnel or piece of paper made into a powder funnel placed through the neck that had the thermometer in it. Keep the stem of the funnel below the greased joint so that none of the particles of aluminum chloride stick to the joint. The aluminum chloride should be taken from a container that has been opened that day. It should be weighed and transferred as rapidly as possible so that it has little opportunity to react with any atmospheric water. Return the thermometer to the flask as quickly as possible.

Cool the benzene-aluminum chloride to 7 to 10° by putting an ice bath around it. Put 7 ml of *t*-butyl chloride into the separatory funnel[1] and swirl to insure mixing. Run about 15 *drops* of *t*-butyl chloride solution into the flask and swirl the flask by applying a gentle rotary motion to the ring stand. Add more ice to the cooling bath and attempt to maintain the temperature of the mixture between 10 and 14°.

Once the evolution of hydrogen chloride begins, add the *t*-butyl chloride a drop at a time over about 15 to 30 min. Attempt to keep the temperature *in the flask* between 8 and 10° by addition of ice and if necessary, by

[1] It is usually called a *dropping funnel* when used like this.

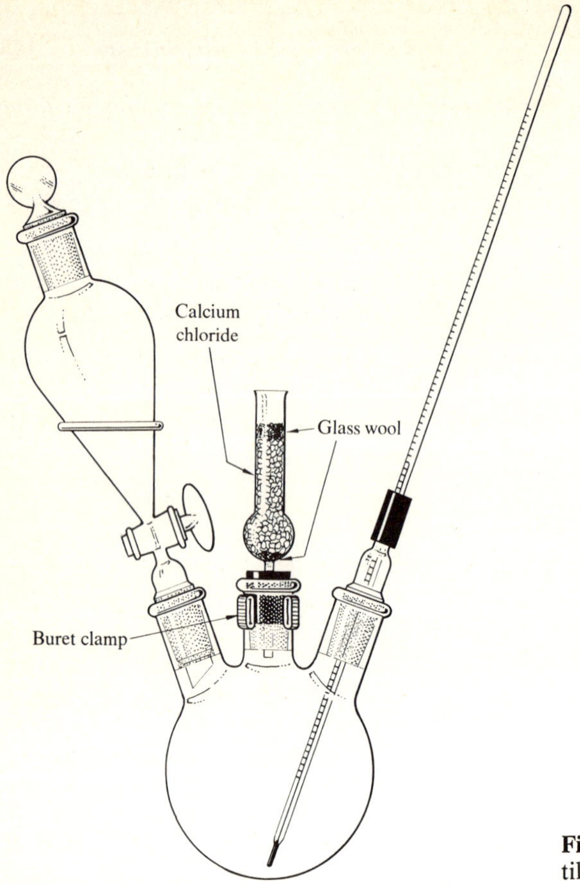

Figure 15-2 Drying tube in place of distillation assembly.

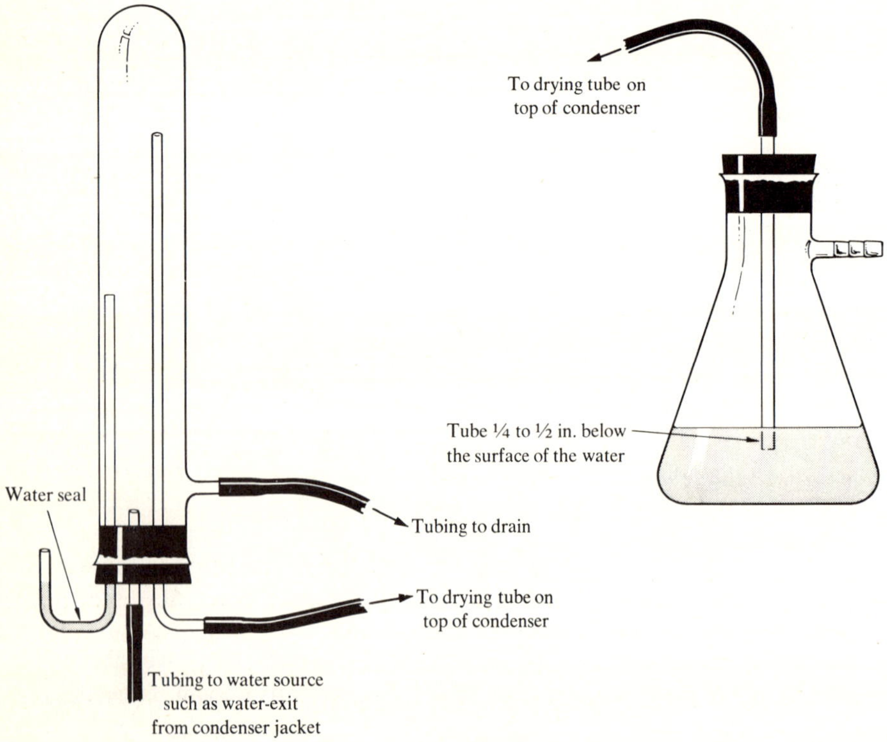

Figure 15-3 Traps for toxic and corrosive vapors.

decreasing the rate of addition of *t*-butyl chloride to the reaction mixture. Swirl and shake the mixture while the *t*-butyl chloride solution is being added.

Allow the mixture to stand about 5 min; shake it occasionally. Remove the cooling bath and shake the mixture occasionally over a 10-min period. Remove the separatory funnel and add 25 g of ice to the mixture. Swirl the flask and allow the ice to melt; then add 30 ml of water.

Problem 15-8. (a) Give two reasons why the mixture is chilled before the addition of *t*-butyl chloride. (b) Give two reasons why the mixture is not cooled below 6° before some *t*-butyl chloride is added.

Problem 15-9. (a) What would happen if after the addition of the *t*-butyl chloride the mixture were put in the desk until the next laboratory period instead of being quenched with ice?

Problem 15-10. What is the purpose of quenching the mixture with ice? Why is ice used instead of water?

Isolation and Purification

Replace the dropping funnel with a stopper. Adapt the condenser for distillation and assemble the apparatus shown in Figure 15-4. Lead the tubing from the receiver to the desk hood or a funnel-water aspirator type

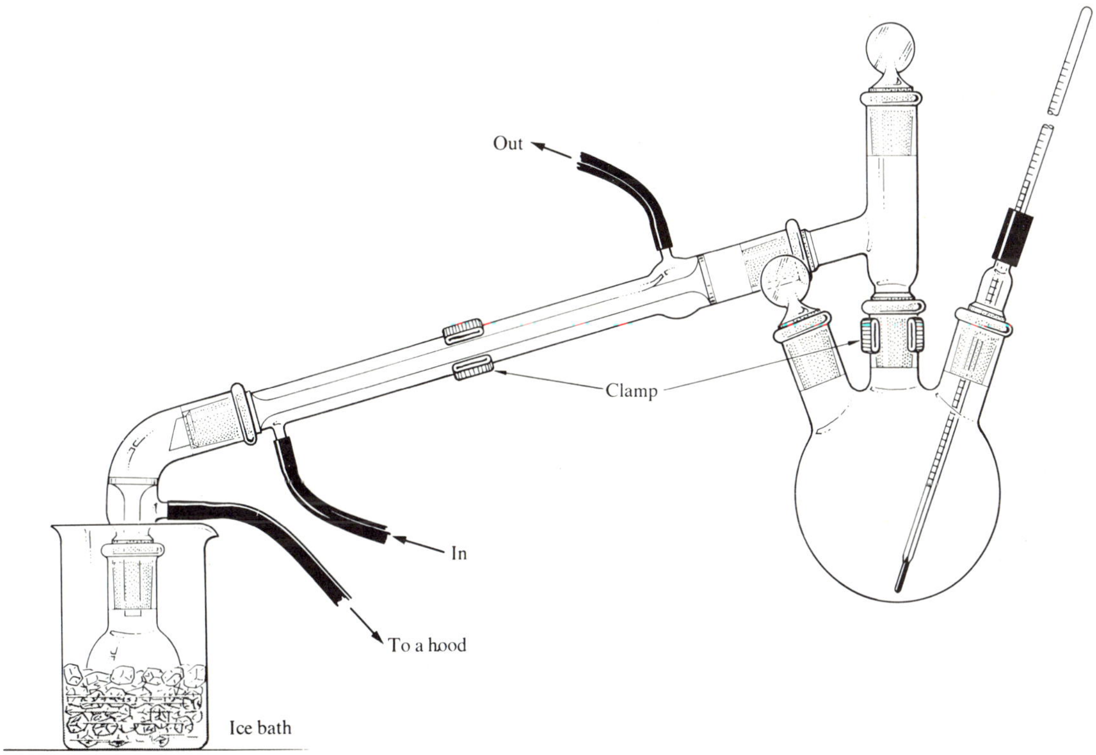

Figure 15-4 Distillation apparatus for removal of benzene.

hood shown in Chapter 12. Place ice water in the beaker around the receiver. Add two boiling chips and distil the mixture until only about 15 ml of water remain in the flask; however, if organic material is still distilling, add more water and continue distilling until no more organic material condenses.

Problem 15-11. Use the appropriate tables in Appendix I. How much water is needed to steam distil all the benzene remaining after the reaction? How much more water is needed to steam distil all the *t*-butylbenzene? What factor might make these calculations in error?

Transfer the distillate to a separatory funnel and remove the aqueous layer. Extract the organic layer with three 40-ml portions of water. Use the first two portions of water to rinse the organic drops from the condenser and receiving flask into the separatory funnel. If there are any crystals in the condenser or adapters, collect them and save them. Add calcium chloride granules to the separatory funnel. Shake the funnel for a few minutes. Dry the stem of the funnel with a pipe cleaner and drain the organic layer into a 100-ml round bottom flask.

Place the flask in a simple distillation apparatus (Fig. 3-1). Collect the fraction that distills at 163 to 172° in a tared yield bottle; label it and hand it in to your instructor. Save the residue in the flask for recovery of *p*-di-*t*-butylbenzene.

p-Di-*t*-Butylbenzene (Optional)

Wash the residue from the three-necked flask into a separatory funnel with 40 ml of water and 30 ml of benzene. Separate the two layers and wash the organic layer with three 40-ml portions of water. Add some granules of calcium chloride to the benzene and shake the separatory funnel for a few minutes. Drain the benzene into the 100-ml flask that contains the residue from the *t*-butylbenzene distillation. Add any crystals collected from the condenser or adapters after the steam distillation. Optional: stir the solution well for a few minutes and transfer 8 to 10 ml to a vial and stopper. This will be used for analysis when Chapter 18 is studied.

Distil the mixture slowly and collect the fraction that boils from 190 to 243° in a test tube. Stir the distillate with about twice its volume of methanol while holding the test tube in an ice bath. Collect the solid by filtration and wash with a small amount of 10% water in methanol solution.

Crystallize the precipitate from the minimum amount of methanol. Allow the hot mixture to cool slowly to room temperature and then chill in an ice bath before filtering. Wash the precipitate with a 25% water in methanol solution. Dry and weigh the precipitate. Determine its melting point and hand it in to your laboratory instructor in a tared yield bottle.

EXERCISES

15-1. Nine grams of water codistills with 40 g of *y*, an unknown compound, at 90° and atmospheric pressure. What is the vapor pressure of *y* at 90°? What is the molecular weight of *y*?

15-2. (a) In the general Friedel-Crafts alkylation

$$\mathrm{ArH + RX \xrightarrow{AlCl_3} ArR + HX}$$

what are the various by-products and the routes to them? (b) Which ones are not likely to occur in the synthesis of *t*-butylbenzene? (c) Of the side reactions that can occur in the synthesis of *t*-butylbenzene what procedures in this experiment minimize their occurrence?

16

Bromination of Benzene

$$C_6H_6 + Br_2 \xrightarrow{FeBr_3} C_6H_5Br + p\text{-}C_6H_4Br_2 + HBr$$

C_6H_6	Br_2	C_6H_5Br	p-$C_6H_4Br_2$
B.p. 80.1°	B.p. 59°	B.p. 156°	M.p. 87.5°
D_4^{20} 0.879	D_4^{20} 3.12	D_4^{20} 1.495	B.p. 219°

INTRODUCTION

Good yields of monohalogenated derivatives are obtained by electrophilic halogenation of aromatic hydrocarbons. Polysubstituted by-products can often be isolated. Steam distillation may be used to isolate two or more products in succession from a reaction mixture if these products have sufficiently different vapor pressures below 100°.

EXPERIMENTAL PROCEDURE

Assemble the apparatus shown in Figure 16-1; use a 100-ml flask. Lead the tubing from the condenser to the water trap (Fig. 15-3) or a desk hood. Put 18 ml of benzene and two small nails in the flask. Place 10 ml of chilled bromine in the separatory funnel.

Heat the benzene to boiling. *Remove the heat source.* Begin adding the bromine a drop at a time. Adjust the rate of addition so there is a vigorous but not violent reaction. After the bromine has been added and the reaction subsides, reflux the mixture until all the red bromine fumes disappear and then heat for an additional 10 min. Allow the mixture to cool and transfer it to a separatory funnel. Wash with 40 ml of 5% sodium hydroxide solution and two 40-ml portions of water. Which layer is aqueous? Do not throw away the product.

Transfer the product to the 100-ml flask in the steam distillation apparatus shown in Figure 16-2. If there are no steam lines in the laboratory, assemble a steam generator (Fig. 16-3). Steam distil the mixture by passing steam into the 100-ml flask. If this flask becomes half filled with water, heat until most of the water is removed. Note the temperature after about 5 ml of water has distilled.

Continue distilling until a few yellow crystals appear in the condenser. Change the receiver and continue steam distilling for 20 min while purifying the bromobenzene. If a significant amount of crystalline p-dibromobenzene collects in the condenser at the end of the second distillation, shut off the cooling water and pass steam through the condenser until the crystals melt and flow into the receiving flask.

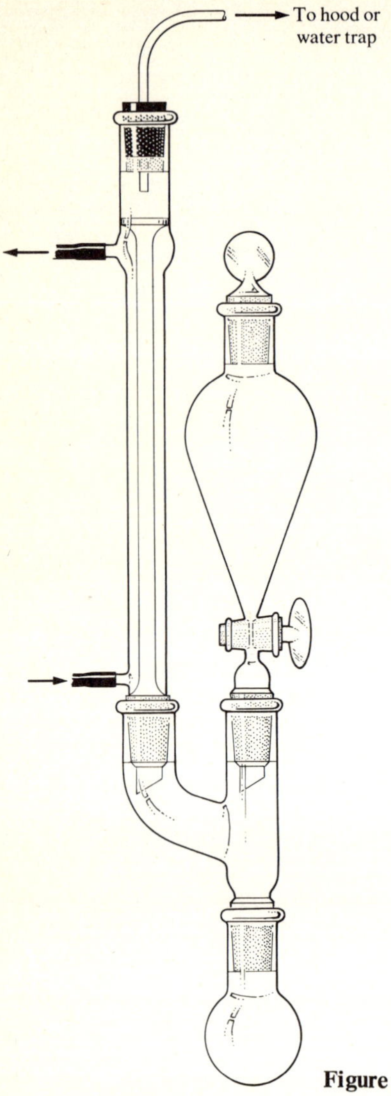

Figure 16-1 Reflux and addition apparatus.

Pour the first distillate into the separatory funnel and separate the two layers. Which layer is the product? Filter the bromobenzene into a 25-ml flask through a funnel with a small amount of magnesium sulfate in it. Distil and collect the fraction that boils from 150 to 160° in a tared yield bottle.

Collect the crystalline product from the second steam distillation by filtration. Recrystallize the *p*-dibromobenzene from 95% ethanol. Dry the product and weigh it in a tared yield bottle.

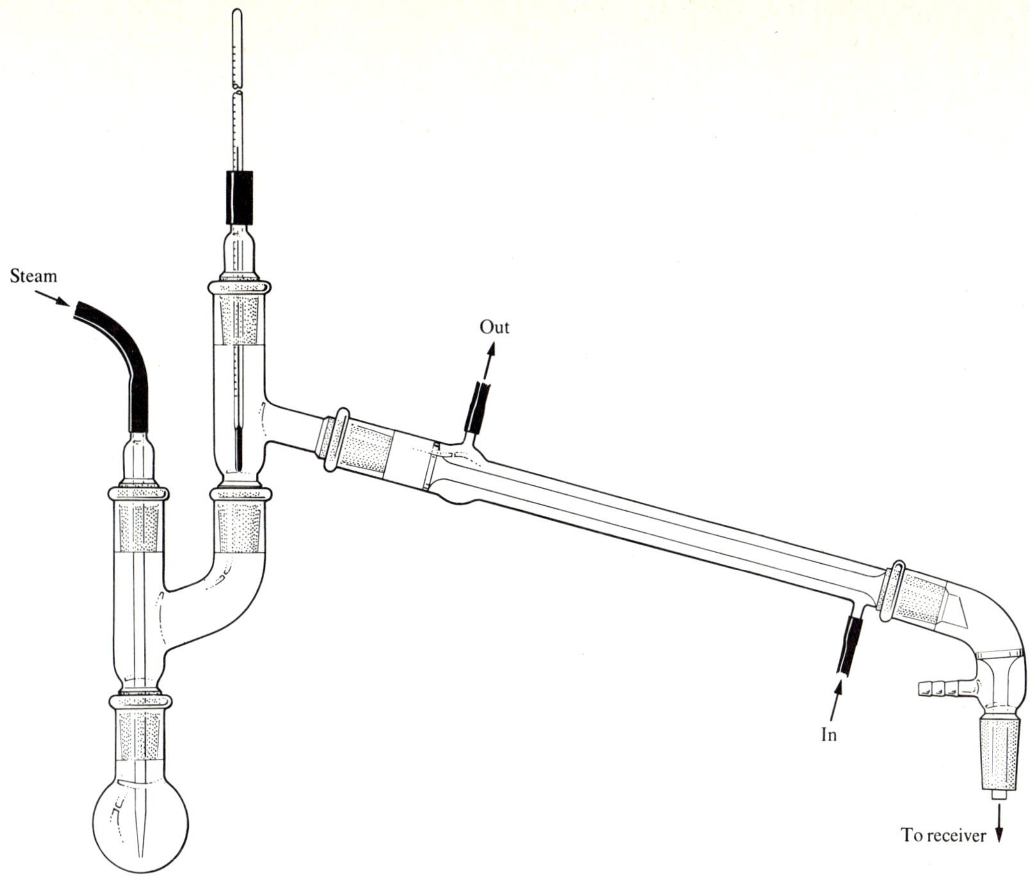

Figure 16-2 Steam distillation apparatus.

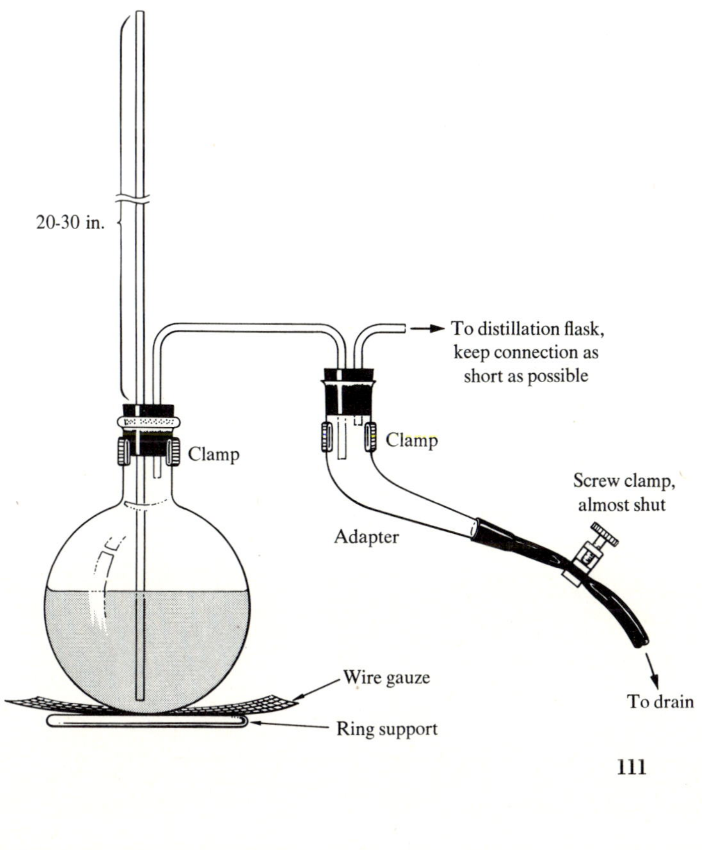

Figure 16-3 Steam generator with watertrap.

EXERCISES

16-1. Compare the steam distillation apparatus in this experiment with that described in Chapter 15. In what situations is one preferable to the other?

16-2. What contaminants are likely to accompany the *p*-dibromobenzene? What step in the purification removes these contaminants?

16-3. From the temperature of the steam distillation of bromobenzene estimate: (a) the vapor pressure of the bromobenzene at the boiling point of the mixture, (b) the amount of water that was necessary to codistil with the amount of bromobenzene obtained as the final product.

17

Friedel-Crafts Acylation

Synthesis of Propiophenone

$$C_2H_5COCl + AlCl_3 \rightarrow C_2H_5CO^+ + AlCl_4^-$$

B.p. 80°
D_4^{20} 1.065

$$C_6H_6 + C_2H_5\overset{+}{C}O \rightarrow C_6H_5\begin{matrix} / \, COC_2H_5 \\ \backslash \, H \end{matrix} \rightarrow C_6H_5COC_2H_5 + H^+$$

B.p. 80° — B.p. 219°
D_4^{20} 0.879 — D_4^{20} 1.01

INTRODUCTION

Friedel-Crafts acylation is more easily controlled than the corresponding alkylation reaction. Friedel-Crafts acylation is not accompanied by polysubstitution or rearrangement. As with alkylation every precaution must be taken to use anhydrous reactants and to keep the reaction dry.

EXPERIMENTAL PROCEDURE

Assemble a distillation apparatus from a 500-ml three-necked flask, separatory funnel (stopcock closed), and condenser as shown in Figure 15-1. Add 100 ml of benzene and a boiling chip to the flask. Distil 15 ml of benzene. Allow the mixture to cool a few minutes and remove the condenser and distilling head, covering the exposed neck with a drying tube (Fig. 15-2) while rinsing the condenser with 5 ml of acetone followed by two 5-ml washes of benzene.

Replace the drying tube with the condenser set for reflux. Put the drying tube in the top of the condenser (Fig. 13-1). Attach a piece of tubing to the drying tube and lead it to a good desk hood. If a good hood is not available, connect the tubing to a combination air and water trap (Fig. 15-3). If a water trap is used, do not connect it until the first 5 drops of propionyl chloride solution have been added. With a *dry* pipet attached to a vacuum line, rubber bulb, well trapped aspirator, or any vacuum source (except your mouth) quickly transfer 20 ml of benzene to the separatory funnel.

Warning! Aluminum chloride is a very strong acid and should be handled very carefully. See Chapter 1.

Add 40 to 45 g of *anhydrous* aluminum chloride to the benzene with the aid of a powder funnel or piece of paper made into a powder funnel placed through the neck that had the thermometer in it. Keep the stem of the funnel below the greased joint so that no particles of aluminum chloride stick to the joint. The aluminum chloride should be taken from a container that has been opened that day. Weigh and transfer the aluminum chloride as rapidly as possible so that it has little opportunity to react with any atmospheric water. Return the thermometer to the flask as quickly as possible.

Warning! Propionyl chloride is an acid chloride. Acid chlorides have strong acid properties. Low molecular weight acid chlorides are strong lachrymators and react quickly with water. Every possible precaution should be taken to keep the propionyl chloride from evaporating into the air and to keep it from reacting with water. Measure and transfer the propionyl chloride as rapidly as possible to keep evaporation and reaction with atmospheric moisture to a minimum.

Put 27 ml of propionyl chloride into the dropping funnel and swirl to insure mixing. Run about 8 *drops* of propionyl chloride solution into the flask and swirl the flask by applying gentle rotary motion to the ring stand. Once the evolution of hydrogen chloride begins, add the propionyl chloride solution a drop at a time over 20 to 30 min. Use a cooling bath of ice and water to prevent the evolution of hydrogen chloride from becoming too vigorous and to maintain the temperature of the mixture below 70°.

> **Problem 17-1.** Why is a temperature higher than was allowed in Friedel-Crafts alkylation permissible here?

When the addition of propionyl chloride is complete, reflux the mixture for 20 min. Cool the flask with the cold water bath used earlier. When the temperature of the flask is below 35°, pour the contents onto a mixture of about 80 g of ice and 15 ml of concentrated hydrochloric acid in an 800-ml beaker and stir with a glass rod. If possible, do this in a hood.

Vacuum filter the mixture after all the ice has melted. Rinse the reaction flask with 10 ml of benzene; use this benzene to rinse the 800-ml beaker and then to wash the residue in the Büchner funnel. Pour the filtrate into the large separatory funnel. Separate and wash the organic layer with 40-ml portions of water, 10% sodium carbonate, and finally water. Dry the organic layer with approximately 10 g of potassium carbonate and filter into a 250-ml round bottom flask.

Using the 250-ml flask, construct the distillation apparatus shown in Figure 3-1. Add a boiling chip and distil. Use a chilled receiver to collect the benzene. When most of the benzene has distilled, transfer it to the waste benzene container before proceeding with the distillation (pp. 2 and 20). Collect the fraction that boils at 205 to 222° in a tared yield bottle and hand it in to your instructor.

EXERCISES

17-1. This synthesis could be one step in the synthesis of 1-phenylpropane. What would the other step(s) be? Why would this method be preferred to the synthesis of 1-phenylpropane by the aluminum chloride-catalyzed alkylation of benzene with propyl chloride?

17-2. Much more aluminum chloride was used in the synthesis of propiophenone than was used in the synthesis of *t*-butylbenzene. Why?

18

Infrared Absorption Spectroscopy

INTRODUCTION

Infrared spectra can readily be obtained by modern instruments. The basic principles and use of these instruments and their application to analysis of reaction products are described here. This chapter and further discussions on absorption spectroscopy in this text emphasize experimental techniques. Some discussions of interpretation and structure-spectra correlation are included when they are very important to the experiment or are not discussed in most of the lecture texts. The spectroscopy sections of this text are intended to be used with lecture texts having discussions on absorption spectroscopy. The student should use the lecture text and supplemental reading for theory and interpretation and use this text for experimental information, applications, and selected aspects of interpretation.

Roger Adams, a distinguished organic chemist active in both research and teaching for the greater part of this century, said in 1966, "The most salient breakthrough for the natural product investigator came with the perfection of infrared spectrophotometers." [1] In addition to its use in the study of natural products, infrared spectroscopy has been extremely useful, if not indispensible, to the modern development of most other branches of organic chemistry. Nobel laureate Robert B. Woodward wrote, ". . . no single tool has had a more dramatic impact upon organic chemistry than infrared measurements . . . simply operated machines for the determination of infrared spectra has permitted a degree of immediate and continuous analytical and structural control in synthetic organic work which was literally unimaginable fifteen years ago. The power of the method grows with each day, and further progress may be expected for a long time to come. None the less, its potentialities are even now greater than many realize. . . ." [2]

Infrared spectroscopy has been so useful to organic chemistry because with proper techniques good infrared spectra can be obtained from almost any organic compound, in any physical state, in any quantity greater than 10 micrograms (and sometimes from even submicrogram quantities). Although usually the purer the compound the better the spectrum, with proper techniques useful structural information can often be obtained from the spectra of compounds that are grossly impure or spectra of mixtures of many compounds. Poor techniques usually give poor spectra or sometimes useless or misleading spectra. Efficient, effective operation of the spectrometer is one of the techniques needed to obtain good spectra. The more that is understood about the construction of a spectrometer, the more effectively it can be used.

SIMPLE SINGLE BEAM SPECTROMETERS

Absorption spectra can be obtained from a relatively simple instrument such as is shown in Figure 18-1. Radiation is produced by the source, usually a heated ceramic rod or wire. A condensing mirror (or lens), M_1, focuses as much radiant energy as possible through the sample and on the entrance slit, S_1.

The sample is placed in the beam from which it selectively absorbs certain frequencies of radiation.

[1] Fourth International Symposium on the Chemistry of Natural Products, Stockholm, 1966.

[2] *Perspectives in Organic Chemistry*, Sir Alexander Todd, ed., Interscience Publishers, Inc., New York, 1956.

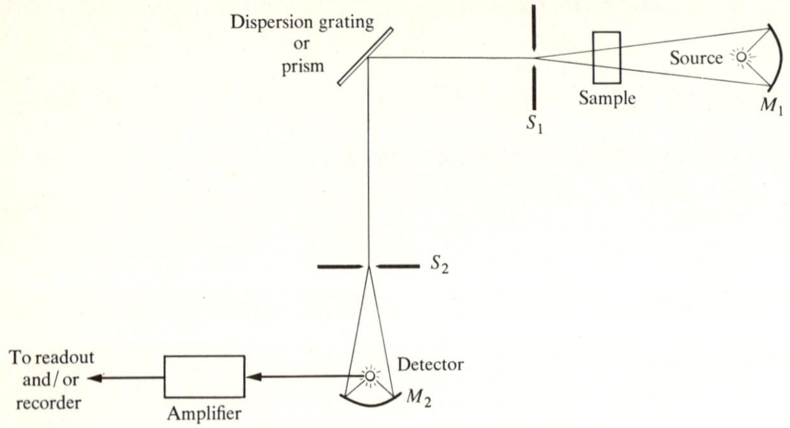

Figure 18-1 Schematic of a simple absorption spectrometer.

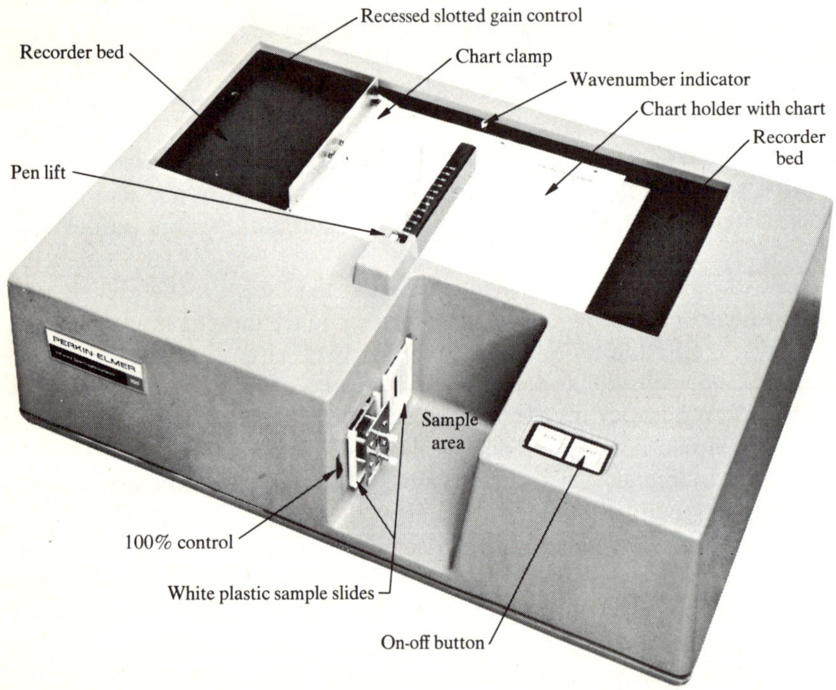

Figure 18-2 A Perkin-Elmer Model 700 infrared spectrophotometer, with a sealed cell in the sample beam.

The function of the entrance slit is to admit to the monochrometer the width of radiation that exactly fills the exit slit. For maximum resolution and passage of maximum energy, the entrance slit width is almost exactly the exit slit width. The only difference is a small correction for characteristics of the optical system.

From the entrance slit the radiation reaches the dispersing element which may be a diffraction grating or prism (one very low resolution[3] instrument uses a filter wheel). The grating or prism disperses (i.e., separates) the continuous radiation from the source into its frequencies.

The exit slit selects the range of these frequencies that fall on the detector. The wider the exit slit, the wider the frequency range of the radiation that falls on the detector. Good resolution requires that the exit slit be as narrow as the need for radiant energy permits. The wider the exit slit the more radiant

[3] *Resolution* is the separation of adjacent frequencies. The higher the resolution the greater the separation, the more spectral detail that can be seen, and, all else being equal, the better the spectrum.

energy that reaches the detector. Since infrared spectrometers are energy limited, the slit widths are a compromise between the narrowness required for the best resolution and the width required for high radiant energy to produce a strong signal at the detector.

The mirror M_2 gathers as much energy as possible from the exit slit and focuses it on the detector. The detector is a transducer that changes the signal carried by the radiant energy to an electrical signal. Most infrared instruments use a thermocouple as a detector. In all modern spectrometers the electrical signal is amplified, and the amplified signal is registered on a meter, other suitable readout, or a recorder. Almost all modern infrared spectrometers use a recorder.

The movement of the abscissa (x-axis) of the recorder is coordinated, often by a direct mechanical

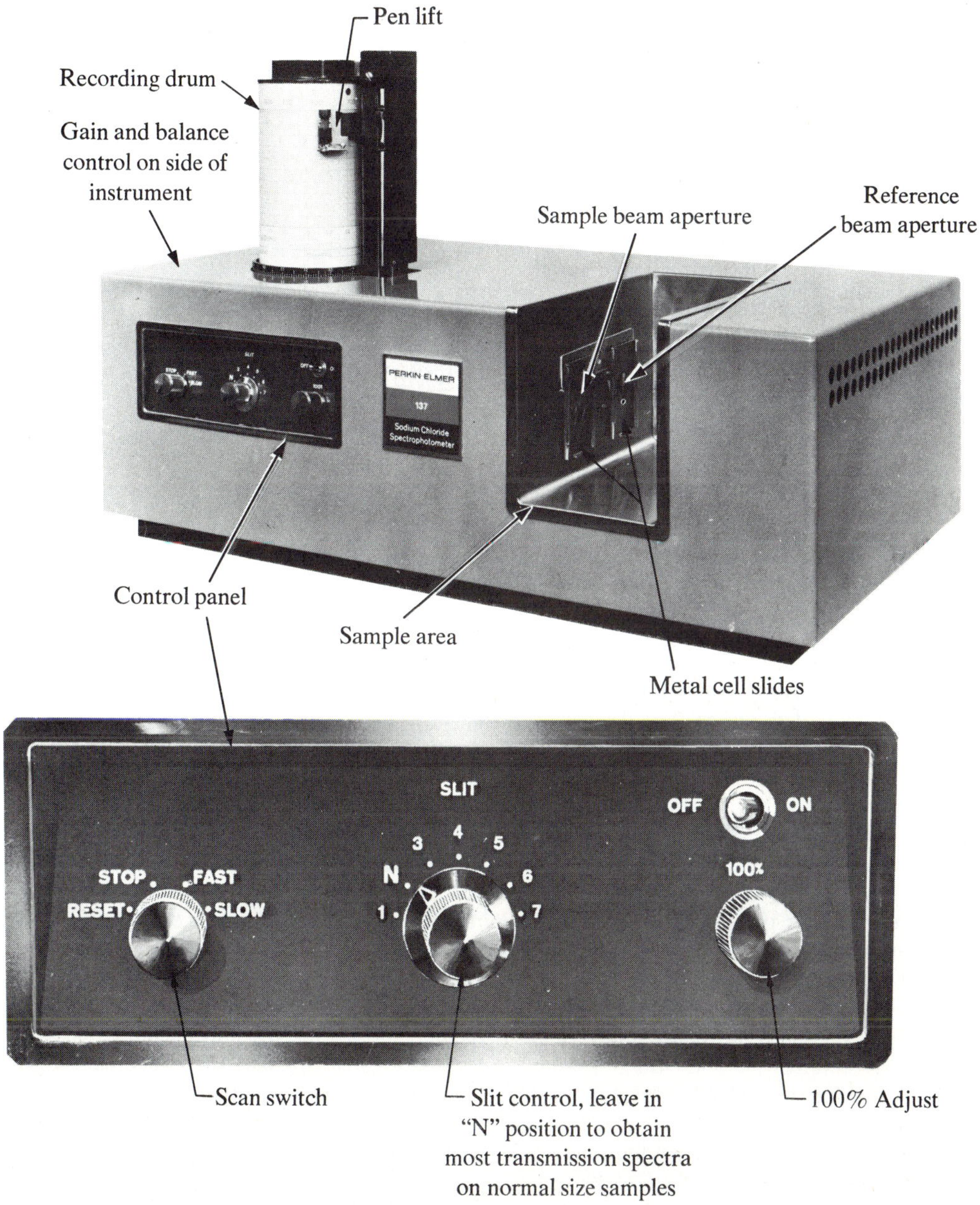

Figure 18-3 A Perkin-Elmer Model 137 infrared spectrophotometer.

linkage, to the prism or grating. In this way the same frequencies can always be recorded at the same chart position. This allows the use of preprinted chart paper. The pen moves along the ordinate (y-axis) and records the amount of energy the sample transmits (i.e., does not absorb). Typical infrared spectrometers likely to be encountered (but considerably more complicated than the one shown in Fig. 18-1) in a student organic chemistry laboratory are shown in Figures 18-2 and 18-3. An infrared spectrum of hexane is shown in Figure 18-4.

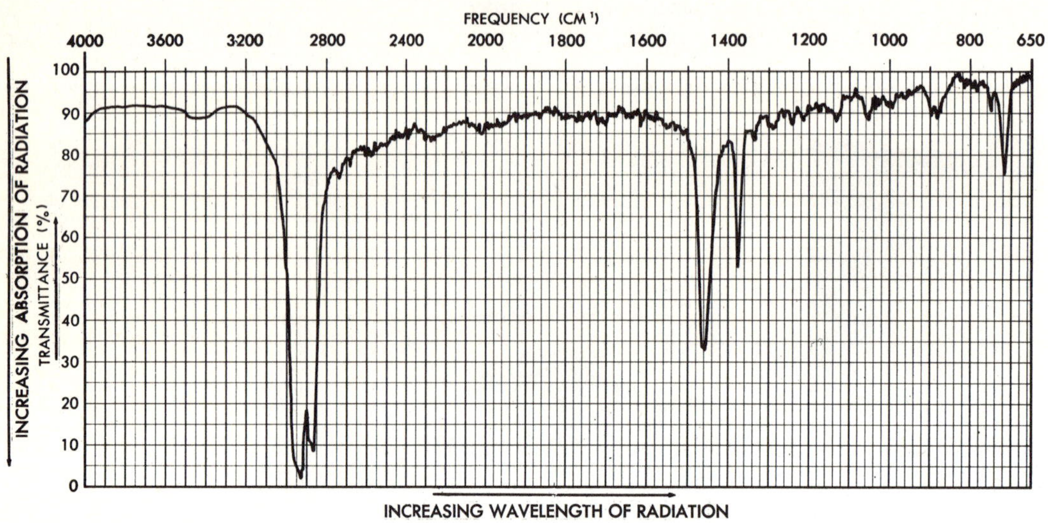

Figure 18-4 Infrared spectrum of hexane, neat, in a 0.015-mm cell.

OPTICAL NULL, DOUBLE BEAM SPECTROPHOTOMETER

General Description

The spectrometer described above and shown in Figure 18-1 has several virtues, principally its simplicity, and many deficiencies. The deficiencies include poor reproducibility of the absorption band intensities, base line drift, and extraneous bands in spectra that are not from absorption by the sample. The causes of these are due mainly to variations in the source intensity and absorption by compounds in the atmosphere, especially carbon dioxide and water. Random signals (usually called *noise*) are introduced into the system principally by thermal noise of the detector and to a lesser extent by any electrical component. Other small deviations may be caused by nonlinear response of the dispersing unit, focusing mirrors, and any other components in the optical system.

To compensate for these effects most modern infrared spectrophotometers are optical null, double beam instruments. All the spectra in this book including Figure 18-4 were obtained on double beam infrared spectrophotometers.

A schematic of a double beam, optical null infrared spectrophotometer is shown in Figure 18-5. The output from the source is divided and focused by the mirrors M_R and M_S into reference and sample beams. These go through the reference and sample areas (see Figures 18-2 and 18-3). The sample beam goes through the sample, and the reference beam goes through the reference material, if present. The chopper is a segmented wheel with a mirrored surface (Fig. 18-6) that alternately reflects the sample beam and transmits the reference beam through the entrance slits. The alternating frequency between the reference and sample beams occurs 6 to 25 times per second depending on the instrument and is a constant value for each instrument. From the chopper to the detector the reference and sample beam use the same path but occupy it at different times.

The detector is connected to an amplifier which will amplify only alternating current of the sar

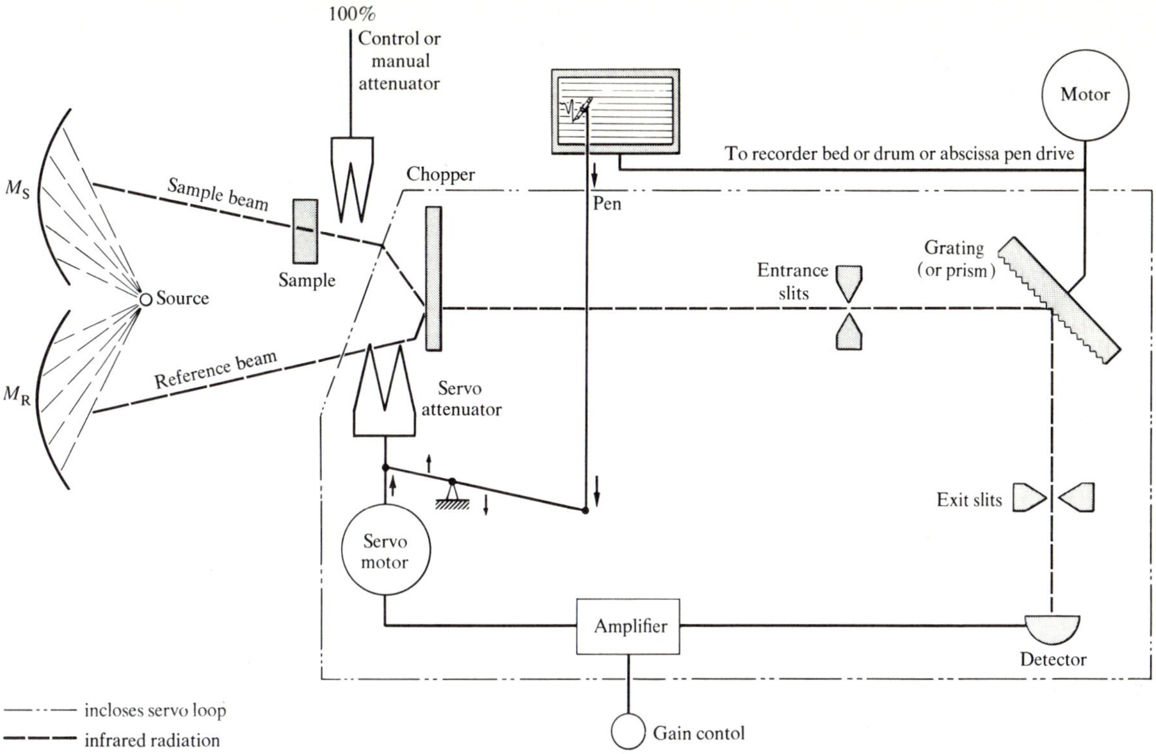

Figure 18-5 Schematic of an optical null, double beam infrared spectrophotometer.

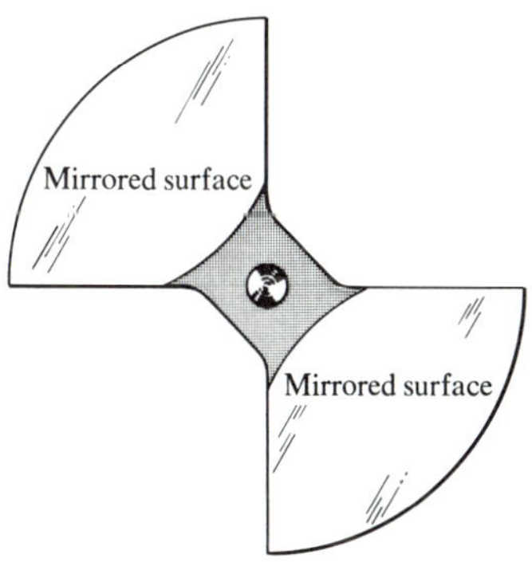

Figure 18-6 A chopper from an optical null, double beam infrared spectrophotometer. Since every turn of the wheel chops each beam twice, the frequency of the alternation of the two beams is twice the frequency of rotation of the chopper.

frequency as the alternation between reference and sample beams. The amplifier drives a motor which controls the position of the pen on the chart and an optical attenuator in the reference beam. The pen and the optical attenuator are synchronized to move together.

Assume a sample of hexane is in the sample beam and nothing is in the reference beam. As the grating (or prism) moves and diffracts frequencies from 4000 cm^{-1} to 3000 cm^{-1} through the exit slit, neither the reference nor the sample beams undergo any changes in attenuation, and they alternately fall on the detector at the same high intensity. This signal from 4000 cm^{-1} to approximately 3000 cm^{-1} can be represented as shown in Figure 18-7a. The detector transduces this heat energy into a relatively high direct electric voltage. Since the amplifier passes only an alternating voltage, nothing happens to the servo motor driven by the amplifier.

When the grating (or prism) diffracts radiation of approximately 3000 cm^{-1} through the exit slit,

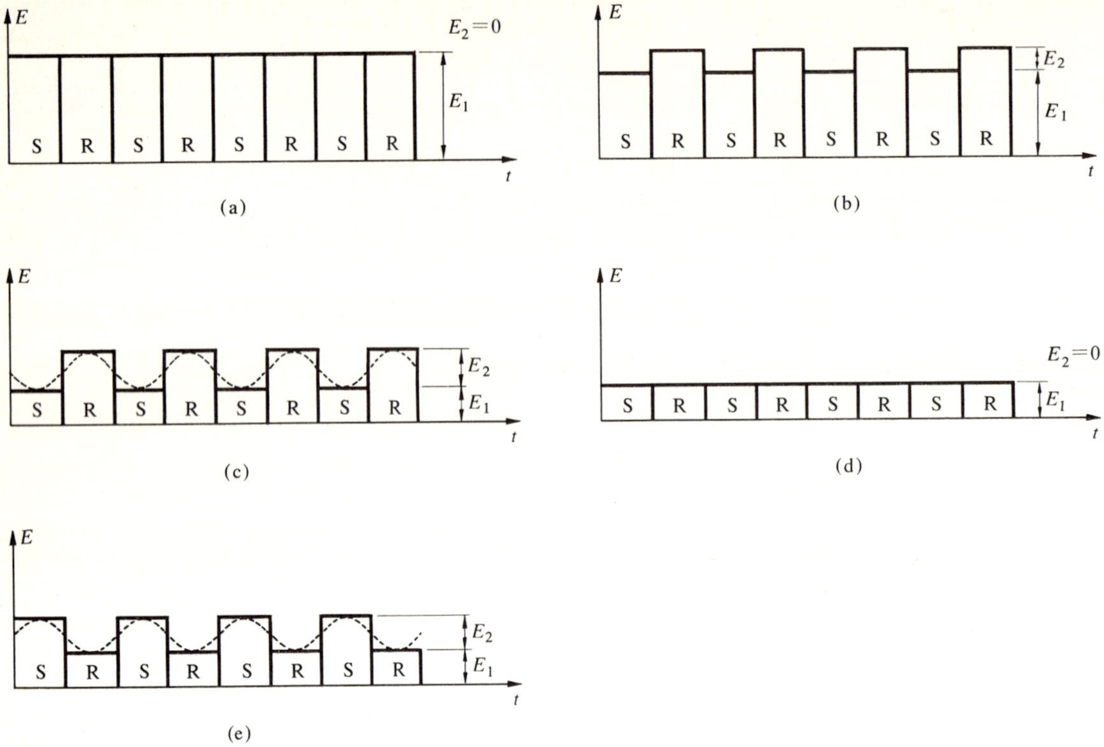

Figure 18-7 Sample and reference beam intensities.

the sample beam radiation reaching the detector is attenuated because of absorption of this frequency by the sample. The radiant energy reaching the detector can be represented by Figure 18-7b. The detector transduces this into an electrical signal which is comprised of two voltages: E_1, a direct current which would be the same as the voltage produced if the unchopped sample beam were allowed to fall steadily on the detector, and E_2, an alternating voltage which is the difference between the voltages produced by the reference beam and the sample beam. The alternating voltage is the only signal amplified. The signal is also rectified, filtered, and then converted into 60 cps AC. Students without the requisite electronic background can ignore these later transformations and still understand how an optical null instrument operates. This amplifier signal drives the motor which moves both the optical wedge or attenuator into the reference beam and the pen on the chart paper. The speed of the servo motor and, therefore, the response of the pen and optical wedge are dependent upon the voltage of the amplified AC signal. The higher the voltage, the faster the pen and wedge respond. As the optical wedge is driven into the reference beam, it blocks part of the beam and attenuates its radiant energy (Fig. 18-7 b and c). The optical wedge continues to be driven into the reference beam until its radiant energy is the same as the radiant energy of the sample beam (Fig. 18-7d). At this point no alternating voltage reaches the amplifier and the servo motor stops.

At approximately 2920 cm^{-1} the sample absorbs less than it did at a slightly higher frequency. As the grating (or prism) turns and frequencies slightly lower than 2925 cm^{-1} are focused through the exit slit and onto the detector, the sample beam has more energy than the reference beam (Fig. 18-7e). The detector transduces the difference between the sample and the reference beams into an AC voltage. The phase is opposite to the AC voltage produced when the sample beam is weaker than the reference beam. The servo motor is phase-sensitive, and when it receives this amplified voltage it withdraws the optical wedge from the reference beam and moves the pen back up the chart paper. The process is repeated for other absorptions. Figures 18-7 and 18-8 show the relative energy of the sample and reference beam and the relative position of the optical attenuator in the reference beam for different absorptions by the C—H stretching vibration in hexane (Fig. 18-9). The entire action of the sample, optical wedge, servo motor,

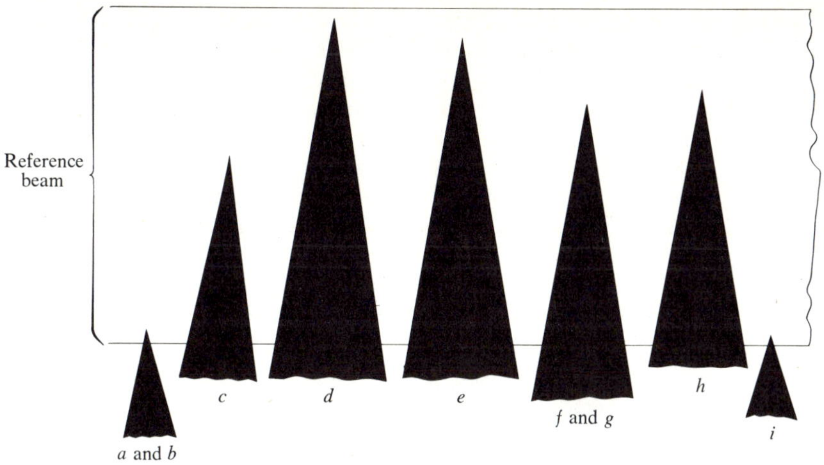

Figure 18-8 Position of the optical wedge in the reference beam at different points of the carbon-hydrogen stretching absorptions of hexane.

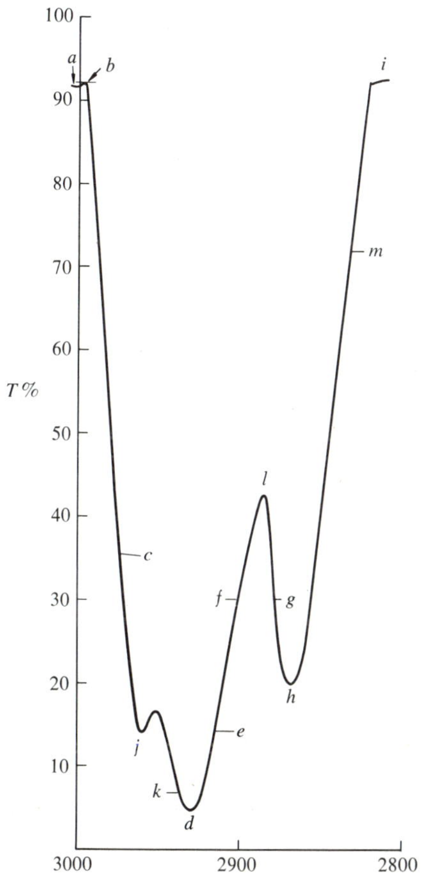

Figure 18-9 Carbon-hydrogen stretching absorptions of hexane.

detector, and amplifier[4] is such as always to keep the sample beam energy and reference beam energy as nearly equal as possible. The optical null system does this by detecting an error signal (the difference between the two beams), amplifying this signal, and using it to drive the optical wedge which corrects

[4] All these components and the two beams comprise what is known as a servo loop.

this error. The pen is attached to or synchronized with the optical wedge and records the correction necessary. Since the correction necessary is the same as the difference between the two signals, the pen records the transmittance (or absorbance) difference of the two beams before attenuation of the reference beam. Absorptions by the sample will, therefore, be recorded but absorptions by compounds in the atmosphere will not be recorded provided the path lengths of the two beams through the atmosphere are the same. Since they have equal effect on both beams, variations of the source will not be recorded; neither will nonlinearity of any of the components in the path shared by both beams (from the chopper to the detector). The spectrum is, therefore, free of all or at least most of the extraneous peaks that are found on single beam instruments. In addition, signals from frequencies significantly different from the frequency of the alternating beams are not amplified so spectra from a double beam instrument have less noise than spectra from a similar single beam instrument of the type shown in Figure 18-1.

Problem 18-1. Draw the relative heights of the sample and reference beams and position of the attenuator in the reference beam for points *f-i* in Figure 18-9. Use the same formats that were used in Figures 18-8 and 18-7.

Problem 18-2. With the spectrophotometer as shown in Figure 18-7a, what would happen to the pen if an opaque object were placed a little way into the reference beam? What would happen to the optical attenuator? The energy of the reference beam? The energy of the sample beam? The AC and DC voltages produced by the detector?

OPERATION OF OPTICAL NULL, DOUBLE BEAM INFRARED SPECTROPHOTOMETER

Use of the 100 Percent Control

To compensate for small amounts of vignetting (blocking of portions of the beam) of the sample beam and different opacities of material placed in the sample and reference beams, an optical attenuator (Fig. 18-5) is placed in the sample beam. This control is shown on the two instruments in Figures 18-2 and 18-3. It is manually adjusted and has no part in the servo loop except indirectly. That is, it affects the servo loop only by causing changes in the sample beam which cause a complimentary change in the position of the optical attenuator in the reference beam.

In general, the 100% control should be set so that the pen begins at 90 to 95% transmittance. If the 100% control is set too low (Fig. 18-10a), a significant amount of contrast is lost and a considerable amount of energy is removed unnecessarily from the sample beam and consequently from the reference beam. This could cause an unfavorable signal-to-noise ratio and loss of spectral detail. If the 100% control is set too high (Fig. 18-10b), bands with weak absorptions are likely to be missed. In addition, most infrared spectrophotometers will be damaged if they are operated very long with the 100% control set too high.

Sometimes a 90 to 95% setting is too high because of excessive scattering of light at only the higher frequencies. The 100% line will then increase as the spectrum is scanned and cause loss of information at the lower frequencies (Fig. 18-11a). When this occurs, the scan will need to be started at transmissions as low as 50 to 60%. If this is necessary, the 100% control may be opened far enough to cause the pen to touch or even exceed the 100% line for a short distance (10 to 20 cm^{-1}); see Figure 18-11b.

Perhaps the easiest way for the beginner to adjust the 100% control is to set it for 100% transmission at 4000 cm^{-1} (although any frequency will do) with nothing in either beam. If the base line is too low (less than 90% over the entire spectrum), another spectrum can be obtained with a higher base line setting.

Occasionally a sample will be so opaque or so much vignetting may occur that even with the 100% control opened all the way the maximum transmittance may be very low. If this happens everything possible should be done to increase the transmittance of the sample and to be certain it is so placed in the sample beam to allow maximum transmission of radiant energy. If after these steps the maximum transmission is still low, the spectrum will have to be accepted with this deficiency. (See, however, Appendix III, Advanced Infrared Techniques.)

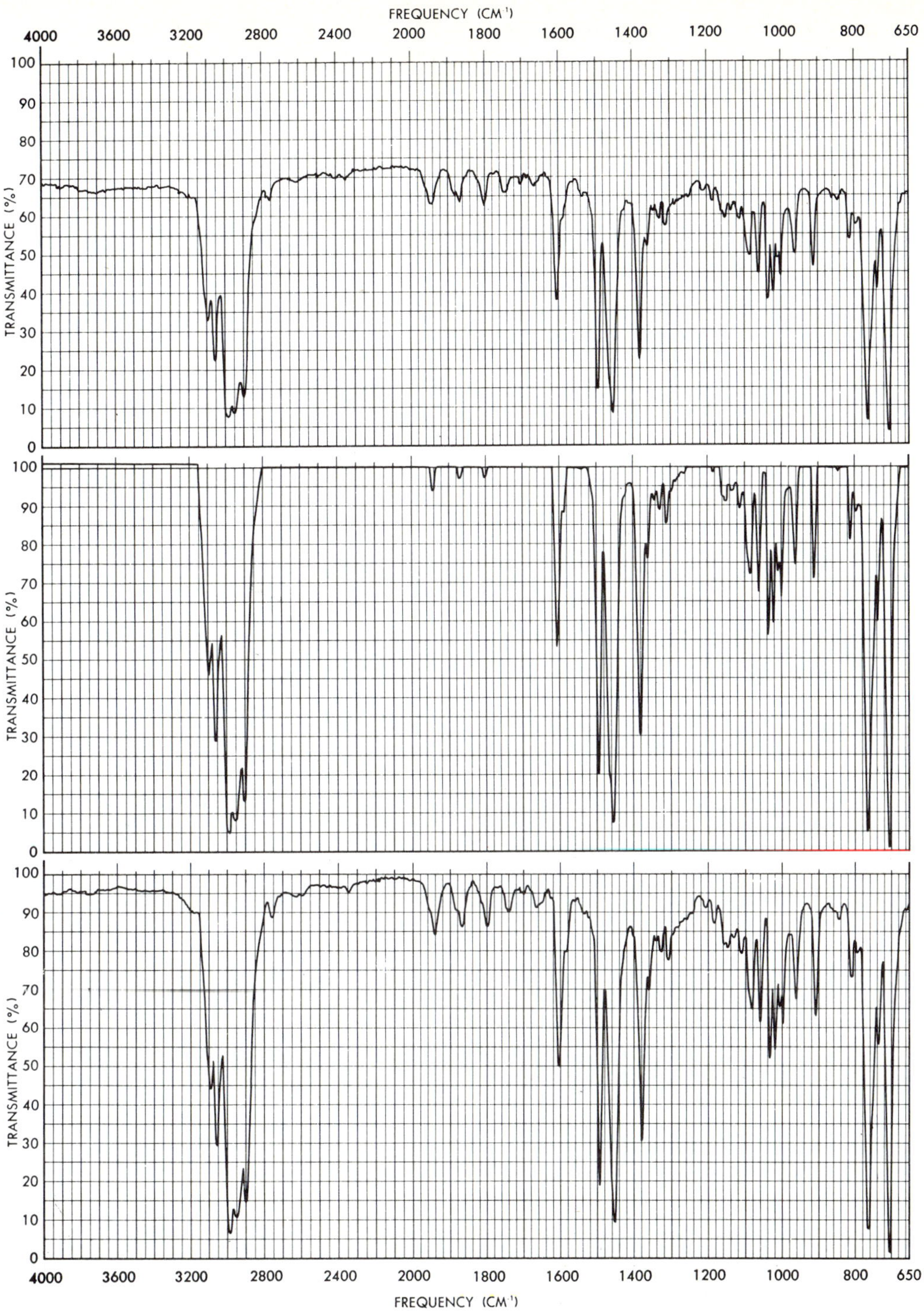

Figure 18-10 (a) 2-Phenylbutane, 100% control set much too low. Note loss of contrast. (b) 2-Phenylbutane, 100% control set much too high. Note loss of spectral information. Overtone and combination bands between 2000 and 1630 cm^{-1} are almost completely obscured. Compare with part c. (c) 2-Phenylbutane, 100% control has optimum adjustments. Note all weak bands are shown. Detail between 2000 and 1620 cm^{-1} is well defined.

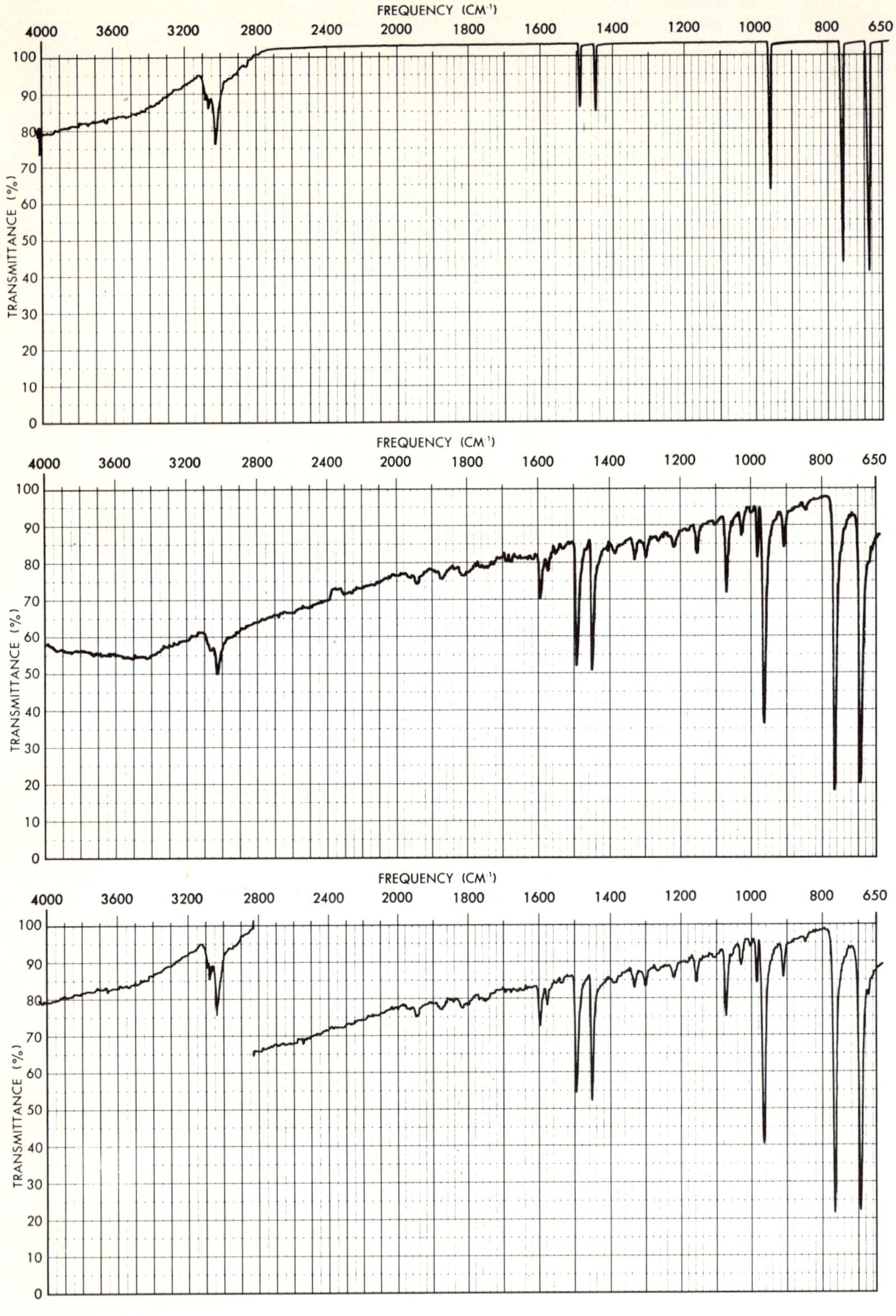

Figure 18-11 Spectra of *trans*-stilbene, 1.2 mg in 296 mg KBr. (a) 100% control set for 69% at beginning. This is too high because of sloping I_0; spectral information lost below 2000 cm^{-1}. (b) Correct setting of 100% control, small amount of pen travel above 100% in 800 cm^{-1} region is not unsatisfactory since no spectral information is lost. (c) Optimum 100% control setting. Maximum contrast above 2400 cm^{-1}, then pen is lifted, scan button pressed to stop chart movement, chart moved back slightly, 100% control adjusted and scan continued.

Use of the Gain Control

The amount of amplification available from an electronic apparatus is often called *gain.* Common examples of gain controls are the volume controls of radios, public address systems, or television sets. All infrared spectrophotometers have a gain control. On some spectrophotometers it is not easily accessible, such as on a Perkin-Elmer Model 700 where it can be reached only with a long screw driver through a hole in the top panel (Fig. 18-2). On older Infracords it was on the back of the instrument. On many spectrophotometers, however, it is readily available on the side (Fig. 18-3) or front of the instrument.

The amount of power in a servo loop such as is found in optical null, double beam spectrometers must be within certain fairly narrow limits or the servo loop will not operate properly. If enough power is not available, the pen response and movement of the optical wedge in the reference beam is sluggish, a lot of spectral detail is missed, and absorption bands appear at frequencies lower than their actual values (Fig. 18-12a). Too much energy will cause the system to oscillate. The energy available is determined by the total optical and electrical energy of the servo loop. The optical energy is determined by the source intensity, sample transparency, and slit width. The only controllable variable that effects the electrical energy is amplifier gain. For a given sample the only variables that can be used to adjust the total power of the servo loop on most infrared spectrophotometers that are found in the beginning organic chemistry laboratory is the gain.[5]

If the gain control is set too low, the pen response will be sluggish and detail will be lost (Fig. 18-12a). This might be compared to a radio with the volume set low where the program is indistinct and many of the softer sounds are lost.

If the gain control is set too high, the pen will have too much overshoot[6] causing exaggeration of the absorption bands and high noise (random signals generated principally in the detector) will make the detection of weak bands and the exact location of all bands difficult (Fig. 18-12b). The noise from even moderately high gain settings can cause errors in band intensities and should not be used for quantitative analysis (Chapter 20). Too high a gain might be compared with too high a volume setting on a radio where the static (random noise) is amplified excessively. Sometimes because of an opaque sample or vignetting, the radiant energy is so low that the electrical energy must be increased considerably by a very high gain setting. In such cases a high gain setting is needed and while the noise is more than is desirable for even qualitative analyses the pen does not have excessive overshoot. This can be compared to tuning in a distant or weak station on a radio. If the station is to be heard at all, the volume control must be turned high, but of course the static will be high also.

If the gain control is very high (or even moderately high when the radiant energy is high), the servo loop will have too much energy (Fig. 18-13). When a random noise signal or an absorption band causes one of the beams to be attenuated (for this discussion and Figs. 18-15 and 18-16, we will assume it is the sample beam), the servo motor drives the optical attenuator (and pen) to equalize the energy in the two beams. Because, however, of excess energy in the servo loop the optical attenuator is driven considerably farther than is needed to attenuate the energy of the reference beam (Fig. 18-16e and g). The signal from the amplifier stops and reverses and now drives the servo motor hard in the other direction. Because of the high gain the wedge and pen are driven much farther than the equilibrium position (line *A*, Fig. 18-15b) so that the reference beam is now stronger than the sample beam (point *V*′: Figs. 18-15b and 18-16h). The oscillation continues until the gain or the radiant energy is decreased.

Spectra with such oscillations (as in Fig. 18-13) are completely unacceptable. For qualitative analysis, spectra with high noise levels are marginally acceptable if they cannot be obtained without the high noise. Oscillations of low amplitude can be easily distingusihed from high noise by the regular periodic, usually sinusoidal character of the signal (Fig. 18-17, also Figs. 18-13 and 18-15b). Noise is completely random with no resemblance to any definite wave form (Fig. 18-18, also Fig. 18-12b).

Problem 18-3. The oscillations in all servo systems or feedback loops are due to too much power in the loop. A public address system with the volume up too high will emit a howl or squeal. It is started by the pickup of random noise by the microphone. Complete the analysis for the generation of this sound and compare it to the oscillations produced by an optical null, double beam spectrophotometer with too high a gain setting. Why does turning down the volume on the public address system stop the feedback squeal instead of just decreasing it?

[5] All of the better medium priced instrument currently in production also have variable slit programs. The discussion in Appendix III will aid in obtaining optimum settings of this control.

[6] *Overshoot* can be defined as the excess travel of the optical wedge beyond that required to equalize the energy in the two beams. See Figure 18-14 for the effect of this on pen movement.

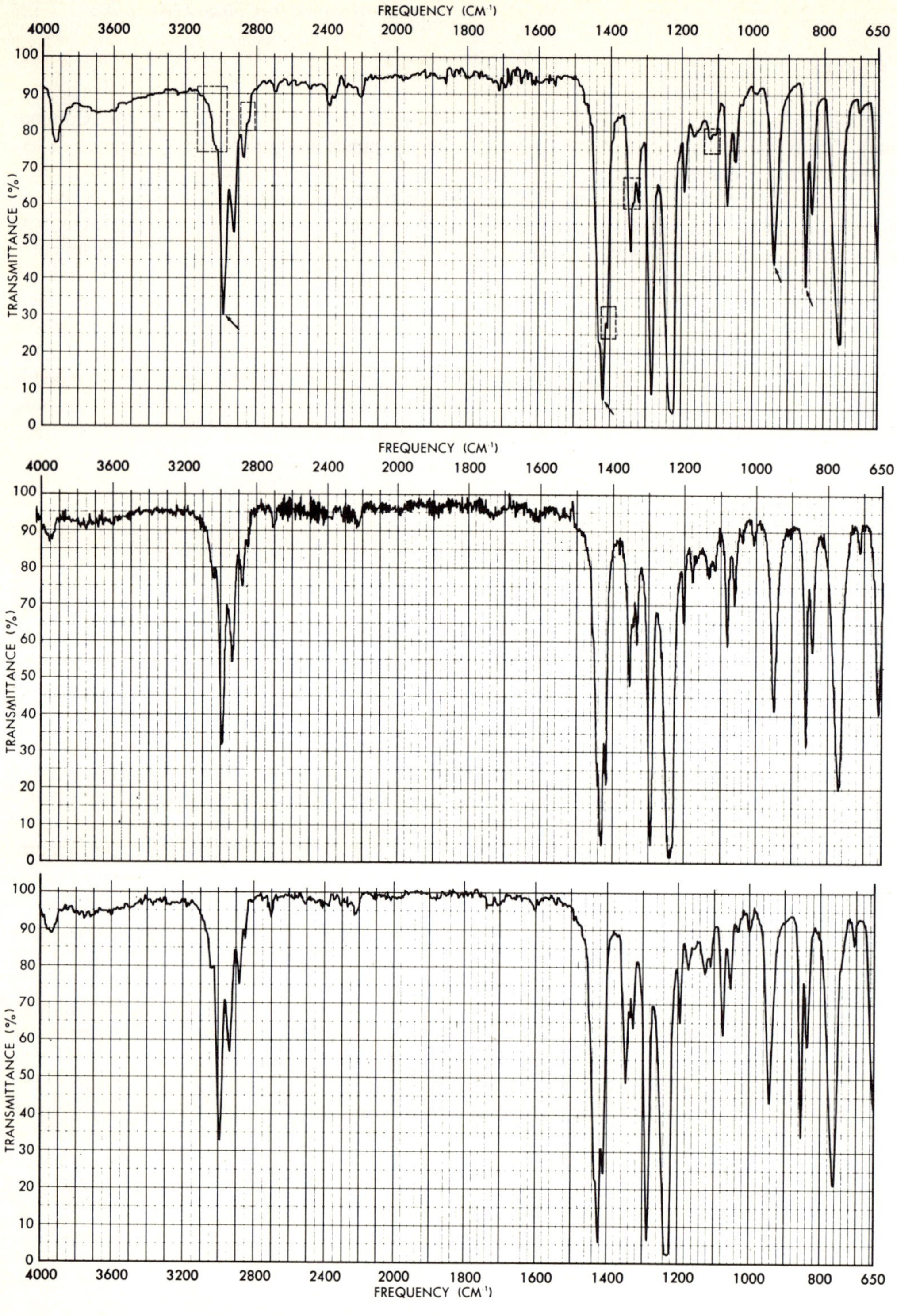

Figure 18-12 (a) Infrared spectrum of 1,3-dibromopropane in a 0.025-mm cell. Gain is too low. Compare with spectrum obtained with optimum gain (see part c). Especially compare boxed areas, which show loss of spectral detail, and arrows, which indicate a small frequency error, with same portions of part c. (b) Infrared spectrum of 1,3-dibromopropane in a 0.025-mm cell. Gain is too high. Note excess noise, especially between 2800 to 1600 cm^{-1}. (c) Infrared spectrum of 1,3-dibromopropane in a 0.025-mm cell. Gain is either optimum or possibly just a little high.

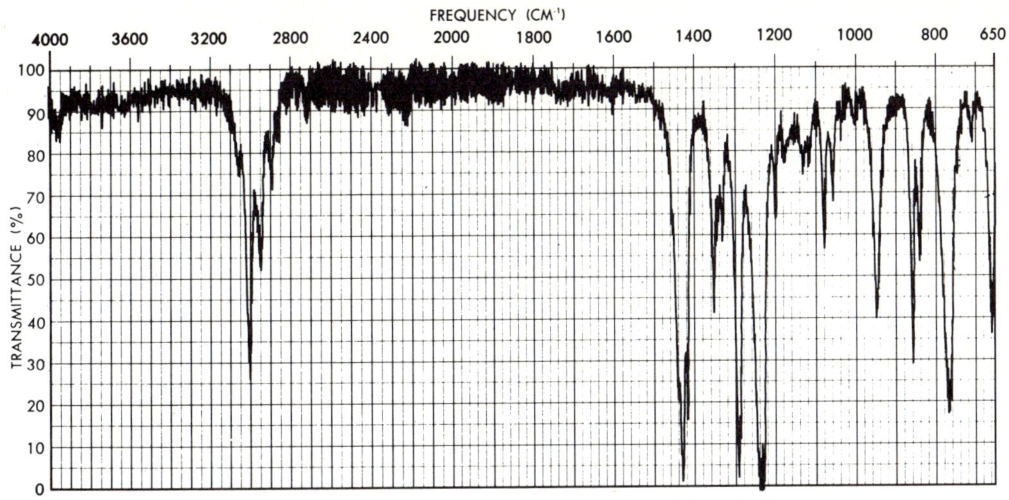

Figure 18-13 Infrared spectrum of 1,3-dibromopropane. Gain is far above optimum setting. Note severe oscillations.

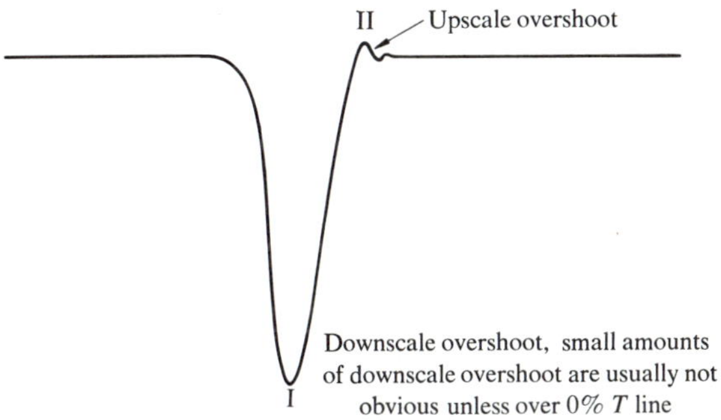

Figure 18-14 Absorption band with a small amount of overshoot.

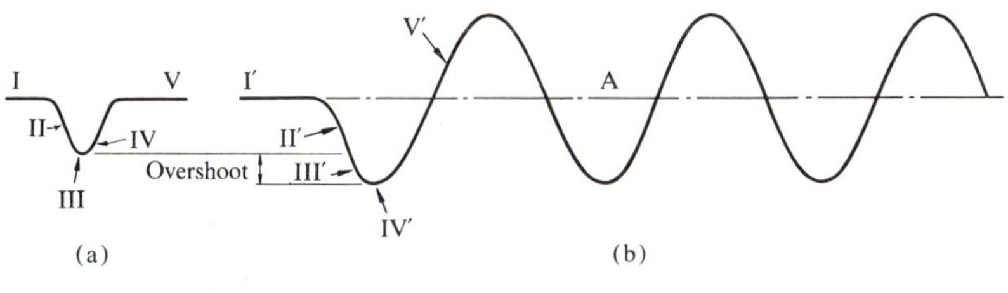

Figure 18-15 Trace from an infrared spectrophotometer recording a small signal. (a) Gain set correctly. (b) Gain set much too high. Identical Roman numerals indicate identical times on two traces and in Figure 18-16.

The gain should be checked at the beginning of each day, after an adjustment on another control that affects the servo loop, and for each sample of significantly different opacity from the previous one. These last two conditions are most often encountered when using micro, reflectance, or other special sampling techniques. Unless the energy in the servo loop is obviously too high or too low, it is necessary to check the gain only once when obtaining spectra on a series of samples with similar opacities and absorptions, such as a group of samples from the same synthesis (as from an organic chemistry class or an industrial process).

The gain should be adjusted with the sample in the beam so that there is a small amount of overshoot and the pen response is fast enough to give sufficient spectral detail. The gain should be set at a frequency

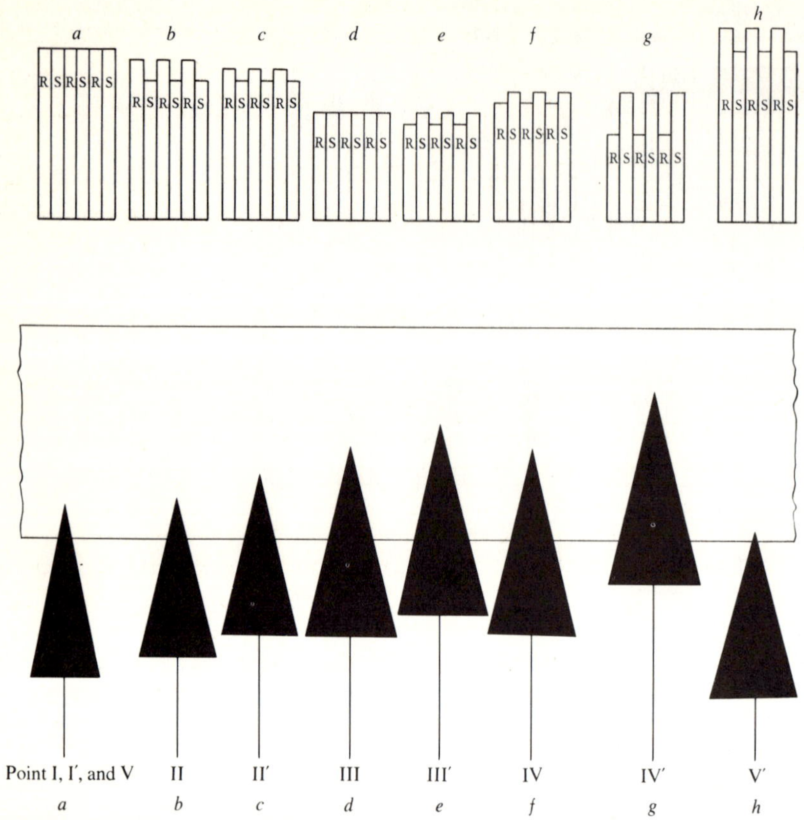

Figure 18-16 Position of optical wedge and relative strengths of sample and reference beams for various points shown in Fig. 18-15.

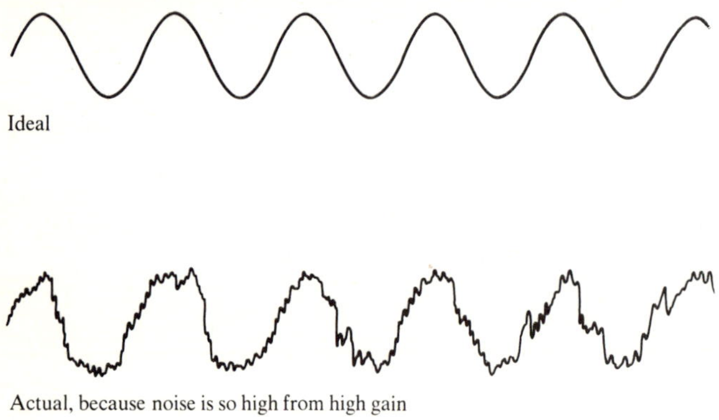

Figure 18-17 Oscillation from an optical null, double beam spectrophotometer. Abscissa exaggerated for clarity.

Figure 18-18 Enlarged noise trace from optical null double beam spectrophotometer.

where the sample, carbon dioxide (Fig. 18-19), or water vapor (Fig. 18-19) has no significant absorptions; the gain should not be set at a frequency near the limits of a grating or prism. A convenient range for a gain setting is between 2650 and 2450 cm^{-1} unless the sample absorbs over this entire range.

There are two approaches to obtaining a correct gain setting. One is to turn the gain up until the pen oscillates, turn it back until the oscillation just stops, and then turn it back just a little more. This method often gives a higher noise level than necessary and is not recommended for quantitative work. The gain can also be set for a specified percent overshoot; this method often takes a little longer but usually

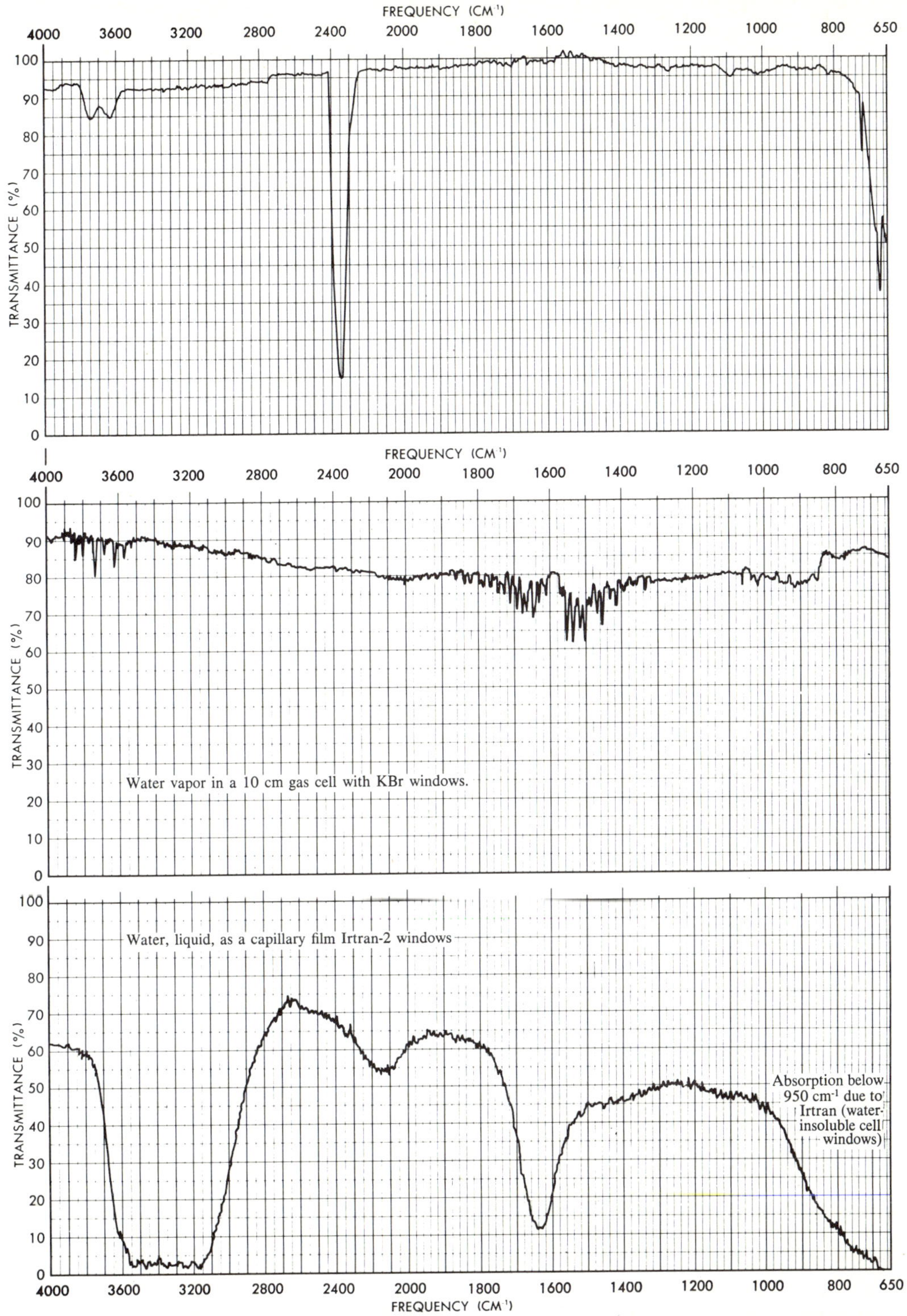

Figure 18-19 Carbon dioxide and water spectra.

gives a more optimum gain setting. For this method set the 100% control so the pen is at 90%, block the sample beam with a finger or piece of cardboard until the pen is at 80%, and quickly remove the obstruction; this produces a 10% transient signal. The pen will rise quickly, overshoot 90% by a small amount, and return to 90%. Adjust the gain so that the pen rises above the equilibrium position (overshoots) by 0 to 1%. Note that a small amount of overshoot is necessary for proper pen response. If it is difficult to observe pen responses of much less than 1%, set the gain so that the overshoot is less than 1% for a 10% transient and greater than 1% for a 20% transient. The Perkin-Elmer Model 700 should be set for 1% or less overshoot on a 15% transient and a 1 to 2% overshoot on a 25% transient.

Alignment of Chart Paper

If a maximum is indicated at a frequency different than that at which the compound absorbs, interpretation of the spectra may be difficult or even give misleading information. Occasionally a modern infrared instrument is out of calibration, but the most frequent cause of a constant frequency shift, at least in student laboratories, is misalignment of the chart paper. Be certain that the chart paper is carefully aligned with the correct mark on the chart drum or recorder bed. Figure 18-20 shows the alignment marks and correct alignment of chart paper for a widely used infrared spectrophotometer. Alignment and placement of chart paper on Perkin-Elmer 137, 237, and 337 infrared spectrophotometers is the same as for the ultraviolet spectrophotometer described in Chapter 26. If one of these instruments with a drum recorder is to be used, see page 206 for instructions.

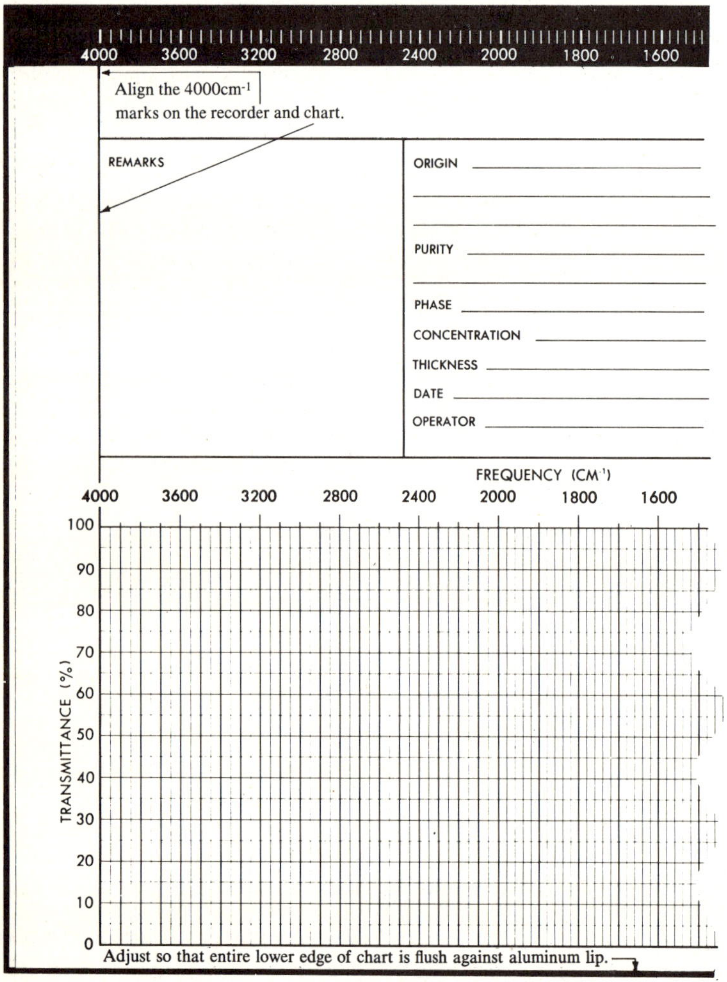

Figure 18-20 Alignment of chart paper on a Perkin-Elmer Model 700.

Instrument Performance Check

To be certain the instrument has not had an unexpected performance change, it should be periodically checked. An instrument that is in constant operation and is used often for quantitative analysis should be checked once or twice a day. Instruments that are used only occasionally need be checked only every other day or as seldom as twice a week. The minimum checking should consist of a polystyrene calibration spectrum, a gain check, and an I_0 line. The first is obtained by placing a polystyrene film in the sample beam, adjusting for optimum gain at 2500 cm^{-1}, and scanning the entire spectrum. This can be used to check wavelength calibration, pen response, and resolution. The I_0 line is obtained by adjusting the pen to approximately 95% transmittance and scanning with nothing in the sample and reference beam. This will give an indication of noise level, excessive deviation of the electronic balance in the amplifier, and any coating on the optical components. These spectra should be compared with spectra obtained when the instrument was installed or most recently serviced. In this way small, gradual, cumulative deteriorations are readily detected.

Proper Care of Infrared Spectrophotometers

All prism infrared spectrophotometers of the correct frequency range to be used with this text have sodium chloride prisms. All infrared spectrophotometers have alkali metal halide windows on the thermocouples. Alkali metal halide cells (for student use usually sodium chloride and occasionally potassium bromide) are used to hold many, if not most, materials in the sample beam. Alkali metal halides are water soluble. Therefore, alkali metal halide spectrometer components and cells are easily damaged by water. Such small amounts and low concentrations of water damage alkali metal halide optical components so readily that even slightly to moderately humid air will etch the components if they are exposed long enough. Never use water or aqueous solutions near an infrared spectrometer or its accessories. When necessary, spectra of aqueous solutions can be obtained with silver halide, polyethylene, or Irtran cells. Even when these are used, care should be taken not to spill any of the aqueous solutions on or near the infrared spectrophotometer.

Never unplug an infrared spectrophotometer or plug it into an AC outlet that is controlled by a switch. Most infrared spectrometers have electrical heaters. These heaters maintain the dispersing unit and other critical components at constant temperature so that frequency calibration of the instrument is independent of room temperature. The heaters also prevent the condensation of atmospheric moisture on water sensitive components.

There are various manipulations that can have the same effect on an infrared spectrophotometer as shifting gears from high to reverse without depressing the clutch on an automobile traveling at 40 mph. These include manually moving the drum, chart holder, or pen cartridge while the instrument is in operation and manually moving the pen along the ordinate (transmission scale) at any time. Do not move the pen manually at any time. Do not move the chart holder on the Perkin-Elmer 700 if the scan button is lit or flashing. Do not move the drum on the Perkin-Elmer 137 unless the scan switch is turned to reset. Do not move the pen carriage on the Beckman IR-5 while the function switch is turned to scan or start.

Do not poke fingers, pencils, screwdrivers, or any other objects through the openings in back of the sample slides. The optical attenuators are immediately inside these openings and are very easily damaged. As was mentioned earlier, do not operate infrared spectrophotometers so that the pen travels above 100% on the transmittance scale.

Do not place pencils or other objects on the recorder bed of a Perkin-Elmer 700. Such objects can jam the chart holder and keep it from moving or can fall into the instrument and damage some of the components.

INTRODUCTORY EXPERIMENTS

Fixed Thickness Sealed Cells

Some of the experiments described in this text specify sealed cells. These cells are often required for quantitative analysis and allow survey spectra of the proper intensity to be obtained more easily. They need a certain amount of care to retain their usefulness. Never touch the flat optical surfaces of the cell windows and keep moisture and all aqueous solutions away from the alkali metal halide windows.

Use a hypodermic syringe either with a Luer lock fitting or a Luer ground glass tip to fill the sealed cell. Remove both Teflon stoppers, place the cell on the desk top and support it at an angle with a pencil or other object (Fig. 18-21a). Draw 0.5 to 1 ml of liquid in the syringe. Fit the syringe on the bottom opening of the cell, slowly fill the cell by depressing the plunger, and verify through the salt window that the liquid is filling the cell and that no bubbles are forming. Less than 0.2 ml is needed to fill a cell up to 0.1 mm thick. Avoid excessive pressure on the syringe as this distorts the cell spacers and enlarges the volume of the cell. Excessive pressure may also rupture the cell and cause a permanent leak. Sometimes very thin cells (less than 0.075 mm) are difficult to fill without excessive pressure. If there is any significant resistance to filling such cells with a slight pressure on the syringe, an empty syringe should be attached to the top port and the plunger pulled while pressure is applied to the other syringe (Fig. 18-21b). This "push-pull" technique is especially useful with slightly

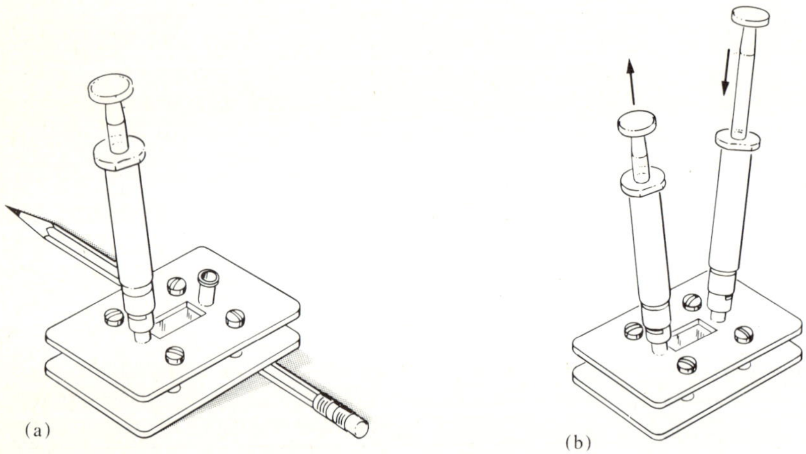

Figure 18-21 (a) Filling a sealed cell. (b) Push-pull technique for filling and cleaning a thin cell. Courtesy of Perkin-Elmer Inc.

viscous liquids. Do not attempt to fill thin sealed cells with highly viscous liquids. Fill the cell until liquid just appears in the top port. If there are any air bubbles, gently tap the cell (the metal frame, not the windows) to remove them. If air bubbles persist, they can usually be removed by gently pushing more liquid through the cell and catching the excess from the top port in a small flask or on a piece of tissue. Do not allow the excess to flow on the outside of the cell windows. If there is not very much of the liquid, the bubbles can be removed by taking the material back into the syringe and refilling the cell.

Lay the cell flat, remove the syringe and place a Teflon plug in the top port. If there is liquid in the bottom port, remove it by touching a piece of tissue to the liquid before putting in the bottom plug. If the plug is inserted while the bottom port is filled with liquid, the resulting hydraulic pressure could distort or rupture the cell.

Clean the cell by flushing several times with a volatile solvent.[7] If the cell is thinner than 0.075 mm, another syringe should be used to supply a vacuum similar to the push-pull technique for filling a thin cell with slightly viscous liquids (Fig. 18-21b). Dry the cell by pushing at least 50 ml of air through with a dry hypodermic syringe. Remove the syringe from the port before pulling back the plunger. An ear syringe will accomplish the drying somewhat faster. If a laboratory vacuum line is available, air can be pulled through the cell. A drying tube should be attached to the other port (Fig. 18-22) in all but the driest environment. Bottled nitrogen (oil free) can also be used to dry the cell. Compressed air should be used only if it is dry and clean. Most sources of labora-

[7] If a solution of a solid has been in the cell, an additional precaution is needed to prevent precipitation of the solid in the cell. Always wash the solution out of the cell with the solvent in which the solid was dissolved. After the wash with the pure solvent, any other liquid can then be used, if necessary, to finish cleaning the cell.

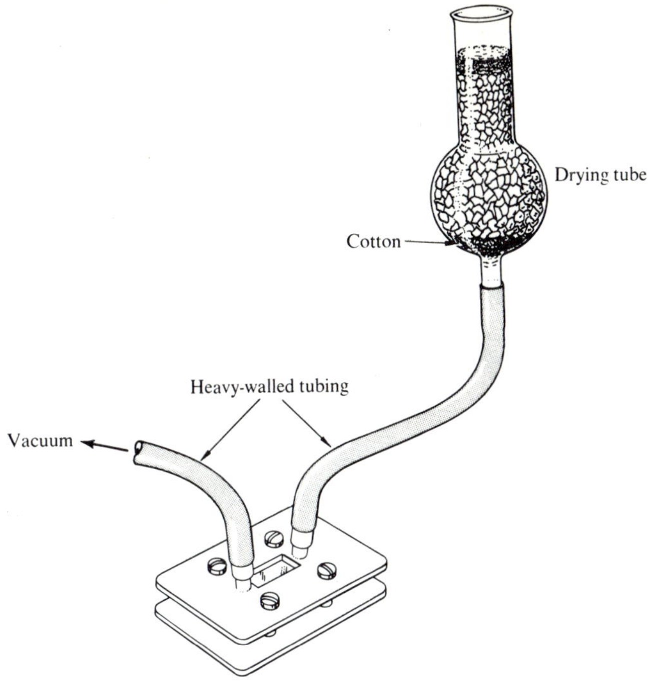

Figure 18-22 Drying a fixed thickness cell with a vacuum.

tory air are both wet and contaminated with oil droplets. Both nitrogen and air must be at very low pressures or the cell will be distorted or opened.

Often the same cell may be used to obtain spectra from several liquids successively. If there is sufficient liquid, the previous material can be removed by flushing the cell two or three times with the sample before filling the cell. When this procedure is used, the cell need not be dried. If spectra are being obtained on solutions of solids, the instructions given in the footnote on page 132 should be carefully followed.

Demountable Liquid Cells

General. Three typical demountable liquid cells are shown in Figures 18-23 to 18-25. They can be used with various size spacers so that cells of any needed thickness can be obtained without maintaining a large inventory of cells. The demountable cell is more easily cleaned and repaired than a sealed cell and its salt plates are more easily polished. Spectra can be obtained on very viscous liquids, greases, oils, films, and gels in demountable cells. Unlike sealed cells, the cell thickness of demountable cells can be made as thin as the spacers available and, if unknown path lengths are acceptable, as thin as can be obtained by squeezing the liquid between two plates (capillary film technique).

The demountable cells often are not as satisfactory for quantitative work (Chapter 20) since they lack both accuracy and reproducibility in the setting of the cell thickness. The particular cells shown in Figures 18-24 and 18-25, however, are adaptable to many quantitative analyses. The nominal thickness for the spacers are not accurate enough for quantitative analysis and change with use. The thickness of the assembled cell is also a function of how tightly the cell is fastened together. If a demountable cell is used for quantitative analysis, the cell thickness must be measured (Appendix III) every time the cell is assembled or an internal standard must be used. Vapors and nonviscous liquids may leak from demountable cells. For normal scan times, however, the leakage is usually only a problem with very volatile liquids. Liquids boiling below 55° should not be used in demountable cells with spacers unless they are dissolved in a less volatile liquid. When the capillary film technique is used, liquids boiling below 80° should not be used. The volatile liquids usually evaporate enough to distort the end of the spectrum. With no spacer liquids boiling below 50° may completely evaporate before the end of the spectrum.

When handling the salt plates or windows for the demountable cells (or any alkali metal halide cell compo-

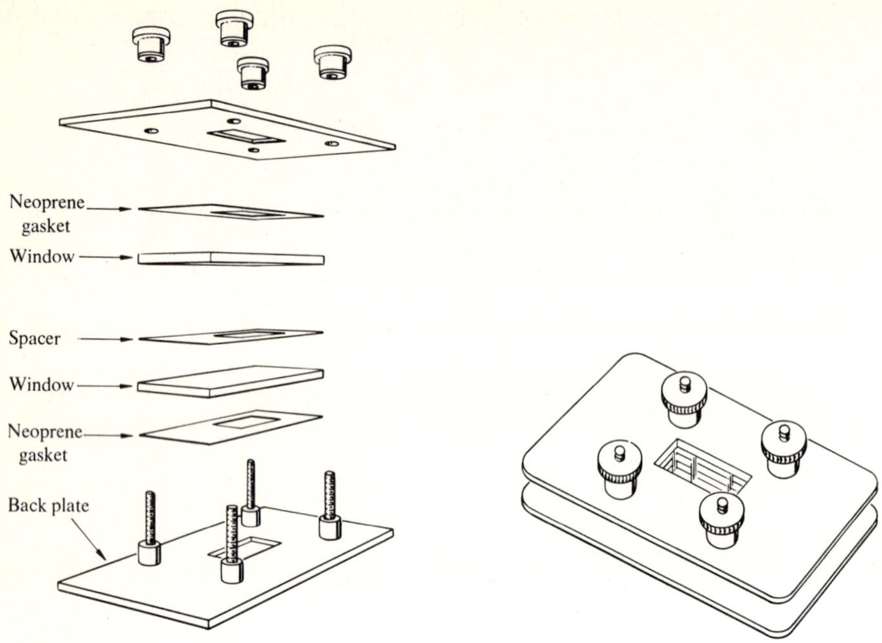

Figure 18-23 Demountable cell without ports. (a) Assembly diagram. (b) Perspective view of assembled cell. Courtesy of Perkin-Elmer Inc.

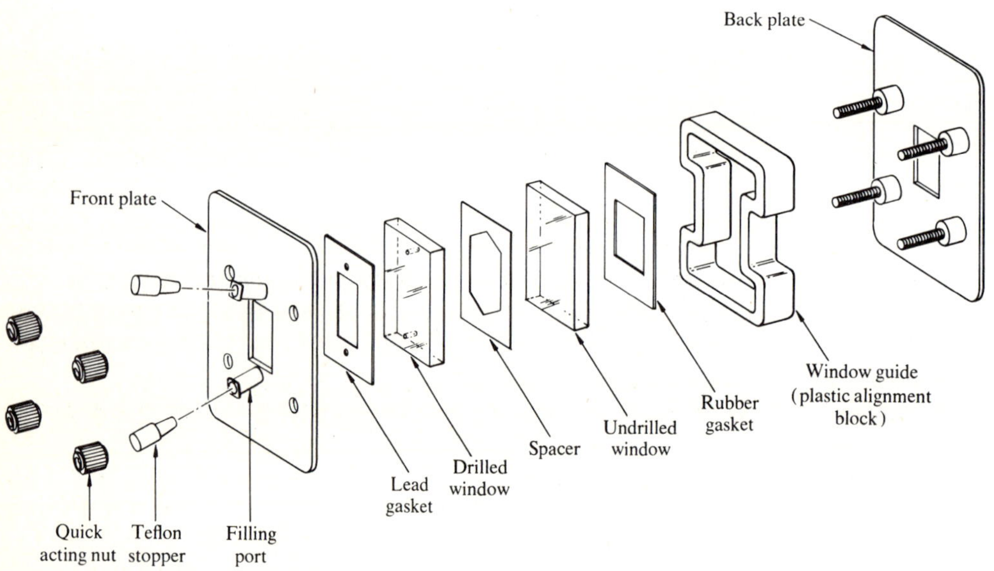

Figure 18-24 Sealed demountable cell or demountable cell with ports. Courtesy of Perkin-Elmer Inc.

nent) do not touch the flat surfaces. These surfaces are easily scratched, and moisture on the skin is sufficient to etch them. If finger cots or rubber or soft cotton gloves are used, the possibility of damaging the plates is decreased. Whether or not these finger covers are used, the salt plates should be touched only on their edges. If gloves or finger cots are not worn be sure your hands are dry and clean.

Demountable Cells Without Injection Ports. With the threaded studs pointing upwards place the back metal

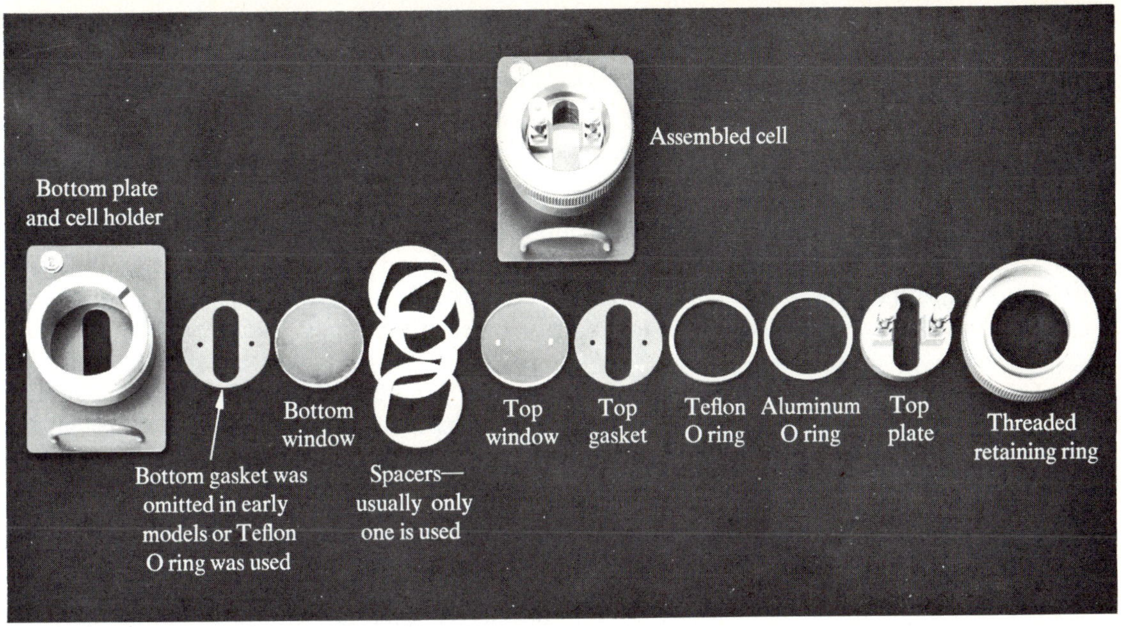

Figure 18-25 Barnes Engineering demountable cell. Courtesy of Barnes Engineering, Stamford, Conn.

plate of the cell (Fig. 18-23) on a flat dry surface. Place one of the sodium chloride windows on the rubber gasket and lay the spacer on the window. If the spacer is not smooth, place another salt plate on and gently press the top window to smooth and align the spacer. Remember to apply this pressure to the edges of the salt window.

Place two or three drops of liquid around and near the center of the window. Put the top window on by touching one of its edges to the corresponding edge on the bottom and slowly lower the other end. If bubbles appear, slightly raise the edge of the top plate nearest the bubble, add another drop or two of liquid near where the bubble was, and slowly lower the top window. If this does not remove the bubbles, open the cell, pour the liquid off, and start over. Align the second rubber gasket on the top window so that the cut-out area matches that of the lower gasket and metal plate. Place the front metal plate on the top of the second gasket so that the studs from the bottom metal plate go through the holes. Spin down the knurled nuts to where they meet a slight resistance and then tighten them evenly, alternately tightening the nuts on opposite corners. Tighten firmly and gently so there are no leaks. **Excessive or uneven tightening will probably break one or both of the salt windows.** For capillary films use the same technique except omit the spacer.

For viscous liquids, greases, gels, and mulls, place a small amount of sample in the center of the bottom window and use the top window to spread it around and then gently press it into a thin film. If there are any bubbles, they usually can be worked out with the top window. Do not use a spacer. Place the gasket and top plate on as instructed above.

Demountable Cells with Ports. The cell shown in Figure 18-24 is the easiest demountable cell developed so far to use. It differs from the cell shown in Figure 18-23 in that:

a. The top gasket is lead.
b. There are ports similar to those on sealed cells.
c. The top window has holes to allow the cell to be used as a fixed thickness cell.
d. There is a plastic alignment block.

This cell is more versatile than the demountable cell described above. The plastic block assures easy and certain alignment of the salt windows, gasket, and spacers and by aiding in the application of even pressure when the nuts are tightened lessens the chances of window breakage. The ports allow this cell (and the Barnes cell described below) to be used as a fixed thickness cell with almost any liquid that is not highly volatile. Once the thickness is measured or once the cell is calibrated with a standard, it can be used as a sealed cell until the cell is opened or the nuts tightened or loosened. When using this cell with films, greases, gels, viscous liquids, and mulls follow the instructions above for the demountable cell. With nonviscous liquids, follow the instructions above for a sealed cell except that this cell may be taken apart for cleaning, if necessary.

Barnes Engineering Demountable Cell. To use this cell follow approximately the same instructions described for assembling and filling the rectangular cells except:

a. There were no bottom gaskets supplied with early production models of this cell.
b. If after assembly the cell is to be used as a fixed thickness cell, use a top window with the requisite holes. These holes *must* be aligned with the spacer, the gasket, and metal top plate if liquid is to be injected into the cell.
c. If a spacer is used, it must *always* be aligned with the thinnest edges at the top and bottom of the cell (Fig. 18-25).
d. Align the gasket so that it can not be seen when the top metal plate is in place. The alignment of the gaskets and spacer is very critical. Misalignment causes vignetting and prevents the cell from being filled through the ports.
e. The entire assembly is held together with a large threaded retaining ring instead of by nuts on threaded studs (Fig. 18-25).
f. Teflon and aluminum O rings fit between the plate and the threaded retaining ring. The Teflon O ring is placed on the plate and the aluminum O ring on the Teflon ring.
g. There is much less possibility of applying uneven pressure on the salt windows with this cell especially if a bottom gasket is used. With nonviscous liquids, however, the range between the pressure necessary to keep the cell from leaking and excessive pressure that cracks windows seems to be narrower for this cell than for the demountable cells with four nuts. Therefore, be very careful of the pressure you apply with the retaining ring.

Spectra of Toluene and Anisole

Using a 0.025-mm fixed thickness cell obtain infrared spectra of toluene and anisole (methoxybenzene). All spectra that you obtain in this course should be placed in a notebook. These spectra along with ones shown in this manual and in the lecture text will be needed as a reference library for future problems and experiments.

> **Problem 18-4.** What is the principal difference in the two spectra? How can this be related to the only difference in structure between the two compounds?

Infrared Spectra of Compounds in Solution

Fixed thickness cells narrower than 0.015 mm are impractical because the narrow cells are difficult to fill and clean. Most polar compounds such as anisole give too strong absorption bands when their spectra are obtained in 0.015-mm cells.

One technique for obtaining usable spectra on such intensely absorbing compounds is to obtain spectra on their dilute solutions. If a nonpolar solvent such as carbon disulfide or carbon tetrachloride is used, the molecules of the compound are surrounded by a medium that is unlikely to affect their vibrational energies. Therefore, infrared spectra of compounds in nonpolar solvents are less likely to show environmental effects than infrared spectra of the compounds in the solid phase, pure liquid phase, or a polar solvent.[8]

Disadvantages in obtaining spectra on compounds in solution are: (a) many compounds are insufficiently soluble in solvents that can be used, and (b) all solvents have significant infrared absorptions of their own. The latter difficulty can be circumvented by obtaining spectra in two solvents that have absorption bands in different regions of the infrared spectrum. Carbon disulfide and carbon tetrachloride are such a pair (Figs. 18-26 and 18-27). It should be remembered that both these solvents are toxic, carbon tetrachloride extremely so, and that carbon disulfide is very flammable.

[8] It has also been stated that if all spectra were obtained in a given nonpolar solvent all spectra could be compared under constant environmental conditions. This is not a very germane argument since most spectra in the various available collections were not obtained from compounds in nonpolar solvents and many compounds are not very soluble in solvents like carbon tetrachloride.

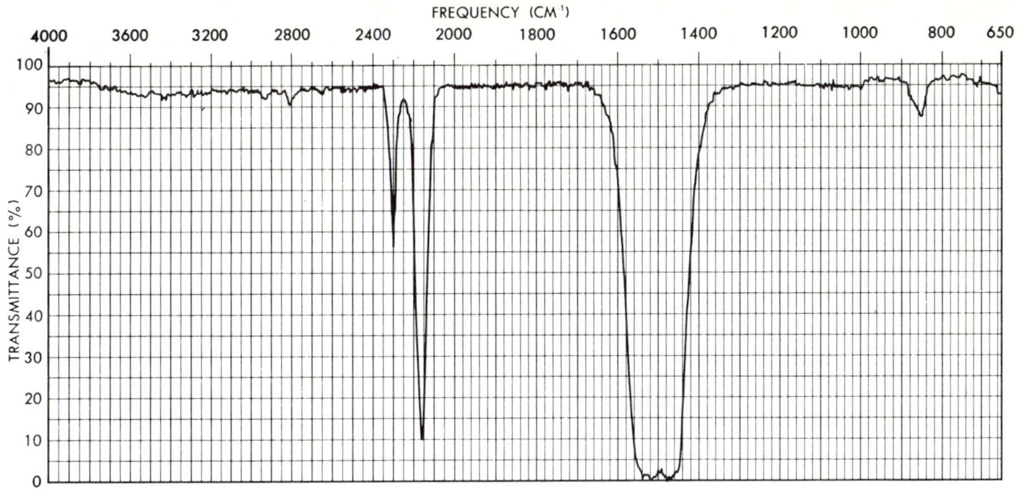

Figure 18-26 Infrared spectrum of carbon disulfide in a 0.05-mm cell.

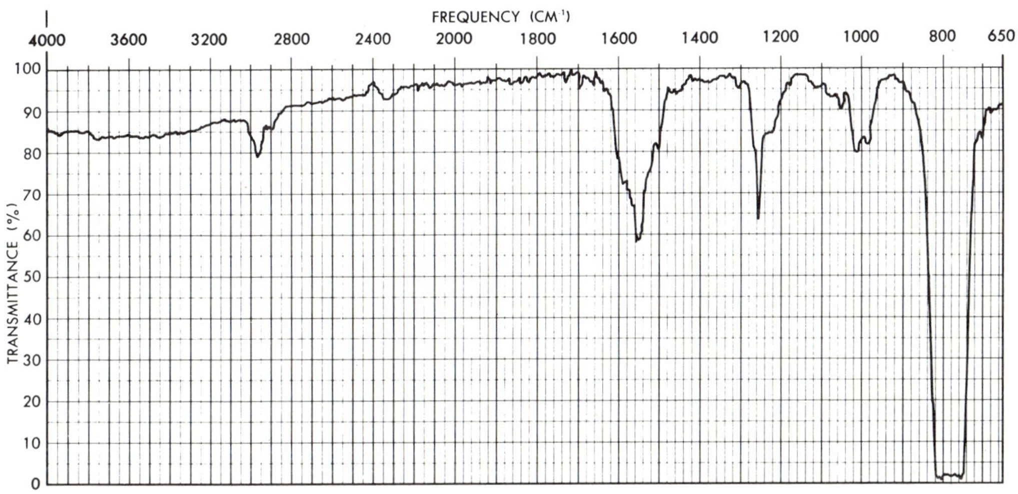

Figure 18-27 Infrared spectrum of carbon tetrachloride in a 0.1-mm cell.

Optimum Intensity

The intensity of a spectrum is optimum when its strongest absorption band or bands lie between 3 and 10% transmittance. Any band that transmits less than 3% of the radiation is likely to be too broad to allow accurate location of the absorption maximum. If the strongest band transmits more than 10%, some of the weaker bands are likely to absorb so little that they cannot easily be seen. When it is important to see the weak absorption bands of polar compounds, it is sometimes necessary to obtain two infrared spectra of a compound: one that has one or more strong bands bottoming at the 0% transmission line to locate the weak bands, and one spectrum that has its strongest absorption band at 5% or greater transmittance to locate the strong bands. Note for instance that the weak combination and overtone bands between 2000 and 1650 cm^{-1} in the spectrum of anisole in the 0.025-mm cell are easily seen.

Problem 18-5. What could be done to obtain a spectrum with more intense bands than one that has been obtained? With less intense bands?

Solution Spectra From an Optical Null, Double Beam Spectrophotometer

Obtain a spectrum of a 2.5% solution of anisole in carbon tetrachloride in a 0.2-mm cell. Compare this spectrum with the one obtained from the pure liquid in a 0.025-mm cell. Note that none of the bands bottom and the frequencies of the absorption maxima that were difficult to locate in the spectrum of the pure liquid are easily located in this spectrum. Note also, however, that the very important out-of-plane bending absorption is completely obscured by the strong absorption of carbon tetrachloride between 720 and 810 cm^{-1}. Also note that the spectrum in the region of 1550, 1200, and 1000 cm^{-1} is slightly distorted because of absorptions by carbon tetrachloride in this area. Note that even the strong C—O stretching band of anisole at 1027 cm^{-1} is changed.

Problem 18-6. If cells of accurately known thicknesses are used for this experiment (or any others with solvent compensation in the reference beam), the slightly thicker cell should be used for the anisole-carbon tetrachloride solution and the slightly thinner cell for the pure carbon tetrachloride. Why?

Fill another 0.2-mm cell with carbon tetrachloride and obtain its spectrum. Return the cell with the anisole-carbon tetrachloride solution to the sample beam cell slide and place the cell with carbon tetrachloride in the reference beam cell slide. Obtain a spectrum.

Problem 18-7. How does the spectrum of pure carbon tetrachloride that was just obtained differ from that shown in Figure 18-27?

Problem 18-8. If the gain was set correctly for obtaining the first spectrum on the carbon tetrachloride-anisole solution, is it necessary to reset the gain when the spectrum is obtained with carbon tetrachloride in the reference beam?

After the chart has been removed and the pen lifted from the recording surface, set the spectrophotometer at 2600 cm^{-1}. Block the sample beam (but not the reference beam) with your hand. When the pen reaches 0% transmittance, determine (with a stop watch, if possible) the number of seconds the pen takes to complete its upscale travel when you quickly remove your hand. Set the spectrophotometer at 1530 cm^{-1}, 1190 cm^{-1}, 955 cm^{-1}, and 795 cm^{-1}, and record the number of seconds the pen takes to travel from 0% transmission to its highest transmission (response time of the pen) for each setting. At 955 cm^{-1} and 795 cm^{-1} block the reference beam with your hand. (Leave cells in both beams for all these timings.) What do you observe?

Problem 18-9. (a) How do you explain the different response times of the pen at the various frequencies? (b) With these response times in mind compare the three spectra of anisole (pure or neat, 2.5% solution uncompensated, and 2.5% solution compensated with solvent in the reference beam) you obtained. In regard to strength of solvent absorption band, what general statement can be made about compensating for solvent absorption with solvent in the reference beam? In regard to thickness of the two cells?

Problem 18-10. (Optional) With the two cells still in the reference beam set the spectrophotometer at 955 cm^{-1}. Turn the gain up to a maximum; if the pen oscillates, decrease the gain until the oscillation just stops and then decrease it just a bit more. Measure the full-scale pen response. Without resetting the gain measure the pen response first at 795 cm^{-1} then at 1190 and 1530 cm^{-1} and finally at 2600 cm^{-1}. Compare it with the pen response times at the lower gain setting. Explain what you observe.

USE OF OUT-OF-PLANE BENDING FREQUENCIES FOR IDENTIFICATION OF UNSATURATED HYDROCARBONS

The hydrogen atoms bonded to sp^2 hybridized carbons have characteristic absorptions from the out-of-plane bendings of these hydrogen atoms. These absorptions are very useful for distinguishing among isomeric alkenes,

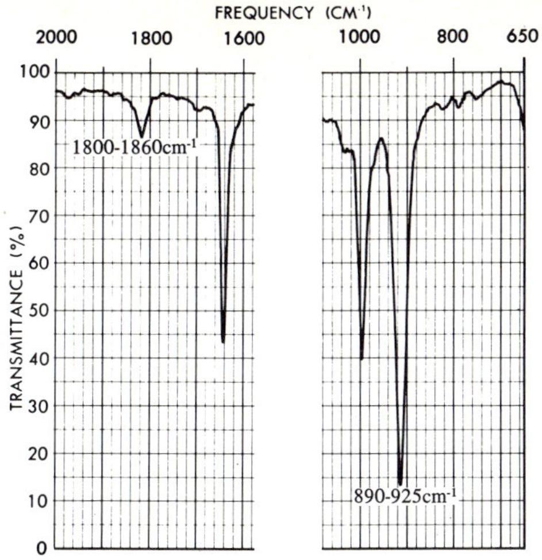

Figure 18-28 Vinyl, $RCH{=}CH_2$, three hydrogen atoms. In Figures 18-28 to 18-45, ignore all absorptions weaker than 10% T between 1100 and 650 cm^{-1} and all marked NC, as these are not characteristic of the functional group and are fingerprint absorptions for the compounds used.

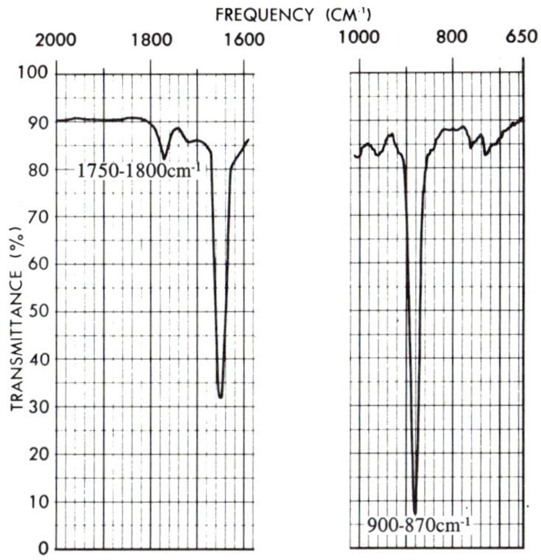

Figure 18-29 Methylene, $R_2C{=}CH_2$, two *geminal* hydrogen atoms.

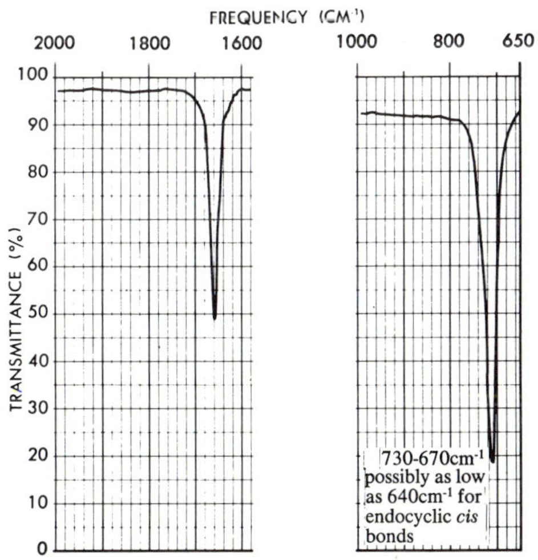

Figure 18-30

Cis, 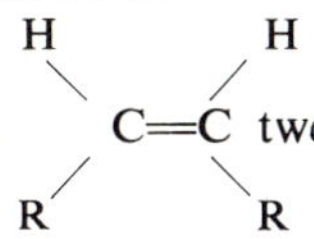 two adjacent *vicinal* hydrogen atoms.

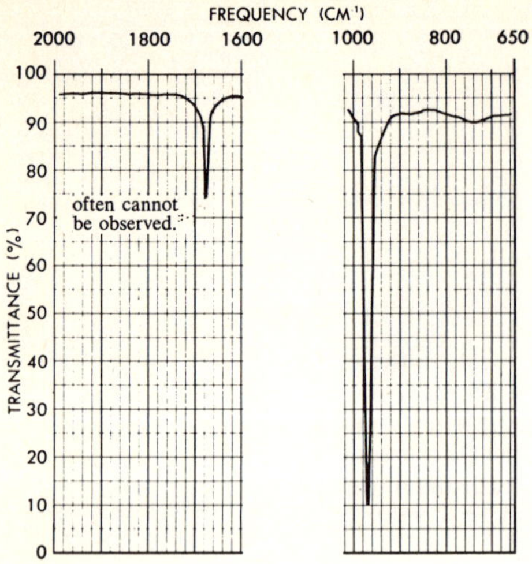

Figure 18-31

Trans, 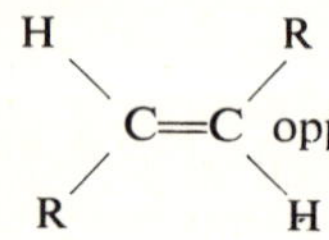 opposite *vicinal* hydrogen atoms.

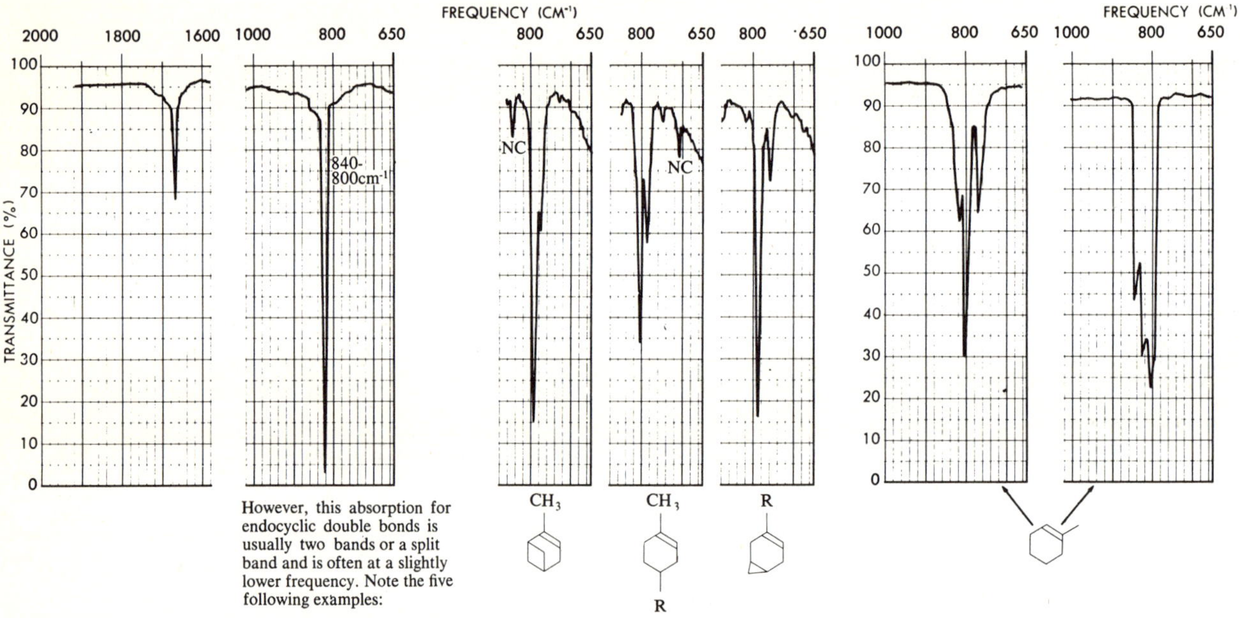

Figure 18-32 Trisubstituted ethylene, one vinyl hydrogen atom.

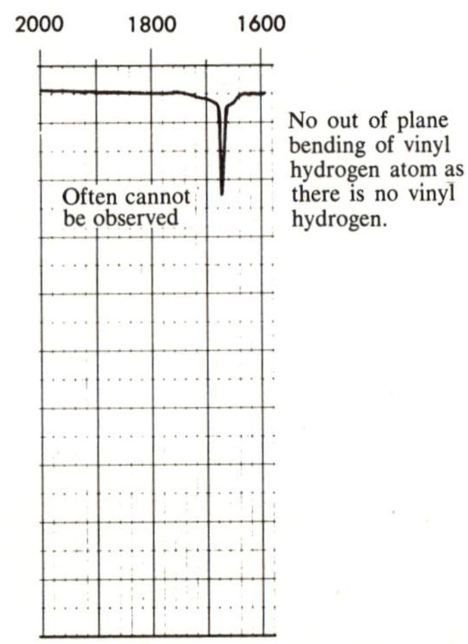

Figure 18-33 Tetrasubstituted ethylene. No vinyl hydrogen atoms.

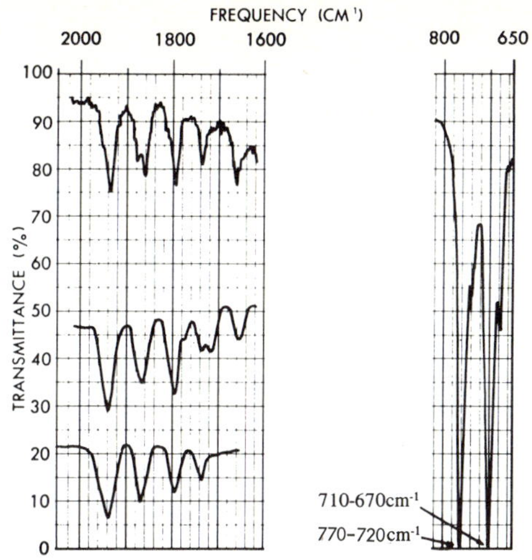

Figure 18-34 Monosubstitution; five adjacent hydrogen atoms.

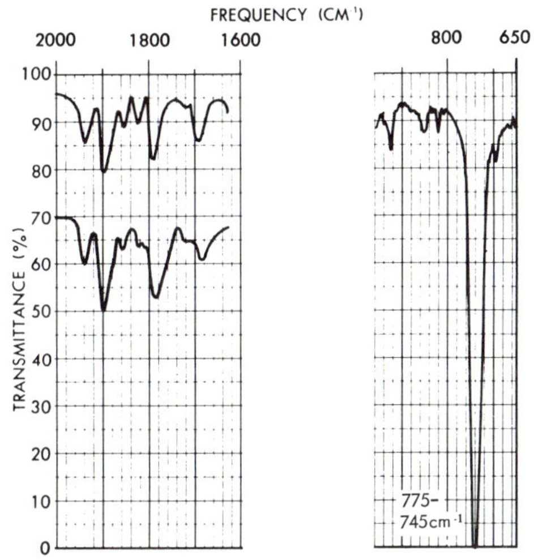

Figure 18-35 *Ortho* substitution; four adjacent hydrogen atoms.

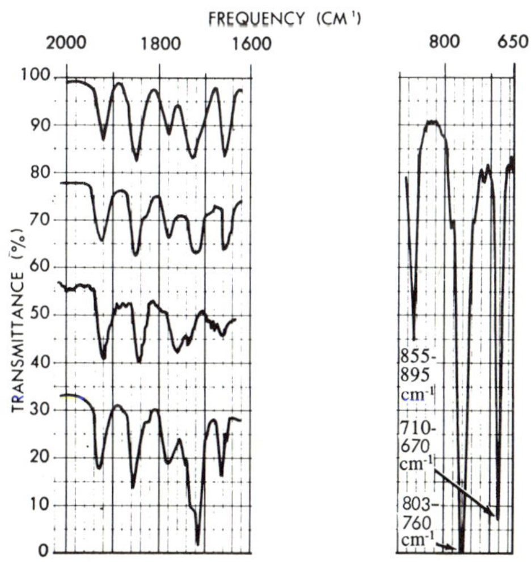

Figure 18-36 *Meta* substitution; one isolated and three adjacent hydrogen atoms.

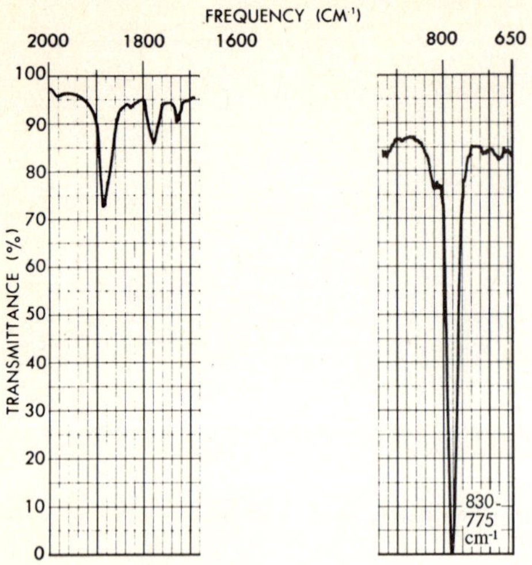

Figure 18-37 *Para* substitution; two adjacent hydrogen atoms.

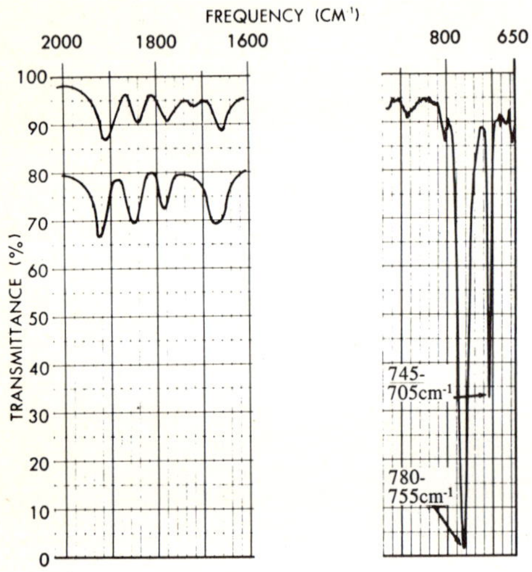

Figure 18-38 1,2,3-Trisubstituted; three adjacent hydrogen atoms.

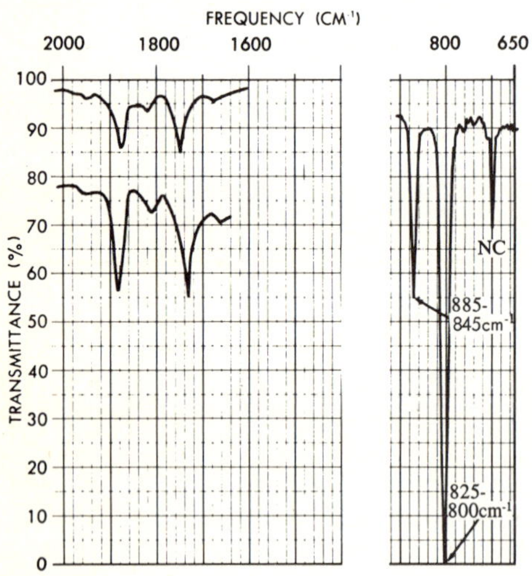

Figure 18-39 1,2,4-Trisubstituted; one isolated and two adjacent hydrogen atoms.

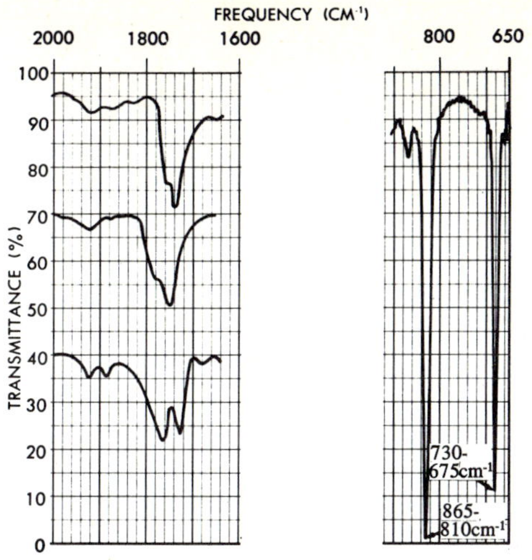

Figure 18-40 1,3,5-Trisubstituted; three isolated hydrogen atoms.

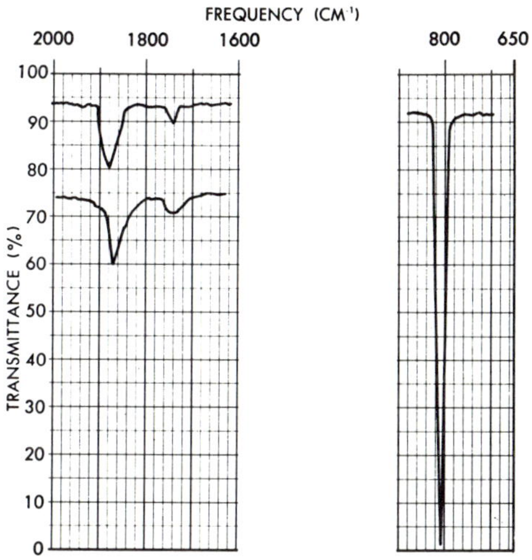

Figure 18-41 1,2,3,4, - Tetrasubstituted; two adjacent hydrogen atoms.

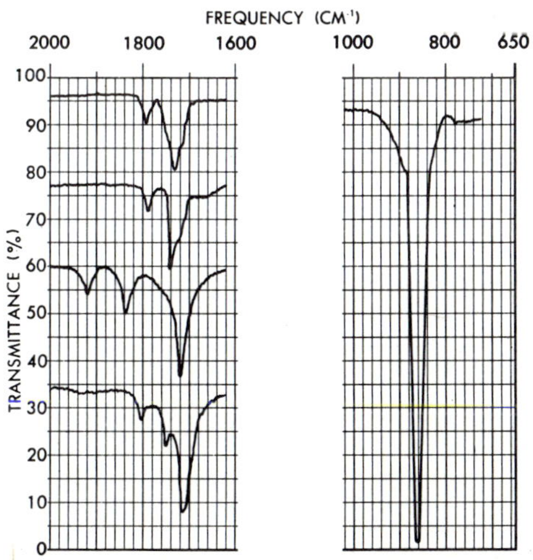

Figure 18-42 1,2,4,5 - Tetrasubstituted; two isolated hydrogen atoms *para* to each other.

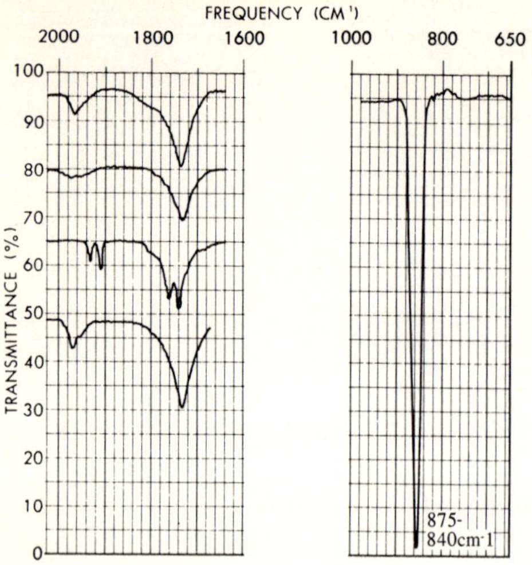

Figure 18-43 1,2,3,5 - Tetrasubstituted; two isolated hydrogen atoms *meta* to each other.

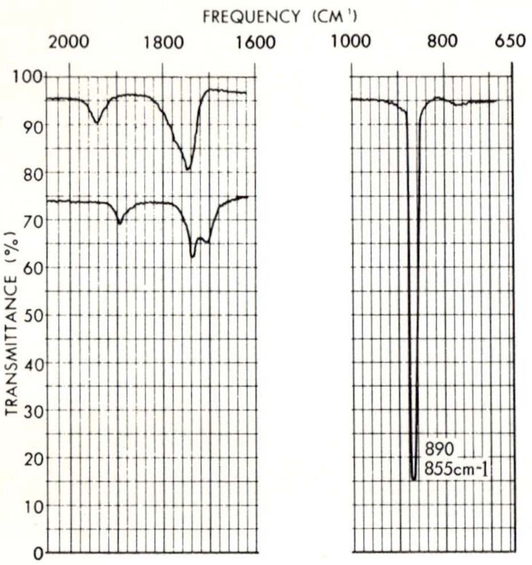

Figure 18-44 Pentasubstituted; one isolated hydrogen atom.

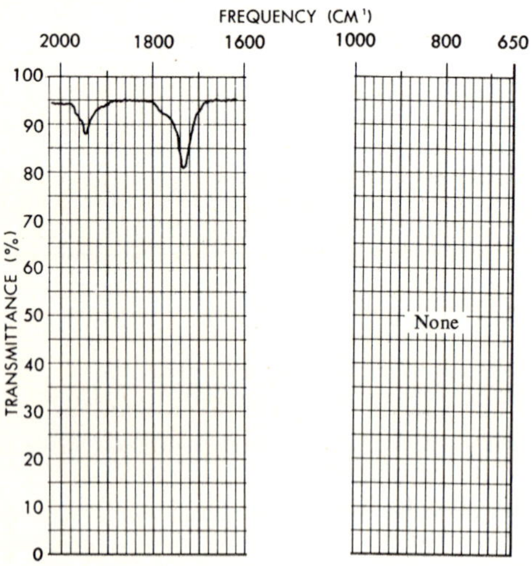

Figure 18-45 Hexasubstituted; no hydrogen atoms on ring.

aromatic hydrocarbons, and many of their derivatives. Figures 18-28 to 18-45 show every possible out-of-plane bending absorption for alkenes, aromatic hydrocarbons, and their derivatives.[9] In addition these figures show the 2000 to 1620 cm^{-1} region which includes the overtone-combination absorptions for both alkenes and aromatic compounds and C=C stretch absorptions for unconjugated alkenes. This region is useful for verification of the information obtained from out-of-plane bending absorptions and for primary diagnosis when other functional groups have strong absorptions obscuring those from the out-of-plane bending or when structural features cause frequency shifts of the out-of-plane bending absorptions.

INFRARED ANALYSIS OF REACTION PRODUCTS FROM CHAPTERS 9 AND 15 (Optional)

Obtain spectra on the reaction products you were asked to save from Chapters 9 and 15. Think carefully about what technique you will use for preparing the sample and obtaining the spectrum. Along with your report of this chapter turn in the spectra you obtain, the conclusions reached about the identity of the products, and a brief but complete explanation of why you chose the sampling techniques you used for obtaining these spectra.

[9] Each of these spectra was obtained on pure liquids except for the solid 1,2,4,5-tetra-, penta-, and hexasubstituted aromatics which were obtained as solutions. The spectra for most of the aromatic absorptions were obtained in a 0.025-mm cell and for all the alkene absorptions in a 0.015-mm cell. The spectra for the three solid polysubstituted aromatic compounds were obtained in solution in a cell of sufficient thickness to place the approximate same number of moles in the sample beam as moles of the pure liquid trisubstituted aromatic compounds in a 0.025-mm cell. An idea of the usual intensities of these absorptions can be obtained from considering the class of compounds, their molecular weight, and the cell thickness. If a spectrum is obtained on a pure liquid and everything else is unchanged, the higher the molecular weight the weaker a specific group absorption is because there are fewer groups to absorb the radiation. The liquids used to obtain the absorptions shown in Figures 18-28 to 18-45 were hydrocarbons with molecular weights between 98 and 158 except for a few of the examples used to obtain the *lower* traces in the 2000 to 1620 cm^{-1} region. Although the relative strength of the various absorptions for a given group of compounds often can be estimated from the cell thickness and percent transmittance of the absorptions shown here, similar compounds of different polarity and similar molecular weight have widely varying absorption. Compare, for instance, the 2000 to 1620 cm^{-1} and 870 to 650 cm^{-1} regions in the spectra of anisole and toluene obtained in a 0.025-mm cell.

For most of the aromatic substitution groupings, several examples of absorptions in the overtone region 2000 to 1620 cm^{-1} are given. The top trace in each example is from the same compound that gave the adjacent out-of-plane bending absorption. The lower traces are from other compounds, some with very polar substituents. These traces show how little the overtone patterns vary for a given type of aromatic substitution and help to define the limits of the variation. These overtone patterns, although highly specific for a given type of aromatic substitution, are often not usable because of other strong absorptions that obscure them or because they are too weak to be seen. Often the latter difficulty can be circumvented by obtaining a second spectrum using a thicker cell. The shapes, relative intensities, and number of the overtone combination absorptions are sometimes more useful than the exact wavenumber of their maxima. These maxima are often shifted more than 10 cm^{-1} (sometimes more than 30 cm^{-1}) to higher frequencies if one or more of the substituents is a strong electron-withdrawing group. This shift should not be confused with the weak to moderate absorptions just above 2000 cm^{-1} for aryl fluorides (2105–2010 cm^{-1}) and aryl ethers (2110–1995 cm^{-1}, usually a doublet). For many nitro compounds the overtone-combination patterns are completely different than those shown here.

The frequency ranges listed at some of the absorption bands in the figures are the amount of variation that may be expected. If no frequency range is given, the location of these absorption bands on almost all compounds known is within 10 cm^{-1} on either side of the one shown. The double-bond stretching absorptions of most alkenes are within 6 cm^{-1} of the absorptions shown. Polar substituents *may* shift the absorptions of the carbon-to-carbon stretch and out-of-plane bending of alkenes to lower frequencies. Several polar substituents may also shift the out-of-plane absorptions of aromatic compounds, especially *para* disubstituted, to lower frequencies.

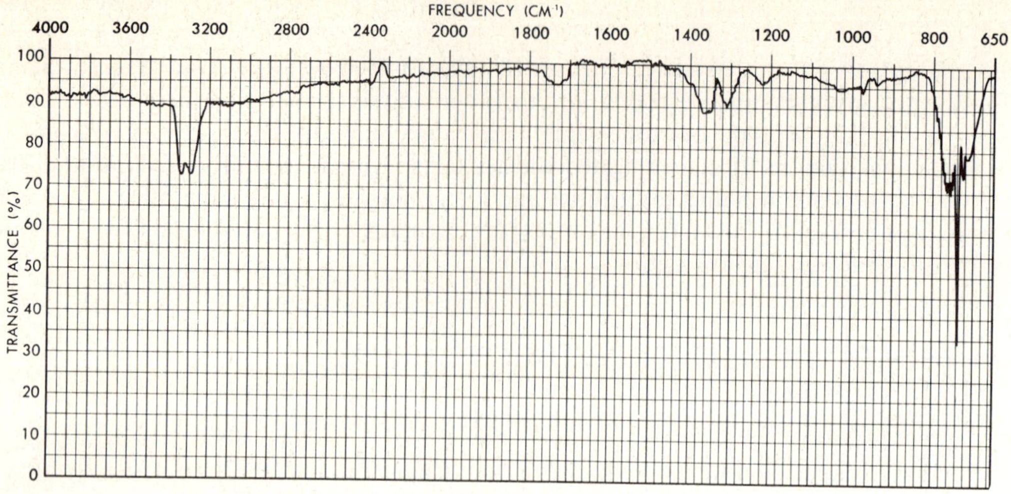

Figure 18-46 Infrared spectrum of acetylene in a 10-cm gas cell.

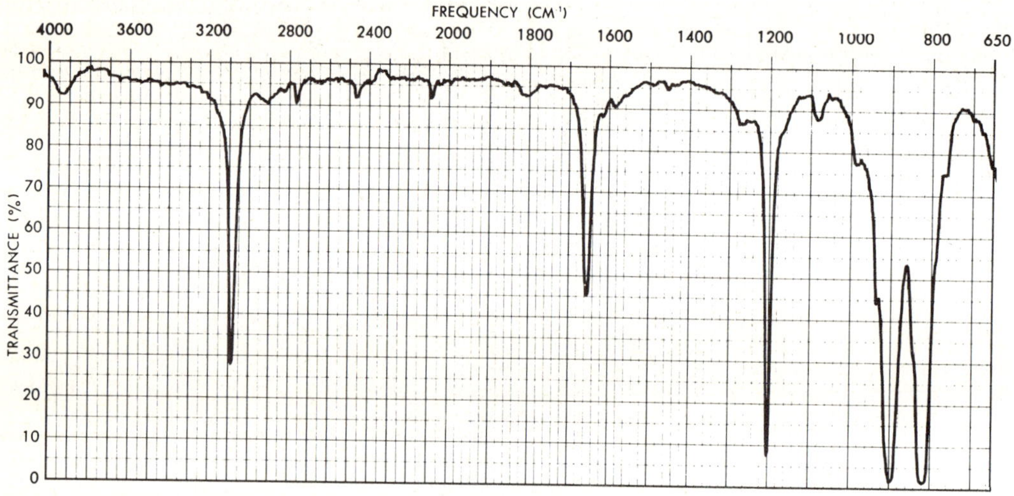

Figure 18-47 $C_2H_2Cl_2$, liquid neat in a 0.015-mm cell.

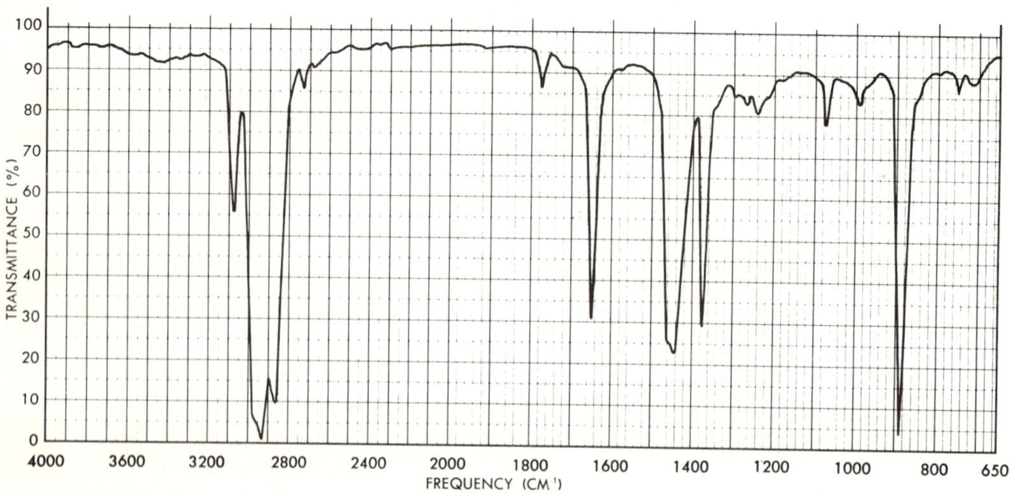

Figure 18-48 C_5H_{10}, liquid neat in a 0.018-mm cell.

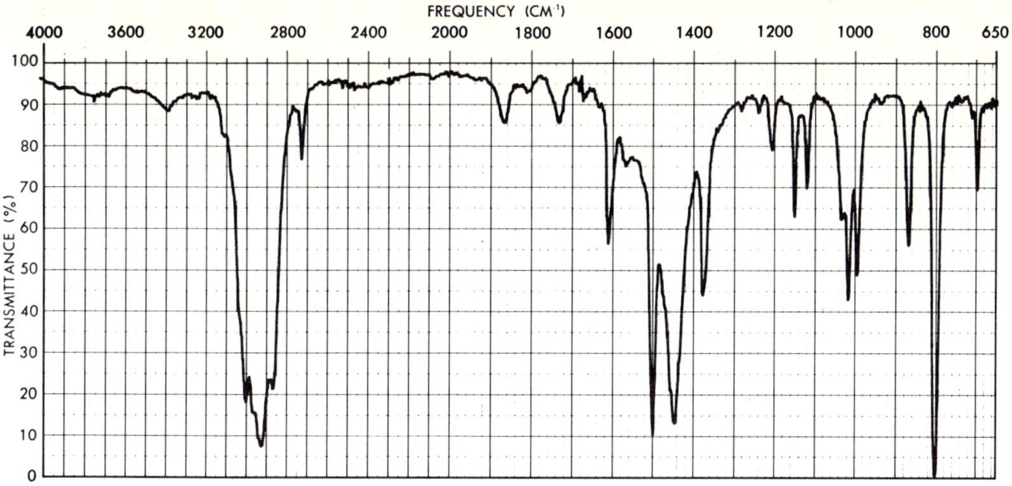

Figure 18-49 C_9H_{12}, liquid neat in a 0.025-mm cell.

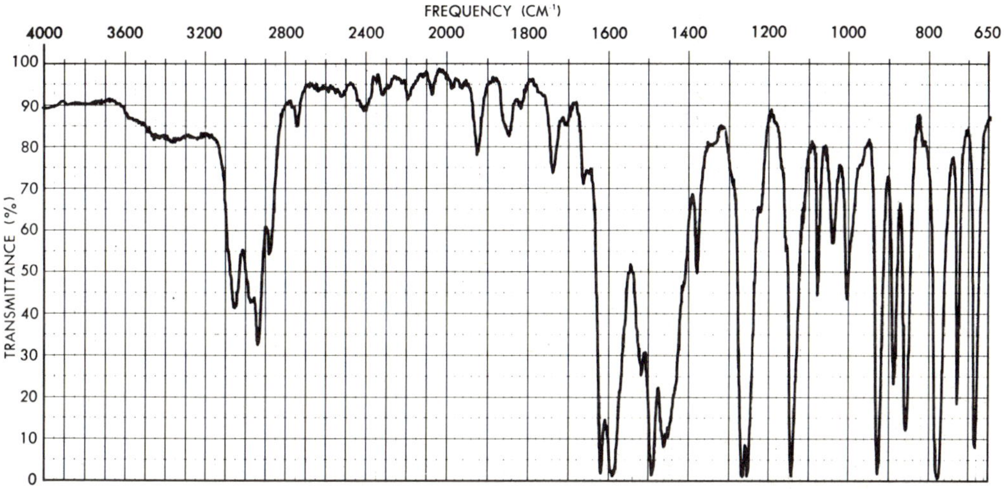

Figure 18-50 C_7H_7F, liquid neat in a 0.020-mm cell.

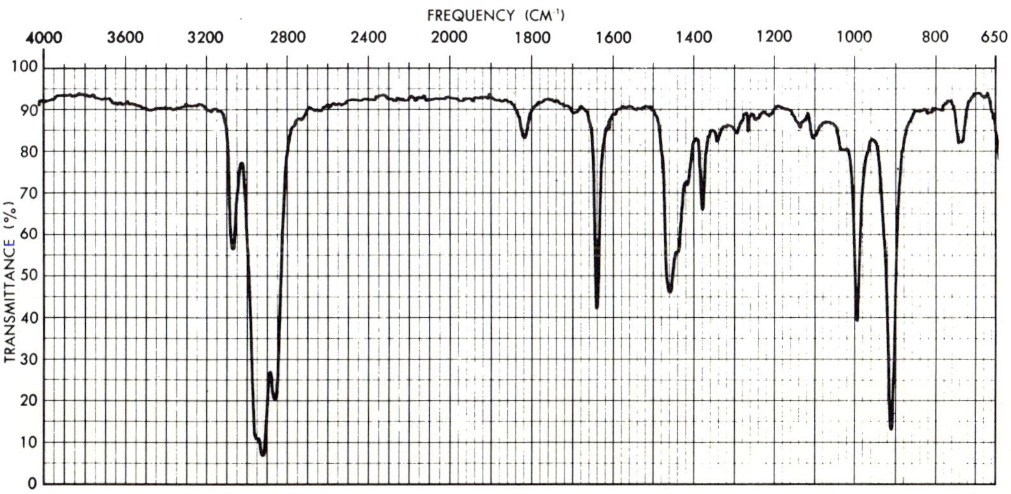

Figure 18-51 C_6H_{12}, liquid neat in a 0.015-mm cell.

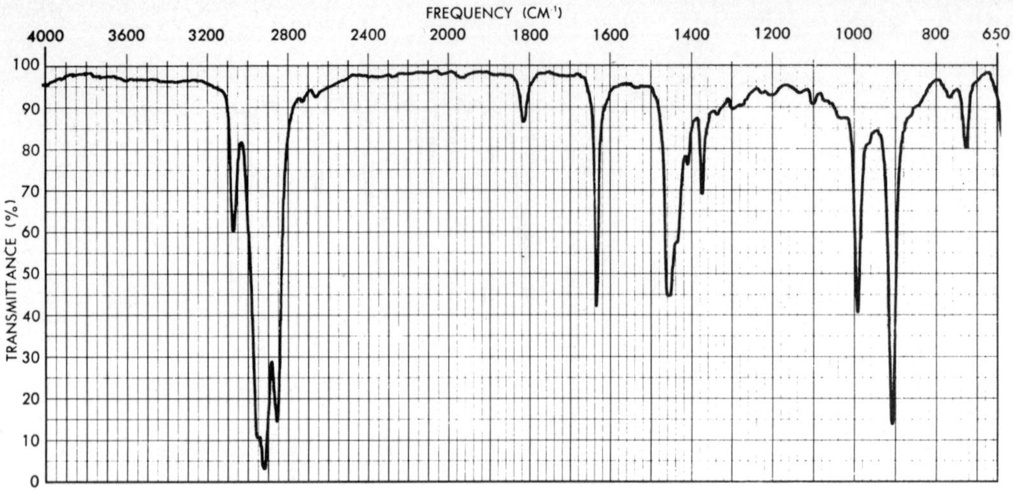

Figure 18-52 C_7H_{14}, liquid neat in a 0.015-mm cell.

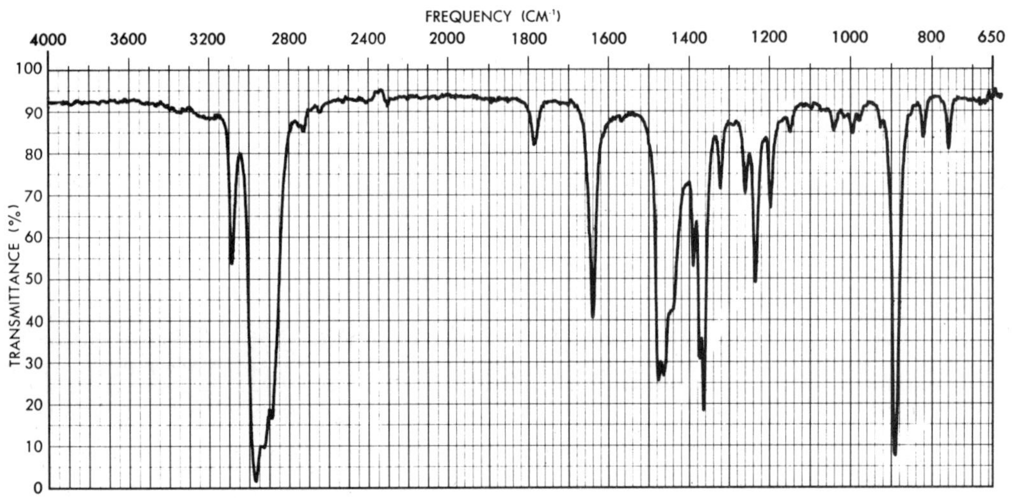

Figure 18-53 C_8H_{16}, liquid neat in a 0.015-mm cell.

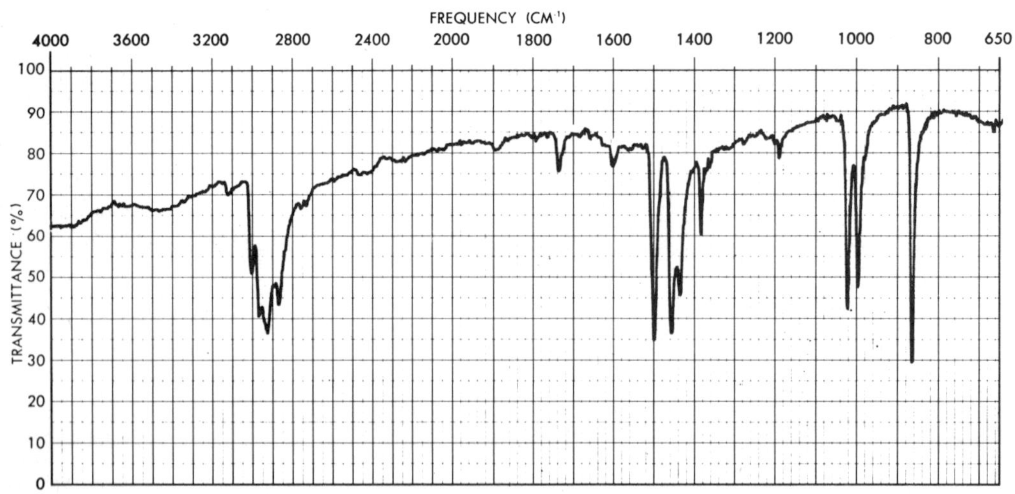

Figure 18-54 $C_{10}H_{14}$, pellet 1.5 mg in 300 mg KBr.

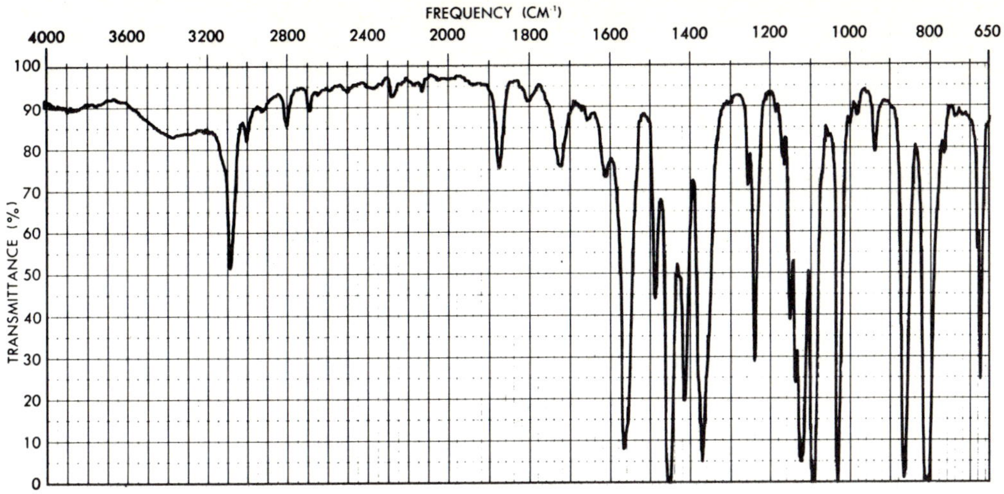

Figure 18-55 1,2,4-Trichlorobenzene, liquid neat in a 0.025-mm cell.

EXERCISES

18-1. Figure 18-46 shows an infrared spectrum of acetylene at 80-mm pressure in an evacuated 10-cm gas cell. (a) Why is there no C≡C stretching absorption band? (b) What is the cause of the "negative" absorption band at 2350 cm^{-1}?

18-2. Identify the compounds from their infrared spectra in Figures 18-47 through 18-54.

18-3. From the experiments in this chapter and your answer to Problem 18-5 summarize the relationship of band intensity, concentration, and path length. How do these relate the band intensity to the number of molecules encountered by the sample beam?

18-4. With optical null, double beam spectrophotometers, what causes the following problems and what could be done to correct them? (a) High noise level. (b) Low transmission where there are no absorption bands. (c) 7% pen overshoot. (d) Oscillation. (e) Sluggish pen response. (f) Frequency calibration off by 4 mm over the entire spectrum. (g) Frequency calibration off by amounts varying from 1 to 7 mm. (h) Poor resolution (i.e., poor separation of adjacent, closely spaced bands).

18-5. The minimum performance check of an infrared spectrophotometer (p. 131) includes obtaining an I_0 line and a polystyrene spectrum and checking the gain and adjusting if necessary. Should these be done in any particular order? Explain your answer.

18-6. Compare the spectrum of 1,2,4-trichlorobenzene (Fig. 18-55) with the 1,2,4-trisubstituted pattern from a hydrocarbon shown in Figure 18-39. Although both spectra were obtained on pure liquids in 0.025-mm cells and the trichlorobenzene (mol. wt. 181.5) has a molecular weight more than 50% higher than the hydrocarbon (mol. wt. 120.2) used to obtain Figure 18-39, the out-of-plane bending and the overtone absorptions of the trichlorobenzene are considerably more intense. Why?

19

The Clemmensen Reduction of Acylbenzene Derivatives

$$C_6H_5\overset{\overset{O}{\|}}{C}R \xrightarrow[\text{HCl}]{\text{Zn(Hg)}} C_6H_5CH_2R$$

INTRODUCTION

A convenient method of synthesizing hydrocarbons is by reduction of their ketone derivatives. This method is especially useful if the requisite ketones are readily available. Acylbenzene derivatives are readily synthesized in good yields by Friedel-Crafts acylation of aromatic hydrocarbons and some of their derivatives. Therefore, acylation and reduction of the ketones obtained are widely applied for introducing an alkyl group onto an aromatic ring.

EXPERIMENTAL PROCEDURE

Synthesis

In a 250-ml round bottom flask add 30 g of zinc wool, 2.0 g of mercuric chloride, 2 ml of concentrated hydrochloric acid, and 35 ml of water. Agitate the mixture for 5 min, allow to settle, and carefully decant the aqueous solution. Immediately add 35 ml of water, 5 ml of glacial acetic acid, and then 35 ml of concentrated hydrochloric acid. Add 15 g of the requisite ketone. Place a condenser on the flask and reflux the mixture for 3 hr. A better yield is obtained if the mixture is refluxed for 5 to 6 hr, and this should be done if the laboratory period is long enough or if the experiment can be carried over to the next period. While the mixture is refluxing, begin the next experiment. Allow the mixture to cool and settle until the next laboratory period even if two laboratory periods were used for refluxing.

Isolation and Purification

Separate the two layers and extract the aqueous layer with 15 ml of hexane. Wash the organic layer from the reaction with 20 ml of 3% aqueous sodium hydroxide solution and then with 20 ml of water. Put the organic layer in a 125-ml Erlenmeyer flask. Use the hexane extract to *successively* extract these two washes. Combine the hexane extract with the organic product in the Erlenmeyer flask and dry the mixture by stirring it with 5 to 10 g of anhydrous magnesium sulfate.

Carefully decant this into a 100-ml flask, using 3 to 4 ml of hexane as a wash to aid the transfer. Remove the hexane by distillation. Distil the residue and collect in a tared yield bottle the fraction that distils 2° on either side of the boiling point given in Table 19-1.

Infrared Analysis of the Product (Optional)

Obtain an infrared spectrum of your product in a 0.025-mm cell. Does the spectrum agree with what you would expect? What is the principal contaminant?

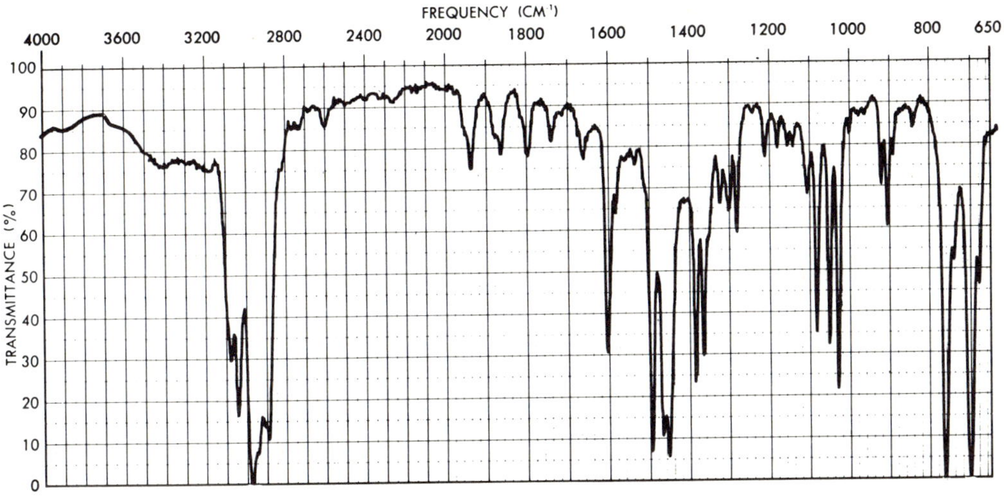

Figure 19-1 Lowest boiling product in Exercise 19-2; liquid neat in a 0.025-mm cell.

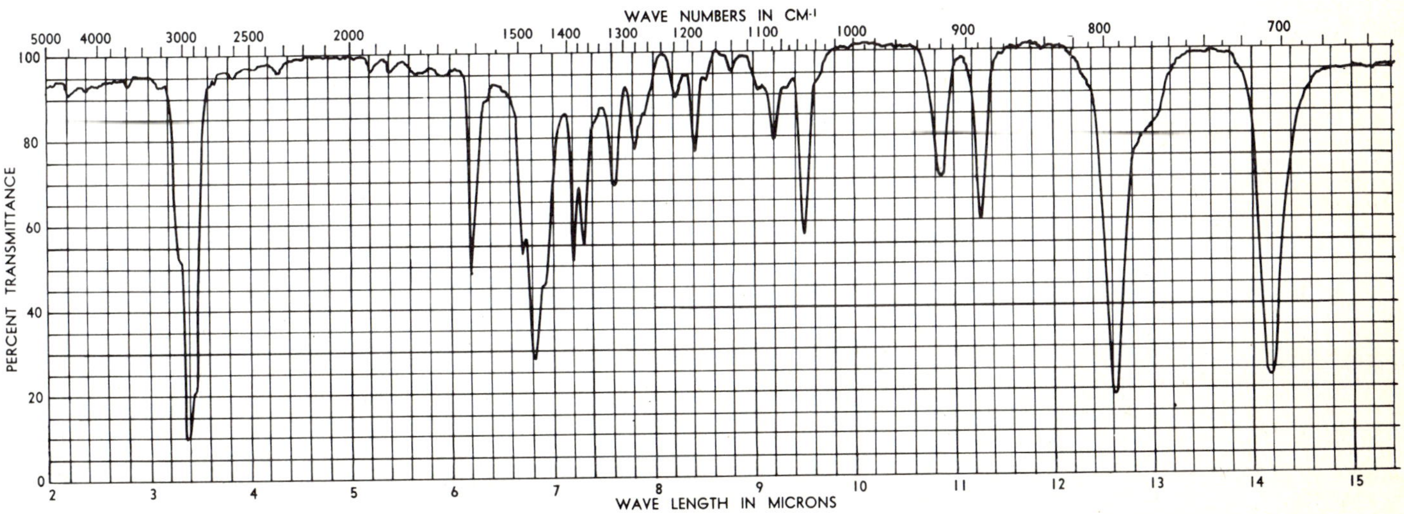

Figure 19-2 Spectrum of a product from reaction described in Problem 19-2. A 0.01-mm cell, b.p. 200–203°. Sadtler Research Laboratories, Inc., Philadelphia, Penna. 19104.

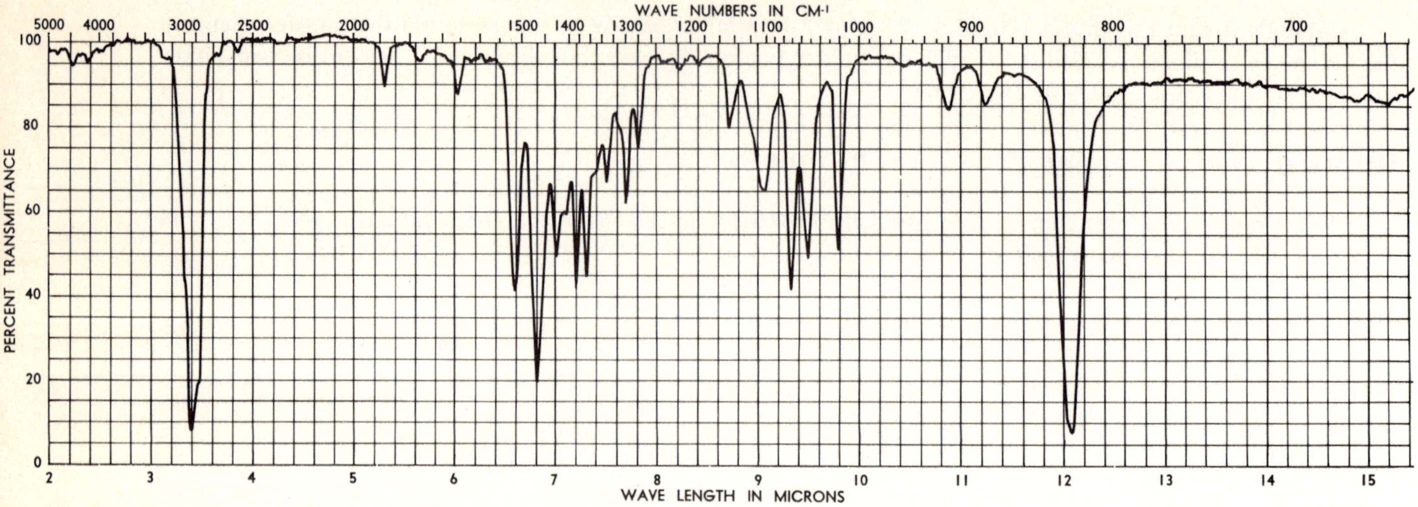

Figure 19-3 Spectrum of a product from reaction described in Exercise 19-2. A 0.01 mm cell, b.p. 200–203°. Sadtler Research Laboratories, Inc., Philadelphia, Penna. 19104.

TABLE 19-1 PHYSICAL PROPERTIES OF ACYLARENES AND THEIR REDUCTION PRODUCTS

Phenone	B.p. (°C)	M.p. (°C)	D_4^{20}	Alkylbenzene	B.p. (°C)	D_4^{20}
Aceto-	203-45	20	1.028	Ethylbenzene	136	0.867
Propio-	218-20	20	1.01	1-Phenylpropane	159	0.862
3-Methylaceto-	219-21		1.007	*m*-Ethyltoluene	161	0.8645
Butyro-	228-30	13		1-Phenylbutane	183	0.860

EXERCISES

19-1 The Clemmensen reduction involves the reaction of three phases—aqueous, organic, and solid (amalgamated zinc). What is the purpose of the glacial acetic acid?

19-2 In an attempt to prepare 1-phenylpropane, 1-chloropropane and aluminum trichloride were stirred with benzene:

$$C_6H_6 + ClCH_2CH_2CH_3 \xrightarrow{AlCl_3} C_6H_5CH_2CH_2CH_3$$

Instead of the desired product the compounds obtained gave the physical properties and spectra shown in Figures 19-1 to 19-3. What were the products? How could 1-phenylpropane be prepared?

20

Quantitative Analysis of a Commercial Xylene by Infrared Spectroscopy and Gas Chromatography

INTRODUCTION

Infrared spectroscopy can be used for quantitative as well as qualitative analysis. Quantitative analysis by both infrared spectroscopy or gas chromatography requires standardization and precise calibration even though the two methods differ greatly. This chapter compares the analyses of a commercial xylene mixture by gas chromatography and infrared spectroscopy.

INFRARED SPECTROSCOPY

Discussion

As was shown in Chapter 18 the absorbance of a given infrared band is determined by the number of molecules the sample beam encounters. This is dependent upon the path length through the sample and the concentration of molecules in the sample. This relation is expressed mathematically by the Beer-Lambert law:

$$\text{Absorbance} = acl = \log \frac{I_0}{I} = \log \frac{1}{T} = \log \frac{100}{\%T}, \tag{20-1}$$

where a = absorption coefficient = absorptivity = ϵ/mol. wt., c is the concentration (g/liter), l the path length (cm; also abbreviated b), I_0 the intensity of incident radiation, I the intensity of transmitted radiation, and

$$T = \text{transmission} = \frac{I}{I_0}. \tag{20-2}$$

If c is in gram moles per liter and l in centimeters, then ϵ is the molar absorptivity or the molar extinction coefficient and is used in place of a. Absorbance is also known as extinction or optical density.

Absorbance is the logarithm of the reciprocal of transmission. Since the pen deflection of most infrared spectrophotometers is linear with transmission of the sample, it is not surprising that most infrared spectra including all those in this text use an ordinate that is linear in transmittance units. For qualitative analysis a transmission scale is superior to an absorbance scale because of the linear relation between the AC voltage from the detector and transmission. Some of the high-priced infrared spectrophotometers have linear absorbance and transmission scales. This requires more complicated and expensive electronics. In infrared spectra with a linear absorbance ordinate most weak to moderate bands are too obscure to be located and the noise level on the strong bands is greatly exaggerated.

As can be seen from Equation 20-1 absorbance is linear with respect to concentration and transmission is not. Absorbance is, therefore, more convenient to use for quantitative analysis than is transmission. Often a paper with a nonlinear absorbance scale on the ordinate can be used. This is not always available for spectrophotometers used in student laboratories so that transmission data must be converted to absorbance data with Equation 20-1 or with the conversion scale on the back cover of this manual. Absorbance data can also be obtained from transmission paper with a ruler calibrated with a logarithmic scale for the size chart paper used. I_0 is the amount of radiation that would be transmitted if the compound being analyzed for were not present, and the other components of the mixture were present.

Selecting the proper I_0 can be a problem if there are bands very close to the one being measured. Figure 20-1 shows some absorption bands with I_0 lines drawn in.

Problem 20-1. Determine I_0 and I for the ten bands (at their absorption maxima) labeled *a* through *j* in Figures 20-2 and 20-3.

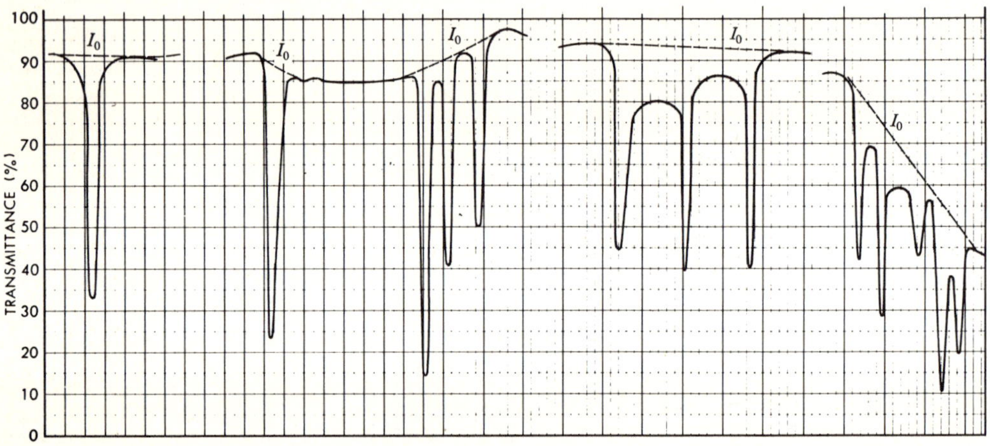

Figure 20-1 Absorption bands with I_0 lines.

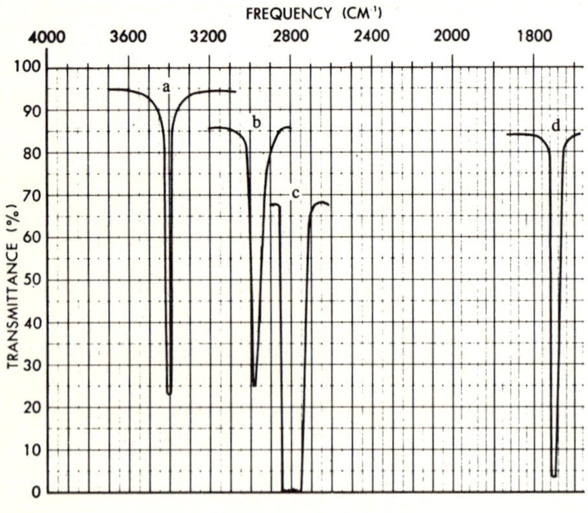

Figure 20-2 Absorption bands. Assume a 0.1-mm cell and a 1 *M* solution in carbon tetrachloride.

Problem 20-2. Calculate the percent transmission and absorbance of the six bands labeled *a* through *f* in Figures 20-2 and -3. From this and the other data available calculate the molar extinction coeffiicient for each of the four bands a — d.

Problem 20-3. Although quantitative analysis often uses absorbance or transmission data obtained from points of the spectrum other than absorption maxima, instrument error is usually smaller if I is obtained from an absorption maximum. Why?

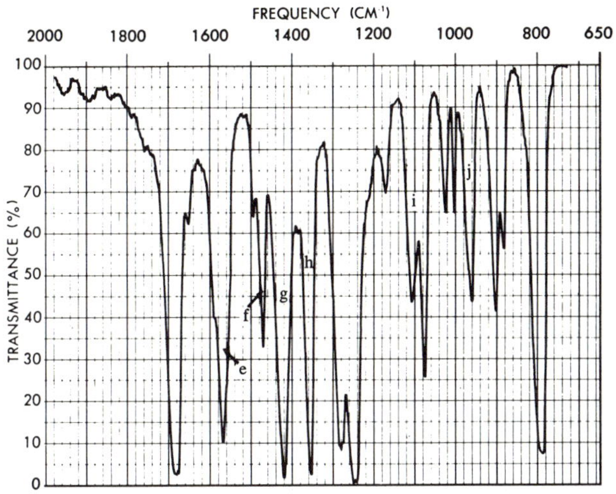

Figure 20-3 Absorption bands.

Problem 20-4. It is preferable that the unknown be so diluted that the point taken for I not be at less than 20% transmittance and not greater than 85%. Explain, considering both instrument limitations and characteristics of the absorbance scale.

Beer's law states that the absorbance of a mixture at a given wavelength is equal to the sum of the absorbances of the mixture's components. For a four-component mixture:

$$A_{\text{mixt}} = A_1 + A_2 + A_3 + A_4 = l(\epsilon_1 c_1 + \epsilon_2 c_2 + \epsilon_3 c_3 + \epsilon_4 c_4). \tag{20-3}$$

Problem 20-5. Figures 20-4 and 20-5 show portions of spectra obtained in solution in 0.1-mm cells. Spectra of carbon tetrachloride and carbon disulfide are shown in Figures 18-26 and 18-27. Data on the compounds in the mixtures are given in Table 20-1. (a) Determine I_0, I, $\%T$, and absorbance for each of the bands in Figures 20-4 and 20-5. (b) Fill in all the blanks in Table 20-1.

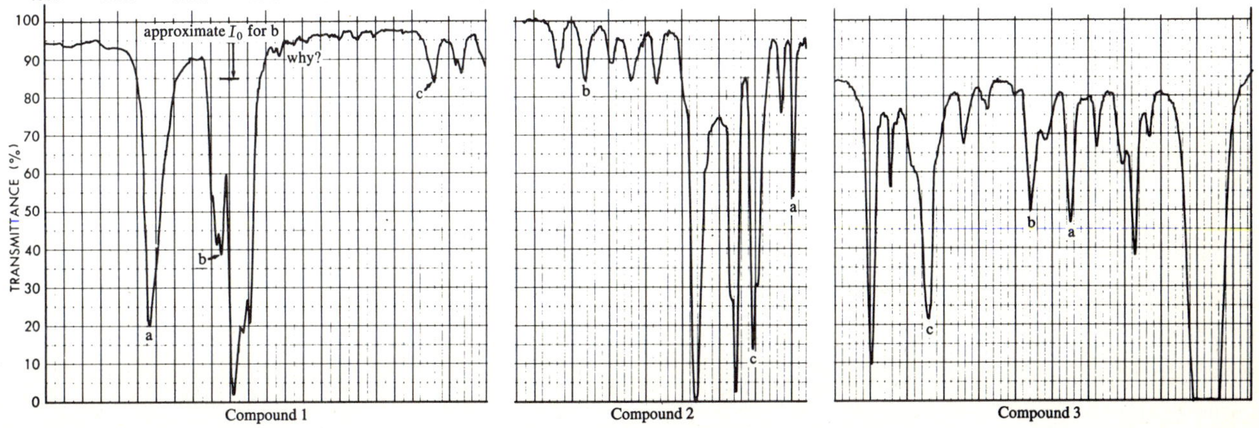

Figure 20-4 Portions of spectra of compounds in carbon tetrachloride solution.

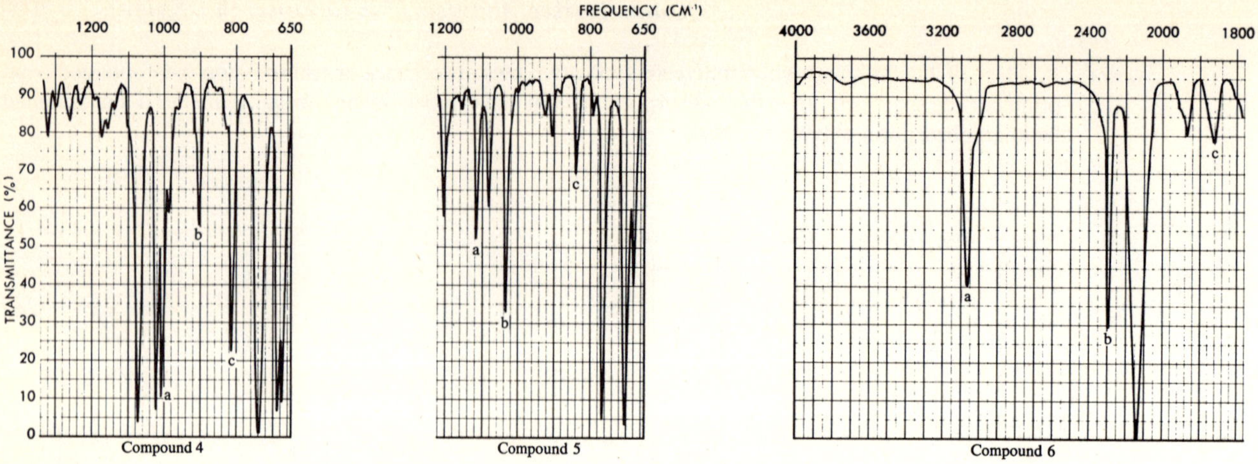

Figure 20-5 Spectra in carbon disulfide (Problem 20-5).

TABLE 20-1 QUANTITIES FOR PROBLEM 20-5

Compound	Band	Mol. wt.	a	ϵ	Conc. (g/liter)
1	a	100			10
1	b				
1	c				
2	a		10	2,000	
2	b				
2	c				
3	a				
3	b			60	10
3	c				
4	a	150		450	
4	b				
4	c				
5	a	200			
5	b				
5	c		10		
6	a			1,000	2
6	b				
6	c				

Analysis of a Commercial Xylene Mixture

Problem 20-6. The spectra of the three xylenes are given in Figure 20-6. Identify each xylene from its spectrum.

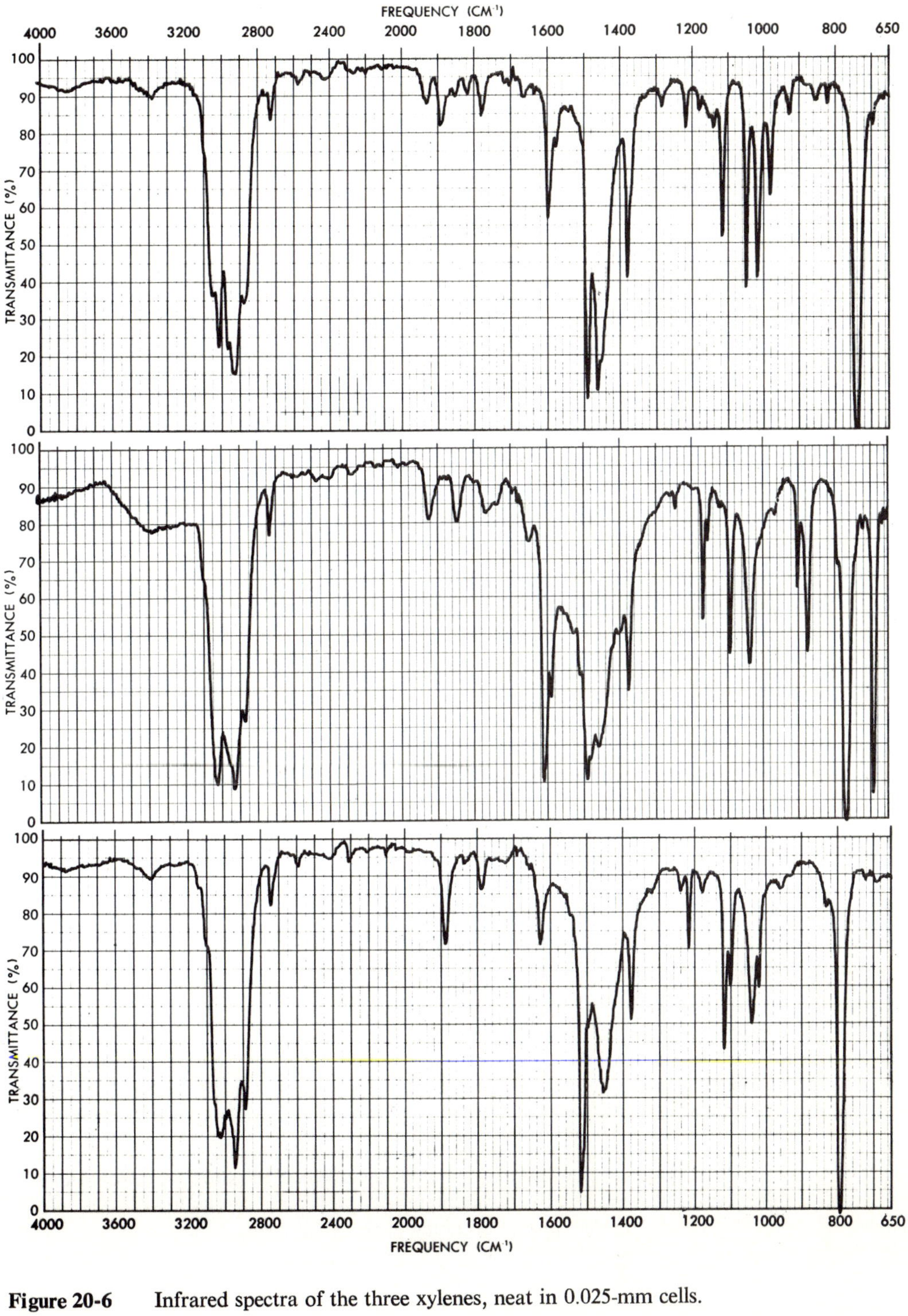

Figure 20-6 Infrared spectra of the three xylenes, neat in 0.025-mm cells.

Calibration. The out-of-plane bending absorption maxima are used as the frequencies for determination of the absorbances. The calibrations and measurement of the unknown are to be done in groups of three students. Each student in the group prepares one of the solutions shown in Table 20-2. Each solution has a total volume of 100 ml with cyclohexane as the solvent.

TABLE 20-2 STANDARD SOLUTIONS FOR ANALYSIS OF XYLENE

Solution	o-xylene (ml)	m-xylene (ml)	p-xylene (ml)
I	1	14	5
II	3	10	7
III	3	4	10

Each student obtains two spectra on his solution in a 0.05-mm cell. Only the region from 800 cm^{-1} to 650 cm^{-1} need be scanned. If the intensities (I) of the spectra you obtain differ by more than 2%, obtain an additional spectrum and if they differ by more than 3%, obtain two additional spectra for a total of four scans. One student of each group should also obtain a full range spectrum of cyclohexane. Use the following frequencies: *p*-xylene, 794 cm^{-1}; *o*-xylene, 741 cm^{-1}; and *m*-xylene, 690 cm^{-1}. Use, however, the maximum intensity or absorbance of each band even if the maximum differs by a few wavenumbers from the above values. If the xylene mixture is also to be analyzed by gas chromatography save the standard solutions for this experiment.

Calculate the absorbance value for each of the three calibration frequencies on every spectrum obtained. Average all the absorbance values for each of the three xylenes. For the solution he prepared the student now has three absorbance values, one for each of the three xylenes. These absorbances are the average of two to four values obtained from the limited range scans.

The data for the group is pooled and one student plots an absorbance-versus-concentration chart for *o*-xylene, one student plots a similar chart for *m*-xylene, and one student plots a chart for *p*-xylene.

Measurement. In a volumetric flask, dilute 20 ml of unknown xylene mixture to 100 ml with cyclohexane. Each student of the group obtains one limited frequency scan (800 to 650 cm^{-1}) of the solution. The absorbance data is pooled and each student averages the three absorbances of the out-of-plane bending bands for all three of the isomers and determines its concentration from the calibration charts. From the concentrations calculate the percent by volume of the three xylenes. The results of the analysis of the xylene mixture should accompany the report for this experiment.

Problem 20-7. In some student laboratories there may be more than one infrared spectrophotometer. It is very important that all the calibration and unknown spectra be obtained on the same infrared spectrophotometer by the three members of each group. Why?

Problem 20-8. It is important that all spectra be obtained in the same cell. Why? If accurately calibrated fixed thickness sealed cells are available, the calibrations and unknown spectra may be obtained with different cells if certain corrections are made. What are these corrections? Assume a 0.053-mm cell was used for the calibration scans and a 0.048-mm cell was used for unknown scans. Calculate the correction necessary.

Problem 20-9. In Chapter 18 it was emphasized that fixed thickness sealed cells are preferred for quantitative analysis. In this experiment, however, the demountable cell with ports can be used if certain precautions are followed. (a) Why is there no loss of accuracy in this experiment if demountable cells are used? (b) What precautions are necessary if they are used?

Analysis for Trace Contaminants (Optional). Obtain two complete spectra (4000–650 cm^{-1}) of the undiluted xylene mixture; one in a 0.025-mm cell and one in a 0.1-mm cell. Do you see evidence for any compounds other than the xylenes?

GAS CHROMATOGRAPHY

Calibration

Like the infrared analysis of the xylene mixture this analysis is done in groups of three students. Use the same solutions for calibration in the gas chromatographic analysis as were used for calibration in the infrared analysis. Divide the data gathering and use in the same manner as described in the section on infrared analysis. Inject separately each of the three xylenes and cyclohexane to be certain of the identity of the peaks and to verify that the xylenes can be separated with the conditions indicated. Inject two sample sizes (10 μl and 6 μl) of the standard solution to obtain more calibration points. As with the infrared analysis use duplicate injections for each size. If the peak heights vary by more than 2%, make three injections, and, if the peak heights vary by more than 3%, make a total of four injections.

After calibration each student in the group has six peak height values which are an average of two to four peak heights. These peak heights are from the three xylenes in the solution which was injected in two different sample sizes. The students pool the data and each student plots a peak height-versus-volume curve for one isomer. It should be remembered that each of the solutions is 80% cyclohexane.

Chromatographic Conditions

Use the isooctyl phthalate column described in Table 8-1 with the same temperature and flow rate. If several chromatographs are in the student laboratory, the same one should be used for chromatography of all standard solutions and the unknown by the three members of each group. If possible, the standard solutions and the unknown should be chromatographed on the same day.

Problem 20-10. What should the injection port and detector temperatures be?

Analysis of Xylene Mixture

Quantitative Analysis. Inject one 5-μl sample of the undiluted xylene mixture. Average the peak heights for each isomer with the peak heights obtained by the other two students in your group. Use this average value to determine the volume of the isomer from the peak height-volume calibration curves.

Detection of Minor Constituents (Optional). Were there any other peaks in the mixture besides those due to the xylenes? From the location of these peaks can you estimate their molecular weight? Note that on most liquid phases including the one used here the retention times are very dependent on the partition coefficient and, therefore, on the structures of the compounds. The molecular weights can be estimated only from relative retention times of similar compounds. The molecular weights are roughly proportional to the log of the retention times for a given group of compounds such as aromatic hydrocarbons, alkanols, and others (review pp. 60–61). For example, an alkane might or might not have a retention time similar to an aromatic hydrocarbon with the same molecular weight. Determine if the minor constituents have the same retention times as various compounds available in the laboratory. Do not randomly inject everything available but select five to seven compounds that might be the minor constituents. Use both the estimation of molecular weights from the retention times and your knowledge of the probable source of the xylene mixture.

Problem 20-11. Are coincident retention times of a standard and an unknown sufficient to establish that the unknown is identical with the standard? If not, what else could be done to confirm the identification?

Problem 20-12. Compare infrared spectroscopy and gas chromatography as to (a) accuracy and precision in analyzing quantitatively a three or more component mixture, (b) detection of minor constituents in a mixture, (c) identification of minor constituents in a mixture, (d) identification of a relatively pure compound, and (e) qualitative analysis of a two-component mixture.

Problem 20-13. Note in this experiment that the composition of the xylene was calculated in percent by volume. How could these figures be converted into percent by weight and mole fraction?

21

Synthesis of Alkyl Halides

INTRODUCTION

Alcohols are the usual starting materials for the laboratory synthesis of alkyl halides. The alcohols are reacted with hydrogen halides or inorganic acid halides to yield the desired alkyl halide:

$$ROH + HX \rightarrow RX + H_2O, \tag{21-1}$$

$$ROH + SOCl_2 \rightarrow RCl + HCl + SO_2, \tag{21-2}$$

$$2ROH + PCl_5 \rightarrow 2RCl + 2HCl + POCl_3. \tag{21-3}$$

Primary alcohols that are not allylic or benzylic react with hydrogen halides by an *S*N2 mechanism; all other alcohols react with hydrogen halides *via* an *S*N1 mechanism.

EXPERIMENTAL PROCEDURE

Synthesis of *n*-Butyl Bromide, an *S*N2 Reaction

$$\underset{\substack{\text{B.p. } 118^\circ \\ D_4^{20}\ 0.810}}{CH_3CH_2CH_2CH_2OH} + H^+ \rightarrow n\text{-Bu}\overset{+}{O}H_2 \xrightarrow{Br^-} \underset{\substack{\text{B.p. } 101.5^\circ \\ D_4^{20}\ 1.275}}{CH_3CH_2CH_2CH_2Br} + H_2O \tag{21-4}$$

Weigh carefully 30 g of ice in a 50-ml beaker. Put the ice and any water that melted from the ice after weighing into a 250-ml round bottom flask. Be sure that a total of 30 g of liquid and solid water is in the flask. Place the flask in an ice bath and slowly add 35 ml of concentrated sulfuric acid, and then add 20 ml of 1-butanol. Stir the mixture well with a glass rod during both additions. Remove the ice bath. With a mortar and pestle crush 44 g of sodium bromide dihydrate into a fine powder. Add the powder to the flask in small portions while stirring the mixture and mashing the added sodium bromide so that it does not agglomerate into lumps. Add a boiling chip.

Set a condenser on the flask for reflux (Fig. 10-1) and lead a rubber tube from a gas-tight connection on the top of the condenser to a hood or an air and water trap (Fig. 15-3). Gently heat the mixture and stir by applying a gentle rotary motion to the ring stand. Continue the heating and swirling until all or almost all the sodium bromide has dissolved. Gently reflux the mixture for 1 hr 40 min. During this time begin the synthesis of the *t*-butyl chloride (part 2). If the optional gas chromatographic analysis is assigned, read that section before proceeding further.

Allow the mixture to cool 5 min. Disconnect the condenser and construct the distillation apparatus shown in

Figure 3-1. Add 20 ml of water and a fresh boiling chip and distil for 20 min. Then examine the distillate every 4 to 5 min by collecting a few milliliters of distillate in a separate small tube and examining it for an organic phase. Stop the distillation when no more organic liquid is distilled.

Combine *all* the distillate in a separatory funnel, add a few pieces of ice and 30 ml of water, shake vigorously, allow to settle, and separate. Place 4 g of ice in a 25-ml Erlenmeyer flask in an ice bath, add 10 ml of concentrated sulfuric acid, and stir with a glass rod. Extract the butyl bromide with this 75% sulfuric acid. Wash the butyl bromide with 20 ml of cold water and 20 ml of cold 5% sodium bicarbonate. *Caution:* review pages 37 and 38.

Problem 21-1. Which layers are the butyl bromide in the above five separations? Be certain of the answer before you discard the nonorganic layers.

Problem 21-2. What is the purpose of the acid extraction?

Dry the butyl bromide by stirring it in a 25-ml Erlenmeyer flask with 5 to 6 g of anhydrous calcium chloride. Decant the dried *n*-butyl bromide into a clean, dry 25-ml distilling or round bottom flask. Add a boiling chip, distil, and collect the fraction that boils at 98 to 103° in a tared yield bottle.

Problem 21-3. Sometimes a fraction is obtained that boils at 70 to 75°; what is this fraction? What could be done to this fraction to obtain a better yield of product? If you obtained such a fraction, verify your answer before proceeding further.

Weigh the product, calculate the % yield, label the bottle, and hand in the product to your laboratory instructor.

Synthesis of *t*-Butyl Chloride, an SN1 Reaction

$$(CH_3)_3COH \xrightarrow[H_2O]{} (CH_3)_3C^+ \xrightarrow{Cl^-} (CH_3)_3CCl \qquad (21\text{-}5)$$

$(CH_3)_3COH$: B.p. 83°, M.p. 25.5°, D_4^{26} 0.779

$(CH_3)_3CCl$: B.p. 51°, D_4^{20} 0.846

Pour 60 ml of concentrated hydrochloric acid that has been cooled to 0 to 5° and 20 ml of *t*-butyl alcohol into a separatory funnel. Shake every few minutes over a 25-min period. Allow the mixture to stand about 10 min. If the optional gas chromatographic analysis is assigned, read that section before proceeding further. Draw off the bottom aqueous layer, and wash with 15 ml of cold water, 15 ml of cold 5% sodium bicarbonate solution, and 15 ml of cold water.

Add 4 to 5 g of anhydrous calcium chloride to the separatory funnel and shake the funnel a few times. Carefully drain off the *t*-butyl chloride into a 50-ml round bottom or distilling flask. Remove the top from the funnel and open the stopcock about one fourth to one half so that all the *t*-butyl chloride drains and leaves the calcium chloride in the funnel.

Distil and collect the fraction that boils at 48 to 53° in a tared yield bottle. Weigh the product, calculate the % yield, label the bottle, and hand in the product to your laboratory instructor.

Problem 21-4. What is the most likely side reaction? What steps are taken to prevent this?

Problem 21-5. What reactions are likely to occur in the isolation and purification of the *t*-butyl chloride? What is done to minimize these reactions? What is done to keep the products of the side reactions from contaminating the *t*-butyl chloride?

Analysis of the Reaction Products by Gas Chromatography

Obtain a 5-μl sample of each of the reaction products before any washing or purification. Chromatograph each sample on the Apiezon column described in Table 8-1. Use also the temperature and flow rate described in Table 8-1. The attenuator should be turned down to its 1-X position or until the noise level and base line drift exceed 3 %. Be prepared to attenuate the signal (pp. 64–65) as some of the later peaks come through.

From the compounds available in the laboratory and from the peak height data and retention times supplied by your instructor identify the contaminants. Does the *S*N1 or the *S*N2 synthesis give more by-products? Explain the origin of all the contaminants in the crude material.

Problem 21-6. (Optional) From your estimation of the amounts of compounds found by gas chromatography can you account for all the starting material? Are much of the products lost on purification? Can you suggest any improvements to increase the yields and purities of the halides?

22

Kinetics of Displacement Reactions

$$CH_3CH_2CH_2I + HO^- \rightarrow CH_3CH_2CH_2OH + I^-$$

$$(CH_3)_3CCl + HO^- \rightarrow (CH_3)_3COH + Cl^-$$

INTRODUCTION

The rates of organic reactions and the effects of concentration of the reactants on the rates can give useful information about the mechanisms of reactions. Rates of reactions can be determined by a variety of methods. Since small changes in temperature have such a marked effect on reaction rates, precise temperature regulation is required for all kinetic studies.

DISCUSSION

Many organic reactions are ionic displacement reactions of the type

$$Y^- + RX \rightarrow RY + X^-,$$

$$Y\colon + RX \rightarrow RY^{\oplus} + X^-,$$

$$Y^- + RX^{\oplus} \rightarrow RY + X.$$

In many of these displacement reactions the rate of the reaction is proportional to the concentration of both reactants because both reactants are needed in the rate-determining step. For example, in the Williamson synthesis of *n*-propyl phenyl ether both the phenoxide ion (Y^-) and 1-bromopropane (RX) collide in the rate-determining step

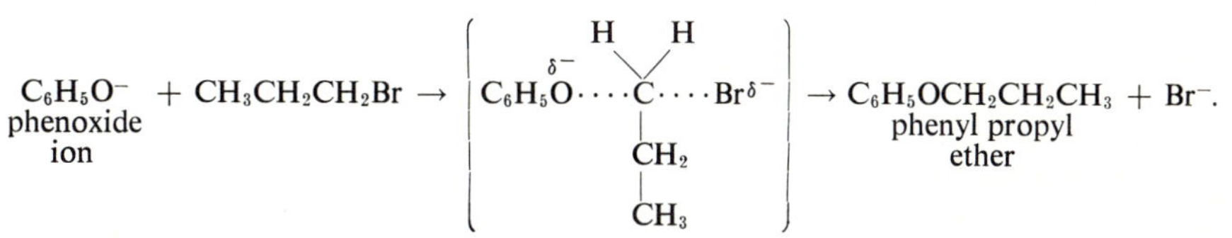

The kinetic equation for such a reaction is

$$\text{Rate} = k[C_6H_5O^-][n\text{-PrBr}], \tag{22-1}$$

or more generally,

$$\text{Rate} = k[Y][RX], \tag{22-2}$$

where k is a rate constant that is independent of concentration and varies with the temperature, solvent, and the composition of Y and RX.

The rate can be expressed as the rate of disappearance of one or both of the reactants or the rate of appearance of one or both of the products. The rate of the reaction at any one time is

$$\frac{-d(a - x)}{dt} = k(a - x)(b - x) = \frac{dp}{dx}, \tag{22-3}$$

where a is the initial concentration of Y, b the initial concentration of RX, p the amount of product formed, and x the amount of reactant used.

Since a is a constant

$$\frac{dx}{dt} = k(a - x)(b - x). \tag{22-4}$$

If the time interval is sufficiently small

$$\frac{\Delta x}{\Delta t} = k(a - x)(b - x), \tag{22-5}$$

where Δx is the change in concentration over the time interval Δt.

In this experiment alkyl halides are hydrolyzed in basic aqueous ethanol. The rate of disappearance of one reactant, the base, is used to measure the rate of reaction.

EXPERIMENTAL PROCEDURE

This experiment is done in pairs. Each student uses his data and the data obtained by his partner. Student 1 prepares a mixture of 33 ml of distilled water and 140 ml of 95% ethanol. Student 2 prepares a mixture of 12 g of potassium hydroxide in 85 ml of this ethanol-water solution. Student 1 uses 1-iodopropane for the alkyl halide and student 2 uses *t*-butyl chloride. Accurately weigh (to ±0.01 g) two 4-g portions of alkyl halide into two 25-ml volumetric flasks. Place these flasks in a constant temperature bath held at ±0.3° approximately 2° above room temperature. Place the ethanol solution and potassium hydroxide solution in separate, tightly sealed flasks in the water bath. Allow 5 to 7 min with occasional swirling for the mixtures to reach the temperature of the bath. Into one flask containing the halide accurately pipet 10 ml of the potassium hydroxide solution. Immediately add aqueous ethanol to bring the volume to exactly 25 ml. Mix thoroughly and quickly and without delay remove and titrate a 1-ml sample. Return the flask to the constant temperature bath at once.

Once the potassium hydroxide solution and alkyl halide are mixed, the rest of the steps must follow very quickly. In the rate equation the quantity b is assumed to be the amount of alkyl halide placed in the flask. If there is any significant time lag between mixing the reactants and titrating the sample, b will be less than the amount of halide added. There can be no similar error in a since it is calculated from the first titration.

Into another flask containing the alkyl halide accurately pipet 20 ml of the potassium hydroxide solution. Immediately add enough aqueous ethanol to bring the mixture to 25 ml, shake the flask briefly and vigorously. Quickly remove and titrate a 1-ml sample. Return the mixture to the constant temperature bath at once. Do the above steps and the titrations described below as quickly as possible. Since you will be working with two flasks, stagger the starting times on each one so that you will not be forced to attempt two titrations at once. The second propyl iodide reaction should be started 5 to 7 min after the first, and the second *t*-butyl chloride reaction should be started about 35 min after the start of the first *t*-butyl chloride reaction. Gently shake each flask in the constant temperature bath as often as once a minute if time permits.

Thirty to sixty seconds before the time intervals shown in Tables 21-1 and 21-2 pipet 1-ml samples from each of the reaction flasks. At the exact time intervals shown on the tables, transfer these samples to a 250-ml Erlenmeyer flask containing 2 to 3 drops of phenolphthalein solution and 30 ml of distilled water. Time zero is when the first sample is transferred to the 250-ml flask just before the reaction flask was returned to the constant temperature bath. If the samples are not transferred at the exact time shown in Tables 22-1 and 22-2, use the time the sample was transferred in your own tables and make the appropriate corrections of Δt.

As soon as each sample has been transferred, titrate it with 0.1 *N* hydrochloric acid as quickly as possible. The end point occurs when the last faint pink color just disappears and the solution turns completely colorless. Accurate determination of the end point is easier if the standard acid solution is added very slowly and the solution stirred well with a glass rod as the pink color begins to disappear. The color change appears sharper if observed against a white background.

TABLE 22-1 KINETIC DATA FOR 1-IODOPROPANE HYDROLYSIS

Sample number	Time (min)	0.1 N HCl (ml)	x (mol/liter)	(a - x) (mol/liter)	(b - x) (mol/liter)	Δt	Δx	$\frac{\Delta x}{\Delta t}$	Average $\frac{\Delta x}{\Delta t}$	k
1	0		0							
						10				
2	10									
						10				
3	20									
						20				
4	30									
						20				
5	60									
						30				
6	90									
						30				
7	120									

TABLE 22-2 KINETIC DATA FOR t-BUTYL CHLORIDE HYDROLYSIS

Sample number	Time (min)	0.1 N HCl (ml)	x (mol/liter)	(a - x) (mol/liter)	(b - x) (mol/liter)	Δt	Δx	$\frac{\Delta x}{\Delta t}$	Average $\frac{\Delta x}{\Delta t}$	k
1	0		0							
						5				
2	5									
						5				
3	10									
						5				
4	15									
						10				
5	25									
						15				
6	40									
						20				
7	60									

Make two tables similar to Table 22-1 or 22-2 depending on which halide you hydrolyzed. Record the times the samples were taken and the amounts of standard acid solution required to titrate each sample.

GAS CHROMATOGRAPHY (Optional)

After the last sample is taken, pour the contents of the reaction flask into a 100-ml beaker containing approximately 5 g of crushed ice and concentrated hydrochloric acid. Use 4 ml of acid for the flask that had 10 ml of potassium hydroxide solution and 10 ml of acid for the flask that had 20 ml of potassium hydroxide solution. Stir for a few minutes and then extract with 5 ml of *p*-cymene. Inject 20 μl of this solution into a gas chromatograph. Use the Apiezon column and the conditions described in the third line of Table 8-1. Determine retention times and approximate peak height concentration data from appropriate compounds available in the laboratory. From this and any similar data furnished by the laboratory instructor determine the identity and concentration of the compounds in the reaction mixture. Compare the gas chromatographic results for the reaction of the two concentrations of base with the halide you studied and the results of your partner.

Problem 22-1. (a) How do you account for the various products? (b) Are there any compounds missing that you would expect to find? If so, where did they go? (c) How do you account for the different *relative* concentrations of compounds obtained by using different concentrations of base?

Problem 22-2. (a) Why not just inject the reaction mixture directly instead of injecting an organic extract? (b) Why was *p*-cymene used to extract the aqueous solutions instead of benzene, chloroform, or ethyl ether?

Problem 22-3. Why would a Carbowax column be unsatisfactory for the analysis of the extracts?

CALCULATION OF RATE CONSTANTS FROM THE TITRATION DATA

Using the titration data from your tables, calculate x, $a - x$, $b - x$ for each sample. Calculate Δx and $\Delta x/\Delta t$, the eighth and ninth columns; note that these are for the times between the samples. The value of $\Delta x/\Delta t$ for the times the samples were taken (second to last or tenth column in the tables) is obtained by averaging the $\Delta x/\Delta t$ values in the ninth column of the tables. That is, $\Delta x/\Delta t$ for sample 2 is obtained by averaging the first and second $\Delta x/\Delta t$ in the ninth column, $\Delta x/\Delta t$ for the third sample is obtained by averaging the second and third numbers in the ninth column, etc. Using Equation 22-5, calculate k for samples 2 to 6 inclusive and record the results in the last column of the tables. Compare your results with your laboratory partner and discuss the two sets of results with him.

Problem 22-4. What is the effect of the concentration of base on the rates of reactions of the two alkyl halides?

Problem 22-5. The value k should be a constant; why does it vary as much as it does?

With one of the halides the principal variation of the rate constants you calculated is due to the error in using Equation 22-5 which is very accurate only for small time differences (Δt). Equation 22-4 can be integrated to Equations 22-6:

$$\frac{1}{a-b}\ln\frac{b(a-x)}{a(b-x)} = kt, \qquad \text{(22-6a)}$$

$$\frac{2.303}{a-b}\log\frac{b(a-x)}{a(b-x)} = kt. \qquad \text{(22-6b)}$$

The validity of these equations is not dependent on any short time interval and can be used to calculate k at any time except for $t = 0$.

Problem 22-6. (Optional for those with sufficient mathematical background) Derive Equation 22-6 from Equation 22-4.

Recalculate k for samples 1–6 and calculate k for sample 7 for your reaction with both concentrations of base using Equation 22-6b.

Problem 22-7. The data on one of the alkyl halides does not fit Equation 22-4, 22-5, or 22-6 because the displacement reactions of that halide do not generally follow second order kinetics. Which halide is it? What order rate equation should be used? Derive the proper rate equation similar to 22-5 and calculate the rate constants for samples 2 to 6 inclusive for its reactions with both concentrations of base.

EXERCISES

22-1. (a) Could the rates of these reactions be studied with infrared spectroscopy instead of titrations? If so, how? Be specific about frequencies, standardization, scans, cell materials, and other details. (b) If a constant temperature cell were available, how could this be used to obtain better data with less work?

22-2. (a) Could gas chromatography be used instead of titrations to determine the rate of the reaction? (b) What might prevent accurate quantitative analysis of the mixture from the reaction flask? (c) What could be done to overcome these problems? (d) Outline a complete procedure for determining the rates of the reactions by gas chromatography.

22-3. In the study of these reactions the possibility that the alkyl halide might react significantly before the first titration is a potential source of error. How could this source of error be completely eliminated without obtaining any more experimental data?

23

Properties and Reactions of Organic Halides

Organic Qualitative Analysis II

INTRODUCTION

Chlorine, bromine, and iodine can be easily identified in organic compounds by sodium fusion and inorganic anion analysis of the products of the fusion. Halogens in organic compounds give characteristic, readily observable reactions to alcoholic silver nitrate and sodium iodide in acetone. Because of these tests haloorganic compounds are usually readily identified.

PHYSICAL PROPERTIES

Boiling Point

As with other organic compounds the boiling points of organic halides can be correlated with their molecular weights (p. 19). The correlation is very close only if halides of a given type (e.g., straight-chain primary chlorides, straight-chain secondary bromides, branched primary bromides, etc.) are compared with each other. Such correlation gives a rough idea of an organic halide's molecular weight from its boiling point.

A chlorine atom has about the same effect on boiling point as an equal weight of methylene groups (CH_2). The other halogens have a smaller effect on boiling point than an equal weight of methylene groups. A bromine atom has about the same effect on boiling point as three methylene groups. In other words, a monobromide has about the same boiling point as the compound that would be obtained by replacement of the bromine atom with a propyl group. An iodine atom has about the same effect on boiling point as four methylene groups (Table 23-1 and Fig. 23-1). Fluorine atoms have a variable effect on boiling points. Often replacement of a hydrogen atom by a fluorine atom has little effect on the boiling point as with the aromatic compounds shown in Table 23-2 even though there is an increase in molecular weight. For some aliphatic compounds replacement of a hydrogen atom by a fluorine atom gives a 25 to 50° increase in the boiling point. If, however, other halogen atoms are present, the replacement of a hydrogen atom by a fluorine atom may greatly increase the boiling point, have almost no effect on the boiling point, or decrease it significantly (Table 23-3). Possible explanations for some of these effects will be studied in Appendix V.

TABLE 23-1 STRUCTURES OF HYDROCARBONS, ORGANIC BROMIDES, AND ORGANIC IODIDES OF SIMILAR BOILING POINT

$RCH_2CH_2CH_2CH_3$	B.p. (°C)	RCH_2Br	B.p (°C)	RI	B.p. (°C)	$Br(CH_2)_{n+1}Br$	B.p. (°C)	$Br(CH_2)_nI$	B.p. (°C)
Pentane	36	Ethyl bromide	38	Methyl iodide	43				
2-Methylpentane	60	Isopropyl bromide	60						
Hexane	68	1-Bromopropane	71	Ethyl iodide	72				
2,2-Dimethylpentane	79	*t*-Butyl bromide	72						
2-Methylhexane	90	Isobutyl bromide	91	Isopropyl iodide	89				
Heptane	98.4	*n*-Butyl bromide	100	Propyl iodide	102	Methylene bromide	98		
Octane	125	*n*-Amyl bromide	129	*n*-Butyl iodide	130	Ethylene bromide	130	Bromoiodo–methane	130
Nonane	150	*n*-Hexyl bromide	157	*n*-Amyl iodide	156	1,3-Dibromo-propane	165	Bromoiodo–ethane	163
Propylbenzene	158	Bromobenzene	156						
Butylbenzene	182	Benzyl bromide	198	Iodobenzene	188				

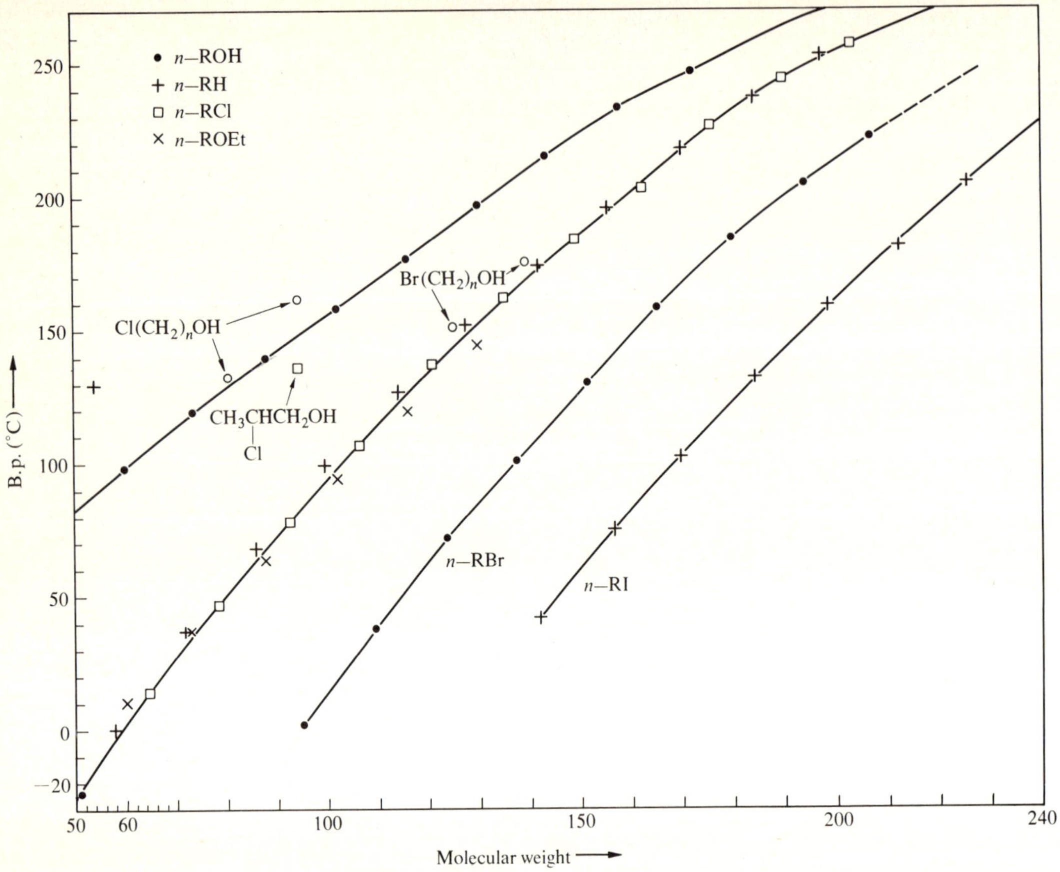

Figure 23-1 Boiling points of alkanes, alkanols (Chapter 28), and some of their halogen derivatives as a function of molecular weight.

TABLE 23-2 COMPARISON OF BOILING POINTS OF AROMATIC FLUORINE COMPOUNDS WITH PARENT COMPOUNDS

	B.p.(°C)		B.p.(°C)
Benzene	80	Fluorobenzene	84.6
Toluene	111	o-Fluorotoluene	114
		m-Fluorotoluene	115
		p-Fluorotoluene	117
Chlorobenzene	132	o-Fluorochlorobenzene	135
		m-Fluorochlorobenzene	124.5
		p-Fluorochlorobenzene	131
Bromobenzene	157	p-Fluorobromobenzene	152
Iodobenzene	183	p-Fluoroiodobenzene	183

TABLE 23-3 COMPARISON OF BOILING POINTS OF ALIPHATIC FLUORINE COMPOUNDS WITH PARENT COMPOUNDS

	B.p. (°C)		B.p. (°C)
Methane	-169.5	Fluoromethane	-142
		Difluoromethane	-51.6
		Trifluoromethane	-82.2
		Tetrafluoromethane	-184
Chloromethane	-24	Fluorochloromethane	-9
		Difluorochloromethane	-31
		Trifluorochloromethane	-81
Bromomethane	3.6	Fluorobromomethane	19
		Trifluorobromomethane	-69
Iodomethane	42.5	Fluoroiodomethane	53.4
		Difluoroiodomethane	21.6
Ethane	-88.6	Fluoroethane	-38
		1,1-Difluoroethane	-25
		1,1,1-Trifluoroethane	-47
		1,1,1,2-Tetrafluoroethane	-21
		Hexafluoroethane	-79
Chloroethane	13	2-Fluoro-1-chloroethane	53
Bromoethane	38.4	2-Fluoro-1-bromoethane	70.5
Propane	-44.5	1-Fluoropropane	-2.5
		2,3 Difluoropropane	0.5
Butane	-0.5	1-Fluorobutane	32

Solubility

Halogen derivatives of the hydrocarbons are similar in solubility to the hydrocarbons from which they are derived. The halogen derivatives of hydrocarbons are insoluble in water. They usually show the same solubility in concentrated mineral acids as the parent hydrocarbons unless an aromatic system has been deactivated to electrophilic reactions by several halogen atoms, e.g., *m*-xylene has borderline solubility in concentrated sulfuric acid, but 4,6-dichloro-1,3-dimethylbenzene is completely insoluble in this acid.

Usually the more halogen atoms a compound has, the less soluble it is in other nonhalogenated solvents. This is more noticeable with compounds that have many fluorine atoms than with compounds with other halogens. Many polyfluoro compounds are soluble in other polyfluoro compounds but insoluble in most other solvents.

Problem 23-1. Note in Table 23-4 that the boiling points of vinyl halides, $-\overset{|}{C}=\overset{|}{C}-X$, are considerably below those of the alkyl halides, while other unsaturated halides have boiling points very close to those of the corresponding alkyl halides and most alkenes have boiling points close to those of the corresponding alkanes. How can this be explained? Hint: see the discussion of dipole moments of these compounds in your lecture text.

TABLE 23-4 COMPARISON OF BOILING POINTS OF UNSATURATED COMPOUNDS WITH SATURATED COUNTERPARTS

Unsaturated compound	B.p. (°C)	Saturated compound	B.p. (°C)	B.p. sat Δ B.p. unsat. (°C)	Aver. diff
$H_2C{=}CHCl$	-14	CH_3CH_2Cl	13	27	
$CH_3CH{=}CHCl$	37	$CH_3CH_2CH_2Cl$	47	10	
$H_2C{=}CHBr$	16	CH_3CH_2Br	38	22	19.2
$CH_3CH{=}CHBr$	60	$CH_3CH_2CH_2Br$	71	11	
$CH_2{=}CHI$	56	CH_3CH_2I	72	26	
$CH_2{=}CH\text{-}CH_2Cl$	45	$CH_3CH_2CH_2Cl$	47	2	
$CH{=}CH(CH_3)CH_2Cl$	72	$(CH_3)_2CHCH_2Cl$	69	-3	
$CH_3CH_2CHCl\text{-}CH{=}CH_2$	94	$CH_3CH_2CHClCH_2$ CH_3	97	3	0.2
$CH_2{=}CH\text{-}CH_2Br$	71	$CH_3CH_2CH_2Br$	71	0	
$CH_2{=}CH\text{-}CH_2I$	103	$CH_3CH_2CH_2I$	103	0	
$H_2C{=}CHCH_2CH_2$	-6	$CH_3CH_2CH_2CH_3$	0	6	
cis-$H_3C\text{-}CH{=}CH\text{-}CH_3$	4	$CH_3CH_2CH_2CH_3$	0	-4	
$H_2C{=}CHCH_2CH_2CH_3$	30	$CH_3(CH_2)_3CH_3$	36	6	0.66
cis-$H_3C\text{-}CH{=}CH\text{-}CH_2CH_3$	37	$CH_3(CH_2)_3CH_3$	36	-1	
$H_2C{=}CH\text{-}CH_2CH_2CH_3$	64	$CH_3(CH_2)_4CH_3$	68	2	
$CH_2{=}C(CH_3)_2$	-7	$(CH_3)_2CHCH_3$	-12	-5	

ELEMENTAL ANALYSIS

Chlorine, bromine, iodine, and a few other elements are easily detected in organic compounds. Since the knowledge of the presence of these elements in an unknown compound aids greatly in the identification of the unknown, elemental analysis is one of the first steps in the identification of organic unknowns.

Beilstein's Test for Halogen

Heat a copper wire having a small loop on the end in a burner flame until the flame is no longer green. Allow the wire to cool and wet the loop with a drop of the compound that is to be examined. Heat the wet wire in the edge of the burner flame. The presence of halogen is indicated by a green or blue-green flame. Some of the low boiling halogen compounds may occasionally give negative tests because they vaporize completely before the wire becomes sufficiently hot to cause decomposition. Usually, however, this test is too sensitive and gives positive tests for compounds that contain only traces of organic halides as impurities. Such false positives are not much of a problem since it is always necessary to do a sodium fusion to identify which halogen is present. The absence of halogen upon sodium fusion when the Beilstein test is positive indicates that the unknown is a nonhalogenated compound with a small amount of organic halide as an impurity.

Sodium Fusion

Danger! Sodium is a very reactive and dangerous metal. It reacts violently with water and with other compounds contaminated by large amounts of water. Sodium also reacts violently with many organic compounds, especially when sodium is hot as in this test. Always use extreme care when doing sodium fusions. Always wear eye protection when you or your neighbor are doing sodium fusions. Be sure to quench any unused pieces of sodium with ethanol and wash the resulting strongly basic solution down the drain with excess water.

Add a *small* drop or about 3 mg of unknown to a *dry*, 5-cm *Pyrex* test tube. With filter paper quickly dry a piece of shiny sodium metal about the size of a large pea (5 × 5 × 5 mm) and add to the test tube. Gently warm the test tube and allow it to stand for 2 min. Heat the tube until sodium vapors rise in the test tube. Remove the tube from the flame and quickly add about 7 mg (2 *small* drops if a liquid) of unknown.[1] Try to drop the unknown directly on the molten sodium and not on the side of the test tube. Heat again until the sodium vapors rise in the test tube and add about 7 mg of the unknown as before. Heat slowly until the sodium vapors begin to rise again. Then heat the test tube to a red heat for at least 3 min and allow it to cool to room temperature.

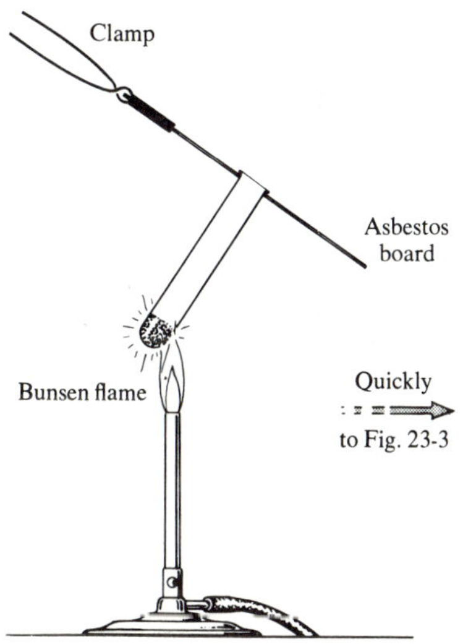

Figure 23-2 Final heating of sodium fusion product.

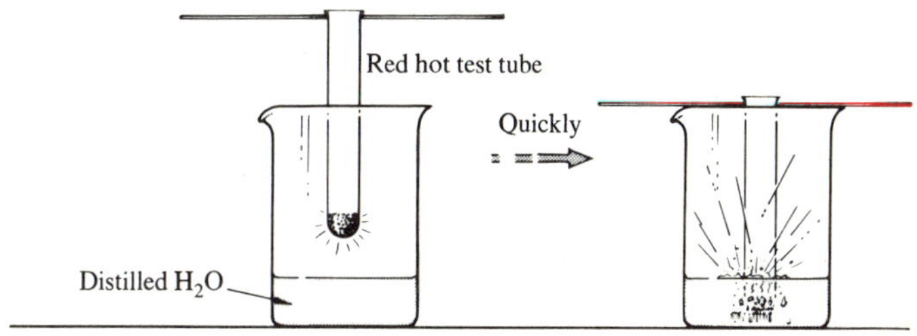

Figure 23-3 Safe quenching of sodium fusion product.

Place the tube in a hole in the center of a 2 in. × 2 in. clean asbestos board. The size of the hole should be such that the test tube fits very snugly. Add 1 ml of ethanol to the test tube. Hold the test tube in the flame by the asbestos board which is secured by a clamp (Fig. 23-2). The excess ethanol is burned while the test tube is heated to a dull red color. Plunge the red hot test tube into 15 ml of distilled water in a 50-ml beaker. If this is done quickly and correctly (Fig. 23-3), the tube shatters from thermal shock, and the asbestos board serves as a shield to prevent small pieces of glass and other material from flying out of the beaker.

Crush the pieces of glass test tube and break up any lumps of carbonaceous material with a glass rod. Heat the solution to boiling, and gravity filter the hot solution. Wash the beaker with two 4-ml portions of hot distilled water and then using this same water, wash the filter paper and the residue.

Combine the washes with the filtrate. The filtrate should be clear and colorless. A slight yellow color, however, is satisfactory. If the filtrate is very dark, the sodium fusion should be repeated as the dark color indicates that the fusion may not have been done well enough to ensure that all the elements present will be detected, and the dark color will mask some of the tests.

[1] At this time or on the first heating, a small explosion will occur with some organic compounds. If this occurs, gently boil about 0.4 g of the unknown with 0.4 ml of glacial acetic acid. After five minutes heat the mixture until the acetic acid is vaporized. The *entire* residue should be subjected to a sodium fusion as described above.

Analysis for Halogen

With a few drops of dilute nitric acid, acidify 6 drops (about 1.2 ml) of the fusion filtrate in a 7.5-cm test tube and add 3 drops of 5% silver nitrate solution. A white or yellow precipitate indicates the presence of halide ions. If the precipitate is yellow, it indicates that the precipitate is all or part silver iodide. A faint yellow color indicates the precipitate is all or part silver bromide or part silver iodide.

Acidify 8 ml of fusion filtrate with 10% sulfuric acid, boil for 2 min, and cool to room temperature. Add 2 drops of carbon tetrachloride and 3 to 4 drops of 10% sodium nitrite solution and shake the test tube. A purple color in the carbon tetrachloride solution indicates a positive test for iodine.

If iodine is present, treat the entire acidified fusion filtrate with 1 g of sodium nitrite and 2 ml of carbon tetrachloride. For later unknowns only half of the fusion filtrate will be so treated; the other half will be used for tests for nitrogen and sulfur. Remove the carbon tetrachloride with a capillary pipet. Repeat this treatment until the carbon tetrachloride no longer shows any purple iodine color. Boil the solution for a minute and then cool.

To 5 ml of the aqueous solution (fusion filtrate if no iodine is present or solution from sodium nitrite treatment if iodine is present) add 2 drops of commercial laundry-type bleach (such as Chlorox). If the solution is not acidic to pH paper after addition of the bleach, add dilute nitric acid a drop at a time until it is acidic and then add one extra drop. Add 0.5 ml of carbon tetrachloride. A brown color in the carbon tetrachloride indicates bromine. Allow time for the bromine color to develop. If no bromine is present, omit the following treatment with potassium persulfate.

To the remaining aqueous solution add 40 ml of water, 2 ml of concentrated nitric acid, 0.6 g of potassium persulfate, and a boiling chip and boil carefully until the solution has been concentrated to about 30 ml (about 7 min of boiling). Cool the mixture to room temperature. Add silver nitrate reagent to the solution. A white precipitate indicates chlorine. If no other halogen is present, a white precipitate from the original test of fusion filtrate with silver nitrate indicates that the halogen is chlorine (Fig. 23-4).

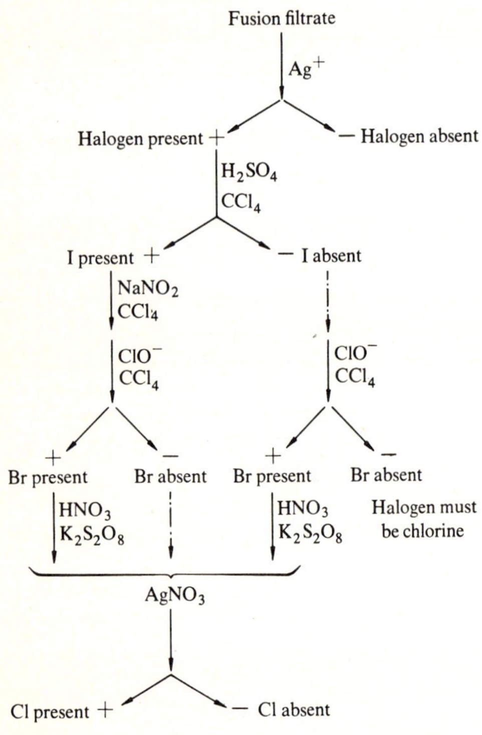

Figure 23-4 Schematic of test of the fusion filtrate for halogen.

CLASSIFICATION TESTS

All compounds that contain chlorine and/or bromine should be tested with sodium iodide in acetone. All compounds that contain chlorine, bromine, and/or iodine should be tested with alcoholic silver nitrate. The tests should not be done on compounds that show no halogen on elemental analysis. Review the portion of your lecture text that discusses *S*N1 and *S*N2 reactions.

Sodium Iodide in Acetone, An *S*N2 Reaction

Discussion.

$$RCl + NaI \xrightarrow{\text{acetone}} RI + NaCl\downarrow \tag{23-1}$$

$$RBr + NaI \xrightarrow{\text{acetone}} RI + NaBr\downarrow \tag{23-2}$$

This test is based on the fact that this is an *S*N2 reaction (Fig. 23-5), that sodium chloride and bromide are insoluble in acetone, and that sodium iodide is soluble in acetone. Iodide ion is a very strong nucleophilic reagent, and compared to water and alcohol, acetone has a low dielectric constant. A medium of low dielectric constant promotes this type of *S*N2 reaction and discourages the corresponding *S*N1 reaction. All these factors assure that the bimolecular mechanism predominates in reactions with this reagent regardless of the structure of the organic halide. Therefore, relative reactivity for various halides with this reagent is methyl > primary > secondary > tertiary > aryl and vinyl. This applies whether or not the primary, secondary, or tertiary halide also happens to be benzylic or allylic. In both *S*N1 and *S*N2 reactions the reactivity of the displaced halogen atom parallels decreasing bond energy and increasing polarizability of the halogen atom. That is, for all nucleophilic displacement reactions the relative rates are RI > RBr > RCl > RF.

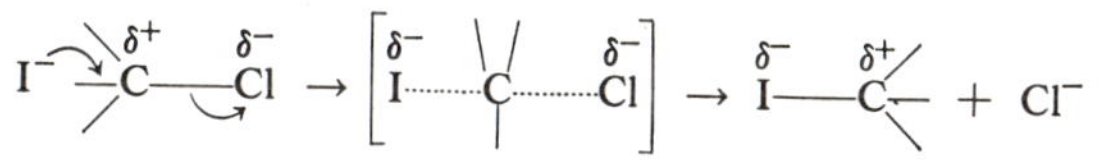

Figure 23-5 An *S*N2 reaction.

Primary bromides give a white precipitate of sodium bromide within 4 min at room temperature. Secondary and tertiary bromides and primary and secondary chlorides react within 8 min at 50°. Tertiary chlorides react only if kept at 50° for twenty-four hours or longer. Aryl and vinyl halides do not react even if left for several days (unless strong electron-withdrawing groups are in the proper positions). Read your lecture text for elaboration on this.

Cyclopentyl halides are slightly more reactive than their secondary and tertiary acyclic counterparts and give precipitates in approximately the same amount of time as similar acyclic halides as outlined above. Cyclopropyl, cyclobutyl, and cyclohexyl halides react more slowly than acyclic secondary halides because they are unable to accommodate the *S*N2 transition state without causing steric interactions such as internal (I) strain. Therefore, with the possible exception of cyclohexyl bromide which may give a small amount of precipitate, none of the cyclopropyl, cyclobutyl, and cyclohexyl halides gives a precipitate with the sodium iodide-acetone reagent at 50° in the allotted time. Compounds with two or more halogen atoms on the *same carbon* atom are less reactive to this reagent than similar compounds with only one halogen on a carbon atom. Chloroform, carbon tetrachloride, 1,1,1-trichloroethane, 1,1-dichloropropane, and similar compounds do not react with sodium iodide in acetone.[2]

[2] The low reactivity of halogens on the same carbon atom with other halogen atoms has been attributed to hyperconjugation as:

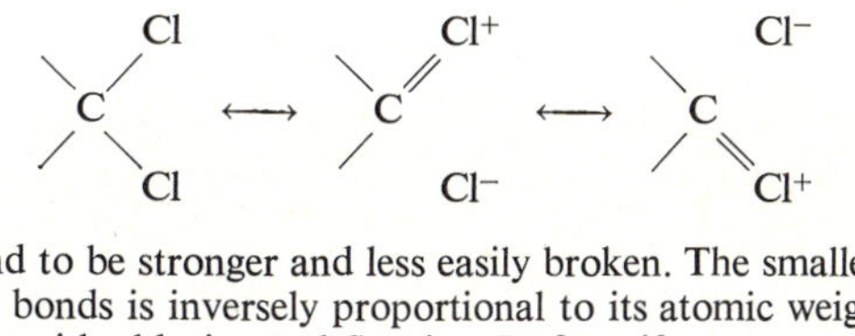

This would cause the carbon-halogen bond to be stronger and less easily broken. The smaller the halogen atom, the more marked the effect because a halogen's ability to form bonds is inversely proportional to its atomic weight. Therefore, this effect is not significant with bromine and iodine but is important with chlorine and fluorine. In fact, if two or more fluorine atoms are on the same carbon atom, they are usually inert to both *S*N1 and *S*N2 reactions.

Some halogen compounds of more complicated structure undergo other reactions in addition to or instead of the usual *S*N1 and *S*N2 reactions. These reactions frequently release iodine, and the resulting solution and precipitate will be reddish-brown:

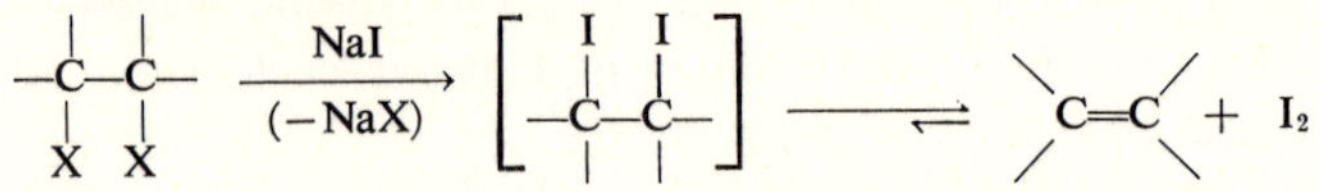

(23-3)

Vicinal dihalides undergo a complex displacement-elimination reaction and give free iodine. The *relative* reactivity of vicinal dihalides towards alcoholic silver nitrates is the same as that for the monohalides (Table 23-5). Triarylmethyl halides immediately give a reddish-brown precipitate. Some polybromo compounds as bromoform and 1,1,2,2-tetrabromoethane give a reddish-brown precipitate within 8 min at 50°. Carbon tetrabromide gives a reddish precipitate within 4 min at room temperature.

TABLE 23-5 RELATIVE REACTIVITIES OF MONOHALIDES AND VICINAL DIHALIDES

$BrCH_2CH_2Br$	ppt. within 2 min at 25°
$CH_3(CH_2)_3Br$	ppt. within 2 min at 25°
$BrCH_2CH_2Cl$	ppt. within 3 min at 25°
$CH_3(CH_2)_3Cl$	ppt. within 4 min at 50°
$ClCH_2CH_2Cl$	ppt. within 4 min at 50°

Experimental Procedure. Dissolve 8 g of sodium iodide in 50 ml of dry acetone and store in a dark bottle away from heat and light. Before use, the solution should be shaken and examined for free iodine. If the reddish-brown color of iodine has developed, the solution should be discarded. A yellow color is normal and does not indicate decomposed reagent.

Dissolve about 0.05 of solid unknown in the minimum volume of dry acetone in a dry test tube. Add 3 drops of this solution or 1 drop of liquid unknown to 1 ml of the reagent. Shake and allow to stand for 4 min at room temperature. If no visible reaction has occurred, place the test tube in a water bath at 50° for 8 min, cool to 20°, and observe whether or not any precipitate has formed. Try this test on ethyl bromide, 2-bromobutane, *t*-butyl chloride, 2-chlorobutane, benzyl chloride, chlorobenzene, and chloroform.

Silver Nitrate in Ethanol, An SN1 Reaction

Discussion.

$$RX + AgNO_3 \rightarrow RONO_2 + AgX \qquad (23\text{-}4)$$

The ease with which an aliphatic halide reacts with alcoholic silver nitrate depends on the ease with which it undergoes an *S*N1 reaction. Usually only *S*N1 reactions occur with this reagent because the nitrate ion is a very poor displacing agent and shows only slightly more tendency than water to act as a nucleophile and because the silver ion helps to pull the halide ion from the carbonium ion. The solvent, 95% ethanol, which is fairly polar and has a moderately high dielectric constant is also more favorable to an

S_N1 reaction than most other organic solvents such as acetone. The reactions that occur with a typical halide are:

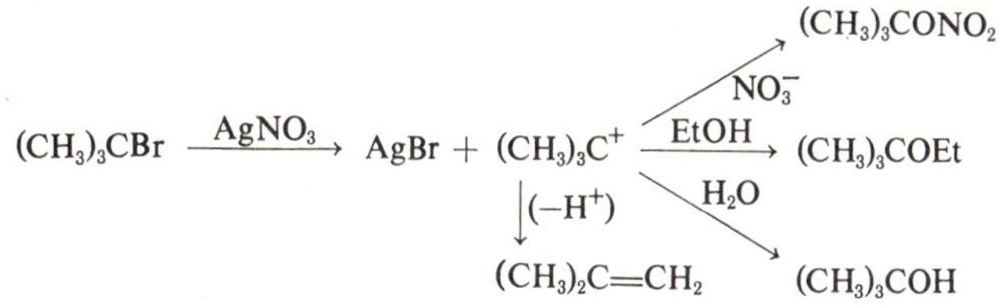

(23-5)

High dielectric constant media promote S_N1 reactions of organic halides (or any S_N1 reaction where an uncharged group ionizes to become an anion) because the transition state of these reactions involves the creation and separation of unlike charges. A solvent of high dielectric constant stabilizes the separation charges in this reaction just as a medium of high dielectric constant between plates of a capacitor stabilizes the separation of charges on that capacitor. See Appendix II. The stabilization of the transition state of reactions similar to those shown in Figure 23-6 by polar or high dielectric solvents is due to the development of ion-dipole attractions in the transition state which do not exist when the reaction is attempted in nonpolar solvents.

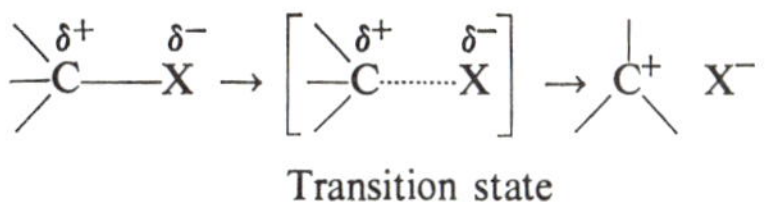

Figure 23-6 S_N1 reaction rate-determining step.

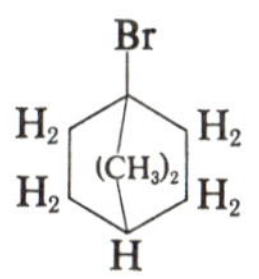

Figure 23-7 Apocamphyl bromide.

The reactivity of organic halides in an S_N1 reaction is allyl, benzyl > tertiary > secondary > primary > methyl > vinyl, aryl. This accounts for the fact that the following give a precipitate within five minutes at room temperature: benzyl, allyl, and tertiary halides; alkyl iodides; monobromoalkanes; 2,4,6-trinitro aryl halides. The following give a precipitate only if boiled with alcoholic silver nitrate: primary and secondary chlorides, 2,4-dinitrophenyl and similar aryl halides, *gem*-dibromides, and *gem*-tribromides. Another group of halides gives no precipitate with alcoholic silver nitrate even when boiled: vinyl halides, aryl halides (unless electron-withdrawing groups are present), alkyl *gem*-dichlorides, alkyl 1,1,1-trichlorides, and carbon tetrachloride. Even though *gem*-dibromo and 1,1,1-tribromoalkanes are less reactive than their monobromo counterparts and give a white precipitate only if boiled, carbon tetrabromide gives a precipitate with alcoholic silver nitrate at 25°.

Problem 23-2. As with the sodium iodide and acetone reagent, alcoholic silver nitrate does not react with compounds with two or more chlorine atoms on the same carbon atom unless the chlorine atoms are on a benzylic or allylic carbon atom. That is, $C_6H_5CCl_3 > C_6H_5CHCl_2 > C_6H_5CH_2Cl$ and $RCH_2Cl > RCHCl_2 > RCCl_3$ in rates of S_N1 reactions. Suggest a reason for the difference in the aliphatic and benzylic series.

Experimental Procedure. Dissolve 2 g of silver nitrate in 100 ml of warm ethanol. Keep the solution in a dark bottle away from heat and light. If the solution becomes gray or brown (colloidal silver) or if more than a very small amount of black silver precipitates, the solution should be discarded.

If the unknown is a solid, dissolve about 0.05 g in the minimum amount of warm ethanol. Allow to cool and add this solution or 1 drop of liquid unknown to 2 ml of the silver nitrate solution. Allow to stand for five minutes and, if no precipitate forms, heat the mixture to boiling and observe. The possibility that a precipitate is the silver salt of an organic acid is eliminated by adding 3 drops of dilute nitric acid which will dissolve these silver salts and not silver halides.

Organic compounds unless very carefully purified usually contain small amounts of impurities which are often isomers. For instance, isobutyl chloride may contain traces of *t*-butyl chloride, and *n*-butyl bromide may contain traces of *sec*-butyl bromide. In both of these examples significant amounts of the impurity could give erroneous results with the alcoholic silver nitrate test. Therefore, in tests in which the unknown gives markedly less precipitate than a known reactive compound, there should be a rough check on percent yield. The molecular weight can be estimated from the physical properties, solubility analysis, and elemental analysis. The weight of unknown to give a certain weight of *dried* silver halide should be determined. If according to these figures the yield is less than 20%, the test should be considered negative. Of course, this precaution applies not only to the silver nitrate test but also to other tests such as sodium iodide in acetone and the iodoform (Chapter 28) tests in which there can be a semiquantitative estimation of yields when the results of the tests are borderline. A mixture containing a small amount of reactive halide impurity can be distinguished from a slowly reacting halide by filtering the test mixture and adding more silver nitrate solution to the filtrate. A precipitate upon this second addition of silver nitrate indicates a positive test. Try this test on butyl bromide, 2-chlorobutane, chlorobenzene, chloroform, and *t*-butyl chloride.

Problem 23-3. Neopentyl chloride, $(CH_3)_3CCH_2Cl$, gives a negative test with sodium iodide in acetone at 50° even though it is a primary chloride. (a) Why? (b) How would neopentyl bromide be expected to react with sodium iodide in acetone? (c) How would neopentyl chloride and neopentyl bromide be expected to react with alcoholic silver nitrate?

Problem 23-4. *S*N1 reactions of alcohols occur in the following steps:

$$ROH + H^+ \rightarrow \overset{+}{R}OH_2 \xrightarrow[(-H_2O)]{} R^+ \xrightarrow{X^-} RX. \qquad (23\text{-}6)$$

Remember that the formation of the carbonium ions is the rate-determining step in an *S*N1 reaction. What would the effect be on the rate of reaction if the solvent were changed from one of low dielectric constant to one of high dielectric constant, all other paramenters being held constant? (Hint: draw the transition state of the rate-determining step.)

Problem 23-5. Although it is a tertiary bromide, apocamphyl bromide (Fig. 23-7) does not react with alcoholic silver nitrate even when warmed for an hour or longer. Why? Would apocamphyl bromide be expected to react with sodium iodide in acetone?

INFLUENCE ON OTHER CLASSIFICATION TESTS BY HALOGEN ATOMS

$H_2SO_4 \cdot SO_3$

Many Class I (sulfuric-acid insoluble) halogen compounds react with fuming sulfuric acid (Chapter 14) in reactions other than aromatic sulfonation. Some bromides and iodides give free halogen. Vicinal dihalides react with fuming sulfuric acid to give a dark, polymeric material. If one or both of the halogen atoms are bromine or iodine, free halogen is liberated. One sequence of reactions that occurs is dehydrohalogenation followed by acid-catalyzed polymerization of the resulting vinyl halide.

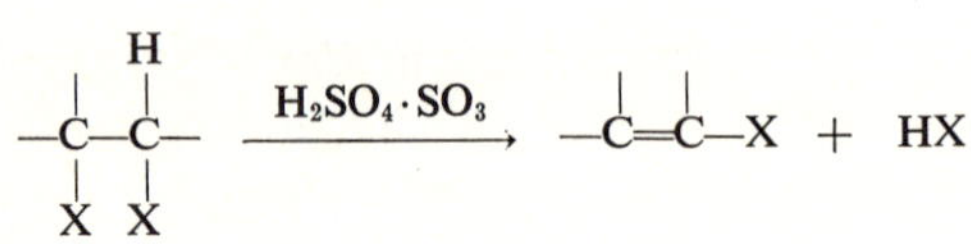

(23-7)

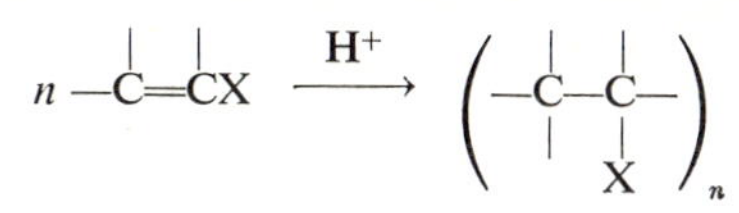

(23-8)

and if X = Br or I,

$$H_2SO_4 + 2HX \rightarrow X_2 + SO_2 + 2H_2O. \qquad (23\text{-}9)$$

$AlCl_3$ and $CHCl_3$

Some aliphatic bromine compounds give a yellow color and many aliphatic iodine compounds give a purple color with this reagent (Chapter 14).

Br_2 in CCl_4

The accumulation of halogen atoms and other electron-withdrawing groups on the vinyl carbon atoms causes the double bonds to which these groups are attached to be less reactive to electrophilic addition. For example, $C_6H_5CCl{=}CClC_6H_5$ does not decolorize bromine in carbon tetrachloride (Chapter 14).

UNKNOWN

The second unknown can be any compound from the tables at the end of Chapter 14 and Tables 23-6 through 23-15. The boiling and/or melting point should be determined and any necessary purification should be done. The elemental analysis is the next step. This is followed by a solubility test in concentrated sulfuric acid. All these indicate which classification tests will give useful information. From the classification tests and the other data the compounds can be identified. Do the following problems before attempting your unknown.

EXERCISES

23-1. In some tables of organic halides the boiling point listed for 1-iodooctane is 225° and for 1-iodononane is 220°. (a) Is this the order you would expect? (b) If one of these were in error, which one would be incorrect? (c) Why? (d) What would you expect the correct boiling point to be?

23-2. A bromine-containing compound is insoluble in concentrated sulfuric acid and boils at 108.5 to 111°. What compounds could it be? What classification test would distinguish among the various possibilities?

23-3. A chlorine-containing compound is insoluble in concentrated sulfuric acid and boils at 158 to 160°. What possible compounds could it be? What classification test would distinguish between them?

TABLE 23-6 ALIPHATIC CHLORIDES (LIQUIDS)

Compound	Physical properties B.p. (^{o}C)	D_4^{20}	n_D^{20}	Derivatives Anilide M.p. (^{o}C)	α-Naphthalide M.p. (^{o}C)
Isopropyl chloride	36.5	0.859	1.378	103	
Methylene chloride	41	1.336	1.4237		
Allyl chloride	44.5	0.940	1.416		
1-Chloropropane	45.5	0.889	1.388	92	121
t-Butyl chloride	51	0.846	1.386	125	147
1,1-Dichloroethane	57	1.175	1.4164		
2-Chloro-1,3-butadiene	59	0.9584	1.458		
Chloroform	61	1.489	1.446		
2-Chlorobutane	68	0.874	1.397	108	129
3-Chloro-2-methylpropene	72	0.9475	1.434		
1,1,1-Trichloroethane	74	1.349	1.4380		
Carbon tetrachloride	77	1.695	1.4630		
1-Chlorobutane	78	0.886	1.4021	63	112
1,2-Dichloroethane	84	1.256	1.4443		
Neopentyl chloride	85	0.879		130.5	
2-Chloro-2methylbutane	86	0.865	1.405	92	138
3-Chloro-1-pentane	93.5	0.8978	1.4254		
3-Chloro-2-methyl-2-butene	94	0.923	1.4320		
1,2-Dichloropropane	96	1.155	1.4388		
3-Chloropentane	97	0.872	1.4082	126	118
1-Chloropentane	106	0.882	1.4121	96	112
1-Chloro-2-pentene	109.4	0.909	1.436		
3-Chloro-2-methylpentane	112	0.863	1.4127	73.5	117
1,1,2-Trichloroethane	114	1.443	1.4707		
		1.121			
Cyclopentyl chloride	114.5	1.005	1.4510		
3-Chlorohexane	123	0.864	1.4163		
1,2-Dichlorobutane	124		1.4401		
1,3-Dichloropropane	125	1.186	1.449		
3-Chloro-2,2,3-trimethylbutane	133				
1,3-Dichloro-2-methylpropane	135.5	1.126	1.4623		
1-Chlorohexane	133	0.878	1.4201	69	106
Chlorocyclohexane	143	0.989	1.4620	146	188
1,1,2,2-Tetrachloroethane	146	1.600	1.4942		
1-Chloro-3,4-dimethylpentane	152	0.883	1.4299	80.5	
1,2,3-Trichloropropane	158	1.417	1.4585		
1-Chloroheptane	159	0.877	1.4261	57	95
Benzyl chloride	179.4	1.20		117	166
1-Chlorooctane	182	0.875	1.431	57	91
β-Chlorostyrene	198	1.109	1.573		
1-Chlorononane	202	0.870	1.434		
p-Chlorobenzyl chloride	213.5			166	
1-Chlorodecane	223	0.868	1.437		
1-Chloroundecane	241	0.868	1.440		
1-Chlorododecane	243.5	0.8673	1.4435		
1-Chlorohexadecane	289d				

TABLE 23-7 AROMATIC CHLORIDES (LIQUIDS)

Compound	Physical properties				Derivatives		
					nitration product		oxidation product (arom. acid)
	b.p. (°C)	m.p. (°C)	D_4^{20}	n_D^{20}	m.p. (°C)	position	m.p. (°C)
Chlorobenzene	132		1.107	1.5250	52	2,4	
2-Chlorotoluene	159		1.082	1.5241	63	3,4	141
1,3-Dichlorobenzene	173		1.288	1.5460	103	4,6	
1-Chloro-3-ethylbenzene	184		1.053	1.5199			158
2-Chloro-1,4-dimethylbenzene	184.5	2	1.058		77 101	5 5,6	320
o-Chlorostyrene	189		1.100	1.5649			141
3-Chloro-1,2-dimethylbenzene	190						186.5
β-Chlorostyrene	198		1.109	1.573			121
2,5-Dichlorotoluene	199	4.3	1.2531		50.5	4	154
3,4-Dichlorotoluene	209		1.2526	1.5471	63	6	206
p-Chlorobenzyl chloride	213.5	29					242.5
3-Chloro-4-isopropyl-1-methylbenzene	217		1.018	1.5171	102.5	2,6	320
1-Chloronaphthalene	259		1.191	1.633	180	4,5	
2-Chlorobenzotrichloride	264	29					
3-Chlorobiphenyl	284.5	16			202.5	4,4	158

TABLE 23-8 ORGANIC CHLORIDES
(SOLIDS)

Compound	Physical properties		Derivatives		
			nitration product	oxidation product (arom. acid)	
	m.p. (°C)	b.p. (°C)	m.p. (°C)	position	m.p. (°C)
p-Chlorobenzyl chloride	29	213.5			242.5
1,1,1,2,2,3,3-Heptachloropropane	29.4	247.5			
2-Chlorobenzotrichloride	29	264			142
2,4,6-Trichlorotoluene	34		53 179	3 3,5	160.5
2-Chlorobiphenyl	34	273			141
2,3,4-Trichlorotoluene	41	231	60 140.5	1 or 6 1,6	186.5
1,2,3,4-Tetrachlorobenzene	44.5	254	64.5 151	5 5,6	
3,4,5-Trichlorotoluene	44.5	244	81.5 163.5	2 2,6	203
2,3,5-Trichlorotoluene	45.6	232	58.5 149.5	4 or 6 4,6	162
1,2,3,5-Tetrachlorobenzene	51	246	40.5 161.5	4 4,6	
1,2,3-Trichlorobenzene	53	218.5	56 92.5	4 4,6	65
1,4-Dichlorobenzene	53	173	54	2	
4,4'-Dichlorodiphenylmethane	55	337	198.5	3,3'	
2-Chloronaphthalene	56	265	173	1,8	
1,3,5-Trichlorobenzene	63	208	68	2	
1,7-Dichloronaphthalene	63.5	286	138.5		
1-Chloro-4-bromobenzene	67.4	197	72	2	
1,4-Dichloronaphthalene	68	287	92	8	
Octachloropropane	160	271			
Hexachloroethane	187	sub.			

TABLE 23-9 ALIPHATIC BROMIDES
(LIQUIDS)

Compound	Physical properties			Derivatives	
	B.p. (°C)	D_4^{20}	n_D^{20}	anilide M.p. (°C)	α-naphthalide M.p. (°C)
Ethyl bromide	38	1.460	1.4252	104	126
2-Bromopropane	60	1.314	1.4250	103	
1-Bromopropane	71	1.353	1.4341	92	121
Allyl bromide	71	1.398	1.4655	114	
t-Butyl bromide	72.5	1.211	1.4283	128	147
Isobutyl bromide	91	1.253	1.4351	109	125
Dibromomethane	98.5	2.496	1.5380		
1-Bromobutane	101	1.274	1.4403	63	112
2-Bromo-2-methylbutane	108	1.196	1.442	92	138
1,1-Dibromoethane	112	2.055	1.5128		
2-Bromopentane	116	1.212	1.442	93	103
1-Bromo-3-methylbutane	120.5	1.213	1.442	108	111
1-Bromopentane	129	1.219	1.445	96	112
1,2-Dibromoethane	132	2.179	1.5379		
2-Bromohexane	146	1.166	1.4852	91.5	
1,2-Dibromo-2-methylpropane	149	1.785	1.512		
Bromoform	151	2.890	1.598		
1-Bromohexane	156	1.175	1.448	69	106
2,3-Dibromobutane	157	1.792	1.515		
Cyclohexyl bromide	165	1.336	1.495	146	188
1,2-Dibromobutane	166	1.820	1.515		
1,3-Dibromopropane	167.5	1.982	1.5231		
1-Bromoheptane	179	1.140	1.451	57	95
1,1,2-Tribromoethane	189	2.621	1.5933		
Benzyl bromide	198	1.438	1.4388	117	166
1-Bromooctane	202	1.112	1.453	57	91
β-Phenylethyl bromide	218	1.359	1.556	97	
1,2,3-Tribromopropane	220	2.402	1.582		
1,1,2,2-Tetrabromoethane	244	2.967	1.638		
1,8-Dibromooctane	271	1.461	1.499		
1,9-Dibromononane	287	1.416			

TABLE 23-10 AROMATIC BROMIDES
(LIQUIDS)

Compound	Physical properties			Derivatives		
				nitration product		oxidation product (arom. acid)
	b.p. (^{o}C)	D_4^{20}	n_D^{20}	m.p. (^{o}C)	position	m.p. (^{o}C)
Bromobenzene	156	1.494	1.560	71	2,3	
2-Bromotoluene	162	1.425	1.4388	82	3,5	150
Benzyl bromide	198	1.438				121
o-Bromoethylbenzene	199	1.355	1.5486			150
o-Bromostyrene	210	1.4160	1.5927			150
1-Bromo-2-isopropylbenzene	210	1.3020	1.5408			150
1,3-Dibromobenzene	218.5	1.952	1.606	61	4	
β-Phenylethyl bromide	218.5	1.359	1.556			121
2-Bromocymene	234	1.267		97		
1-Bromonaphthalene	281	1.484	1.658	85	4	
2-Bromobiphenyl	297					150

TABLE 23-11 ORGANIC BROMIDES
(SOLIDS)

Compound	Physical Properties		Derivatives		
			nitration product		oxidation product (Arom. acid)
	m.p. (^{o}C)	b.p. (^{o}C)	m.p. (^{o}C)	position	m.p. (^{o}C)
2-Bromo-2,3-dimethylbutane	24.5	134			
4-Bromotoluene	28.5	184			251.5
p-Bromobenzyl bromide	41				251.5
2-Bromonaphthalene	59	281			
1,2-Dibromonaphthalene	67				196
1,2,2,4-Tetrabromobutane	72.5				
1,4-Dibromonaphthalene	82				135
1,4-Dibromobenzene	89	219	84	2,5	
4-Bromobiphenyl	89	310			251
1,3,5-Tribromobenzene	120	271			
4,4'-Dibromobiphenyl	164				251
1,2,4,5-Tetrabromobenzene	180		168	3	

TABLE 23-12 ORGANIC IODINE COMPOUNDS (LIQUIDS)

Compound	Physical properties			Derivatives	
	b.p. (°C)	D_4^{20}	n_D^{20}	Anilide m.p. (°C)	α—naphthalide m.p. (°C)
Methyl iodide	43	2.282	1.532	114	160
Vinyl iodide	56			104	
Ethyl iodide	72	1.940	1.514	104	126
2-Iodopropane	90	1.703	1.499	103	
1-Iodopropane	102.5	1.743	1.505	92	121
Allyl iodide	103	1.777	1.578	114	121
Isobutyl iodide	120	1.602	1.496	109	125
1-Iodobutane	130	1.616	1.499	63	112
2-Iodopentane	142	1.510	1.496	93	
1-Iodopentane	155	1.512	1.496	96	112
Cyclopentyl iodide	166.5	1.710	1.5447		
Cyclohexyl iodide	179	1.616		146	188

Compound	b.p. (°C)	D_4^{20}	n_D^{20}	Nitration product m.p. (°C)	position	Oxidation product (arom. acid) m.p. (°C)
Iodobenzene	186.5	1.831	1.620	171	4	
3-Iodotoluene	204	1.698		108	4,6	187
1-Iodoheptane	204	1.373	1.489			
2-Iodotoluene	211	1.698		103	6	162
1-Iodooctane	255.5	1.330	1.489			
4-Iodo-1-isopropylbenzene	237					
1-Iodonaphthalene	305					

TABLE 23-13 ORGANIC IODINE COMPOUNDS
(SOLIDS)

Compound	Physical properties		Derivative nitration product	
	M.p(°C)	B.p. (°C)	M.p. (°C)	Position
1-Iodohexadecane	22			
Benzyl iodide	24			
4-Iodotoluene	35	211		
5-Iodo-1, 2, 4-trimethylbenzene	37	257		
1, 3-Diiodobenzene	40	285		
2-Iodonaphthalene	55	309		
1, 1-Diiodoethane	81			
4-Iodobiphenyl	114	320d		
Iodoform	119			
1, 4-Diiodobenzene	129	289	171	2, 5

TABLE 23-14 ORGANIC HALIDES CONTAINING TWO OR MORE DIFFERENT HALOGEN ATOMS
(LIQUIDS)

Compound	Physical properties			
	B.p. (°C)	M.p. (°C)	D_4^{20}	n_D^{20}
Chlorobromomethane	68.5		1.99	1.488
1-Chloro-1-bromoethane	83		1.65	
1-Chloro-2-bromoethane	106.5	-16.6	1.688	1.4917
Chloroiodomethane	109		2.49	1.5822
1-Chloro-1-iodoethane	118		2.053	
1-Chloro-1, 1-dibromoethane	122		2.12	
Dichlorodibromomethane	135	22	2.43	
1-Chloro-2-iodoethane	140	-15.6	2.123	
1-Chloro-3-bromopropane	144.6		1.594	1.4861
1-Chloro-3-bromobenzene	195.5	-21	1.630	1.576
Chlorodiiodomethane	200d	-4	3.172	

TABLE 23-15 ORGANIC HALIDES CONTAINING TWO OR MORE DIFFERENT HALOGEN ATOMS (SOLIDS)

Compound	Physical properties		Derivative Oxidation product (arom. acid)
	M.p. (°C)	B.p. (°C)	M.p. (°C)
Dichlorodibromomethane	22	135	
1-Chloro-1,1,2,2-tetrabromoethane	32.5		
_p_-Bromobenzyl bromide	41		252
_p_-Chlorobenzyl bromide	51		242
Chlorotribromoethane	55	160	
Bromodiiodomethane	60		
1-Chloro-4-bromobenzene	67.4	197	
Chloropentabromoethane	170d		

24

The Grignard Reaction

Synthesis of 3-Methyl-3-heptanol

$$CH_3(CH_2)_3Br + Mg \xrightarrow{Et_2O} CH_3(CH_2)_3MgBr$$

B.p. 102°
D_4^{20} 1.277

$$CH_3(CH_2)_3MgBr + CH_3COCH_2CH_3 \rightarrow CH_3(CH_2)_3(CH_3)\underset{\substack{|\\ C_2H_5}}{C}O^-Mg^{++}Br^-$$

(for $CH_3COCH_2CH_3$) B.p. 79.6°
D_4^{20} 0.805

$$CH_3(CH_2)_3(CH_3)\underset{\substack{|\\ C_2H_5}}{C}OMg^+Br^- + H_3O^+ \rightarrow CH_3(CH_2)_3(CH_3)\overset{\substack{H\\ O\\ |}}{C}C_2H_5 + Mg^{++}Br^- + H_2O$$

(for the product) B.p. 162°
D_4^{20} 0.822

INTRODUCTION

The Grignard reaction can be used to synthesize most monofunctional alcohols and many polyfunctional alcohols. In this experiment a tertiary alcohol is synthesized from methyl ethyl ketone and *n*-butylmagnesium bromide. It is very important to keep the reactants and the apparatus dry to ensure that the reaction occurs and that the maximum yield of pure product can be obtained.

EXPERIMENTAL PROCEDURE

Danger! Ether is a very flammable solvent. There should be no flames in the laboratory when this large a quantity of ether is being used. Steam or heating mantles should be used as heat sources. If these are not available, hot water may be used as the heat source. Hot plates with enclosed heating coils may be used to heat large beakers or pans of water in a section of the laboratory remote from where the ether is being used. The hot plates should be set on "low." If only burner flames are available as a primary heat source, the water should be heated in a room close to but separate from the laboratory.

Using a 250-ml round bottom flask, construct a reflux and addition apparatus similar to that shown in

Figure 16-1 except that a calcium chloride tube should be placed in the top of the condenser instead of a cork and tube leading to a hood. Rub 5 g of clean, dry magnesium turnings with steel wool and put them in the flask.

A method of ensuring a dry reaction flask is to heat the flask and magnesium. Disconnect the flask containing the magnesium from the Claisen head and heat the flask directly with an air-starved, yellow Bunsen flame (the hotter, blue flame may crack the flask) for a few minutes. Reconnect the hot flask to the apparatus. *Extinguish the Bunsen flame* and allow the flask to cool. The entire class must do this *before any* of the ether is opened. Any late arrivals will have to omit this heating.

Dissolve 20 ml of *n*-butyl bromide in 50 ml of *anhydrous* ether. Since ether rapidly absorbs moisture from the air, do not use ether from a container that has been opened before the day of this synthesis. Place the ether solution in the dropping funnel and stopper the funnel. Add a crystal of iodine to the flask. Prepare an ice water bath; this is to be placed around the reaction flask to moderate the reaction if it proceeds too rapidly and the ether boils too vigorously.

Run 7 to 10 ml of the *n*-butyl bromide solution onto the magnesium. The reaction starts in a few minutes if all traces of moisture have been kept from the reaction mixture. Cool the reaction if the mixture boils violently. If any product at all is to be produced by this synthesis, it is very important that this reaction occurs before any more solution is added. If the *n*-butyl bromide does not react, heat the flask with a hot water bath. If even this does not start the reaction, notify the instructor and he will add a small amount of methylmagnesium iodide.

Add 40 ml more of anhydrous ether to the dropping funnel and briefly and vigorously stir the mixture with a glass rod. When reaction of the magnesium with the first 7 to 10 ml of butyl bromide has moderated, allow the butyl bromide-ether solution to drip slowly into the flask. Set the rate of addition so that heat from the reaction causes moderate refluxing of the ether. Stir the mixture every 2 to 3 min by applying a gentle rotary motion to the ring stand. After all the solution has been added, reflux the mixture for 25 min. Heat gently, otherwise some of the ether will escape from the condenser and be a fire and health hazard. If almost all the magnesium has not reacted, continue refluxing the mixture until it does react.

Problem 24-1. What side reactions may occur before the methyl ethyl ketone is added? What is done in this experiment to prevent these reactions? What else could be done to prevent these side reactions?

Problem 24-2. Other than increasing the rate of reaction what is the purpose of refluxing the ethyl ether?

In the next step low temperature, slow addition of ketone, and effective mixing are essential for a good yield of product. Pour most of the water from the ice water cooling bath and add some calcium chloride and more ice. Be sure that the stopcock is closed on the separatory funnel. Place the ice-bath around the reaction flask. Pour 19 ml of methyl ethyl ketone and 35 ml of ether into the dropping funnel, mix briefly and vigorously with a glass rod, and stopper the funnel. Run this solution into the cold Grignard mixture; adjust the rate of addition so that there is a rapid flow of *individual drops.* Agitate the flask by applying a gentle rotary motion to the ring stand. The motion should be vigorous enough to cause all or almost all of the white salt that precipitates to dissolve. When all the solution has been added, remove the cooling bath and replace it with a heating bath that is *initially* at 32 to 35°. Agitate the mixture every 3 to 4 min for the next 15 min.

Put approximately 60 ml of water, 60 g of crushed ice, and 20 ml of glacial acetic acid into the large separatory funnel. Pour the reaction mixture into the separatory funnel. Rinse the flask with two 20-ml washes of 10% aqueous acetic acid and pour the rinses into the separatory funnel. Gently shake the mixture. The magnesium alkoxide is hydrolyzed very rapidly by water. The acetic acid is added to produce water-soluble magnesium acetate rather than a voluminous precipitate of magnesium hydroxide.

As soon as the ice has melted, shake the funnel vigorously and briefly and separate the two layers. It is especially important in this experiment that the separations be as sharp and effective as possible; review page 38. Save the aqueous layer until it is extracted with ether. Wash the ether layer with 50 ml of 5% sodium bisulfite solution, 50 ml of 5% sodium carbonate solution, and 100 ml of saturated sodium chloride solution. Collect each aqueous layer in a separate container and do not discard until each has been extracted with ether.

Add about 10 g of anhydrous potassium carbonate to a 250-ml Erlenmeyer flask and transfer the ether solution to this flask. Stir with a glass rod and stopper the flask. Rinse the reaction flask with 40 ml of ether (does not have to be anhydrous) and pour the wash into the separatory funnel. Use this to extract successively the aqueous bisulfite, carbonate, and chloride solutions that were used to wash the ether solution. Combine this ether wash with the ether being dried by potassium carbonate. Stir the mixture occasionally for 20 min or longer while disassembling and cleaning your apparatus and constructing the distillation apparatus shown in Figure 5-5. Use the *dried* 250-ml round bottom flask.

Place a boiling chip in the 250-ml flask and decant the ether solution from the potassium carbonate into it. Heat the mixture with hot water or steam until no more liquid distils. If a water bath is used, change it so it is at least 75° just before you stop the distillation. Cool the flask in an ice bath. If any water is evident (two liquid phases or a cloudiness), add 2 to 4 g of potassium carbonate. Pour the liquid (decant carefully from the potassium carbonate if it was used) into a 50-ml round bottom flask with a boiling chip in it. Use 10 ml of distillate from the first distillation to rinse any material remaining in the 250-ml flask into the 50-ml flask. Place the 50-ml flask under the fractionating column and heat again with steam or a water bath that is at least 75° until no more of the mixture distils. Dispose of the distillate into the waste ether containers.

The mixture will now need to be heated to a higher temperature. If heating mantles are not used, **do not proceed further until everyone in the laboratory has completed the ether distillation,** and all the ether distillates are in the used ether containers. Distil the mixture and collect the fraction that boils at 157 to 165° in a tared yield bottle. Weigh the product, calculate the % yield, label the bottle, and hand it in to your instructor.

GAS CHROMATOGRAPHY OPTION

Chromatograph 10 μl of (a) the purified product, (b) the material in the 50-ml flask after the ether is removed and just before the final high temperature fractional distillation, (c) any material that distils through the fractionating column below 157° during the final high temperature distillation, and (d) the last drops of ether that distilled from the 250-ml flask. Use the Apiezon column and the conditions described in Table 8-1. If another gas chromatograph with a Carbowax column is available, chromatograph the above four fractions on this column also. Use the conditions described in Table 8-1 except set the column even temperature at 85° (for the Carbowax column only). From retention times and peak height data supplied by the instructor and obtained on standards you chromatograph, identify all the compounds in each fraction and determine the relative amount of each.

EXERCISES

24-1. What by-product is likely to be produced in the isolation and purification of the product? What steps in the procedure minimize the formation of this compound?

24-2. What other by-products are likely? Write equations for their formation.

24-3. The product actually consists of two compounds. What are they and how much of each did you isolate?

25

Adsorption Column Chromatography

Acylation of Ferrocene

$$(C_5H_5)_2Fe + (CH_3CO)_2O \xrightarrow{H_3PO_4} C_5H_5FeC_5H_4COCH_3$$

M.p. 173° B.p. 136.5°, D_4^{20} 1.082 M.p. 85–86°

INTRODUCTION

In Chapter 8 separation and analysis by chromatography was introduced, and the primary concern was with gas-liquid chromatography or gas partition chromatography. This chapter introduces liquid-solid adsorption chromatography.

Many compounds that show aromatic properties are not closely related to benzene. Their main resemblance to benzene is that they have clouds of 6 delocalized π electrons. One such compound is ferrocene (Fig. 25-1).

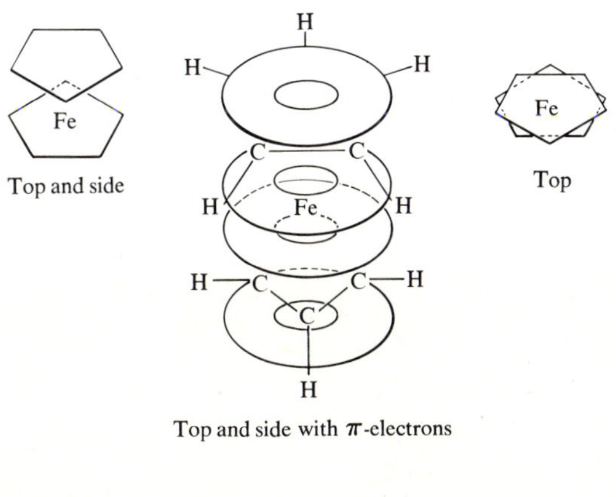

Top and side   Top and side with π-electrons   Top

Figure 25-1 Three views of ferrocene.

Like other aromatic compounds ferrocene undergoes electrophilic substitution. One of these substitutions, Friedel-Crafts acylation with acetic anhydride, is demonstrated in this chapter.

DISCUSSION OF CHROMATOGRAPHY

Description

Compounds can be separated and purified and mixtures analyzed by selective adsorption. One method of accomplishing this is to pack a column or tube with an appropriate adsorbent (material with a large surface area). Table 25-1 gives some typical adsorbents in order of their increasing adsorbing ability. The order may be changed slightly when different groups of compounds are adsorbed.

TABLE 25-1 ADSORBENTS

Carbon black*	Very strong adsorbent
Magnesium silicate	
Alumina, Al_2O_3	
Silica gel	
Aluminum hydroxide, $Al(OH)_3$	decreasing
Calcium carbonate	absorbent
Calcium sulfate	power
Talc	
Sugar	
Powdered cellulose	weak adsorbent

*A weaker adsorbent for compounds that can hydrogen-bond. Adsorbent power more dependent upon method of activation and compounds being chromatographed than other adsorbents.

The substance to be separated is deposited on the adsorbent. Solvents of increasing eluant power are passed through the column, and the compounds are eluted down the column. The least strongly adsorbed compound is eluted first; the most strongly adsorbed compound is eluted last. The more polar the compounds and the stronger their hydrogen bonds with the adsorbent, the more strongly they are adsorbed Table 25-2 gives the order of elution of groups of compounds from silica.

TABLE 25-2 APPROXIMATE ORDER OF ELUTION OF COMPOUNDS*

Alkanes	most quickly eluted
Alkenes	
Aromatic hydrocarbons	
Ethers	
Esters	
Ketones	
Aldehydes	increasing
Alcohols	elution
Amides except N, N-disubstituted	time
Diols	
Carboxylic acids	
Ketocarboxylic acids	
Hydroxycarboxylic acids	least quickly eluted

*From adsorbents given in Table 25-1 except for carbon black.

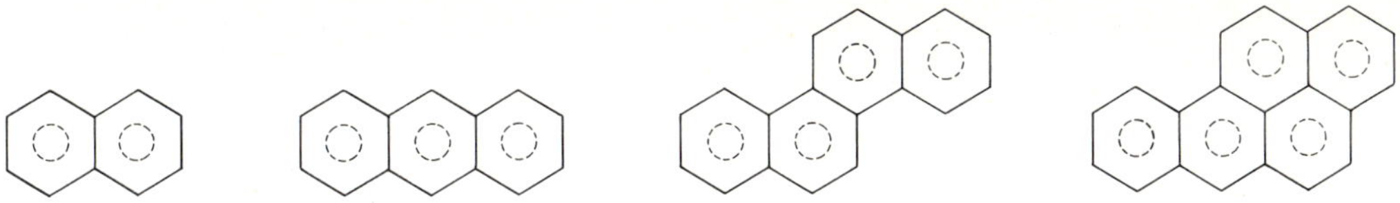

Naphthalene | Anthracene | Chrysene | 3,4-Benzpyrene

Figure 25-2 Structure of polynuclear aromatic hydrocarbons.

Problem 25-1. Arrange the following solid aromatic hydrocarbons in order of their elution from alumina: anthracene, chrysene, hexamethylbenzene, naphthalene, 3,4-benzpyrene. See Figure 25-2 for their structures.

Problem 25-2. As can be seen from the footnotes to Tables 25-1 and 25-2, carbon black shows markedly different properties than the other adsorbents. Suggest an explanation for this.

The solvents displace the adsorbed molecules by forming stronger attractions with the adsorbent than the molecules they displace.[1] The elution power of a solvent is a function of its polarity and on most adsorbents, such as aluminum oxide and magnesium silicate, of its hydrogen-bonding ability. The more polar the solvent and the more it can hydrogen-bond with the adsorbent, the stronger its eluting power.

TABLE 25-3 ELUENT POWER OF SOLVENTS

Solvent	
Petroleum ether, ligroin*	low polarity, weak eluents
Cyclohexane	
Carbon disulfide	
Carbon tetrachloride	
Benzene	
Methylene chloride	
Chloroform	
Ethyl ether	
Tetrahydrofuran	increasing
Ethyl acetate	polarity
Methyl ethyl ketone	and
Acetone	H-bonding
1-Butanol	
1-Propanol	
Ethanol	
Methanol	
Phenol	
Water	
Acetic acid	
Pyridine	high polarity, strong eluents ↓

*Increases with boiling point of fraction used.

Problem 25-3. Arrange the following series of solvents in order of their increasing eluting power from alumina: (a) methylene chloride, carbon tetrachloride, chloroform; (b) methanol, ethanol, 1-propanol, 1-butanol; (c) water, acetone, cyclohexane, acetic acid, benzene, methanol. Verify your answers in Table 25-3. Explain your answer to part a.

[1] There is much more solvent than compound being separated. Because of mass action a solvent may displace a compound even though the solvent has *slightly* weaker attractive forces for the adsorbent than the displaced compound. In such a situation the compound is eluted down the column very slowly.

Problem 25-4. Clarification (pp. 81, 83, and 86) has been compared to a single plate adsorption chromatography. Explain.

Reproducibility

Because of the strong displacing power of polar solvents very small amounts of polar contaminates in nonpolar solvents can elute many compounds that would not be eluted by pure nonpolar solvents. For this reason solvents should be free of even traces of compounds of significantly higher polarity. Contamination of benzene by as little as ½% of ethanol or methanol would drastically increase its eluting power for many compounds. The amount of toluene present contaminating benzene could be as much as 5% without significantly affecting the eluting power of benzene. Contamination by water also greatly increases the eluting power of nonpolar and moderately polar solvents. Contamination of solvents by water is probably the principal reason for nonreproducibility of some column adsorption chromatograms.

If adsorption chromatographies are to be reproducible, all water must be kept from the adsorbent and solvents or a known, reproducible amount must be present. If the adsorbent is contaminated with polar compounds, such as water, these polar contaminants are strongly adsorbed. Other less polar molecules can not be adsorbed where the polar contaminants are adsorbed and are eluted past them.

Efficiency

Approximately 30 g of adsorbent is required for each gram of material to be chromatographed. A slightly lower ratio of adsorbent to compound can be used if the compounds to be separated are widely different in their polarities, and a much larger ratio is needed to separate compounds that have almost identical attractions for the adsorbent. The column becomes overloaded and the efficiency of separation is greatly reduced if there is not enough adsorbent to keep the different compounds in separate zones.

A long thin column has more theoretical plates and is more effective at separation provided the ratio of adsorbent to compounds to be separated is sufficiently high. A short wide column allows faster separation and the use of a somewhat lower ratio of adsorbent to compounds to be separated. The column should, of course, be long enough to separate the compounds being chromatographed.

Gas bubbles and cracks in the column greatly reduce efficiency because these voids bypass significant amounts of adsorbent and cause channeling. Since the voids are almost never horizontally uniform, they also cause uneven flow down the column, and this often causes considerable overlap of the compounds being separated. Bubbles and cracks can be minimized by proper packing procedures, one of which is given in the experimental section of this chapter.

Large variations in temperature may cause small but significant differences in the ease of elution of many compounds. More serious, variations in temperature may cause some of the solvent to vaporize in the column and form gas bubbles. To avoid this a solvent should not be used (a) in any concentration if its boiling point is not at least 3° above room temperature, (b) in a concentration of greater than 10% if its boiling point is not at least 5° above room temperature, and (c) in a concentration of greater than 30% if its boiling point is not at least 10° above room temperature.

Comparison with Gas Chromatography

The number of theoretical plates possible in liquid-solid adsorption column chromatography is not nearly so high as the number possible with gas chromatography. Comparable separations usually take significantly longer with column chromatography. Liquid-solid column chromatography can not be used with so small quantities as gas chromatography nor can it be used to separate gases and volatile liquids.

Even with these disadvantages many mixtures can be separated only by or better by liquid-solid column chromatography because:

a. Nonvolatile compounds and compounds that decompose at moderately elevated temperatures can be easily separated by liquid-solid adsorption chromatography.
b. Column chromatography can readily be used to separate large quantities.
c. The equipment for column chromatography is considerably less expensive than that required for gas

chromatography. Some new techniques use equipment that is comparable with GLC instruments in both cost and speed.

Comparison with Recrystallization

Column chromatography is often used for purifications that might also be attempted by recrystallization. Column chromatography is usually the better choice unless the compounds have widely different solubilities or a significant loss in the recovery of the compounds can be tolerated. The superiority of chromatography is due not only to the large number of possible theoretical plates that can be utilized without the expenditure of considerable effort but also to the high selectivity in adsorption and elution of many compounds that have similar solubilities. This selectivity can be changed completely or modified only slightly by varying adsorbents or eluting solvents. Note the similarity of this comparison to that between gas chromatography and fractional distillation (p. 54).

EXPERIMENTAL PROCEDURE

Acetylation of Ferrocene

In a 25-ml round bottom flask add 4.0 g of ferrocene, 12 ml of acetic anhydride, and 2.5 ml of 85% phosphoric acid.[2] Equip the flask with a reflux condenser with a calcium chloride drying tube in the top. Heat the mixture on a steam bath for 10 min. During this heating period and other waiting periods prescribed in this section prepare the chromatography column as described below.

Pour the mixture onto 50 g of ice in a 100-ml beaker in an ice bath. Rinse the flask with 5 ml of ice water; pour the rinse into the beaker. Stir a few minutes with a glass rod, cautiously add 5 pellets of sodium hydroxide, and then bring the mixture to pH 7 (Hydroin paper) by adding solid sodium bicarbonate (be careful of foaming). Allow the mixture to stand in the ice bath for 30 to 40 min and suction filter. Discard the filtrate and dry the precipitate in the funnel by pulling air through it for 15 min.

Problem 25-5. What does this dried product contain?

Problem 25-6. What is the order of elution of the compounds given in the answer to Problem 25-5?

Preparation of a Chromatography Column

There should be no flames in the laboratory while preparing the columns, chromatographing the reaction mixtures, and evaporating the eluants. If steam and heating mantles are not available, use the heat sources described on page 188.

In a column approximately 2 cm in diameter and 32 to 45 cm[3] in length place a small amount of glass wool. If the bottom of the column has a small fritted disk, the glass wool should be omitted. Close the bottom of the column. If the column is not fitted with a stopcock, a piece of tubing with a clamp can be used. Fill the column with petroleum ether (b.p. 40 to 60° at 760 mm) or ligroin (b.p. 35 to 60° at 760 mm).[4] Add about ½ in of clean sand.[5] Align the column parallel to the earth's gravitational pull. Use a plumb line if necessary. Smooth

[2] If only a 50-ml buret is available for use as a column, the following quantities should be used: ferrocene, 3 g; acetic anhydride, 9 ml; phosphoric acid, 2 ml; and crushed ice, 40 g.

[3] If such a column is not available, a 50-ml buret can be used. Chromatography in this narrower, higher column requires more time than the column suggested above and not as much material can be chromatographed.

[4] If petroleum ether or ligroin in these boiling ranges is unavailable or if altitude or room temperature contraindicates use of solvents with such a low boiling range (p. 194), one of the following commercially available alkanes or alkane mixtures should be used. Wherever petroleum ether is called for in this chapter, the alkane solvent selected from this list should be used. Listed in order of preference, they are: hexane, cyclohexane, ligroin (b.p. 66 to 75°), naphtha solvent (b.p. 60 to 90°), heptane, and petroleum ether (b.p. 60 to 110°). If technical grade ligroin is used, wash it with aqueous permanganate and dry. If any of the last three is used and the only heat source available is remotely heated hot water, vacuum distillation (Chapter 29) will be necessary to remove the solvent from the eluted compounds.

[5] Omit the sand on the *bottom* of the column if a Teflon plug or valve is used to control the flow at the bottom of the column.

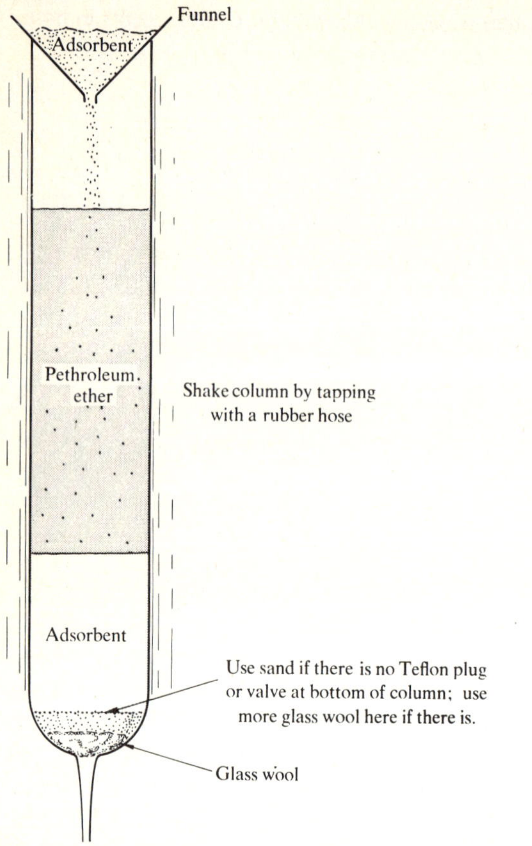

Figure 25-3 Filling a chromatography column with adsorbent.

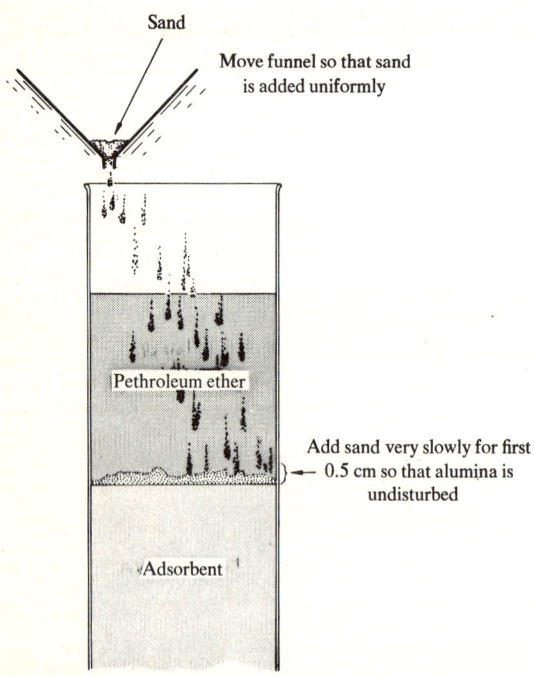

Figure 25-4 Placement of sand on top of an alumina column.

the surface of the sand by tapping the column with a piece of vacuum tubing. Tap first one side and then the other in the middle of the column. Be careful not to change the vertical alignment of the column.

Avoid all excessive exposure of alumina to air. Keep the bottles tightly capped except when actually removing adsorbent. Place a stemless funnel in the top of the column. Slowly add 140 g (115 g if a 50-ml buret is used) of acid-washed alumina to the column through the funnel (Fig. 25-3). While the alumina is going into the column, tap the column with the rubber tubing as described above. Allow the petroleum ether to flow slowly out of the column in a rapid stream of drops; however, maintain the level of petroleum ether well above the top of the settling alumina by adding more petroleum ether when necessary. The petroleum ether that flows from the bottom of the column during this packing should be reused. To avoid the development of cracks never allow any portion of the alumina to dry. Always keep a generous layer of solvent or wet sand on top of the alumina while preparing the column and during the chromatography.

When all the alumina has been added, fill the column to within 3 cm of the top with petroleum ether and close the stopcock. Gently stir the top and allow the alumina to settle so that its top forms as a planar horizontal surface. Slowly add sand a few grains at a time (Fig. 25-4) so that the top of the alumina is not disturbed. When approximately 0.5 cm of sand is on top of the alumina, the sand can be added more rapidly without disturbing the alumina. Add sand until there is a 2 to 3 cm layer on top of the alumina. Seal the top of the column with a *cork* stopper. If at any time during the chromatography it is necessary to leave the column, close the stopcock or clamp, place a 250-ml beaker under the column, fill the column almost full with the solvent being used, and tightly stopper it.

Chromatography of the Reaction Mixture

Dissolve as much of the dried reaction mixture as possible in 15 ml of warm petroleum ether in a 25-ml Erlenmeyer flask. Drain the column until only the bottom two-thirds of the top *sand* layer is wet with solvent. Pour[6] the petroleum ether solution on the column. During the transfer allow the column to drain so that the bottom half of the top sand layer never dries and the solution of the reaction mixture never is more than 0.5 cm above the sand layer (Fig. 25-5). When all the petroleum ether solution has been transferred, dissolve the petroleum ether-insoluble residue in the *minimum* amount of warm benzene (6 to 10 ml). Transfer the benzene solution to the chromatography column using the technique described above. Rinse first the flask (and then the pipet if it was used, with the same wash) with 5 ml of warm petroleum ether. Transfer the wash to the sand on the column.

Add another 1 to 1.5 cm of sand to the top of the column. Fill the column with petroleum ether and adjust so that 1 to 1.5 drops a second flow from the bottom. Add a 1-hole stopper, a ring support, and a separatory funnel to the column to construct the apparatus shown in Figure 25-6. Add 100 ml of petroleum ether to the separatory funnel. Open the stopcock on the separatory funnel, and use a cork with a slit in it[7] on top of the separatory funnel. The first 20 ml (and only the first 20 ml) of petroleum ether that is collected after the reaction mixture is placed on the column should be used to make the first ethyl ether-petroleum ether mixture.

Adjust the column stopcock so that an almost continuous stream of drops flows rapidly from the bottom of the column. If this does not occur even when the clamp or stopcock is wide open, increase the head of the solvent to increase the pressure. Do this by raising the tip of the separatory funnel until it is just below the bottom of the cork on the column. Elute with petroleum ether until the unreacted ferrocene (yellow band) begins to come off the column. Collect all the ferrocene eluant in a round bottom flask.

When the ferrocene begins to come off the column, change the solvent to 50 ml of 2% ethyl ether–98% petroleum ether.[8] Each time the eluant is changed, the remaining solvent in the column should be drained to

[6] If a 50-ml buret is used the solvent should be transferred with a capillary-tipped pipet.

[7] If the cork on top of the column is not air tight, the separatory funnel should be tightly stoppered with a ground-glass stopper to prevent the column from overflowing.

[8] If altitude or room temperature does not allow the use of ethyl ether (b.p. 34.6°), dried tetrahydrofuran (b.p. 65°) should be used instead.

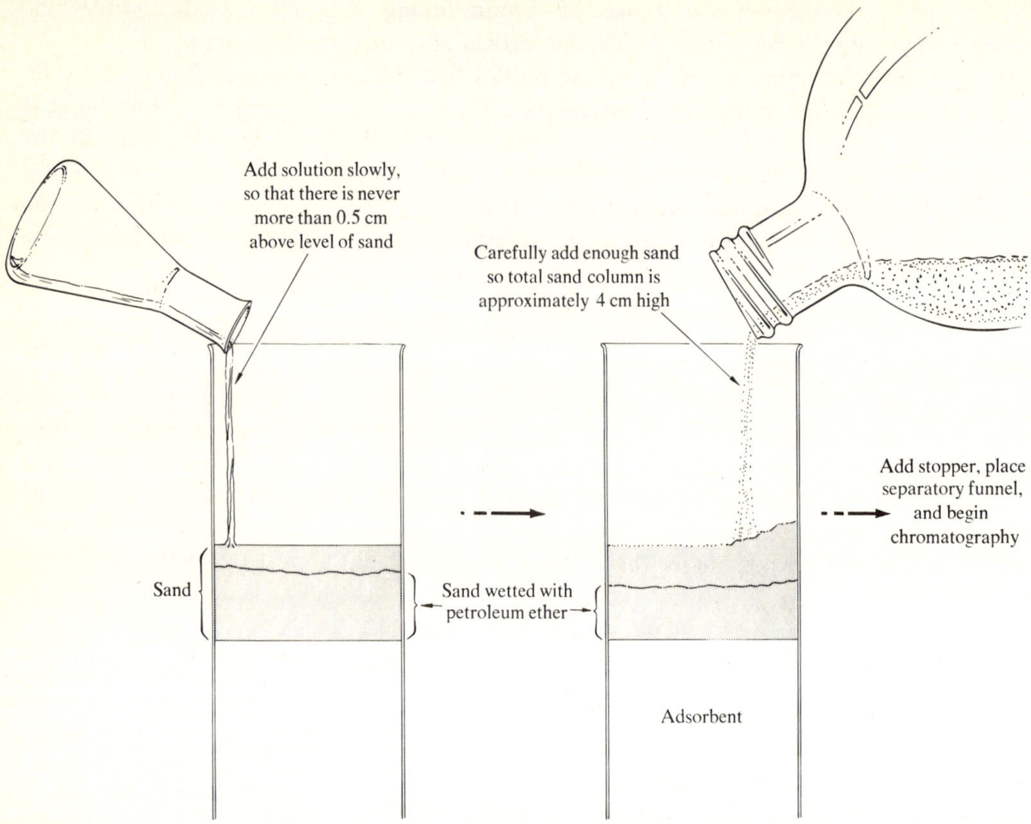

Figure 25-5 Transfer of the reaction mixture to the column.

just below the top of the sand. Continue eluting the column with 25 ml each of the following solvent mixtures until the acetylferrocene (orange-red band) begins to move down the column moderately fast:

ethyl ether	petroleum ether
4%	96%
10%	90%
20%	80%
50%	50%
100%	

Continue eluting with the solvent mixture of lowest ether concentration that moves the acetylferrocene. Collect all the acetylferrocene solution in a round bottom flask. The exact composition of solvent that elutes the acetylferrocene is dependent upon the amount of water in the ether and the method of activating and acid-washing the alumina. Be careful not to use a higher concentration of ethyl ether than is necessary to elute the acetylferrocene reasonably rapidly; too high a concentration of ethyl ether may elute the brown resinous by-products and contaminate the acetylferrocene with these materials.

Problem 25-7. Why is it very important not to use excess benzene when transferring the reaction product to the alumina column?

Evaporate the solvents from the ferrocene and acetylferrocene. Dry the two compounds by crushing and mashing them *separately* on a porous plate and determine their melting points. Transfer the two compounds to *separate* tared yield bottles and determine both % yield of acetylferrocene and % recovery of ferrocene. Hand in both the product and recovered starting material in separate bottles to your instructor.

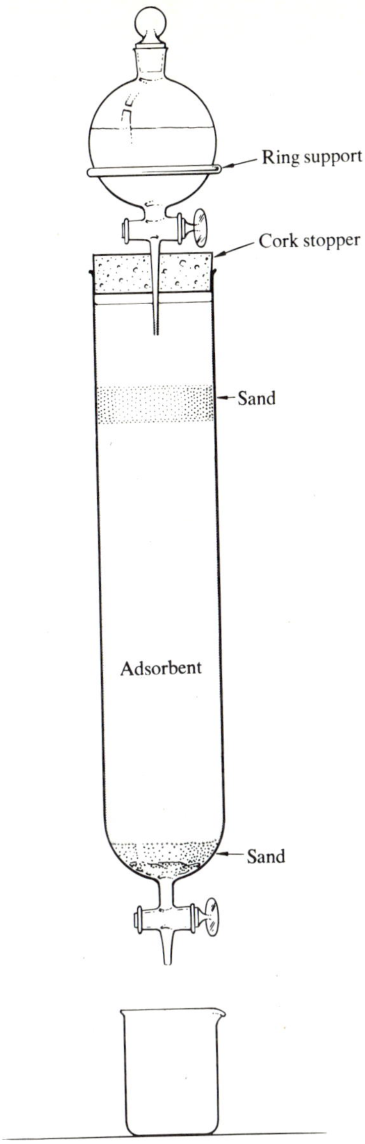

Figure 25-6 Completed chromatography column with a supply of eluant.

EXERCISES

25-1. Compare the conditions of this reaction with those for acylation of benzene (Chapter 17). Is ferrocene more or less reactive than benzene towards Friedel-Crafts acylation?

25-2. There are three and only three possible diacetylferrocenes. Draw their structures. (a) What does the number of isomers indicate about the relative orientations of the two cyclopentadienyl rings? (b) Which diacetyl isomer(s) would be expected from Friedel-Crafts diacetylation of ferrocene?

25-3. (a) Would significant amounts of diacetylferrocene be expected from the reaction described in this chapter? (b) If a diacetyl derivative were produced, when would it be eluted from the alumina column?

25-4. Drawn two schematics of a liquid-solid chromatography similar to those shown in Figures 8-1 and 8-2 for GLC. Label all components as was done in the schematics in Chapter 8 and indicate the principal differences and similarities between liquid-solid adsorption chromatography and GLC.

26

Separation of Natural Products by Thin-Layer Chromatography

Identification of Plant Pigments by their Electronic Spectra

INTRODUCTION

Thin-layer chromatography (TLC), a relatively new diagnostic technique, can be used to separate small quantities with simple apparatus. Ultraviolet absorption spectroscopy, another modern technique, is very useful for the analyses and structural determination of organic compounds. Although ultraviolet spectroscopy is not as universally applicable and does not usually give as much structural information as infrared spectroscopy, there are many organic applications where ultraviolet spectroscopy is more useful than any other spectroscopic technique. Some of the principal applications of ultraviolet spectroscopy include qualitative analysis and structure determination of organic compounds with conjugated electron systems, detection of very small amounts of compounds that absorb in the ultraviolet, spectra studies in aqueous solutions, and very precise quantitative analysis of compounds that have ultraviolet or visible absorptions.

Green plant leaves contain many pigments of widely different structures. Like all colored organic compounds these pigments have extensively conjugated electron systems and have characteristic absorption bands of very high extinction coefficients in the ultraviolet and visible regions of the electromagnetic spectrum.

THIN-LAYER CHROMATOGRAPHY

General

Thin-layer chromatography is usually an analytical adsorptive technique (p. 49), although sometimes it involves partition. The adsorbent, usually silicic acid or silica gel (often mixed with a binder) or sometimes alumina or other material is spread in a thin layer on a glass plate or other inert support. The adsorbent is usually activated by heat after application. The material to be separated is applied in small spots on the adsorbent. The plate is placed in a chamber and the material eluted with a solvent or solvent mixture. The same principles of adsorption chromatography that were discussed in Chapter 25 are applicable to this technique.

Comparison with Other Chromatographic Techniques

Because of its similarity to column chromatography and its rapidity, thin-layer chromatography is often used to determine the optimum adsorbent and eluant for column chromatography. It is useful for separating smaller quantities than are conveniently separated by column chromatography but cannot be used to separate nearly as large quantities of material as column chromatography can. Thin-layer chromatography is faster than column chromatography. Usually the resolution is poorer with thin-layer chromatography because of experimental difficulties in handling long slides. Unlike column chromatography, thin-layer chromatography does not permit the easy changing of solvents for elution. Accurate temperature control and use of high and low temperature is more easily achieved with thin-layer chromatography than with column chromatography.

Thin-layer chromatography requires much less expensive apparatus than gas chromatography and can be used with nonvolatile compounds. It cannot be used to separate volatile materials. Quantitative analysis by thin-layer chromatography is not nearly as precise or accurate as by gas chromatography. Thin-layer chromatography is not so adaptable to detecting very small quantities (submicrogram) of material or to preparative separations as gas chromatography is. Gas chromatography is capable of much higher resolution than can be obtained with thin-layer chromatography.

Experimental Procedure

Extraction of Plant Pigments. Crush 8 to 10 g of spinach or other green leaves with a mixture of 5 ml of acetone and 10 ml petroleum ether. Pour the solution into a separatory funnel and extract by swirling (do not shake, as this subjects the plant pigments to unnecessary air oxidation) with two 20-ml portions of cold, recently boiled water. If nitrogen is available, the plant leaves are preferably crushed and extracted in a beaker and separatory funnel filled with nitrogen.

Drain the organic phase into a 25-ml Erlenmeyer flask that has 3 to 4 g of sodium sulfate in it. Stir the mixture with a glass rod. If nitrogen is available, concentrate the solution to about 40% of its original volume by forcing a stream of nitrogen on it (Fig. 26-1). Stopper the mixture. If it is not to be chromatographed immediately, fill the flask with nitrogen and store it in a refrigerator or cold room until it is needed.

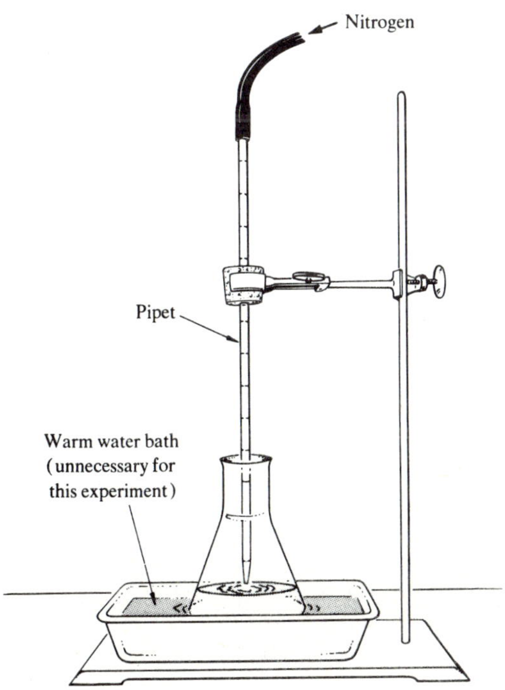

Figure 26-1 Evaporation of solvent from air-sensitive materials by a stream of nitrogen impinging on the surface. Note that the flow of the gas stream is carefully adjusted to prevent splattering of the solvent.

Preparation of Thin-Layer Strips. Heat 80 ml of water and 1 g wheat starch on a water bath at 65 to 70°; stir the mixture until all the wheat starch dissolves. Allow the mixture to cool to room temperature and slowly add with vigorous stirring 40 g of 100 mesh silicic acid (analytical grade). When approximately half the silicic acid has been blended with the mixture, add another 25 ml of water and continue to stir in the rest of the silicic acid. Pour the mixture onto the glass strips (250 × 30 mm strips are recommended but smaller sizes can be used if a loss in resolution is acceptable). Use a stirring rod to spread the slurry while gently tapping the sides. If available, use a wet film applicator to smooth the adsorbent layer and adjust its thickness to 1 to 1.5 mm.

The glass strips with the thin layer of adsorbent should be handled very carefully so that the adsorbent layer is not disturbed. Allow the glass strips to stand for 25 to 30 min. Place them in an oven at 75° for 1 hr and then at 110° for 2 hr; this is in addition to the time required for the temperature to increase from 75 to 110°. Remove the strips from the oven while they are still hot and store them in a desiccator until they are used. It is important that this procedure be followed as closely as possible to ensure proper activation of the adsorbent and to prevent the formation of bubbles and other defects in the thin layer.

Chromatography of the Plant Pigments. As soon as the glass plates are removed from the desiccator, they should be spotted and placed in the chromatographic chamber as quickly as possible to minimize deactivation of the silicic acid and oxidation of the plant pigments. Spot the solution of plant pigments on two points 1.0 cm apart on a line 1.5 to 1.8 cm from the end of the glass plate and equidistant from either side (Fig. 26-2). Use a micro or capillary pipet to spot 8 small drops of the pigment solution in each position. Allow the solvent from each drop to evaporate before adding the next. If a glass plate narrower than 30 mm is used, spot only one point with 10 to 12 small drops 1.5 to 1.8 cm from the end and equidistant from the sides. If the leaf extract was concentrated under a stream of nitrogen, half the number of drops specified above may be used.

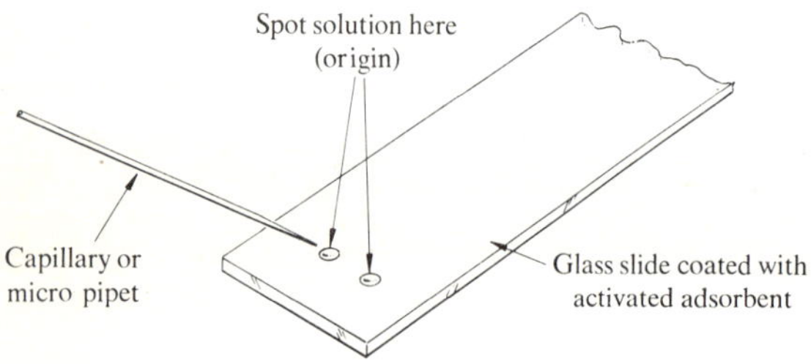

Figure 26-2 Spotting of plant pigments on a glass slide.

Figure 26-3 Test tube developing chamber with thin-layer strip and wick.

A 300 × 38 mm test tube is a convenient developing chamber. If the glass strip is smaller than 200 × 25 mm, a smaller test tube can be used. The developer (i.e., eluting solvent) is a mixture of 10 ml of isooctane, 2 ml of acetone, and 2 ml of anhydrous ethyl ether. Enough of the eluting solvent is placed in the test tube to come within 3 to 4 mm of the bottom edge of the spot. The solvent should under no condition touch the spot. Carefully place the spotted plate in the test tube. Fold a piece of filter paper for use as a wick. Fill the test tube with nitrogen if it is available. Adjust the filter paper so that it touches the top of the silicic acid, folds over the lip of the test tube, and can be fastened in place with the stopper (Fig. 26-3). Since some of the pigments are not separated by the time the solvent front reaches the top of silicic acid, it is necessary to remove some solvent from the silicic acid layer. The wick does this by taking solvent from the silicic acid by capillary action and transferring it to the outside where it evaporates.

Shield the stoppered test tube from light and drafts. After 2 hr look at the chromatogram and look again every 10 to 15 min. Remove the chromatogram from the tube when the upper edge of the top yellow spot is between 5 and 10 mm from the paper wick. This usually takes about 3 hr.

Note the colors of each of the spots and their distance from the origin. If ultraviolet spectra are to be obtained, proceed as soon as possible with the instruction under the experimental procedure for the identification of plant pigments by their electronic absorption spectra in the following section. If a delay of more than 5 min is necessary, store the developed chromatograms in the dark under a layer of nitrogen.

Problem 26-1. Usually the carotenes (the top yellow spots) are not completely resolved by the above procedure. Suggest a modified procedure that would give better resolution of the carotenes. Verify your answer, and, if your instructor indicates, try this procedure on the green leaf extract.

Problem 26-2. Before determining the identity of the plant pigments in each spot, make as many rational judgments as possible about their structure from information you have obtained *so far* in this experiment.

Separation of Other Plant Pigments (*Optional*). Repeat the above procedures on crushed tomato pulp, carrot scrapings, or geranium flower petals.

ELECTRONIC ABSORPTION SPECTROSCOPY

Introduction

Molecules absorb radiant energy in the visible-ultraviolet region of the spectrum. The absorption occurs when the incident radiation has the same energy as that required for an electronic transition of the molecules. Since most of these electronic transitions are in the ultraviolet region, ultraviolet-visible or electronic spectra is often called ultraviolet or UV spectra.

Most commercial ultraviolet recording spectrophotometers have a lower wavelength range from 190 to 210 $m\mu$. Even if the low wavelength range were not instrument limited, obtaining spectra below 190 $m\mu$ would be plagued with many problems such as solvent and atmospheric absorptions. The principal electronic transitions at wavelengths above 200 $m\mu$ are those involving unbonded or *pi* electrons. The electronic transitions of most nonconjugated *pi* bonds are at wavelengths below 200 $m\mu$. Electronic transitions of conjugated *pi* systems such as —C=C—C=C—, —C=C—C=O (with single bonds drawn upward from each carbon) are above 215 $m\mu$. The more extensive the conjugation is the more absorption maxima, the greater their intensities, and the longer their wavelengths. Space in this laboratory manual does not permit an extensive discussion on the theoretical and interpretive aspects of electronic absorption spectroscopy. Read your lecture text or one of the small monographs on this subject for theoretical considerations and structure correlations of electronic spectra.

Instrumentation

Many of the same principles discussed in Chapter 18 on infrared spectroscopy are also applicable to ultraviolet spectroscopy. There are, however, many differences. One of the principal ones is that in ultraviolet spectroscopy there is not nearly as much energy limitation as in infrared spectroscopy. The hydrogen or deuterium and tungsten lamps that are used as sources in ultraviolet spectrophotometers emit considerably more radiant energy than the heated ceramic rods and other sources used in infrared spectrophotometers. In addition, the electronic energy transitions are usually very much larger than vibrational energy transitions. This means that conservation of energy is not as critical in ultraviolet spectrophotometers. For example, the slit width can be optimized for factors other than letting through energy. Ultraviolet spectroscopy is potentially more accurate and more precise for quantitative analysis than infrared spectroscopy. This is not necessarily always realized in practice. If the compounds have the requisite absorption band, smaller amounts can be detected and often identified by ultraviolet spectroscopy than by infrared spectroscopy.

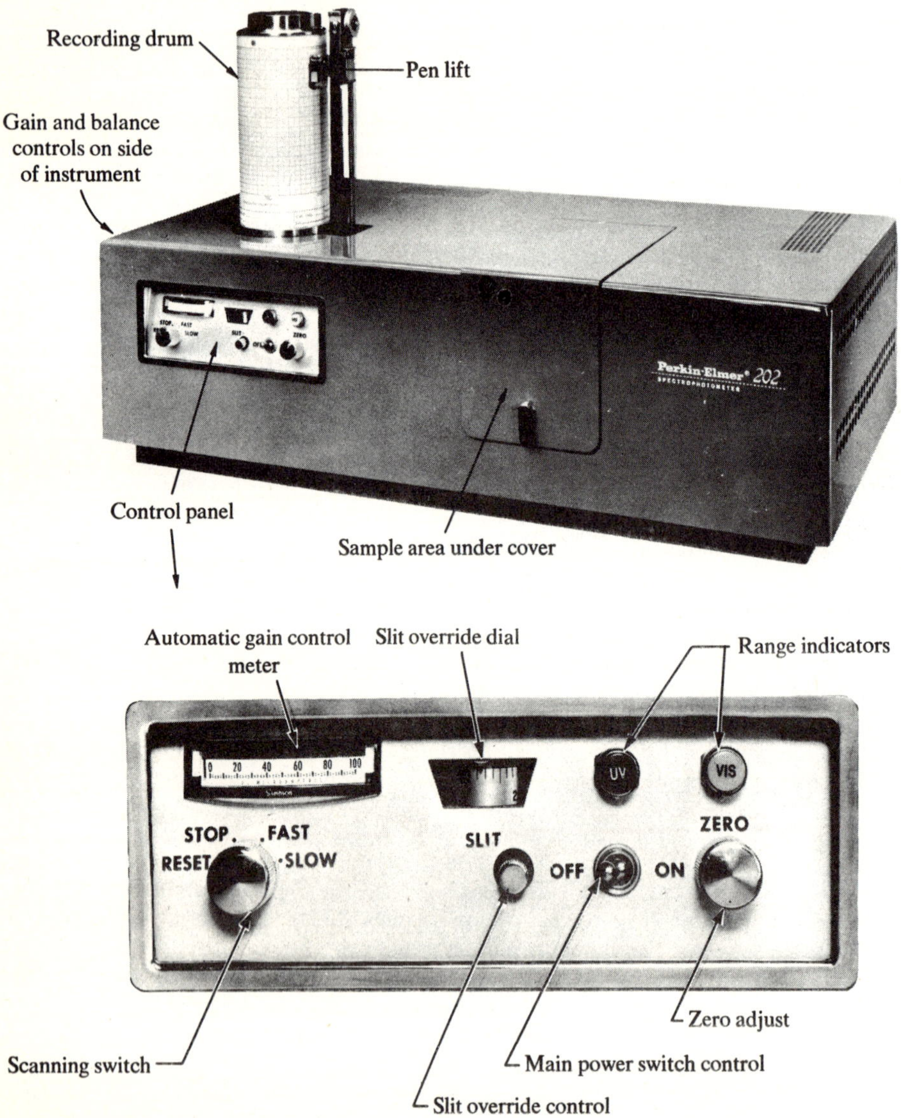

Figure 26-4 An ultraviolet-visible spectrophotometer and its operating controls. Courtesy of Perkin-Elmer, Inc.

There are more basic designs of ultraviolet spectrophotometers in use than of infrared spectrophotometers. Almost all recording infrared spectrophotometers in use today (the author estimates at least 99.5%) are double beam, optical null instruments. Although the recording ultraviolet spectrometer chosen as an example for this manual is also an optical null, double beam instrument (Fig. 26-4), other basic designs are also in common and perhaps even more prevalent use. These include electrical null, ratio recording, and single beam. Some single beam instruments can function as double beam instruments by rapidly exchanging the sample and reference material and recording the difference signal. It is beyond the scope of this manual to discuss each design and its relative merits.

Another difference is that the ordinate pen response on most ultraviolet spectrophotometers is linear in absorbance units (p. 153), although some ultraviolet spectrophotometers can record linear in transmittance units also. A linear absorbance scale is very convenient because these spectrophotometers are used so much for quantitative analysis. In addition a logarithmic scale can better accommodate most of the maxima of a complex spectrum on a single chart and show the difference in absorbances of the intense peaks. When, however, a linear absorbance scale is used to obtain spectra on compounds that have both very intense and very weak absorbances, it is necessary to obtain spectra on at least two concentrations. One is dilute enough so that the very intense peaks are not off scale and the other is concentrated enough so that the weak peaks can be seen. The high source energy permits operation of the instrument with a high signal to noise ratio; therefore, the use of a linear absorbance scale does not cause excessive noise to be shown on the absorption maxima.

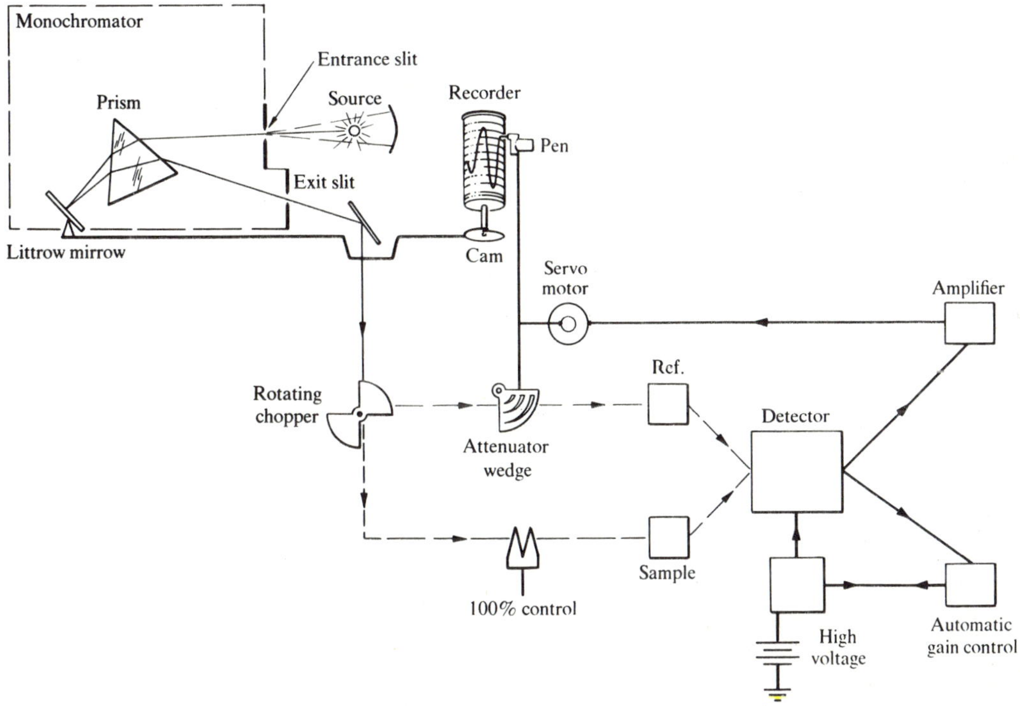

Figure 26-5 Schematic of an optical null, double beam ultraviolet-visible recording spectrophotometer.

Problem 26-3. Figure 26-5 shows a schematic of an ultraviolet-visible recording spectrophotometer. Compare with Figure 18-5. Other than the difference in the dispersing units of the two instruments and automatic gain on the ultraviolet spectrophotometer, what is the principal difference and how does this relate to some of the principal differences between infrared and ultraviolet spectroscopy?

Sampling Procedures

Most ultraviolet cells are made from quartz which is transparent to ultraviolet and visible radiation. This simplifies sampling procedures as quartz is insoluble in almost all common solvents and is very inert. Ultraviolet spectra are often obtained in aqueous solutions not only because cells insoluble in water are readily available but mainly because water has no electronic absorption bands above 200 mμ.

Operation of An Ultraviolet Spectrophotometer

Figure 26-4 shows the instrument and its various controls. The gain control is on the back of the instrument. Because of the automatic gain circuit the adjustment of the gain control with this instrument is not as critical as the adjustment of the gain control on infrared spectrophotometers. It does not need to be adjusted when spectra are obtained on samples of different opacities. It is probably best set once a week by the laboratory instructor and not disturbed the remainder of the time.

Warning. Do not move the chart drum or attempt to remove or adjust the chart paper unless the scanning switch is turned to Reset. To insert chart paper remove the drum by lifting it straight up. Lift the edge of metal paper holder by pressing down the end of curved metal lever shown in Figure 26-6. Place the short wavelength end of the chart paper (of proper range) under the metal paper holder. Partially wrap the paper around the chart drum with the top of the chart paper flush against the upper rim of the drum. Align the index lines on the drum and the chart paper. Insert the long wavelength edge of the chart into the other paper holder, smooth

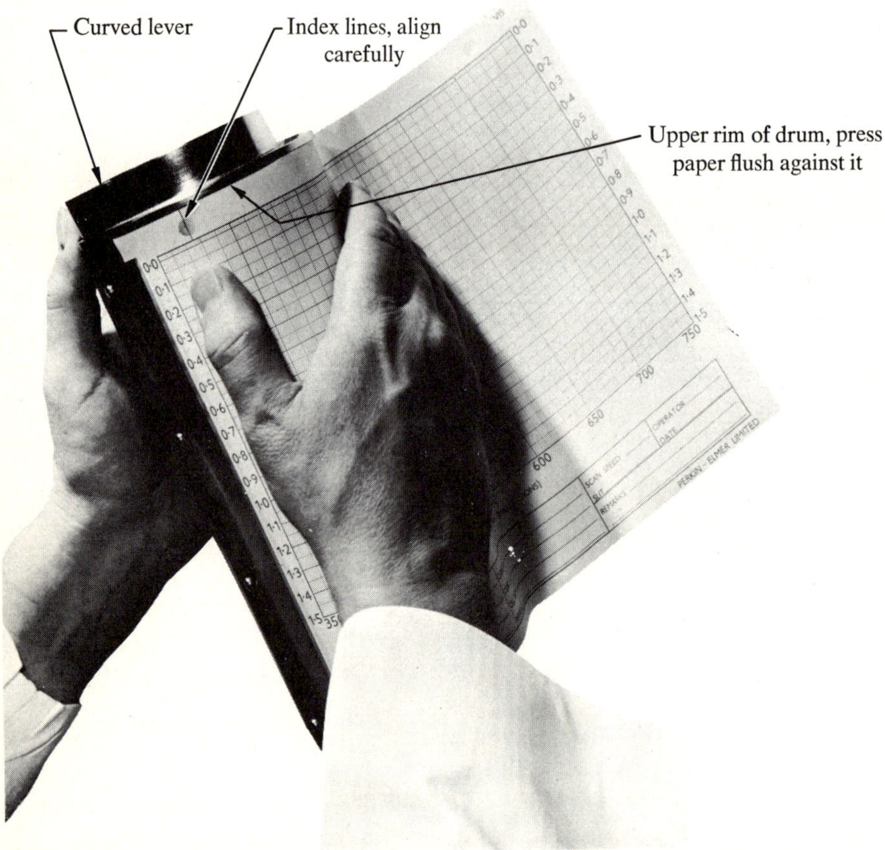

Figure 26-6 Placement and alignment of paper on the recorder drum. Courtesy of Perkin-Elmer, Inc.

the paper so that it is taut, and fasten the long wavelength end. If there are any wrinkles in the chart paper or if the index mark of paper has become misaligned with the index mark on the drum, make the appropriate corrective adjustments. Be certain the position selector in the base of drum is on the N (or normal) position (Fig. 26-7). Align the drum vertically so that the engaging pin fits in the hole in the base of the drum and lower the drum.

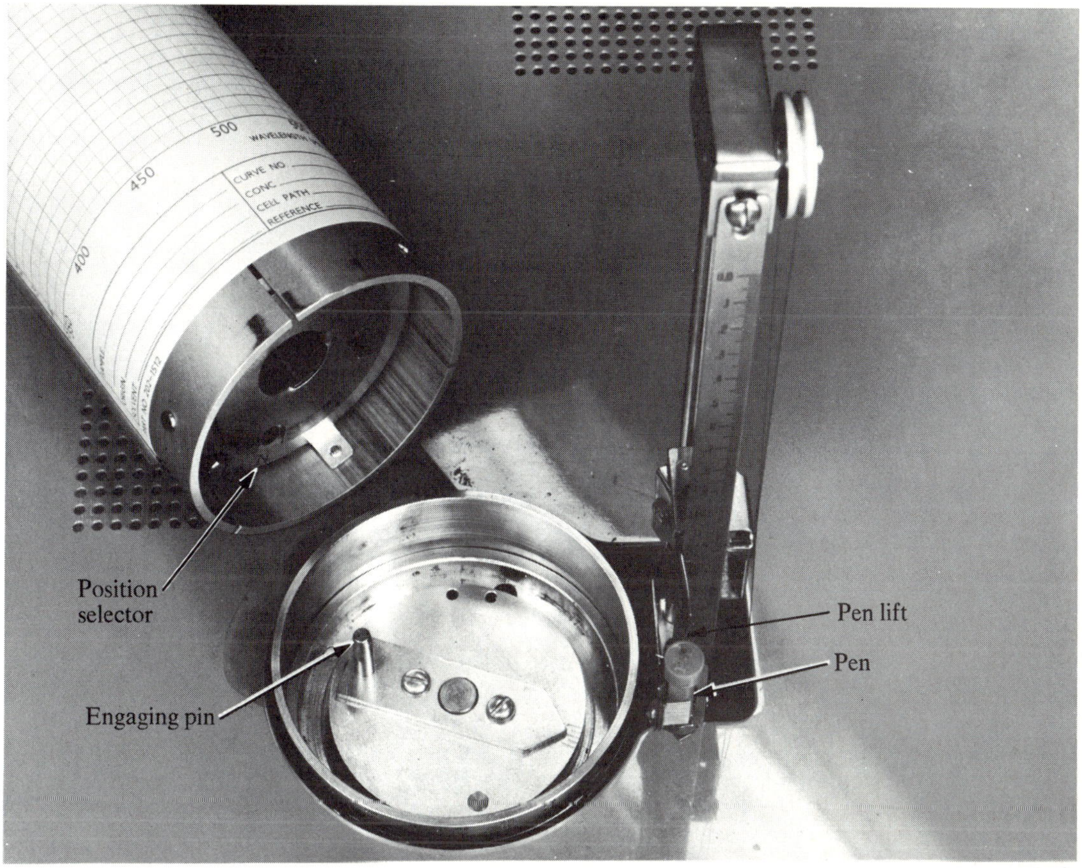

Figure 26-7 Position selector in base of recorder drum. Courtesy of Perkin-Elmer, Inc.

For the identification of plant pigments in this experiment spectra in both the ultraviolet and visible should be obtained. Make sure the correct chart paper is used for the range in which the instrument is operating. Note the range indicators on the front panel. To change from the ultraviolet to the visible range rotate the drum clockwise by hand (**scan switch on Reset!**) until the paper holder passes the wavelength indicator. A "click-bong" noise and lighting of the VIS indicator shows the range has been changed. To change from visible to ultraviolet range rotate the drum counterclockwise; a noise and lighting of the "UV" indicator shows the range has been changed.

If the slit override dial shows any setting other than 25, set the slit override control so that the dial reads 25. The zero adjust is the same as the 100% control on infrared spectrophotometers, and it is adjusted in the same way (pp. 122–124). The control is easily damaged if it is forced beyond its limiting position in either direction. **Do not force this control.** To obtain a spectrum, set the cell in the sample compartment, open the shutter and lock the sample compartment cover by turning the shutter control clockwise, push the pen against the paper, and turn the scan switch to Slow.

Identification of Plant Pigments by Their Electronic Absorption Spectra

Discussion. Ultraviolet spectra are most usually used to identify gross structural features such as the number of aromatic rings, presence or absence of heteroatoms, the presence, type, and length of conjugated systems, etc. Because of the many absorption maxima of plant pigments electronic absorption spectra

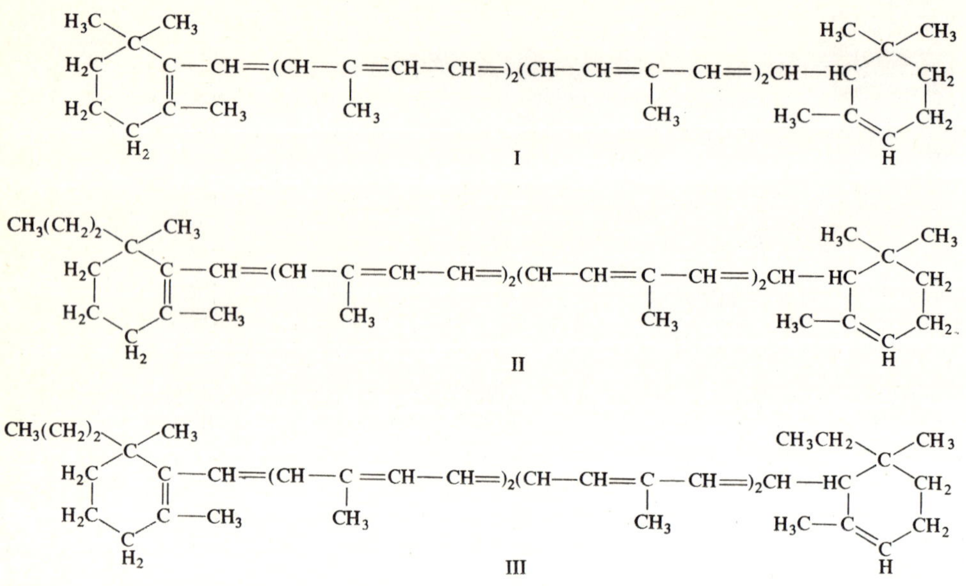

Figure 26-8 Structure of some carotenes.

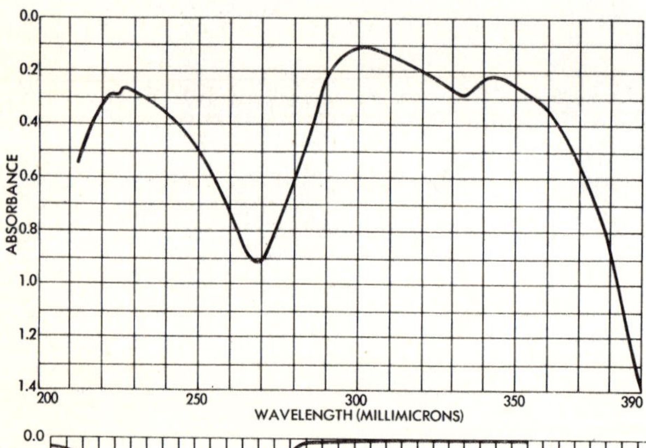

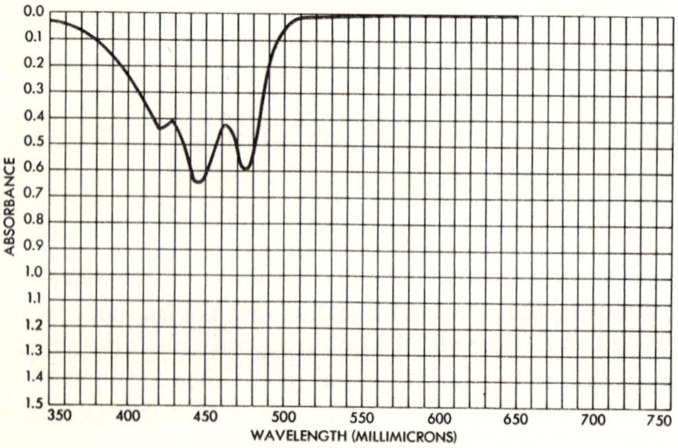

Figure 26-9 Ultraviolet spectrum of carotenes in Figure 26-8.

can be used to fingerprint these compounds. Many small differences in structure are not detectable by ultraviolet spectroscopy. For instance, there is no significant difference[1] in the ultraviolet spectra of carotenes I, II, and III (Fig. 26-8); all show the spectrum given in Figure 26-9. If one of the plant pigments has a spectrum identical with Figure 26-9, assume it is due to carotene I because I has been found in plants and II and III have not. If this investigation were part of some original research, it would be necessary to verify the identification of I by chemical or other spectral methods.

Experimental Procedure. Every possible precaution should be taken to prevent oxidation and photodegradation of the plant pigments. As soon as solvent has evaporated from the plate, the plant pigment spots should be removed and extracted unless the plate can be stored in the dark under an atmosphere of nitrogen.

Problem 26-4. Why is it important that all the solvent evaporate before the spots are removed and extracted?

With a spatula transfer each spot and a little of the adjacent adsorbent into a small test tube; use a separate test tube for each spot. Add 5 ml of solvent mixture. The solvent mixture for the top yellow spots should be made from equal volumes of spectral grade ethyl ether and isooctane; for all the other spots use a mixture of equal amounts of spectral grade methanol and isooctane. Mix the contents of the test tube well with a glass rod and allow to settle and then pour through a cotton plug in a small glass funnel into a quartz UV cell, or centrifuge and carefully decant the supernatant liquid into a quartz cell.

Obtain both visible and ultraviolet spectra on each of the extracted pigments. Use the appropriate solvent mixture in the reference beam. While obtaining a spectrum on one pigment extract the next. If it is necessary to wait more than 5 min between extraction of a pigment and placement in the spectrophotometer, put the extract in a cold, dark place.

If the solution is too contentrated as indicated by one or more peaks that flatten against the 1.5 absorbance line, obtain another spectrum on a sample diluted five times with the appropriate solvent mixture. If only peaks in one range are too strong, it is unnecessary to obtain a spectrum on the diluted sample for the range that had no peaks against the 1.5 absorbance line.

Compare the spectra you obtained with the plant pigment electronic absorption data and spectra in Appendix VI. Identify as many of the pigments as possible.

EXERCISES

26-1. Why is it necessary to elute thin-layer plates inside a sealed container?

26-2. The complete procedure using both parts of this chapter is for separation and identification (qualitative analysis) of plant pigments.

a. How could this method be used for quantitative analysis to determine the relative amounts of the various plant pigments? What might be the largest sources of error in such an analysis?
b. How could this method be adapted to determine the absolute amounts of plant pigments in the leaves?
c. It was mentioned on page 201 that thin-layer chromatography was not as accurate or precise as gas chromatography for quantitative analysis. Would this generality apply to the methods proposed in the answers for parts a and b of this problem?

26-3. It is important that the exposure of the plant pigments to air and light be kept at a minimum at all times, but for the qualitative identification of the plant pigments it is especially important that exposure between separation by TLC and identification by their electronic absorption spectra be kept to a minimum. Why?

[1] If spectra were obtained on equal concentrations by weight, the spectrum of I would be slightly more intense because the molar extinction coefficients would be identical and I has a lower molecular weight.

27

Infrared Studies of Hydrogen Bonding of Alcohols

INTRODUCTION

This experiment demonstrates the effect of environment on infrared spectra, particularly the effect of hydrogen bonding on hydroxyl stretching vibrations. The demonstration of the effect of concentration on hydrogen bonding requires that spectra be obtained on different concentrations with various path lengths. A variable path length cell is a convenient accessory for obtaining spectra on liquids in different path lengths. Less conveniently, different path lengths can also be obtained with demountable cells using various size spacers.

DISCUSSION[1]

The hydroxyl groups of alcohols and other functional groups comprised of a hydrogen atom on a strongly electronegative atom undergo hydrogen bonding. When the hydrogen atom is hydrogen bonded, the force constant of the covalent bond between the proton and electronegative element is lowered slightly. There-

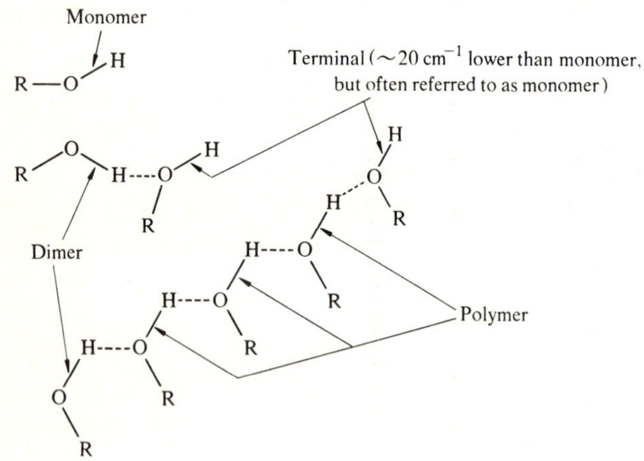

Figure 27-1 Hydrogen bonding of alcohols.

fore, hydrogen bonding causes the hydrogen-oxygen stretching absorption to be at lower frequencies than frequencies of the unassociated groups. As a result alcohols show three hydroxyl stretch absorptions. One is from the monomer whose hydroxylic protons are not hydrogen bonded. This absorption is difficult

[1] It is assumed that the student has thoroughly studied the section in his lecture text on hydrogen bonding before reading this chapter. See also page 214.

to distinguish from the absorption of the terminal hydroxyl with the nonhydrogen-bonded proton that occurs on the end of all associated groups of alcohols as it is usually weaker and only 20 cm^{-1} above it. In this text both of these absorptions are referred to as monomer bands (Fig. 27-1). Another absorption is from the hydroxyl stretch vibration of protons hydrogen bonded to an alcohol and covalently bonded to an oxygen atom that is not hydrogen bonded. This occurs on the other end of associated groups of alcohols and is often called a dimer hydroxyl. The third or polymeric absorption band is from the stretch frequency of hydroxyls that have both the hydrogen and oxygen hydrogen bonded. This last type can occur only when three or more molecules are associated by hydrogen bonding.

In studying association it is often indispensible to obtain data from different concentrations. For a given path length useful infrared absorption bands can be obtained only from relatively narrow ranges of concentration. If, however, the path length is varied, spectra can be obtained on a wide range of concentrations.

Problem 27-1. What law justifies this last statement?

Problem 27-2. Assume a compound has an absorption at 1400 cm^{-1} and its absorption coefficient does not vary with concentration. If a 10% solution (100 g/l) in CCl_4 in a 0.05-mm cell has a transmittance of 20%, what is the absorbance and absorption coefficient? What path length cell would be required to give the same transmittance with a 5% concentration? With a 2% concentration? Assume an I_0 of 100% *T*.

The path lengths can be changed by using the variable path cell described in the following section or by using one of the two demountable cells with ports described on pages 135–136. Up to four spacers can be used in the demountable cells if spacers thick enough to give the specified path lengths are not available. If several spacers are used, extra care is needed in tightening the cells to prevent leakage.

EXPERIMENTAL PROCEDURE

Variable Path Cell

This cell is a precision made accessory (Fig. 27-2). If it is to retain its usefulness, certain precautions are necessary:

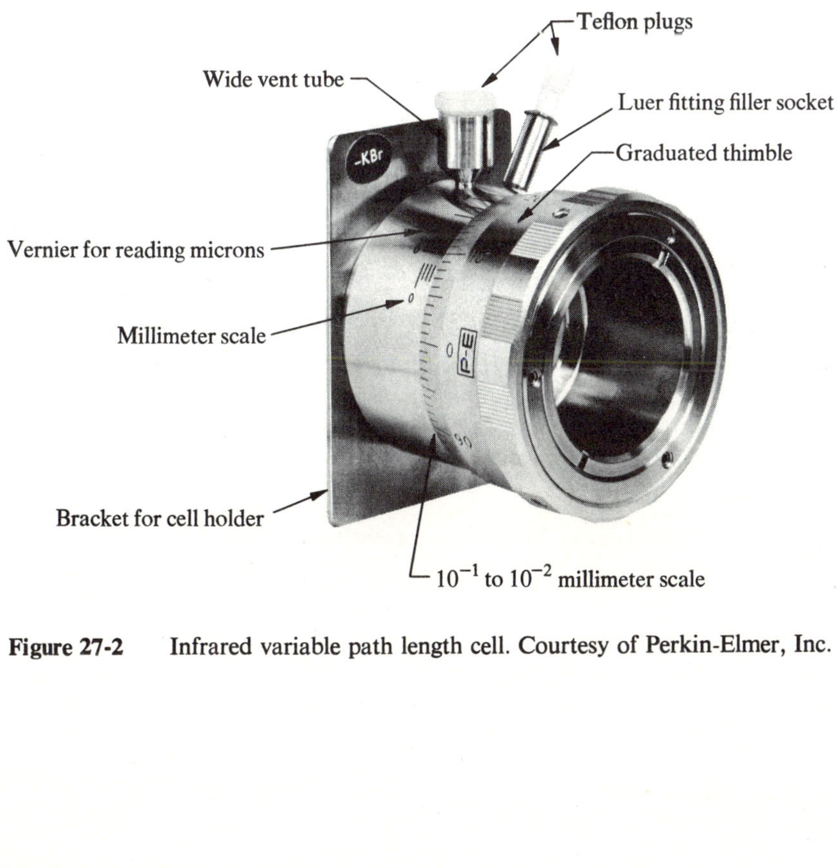

Figure 27-2 Infrared variable path length cell. Courtesy of Perkin-Elmer, Inc.

a. Do not change the path length if the cell temperature is above 30°.
b. Do not change the path length with both plugs in.
c. Never heat variable path cells above 35°.
d. Do not fill variable path cells with solutions that may precipitate solids or become very viscous if a small portion of the solvent evaporates. Do not put materials in the cell that are likely to polymerize.
e. Avoid spilling solvents on the calibrated scales or the dark filler will be washed away.
f. If solvent spills on the outside of the cell, allow it to run off or evaporate; do not attempt to wipe it off.
g. Do not attempt to set the path length smaller than 0.030 mm or greater than 6.4 mm.
h. Exercise the usual care in working with alkali metal halide windows.

Remove both stoppers. Set the cell at the path length desired using the vertical markings on the body assembly to read the path length in millimeters, the scale on the thimble to read 10^{-1} to 10^{-2} mm, and the vernier on the stationary body assembly to obtain the nearest micron (10^{-3} mm). To fill the cell use a hypodermic syringe on the Luer lock fitting or carefully pour liquid into the wide vent tube. Look through the cell to see if there is an air bubble. If there is, add more liquid and work the bubble out by rocking the cell through a 30° arc with the ports pointed up. Both Teflon plugs should be off. The path length can be altered once liquid is in the cell if one of the ports is open. If the path is shortened considerably when the cell is filled, remove the excess liquid from the wide vent tube with a syringe rather than letting it flow over the cell. If the cell is lengthened considerably after filling, more liquid will have to be added to remove any air bubble that forms. The final adjustment of the cell is made while it is in the spectrophotometer with one of the stoppers off.

Problem 27-3. What are the path lengths of the two cells shown in Figure 27-3?

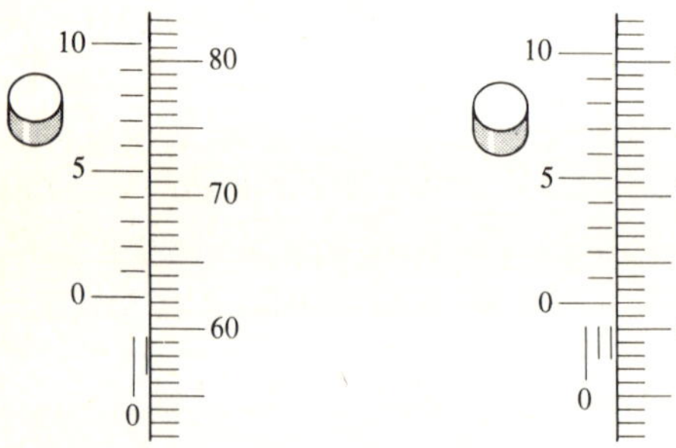

Figure 27-3 Scale readings on variable path cell (Problem 27-3).

Clean the cell immediately after use. Pour the liquid out the top. Close the cell to 0.1 mm. Wash the cell several times with a volatile solvent. Drain the cell as well as possible after each wash. If, however, a solution of a solid compound has been in the cell, follow the precautions in the footnote on page 132. Place the cell in the sample beam of an infrared spectrophotometer and push dry air through it with a hypodermic or ear syringe.

Effects of Concentration on Hydrogen Bonding

Prepare 12 (3 ml in 25 ml), 3, 1.2, 0.6, 0.3, 0.1, and 0.03% by volume solutions of methanol in carbon tetrachloride. Obtain complete spectra (4000 to 650 cm^{-1}) on the 12% solution in a 0.05-mm cell, on the 0.1% solution in a 6.0-mm cell, and on pure methanol as a capillary film. Obtain spectra from 4000 to 3000 cm^{-1} on the 3% solution in a 0.2-mm cell, on the 1.2% solution in a 0.5-mm cell, on the 0.6% solution in the 1.0 mm cell, on the 0.3% solution in a 2.0-mm cell, on the 0.1% solution in a 6.0-mm cell, and on the 0.03% solution in a 6.4-mm cell.[2]

[2] If neither a variable path cell nor the necessary size spacers are available, the very dilute solutions may have to be omitted and cell thicknesses other than the ones discussed above may have to be used. If so, change the concentrations using Equation 20-1 so there are always approximately the same number of methanol molecules in the sample beam. For example, an 0.8-mm cell needs a 0.75% by volume solution. If the thickest cell available were only 0.8 mm, 4000 to 3100 cm^{-1} scans on 0.3% and 0.15% by volume methanol in carbon tetrachloride solutions should also be obtained.

Problem 27-4. What is the effect of concentration on the intensities of the various hydroxyl stretching frequencies?

Solvent Effects on Hydrogen Bonding

Obtain spectra on 12% by volume solutions of methanol in dioxane and in chloroform in a 0.05-mm cell. Obtain spectra on pure dioxane and pure chloroform in a 0.05-mm cell.

EXERCISES

27-1. Explain any differences in the 4000 to 3100 cm^{-1} region of the spectra of 12% methanol solutions in carbon tetrachloride, chloroform, and dioxane.

27-2. Are there differences in the 3100 to 650 cm^{-1} region in the methanol absorption bands obtained in carbon tetrachloride, dioxane, or chloroform solutions or from pure methanol as a capillary film? Why?

27-3. From the spectra you have obtained determine the hydroxyl stretching frequencies of the "monomer," "dimer," and "polymer" of methanol. Include the frequencies and your rational for determining them in your report for this experiment.

27-4. A variable path cell is also very useful as a reference beam cell when solvent compensation is desired (pp. 138 and 244). Why is it more useful than other cells for this purpose? Devise a method for utilizing it for solvent compensation using as an example a methanol in carbon tetrachloride solution in a 0.05-mm cell. Be specific about path length, frequencies, and other details.

27-5. Could the information you just obtained be used to calculate an equilibrium constant for the association of methanol? If so, how? If not, what other information would be needed? Be specific about frequencies, concentrations, and calculation of various values of Equation 20-1.

27-6. Indicate the relative order of bond strength of the hydrogen to oxygen covalent bond in dimer, monomer, polymer, and terminal methanol (Fig. 27-1).

28

Properties and Reactions of Alcohols

Organic Qualitative Analysis III

INTRODUCTION

The boiling point, melting point, and solubility of a compound are determined by its structure and molecular weight. Structure can be used to estimate these physical properties, and physical properties can be used to help determine structure.

The solubility of organic compounds in different solvents can be used to classify organic compounds into a few general classes. From this classification and other physical properties information can be obtained as to which classification tests may be useful for identification of an organic compound.

Like all functional groups the alcohol group undergoes characteristic chemical reactions. Some of these reactions are readily observable and can be used to identify the alcohol group or to distinguish among the different classes of alcohols.

PHYSICAL PROPERTIES

Solubility in Water

Polar Attractions and Effects in Homologous Series. Water solubility serves to distinguish low molecular weight alcohols and other low molecular weight compounds containing oxygen, nitrogen, or sulfur functional groups from other compounds. With the techniques and limitations described in the experimental section, the borderline of water solubility for monofunctional alcohols is between those alcohols with 4 and 5 carbon atoms.[1]

Many organic polar compounds dissolve in water because of the dipole-dipole attractions between water and the solute molecules (Fig. 28-1). Hydrogen bonding is the type of dipole-dipole interaction responsible for the most important attractions between water and water-soluble, oxygen and nitrogen-containing, nonionic compounds. This type of dipole-dipole attraction is stronger than other dipole-dipole interactions because the hydrogen atom is so small it allows a closer approach of the negative pole (usually a fluorine, oxygen, or nitrogen atom). Thus water is a good solvent for polar organic substances.

[1] Aldehydes and ketones show approximately the same water solubility as alcohols with the same number of carbon atoms, acids and amines are somewhat more soluble than their alcohol analogs, and esters and ethers are slightly less soluble than alcohols with the same number of carbon atoms. Later chapters discuss these solubility relationships further.

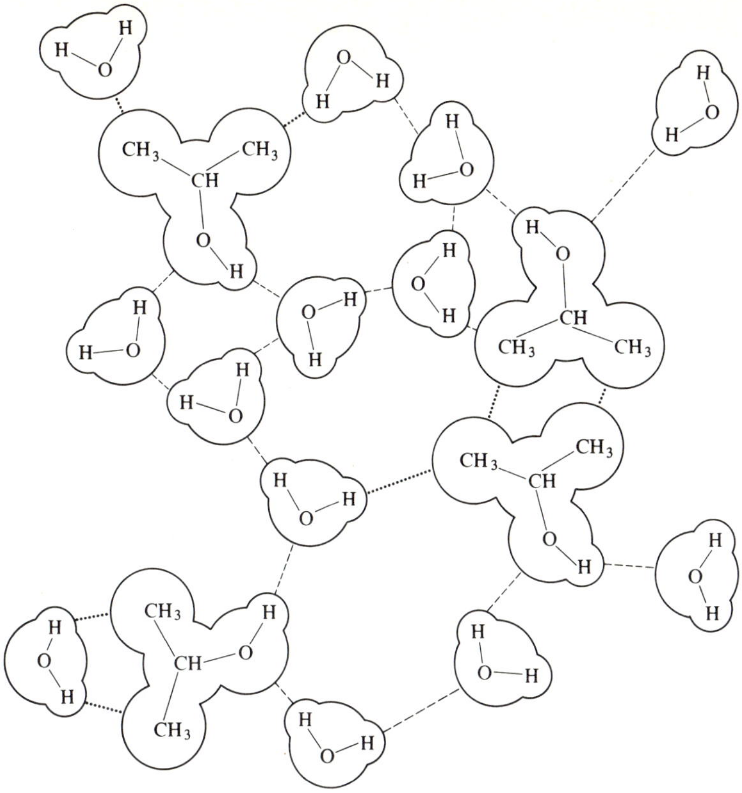

Figure-28-1 Hydrogen bonding between water and 2-propanol molecules in solution. Dashed lines indicate the relatively strong hydrogen bonds and the dotted lines indicate the much weaker van der Waals forces and dipole-induced dipole attractions.

Problem 28-1. List 4 polar solvents and 6 nonpolar solvents that you have used.

Many, if not most, organic compounds are composed of molecules with a polar and a nonpolar part. The solubility of these compounds depends not only on their molecular weight but also on the proportion of polar and nonpolar moieties. As a homologous series, such as alcohols, is followed to increasingly higher molecular weights, the relative hydrocarbon portion of the molecules increases and the properties of the compounds become more like those of the parent hydrocarbons (see Table 28-1). This causes water solubility to decrease and solubility in nonpolar solvents, as benzene and ether, to increase. The larger the alkyl radical, the more important the London dispersion forces are and the less important the dipole attractions. In an alcohol such as 1-tetradecanol the intermolecular attractions are mainly from weak London dispersion forces. Such molecules can be attracted to water only by dipole-dipole attractions at the relatively small alcohol group and by weak dipole-induced dipole attractions over the rest of the molecule (Fig. 28-2). Not enough energy can be supplied by these interactions to break hydrogen bonds between enough water molecules to allow the long tetradecyl group to slip in.

Effect of Aromatic Structure. A phenyl group on most oxygen and nitrogen functional compounds has about the same effect on *water* solubility as 3 to 4 carbon atoms or approximately 3½ CH_2 groups. See the discussion below on possible effects of cyclization. Phenol (see Table 28-1) is intermediate in solubility between 1-propanol and 1-butanol; benzyl alcohol is intermediate in solubility between 1-butanol and 1-pentanol; 2-phenylethanol is intermediate in solubility between 1-pentanol and 1-hexanol. Similar relations are observed in other groups of compounds; butylamine ($CH_3CH_2CH_2CH_2NH_2$) and benzylamine ($C_6H_5CH_2NH_2$) are completely miscible with water in all proportions; and 2-phenylethylamine is intermediate in solubility between butylamine and pentylamine.

Aromatic compounds are more soluble in water than similar aliphatic compounds with the same number of carbon atoms because: (a) the aromatic compounds are more compact. This can account for the higher water solubility of only liquid aromatic compounds. (b) The *pi* clouds of aromatic systems are more easily moved or *polarized* by polar or ionic charges than the electrons in aliphatic molecules

TABLE 28-1 PHYSICAL PROPERTIES OF HOMOLOGUS ALCOHOLS

Alcohol	Solubility (g alcohol/ 100 g H_2O)	D_4^{20}	B.p. (°C)	Parent hydrocarbon sol, (g/100g of H_2O)	B.p. (°C)	Difference in between RH and ROH
Methanol	completely miscible	0.7915	64.5		-162	226.5
Ethanol	completely miscible	0.7894	78.3		-88.5	166.8
1-Propanol	completely miscible	0.8036	97		-42	139
1-Butanol	7.9 "soluble"	0.8096	118		0	118
1-Pentanol	2.3 "insoluble"	0.8148	138	0.040	36	102
1-Hexanol	0.6 insoluble	0.8189	156		69	87
1-Heptanol	0.20 insoluble	0.8224	176	0.006	98	78
1-Octanol	0.05 insoluble	0.8249	195	0.0025	126	69
1-Nonanol	insoluble	0.8271	215		151	64
1-Decanol	insoluble	0.8292	232		174	58
1-Undecanol	0.019insoluble	0.8299	243		195.5	47.5
1-Dodecanol	0.017 insoluble		257		215	43.0
1-Tridecanol	0.016 insoluble				234	
1-Tetradecanol	0.014 insoluble				252	
1-Pentadecanol					266	
Phenol (C_6H_5OH)	9.3 soluble		182		80.1	102
Benzyl alcohol	4 borderline solubility	1.0454	205		110.6	84.4
2-Phenyl-1-ethanol	1.6 insoluble	1.0247	221		136.2	85
3-Phenyl-1-propanol	insoluble	1.0079	236		159.2	77

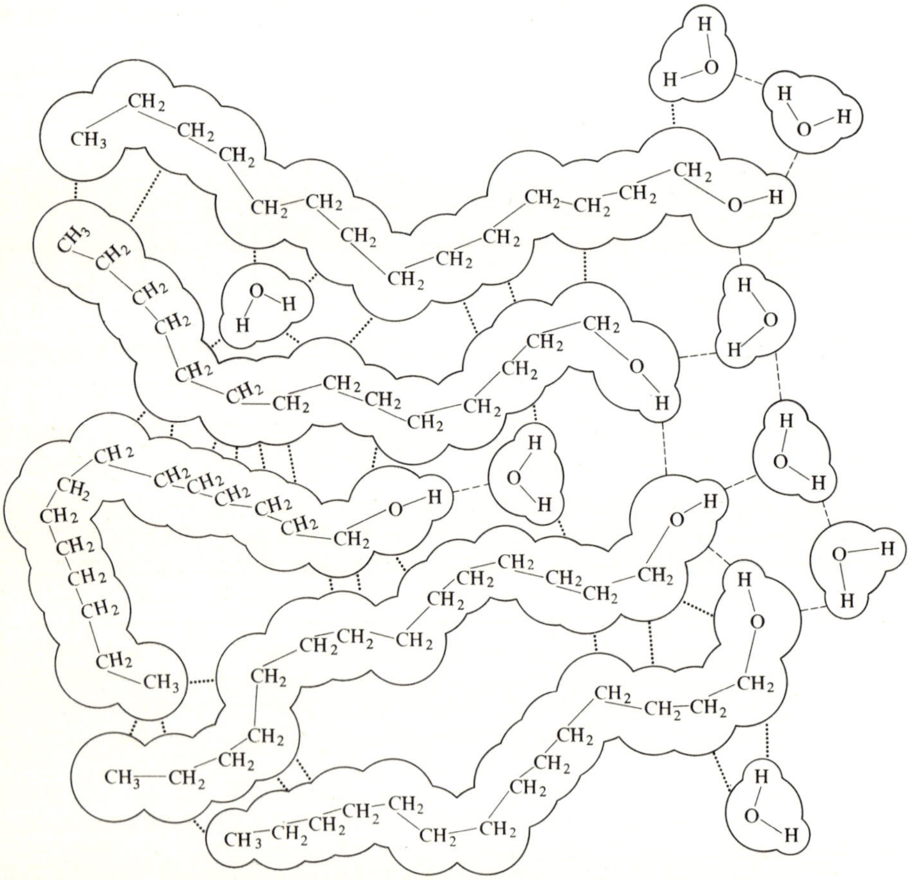

Figure 28-2 Attractions between water and tetradecanol in liquid phase (above 38 °C). Dashed lines indicate the relatively strong hydrogen bonds and the dotted lines indicate the much weaker van der Waals forces and dipole-induced dipole attractions.

are so that the induced dipoles are larger and the dipole-induced dipole attractions are stronger. (See footnote on page 221 for a further discussion of these factors.)

Effect of Halogens. A haloalcohol is less soluble than is the alcohol from which it is derived. The water solubility of haloalcohols is in the general order H-R-OH > Cl-R-OH > Br-R-OH > CH_3-R-OH > I-R-OH, where R is the same structure in all five groups of compounds. This rule may not hold if the halogen atoms are adjacent to an oxygen functional group because the inductive effect of the halogen may change the polarity of the group (or cause other phenomena) compared to its polarity when the halogen is absent. The effect of a halogen atom on solubility is small; a chlorine or bromine atom lowers water solubility less than a methyl group does (Table 28-2). Most halogen derivatives of the alcohols are considerably more soluble in water than their molecular weights would indicate. This effect occurs in other groups of compounds and is especially marked where a halogen atom is close to an oxygen-functional group.[2]

TABLE 28-2 WATER SOLUBILITY OF ALKANOLS AND HALOALKANOLS

Haloalkanol	Mol. wt.	Solubility ($g/100g. H_2O$)	Similar alkanol	Mol. wt.	Solubility ($g/100g\ H_2O$)
$ClCH_2CH_2OH$	80.5	∞	$CH_3(CH_2)_3OH$	74	7.9
$ClCH_2CH(OH)CH_3$	94.5	∞	$CH_3(CH_2)_2CH(OH)CH_3$	88	4.8
$Cl(CH_2)_3OH$	94.5	50	$CH_3(CH_2)_4OH$	88	2.3
$(CH_3)_2C(OH)\text{-}CH_2Cl$	108.5	6.6	$(CH_3)_2C(OH)\text{-}CH_2CH_2CH_3$	102	< 1.3
$Br(CH_2)_3OH$	139	16.6	$CH_3(CH_2)_8OH$	144	< 0.05

Effect of Chain Branching. Chain branching usually makes organic molecules more compact. If one molecule is more compact than the other and *all other factors are constant*, the more compact one usually has a lower boiling point. If the compound is a liquid, it has a higher solubility (Table 28-2). The more compact a structure, the less the attraction the molecules have for each other in the liquid phase because compact molecules have less surface for polar and London dispersion forces to act upon. (Review Chapters 2, 6, and 12.)

Regularity, symmetry, and compactness in molecules permit stronger and more compact crystal structures and, consequently, higher lattice energies. These higher intracrystalline attractive forces cause the compounds whose molecules are more branched and compact to have lower solubilities (if solids) in all solvents and higher melting points than their less compact isomers do. For instance, benzene melts much higher than does toluene even though benzene has a lower molecular weight; *para*-disubstituted benzenes usually melt higher and, if solids, are less soluble than their *ortho* and *meta* isomers. Compact molecules usually melt higher and boil lower than their less compact isomers do, i.e., neopentane melts higher but boils lower than any of its isomers, and *t*-butyl alcohol melts higher and boils lower than any of the other 4 carbon atom alcohols (Table 28-3).

Problem 28-2. From the tables in other chapters in this text and in chemical handbooks find 3

[2] This may be due to at least two causes: (a) The halogen atoms are very dense (see discussion on densities in the next section). As a result the water molecules are not separated so much when a molecule bearing a halogen atom slips between them as they are by a molecule of similar weight but having no halogen atoms. This is similar to one of the effects of aryl rings on water solubility. (b) Another effect is that the halogen atom by withdrawing electrons makes the oxygen atom negative and thereby able to form stronger hydrogen bonds with water molecules.

TABLE 28-3 EFFECTS OF BRANCHED AND CYCLIC STRUCTURES ON PHYSICAL PROPERTIES

Compound	B.p. (°C)	M.p. (°C)	Water solubility ($g/100g\ H_2O$, 20°C)	D_4^{20}
Cyclopentane	50	-94		0.746
Pentane	36	-130	0.04	0.626
Isopentane	28	-160		0.620
Neopentane	9.5	-17		
Cyclohexane	80.7	6.5		0.778
Hexane	68.3	-95		0.659
3-Methylpentane	63	-118		0.664
2-Methylpentane	60.3	-154		0.653
2,3-Dimethylbutane	58	-129		0.662
2,2-Dimethylbutane	50	-98		0.649
1-Butanol	117	-90	7.9	0.8096
Isobutyl alcohol	108	-108	10.0	0.8020
2-Butanol	99.5	-114	12.5	0.8069
t-Butyl alcohol	83	25.5	∞	0.789
Cyclopentanol	140.8			0.9469
1-Pentanol	138	-78.5	2.3	0.8148
3-Methyl-1-butanol	132	-117	2.8	0.8092
2-Methyl-1-butanol	128.7		3.2	0.8193
2-Pentanol	119		4.8	0.8092
3-Pentanol	116	-9	5.3	0.8204
3-Methyl-2-butanol	113.5		6.0	0.8180
2,2-Dimethyl-1-propanol	113.5	52.5	3.7	0.812
2-Methyl-2-butanol	102.4	-9	12.2	0.8090

series of 3 or more isomers in which the highest melting member is also the lowest boiling one. Indicate what you would expect the *relative* solubility of each isomer to be in water, in hexane, and dibromodifluoromethane.

Boiling Points and Melting Points

Polar Attractions and Effects in Homologous Series. Because of the strong polar attractions that are the hydrogen bonds, alcohols melt and boil higher than other compounds of similar weight that we have studied so far. As with all homologous series the boiling points of alcohols increase as the number of methylene groups increases. For monofunctional alcohols the difference in the boiling point between the alcohol and the parent hydrocarbon decreases as the molecular weight increases (last column of Table 28-1).

Effect of Halogen Atoms. Halogen atoms have approximately the same effect on boiling points of alcohols and other polar compounds as they do on the boiling points of hydrocarbons (pp. 168–170). Chlorine atoms have the same effect as an equal weight of methylene groups (top curve, Fig. 23-1), and bromine and iodine increase the boiling point less than an equal weight of methylene groups. Sometimes, however, a fluorine atom or two or more other halogen's *alpha* to a polar functional group cause the compound to boil and melt higher than would be indicated by molecular weight considerations. This is due to the inductive effect of the halogens increasing the polarity of the functional group. This is also one of the reasons that most of these compounds have higher water solubilities than molecular weight considerations would indicate. See page 217 and Figure 28-3.

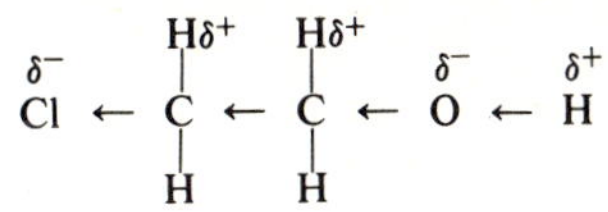

Figure 28-3 Inductive effect in 2-chloroethanol.

Effect of Branched and Cyclic Structures. The previous section on solubility indicated that liquids with highly branched molecules are more soluble in water and other solvents and lower boiling than their unbranched isomers. On the other hand solids with highly branched structures are less soluble in water and other solvents and are higher melting than their unbranched isomers. The phenomena of the relatively high melting points and low boiling points of highly branched molecules such as neopentane and *t*-butyl alcohol can be accounted for by comparing their alignment, organization, and order in the solid and liquid states. Perfect or nearly perfect orientation of straight-chain compounds as *n*-pentane occurs only in the solid state (Fig. 28-4). When the solid melts, the straight-chain compounds become

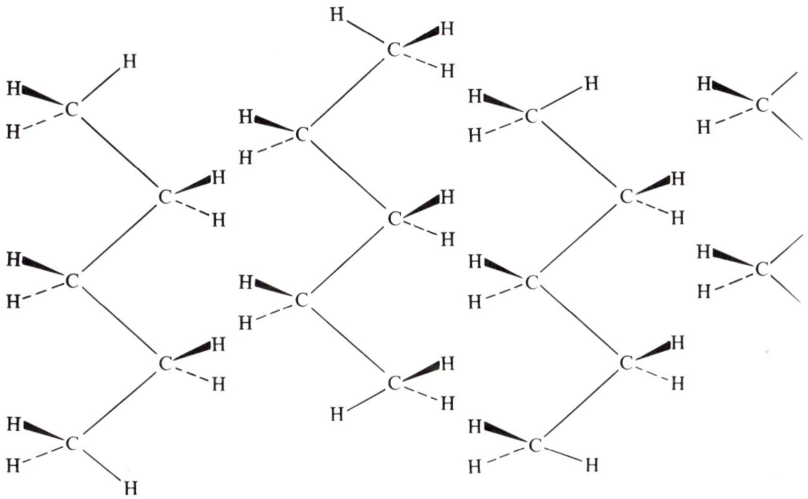

Figure 28-4 Crystalline pentane, an example of a perfectly aligned straight-chain compound.

very disordered (Figs. 28-5 and 2-7); however, because of the straight-chain molecules' length and flexibility, the randomness and disorder do not cause a large reduction in intermolecular attractions. If the molecule bends away from a second molecule, it will interact with a third molecule (or perhaps another part of the second molecule) that probably will be in one of the orientations necessary to have the same or greater intermolecular attractive forces. This could be compared to the difference between uncooked, rigid spaghetti and cooked spaghetti. The cooked spaghetti is completely randomly arranged; however, there is just as much (if not more) contact between the different pieces in the cooked as in the boxed uncooked spaghetti.

Highly branched molecules do not have so much chance to bend and coil as their straight-chain counterparts. Some, such as *t*-butyl alcohol and neopentane, have *no* coiling or significant bending. For instance, in solid neopentane the 4 methyl groups are oriented toward 4 methyl groups of other molecules (Fig. 28-6). The crystal structure somewhat resembles that of the diamond except the forces holding the units together are much weaker in solid neopentane. Any disorientation of this structure will cause a considerable drop in intermolecular attractions. Conversely, highly branched molecules such as neopentane freeze or crystallize so much more easily than their straight-chain counterparts because it is very easy for them to assume the orientation necessary for the maximum intermolecular

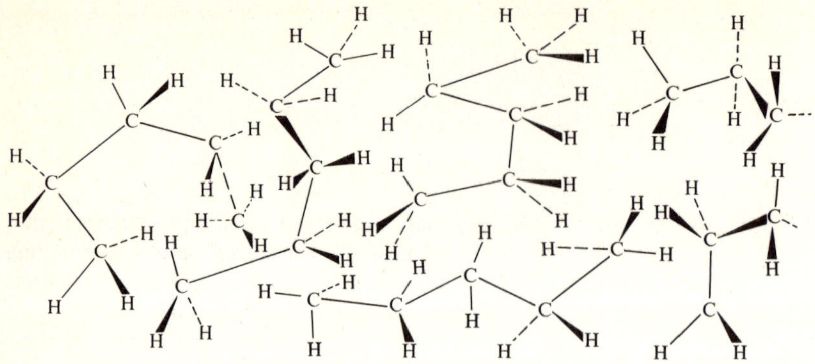

Figure 28-5 Liquid pentane, an example of straight-chain compounds nonaligned and randomly arranged.

attractions of the solid state once they are in the close contact of the liquid phase. If neopentane in the liquid state has maximum intermolecular attraction with 4 methyl groups on neighboring molecules (circled area in Fig. 28-6), there is formed a small segment of crystal. If this occurs with several adjacent molecules, a small crystallite is formed.

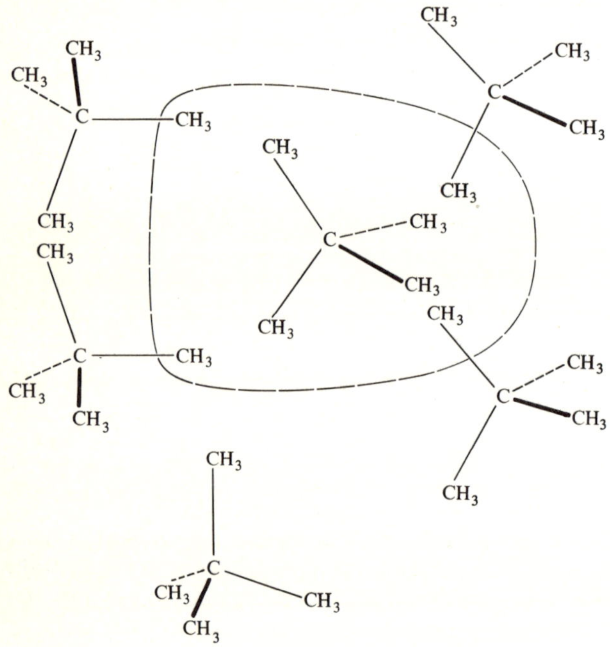

Figure 28-6 Crystalline neopentane.

In a solid the molecules are very close and there is permanent regular structure; in a liquid the molecules are very close but any intermolecular structure is not permanent; and in gases the molecules are relatively far apart and there is no intermolecular structure. Therefore, it is easy to see why molecules such as neopentane solidify easily. When put close together, i.e., condensed to a liquid, they will more or less fall into the correct orientation required for a crystalline solid.

The difference between the ease of orientation or organization of straight-chain molecules and highly branched molecules may be compared to the ease of orienting spherical meatballs (neopentane molecules) and strands of spaghetti (*n*-pentane molecules). All that is necessary to get the meatballs or other spheres in their most closely packed orientation is simply to pour them into a container or even in a pile. There is no randomness possible in the orientation of spheres. The more molecules approach the symmetry and compactness of spheres the less randomness possible. To regularly organize spaghetti or long molecules like pentane requires the considerable effort of lining up each one and placing it parallel to the next.

Notice in Table 28-4 that compounds with a small amount of branching (see isopentane, 2-butanol, and 2-methyl-1-butanol) have lower boiling points, higher solubility (whether a solid or liquid), and perhaps unexpectedly lower melting points than their straight-chain isomers. A *small amount* of branching causes less probability of maximum orientation in the solid and less intermolecular interaction in both the solid and the liquid. The extra piece prevents close alignment of the parallel chains and causes little increased interaction. Compare this with small pieces of spaghetti spliced onto the long pieces; it would require more effort to align the branched pieces with other branched pieces than it does with normal spaghetti.

TABLE 28-4 COMPARISON OF PHYSICAL PROPERTIES OF ANALOGOUS STRAIGHT-CHAIN, BRANCHED-CHAIN, CYCLIC, AND AROMATIC COMPOUNDS

Compound	B.p. (°C)	M.p. (°C)	D_4^{20}
Pentane	36.3	-129.7	0.629
Cyclopentane	49.5	-93.3	0.745
Hexane	69	-95	0.659
Methylcyclopentane	72		0.746
Cyclohexane	81	6.5	0.778
Benzene	80.1	5	0.8790
2-Methylhexane	90.0	-118.2	0.679
Heptane	98.4	-90.6	0.684
Methylcyclohexane	100.9	-126.3	0.769
Toluene	110.6	-95	0.8670
Cycloheptane	119.0	-12	0.810
3-Methyl-1-butanol	132	-117	0.8092
1-Pentanol	138	-78.5	0.8148
Cyclopentanol	140		0.9469
4-Methylpentanol	152.5		0.8131
1-Hexanol	156	-47	0.8189
Cyclohexanol	160	25.5	0.926
Phenol	182.5	42	1.069

Note in Table 28-3 that cyclic compounds have higher boiling points even though they are more compact than their straight-chain counterparts. This is probably because the cyclic compounds are planar or nearly so. This planarity allows the molecules to be closer together than their branched or straight-chain analogs; the intermolecular forces, therefore, act over a smaller distance and the attractive forces are greater than where such approach of molecules cannot occur. This hypothesis is supported by the fact that the densities of cyclic compounds are significantly greater (about 20%) than their open-chain counterparts (Table 28-4). Obviously the molecules of the cyclic compounds are closer together.[3]

[3] A second, more tenuous observation is that cyclic compounds whose molecules have all of their bulk coplanar or nearly so (Fig. 28-7) often have boiling points considerably higher than all their straight-chain counterparts. On the other hand, cyclic compounds which have a significant proportion of their bulk noncoplanar usually have boiling points only slightly above some of their straight-chain counterparts (Table 28-4). The most striking example of this is in the aromatic series where all the ring carbon atoms and all the atoms attached to the ring must be coplanar. The aromatic compounds are considerably denser than their acyclic and even their cyclic counterparts and usually have higher boiling and melting points than these aliphatic analogs. The difference in these physical constants, however, may be due largely to the fact that the easily polarized *pi* electron cloud causes greater dipole-induced dipole and induced dipole-induced dipole (London dispersion forces) interactions.

At least this picture is consistent with the water solubility of aromatic compounds. That is, an aromatic system has the same effect on water solubility in a series of compounds with oxygen or nitrogen functional groups as only about half the number of aliphatic carbon atoms in straight-chain molecules or about two-thirds the number of aliphatic carbon atoms in branched molecules. This could be accounted for by the water molecules' being able to pack more closely with the flat aromatic molecules. This closeness would allow

As would be expected, cyclic compounds melt higher than their acyclic counterparts. Because cyclic structures can assume fewer conformations, there is less randomness possible and, therefore, they have a smaller liquid temperature range than their acyclic counterparts. Of course, some noncyclic compounds such as neopentane are so symmetrical that they can assume oriented relationships more easily than similar cyclic compounds and have higher melting points.

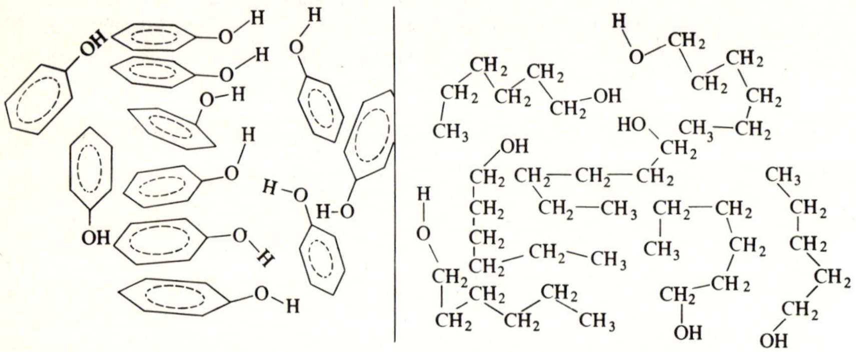

Figure 28-7 Comparison of the more closely packed cyclic molecules with the analogous straight-chain molecules.

Hydrocarbons were chosen for most of the examples above because the attractions are pure London dispersion forces which are more or less uniform over the entire surface of the molecules. These rules, however, still apply with additional complications when the forces are the more intense and localized polar attractions. This is demonstrated by the alcohols in Tables 28-3 and -4. All rules discussed above are more reliable with liquids than with solids because the intracrystalline forces of solids often show considerable variation with subtle changes of structure, configuration, and conformation.

DENSITY

A general idea of the density of an organic compound can be obtained by carefully observing its behavior during the solubility tests which are done during the systemic analysis of all unknowns. A liquid is denser than the solvent in which it is being tested if it sinks to the bottom, less dense if it floats to the top, and approximately the same density if it stays dispersed as fine bubbles and settles to the top or bottom only after standing for more than a minute.

All hydrocarbons and all water-insoluble, monofunctional oxygen derivatives of *aliphatic* hydrocarbons are lighter than water. The specific gravity of organic halogen compounds increases in the order RH < RF < RCl < RBr < RI. All organic bromides and iodides are denser than water. Organic halogen compounds are more dense than the corresponding nonhalogenated compounds because the halogen atoms have very high mass to volume ratio compared with most organic moieties.[4]

Most aromatic compounds are considerably denser than their aliphatic counterparts. *All* aromatic halides are denser than water. This is true even if the halogen atom is not attached directly to the aromatic

---

fewer dipole-dipole attractions between water molecules to be broken that were not compensated for by dipole-dipole attractions between the water molecules and the functional groups. Also, since the easily polarized *pi* electrons of the aromatic ring are so exposed, fairly strong dipole-induced dipole attractions can exist to compensate partially for the interrupted dipole-dipole attractions between the water molecules.

[4] The density of an organic moiety can be obtained by dividing the volume of the group into the sum of the atomic weights of the atoms in the group. The volume of the group, or at least the space it occupies, can be approximated by assuming that it is a sphere with a radius the same as the van der Waals radii of the group in question. In mass units per cubic angstrom (10^{-24}cc) the following group densities are obtained: H, 0.14; CH_3, 0.223; F, 1.84; Cl, 1.49; Br, 2.57; I, 3.09.

ring, e.g., benzyl chloride, D_4^{20} 1.1026. The combination of an aromatic group and an alcohol brings the density only up to (and usually slightly over) that of water (Table 28-5). All monochloro and monofluoro derivatives of the aliphatic hydrocarbons are less dense than water. A cyclic structure, however, leads to a more compact molecule, and cyclic compounds are more dense than their noncyclic counterparts although most saturated, cyclic compounds are not so dense as similar aromatic compounds. For example, saturated, cyclic chlorides have approximately the same density as water, e.g., cyclopentyl chloride, D_4^{20} 1.005, and cyclohexyl chloride, D_4^{20} 0.989.

TABLE 28-5 DENSITIES OF AROMATIC ALCOHOLS

Alcohol	D_4^{20}	Alcohol	D_4^{20}
$C_6H_5CH_2OH$	1.045	$C_6H_5CHOHCH(CH_3)_2$	0.985
m-$CH_3C_6H_4CH_2OH$	0.918	$C_6H_5CHOHCH_2CH_2CH_3$	1.020
$C_6H_5(CH_2)_2OH$	1.025	$C_6H_5C{\equiv}CCHOHCH_3$	1.037
$C_6H_5(CH_3)CHOH$	1.013	$C_6H_5C{\equiv}CCHOHCH_2CH_3$	1.015
$C_6H_5(CH_2)_3OH$	1.008	$C_6H_5CHOH(CH_2)_5CH_3$	0.946
$C_6H_5CHOHCH_2CH_3$	1.001		

The greater number of halogen atoms a compound contains, the greater its density. Every organic compound with two or more halogen atoms is denser than water. Similarly, the combination of a chlorine atom and an oxygen-functional group causes the compound to be denser than water, e.g., ethylene chlorohydrin, D_4^{20} 1.213.

Most compounds that do not dissolve in or react with concentrated sulfuric acid or 85% phosphoric acid are less dense than these acids (densities about 1.84). Iodides and polyhalides are among the few types of organic compounds that may be denser than and insoluble in concentrated sulfuric and 85% phosphoric acids (Fig. 28-8).

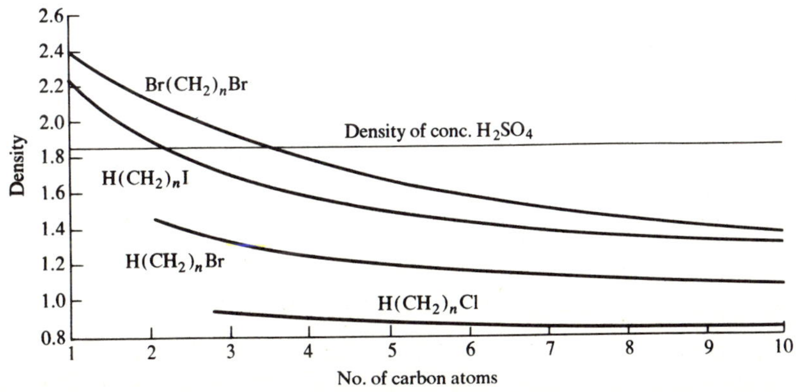

Figure 28-8 Densities of haloalkanes as a function of the number of carbon atoms.

SOLUBILITY IN CONCENTRATED ACIDS

H_2SO_4

Almost all compounds containing oxygen dissolve in or react with concentrated sulfuric acid. They behave as weak bases because of the unshared electron pairs on the oxygen atoms. These ionic protonated alcohols are soluble in the excess concentrated sulfuric acid:

$$H_2SO_4 + ROH \rightarrow \underset{\text{oxonium ion}}{\overset{+}{ROH_2}} + HSO_4^-. \qquad (28\text{-}1)$$

Organic, oxygen-containing compounds can react in other ways with the concentrated sulfuric acid. Primary and secondary alcohols often react to give the highly polar acid sulfate esters that are soluble in concentrated sulfuric acid:

$$RCH_2OH + 2H_2SO_4 \rightarrow RCH_2OSO_3^- + H^+ + H_3O^+ + HSO_4^-, \qquad (28\text{-}2)$$

$$R_2CHOH + 2H_2SO_4 \rightarrow R_2CHOSO_3^- + H^+ + H_3O^+ + HSO_4^-. \qquad (28\text{-}3)$$

Triarylcarbinols and some other compounds dissolve in concentrated sulfuric acid to produce stable carbonium ions:

$$Ar_3COH + 2H_2SO_4 \rightarrow Ar_3C^+ + H_3O^+ + 2HSO_4^-. \qquad (28\text{-}4)$$

Many secondary and most tertiary alcohols are dehydrated by sulfuric acid at room temperature to give alkenes. The excess sulfuric acid usually catalyzes the carbonium ion polymerization of these alkenes to insoluble, high molecular weight polyalkenes:

$$\underset{\displaystyle OH}{R_2C{-}CH_2R} \xrightarrow{H^+} \underset{\displaystyle \overset{}{OH_2^+}}{R_2C{-}CH_2R} \xrightarrow[(-H_2O)]{} R_2\overset{+}{C}{-}CH_2R \rightarrow R_2C{=}CHR + H^+ \qquad (28\text{-}5a)$$

$$R_2C{=}CHR \xrightarrow{H^+} R_2\overset{+}{C}{-}CH_2R \xrightarrow{R_2C=CHR} RHC{-}\underset{R}{\overset{R}{C}}{-}\underset{R}{\overset{H}{C}}{-}\underset{R}{\overset{R}{C^+}} \rightarrow \text{etc.} \qquad (28\text{-}5b)$$

Benzyl type alcohols usually undergo Friedel-Crafts reactions to give insoluble, orange condensates. The average composition of the condensate from benzyl alcohol corresponds roughly to

$$C_6H_5CH_2(C_6H_4CH_2)_{10}OH.$$

$$C_6H_5CH_2OH + H_2SO_4 \rightarrow C_6H_5CH_2^+ + H_3O^+ + 2HSO_4^- \qquad (28\text{–}6)$$

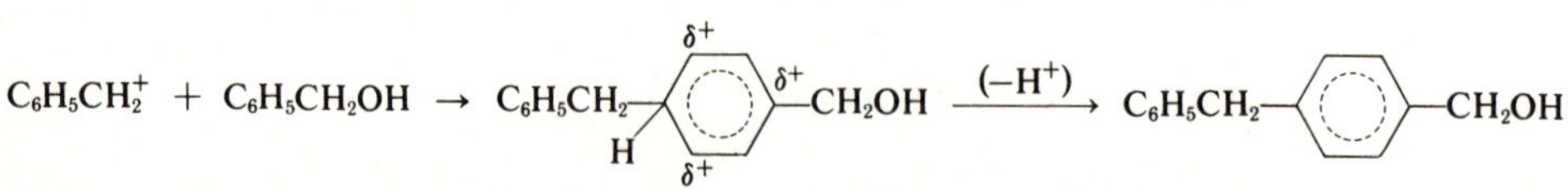

(28–7)

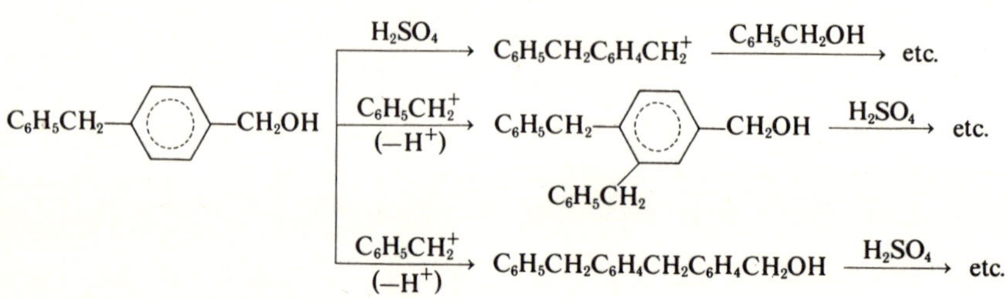

(28–8)

Since there are so many carbon atoms per hydroxyl group, the product is insoluble in concentrated sulfuric acid.

Problem 28-3. How would the following alcohols be expected to react with concentrated sulfuric acid: (a) *o*-$C_6H_4CH_2OH \cdot CO_2CH_3$, (b) *p*-$ClC_6H_4CH_2OH$, (c) 2,4,6-trichlorobenzyl alcohol?

As with the hydrocarbons all of the reactions described in the above paragraph should be considered a soluble test. If, however, any insoluble products or colors form, they should be noted, as such reactions may give further clues to the identity of the compound.

85% H_3PO_4

Phosphoric acid is used to distinguish low molecular weight, oxygen-containing compounds from the rest of the compounds that are soluble in concentrated sulfuric acid: most alkenes, reactive aromatic hydrocarbons, and high molecular weight, oxygen-containing compounds.

The solvent action of the 85% phosphoric acid is due almost entirely to the formation of organic ions from protonated, oxygen-functional groups. Unlike the reactions with concentrated sulfuric acid there is almost no polymerization, color formation, heat or gas evolution, or other reaction accompanying solution in 85% phosphoric acid. Alcohols and other groups with oxygen functions dissolve if they have fewer than 9 or 10 carbon atoms.[5]

Caution! Even though 85% phosphoric acid has a less drastic action than sulfuric acid on most organic compounds, it is a very corrosive and dangerous substance and should be treated with the same respect shown all acids.

Problem 28-4. A few alkenes such as the pentylenes dissolve in phosphoric acid. If some alkenes were to dissolve in phosphoric acid, why could it be predicted that the pentylenes would dissolve instead of most of the other alkenes given in Table 14-2?

SOLUBILITY CLASSIFICATION

Discussion

The solubility of organic compounds in water, acids, and other solvents (Chapters 32, 41, and 44) can be used to classify organic compounds so that they may be identified more easily. The solubility classification needed for unknowns assigned with this chapter is given in Figure 28-9.

Experimental Procedure

Put 3 ml of solvent (acid, water, or aqueous solution) in a test tube, and slowly add 0.15 to 0.20 g of unknown (0.2 ml of a liquid) in small portions. Make sure all solids are finely powdered so that a slowly dissolving solid will not be recorded as insoluble. *Agitate vigorously* with a clean *glass rod* after the addition of each portion of unknown. If any portion of the unknown fails to dissolve or react after 1 min of agitation, do not add any more unknown and record the sample as insoluble in that particular solvent. If the unknown dissolves, record any further evidence of reaction such as color change, polymer formation, or other details. If the unknown is a liquid or a solid melting below 35°, wait at least 3 min after the last agitation and examine closely to make sure that two phases have not separated.

Test the water solubility of isopropyl alcohol, benzene, cyclohexene, 2-pentanol, benzyl alcohol, 1-decanol, 1-octadecanol, and 2-pentene. All those that are soluble, record as Class S compounds. Test the solubility of all the water-insoluble compounds in concentrated sulfuric acid. Record all those that are insoluble as Class I

[5] For a few groups of compounds such as ethers and esters of dicarboxylic acids, the borderline of solubility in phosphoric acid occurs at about seven carbon atoms. There is, however, a considerable number of exceptions to these rules.

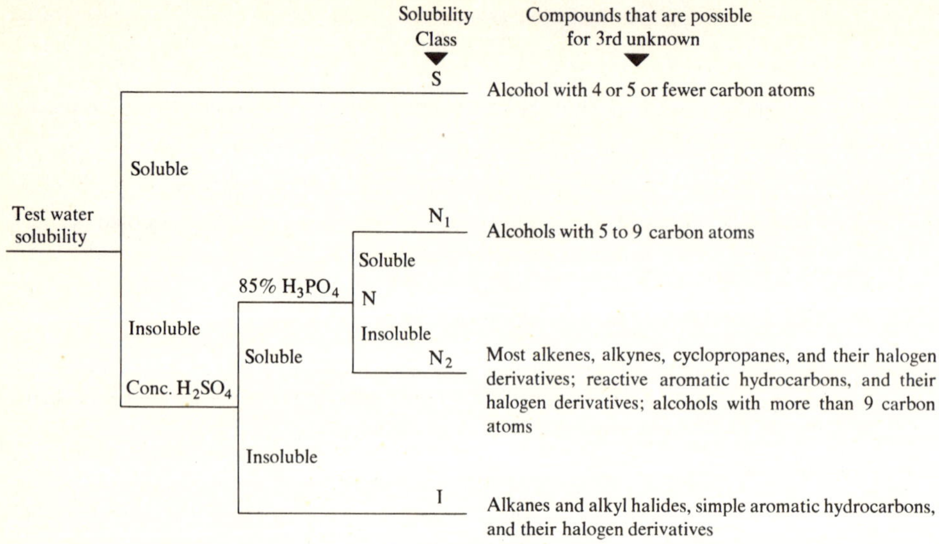

Figure 28-9 Solubility classification for unknowns.

compounds. Test the solubility of all the sulfuric acid-soluble compounds in 85% phosphoric acid. Record those that are soluble in phosphoric acid as Class N_1 compounds and those that are insoluble as Class N_2 compounds. Refer to Figure 28-9.[6]

The observation of two phases is often difficult when two colorless liquids are used. Shake the mixture briefly but vigorously. The mixture will become cloudy if two phases are present.

> **Problem 28-5.** Occasionally even this precaution will not enable you to distinguish two phases. When you are unable to do so, what properties of the two must be the same?

This technique of adding unknown to solvent has the advantage of conserving unknown for the next test if the compound proves to be insoluble. It has, however, the disadvantage of producing an occasional violent reaction when certain compounds are added to sulfuric acid in this manner. **Always wear eye protection when testing solubilities in acids, and always keep the test tube pointed away from you and others.**

If a *solid* seems to be insoluble in *water*, it is permissible to warm the mixture to about 50° in a water bath.[7] It is *very important* to allow the solution to return to room temperature before concluding that the solid is soluble. Take the usual steps to make sure there is no supersaturation. Agitate the solution, scratch the inside test tube wall, or add a small seed crystal. Do not cool the solution below 20°. Why? If the sample of unknown is small, any portions used in the solubility test with water that did not dissolve may be recovered and reused.

It is important to measure the quantities accurately. Measure liquids, both solvents and unknowns, with a graduated pipet and weigh solids on a balance to the nearest 0.02 g. Unless fairly careful control is kept over amounts used, compounds may be recorded in the incorrect solubility class. For instance, 25% excess of 1-butanol and 20% too little solvent would cause 1-butanol to appear water insoluble and to be classified erroneously as an N_1 compound.[8] If benzyl alcohol were the unknown and 35% too little of the unknown *or* 50% too much solvent were used, the benzyl alcohol would appear to be in Class S rather than Class N_1.

[6] The symbols S for water-*s*oluble compounds, I for sulfuric acid-*in*soluble or inert compounds, N for neutral compounds soluble in sulfuric acid, and similar symbols introduced are useful memory aids and shorthand but should not be memorized for their sake alone. Low molecular weight, oxygen-functional compounds should be thought of as water soluble rather than Class S unless, of course, it is understood that Class S means water soluble. If the symbols are memorized and solubility relations not understood, the symbols will be only a useless crutch that retard the laboratory work instead of being the aid they are intended.

[7] When solubility in ether is determined (Chapter 32), the unknown may be boiled in the ether for a few minutes. However, heat should not be used with any of the other solvents or solutions used for solubility tests as irreversible reactions may occur.

[8] If the correct amounts were used in some of the subsequent solubility tests, the unknown could appear to be in solubility Classes A or B. See Chapters 32 and 44.

CLASSIFICATION TESTS

Sodium

Discussion. Two general groups of organic compounds are acidic enough to release hydrogen gas when treated with sodium. Compounds which have a hydrogen atom on an electronegative atom, as —SH, $-\overset{|}{N}H$, —OH, comprise the larger of these two groups. The only class of compounds studied thus far that meets this requirement is the alcohols:

$$2ROH + 2Na \rightarrow 2R{-}O^- Na^+ + H_2\uparrow \qquad (28\text{-}9)$$

The other group of organic compounds that react this way consists of compounds that lose hydrogen from a carbon atom to give a relatively stable carbanion. The only class of compounds studied thus far that meets this requirement is the 1-alkynes although there are other important types of organic compounds that reach this way:

$$2RC{\equiv}CH + 2Na \rightarrow 2RC{\equiv}\bar{C}\ Na^+ + H_2\uparrow \qquad (28\text{-}10)$$

This reaction should be interpreted very carefully because:

a. So many organic compounds contain water as an impurity. This is especially true if the compounds show significant mutual solubility with water.
b. Sodium that has been cut or allowed to stand for more than a few minutes in moist air adsorbs moisture on its surface and does not react with a good portion of this adsorbed moisture. When this sodium is placed in a completely anhydrous, nonacidic solvent such as dry benzene, the adsorbed moisture reacts with the sodium to evolve hydrogen. This gas evolution is not vigorous and usually of short duration because of the small amount of water involved.
c. The rate of hydrogen evolution decreases with increasing molecular weight of acidic compounds. With monofunctional alcohols the test becomes difficult to interpret if the alcohol has more than nine carbon atoms, and almost useless if it has more than 14 carbon atoms. The rate of evolution of hydrogen for these compounds is so small as to be mistaken easily for a nonacidic compound containing a small amount of water.

Difficulty (a) can be surmounted by thoroughly drying the unknown. Difficulty (b) can be overcome partially by keeping the exposure of the sodium to the atmosphere at an absolute minimum and by cutting the sodium with a knife or spatula that just has been *warmed briefly* in a burner flame. **Be sure that the blade is not so hot that it will ignite the sodium.** Difficulty (c) can be overcome partially by taking the precautions outlined above for the other two and, if the gas evolution at the beginning is slow, by observing whether or not it persists longer than three minutes.

Problem 28-6. If the test is thought to be positive, a small amount of ether can be added to the reacted mixture. A precipitate will verify the test. What is the precipitate? How does it verify the test?

Experimental Procedure. Dry the sample in a test tube. If a solid, it can be dried by heating to 120° for 30 min in an oil bath. If a liquid, it can be dried by stirring with about one-third its weight of anhydrous sodium sulfate.[9] In a *dry* 7.5-cm test tube place 1 ml of the dried compound. Add about 0.2 g of sodium (0.1 g or less if the compound boils above 180°) as two or three small pieces. **Warning! Sodium is very reactive! Quench all residues and scraps of sodium with excess ethanol and wash down the drain the resulting strongly basic solution (after all the sodium has reacted) with excess water. Always wear eye protection when working with sodium.**

If the unknown is a solid, dissolve 0.5 to 0.8 g in 5 ml of dry benzene. Warm, if necessary, to effect solution. Add about 0.1 mg or less of sodium. If the evolution of hydrogen is feeble but persists for at least 4 min, record the test as positive. If no hydrogen is being evolved, gently shake the tube and observe 1 min later. If

[9] It is assumed that the unknown has been purified by distillation or recrystallation if the physical constants indicate an impure compound.

hydrogen is now being evolved, observe again in 3 min. If the evolution persists, record the test as positive. For any compound that gives a positive test wait until the reaction stops. Cool to room temperature and look for the appearance of a precipitate to verify the positive test. If no precipitate appears add some ether. Try this test on isopropyl alcohol, 1-pentanol, 1-octanol, 1-dodecanol, acetone, tetrahydrofuran, and phenylacetylene. Be sure to use the special samples that have been prepared for the test by the laboratory instructor.

Problem 28-7. Why should calcium chloride not be used as a drying agent for this test?

Acid Chlorides

Discussion. Because of the occasional difficulties in interpreting the acid chloride and sodium reactions, all Class N and neutral Class S compounds should be tested with both acid chlorides and sodium. The final decision as to whether or not an alcohol group is present should be made after the results of both these tests are considered along with the physical constants and behavior during the solubility tests.

Alcohols react with acid chlorides to give esters. These products usually appear distinctly different from the parent alcohol.

$$RCH_2OH + C_6H_5COCl \rightarrow C_6H_5CO_2CH_2R + HCl \qquad (28\text{-}11)$$

$$R_2CHOH + CH_3COCl \rightarrow CH_3CO_2CHR_2 + HCl \qquad (28\text{-}12)$$

$$R_3COH + CH_3COCl \rightarrow R_3CCl + CH_3CO_2H \qquad (28\text{-}13)$$

or

$$R_3COH + CH_3COCl + R_3N \rightarrow CH_3CO_2CR_3 + R_3NH^+ + Cl^- \qquad (28\text{-}14)$$

The esters are usually fragrant smelling liquids or crystalline solids. Primary and secondary alcohols of low molecular weight react very vigorously and exothermally with acetyl chloride. All alcohols except methanol and ethanol give esters that are water insoluble whereas all alcohols with 4 and some with 5 carbon atoms and many polyfunctional alcohols are water soluble. Tertiary alcohols react with acetyl chloride to give tertiary chlorides. If a small amount of certain amines such as pyridine and N,N-dimethylaniline are added, even tertiary alcohols will form esters. Since tertiary chlorides differ from alcohols in odor, density, and other easily recognized properties, it is not always necessary to make the ester to establish that a tertiary alcohol is present.

Experimental Procedure. **Be careful! Wear eye protection when doing this experiment. The reaction may be violent. Acid chlorides are usually strong lachrymators. Any excess of these acid halides and residues from the reactions should be treated carefully with excess ammonium hydroxide and washed down the drain. The precautions and antidotes for these compounds are the same as for strong acids.**

Acetyl Chloride: put 0.5 ml of the compound to be examined in a 7.5-cm test tube, carefully add 5 drops (about 0.25 ml) of acetyl chloride, and note what happens. Shake the tube, allow it to stand for a few minutes, and pour into 4 ml of 5% sodium bicarbonate solution in a 10-cm test tube. Try this test on ethanol, *t*-butyl alcohol, 2-pentanol, cyclohexene, water, and 1-octanol. Repeat the test on *t*-butyl alcohol, 2-pentanol, and cyclohexene, but add 4 drops of pyridine to the compound before adding the acetyl chloride.

Benzoyl Chloride (Schotten-Baumann reaction): put 1 ml of the compound to be examined in a 7.5-cm test tube, add 5 ml of 5% aqueous sodium hydroxide and 8 drops (about 0.4 ml) of benzoyl chloride, and shake the stoppered tube vigorously. Try this test on ethanol, 2-pentanol, 1-butanol, 1-octanol, and cyclohexene.

Problem 28-8. What is the purpose of the sodium hydroxide solution in the Schotten-Baumann reaction?

The Lucas Test, A Carbonium Ion Reaction

Discussion. This test is dependent upon the rate of ionization of the acidified hydroxy group, $\overset{+}{-OH_2}$ or $\underset{^{+}Zn}{\underset{|}{-OH}}$, from the organic moiety to give a carbonium ion.

$$R{-}OH \xrightarrow[HCl]{ZnCl_2} R{-}\overset{\overset{H}{|}}{\underset{+}{O}}{-}Zn^+ \rightarrow \left[\begin{matrix}\delta^+ \ \delta^+ \\ R{-}{-}{-}OH \\ | \\ ^{+}Zn\end{matrix}\right] \rightarrow R^+ + Zn^{++} + OH^- \qquad (28\text{-}15a)$$

$$R{-}OH \xrightarrow[HCl]{ZnCl_2} ROH_2 \rightarrow \left[\begin{matrix}\delta^+ \ \delta^+ \\ R{-}{-}OH_2\end{matrix}\right] \rightarrow R^+ + H_2O \qquad (28\text{-}15b)$$

$$R^+ + Cl^- \rightarrow R{-}Cl \qquad (28\text{-}15c)$$

The rate of the reaction is dependent on the stability of the transition state of the rate-determining step $\left[\begin{matrix}\delta^+ \quad \delta^+ \\ R{-}{-}{-}{-}OH_2\end{matrix}\right]$ or $\left[\begin{matrix}\delta^+ \quad \delta^+ \\ R{-}{-}{-}{-}OH \\ | \\ ^{+}Zn\end{matrix}\right]$. Since the transition state resembles the carbonium ion especially in the development of the positive charge, structural features that stabilize the carbonium ion also stabilize the transition state and cause an increase in the rate of reaction. The different rates of reaction with the Lucas reagent are used to distinguish primary, secondary, and tertiary alcohols from one another. The stability of the carbonium ions is $3° > 2° > 1° > CH_3^+$; therefore, this is the same sequence of reaction rates of alcohols from which the carbonium ions are derived.

The visible change indicating a reaction has taken place is the separation of the insoluble organic chloride from the aqueous phase. Tertiary alcohols react almost instantly as shown by an immediate white, cloudy suspension. A second phase of organic chloride separates within 5 min. Because of the stability of benzylic and allylic type carbonium ions, alcohols with the structures $Ar{-}\overset{|}{\underset{|}{C}}{-}OH$ and $-\overset{|}{C}{=}\overset{|}{C}{-}\overset{|}{\underset{|}{C}}{-}OH$ give an insoluble chloride as rapidly or more rapidly than the tertiary alcohols.

Problem 28-9. Allyl alcohol forms carbonium ions immediately with the reagent but does not give evidence of reaction for at least 7 min and often never shows any sign of reaction. This is despite the fact that most other alcohols with the same basic structure, $-\overset{|}{C}{=}\overset{|}{C}{-}\overset{|}{\underset{|}{C}}{-}OH$, give an instant cloudiness, e.g., cinnamyl alcohol ($C_6H_5CH{=}CH{-}CH_2OH$). Why? What could be done to cause the allyl chloride to form a second phase?

Secondary alcohols produce a milky suspension within 5 min of their addition to the Lucas reagent and a second phase within 10 min. Primary alcohols other than the allylic and benzylic type show no sign of reaction after an hour or longer.

The times given were measured when the reaction mixture was at a temperature of 25 to 27 °C. Temperatures considerably above or below this cause a marked change in the rate of reaction, especially in the reactions of secondary alcohols. Near room temperature a 10° increase in temperature approximately doubles the rate of this type of reaction. Temperatures as low as 20 °C (68 °F) and as high as 30 °C (86 °F) are common in many laboratories, and extremes of 18 °C (65 °F) and 36 °C (96.8 °F) occasionally occur. Such deviations in temperature are not likely to go unnoticed by the student. When such extremes occur, the alcohols should be tested in a water bath at 26° or at least the possibilities of the influence on rate should be considered in the interpretation of the data.

The consideration of the effect of extremes in the laboratory temperature is important not only for the Lucas test but also for all classification tests that depend on differences in reaction rate to classify functional groups. (See Chapters 23 and 61.) With the Lucas test at low temperature the only problem is that secondary alcohols react more slowly than usual. There is not much chance for misinterpretation in this case because primary alcohols react extremely slowly and tertiary, benzylic, and allylic alcohols react very rapidly at any temperature likely to be encountered in the laboratory. At the higher temperatures some secondary alcohols react rapidly enough to make their differentiation from tertiary alcohols more difficult. This difficulty can be overcome by conducting the test in a beaker of water maintained at 26° by the addition of small pieces of ice.

In addition, if the alcohol is sufficiently soluble in concentrated hydrochloric acid,[10] it can be tested by dissolving it in concentrated hydrochloric acid. Tertiary alcohols react almost immediately and form a separate phase on top of the acid within 3 min. Secondary alcohols show no sign of reaction within 10 min.

Problem 28-10. At sufficiently low temperatures allyl alcohol may *appear* to react more rapidly than it does at higher temperatures. Why?

By increasing the acidity of the medium, the zinc chloride serves two purposes in this test. It increases the rate of formation of the carbonium ions so that the reaction with secondary alcohols can be observed within a convenient time period. It also increases the solubility of the alcohol so that alcohols with as many as 6 carbon atoms (up to 10 carbon atoms if an aromatic ring is present or if the alcohol is polyfunctional) dissolve in Lucas reagent. This increase in solubility is due to the formation of oxonium ions which are soluble in the polar medium of the Lucas reagent. The dissolving of water-insoluble compounds by this reagent occurs because of the same basic reasons that certain alcohols dissolve without further reaction in sulfuric acid and that low molecular weight alcohols dissolve in phosphoric acid. The test is difficult to interpret if the alcohol is insoluble in the reagent. Why?

Unless the purity of the unknown is absolutely certain, only the appearance of a second phase on top of the aqueous solution should be considered a definite positive test. This is because small amounts of impurities may produce cloudiness before the pure compound could show any sign of reaction. For instance, 1-butanol with a 20% impurity of 2-butanol gives a slight cloudiness within 6 min and 2-butanol with a 5% impurity of *t*-butyl alcohol gives a slight cloudiness within 30 sec. In neither of these two cases does a distinct upper layer appear before it would with the pure compound.

Since the test depends upon the appearance of a second phase, the test is most easily interpreted if the alcohol is soluble in the reagent, i.e., all aliphatic alcohols with 5 or fewer carbon atoms, some branched hexyl alcohols, aromatic alcohols with 9 or fewer carbon atoms, and some polyfunctional aliphatic alcohols with 6 or more carbon atoms. It is, however, possible to get some indication of the type of alcohol group present even if the alcohol does not dissolve completely in the reagent. Thus, both 1-hexanol and 3-heptanol fail to give a homogenous solution with Lucas reagent, but after the samples have been agitated for a minute and allowed to stand for several more minutes, only the lower phase in the tube containing the secondary alcohol has the characteristic milky appearance of a positive test. Of course, such results should be accepted with reservations.

Experimental Procedure. Lucas reagent is made by dissolving 136 g of *anhydrous* zinc chloride in 105 g of concentrated hydrochloric acid. The reagent is brought to 26° before using. Quickly add 9 ml of the reagent to 1 ml of the alcohol in a *dry* test tube. Stopper and shake the tube vigorously but briefly. Note the time required for the appearance of a top layer. If nothing appears after 25 min (40 min if the temperature of the environment of the reaction is at 20°), record the test as negative and discard the solution. If the alcohol is not completely soluble in Lucas reagent, shake the stoppered tube vigorously for 30 sec and observe as above. Try this test on 1-butanol, 2-butanol, *t*-butyl alcohol, 3-heptanol, and benzyl alcohol.

To 1 ml of alcohol in a test tube add 9 ml of concentrated hydrochloric acid. Shake the tube vigorously, allow to stand, and observe carefully what happens during the first 4 min after this addition. Try this test on 2-butanol, *t*-butyl alcohol, and allyl alcohol.

Problem 28-11. Why does the secondary alcohol 1-chloro-2-propanol not give a visible reaction with Lucas reagent within 30 min? How would 1,3-dibromo-2-propanol react?

Problem 28-12. Although isopropyl alcohol gives a milky solution within 6 min, it often fails to give a second phase. Why? (Hint: see Table 23-6.) How could you be sure to observe the top layer when isopropyl alcohol is tested?

[10] All secondary and tertiary alcohols that boil below 141° are sufficiently soluble in the acid to use this test.

CrO_3 in Aqueous H_2SO_4

Discussion. Primary and secondary alcohols can be oxidized readily and without breaking the carbon skeleton.

$$3RCH_2OH + \underset{\text{orange}}{4CrO_3} + 12H^+ \rightarrow 3RCO_2H + \underset{\text{green}}{4Cr^{+++}} + 9H_2O \qquad (28\text{-}16)$$

$$6R_2CHOH + \underset{\text{orange}}{4CrO_3} + 12H^+ \rightarrow 6R_2CO + \underset{\text{green}}{4Cr^{+++}} + 9H_2O \qquad (28\text{-}17)$$

Tertiary alcohols are fairly resistant to oxidation because they have no hydrogen on the alpha carbon atom. Primary or secondary alcohols react with chromic oxide and sulfuric acid within *3 sec* at 24° to give a green precipitate which causes the solution to become opaque and a green color to replace the orange. With tertiary alcohols or most nonhydroxylic organic compounds the solution *may* become slightly opaque, but the solution does not turn green at room temperature.

This oxidation test is relatively insensitive to impurities. Tertiary alcohols that are contaminated with as much as 20% of a secondary alcohol do not give the *immediate* opacity and *green color* that is considered a positive test. Although with this much impurity, the solution becomes opaque within 30 sec and a green color appears within 2 min.

This test can be applied universally to all alcohols, even those of high molecular weight. Carbon to carbon double and triple bonds do not cause cloudiness before 30 sec, and a green color does not develop until later.[11] Thus, compounds as $CH_3CH{=}CHCH_2\underset{|\atop OH}{C}(CH_3)_2$ do not give false positives for primary and secondary alcohols.

Experimental Procedure. Dissolve 1 g of chromic oxide in 1 ml of concentrated sulfuric acid and then dilute in 3 ml of water. **Be careful! Wear eye protection!** Dissolve about 0.01 g or 1 drop of the alcohol to be tested in 1 ml of pure acetone. The acetone should be distilled from potassium permanganate to remove any easily oxidizable impurities. Add 1 drop of the chromic oxide reagent to the unknown solution. Shake the tube vigorously for 2 sec. Note the time required for any color change or precipitate. Try this test on benzyl alcohol, 1-octadecanol, 1-butanol, 2-butanol, and *t*-butyl alcohol. Be sure to use a clean, *dry* test tube.

Interference with the Baeyer Test for Unsaturation

Sometimes saturated primary and secondary alcohols decolorize neutral, dilute potassium permanganate (Chapter 14). This is due to the presence of easily oxidizable impurities, such as aldehydes, in the alcohol. Reasonably pure, *saturated* alcohols do not decolorize dilute, cold, *neutral* permanganate solution for at least 5 min. Disregard any statements in your lecture text to the contrary.

The Iodoform Test

Discussion. Secondary methyl carbinols can be distinguished from other secondary alcohols by use of the iodoform test.

$$R\underset{}{\overset{OH\atop |}{C}}HCH_3 + 4NaOI \rightarrow RCO_2^- + HCI_3\downarrow + 4Na^+ + I^- + 2HO^- + H_2O \qquad (28\text{-}18)$$

Another group of compounds that gives positive tests is the iodomethylcarbinols, $R\overset{OH\atop |}{C}HCH_2I$ and $R\overset{OH\atop |}{C}HCHI_2$. A yellow precipitate of iodoform indicates a positive test. If verification is needed, the melting point of the precipitate can be determined (CHI_3, m.p. 120 to 121°). Occasionally the iodoform forms an oil instead of a crystalline solid. If so, crystallization can be induced by the usual methods (Chapter 12).

[11] Impure alkenes, however, may give positive tests. This is probably due to the presence of hydroperoxides.

Experimental Procedure. Dissolve 5 g of iodine in a solution of 10 g of potassium iodide and 40 ml of distilled water. Store away from heat and light in a brown glass bottle. Dissolve about 0.1 g of compound to be tested in 2 ml of water in a 15-cm test tube. If the compound is insoluble in water, dissolve in a maximum of 4 ml of dioxane. Add 3 ml of 10% sodium hydroxide solution. Add the iodine-potassium iodide solution drop by drop while stirring until a slight excess of iodine is present as shown by the persistence of the dark iodine color. Place the test tube in a water bath at 60°. If the iodine color disappears, add more iodine-potassium iodide solution drop by drop until the dark color persists for at least 2 min at 60°. Add 10% sodium hydroxide solution a drop at a time with agitation until the dark color disappears. Remove the tube from the water bath and add 10 to 15 ml of water, taking care not to overfill the test tube. If the test is positive, the iodoform precipitates at this point. Try this test on 2-propanol, 2-pentanol, and 3-pentanol.

Problem 28-13. What simple primary alcohol gives a positive iodoform test?

UNKNOWN

Hand in a clean dry test tube; you will receive your third unknown in it. Your third unknown can be any compound from the tables at the end of this chapter (Tables 28-6 through 28-8, except compounds marked with an asterisk) *and* any of the compounds from the tables at the end of Chapters 14 and 23.

The recommended procedure for identification is to determine first the melting point and/or boiling point. If the physical constants indicate that the unknown is impure, it should be purified. The solubility should be determined by the methods outlined beginning on page 225. The results of this determination (Fig. 28-9) indicate which classification tests from Chapters 14, 23, and 28 should be done. Many or all of these classification tests may be omitted if instrumental techniques are used. The physical constants and the results from the solubility and classification tests should indicate unambiguously which compound in the tables is the unknown. Do the following problems before attempting your unknown. Assume the compounds listed in the tables at the end of this chapter and Chapters 14 and 23 are the only solutions to the problems.

EXERCISES

28-1. A water-soluble compound has a boiling range of 96 to 97°; what are the possible structures? What three tests could you use to distinguish among the possibilities? Which one of the three tests would be the least ambiguous and the least subject to interference by impurities?

28-2. A Class N_2 compound has a boiling range of 212 to 213.5° and reacts with sodium to give a few gas bubbles. It gives no apparent reaction with acetyl chloride but it decolorizes cold, dilute permanganate and bromine in carbon tetrachloride. What is the compound?

28-3. A Class N_1 compound has a boiling range of 160 to 161°. Which compounds are possible? What tests would positively identify this unknown?

28-4. A Class S compound (I) reacts with sodium to give hydrogen and reacts with acetyl chloride to give a water-insoluble liquid denser than water. I contains bromine and boils at 148 to 149.6°. When an alkaline solution of I is treated with iodine-potassium iodide reagent, a yellow solid (II) precipitates which melts at 58 to 59.6°. Identify I and II.

28-5. A Class S compound (III) that contains chlorine evolves hydrogen when treated with sodium and boils at 126.8 to 128°. When an alkaline solution of III is treated with iodine-potassium iodide solution, a colorless dense liquid (IV) separates. All attempts to induce crystallization of IV fail. IV boils at about 200°. Identify III and IV.

28-6. Which other compounds in Table 28-8 would give pseudoiodoform tests similar to compounds I and III above?

TABLE 28-6 ALCOHOLS
(LIQUIDS)

Alcohol	B.p. (°C)	M.p. (°C)	Derivatives m.p. (°C) 3,5-dinitro-benzoate	phenyl urethane	α-naphthyl urethane
Methyl alcohol	65		107	47	124
Ethyl alcohol	78		93	52	79
Isopropyl alcohol	83		122	88	106
t-Butyl alcohol	83	25.5	142	136	101
Allyl alcohol	97		48	70	109
n-Propyl alcohol	97		74	51	80
sec-Butyl alcohol	99	-114	75	65	97
2-Methyl-2-butanol	102	-8.5	117	42	71
Isobutyl alcohol	108	-108	86	86	104
3-Methyl-2-butanol	113	-9	76	68	112
n-Butyl alcohol	118	-90	64	57	71
*3-Pentanol	116		97	48	71
2-Pentanol	119		61		76
*2-Methoxyethanol	125				113
1-Chloro-2-propanol	127		77.5		
2-Methyl-1-butanol	129		70	31	82
*Isoamyl alcohol	130	-117	62	55	67
4-Methyl-2-pentanol	131.5		65	144	68
*2-Ethoxyethanol	135		75		67.4
1-Pentanol	138	-78.5	46	46	68
Cyclopentanol	140		115	132	118
3-Ethyl-3-pentanol	142				
*Acetoin (3-hydroxybutanone)	145	-72			
*Hydroxyacetone (1-hydroxy-propanone)	146	-71			
*4-Heptanol	156	-41.5	64		80
1-Hexanol	156	-47	58	42	59
Cyclohexanol	160	25.5	112	82	128
2 Heptanol	160		49		54
*2-Methylcyclohexanol	165	21	115	103	155
*Furfuryl alcohol	170		80	45	129
*Diisobutylcarbinol	172			62	
*Pinacol	172			215	
4-Methylcyclohexanol	174		130	125	160
1-Heptanol	176.5	-34	47	65	62
*dl-2,3-Butanediol	176.7	7.6			
*Tetrahydrofurfuryl alcohol	178		84	61	
*2-Octanol	179		32	114	63.5
*meso-2,3-Butanediol	182	34.4		201	
*Propylene glycol (1,2-pro-panediol)	188			153	
*Butyroin	190				
1-Octanol	192	-16	61	74	66.5
*Ethylene glycol	198	-12.6	169	157	176
Linalool	199			65	53
1- Phenylethanol	203	20	95	94	106
Benzyl alcohol	205	-15.3	112	78	154
1-Nonanol	214		52	62	65
*1,3-Propanediol	215	-30	178	137	164
m-Tolylcarbinol	217				116

*Will not be given as an unknown until Chapter 61 is assigned.

TABLE 28-6 (cont.)

Alcohol	B.p. (°C)	M.p. (°C)	Derivatives, m.p. (°C) 3,5-dinitro-benzoate	phenyl urethane	α-napthyl urethane
2-Phenyl-1-ethanol	221	-25.8	103	79.5	119
Citronellol	222				
Geraniol	229		62.5		47.5
1-Decanol	232		57	60	71
3-Phenyl-1-propanol	235		92	47	
*Diethylene glycol	245		149		
Cinnamyl alcohol (3-phenyl-2-propen-1-ol)	257	33	121	91	114
*Glycerol	290d		188(tri)	180(tri)	191(tri)
p-Methoxyphenylmethylcarbinol	310d			82.5	

TABLE 28-7 ALCOHOLS
(SOLIDS)

Alcohol	M.p.(°C)	B.p.(°C)	Derivatives, m.p.(°C) 3,5-dinitro-benzoate	phenyl urethane	napthyl urethane
1-Phenylethanol	20	202	94	92	106
2-Methylcyclohexanol	21	167.4	115	105	155
1,1-Dimethyl-2-phenyl-ethanol	24	216			
1-Dodecanol	24.5	257	60	74	80
p-Methoxybenzyl alcohol	25	259		93	
Cyclohexanol	25	161	112	82	128
t-Butyl alcohol	25.5	82.5	142	136	101
Cinnamyl alcohol	33	257	121	91	114
o-Tolylcarbinol	35.5	219		79	
trans-9-Octadecen-1-ol	36	333		57	71
1-Tetradecanol (myristyl alcohol)	39	278	67	73	82
*Pinacol (tetramethyl glycol)	45	172			
1-Hexadecanol (cetyl alcohol)	50	295	66	73	82
2,2,6-Trimethylcyclohexanol	51	187			
trans-3,3,5-Trimethylcyclo-hexanol	52	196			
1-Heptadecanol	54	310	121.5		88.5
1,2-Diphenylethanol	67				
*Nitrobenzyl alcohol	74	270			
10-Nonadecanol	85			96	54
**o*-Hydroxybenzyl alcohol (saligenin)	86				
**d*-Sorbitol	90				
**meso*-Erythritol	121	330			
**dl*-Benzoin	136	344		165	140
Cholesterol	148.5			168	176
Triphenylcarbinol	162	380			

*Will not be given out as an unknown until Chapter 61 is assigned.

TABLE 28-7 (cont.)

Alcohol	M.p. (°C)	B.p. (°C)	Derivatives, m.p. (°C) 3,5-dinitrobenzoate	phenyl urethane	napthyl urethane
Ergosterol	165		202	185	202
*d-Mannitol	166			303	
d-Borneol	209	212	155	139	131
*meso-Inositol	223		86		

TABLE 28-8 HALOALCOHOLS (LIQUIDS)

Haloalcohol	B.p. (°C)	M.p. (°C)	D_4^{20}	Derivatives, M.p. (°C) 3,5-dinitrobenzoate	phenyl-urethane	naphthyl-urethane
1-Chloro-2-propanol	127		1.115	77		
1-Chloro-2-methyl-2-propanol	128	-20	1.061			
2-Chloro-1-propanol	133.5		1.103	76		
3-Chloro-2-butanol	137		1.105			
2-Iodoethanol	176d		2.197			
2-Bromoethanol	150d		1.772			86
Trichloroethanol	151	19		142	87	120
2-Bromo-2-propen-1-ol	154		1.597			
3-Chloro-1-propanol	161		1.311		38	78
1,3-Dichloro-2-propanol	176				73	115
3-Bromo-1-propanol	176d		1.571			73
2,2-Dibromoethanol	181					
2,3-Dibromo-1-propanol	219d				84	
2,3-Dichloro-1-propanol	182				73	93

29

Synthesis of Alcohols by Borohydride Reduction

Vacuum Distillation

INTRODUCTION

Sodium borohydride can be used to reduce aldehydes and ketones. By use of Lewis acids and sodium borohydride, esters can be reduced to primary alcohols.

DISCUSSION

Lithium aluminum hydride is a very versatile reducing agent. It can be used to reduce any compound containing the carbonyl ($\text{—}\overset{O}{\overset{\|}{C}}\text{—}$) group, including carboxylic acids, RCO_2H, and carbon dioxide, to the corresponding alcohol. Lithium aluminum hydride also reduces alkyl bromides, alkyl iodides, and some alkyl chlorides. Lithium aluminum hydride, however, is difficult and hazardous to handle because it reacts violently and exothermically with water. It is usually used in anhydrous ether, which presents an additional fire hazard.

Sodium borohydride reacts rapidly with water only at low pH, and the reaction is not so violent as the reaction of lithium aluminum hydride with water. In a neutral or basic solution alkali metal borohydrides react with water very slowly or not at all. Sodium borohydride is not so powerful a reducing agent as lithium aluminum hydride. By itself sodium borohydride does not reduce esters, acids, or alkyl halides:

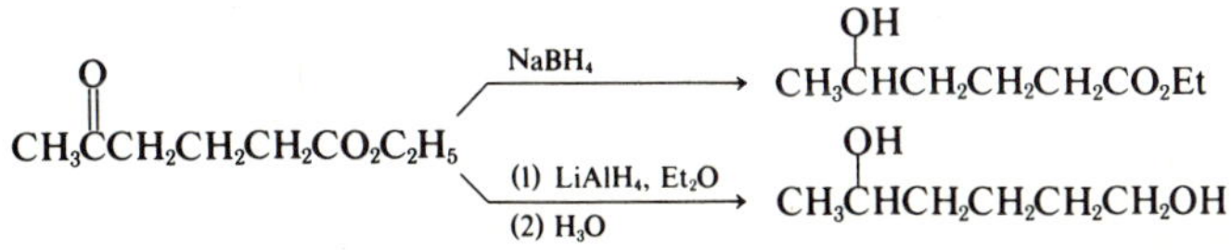

(29-1)

Sodium borohydrides can, however, be used to reduce esters, acids, and amides if Lewis acids are present.

$$RCO_2Et \xrightarrow[AlCl_3]{NaBH_4} RCH_2OH \qquad (29\text{-}2a)$$

$$RCO_2H \xrightarrow[AlCl_3]{NaBH_4} RCH_2OH \qquad (29\text{-}2b)$$

Sodium borohydride can not directly reduce carbon to carbon double bonds.

$$(CH_3)_2C{=}CH\overset{O}{\overset{\|}{C}}CH_3 \xrightarrow{NaBH_4} (CH_3)_2C{=}CH\overset{OH}{\overset{|}{C}}HCH_3 \qquad (29\text{-}3)$$

Lithium aluminum hydride reduces carbon to carbon double bonds only if they are allylic to oxygen functional groups; other carbon to carbon double bonds are not affected.

$$C_6H_5CH{=}CHCHO \xrightarrow[(2)\ H_3O^+]{(1)\ LiAlH_4,\ Et_2O} C_6H_5CH_2CH_2CH_2OH \qquad (29\text{-}4)$$

$$CH_3CH{=}CHCH_2CH_2CHO \xrightarrow[NaBH_4]{(1)\ LiAlH_4,\ Et_2O;\ (2)\ H_3O} CH_3CH{=}CHCH_2CH_2CH_2OH \qquad (29\text{-}5)$$

In the presence of acid and a platinum catalyst sodium borohydride does reduce carbon to carbon double bonds. It is not, however, a *direct* reduction. The sodium borohydride decomposes in the acid solution to generate hydrogen *in situ*. The double bond is then reduced by conventional catalytic hydrogenation.

$$NaBH_4 + H_3O^+ + 2H_2O \longrightarrow Na^+ + B(OH)_3 + 4H_2 \qquad (29\text{-}6)$$

$$RCH{=}CHR + H_2 \xrightarrow{Pt} RCH_2CH_2R \qquad (29\text{-}7)$$

Certain aldehydes such as CCl_3CHO (chloral) and CF_3CHO readily form hydrates [$CCl_3CH(OH)_2$, chloral hydrate, and $CF_3CH(OH)_2$]. These hydrates, like the parent aldehydes, are reduced to primary alcohols by sodium borohydride.

SYNTHESIS OF 2,2,2-TRICHLOROETHANOL

$$CCl_3CH(OH)_2 \xrightarrow{NaBH_4} CCl_3CH_2OH \qquad (29\text{-}8)$$

B.p. 96°d
M.p. 52°

B.p. 151.5°
M.p. 19°
D_4^{20} 1.555

Dissolve 12.5 g of chloral hydrate (**warning: toxic!**) in 15 ml of water in a 100-ml round bottom flask. With this flask, a Claisen adapter, thermometer, and dropping funnel construct the apparatus shown in Figure 29-1. Put the apparatus in an ice bath. Note that the cork holding the thermometer is held *above* the joint. *Do not place the cork in the joint* because a pressure buildup would transform such an apparatus into a glass bomb. Be sure the mercury bulb is below the surface of the liquid but not touching the bottom of the flask.

Dissolve 0.9 g of sodium borohydride in 15 ml of cold water. Pour the solution into the dropping funnel. Add the borohydride solution at a rate that allows the heat of reaction to maintain a temperature of 20 to 28°. While adding the borohydride, agitate the mixture by applying a gentle rotary motion to the ring stand. When addition is completed, remove the ice bath and allow the mixture to stand for 20 min. Stir every 3 to 4 min. Return the ice bath and add 5 ml of 10% hydrochloric acid.

Pour the mixture into a separatory funnel and extract it with 15 ml of ether. Wash the ether with 20 ml of saturated sodium chloride solution. Drain the ether into a 50-ml Erlenmeyer flask containing approximately 5 g of magnesium sulfate. Repeat the extraction with an additional 15 ml of ether. Use the same 20 ml of saturated sodium chloride solution to wash the second ether extract. Add the second ether extract to the first. Stir

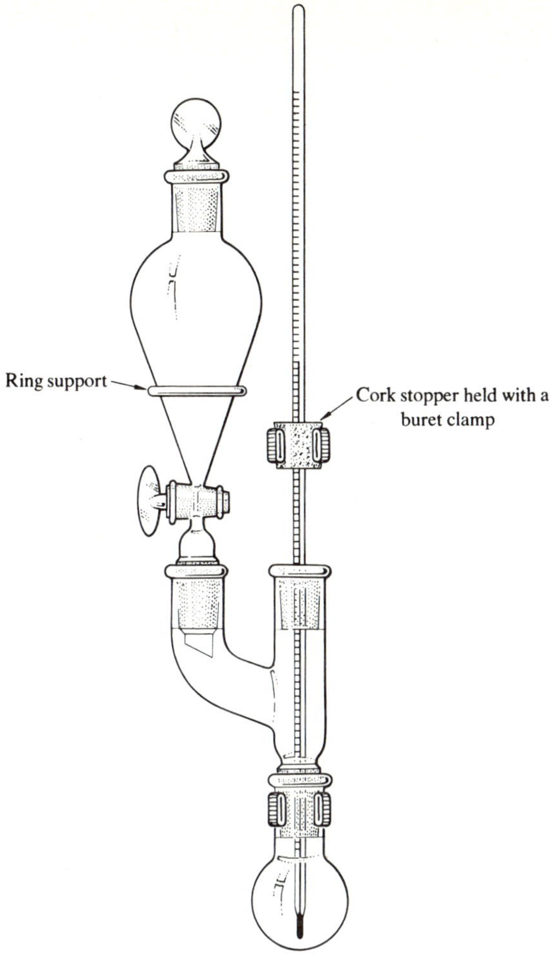

Figure 29-1 Reaction flask with dropping funnel and adapters.

the ether solution and allow the magnesium sulfate to settle. Decant the ether into a 50 ml round bottom or distilling flask. Add a boiling chip and assemble the apparatus shown in Figure 3-1.

Remove the ether by distillation. **Remember ether is a fire hazard. There should be no flames in the area while the ether is being distilled. Chill the receiver flask in an ice bath.** Distil the residue and collect the fraction that boils at 149 to 154° in a tared yield bottle. Weigh the product, calculate the % yield, label the bottle, and hand in the product to your laboratory instructor.

SYNTHESIS OF BENZYL ALCOHOL

$$\underset{\substack{\text{B.p. } 213^\circ \\ D_4^{20}\ 1.047}}{C_6H_5CO_2Et} \xrightarrow[AlCl_3]{NaBH_4} \underset{\substack{\text{B.p. } 205^\circ,\ 113^{\circ 30mm},\ 100^{\circ 15mm} \\ \text{M.p. } -15^\circ \\ D_4^{20}\ 1.045}}{C_6H_5CH_2OH} \qquad (29\text{-}9)$$

Vacuum Distillation

A liquid boils when its vapor pressure equals the pressure of the environment (Chapter 2). A liquid boils at a lower temperature when the pressure over it is reduced. A reduction in pressure can be obtained by applying a vacuum from a water aspirator, mechanical pump, or other source.

Danger! When a vacuum is applied to the inside of a container, the difference between the pressure inside and outside the container causes a stress on the container. If the container is glass, this stress could cause the container to shatter or implode. Never apply a vacuum to thin walled, flat bottom glass containers such as Erlenmeyer or Florence flasks. Never apply a vacuum to a glass wall that has a crack or star on it. **Always wear eye protection when conducting a vacuum distillation.**

Problem 29-1. What is the vapor pressure of benzyl alcohol at 113°?

Experimental Procedure

Pour into a 500-ml three-neck flask 40 ml of 1,2-dimethoxyethane (glyme) that has been dried over anhydrous magnesium sulfate for at least 24 hrs. Add 2.0 g of sodium borohydride. Place a thermometer in one side neck, a dropping funnel in the other, and a calcium chloride drying tube in the center neck as in Figure 15-2. Add 6.6 g of ethyl benzoate to the flask. Place an ice bath around the flask.

Dissolve 2.4 g of aluminum chloride in 30 ml of glyme in a 50-ml Erlenmeyer flask cooled in an ice bath. Stir the mixture with a glass rod. **Danger:** see comments on pages 97 and 105. Pour the aluminum chloride solution into the dropping funnel. Add the aluminum chloride solution to the flask while stirring the mixture by applying a gentle rotary motion to the ring stand. Adjust the rate of addition and the stirring so that the heat of reaction maintains the mixture at 60 to 90°. After the addition is complete and the temperature has cooled to 50°, heat the mixture on a hot water bath to 90 to 100° for 30 min.

Problem 29-2. In the first experiment described in this chapter sodium borohydride was used in a water solution. Why is it very important to keep all moisture out of this reaction?

Cool the mixture in an ice bath, add 130 to 150 g of crushed ice and stir. Add a few pieces of ice to 30 ml of 10% hydrochloric acid. Pour the acid into the reaction flask and stir until all the metal salts dissolve. While stirring remove and wash the dropping funnel. Pour the mixture into the large separatory funnel and extract it with three 20-ml portions of chloroform. Use each portion of chloroform to rinse the reaction flask before using it to extract the reaction mixture.

Drain each chloroform extract into the cleaned, small separatory funnel that was used as a dropping funnel. Wash each chloroform layer with the same 20 ml of water. Do not remove the water from the small separatory funnel during these washes. Drain the first two washed extracts into a 125-ml Erlenmeyer flask with 4 to 5 g of magnesium sulfate in it. Stir and allow to stand while extracting the mixture for the last time and while rinsing this last extract. Place a boiling chip in a 100-ml round bottom or distilling flask and filter the combined first two extracts into it. Drain the third extract into the 125-ml Erlenmeyer flask and add 2 to 3 g more of magnesium sulfate. Stir the mixture and filter it into the 100-ml round bottom flask.

Problem 29-3. Why is the last chloroform extract dried and washed separately from the first two?

Problem 29-4. Why is chloroform used instead of ethyl ether to extract the mixture?

Construct the apparatus shown in Figure 29-2. Use the flask containing the boiling chips and chloroform as the distilling flask. Distil the mixture at atmospheric pressure until the oil bath reaches 110°. Turn off the flame and add a few more boiling chips. Turn on the vacuum and slowly raise the temperature of the oil bath. After the first 4 to 6 drops have collected in the receiving flask, change receiving flasks and distil until about 1 ml of liquid remains in the distilling flask. The liquid will boil at 100 to 135° (head temperature). The exact boiling point will be determined by the amount of vacuum which is determined by the rate of flow and temperature of the water, leaks in the system, and other factors.

Pour the benzyl alcohol into a tared yield bottle. Weigh the product, calculate the % yield, label the bottle, and hand in the product to your laboratory instructor.

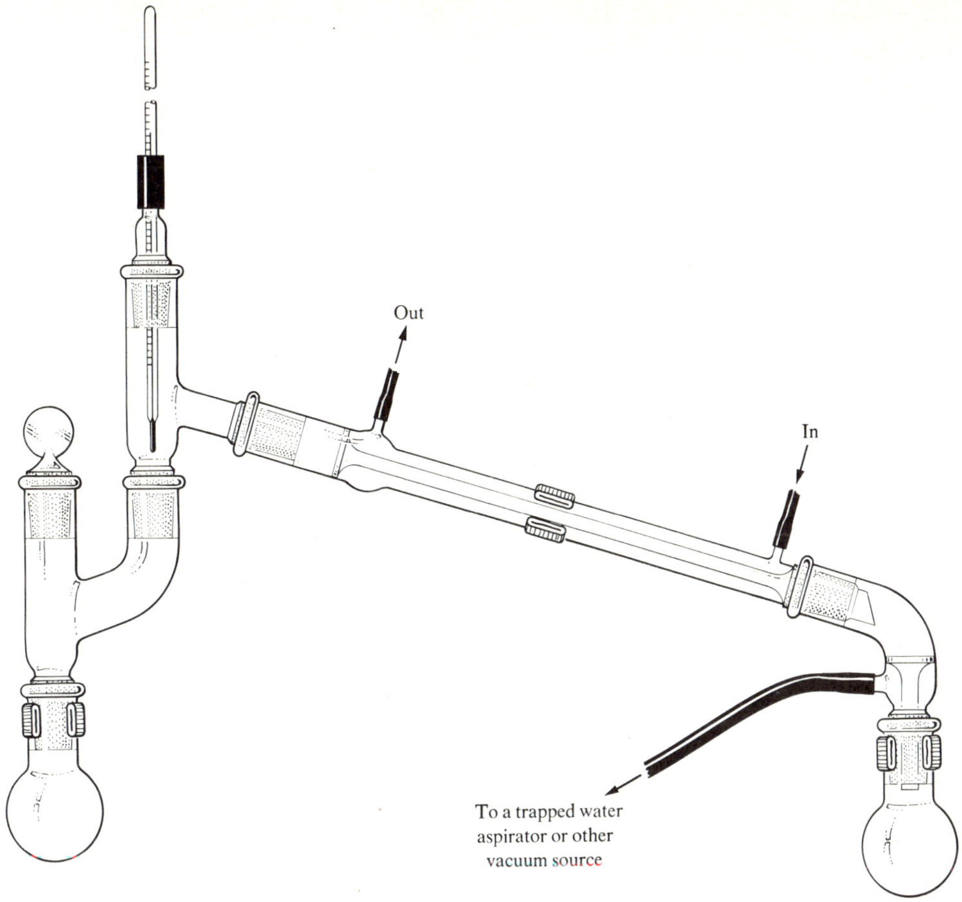

Figure 29-2 Vacuum distillation apparatus.

INFRARED ANALYSIS (Optional)

Obtain infrared spectra of (a) your purified product, (b) any material that distilled between 100 and 148° and above 154° (keep separate) if you synthesized the trichloroethanol, (c) the first few drops that collected as the first (discarded) fraction in the final purification of benzyl alcohol if you synthesized this compound, and (d) any residue in flask after the fractional distillation of the benzyl alcohol. All spectra should be obtained in 0.05-mm cells so that minor constituents can be seen more easily. In addition, obtain spectra on the *pure* compounds as capillary films so that the maxima of strong absorptions can be accurately determined. Correlate as many of the absorptions as possible with the structure of the products. Which absorptions are from impurities? What are the impurities?

30

The Williamson Synthesis of Methyl Butyl Ether

An S_N2 Reaction

INTRODUCTION

Unsymmetrical ethers can be made by the Williamson synthesis. By judicious selection of reactants a large variety of pure ethers can be made in good yield by use of this synthesis.

$$\underset{\substack{\text{B.p. } 65^\circ \\ D_4^{20}\ 0.7914}}{2CH_3OH} + 2Na \rightarrow 2CH_3O^- Na^+ + H_2\uparrow$$

$$CH_3O^- + \underset{\substack{\text{B.p. } 101.5^\circ \\ D_4^{20}\ 1.276}}{CH_3(CH_2)_3Br} \rightarrow \underset{\substack{\text{B.p. } 71^\circ \\ D_4^{20}\ 0.744}}{CH_3(CH_2)_3OCH_3} + Br^-$$

EXPERIMENTAL PROCEDURE

Carefully add 2.5 g of sodium to 30 ml of dry methanol[1] in a 100-ml round bottom flask. Extreme caution must be used in handling the sodium. The work should be planned so that the sodium is cut, wiped clean, weighed, and added to the methanol as quickly as possible to prevent the sodium from reacting with moisture from the air. The general precautions including *eye protection* described in the use of sodium in Chapter 23 should be followed whenever sodium is used. Because of the evolution of hydrogen the reaction with sodium should be done in a hood.

When the sodium finishes reacting with the methanol, cool the mixture and add 10 ml (12.8 g) of 1-bromobutane (*n*-butyl bromide) and connect the flask at once to a reflux condenser fitted with a drying tube. Swirl the mixture until the reactants are thoroughly mixed. After the mixture seems to stop reacting, reflux it for 1 hr 45 min.

Problem 30-1. Up to this point every precaution should be taken to keep moisture from the mixture. Why?

[1] Reagent grade (A.C.S.) methanol is usually dry enough for this purpose as it is supposed to contain less than 0.15% water. If the bottle has been opened previously or if a poorer grade of methanol is used, barium hydroxide should be shaken with the methanol, the mixture allowed to stand overnight, and the methanol decanted from the drying agent. Calcium chloride should not be used as a drying agent as it complexes with alcohols.

Allow the mixture to cool no longer than 5 min. Remove the drying tube, wipe the joint clean, and add 25 ml of water through the top of the condenser. Agitate until all the inorganic residue dissolves. Connect a distillation head to the flask and place the condenser for distillation. Those not using ground glass equipment should connect the condenser to the flask with two corks and a short piece of glass tubing (Fig. 30-1). Distil until about 15 ml of water solution remains in the flask.

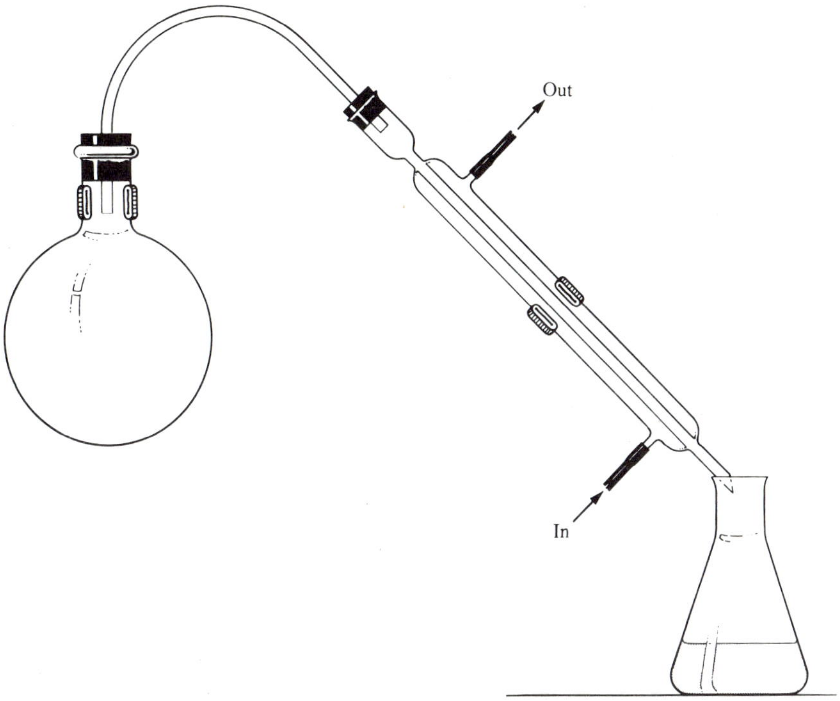

Figure 30-1 Distillation apparatus using a short-necked, round bottom flask.

Problem 30-2. (a) What by-products are likely to be formed in this reaction? (b) What other contaminants might there be? (c) What precautions in the above procedure are designed to keep these by-products to a minimum? (d) Can you suggest any additional precautions to minimize the formation of by-products? (e) Are the by-products likely to interfere with purification of the product?

Problem 30-3. Methanol and methyl butyl ether form an azeotrope boiling at 56° (65% methyl butyl ether and 35% methanol). Suggest a method that would separate the methanol from the product *quickly* and *easily* with the equipment and material available to you.

Check your answer to Problem 30-3 in the answer section of the book and *follow the procedures described there*. Then distil the purified liquid through the fractionating column (Fig. 5-5) and collect the fraction boiling at 66 to 76°.

Problem 30-4. A lower boiling fraction *may* also be obtained at this point. What is its composition? If the fraction contains any methyl butyl ether, how could it be recovered?

If a lower boiling fraction is obtained, follow the procedure in the answer to Problem 30-4 before proceeding further. Redistil the fraction that boils at 66 to 75° through the fractionating column. Collect the fraction that boils at 68.5 to 72° in a tared yield bottle.

Problem 30-5. How could the following be synthesized from compounds of 6 carbon atoms or fewer? (a) $(CH_3)_2CHCH_2CH_2OEt$, (b) $CH_3CH_2C(CH_3)_2OEt$, (c) $C_6H_5O—CH(CH_3)_2$, (d) $(CH_3)_3$-C—O—$C(CH_3)_3$, (e) $C_6H_5OC(CH_3)_3$.

INSTRUMENTATION OPTIONS

Infrared

Standard transmission spectra. Obtain spectra using 0.015-mm (if not available, use 0.025-mm) and 0.2-mm cells on (a) the liquid obtained from the purification described in the answer to Problem 30-3 before it is fractionally distilled, (b) the final purified methyl butyl ether, (c) the material that boils below 66° that collects as the first fraction from the fractional distillation, (d) the material that boils above 76° that remains in the flask as a residue from the fractional distillation. Assign as many of the absorption bands as possible assuming a pure product. What absorptions can not be accounted for? Which of the four fractions contain the extra absorptions? From your knowledge of the reactants, the likely by-products, and the spectra, determine which fractions contain which impurities.

Problem 30-6. Note that the spectra obtained with a 0.015-mm cell are sufficiently intense to identify the product. What is the purpose of obtaining the spectra in the 0.2-mm cells?

Differential analysis. Use two 0.2-mm cells that have the same or almost the same thickness. If they do not have the same cell thickness, use the one that is slightly thinner for the reference beam. Fill the cell for the reference beam with the purified methyl butyl ether (liquid b). With this cell in the reference beam obtain spectra on liquids (a), (c), and (d) above. Since there is relatively pure methyl butyl ether in the reference beam, an identical amount of methyl butyl ether in the sample beam causes no pen deflection. Small amounts of impurities can be detected that might be masked by weak or medium absorptions of the methyl butyl ether.

Problem 30-7. Note that if there is a sufficient difference in the path length of the two cells some of the medium absorptions of methyl butyl ether will be evident. Why are the strong absorptions not seen?

Problem 30-8. Note the appearance of the differential spectrum from 2970 to 2860 cm^{-1}, from 1160 to 1070 cm^{-1}, and near 1460 cm^{-1}. Can you conclude that the contaminates have no absorptions in these regions? With differential spectra can there be any information in regions where the major components absorb strongly?

Problem 30-9. (a) Why is it important to put the slightly thinner cell in the reference beam? (b) There are at least three reasons why negative (i.e., toward higher transmittance) absorption peaks might appear in an infrared spectrum. Name them.

Gas Chromatography

Chromatograph 10 μl of liquids (a), (b), (c), (d) (specified in the instructions in the infrared section above) on the Carbowax column. Use the conditions specified in Table 8-1. If a second instrument is available, chromatograph 20 μl of liquids (a) and (c) on the Apiezon column described in Table 8-I. Use the conditions described in the second footnote on p. 61. Chromatograph as many of the possible contaminants that you have available. From peak height and retention time data and the amounts injected report the identity and approximate amounts of each of the compounds in the four liquid samples analyzed.

EXERCISES

30-1. (a) Are coincident retention times on one column sufficient for unambiguous identification of completely unknown materials? (b) For identification of compounds in a mixture about which some information is known such as the product from the reaction described in this chapter? (c) Answer (a) and (b) again for coincident retention times on *two* different columns.

30-2. Compare the results of the chromatograms and the infrared spectra. (a) From which is it easier to identify the methyl butyl ether? (b) To identify the impurities? (c) To determine the quantities of the product and the impurities?

31

Synthesis of Carboxylic Acids by Oxidation of Alcohols

Synthesis of Isobutyric Acid

$$3(CH_3)_2CHCH_2OH + 4MnO_4^- \rightarrow 3(CH_3)_2CHCO_2^- + 4MnO_2\downarrow + HO^- + 2H_2O$$

B.p. 108°
D_4^{20} 0.802

$$(CH_3)_2CHCO_2^- + H^+ \rightarrow (CH_3)_2CHCO_2H$$

B.p. 155°
D_4^{20} 0.948

INTRODUCTION

Primary alcohols are easily oxidized to the corresponding carboxylic acid. If the alcohol has no easily oxidized groups, such as olefinic double bonds, the yield of carboxylic acid is usually very high. When the primary alcohol is available, oxidation is often the method chosen for preparation of a carboxylic acid.

EXPERIMENTAL PROCEDURE

In a 500-ml Erlenmeyer flask place 40 ml of water, 2.5 g of sodium hydroxide, and 20 ml of isobutyl alcohol. Stir the contents a few minutes. Put a thermometer in the flask. Dissolve 44 g of potassium permanganate in 360 ml of water in the large separatory funnel and allow it to run slowly into the alcohol mixture. Stir the contents and adjust the rate of addition to maintain the temperature of the reaction below 40°. When the deep purple of the permanganate persists, stop the addition and stir the mixture for a few minutes. If the purple disappears after stirring, keep the flask in the ice bath and continue adding the permanganate solution. Once the purple color persists after stirring, add 5 to 10 ml more of the permanganate solution. Allow the mixture to stand for 40 min and stir it occasionally. If the purple color disappears during this time, add another 15 to 20 ml of the permanganate solution and stir the reaction mixture.

Gravity filter the mixture through fluted filter paper. When 80 to 100 ml has filtered, transfer the filtrate to a 500-ml round bottom or distilling flask with a boiling chip in it and begin distilling the solution while the remainder of the mixture is filtered. As more of the mixture filters, add it to the 500-ml distilling flask. Add a fresh boiling chip after each addition. Wash the precipitate with two 25-ml portions of water. Transfer the washes to the distilling flask and continue the distillation until only 30 to 40 ml of liquid remains.

Disassemble the distillation apparatus and place the flask in an ice bath. Add a few pieces of ice and then add 15 ml of concentrated hydrochloric acid. Transfer the acidified mixture to a separatory funnel and extract with five 15-ml portions of chloroform. Use the first three portions of chloroform to rinse the 500-ml flask before using them to extract the acidified mixture. Combine the extracts in a 100-ml Erlenmeyer flask and add 5 to 6 g of calcium chloride; stir this occasionally over a 10-min period.

Decant the chloroform solution into a 100-ml round bottom or distilling flask. Add a boiling chip. With a short condenser and the appropriate adapters construct a distillation apparatus. Distil and collect the fraction that boils at 152 to 156° in a tared yield bottle. Read the following section if assigned. Weigh the product, calculate the % yield, label the bottle, and hand in the product to your instructor.

INFRARED SPECTRA (Optional)

Collect the fraction that boils between 95 to 149°; obtain a spectrum on it in a 0.025-mm cell. What is the composition of this fraction? Obtain spectra on the product in a 0.025-mm cell and between two salt plates as a capillary film. Are there any contaminants? If so, what are they?

32

Properties and Reactions of Carboxylic Acids and Their Salts

Organic Qualitative Analysis IV. Preparation of Derivatives

INTRODUCTION

The presence of the carboxylic acid group in an organic compound is easily established by its acid reactions. As was seen for other groups of organic compounds, physical properties are very useful for identification of carboxylic acids.

Ethyl ether and 5% aqueous sodium hydroxide are useful solvents in aiding in the classification of organic compounds. To establish unambiguously the identity of an organic unknown at least one solid derivative of known melting point should be synthesized.

PHYSICAL PROPERTIES

Boiling Points of Carboxylic Acids

The boiling points of carboxylic acids increase with increasing molecular weight in the same more or less regular manner as shown by other groups of organic compounds (Chapters 23 and 28). Similarly to these other groups of compounds, the boiling points decrease as branching is increased in a series of isomers (Fig. 32-1). Cyclic carboxylic acids also boil at slightly higher temperatures than do their straight-chain counterparts. Review pages 218–222.

Carboxylic acids boil at higher temperatures than do alcohols of similar structure and molecular weight (Fig. 32-2). That is, a carboxylic acid boils at almost exactly the same temperature as does the analogous alcohol with two more methylene groups (an increase of 28 in molecular weight). The high boiling points of carboxylic acids are a result of the dimerization of the acids (Fig. 32-3). Since the dimer is held together by *two* hydrogen bonds, carboxylic acids exist mainly as the dimer in solid, liquid, *and vapor* states.

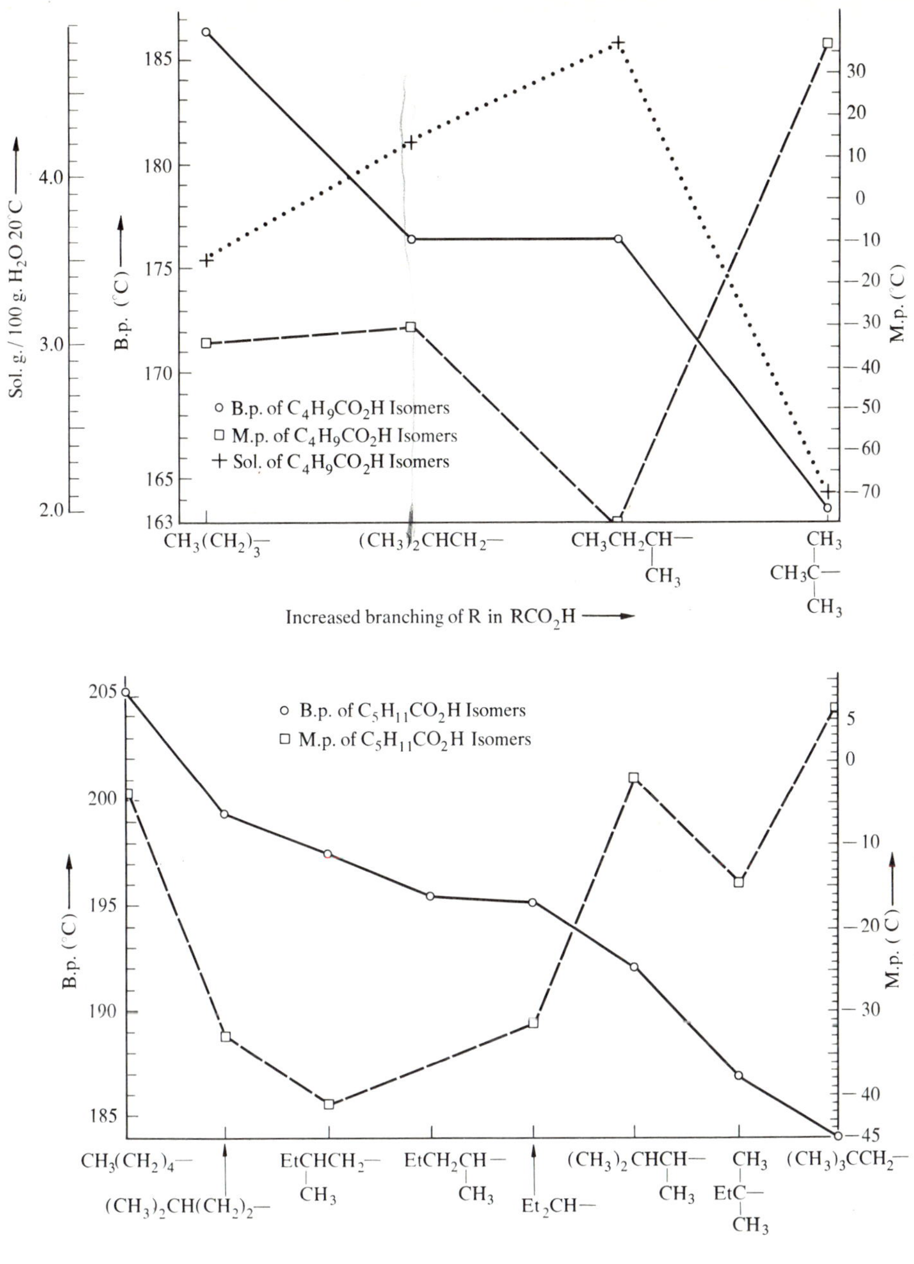

Figure 32-1 Boiling points, melting points, and water solubility as a function of branching of molecules for isomers of $C_4H_9CO_2H$ and $C_5H_{11}CO_2H$.

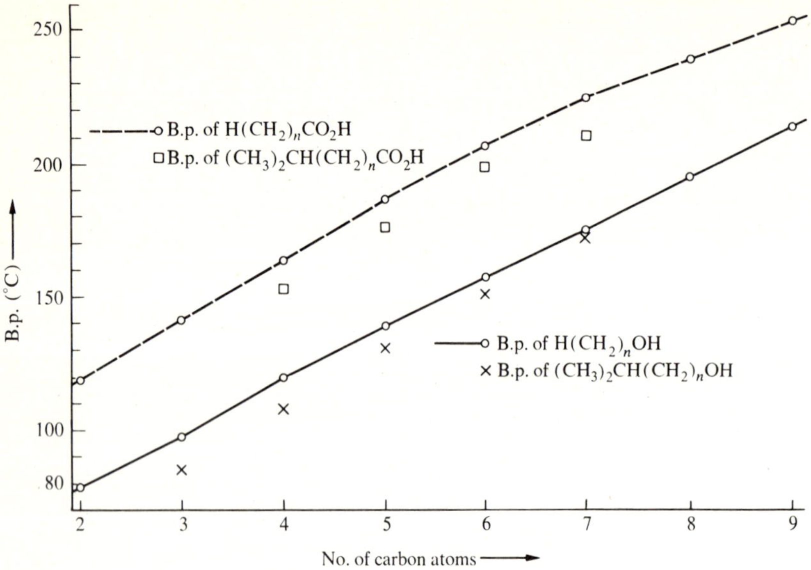

Figure 32-2 Boiling points of homologous alcohols and carboxylic acids.

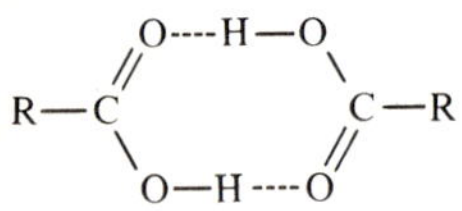

Figure 32-3 Carboxylic acid dimer.

Water Solubility of Carboxylic Acids

The water solubilities of carboxylic acids show the same variations with molecular weight and the same variations between branched and straight-chain compounds (Fig. 32-1), between cyclic and non-cyclic compounds, and between aromatic and nonaromatic compounds that are shown by other mono-functional organic compounds. Carboxylic acids are somewhat more soluble than are alcohols with an equal number of carbon atoms (Table 32-1). This relatively high solubility is due to two factors, the first being the principal reason. (a) The —CO_2H group is more polar than is the —CH_2OH group; as a result, stronger dipole-dipole attractions exist between the carboxy group and the water molecules. (b) A small portion of the acid molecules are ionized in dilute, aqueous solution. Such ions are very "soluble" in water because of the relatively strong ion-dipole attractions.

Problem 32-1. Of all the acids in Table 32-1 only 2,2-dimethylpropanoic (trimethylacetic) acid and decanoic acid are less soluble than the analogous alcohols. Suggest an explanation for this. (Hint: see the other physical constants for these acids from the tables at the end of this chapter.) Would benzoic acid be expected to have a higher or lower water solubility than benzyl alcohol? Benzoic acid has a water solubility of 0.25 g/ 100 g. of water at 20°; see Table 28-1 for the water solubility of benzyl alcohol.

TABLE 32-1 COMPARATIVE WATER SOLUBILITIES OF ANALOGOUS ALCOHOLS AND CARBOXYLIC ACIDS

Alcohol	Solubility (100g H_2O, 20°C)	Acid	Solubility (g/100g H_2O, 20°C)
1-Butanol	7.9	Butanoic acid	∞
1-Pentanol	2.3	Pentanoic acid	3.5
3-Methyl-1-butanol	2.8	3-Methyl-1-butanoic acid	4.2
2-Methyl-1-butanol	3.2	2-Methyl-1-butanoic acid	~4.8
2,2-Dimethyl-1-propanol	3.7	2,2-Dimethyl-1-propanoic acid	2.1
1-Hexanol	0.6	Hexanoic (caproic) acid	1.1
1-Heptanol	0.2	Heptanoic acid	0.27
1-Octanol	0.05	Octanoic (caprylic)acid	0.25
1-Decanol	0.02	Decanoic (capric) acid	0.003

Physical Constants of the Salts of Carboxylic Acids

Since salts of carboxylic acids are ionic compounds, they have the physical properties associated with ionic substances. That is, they have high boiling and melting points, high solubilities in polar solvents such as water, and low solubilities in nonpolar solvents such as ether and benzene.

Solubility in Ether

Ether is used as a solvent in the solubility classification scheme to divide water-soluble compounds into two subclasses: ether-soluble (Class S_1) and ether-insoluble (Class S_2). Monofunctional oxygen compounds such as low molecular weight alcohols, acids, and esters are in Class S_1. Ionic compounds, such as salts of carboxylic acids, and polyfunctional compounds, such as hydroxy acids and dicarboxylic acids, are insoluble in ether and usually very soluble in water and, therefore, are in Class S_2.

ACID-BASE REACTIONS

Carboxylic acids are sufficiently acidic to give acid reactions to wet pH or litmus paper. Since carboxylic acids are stronger acids than water and carbonic acid, they react with metal hydroxides, carbonates, and bicarbonates to give metal carboxylates and to liberate water and carbon dioxide:

$$HO^- + RCO_2H \rightarrow RCO_2^- + H_2O, \qquad (32\text{-}1)$$

$$CO_3^{=} + 2RCO_2H \rightarrow 2RCO_2^- + H_2O + CO_2, \qquad (32\text{-}2)$$

$$CO_3^{=} + RCO_2H \rightarrow RCO_2^- + HCO_3^-, \qquad (32\text{-}3)$$

$$HCO_3^- + RCO_2H \rightarrow RCO_2^- + H_2O + CO_2. \qquad (32\text{-}4)$$

The acidic properties of some organic compounds can be used to classify these compounds in the solubility classification scheme. Water-insoluble carboxylic acids dissolve in aqueous solutions of alkali metal hydroxides, carbonates, and bicarbonates because the resulting ionic alkali metal carboxylates are soluble in water.

ACIDITY AND SOLUBILITY CLASSIFICATION

Discussion

All water-insoluble unknowns are tested in aqueous 5% sodium hydroxide. If soluble in this reagent, the compounds are in Class A (for acids); if insoluble, the compounds are tested for solubility in concentrated acids (Chapter 28). All carboxylic acids are found in Classes S_1, S_2, or A. High molecular weight acids (12 or more carbon atoms for straight-chain acids, more than 12 carbon atoms for the branched-chain or cyclic acids) dissolve slowly in aqueous sodium hydroxide and aqueous sodium bicarbonate, forming salts which are similar in composition to the main constituents of nonsynthetic soaps. The resulting mixture is not a clear, true solution but is an opalescent colloid solution that foams when shaken. This behavior is readily recognized after it has been observed a few times. Note this behavior the next time you shower. If a compound does this, it is considered soluble in the aqueous base and in Class A. Any soap formation should be noted because the occurrence or nonoccurrence of such colloids aids in the identification of organic acids.

To verify this classification test or to determine whether or not a compound is partly soluble in the aqueous sodium hydroxide but not in water, the sodium hydroxide can be neutralized with 20% hydrochloric acid.[1] The separation of a second phase indicates that at least some of the compound dissolved.

$$RCO_2^- + H_3O^+ \rightarrow RCO_2H + H_2O \qquad (32\text{-}5)$$

If too much of a compound of borderline water solubility is used in the water solubility test, the compound appears to be water-insoluble. If the correct amount were used in the solubility test with 5% sodium hydroxide, the compound would dissolve and be thought mistakenly to be acidic. Such a compound usually will not separate as a second phase when its basic solution is acidified. If the compound does dissolve in aqueous sodium hydroxide and the basic solution does not give a precipitate with 20% hydrochloric acid, the solubility tests should be repeated with special care devoted to making sure that all quantities are measured accurately.

The reaction of pH paper to water-soluble compounds can be used as a basis for the subclassification of solubility classes S_1 and S_2. If the aqueous solution of an S compound gives an acid reaction to pH paper (pH 4 or less), the compound is in Class S_1a or S_2a. If the solution gives a weakly basic reaction to pH paper (pH 8 to 10), the compound is in Class S_1w or S_2w. If the solution gives a strongly basic reaction to pH paper (pH 10 or greater), the compound is in Class S_1b or S_2b. Water-soluble compounds whose aqueous solutions have a pH between 4 and 8 are in Class S_1n or S_2n. Sometimes the drop of aqueous solution on the pH paper may give a borderline color and make the approximate determination of pH difficult. If this occurs, put a drop of the liquid unknown on wet pH paper. If the unknown is a solid, place a small crystal or two on the pH paper, then put a small droplet of water on the solid. Observe the resulting color change.

All water-soluble carboxylic acids are in Class S_1a or S_2a. A solution of a salt of a strong base and weak acid has a pH greater than 7. This is the reason solutions of alkali metal carboxylates usually have a pH of 8 to 9. Sometimes, however, the salt may contain a significant amount of free acid which lowers the pH of the resulting solution.

Problem 32-2. Solutions of *pure* sodium trichloroacetate have a pH of approximately 7. Why?

Experimental Procedure

Test the water solubility of the following compounds: acetic acid, valeric acid, stearic acid, capric acid, sodium stearate, sodium acetate, sodium benzoate, 1-butanol, 1-hexanol, and ethylene glycol. Use the proportions and directions given on page 225.

For each of the compounds that is *water soluble* place 2 drops of the aqueous solution on pH paper and observe any color change. Test each of the *water-soluble* compounds for solubility in ether, using again the proportions given in Chapter 28. Test each of the *water-insoluble* compounds for solubility in 5% aqueous sodium hydroxide. Save the solution of each compound that dissolves in the base.

[1] This should not be done on Class A(B) or amphoteric compounds. These will be assigned with later chapters.

To each of these basic solutions add 20% hydrochloric acid drop by drop until the pH of each solution is approximately 7. Observe what happens. Add more acid until a pH of 2 is obtained and then add 1 drop more of 20% hydrochloric acid. From the results of this experimental section classify as many as possible of the 11 compounds into solubility classes S_1a, S_1n, S_1w, S_1b, S_2a, S_2n, S_2w, S_2b, and A. Compare your results with Figure 32-4. Always keep in mind what solubilities these symbols indicate. Memorizing the solubility classes for compounds without understanding what solubility and structures these classes indicate is useless.

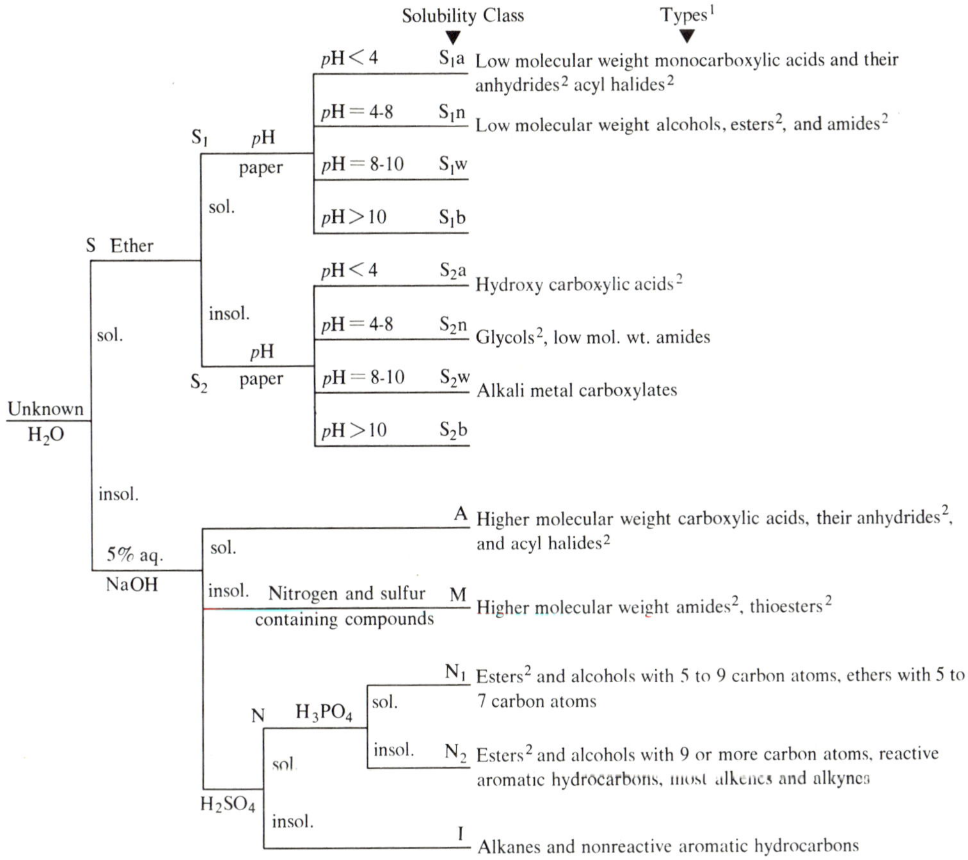

[1]Most halogen compounds are in the same solubility class as are the organic compounds from which they were derived.
[2]Solubility relationships discussed in later chapters.

Figure 32-4 Solubility classification chart. This chart presents an over-all view of the solubility classifictaion. There are some omissions and exceptions. The structure-solubility relations in the text should be thoroughly studied to obtain maximum information from the solubility tests.

SYNTHESIS OF DERIVATIVES

Derivatives in the Organic Qualitative Analysis Scheme

There are approximately 1.5 million known organic compounds, and the number is increasing exponentially. The number of organic compounds possible, while not literally infinite, is extremely large. Therefore, if an unknown organic substance is to be identified in a reasonable amount (or possibly even a finite amount) of time, a logical and systematic method of identification must be used. This systematic method may be based on instrumentation, on a variety of relatively simple physical and chemical tests, or a combination of these. The method introduced in this chapter is based on the second. It is introduced for three reasons: (a) the method illustrates many of the techniques and principles of chemistry to which the beginning student of organic chemistry has been introduced; (b) it allows all beginning students to identify the unknowns quickly and accurately with the techniques, knowledge, equipment, and other resources available to them; (c) instrumental methods are introduced in other chapters.

The importance of a logical and systematic approach can not be emphasized too strongly. Although occasionally a quick identification may be made by superficial examination, the odds against this are very great because of the very great number of organic compounds and because of the frequent alteration of a compound's odor, color, crystalline form, and/or viscosity by small amounts of impurities. For this reason, any attempt to identify a compound by *first* seeing whether or not it is similar to some known sample is usually a waste of time. Of course, once a tentative identification has been made by a logical and systematic procedure, comparison with a known compound of the same postulated structure as that of the unknown is valuable. This is especially true if the comparison is by mixed melting points (Chapter 11) or instrumental methods such as identity of infrared spectra, nuclear magnetic resonance spectra, retention times, and X-ray diffractograms.

The first step in identification of an unknown is the determination of simple physical properties. The most valuable (and the only determinations recommended at this time) are boiling points and melting points. For the more advanced student, densities, refractive indexes, and optical rotations are often valuable. Wide ranges in either the boiling point or melting point are indicative of an impure sample. If the purity of a solid is in question, it should be recrystallized to a constant melting point. A liquid should be purified by distillation. Chapters 3, 5, and 12 should be reviewed and consulted when needed for this part of the procedure.

After the compound has been purified and after the physical constants have been determined, the presence of elements other than carbon, hydrogen, and oxygen is determined. For the unknown assigned with this chapter only the identification of halogens is needed (Chapter 23). It soon will be necessary to analyze for the presence of nitrogen and sulfur (Chapter 35).

The next step in the procedure is to determine the solubility class of the compound. See Figure 32-4 and similar charts in Chapters 41 and 61. The student should realize, however, that some groups and important exceptions are omitted from the charts in the later chapters. A student with a good knowledge of beginning organic chemistry will be able to expand the charts sufficiently to identify any unknown given in this course.

The results obtained by determination of the physical properties, the elements present, and the solubility class indicate which types of compounds the unknown may be, and this indicates which classification or group tests to use. *Group tests* depend upon characteristic reactions of functional groups. These reactions are usually associated with noticeable but sometimes subtle *phase changes* or *color changes* or both. The phase changes involve the separation of one phase into two or the merger of two phases into one. The color changes include the appearance of disappearance of a color or colors or the change from one color to another. A thorough mastery of the subject matter is required to interpret these results and proceed in the most efficient manner. The use of classification tests that cannot possibly yield new information about the unknown is a waste of time, effort, and patience; the omission of tests that would give information about the unknown may also waste time and may result in inability to identify the unknown.

As in Chapters 14, 23, and 28, new classification tests are outlined with each new unknown. Other classification tests may be devised from the material the student has learned in lecture, laboratory, or his own reading. This last method is encouraged, and the student should feel free to consult both the laboratory instructor and the professor about any original tests.

To promote accuracy in the classification and solubility tests the student should do the tests with adequate blanks or controls. Until the student has mastered a test so that both the technique and the interpretation of results are routine (this may never occur for many tests), he should conduct the tests on known compounds that give positive and negative results immediately before or during the same test on the unknown. If the test can give more than two results (as the Lucas test for alcohols), *at least* one representative known for each type of result should be tested. Only in this way can one be sure of the results of classification tests. The importance of testing knowns or using blanks can not be overemphasized. This applies not only for this course but also for all scientific work.

The results of the above tests, if selected judiciously, conducted correctly, and interpreted wisely, will indicate that the unknown is one of several possible compounds. The melting point of a derivative of the unknown will usually indicate which of several possible compounds it is. When the number of unknowns is limited as in this course, the results of the classification tests may indicate that only one compound is possible. Even in this instance it is advisable to make a derivative for verification.

With the first three unknowns the preparation of derivatives was not essential because the number of compounds in the tables was much less than the actual compounds in a given class. Also, because of the relatively small number of organic compounds assigned as possible unknowns in future chapters, your laboratory instructor may tell you that it will be unnecessary to prepare a derivative if you are *absolutely certain* of the identity of the unknown in question. This is done to save time and is not unsatisfactory for these unknowns because of the limited number of possible compounds. It should be remembered, however, that unless the number of possible unknowns is artificially limited or unless instrumental methods are used,[2] a *minimum* of one *well chosen, solid* derivative is *always necessary* for the identification of an organic unknown.

Problem 32-3. Most isomeric alkenes of the *same* carbon skeleton, i.e., differing only in the position of the double bond, boil within 12° of one another. If all compounds possible were listed in Table 14-2, approximately how many *new* compounds that boil between 112 and 125° would be added to that table? To obtain a rough idea draw all isomers of C_8H_{16} with the double bond in the 1 and in the 2 position.

Problem 32-4. There are three isomers of C_9H_{12} listed in Table 14-3. Draw the structures of all other benzenoid isomers of C_9H_{12}. What would be the solubility class of each isomer? The nonlisted isomers boil between 160 and 169°. Assume the limitations on purity of compounds, technique, and apparatus are such that compounds must boil at least 4.5° apart to be positively identified by boiling points. What would be the *greatest* number of these isomers that could be identified by boiling point, solubility classification, and group classification tests? How could the others be identified without instrumental techniques; with instrumental techniques?

Selection of Derivatives

Derivatives that are useful for identification are solids of known melting point into which the unknowns may be converted unambiguously and usually simply. If possible, the derivatives' melting points should be between 55 and 220°; melting points can be most accurately determined in this range. The derivatives should be chosen so that any compounds that may be the unknown give known derivatives that melt at least 5° and preferably 8° apart. If this cannot be done, then two or more derivatives should be made. For example, assume preliminary tests indicate an unknown is a carboxylic acid that melts at 28.5 to 29.7°. A search of the literature indicates the following possibilities:

		Derivatives, m.p. (°C)		
Acid	*M.p. (°C)*	*Anilide*	*Amide*	p-*Bromophenacyl ester*
Undecanoic	29	71	103	68
Cyclohexanecarboxylic	31	142	186	
Capric	31.5	63	108	67

Either the anilide or the amide would distinguish the cyclohexanecarboxylic acid from the other two; however, the anilide would be the preferred derivative because all the anilides melt at least 8° apart, whereas two of the amides melt only 5° apart. The *p*-bromophenacyl ester should not be prepared for this identification because the melting point of this derivative has not been given for cyclohexanecarboxylic acid and the two derivatives listed melt much too close together to make differentiation possible. For some unknowns it may be necessary to prepare more than one derivative although this will not occur in this course as much as when the number of possible compounds is unrestricted.

The derivative should be recrystallized, dried, and/or otherwise purified until a narrow melting range is obtained. If the melting point of the derivative is more than 1.0° higher or 2.5° lower[3] than that of any possible derivative listed, it should be recrystallized until its melting point approaches at least that closely the melting point of a derivative of a possible compound. If this cannot be done, the unknown

[2] The preparation of a derivative *may* not be *absolutely* necessary if complex instrumental methods are available for comparison of the unknown with known compounds, one of which is identical to the unknown. For example, retention time on two different liquid phases identical with those of a known standard is more positive in establishing identity than is the melting point of one well chosen derivative.

[3] Assumes a noncalibrated thermometer is used. If a calibrated thermometer is used, the limits should be 1.5° lower and 0.6° higher.

should be recrystallized until the melting point does not change (constant melting point). If the unknown is one of two or more possible compounds, the selected derivatives of which melt near one another (within 10°), the prepared derivative should also be recrystallized to a constant melting point.

For solid unknowns a mixed melting point can be obtained as a confirmation provided a known sample is available (Chapter 11). If a depression is observed, the compounds are not the same. If no depression is observed, the chances are about nineteen out of twenty that the compounds are identical. Straight-chain alkanes of similar molecular weight are examples of compounds that do not depress one another's melting points.

TABLE 32-2 CARBOXYLIC ACIDS AND DERIVATIVES

Acid	M.p. (°C)	Derivatives, M.p. (°C)				
		amide	anilide	p-nitrobenzyl ester	p-bromophenacyl ester	p-toluidide
2,5-Dimethylbenzoic	132	186	140			
Furoic	133	143	123	134	138	171
Sebacic	133	210	202	73	147	201
Cinnamic	133	148	151	117	145	168
Malonic	133d	170(di)	230(di)	85(di)		255
d- or l-Mandelic	133	122	151	124		
Acetylsalicyclic	135	138	136	90		
β-Campholytic	135	130	104			114
Acetonedicarboxylic	135d		155(di)			
Methylmalonic	137d	217	182	75		
Phenylpropiolic	137.5	102	127	83		

When the results of the analysis are to be reported, the student should have no doubt about the identity of the unknown. If the structure is not certain, further tests should be done and more derivatives prepared until the identity of the unknown is *absolutely certain*.

Problem 32-5. If only the information in Table 32-2 were to be used, how many derivatives and which ones would have to be made to identify positively an acid that melts at 132.8 to 135.2°?

Problem 32-6. The *p*-bromophenacyl derivatives would be useful to distinguish between which pairs of acids in Table 32-2? For the identification of which pairs of acids in Table 32-2 would it be wise not to rely on the amide? On the *p*-nitrobenzyl ester? On the anilide?

Preparation of Derivatives

Amides. Discussion: the most common derivatives used for identification of acids are amides. In one method of synthesis of amides the acid or salt of the acid is converted to the acid chloride by treatment with inorganic acid chlorides such as thionyl chloride, phosphorus pentachloride, and phosphorus trichloride. The acid chloride is then treated with ammonia or an amine:

$$RCO_2H + SOCl_2 \rightarrow RCOCl + SO_2 + HCl, \qquad (32\text{-}6)$$

$$RCOCl + 2RNH_2 \rightarrow R\overset{\overset{\displaystyle O}{\|}}{C}NHR' + R\overset{+}{N}H_3 + Cl^-. \qquad (32\text{-}7)$$

Experimental procedure: **Warning! These inorganic acid chlorides are very corrosive and toxic. They should be treated as strong acids (Chapter 1). They react quickly with moisture. Every possible precaution should be taken to keep moisture away.**

Synthesis of unsubstituted amides. Reflux 1 g of acid with 5 ml of thionyl chloride for 30 min. Pour the reaction mixture **slowly and carefully with stirring (if done too quickly, this can be a very violent reaction)** into 15 ml concentrated aqueous ammonia at 0°. The amides usually precipitate if the mixture is kept at 0°. Filter the aqueous solution from the amide, and crystallize the amide from aqueous ethanol.

$$\underset{\text{acid chloride}}{RCOCl} + 2NH_3 \rightarrow \underset{\text{amide}}{RCONH_2} + \overset{+}{N}H_4 + Cl^- \qquad (32\text{-}8)$$

Synthesis of anilides, I. Heat an alkali metal salt of the acid (0.4 g), 1 ml of aniline (or similar amine), and 0.3 ml of concentrated hydrochloric acid for 1 hr in a test tube in an oil bath at 150 to 160°. Purify the product by the one of the following two methods:

a. If the acid is thought to contain fewer than 6 carbon atoms, extract the crude product from above with 5 ml of boiling 95% ethanol. Pour the extract into 50 ml of hot water, evaporate the resulting solution to a volume of 10 ml, then cool to 0° by placing the flask in an ice bath. Filter the cooled mixture, and recrystallize the precipitated amide from water or aqueous ethanol.

b. If the acid is thought to have 6 or more carbon atoms, powder the crude product and wash with 15 ml of 5% hydrochloric acid and two 10-ml portions of ice water. Extract the residue with 35 ml of boiling 50% ethanol. Filter the extract and cool the filtrate to 0°. Recrystallize the precipitated amide from aqueous ethanol.

$$\underset{\text{sodium carboxylate}}{RCO_2Na} + \underset{\text{anilium chloride}}{Ar\overset{+}{N}H_3\overset{-}{Cl}} \xrightarrow{HCl} \underset{\text{anilide}}{R\overset{O}{\overset{\|}{C}}\text{—}NHAr} + NaCl + H_2O \qquad (32\text{-}9)$$

$$\underset{\text{sodium acetate}}{CH_3CO_2Na} + \underset{\text{aniline}}{C_6H_5NH_2} \xrightarrow{HCl} \underset{\text{acetanilide}}{C_6H_5NH\overset{O}{\overset{\|}{C}}CH_3} + NaCl + H_2O \qquad (32\text{-}10)$$

Synthesis of anilides, II. Mix 1 g of the acid or its alkali metal salt with 3 ml of thionyl chloride in a 10- or 25-ml flask. Reflux the mixture, using a small condenser, for 30 min and cool. Add 1.5 g of aniline (or similar amine) in 30 ml of benzene, and heat the mixture on a steam bath with gentle agitation for 2 min. Agitate the mixture in a small separatory funnel with 10 ml of ice water and remove the aqueous (bottom) layer. Wash the benzene layer successively with 10 ml of 5% hydrochloric acid, 10 ml of 5% aqueous sodium hydroxide, and two 5-ml portions of water. Evaporate the benzene and crystallize the amide from aqueous ethanol.

$$\underset{\text{acid chloride}}{RCOCl} + \underset{\text{aromatic amine}}{2ArNH_2} \rightarrow \underset{\text{anilide}}{R\overset{O}{\overset{\|}{C}}NHAr} + \underset{\text{arylammonium chloride}}{Ar\overset{+}{N}H_3Cl} \qquad (32\text{-}11)$$

$$\underset{\text{butyryl chloride}}{C_3H_7COCl} + \underset{\text{aniline}}{2C_6H_5NH_2} \rightarrow \underset{\text{butyranilide}}{C_3H_7\overset{O}{\overset{\|}{C}}NHC_6H_5} + \underset{\text{anilium chloride}}{C_6H_5\overset{+}{N}H_3Cl^-} \qquad (32\text{-}12)$$

Problem 32-7. What is the purpose of each of the washes described in the above paragraph?

Esters. Discussion: the most commonly used esters for the identification of carboxylic acids are the *p*-nitrobenzyl, phenacyl, *p*-phenylphenacyl, and *p*-halophenacyl esters. These are synthesized by treating the appropriate halide with the salt of the acid:

$$\underset{}{RCO_2^-Na^+} + \underset{\substack{\text{benzyl}\\\text{halide}}}{XCH_2Ar} \rightarrow \underset{\substack{\text{benzyl}\\\text{carboxylate}}}{RCO_2CH_2Ar} + Na^+X^-, \qquad (32\text{-}13)$$

$$RCO_2^+Na^- + \underset{\substack{\text{aracyl}\\\text{halide}}}{XCH_2COAr} \rightarrow \underset{\substack{\text{aracyl}\\\text{carboxylate}}}{RCO_2CH_2COAr} + Na^+X^-. \qquad (32\text{-}14)$$

Experimental procedure: **Caution!** Arylacyl and benzyl halides are lachrymators; tear gas is a phenacyl halide.

Add 10% sodium hydroxide to a solution of 1 g of the acid in 5 ml of water until the pH is approximately 7. If too much sodium hydroxide is added, add a small amount of acid to bring the pH to 7. A neutral solution is very important because phenacyl and benzyl halides are hydrolyzed in alkaline solution. In acid solution a sufficient amount of the sodium salt of the unknown acid will not be present to produce a good yield. If the unknown is an alkali metal salt of the acid, dissolve it in 8 ml of water. If necessary, adjust the pH of the resulting solution to 7 by the addition of 10% hydrochloric acid a drop at a time.

To this solution add 1 g of the benzyl or phenacyl halide in 10 ml of ethanol, and reflux the mixture for 1 hr. (Reflux for 2 hr if the acid is dibasic, and for 3 hr if it is tribasic. Although these last two types of acids are listed in the tables at the end of the chapter, they will not be assigned as possible unknowns until multiple functional compounds are studied in Chapter 61.) If a solid separates before refluxing is complete, add ethanol a drop at a time to the refluxing mixture until the solid redissolves. Allow the solution to cool to room temperature, wash the precipitate with two 5-ml portions of cold water, and recrystallize from ethanol. If no ester precipitates, add water until the ester begins to precipitate, and then reflux the mixture and allow to cool slowly to room temperature. If necessary, cool to 0° to cause precipitation.

Nitro Derivatives. Discussion: now that the principles of using derivatives are understood, the preparation of nitro derivatives may be desirable to verify identities of aromatic compounds. The procedure for synthesis of *m*-dinitrobenzene (Chapter 13) can be used for nitration of less active aromatic compounds like benzoic acid or nitrobenzene. If only a small quantity of unknown is available, all quantities can be reduced to one-third or one-fourth the values stated in Chapter 13 and a 10-cm test tube used. For aromatic hydrocarbons and their monohalo derivatives the procedure described below gives good yields of pure nitro derivatives from small amounts of unknown.

$$ArH + NO_2^+ \rightarrow ArNO_2 + H^+ \qquad (32\text{-}15)$$

Experimental procedure: **Warning! Follow the precautions in Chapter 13 for reactions of nitric acid with aromatic compounds.**

Add 1 g of unknown to 4 ml of concentrated sulfuric acid in a 20-cm test tube. Add 4 ml of concentrated nitric acid, a few drops at a time. Shake the tube after each addition. After all the nitric acid has been added, put the tube in a water bath maintained at 45 to 48° for 5 min. Shake the tube a few times during this warming. Pour the mixture onto 40 g of ice, and filter the mixture through a Hirsh funnel or the smallest Büchner funnel available. Recrystallize the precipitate from methanol. It may be necessary to add a small amount of water to the methanol to obtain a good yield of crystals (Chapter 12).

IDENTIFICATION OF SALTS OF CARBOXYLIC ACIDS

Ignition Tests

Discussion. Often ignition tests give valuable information about the identification of organic unknowns. Aromatic and other compounds with much unsaturation give smoky flames. Carbohydrates char and give a readily identifiable caramel odor. Metal carboxylates are the only compounds that have been studied thus far that leave residues on ignition. The residues from ignition of most metal carboxylates are strongly basic to pH paper. The solid unknowns should be tested by ignition before their melting points are determined.

$$12O_2 + 2C_4H_9CO_2Na \rightarrow Na_2CO_3 + 8CO_2 + 9H_2O \quad (32\text{-}16)$$

Experimental Procedure. Place one drop or a *few small* crystals of the compound on a *clean*, ceramic crucible cover. Put the cover in the edge of a low burner flame. Note whether or not the compound is explosive or flammable. If it burns, note the color and type of flame. Heat gently over the small flame and then heat strongly.

If the compound is a solid, note whether or not and how it melts. Can any new odors be detected? Be careful because sometimes the gases are toxic. Nitriles, for instance, often evolve hydrogen cyanide when heated. If there is any residue, allow the crucible cover to cool. Add a drop of distilled water to the residue. Test the drop with wet pH paper. Add a drop of 20% hydrochloric acid to the residue and water solution and note whether or not a gas is evolved. Wet a platinum wire with the hydrochloric acid solution and do a flame test for any metal ions. Try this test on cyclohexane, *m*-dinitrobenzene, toluene, ethanol, butyric acid, stearic acid, bromobenzene, sodium acetate, calcium acetate or lactate, copper acetate, and ammonium benzoate.

Isolation of Acid from Salt

Discussion. That an unknown is an alkali metal carboxylate can be determined from its solubility class, its high melting point, and its behavior on ignition. Because of their ionic crystal lattice many metal salts of carboxylic acids melt too high to have their melting points determined in a mineral oil bath. The cation can be identified by usual inorganic qualitative analysis schemes. This is unnecessary for unknowns assigned with this course because only the *sodium* carboxylates will be among the possible unknowns. The carboxylate moiety is usually identified by adding excess hydrochloric acid to a concentrated aqueous solution of the salt and identifying the acid.

$$RCO_2^- + H^+ \rightarrow RCO_2H \quad (32\text{-}17)$$

Experimental Procedure. Dissolve 1 to 2 g of acid salt in the minimum amount of water. Add 20% hydrochloric acid to this solution until a pH of 2 or less is obtained and then add 3 ml more of the hydrochloric acid. If the organic acid is a liquid or a water-soluble solid, pour the mixture into a separatory funnel and extract it with four successive 15-ml portions of ether. Extract individually and successively the ether extracts with 10 ml of water and 10 ml of concentrated aqueous sodium chloride solution. Filter the ether layers individually through a funnel packed with the same sodium sulfate. Combine the ether layers and remove the ether by distillation. Purify the acid by distillation if it is a liquid and by recrystallization if it is a solid.

If the acid is an insoluble solid, separate by filtration immediately after adding the dilute acid, wash with small portions of cold water, and recrystallize from ethanol. Identify the acid in the usual manner. Note that the salts may be used to prepare the derivatives directlv

UNKNOWN

This unknown may be any compound that was assigned as a possible unknown when Chapters 14, 23, and 28 were studied, any acid from the tables at the end of this chapter that does not have an asterisk, and the sodium salts of these acids.

A solubility test of S_1a or A is indicative of a carboxylic acid; however, other compounds that will be studied later may also be in these solubility classes. Even so, the physical constants and results of the elemental analysis and solubility classification tests are usually sufficient to establish the presence of a $—CO_2H$ group, and no further tests for this group are needed. All acids should be tested to determine whether or not they are saturated or unsaturated. All halogen-containing compounds, including halo-acids, should be tested with sodium iodide in acetone and alcoholic silver nitrate.

Do the following problems before attempting to solve your unknown.

EXERCISES

32-1. A Class A compound has a boiling point around 204°. What is this compound? How could this be verified?

32-2. A Class S_1a compound that contains chlorine boils at 185 to 186.3°. What is the compound?

32-3. An impure semisolid (Class S_1a) contains bromine and boils at 206 to 207.2°. Which compounds could the unknown be? Which derivative should be prepared to differentiate between the possible compounds?

32-4. A Class S_1a compound contains bromine and melts at 48 to 49°. Which compounds could the unknown be? How would you differentiate among the possible compounds?

32-5. A Class A compound that does not form a soapy solution when shaken with sodium hydroxide melts at 62.7 to 63.5°. It decolorizes both dilute, neutral potassium permanganate solution and bromine in carbon tetrachloride. What is the compound? Where should its amide melt?

32-6. A Class S_2w compound (I) contains no halogen and leaves a residue on ignition which proves to be sodium carbonate. Treatment of a concentrated aqueous solution of I with 20% hydrochloric acid precipitates II which melts at 131.5 to 132.8°. II decolorizes dilute, neutral potassium permanganate but reacts very slowly with bromine in carbon tetrachloride. When I is heated with aniline and concentrated hydrochloric acid, III is produced. III after recrystallization melts at 153.6 to 154.8°. Identify I, II, and III.

32-7. A Class I compound, C_8H_8ClBr, I gives a white precipitate with alcoholic silver nitrate only after boiling. It gives a precipitate with sodium iodide-acetone solution when heated to 50°. I is oxidized with chromium trioxide and sulfuric acid to a Class A compound (II) which melts at 154 to 155°. II reacts with thionyl chloride to give III which reacts with concentrated aqueous ammonia to give IV, a compound melting at 154 to 155°. A mixture of II and IV melts at 139 to 143°. Identify I, II, III, and IV.

32-8 A Class I compound, C_7H_6ICl, I gives an immediate white precipitate with alcoholic silver nitrate and reacts at 50° with sodium iodide-acetone solution to give a white precipitate. I is heated with aqueous potassium hydroxide and potassium permanganate to give a Class S_2w compound (II). When an aqueous solution of II is acidified, a Class A compound (III) is obtained that melts at 263 to 264°. Identify I, II, and III.

32-9. What conditions indicate a derivative should be recrystallized to a constant melting point? What conditions indicate it is not necessary to recrystallize to a constant melting point?

TABLE 32-3 CARBOXYLIC ACIDS
(LIQUIDS)

Acid	B.p. (°C)	M.p. (°C)	Melting points of derivatives (°C) anilide	p-toluide	p-nitrobenzyl ester	p-bromophenacyl ester
Formic	101	8.4	47	53	31	137
Acetic	118	16.6	114	147	78	85
Acrylic	140	13	104	141		
Propionic	140	-20.8	106	124	31	63
Propynoic (Propiolic)	144d	18	87			
Isobutyric	155	-46.1	105	108		77
Butanoic	163	-6	96	73	35	63
*Pyruvic	165d	13.6	104			
cis-2-Buten-1-oic	168	15	102	132		82
2-Methylbutanoic	176	-75	110	93		55
3-Methylbutanoic	176	-30	109	107		68
3,3-Dimethylbutanoic	184	6.7	134	134		105
2-Chloropropanoic	186		92	124		
Pentanoic (Valeric)	186	-34.5	63	73		75

*These acids will not be assigned until polyfunctional compounds are studied (Chapter 61).

TABLE 32-3 (cont.)

Acid	B.p. (°C)	M.p. (°C)	Melting points of derivatives (°C) anilide	p-toluide	p-nitro-benzyl ester	p-bromo phenacyl ester
Dichloroacetic	194	5.5	118	153		99
2-Ethylbutanoic	195	-31.8	126	116		
4-Methylpentanoic	199	-33	111.5	63		77
2-Bromo-2-methylpropanoic	200	48	83	148		
Methoxyacetic	203		58			
n-Caproic	205	-3.5	95	75		72
2-Bromopropanoic	205	24	99			
Ethoxyacetic	206.5	−51	94	32		105
Bromoacetic	208	50	131	91	88	
2-Bromobutanoic	217d		98	92	49	
Heptanoic	223.5	-7.5	70	80.5		72
Cyclohexanecarboxylic	233	31	145			
n-Caprylic	238	16	57	70		67.5
Nonanoic (Pelargonic)	253	12.3	57	84		68
n-Capric	270	31.5	65			67
Undecanoic	275	29	71			68
10-Undecenoic	275	24				

TABLE 32-4 CARBOXYLIC ACIDS (SOLIDS)

Acid	M.p. (°C)	B.p. (°C)	Melting points of derivatives (°C) anilide	amide	p-nitro-benzyl ester	p-bromo-phenacyl ester
3-Bromobutanoic	22	220d		92		
10-Undecenoic	24.5	275		86		
2-Bromopropanoic	25.7	204	99	123		
Undecanoic	28.6	280	71	103		68
Cyclohexanecarboxylic	31	233	145	186		
Capric	31.5	270	69	108		67
*Levulinic	34	246	102	108	61	84
Trimethylacetic (pivalic)	35	164	129	156		76
β-Chloropropionic	41.5	204		101		
Dodecanoic (lauric)	43	299	77	98		76
α Bromoisovaleric	44	230d	116	133		
cis-α,β-Dimethylacrylic	46	185	126	128		
Dibromoacetic	48	233d		156		
2-Phenylpropanoic (hydrocinnamic)	48	280	98	82	36	104
α-Bromoisobutyric	48	200	83	148		
Bromoacetic	50	208	131	91	88	
Trichloroacetic	57.5	197.5	94	141	80	
β-Bromopropionic	62			111		
Palmitic	62.5	222	90	106	42	86
Chloroacetic	63	188	134	121		105
trans-α,β-Dimethylacrylic	64	197	77	76	64	68
α,β-Dibromopropionic	66.5			130		
*Cyanoacetic	66	165d	198.5	119.5		
trans-Crotonic	72	189	118	160	67	96
Phenylacetic	76.5	256	117.5	156	65	89
*Glycolic	79.3		96.7	120	107	138
*α-Hydroxylisobutyric	79	212	136		80	98
β-Iodopropionic	82.5			101		
Iodoacetic	84		143	95		
*Methylmaleic (citraconic)	91		175	187d	71	
*Phthalaldehydic	97					
*Glutaric	97	303	224	175	69	137
Phenoxyacetic	99	285d	99	101		148

*These acids will not be assigned until polyfunctional compounds are studied (Chapter 61).

TABLE 32-4 (cont.)

Acid	M.p. (°C)	B.p. (°C)	Melting points of derivatives (°C) anilide	amide	p-nitro-benzyl ester	p-bromo-phenacyl ester
*l-Malic	100		197	102(mono) 156(di)	124	179
*Citric	100		192	215d	102	148
o-Methoxybenzoic	100	200	62	128.6		113
*Oxalic (hydrated)	101		257(mono) 148(di)	419d	204	242d
o-Toluic	103	260	125	142	91	57
*Pimelic	105	223	155	175		137
*Azelic	106		185	175	44	131
*Ethylmalonic	111		150	214	75	
m-Toluic	113	263sub	125.5	97	86.5	108
*β-Benzoylpropionic	116		150	145.5		
p-Isopropylbenzoic	117.5			133		
*Tropic	117.5			169		
*Benzylmalonic	118d		217	225	119	
*dl-Mandelic	119d		151.5	133.7	123.5	
*m-Nitrophenylacetic	120			110		
Benzoic	122.3	249	160	129.3	89	119
*Trichlorolactic	124		164	96		
*3-Nitrosalicylic(hydrated)	125			145		
*o-Benzoylbenzoic	126		195	165	100	
2,4-Dimethylbenzoic	127		141	180		
*Maleic	130		187	153	89	168
2,5-Dimethylbenzoic	132		140	186		
Tribromoacetic	133	245d		122		
*Furoic (pyromucic)	133	231	123	143	134	138
*Sebacic	133		201	210	73	147
*d- or l-Mandelic	133			122		
trans-Cinnamic	133	300	155	147	116	145
α-Napthylacetic	133		155	181		
*Malonic	133d		224	170	85	
*Acetone-1,3-dicarboxylic	135d		155			
*Acetylsalicylic	135	140d	136	138	90	
Phenylpropiolic	136.5sub		125.5	102	83	
*cis-Pentendioic (Glutaconic)	137		135(mono)			
*trans-Pentendioic	138		167(mono) 228(di)			
*Methylmalonic	138d		182	217	75	
2,6-Dichlorobenzoic	139			202		
*m-Nitrobenzoic	140		153.5	142	141	132
*meso-Tartaric	140			189.5		
*Suberic	140		187	216	85	144
o-Nitrophenylacetic	141			161		
o-Chlorobenzoic	142		117		106	106
*3-Nitrosalicylic	144			145		
Diphenylacetic	148		180	167		
o-Nitrobenzoic	146		155	175	112	107
p-Hydroxyphenylacetic	148			167		
o-Bromobenzoic	150		141	155	110	
*Benzilic	150		175	154	99	152
*Adipic	152		235	220	106	154
p-Nitrophenylacetic	152.5		198	198		207
2,5-Dichlorobenzoic	153			155		
*Citric	153		192	214d	102	148
m-Bromobenzoic	155		146	155	105	
2,4,6-Trimethylbenzoic	155			188		

TABLE 32-4 (cont.)

Acid	M.p. (°C)	B.p. (°C)	Melting points of derivatives (°C) anilide	amide	p-nitro-benzyl ester	p-bromo-phenacyl ester
*Salicylic	158		134	139	97.5	140
m-Chlorobenzoic	158		122	134	107	116
o-Iodobenzoic	162		142	110	111	146
α-Naphthoic	162		161	205		135
*Thiosalicylic	163					
3,4-Dimethylbenzoic	164			130		
*4-Nitrophthalic	165		192	200d		
*Itaconic (methylenesuccinic)	165d		190	192	91	117
*3,5-Dimethylbenzoic	166			133		
*Propane-1,2,3-tricarboxylic	166		252	207d		138
*d l-Dibromosuccinic	167					
*d- or l-Tartaric	169		180	195	163	
*3,5-Dinitrosalicylic	173		140	158	104	153
*Acetylenedicarboxylic	179			294d		
p-Fluorobenzoic	182			154		
2,4-Dinitrobenzoic	183			203	142	158
p-Methoxybenzoic	184		168	162	132	152
β-Naphthoic	185		173	195		
*Acetylanthranilic	185		167	171		
*d-Camphoric	187		203	192	66	
*Hippuric	187		208	183	136	151
*Succinic	188		148(mono) 228.5(di)	242	88(di)	211(di)
*4-Methoxy-3-nitrobenzoic	189		163			
*3,4-Dihydroxybenzoic	199d		166	212	188	
m-Nitrocinnamic	199			196	174	173
2-Chloro-2,5-dinitrobenzoic	199					
*Fumaric	200		314	270	151	197
*m-Hydroxybenzoic	201		155	170	106	176
3,5-Dinitrobenzoic	202		234	183	157	
*Methylfumaric	202		185	176	134	
*d l-Tartaric	204			226	147	
*p-Coumaric (p-hydroxycinnamic)	206			194		
*Phthalic	206d		169	149	155	153
*o-Coumaric	207			209d	152	
*4-Hydroxy-3-methoxybenzoic (vanillic)	207				140	
*p-Hydroxybenzoic	213		202	162	198	191
*2,4-Dihydroxybenzoic	213		127	222	189	
*p-Cyanobenzoic	214		179		189	
*3-Nitrophthalic	218		234	201	189	
2,4,6-Trinitrobenzoic	220d			264d		
*2-Hydroxy-3-naphthoic	222		249	218		
*5-Nitrosalicylic	227		224	225		
*Biphenyl-2,2-dicarboxylic (diphenic)	228		230	212	183	
*o-Nitrocinnamic	240			185	132	141
3,4,5-Trihydroxybenzoic (gallic)	240d		207	245		
*p-Nitrobenzoic	241		217	201	168	137
p-Chlorobenzoic	242		194	179	129	
*Tetrachlorophthalic	250d				180	
p-Bromobenzoic	251		197	189	180	
p-Iodobenzoic	265		210	217	141	147
*p-Nitrocinnamic	285			204	186	191

33

Carbonation of Grignard Reagents

Synthesis of Cyclohexanecarboxylic Acid

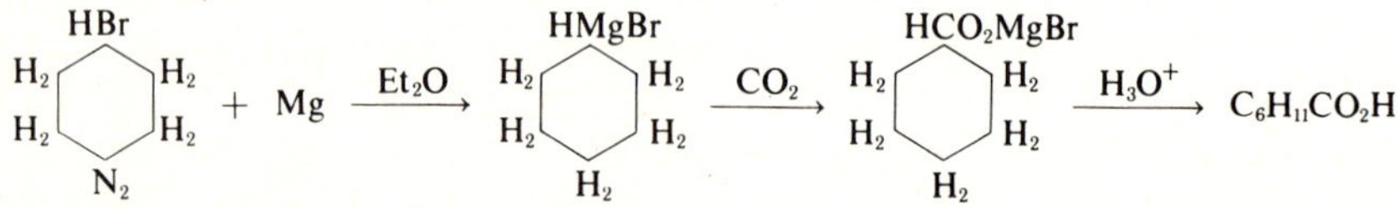

B.p. 165°
D_4^{20} 1.336

B.p. 233°
M.p. 31°

INTRODUCTION

The Grignard reaction can be used to synthesize many carboxylic acids. In this experiment cyclohexylmagnesium bromide is carbonated with dry ice. Most Grignard reagents can be carbonated satisfactorily with dry ice. Use of gaseous instead of solid carbon dioxide requires construction of an elaborate gas generation and purification train. It is very important to keep the reactants and apparatus dry to ensure that the reaction occurs and that the maximum yield of pure product can be obtained.

EXPERIMENTAL PROCEDURE

Danger! Ether is a very flammable solvent. There should be no flames in the laboratory. Follow the safety precautions and the procedure for drying given on pages 188 and 189. **Do not allow dry ice to come in contact with your skin; it can cause severe frost bite.**

Using a 100-ml round bottom flask, construct a reflux and addition apparatus similar to that shown in Figure 16-1 except that a calcium chloride tube should be placed in the top of the condenser instead of a cork and tube leading to a hood. Rub 1.1 g of clean, dry magnesium turnings with steel wool and put them in the flask.

Dissolve 5 ml of cyclohexyl bromide in 20 ml of *anhydrous* ether. Since ether rapidly absorbs moisture from the air, do not use ether from a container that has been opened before the day of this synthesis. Place the ether solution in the dropping funnel and stopper the funnel. Add a crystal of iodine to the flask. Prepare an ice bath; this is to be placed around the reaction flask to moderate the reaction if it proceeds too rapidly and the ether boils too vigorously.

Run 3 to 6 ml of the cyclohexyl bromide solution onto the magnesium. The reaction starts in a few minutes if all traces of moisture have been kept from the reaction mixture. Cool the reaction if the mixture boils violently. If any product at all is to be produced by this synthesis, it is very important that this reaction occurs before any more solution is added. If the cyclohexyl bromide does not react, heat the flask with a hot water bath. If even this does not start the reaction, notify the instructor, and he will add a small amount of methylmagnesium iodide.

Add 20 ml more of anhydrous ether to the dropping funnel and briefly stir the mixture vigorously with a glass rod. When reaction of the magnesium with the first 3 to 6 ml of cyclohexyl bromide solution has moderated, allow the cyclohexyl bromide-ether solution to drip slowly into the flask. Set the rate of addition so that heat from the reaction causes moderate refluxing of the ether. Stir the mixture every 2 to 3 min by applying a gentle rotary motion to the ring stand. After all the solution has been added, reflux the mixture for 25 min. Heat gently, otherwise some of the ether will escape into the room. If almost none of the magnesium has reacted, continue refluxing the mixture until it does react. Remove the heat source while preparing the carbonation flask.

Wipe the frost from a few pieces of dry ice, total weight 40 to 55 g. Crush the dry ice in a clean dry towel with a pestle or small hammer. Immediately transfer the dry ice to a dry 500-ml Erlenmeyer flask containing 15 ml of anhydrous ether. Quickly pour the Grignard mixture into the Erlenmeyer flask. Try to keep any unreacted magnesium in the round bottom flask. Rinse the round bottom flask twice with 10 ml of dry ether. Pour the rinses into the carbonation flask.

Allow the mixture to stand about 5 min after all the dry ice has disappeared. Mix 25 ml of concentrated hydrochloric acid with 20 g of crushed ice. Add 15 to 20 g of ice to the carbonation flask and then add the cold hydrochloric acid. Stir the mixture well with a glass rod. When the resinous mass has dissolved or at least dispersed, pour the mixture into a separatory funnel and shake the funnel. Separate the two layers, extract the aqueous layer with 20 ml of ether (Extraction A), and combine this extract with the first ether layer.

Extract the ether with two 20-ml portions of 10% aqueous sodium hydroxide and 20 ml of water. Combine the three extracts in a 250-ml Erlenmeyer flask. Immediately dispose of the ether in the way the laboratory instructor tells you. Place the flask containing the extracts in an ice bath. Mix 20 ml of concentrated hydrochloric acid with 15 g of ice. Pour the cold hydrochloric acid into the Erlenmeyer flask and stir the mixture. After a few minutes an oil separates. Grind the oil against the flask with a glass rod. Filter after the oil solidifies.

Problem 33-1. Which of the following solvents should be used for crystallizing the product: benzene, cyclohexane, petroleum ether, water, methanol, toluene? Why?

Problem 33-2. Because of its low melting point cyclohexanecarboxylic acid often comes out of solution as an oil. What precautions should be taken to prevent oil formation?

Check your answers to Problems 33-1 and 33-2; then recrystallize and dry the product. Determine the melting point and weigh in a tared yield bottle. Calculate the % yield, label the bottle, and hand in the product to your instructor.

ALTERNATE, SYNTHESIS OF PHENYLACETIC ACID

Follow the directions above except: 1) use 1.2 g of Mg; 2) use 7 ml of benzyl chloride instead of cyclohexyl bromide; 3) extract the aqueous layer after the HCl treatment of the carboration mixture with *two* 20-ml portions of ether (labeled extraction A) and combine these with the first ether layer: 4) after the final neutralization with HCl (second to last paragraph) the grinding is usually unnecessary. The yield can be increased, however, by extracting the acid filtrate with two 15-ml portions of chloroform. Remove the chloroform by distillation and combine any crystals with the product *before* recrystallization. Phenylacetic acid melts at 76.5° and boils at 256°.

EXERCISES

33-1. In Experiment 31 isobutyric acid was made from isobutyl alcohol. How could the following acids be synthesized from isobutyl alcohol? (a) 3-methylbutanoic acid, (b) 5-methyl-2-hexenoic acid, (c) 2,2-dimethylpropanoic acid, (d) *p-t*-butylbenzoic acid.

34

Synthesis of Cyclohexanone

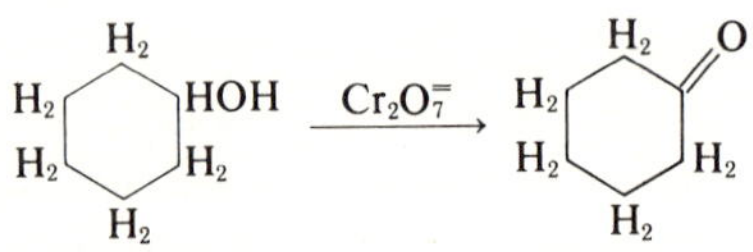

B.p. 161°
M.p. 25°
D_4^{30} 0.94

B.p. 156°
D_4^{20} 0.946

INTRODUCTION

Secondary alcohols are readily oxidized to ketones in high yield with the formation of few by-products. Unlike aldehydes, ketones are usually not oxidized under weak or moderately strong oxidizing conditions.[1]

EXPERIMENTAL PROCEDURE

Pour 52 ml of water in a 250-ml Erlenmeyer flask. Place the flask in an ice bath and add 16 ml of concentrated sulfuric acid while gently stirring with a thermometer. After stirring for 3 to 4 min, add 16 ml of cyclohexanol. Dissolve 16 g of sodium dichromate ($Na_2Cr_2O_7 \cdot 2H_2O$) in 8 ml of water. Slowly add the dichromate solution to the flask while stirring the mixture with the thermometer. The stirring and rate of addition should be adjusted so that the temperature of the mixture is 22 to 30°. Stir 3 to 4 min after all the dichromate solution is added.

Remove the flask from the ice bath; however, keep the ice bath nearby in case it is necessary to cool the reaction again. Allow the mixture to stand for 80 min. Stir the mixture and check the temperature after 5 min; repeat this every 10 min. If the temperature reaches 30°, return the flask to the ice bath and stir until the temperature drops to 20°. Remove the flask from the ice bath and proceed as instructed above. During this period begin the next experiment.

Problem 34-1. Why is it important to keep the temperature of the oxidation below 30°?

Cool the mixture to 0 to 5°. Pour the mixture into a separatory funnel and drain off the bottom layer. Save this layer and all the washes until they have been extracted once with ether. If the gas chromatography option is assigned, read that section before proceeding further. Wash the organic layer with 15 ml of cold 5% sodium bicarbonate solution and 15 ml of ice water. Add 3 to 4 g of calcium chloride to the separatory funnel and shake the funnel. Usually there will be enough water dissolved in the cyclohexanone to dissolve the calcium

[1] The exceptions occur when certain other functional groups are present.

chloride. If this occurs, drain off the aqueous layer and add more calcium chloride. Repeat this until the calcium chloride does not dissolve. Drain the cyclohexanone into a 50-ml round bottom flask, or if standard taper equipment is not being used, drain it into a 50-ml distilling flask. Keep the calcium chloride granules out of the flask.

Put 15 ml of ether (**fire hazard**) into the separatory funnel and successively extract the aqueous acid layer, the 5% sodium bicarbonate washes, the water wash, and the combined calcium chloride solution and solid, Dry the ether layer with calcium chloride and drain it into the flask containing the cyclohexanone. Distil the mixture and collect the fraction that boils at 150 to 157° in a tared yield bottle. Weigh the product, calculate the % yield, label the product, and hand it in to your instructor.

GAS CHROMATOGRAPHY (Optional)

Using the conditions described in Table 8-1, chromatograph on an Apiezon column 5 μl samples of (a) the organic layer before the 5% sodium bicarbonate wash, (b) the purified product, (c) any fractions collected that distilled between 65 and 140°. Determine retention times on any compounds available that you have good reason to think might be present. From this data and any retention time and peak height data that your instructor might supply, determine what contaminants are present and estimate their relative concentrations. What steps in the purification removed which contaminants?

EXERCISES

34-1. Why is cyclohexanol not likely to contaminate the product?

34-2. Sometimes a class N_2 compound, b.p. 82 to 84°, is produced as a by-product. It does not react with sodium and decolorizes dilute neutral permanganate. What is the compound? How is it formed?

34-3. How could 4-cyclohept*en*one be prepared from 4-cycloheptenol?

35

Properties and Reactions of Carboxylic Acid Derivatives

Organic Qualitative Analysis V

INTRODUCTION

Carboxylic acid derivatives can be identified by their physical properties and reactions to various classification tests. The presence of nitrogen and sulfur in a compound is a further aid to identification of the compounds that contain these elements. The ester, amide, and nitrile groups are not quite so quickly identified by chemical tests as the carboxylic acid group.

PHYSICAL PROPERTIES

Boiling and Melting Points

Because acid chlorides, acid anhydrides, and esters cannot undergo hydrogen bonding in pure liquid or solid state, they melt and boil significantly lower than do alcohols and acids of similar molecular weights. The polarity of the carbonyl groups gives rise to considerable dipole-dipole and dipole-induced dipole attractions (Fig. 35-1). As a result monofunctional acid derivatives have boiling points between those of hydrocarbons and alcohols of similar carbon skeleton and molecular weights (Table 35-1).

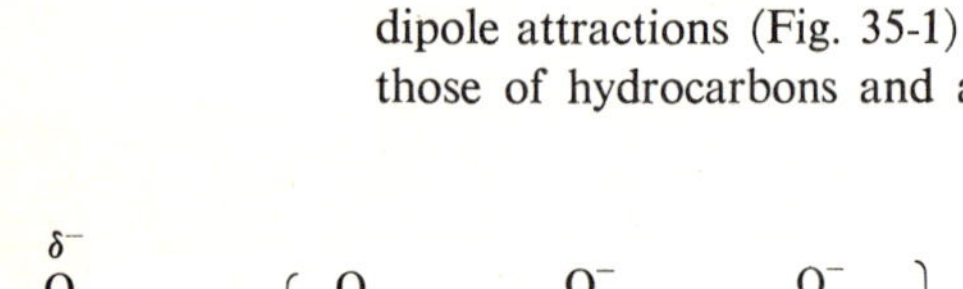

Figure 35-1 Polarity of carbonyl group in acid derivatives.

For molecular weights above 100, hydrocarbons and esters of similar carbon skeletons and molecular weight have similar boiling points. There are two reasons for this. As the molecular weight increases in a homologous series, the fraction of hydrocarbon or nonpolar portion of the molecule increases. Monofunctional acid derivatives thus resemble the hydrocarbon more and more as the molecular weight of the acid derivatives increases. That is, the polarity of the carbonyl group is responsible for a smaller fraction of the total attractive forces in the higher molecular weight acid derivatives. The second reason is that straight-chain acid derivatives resemble branched hydrocarbons because of the carbonyl oxygen; e.g., propyl acetate should be compared with 2-methylhexane, ethyl propionate with 3-methylhexane, and butyl propionate with 3-methyloctane (Fig. 35-2). Even with this latter comparison it is found that esters with more than 8 carbon atoms have boiling points only slightly above those of the branched alkane of similar molecular weight.

TABLE 35-1 PHYSICAL CONSTANTS OF ACID DERIVATIVES COMPARED WITH HYDROCARBONS AND ALCOHOLS

Compound	Mol. wt.	B.p. (°C)	M.p. (°C)	Solubility (g/100 ml H_2O, 20°C)
Pentane	72	36.0	-137.0	0.04
Ethyl formate	74	54.2	79.4	12.0
Methyl acetate	74	57.3	-98.7	32.0
Acetyl chloride	78.5	52.0	-115.0	d
2-Butanol	74	99.5	-89.5	12.5
Hexane	86	68.3	-95.0	0.014
Methyl proprionate	88	79.7		6.5
Ethyl acetate	88	77.2	-86.6	9.2
Propyl formate	88	80.9	-92.9	2.8
Propionyl chloride	92.5	80.0	-94	d
2-Pentanol	88.0	119.9		4.8
3-Pentanol	88.0	116.1	-9.0	5.5
Isopropyl formate	88.0	71.0		2.0
3-Methyl-2-butanol	88.0	113.5		6.0
Heptane	100.0	98.0	-90.6	0.005
2-Methylhexane	100.0	90.0	-118.0	< 0.01
3-Methylhexane	100.0	92	-119	< 0.01
Propyl acetate	102	101.6	-95	1.8
Ethyl propionate	102	99.1	-73.9	2.4
Butyrl chloride	106.5	101.5	-89	d
Acetic anhydride	102	140	-73	d
3-Hexanol	102	135		~ 0.8
2-Hexanol	108	137		~ 0.8
Nonane	128	150		> 0.004
3-Methyloctane	128	142.5		0.004
Propyl butyrate	130	142.7	-95.2	0.18
Butyl propionate	130	146	-89.6	0.17
Hexanoyl chloride	134.6	152.6	-87.0	d
Propionic anhydride	130	168	-45	d
3-Methyl-3-heptanol	130	164		~ 0.1
2-Octanol	130	179	-38.6	0.091

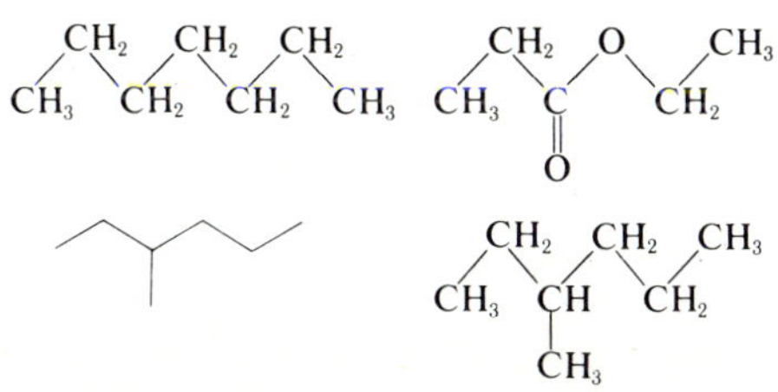

Figure 35-2 Comparison of branching of ethyl propionate and hydrocarbons of similar molecular weight.

Table 35-1 shows that carboxylic anhydrides have boiling points considerably higher than those of esters and acid chlorides of similar molecular weight and carbon skeleton. Acid anhydride molecules have *two* polar carbonyl groups resulting in considerably more dipole-dipole attractions and dipole-induced dipole attractions than exist in monofunctional acid derivatives. Amides also melt and boil higher than do esters and acid chlorides of similar molecular weight (see p. 271).

Solubilities

The borderline of water solubility of monofunctional acid derivatives is at compounds with 4 to 5 carbon atoms. Compared with alcohols having the same number of carbon atoms, amides (except N,N-disubstituted amides) have about the same water solubility, esters and acid chlorides are slightly less soluble, and carboxylic anhydrides are slightly more soluble (Table 35-2). Chain branching, cyclization, aromatic groups, and halogen atoms have the same effects on water solubility of acid derivatives as on the classes of compounds discussed in Chapters 28 and 32.

TABLE 35-2 COMPARISON OF PHYSICAL PROPERTIES OF AMIDES AND ESTERS

Compound	Mol. wt.	B.p. (°C)	M.p. (°C)	Solubility (g/100 ml H_2O, 20°C)
Propionamide	73	213	79	> 20
N-Methylacetamide	73	206	28	> 30
Ethylformamide	73	198		∞
Dimethylformamide	73	175		∞
Ethyl formate	74	54	-80	12
Butyramide	87	216	116	17.4
N-Ethylacetamide	87	205		∞
N,N-Dimethylacetamide	87	166		∞
Ethyl acetate	88	77	-86.6	9.2
Benzamide	121	288	130	1.3
N-Ethylbenzamide	149	299	70.5	0.5
N,N-Dimethylbenzamide	149	272	41.5	10.0
Ethyl benzoate	150	213	-34	0.08

All acyl chlorides except acetyl chloride are insoluble in water solutions. However, they react so rapidly with water or dilute aqueous sodium hydroxide and bicarbonate that only a few high molecular weight acyl halides do not dissolve in these solvents. Therefore, most acyl chlorides are in Class S_1a or A, and some high molecular weight acyl chlorides are in Class N. The lower the molecular weight of the acyl halide, the faster and more violently it reacts with bases. The reaction of lower molecular weight acid halides with water can be dangerous. Acyl chlorides of 6 or less carbon atoms react so rapidly with water they are in Class S_1a.

Acyl anhydrides up to about 6 carbon atoms are in solubility Class S_1a. Acid anhydrides are less reactive than acyl halides and react rapidly with water only if warmed. Since most acid anhydrides react rapidly with aqueous sodium hydroxide or bicarbonate, only a few high molecular weight anhydrides are insoluble in this reagent and are in Class N_2.

Monofunctional esters with less than 5 carbon atoms are in Class S_1n, those with 5 to 9 carbon atoms are in Class N_1, and those with 9 or more carbon atoms are in Class N_2. Esters of a few acids, such as pyruvic and formic, and *low molecular weight* primary alcohols are hydrolyzed rapidly by water at room temperature so that they often appear to be in Class S_1a.

Amides with 1 or 2 hydrogens on the amido-nitrogen are capable of undergoing hydrogen bonding (Fig. 35-3). As a result amides are much higher melting, much higher boiling, slightly more soluble in water, and less soluble in solvents of low polarity than other acid derivatives of the same molecular weight.

Compounds having only 1 amido hydrogen melt and boil lower than those having 2 amido

Figure 35-3 Hydrogen bonding of amides.

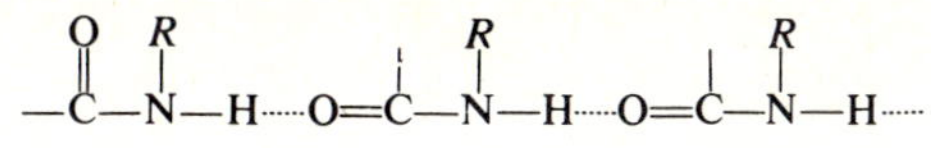

hydrogens (Table 35-2). N,N-Disubstituted amides have no amido hydrogen; therefore, molecules of these compounds cannot undergo hydrogen bonding. Consequently, N,N-disubstituted amides boil and melt considerably lower than do amides of the same molecular weight having 1 or 2 amido hydrogen (Table 35-2). All amides melt and boil significantly above esters of similar molecular weight and structure. Molecules of N,N-disubstituted amides cannot form hydrogen bonds with each other but can with other molecules capable of supplying the hydrogen such as water. Therefore, N,N-disubstituted amido groups have *stronger* attractive forces for *solvent* molecules like water than for other amido groups. This causes N,N-disubstituted amides to be much more soluble in water than are amides having 1 or 2 amido hydrogens.

Problem 35-1. Would you expect N,N-disubstituted amides to be more or less soluble than unsubstituted amides in nonpolar solvents like hexane and benzene?

Acid derivatives like all other compounds containing oxygen are soluble in or react with concentrated sulfuric acid.

SOLUBILITY CLASS M

Most organic nitrogen and sulfur compounds have basic electron pairs similar to those of organic oxygen compounds. These basic electron pairs are able to coordinate with protons in concentrated sulfuric acid to give soluble ionic onium compounds. For this reason almost all compounds containing nitrogen and sulfur are soluble in concentrated sulfuric acid. It would be a waste of time to test the solubility of these compounds in sulfuric acid. All compounds containing nitrogen or sulfur that are insoluble in aqueous solutions (water, dilute sodium hydroxide, and dilute hydrochloric acid) are assigned to a *miscellaneous* solubility class called Class M without further testing. Most unsubstituted carboxamides are in Class M. Figure 32-4 shows the solubility classification scheme with Class M.

Problem 35-2. Predict the solubility classes of the following compounds: methyl acetate, methyl decanoate, methyl pyruvate, *sec*-butyl pyruvate, *p*-nitrobenzoyl chloride, butyric anhydride, N,N-dimethylformamide, N-ethylpentanamide, ethyl formate.

ELEMENTAL ANALYSIS

Since further assigned unknowns may contain nitrogen and sulfur, the presence of these elements in addition to chlorine, bromine, and iodine must now be determined.

Sulfur

Acidify 3 ml of filtrate[1] from the sodium fusion (p. 174) with acetic acid and add 4 drops of 1% aqueous lead acetate. A black precipitate is a positive test for sulfur.

$$S^{=} + Pb^{++} \rightarrow PbS \downarrow \qquad (35\text{-}1)$$

[1] It is very important that the fusion filtrate not be dark so that the tests for nitrogen and sulfur are masked. If the solution is darker than a light yellow or a very light grey, the sodium fusion should be repeated.

Nitrogen

a. Test 1. To 3 ml of the filtrate from the sodium fusion add 5 drops of *freshly* prepared 5% aqueous ferrous sulfate, 5 drops of 10% aqueous sodium hydroxide, and 5 drops of 30% aqueous potassium fluoride. Boil the resulting mixture gently for a few seconds. If sulfur is present, modify the above by adding 10 to 12 drops of ferrous sulfate solution and 1 ml of 10% sodium hydroxide solution. Heat to boiling and *immediately* filter or centrifuge. Add 5 drops of 30% potassium fluoride solution to the filtrate.

Cool the suspension of iron hydroxide and treat with 2 drops of 5% aqueous ferric chloride and sufficient 30% sulfuric acid to dissolve the insoluble iron hydroxides. If cyanide ions are in the filtrate from the sodium fusion, a brilliant blue precipitate of Prussian blue appears at this point.

$$Fe^{++} + 2CN^- \rightarrow Fe(CN)_2 \tag{35-2}$$

$$Fe(CN)_2 + 4CN^- + Fe^{+++} \rightarrow Fe[Fe(CN)_6]^- \tag{35-3}$$

The amount of cyanide ions in the solution may be so small that the ferric ferrocyanide does not appear as a precipitate but as a blue solution or colloidal suspension. A small amount of precipitate is more easily detected by filtering the solution and observing the blue precipitate against the white background of the filter paper.

b. Test 2. Place 1 ml of the filtrate from the sodium fusion in a small evaporating dish and acidify with a few drops of acetic acid. Moisten a piece of filter paper with a mixture of equal parts of cupric acetate solution (0.29 g in 100 ml of water) and an aqueous solution of benzidine acetate. This is made by diluting a saturated solution of benzidine acetate with an equal volume of distilled water. Stick the moist filter paper to the bottom of a watch glass. Place the watch glass with the filter paper into the evaporating dish. If the paper does not develop a blue color immediately, gently warm the dish. A blue color either with or without warming is a positive test.

The test for nitrogen is possibly the most difficult of the procedures required of the beginning organic chemistry student. Nitrogen due to nitro and nitroso groups is usually more difficult to detect than nitrogen due to amine and amide groups. Because of this difficulty the faintest positive test should be considered evidence for the presence of nitrogen. If the first test is positive, the second procedure is not necessary. If, however, the first test is negative, the second test should be done. If both tests are negative, assume there is no nitrogen in the compound unless there are *strong* indications in the classification tests that a nitrogen functional group is present.

Test for Halogen in the Presence of Sulfur or Nitrogen

If sulfur or nitrogen is present in the unknown, a preliminary step to remove sulfide and cyanide is required prior to testing with silver nitrate. If the presence of either or both of these elements is known or even suspected, add 2 ml of dilute nitric acid to the half of the fusion filtrate to be used for the tests for halogens. Add dilute nitric acid a drop at a time until the solution is acidic and then add two more drops of nitric acid. Boil the solution for 3 to 4 min; any nitrogen is removed as hydrogen cyanide and any sulfur is removed as hydrogen sulfide. Proceed with the test for halogens as described on page 174.

REACTIONS OF ACID DERIVATIVES USED AS CLASSIFICATION TESTS

Alcoholic Silver Nitrate

All acyl halides give immediate precipitates with alcoholic silver nitrate and sodium iodide at room temperatures (see Chapter 23). Verify with known acid halides.

The Hydroxamic Acid Test

Discussion. Most acid derivatives react with hydroxylamine and its salts to give hydroxamic acids. This reaction is a specific example of the more general ammonolysis (Eqs. 35-10 and 11). R′ can equal alkyl, aryl, hydrogen, hydroxy, —NHAr, or other groups.

$$\mathrm{RCONH_2 + H_2NOH \xrightarrow{HO^-} R\overset{\overset{\displaystyle O}{\|}}{C}NHOH + NH_3} \quad (35\text{-}4)$$

$$\mathrm{RCOCl + H_2NOH \xrightarrow{H^+} R\overset{\overset{\displaystyle O}{\|}}{C}NHOH + HCl} \quad (35\text{-}5)$$

$$\mathrm{(RCO)_2O + H_2NOH \xrightarrow{H^+} R\overset{\overset{\displaystyle O}{\|}}{C}NHOH + RCO_2H} \quad (35\text{-}6)$$

$$\mathrm{RCO_2Et + H_2NOH \xrightarrow{H^+} R\overset{\overset{\displaystyle O}{\|}}{C}NHOH + EtOH} \quad (35\text{-}7)$$

$$\mathrm{RCCl_3 + H_2NOH + H_2O \xrightarrow{H^+} R\overset{\overset{\displaystyle O}{\|}}{C}NHOH + 3HCl} \quad (35\text{-}8)$$

$$\mathrm{RCN + H_2NOH \xrightarrow{HO^-} R\overset{\overset{\displaystyle NH}{\|}}{C}NHOH + NH_3} \quad (35\text{-}9)$$

$$\mathrm{R\overset{\overset{\displaystyle O}{\|}}{C}Y + H_2NR' \longrightarrow R\overset{\overset{\displaystyle O}{\|}}{C}NHR' + HY} \quad (35\text{-}10)$$

$$\mathrm{R\overset{\overset{\displaystyle O}{\|}}{C}Y + HNR'_2 \longrightarrow R\overset{\overset{\displaystyle O}{\|}}{C}NR'_2 + HY} \quad (35\text{-}11)$$

The results of this test depend upon the formation of a colored chelate or complex of hydroxamic acids with ferric ion. Many compounds, such as phenols, primary and secondary nitro compounds, and β-dicarbonyls, do not yield hydroxamic acids and often give colors with ferric chloride whether or not hydroxylamine is present.

$\mathrm{3RCONHOH + FeCl_3 \rightarrow}$ 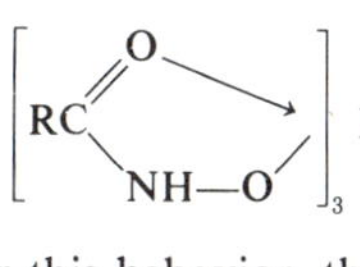 $\mathrm{Fe + 3HCl}$ (35-12)

If the unknown were not checked first for this behavior, these compounds would give false positives and would be a major source of error. Therefore, first test the compound with ferric chloride, and, if a strong persistent color is observed, do not use the hydroxamic acid test.

Experimental Procedure. All acid derivatives except nitriles and some amides: dissolve about 40 mg (a drop if a liquid) of the compound to be tested in the minimum amount of ethanol. Add 1 ml of 1 N hydrochloric acid and then 1 drop of ferric chloride solution. The hydrochloric acid will stop color formation by all enols, nitro compounds, and some phenols. If a persistent and definite color develops (usually red, orange, green, blue, or violet), the following test should *not* be made.

Heat to boiling a mixture of 40 mg of the unknown, 1 ml of 0.5 N ethanolic hydroxylamine hydrochloride solution and 0.2 ml of 6 N sodium hydroxide. Allow to cool for 3 min and add 2 ml of 1 N hydrochloric acid. If the solution becomes cloudy, add 95% ethanol a drop at a time until the cloudiness disappears; do not add more than 3 ml. Add a drop of 5% aqueous ferric chloride solution. If a color forms but does not persist, slowly add the ferric chloride drop by drop until the color becomes definite and persistent. A change from the usual yellow of the hydroxamic acid solution to the distinct deep red to purple of the ferric chelate is a positive test.

The test is positive for carboxylic esters, acid anhydrides, acyl halides, and most aliphatic amides. The color with amides is lighter than for other acyl derivatives and does not appear at all for many carboxamides. Another hydroxamic acid test that is positive for almost all amides and nitriles is described below. Esters of sulfonic and carbonic acids do not give positive tests. Many unpurified carboxylic acids and other organic compounds often give false positive tests because they are contaminated with esters and anhydrides. This contamination is especially likely with the multifunctional carboxylic acids studied in Chapter 61. Formic acid is the only *pure* acid that gives a positive test.

Acid anhydrides and chlorides: to determine if the unknown is an acid halide or anhydride instead of another acid derivative, add about 40 mg of the unknown to 0.5 *N* hydroxylamine hydrochloride solution; then add 2 drops of 2 *N* hydrochloric acid. Allow the mixture to stand a minute and then heat to boiling. Cool and add ferric chloride solution as above. This test is positive for most (but not all) acid anhydrides and acid chlorides but negative (colorless or yellow) for all esters and amides.

Check of molecular weight: the ferric chelates of hydroxamic acids like all monofunctional organic compounds become more soluble in organic solvents and less soluble in aqueous solvents as the number of carbon atoms increases. After a positive color test is obtained as described above, add 1 ml of chloroform and shake the mixture thoroughly. The color will be in the chloroform layer if the unknown is derived from an acid of more than 10 carbon atoms and in the aqueous layer if the unknown is derived from an acid of less than 8 carbon atoms. If the acid has from 8 to 10 carbon atoms inclusive, this part of the test is unreliable. The presence of other functional groups which could radically change the solubility of the chelate also causes this molecular weight verification to be reliable.

Amides and nitriles: before attempting this test dissolve about 30 mg (1 small drop of liquid) of unknown in the minimum amount of propylene glycol and test it with ferric chloride solution. If a persistent red to violet color is obtained, do not complete the test. Dissolve approximately 40 mg of unknown in the minimum amount of propylene glycol in a test tube in a water bath at 70°. Add this solution to 3 ml of a 1 *N* solution of hydroxylamine hydrochloride in propylene glycol. Boil the mixture slowly for 2 to 3 min while stirring with a glass rod to prevent bumping. Cautiously add 5% alcoholic ferric chloride drop by drop (not more than 1 ml). Deep red to violet is a positive test; yellow is a negative test. Consider brown colors or precipitates as neither positive nor negative.

All nitriles and unsubstituted amides give a positive test with the possible exception of highly sterically hindered compounds like 2,6-dialkylbenzamides. Most substituted amides give positive tests if not substituted with large groups, e.g., compounds like benzanilide give negative tests and N-methyl or -ethyl and N,N-dimethyl or -diethylcarboxamides give positive tests.

Anilide Test for Acid Halides

Put 2 drops of aniline in a test tube and add about 40 mg of unknown. (**Caution:** this can be a violent reaction especially with the lower molecular weight acid halides.) If no noticeable reaction occurs, warm the mixture for 2 to 3 min (no longer) in a water bath at 56 to 60°. Add 2 ml of water, acidify with 6 *N* hydrochloric acid, and vigorously shake the test tube.

$$RCOCl + 2C_6H_5NH_2 \rightarrow RCONHC_6H_5\downarrow + C_6H_5NH_3^+\,Cl^- \tag{35-13}$$

A precipitate of the water-insoluble anilide is a positive test. Acid anhydrides will not form significant amounts of the anilide if the above instructions are *closely* followed.

Wash the precipitate successively with 2 ml of cold 6 *N* hydrochloric acid, two 1-ml portions of 6 *N* sodium hydroxide, and two 2-ml portions of cold water. Determine the melting point of the product and recrystallize it from aqueous ethanol until a constant melting point is obtained. This procedure serves two purposes. First, the aniline may be used as a derivative to establish the identity of the unknown. Second, acids may form insoluble aniline salts under the conditions of the test. If the acid and base washes do not dissolve any such insoluble salt, the relatively high melting point of the salt indicates that it is ionic. Try this test on benzoyl chloride, acetyl chloride, acetic anhydride, and acetic acid.

Potassium and Sodium Hydroxide Solutions

Discussion. In addition to its use as a reagent for preliminary solubility examinations, aqueous sodium hydroxide in various concentrations may be used as a classification reagent to detect and help identify ammonium and amine salts (Chapter 44), esters, amides, nitriles, most nitro compounds, aldehydes, and certain polyfunctional compounds (Chapter 61). Sodium hydroxide solution may also be used to hydrolyze amine salts, amides, nitriles, and esters into acids and amines or alcohols so that derivatives may be made of these moieties. If the acid, amine, or alcohol produced is a solid, it may be used as a derivative without further reaction.

$$RCO_2R' + NaOH \rightarrow RCO_2Na + R'OH \qquad (35\text{-}14)$$

$$RCO_2Na + H^+ \rightarrow RCO_2H + Na^+ \qquad (35\text{-}15)$$

Esters. Procedure 1: In a 50-ml flask with a boiling chip reflux 3 g of unknown and 33 ml of 25% sodium hydroxide solution. Continue refluxing at least 5 min after the ester layer disappears. This usually means refluxing 25 min for esters boiling below 100°, about 45 min for most esters boiling between 100 and 140°, and up to 2 hr for esters boiling to 200°. For esters boiling above 200° use Procedure II below. Change the condenser for distillation (Fig. 3-1) and distil about 7 ml. Note if the distillate separates into two layers; if it does not separate, saturate the distillate with potassium carbonate. This causes any water-soluble alcohol to separate into a top layer.

Positive tests for esters are indicated by disappearance of the water-insoluble unknown and if the ester has a fragrant odor, as most low molecular weight esters do, by replacement of this odor with that of the alcohol or a complete disappearance of any odor. If the distillate has separated into two liquid layers, the presence of a volatile or steam-distillable alcohol is indicated. This alcohol can then be identified by the methods outlined in Chapter 28. If no alcohol distils, extract the residue from the distillation with two 10-ml portions of ether. If evaporation of the ether fails to give an alcohol, the ester is probably an internal ester or a lactone. (Lactones will not be assigned as possible unknowns until compounds with multiple functional groups are studied in Chapter 61.)

The aqueous residue from the above distillation (or extraction) is cooled in an ice bath and chilled, concentrated hydrochloric acid is cautiously added to it. Stir the mixture thoroughly, and, if an acid precipitates, filter it. If a water-insoluble liquid forms, separate it from the aqueous layer in a separatory funnel. If little or no insoluble acid forms, extract the acidified aqueous layer with two 10-ml portions of ether. Evaporate the ether to obtain the acid. If the acid is a solid, its melting point after one or two recrystallizations and the physical and chemical properties of the ester and alcohol usually identify the ester. If not, the acid may be identified by the methods outlined in Chapter 32.

Procedure II: In a 25-ml distilling or round bottom flask add 6 ml of diethylene glycol, 0.7 g of potassium hydroxide, and 0.8 ml of water. Dissolve the alkali by heating gently, cool, and add about 1.5 g of ester, mixing thoroughly. With a thermometer, appropriate adapters, and the flask containing the ester construct the apparatus shown in Figure 3-1 or in Figure 7-3. Heat cautiously and agitate the mixture by gently shaking the apparatus.

When only one *liquid* phase (a solid phase may also be present) remains, stop the agitation and heat more vigorously; the alcohol should distil. Stop heating when no more liquid distils or when the temperature reaches 210°. Diethylene glycol boils at 244°, and an alcohol boiling above 210° could not be obtained pure from this hydrolysis by simple distillation.

To the cooled residue (which may be a suspension) add 10 ml of ethanol and 8 ml of water and agitate. If a suspension or precipitate has formed, attempt to get it into solution. If any solid present does not dissolve, add 1 ml of 20% sulfuric acid and agitate; this last addition should be omitted if no solid is present. Either of the following two purifications may be used, or the solution may be divided into two approximately equal portions and both procedures followed.

Purification A (*solid acids only*): Add 10 g of ice to the solution. Then slowly add 20% sulfuric acid until the mixture is very acid to pH paper. Allow to stand for a few minutes and make sure the reaction mixture is at or below room temperature before filtering to collect the precipitated acid. Wash the acid with two 3-ml portions of water and recrystallize it from ethanol to a constant melting point.

Purification B (for any acid for which the melting points of the appropriate derivatives are listed): Acidify the solution with 20% sulfuric acid until a pH of 8.0 to 9.0 is obtained. Allow to stand for 5 min and filter. The filtrate can be treated directly with a *p*-nitrobenzyl halide or *p*-bromophenacyl halide to obtain the corresponding ester derivative. See page 258 for details. In rare instances it may be necessary to prepare two derivatives. In this case the filtrate should be divided into two portions and treated with different reagents.

All esters except sterically hindered ones like *o,o'*-disubstituted benzoic esters are hydrolyzed by one of the above procedures. Compounds that may appear to give false positives are aldehydes and certain polyfunctional compounds, but these are easily distinguished from esters by other tests that will be discussed when these compounds are studied.

Amides and Nitriles. In these equations, R equals H, alkyl, or aryl, and W is a strong electron-withdrawing group.

$$RCONH_2 + NaOH \rightarrow RCO_2Na + NH_3 \quad (35\text{-}16)$$

$$RCONR'_2 + NaOH \rightarrow RCO_2Na + R'_2NH \quad (35\text{-}17)$$

$$RCN + NaOH + H_2O \rightarrow RCO_2Na + NH_3 \quad (35\text{-}18)$$

$$ArNR_2 + NaOH + H_2O \rightarrow \text{no reaction (but see Equation 35-20)} \quad (35\text{-}19)$$

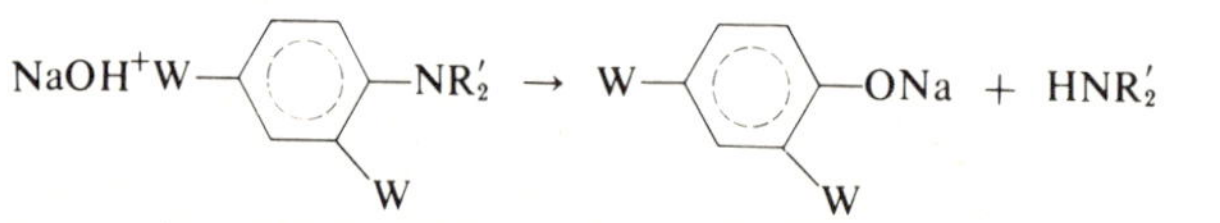

(35-20)

$$R'_3\overset{+}{N}H\ X^- + NaOH \rightarrow R'_3N + NaX + H_2O \quad (35\text{-}21)$$

To about 0.1 g of unknown in a test tube add 3 to 4 ml of 10% sodium hydroxide solution. Shake, cautiously smell, and note whether ammonia or a volatile amine comes off. Volatile amines with 4 or more carbon atoms have strong fishy odors. Amines with 1 to 3 carbon atoms have a combination of strong fishy and ammoniacal odors. The presence of ammonia and volatile amines also can be detected by their action on moist pH paper (strongly basic) and paper wetted with 5% copper sulfate solution (deep blue). If no reaction seems to occur, add a small boiling chip and gently boil for a few minutes. *Remove from the flame;* note the action of the hot vapors on the moist test papers and cautiously note the odor of the vapors. Try this test on acetamide, benzamide, N-methylacetamide, N,N-dimethylbenzamide, N,N-dimethylformamide, and acetonitrile.

If quantities of acids and amines sufficient for identification of the separate moieties are desired, the quantities for the above classification test should be increased by a factor of 8. Put the reactants in the apparatus shown in Figure 3-1 or in Figure 7-3. If there is a possibility the unknown may be a substituted amide, the outlet of the adapter should be below the surface of 35 ml of 10% hydrochloric acid in a 50-ml Erlenmeyer flask. Distil slowly for *at least* 6 min after all the solid has dissolved and an over-all minimum of 10 min. When the flame is removed from the distilling flask, be sure to vent the system to the atmosphere to avoid backing of the solution from the receiver into the rest of the apparatus.

Any amine salt in the distillate can be identified by the Hinsberg test (Chapter 44). If no amine distils and there is a possibility that the unknown is a substituted amide, continue the distillation for an extra 20 min. If still no amine has distilled, cool the residue from the distillation and extract it with four 15-ml portions of ether. Distil the ether from the extracted amine. The amine can be identified as described in Chapter 44 (amines). No N-substituted amide will be given as an unknown until amines are studied. If no amine is found by the above procedure, the unknown probably is not an N-substituted amide unless it is a lactam (cyclic N-substituted amide).

Treat the aqueous residue from the above distillation or extraction like the alkaline residue in Procedure I (p. 275) and identify the acid moiety.

DERIVATIVES

The identity of acid derivatives is confirmed by synthesizing derivatives of the alcohol or amine moieties and synthesizing other derivatives of the acid. See the tables of the compounds in the appropriate chapters for melting points of their derivatives. The anilide test for acid chlorides and the reactions of the alkali metal hydroxides in the preceding pages also should be consulted.

Derivatives of Acid Halides and Anhydrides

These compounds may be treated directly with aniline. (See p. 274 on classification tests.) Anilides of acid anhydrides may be made by using the same procedure except that a drop of pyridine is added to the aniline and the mixture is warmed on a hot water bath for 2 hr instead of 2 min. If the corresponding amide is insoluble in water, it may be made by treating the acid halide or anhydride with an excess of concentrated aqueous ammonia.

$$(RCO)_2O + 2H_2NR' \rightarrow RCONHR' + RCO_2^- + H_3\overset{+}{N}R' \qquad (35\text{-}22)$$

The acid halide or anhydride may also be hydrolyzed to the corresponding acid. If the acid is a solid, it may serve as the derivative. The acid halide or anhydride may be hydrolyzed with dilute aqueous sodium hydroxide and a solid ester derivative obtained by treating the solution as described in Purification B, page 276.

Derivatives of Carboxamides, Nitriles, and Esters

There are two basic methods of making derivatives of amides, nitriles, and esters. The first involves making the derivatives directly from the ester or the amide without hydrolysis. The second method involves hydrolysis of the ester or amide by methods outlined in the previous pages on the use of sodium and potassium hydroxide as a classification reagent. The acids, amines, and alcohols produced by this hydrolysis are then identified.

Acid Hydrolysis of Amides. Many substituted amides and nitriles are hydrolyzed to acids more easily by refluxing with 20% sulfuric acid.

$$RCONR'R'' + H^+ + H_2O \rightarrow RCO_2H + R'R''NH_2^+ \qquad (35\text{-}23)$$

Nitriles may be hydrolyzed to amides by treatment with 90% sulfuric acid at 40° for an hour. The amide is recrystallized from aqueous ethanol and used as a derivative.

$$RCN + H_2O \xrightarrow{H_2SO_4} RCONH_2 \qquad (35\text{-}24)$$

If needed, the amide used to identify the nitrile may be hydrolyzed to the acid for a second derivative.

3,5-Dinitrobenzoates by Ester Interchange. Mix 1.5 g of ester with 1.0 to 1.4 g of 3,5-dinitrobenzoic acid and 2 drops of concentrated sulfuric acid. Reflux the mixture in an oil bath for 45 min. Do not exceed a bath temperature of 160° even if the mixture does not reflux.

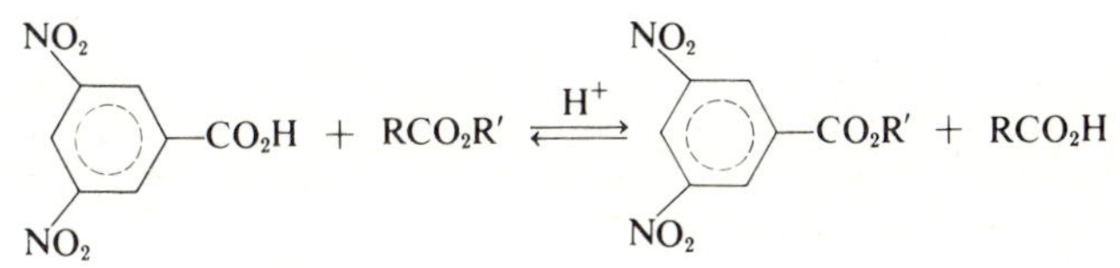

(35-25)

If the 3,5-dinitrobenzoic acid does not dissolve soon after heating, heat for an extra 20 min. Cool to room temperature and mix thoroughly with 25 ml of ether. Extract the ether solution with three 10-ml portions of cold 5% aqueous sodium carbonate and two 10-ml portions of water. Evaporate the ether, dissolve the residue (usually an oil at this point) in 6 ml of hot ethanol, and filter the solution. Add water while stirring until the cloudiness persists. Warm the solution until clear and allow to cool to room temperature. If only a very small amount or no precipitate forms, chill the mixture in an ice bath and stir it, if necessary. Wash the precipitate with cold aqueous ethanol and recrystallize it from this solvent mixture. The second recrystallization is usually much easier than the first.

$$\text{EtMgBr} + C_6H_5NH_2 \xrightarrow{Et_2O} C_6H_5NHMgBr + C_2H_6 \qquad (35\text{-}26)$$

$$2C_6H_5NHMgBr + R'CO_2R \xrightarrow{Et_2O} R'\underset{\underset{\text{OMgBr}}{|}}{C}(NHC_6H_5)_2 + \text{ROMgBr} \qquad (35\text{-}27)$$

$$R'\underset{\underset{\text{OMgBr}}{|}}{C}(NHC_6H_5)_2 + \underset{\text{(excess)}}{2HCl} \xrightarrow{H_2O} R'CONHC_6H_5 + C_6H_5\overset{+}{N}H_3 + Mg^{++} + 2Cl + Br^- \qquad (35\text{-}28)$$

Anilides from Esters. Prepare the ethylmagnesium bromide from 0.25 g of magnesium, 10 ml of *absolute* ether, and 1.15 g of ethyl bromide, using the general procedure for preparing Grignard reagents discussed in Chapters 24 and 33. When all or almost all the magnesium has been dissolved, cool and slowly add 1 g of aniline in 8 ml of ether. After 5 min add about 0.4 g of ester and reflux for 10 min. The importance of keeping moisture (and to a lesser extent carbon dioxide) from the reaction mixture cannot be emphasized too strongly. All procedures described in previous Grignard experiments for keeping the reaction free of carbon dioxide and water should be carefully followed.

Cool, add 10 ml of 6 *N* hydrochloric acid. Pour the mixture into a small separatory funnel and draw off the aqueous (bottom) layer. Wash the ether with 5 ml of 6 *N* hydrochloric acid and with three 5-ml portions of water. Evaporate the ether and dissolve the residue in 5 ml of boiling ethanol. Filter while hot and add water a few drops at a time with stirring to the filtrate. Add the water a few drops beyond the point where a definite cloudiness persists (about 2 to 3 ml of water). Heat the mixture to its boiling point and allow to cool to room temperature. If the precipitate is scant, chill in an ice bath before filtering. Recrystallize the precipitate from 4 ml of ethanol and about 2 ml of water, using the procedure described above. Dry the crystals and determine their melting point.

UNKNOWN

The unknown assigned with this chapter can be any compound from the tables at the end of Chapters 14, 23, 28, 32, and at the end of this chapter (except N-substituted and N,N-disubstituted amides). For some of the acid derivatives the melting points and boiling points of the hydrolysis products are given. By use of these transition points the melting points of solid derivatives of these compounds can easily be obtained from the tables at the end of Chapters 28 and 32. If the hydrolysis products are easily purified solids, then these can be used as the derivatives. The boiling point of the derivative of an unknown should never be relied upon as a final confirmation in the identification of the unknown in the same way that the melting point of a derivative may be relied upon. Do the problems below before attempting to identify your unknown.

EXERCISES

35-1. In the synthesis of solid esters from acid derivatives on page 276 (Purification B) there is an instruction to adjust the pH to 8.0 to 9.0. Why is it important to have the solution neither too acidic nor basic for the synthesis of solid esters?

35-2. Class M compound *A* had a melting point of 93.5 to 96°. It did not decolorize dilute permanganate solution. It reacted with hot sodium hydroxide solution to give vapors that were basic to wet pH paper and turned filter paper wetted with dilute aqueous cupric sulfate a deep blue. When the basic sodium hydroxide solution was acidified, a compound *B* with a melting point of 109 to 111° precipitated. Draw the structures of *A* and *B*.

35-3. Liquid *C* that boiled at 69 to 71° was in Class S_1-N_1. The liquid gave a positive hydroxamic acid test and decolorized dilute permanganate solution. *C* reacted with aqueous sodium hydroxide; the distillate from this reaction was a very volatile compound *D* that decolorized Fehling's solution. Treatment of the sodium hydroxide solution according to Purification B, page 276, with *p*-bromophenacyl bromide gave compound *E* that melted at 82.5 to 84.5°. Identify *C*, *D*, and *E*.

35-4. How can the anilide of ethyl *p*-nitrobenzoate be prepared?

35-5. A Class S_1a compound had a boiling point of 92 to 94° and contained sulfur but no halogens or nitrogen. What is the compound?

35-6. Class M solid *F* had a melting point of 26.8 to 29.4°. When *F* was treated with 90% sulfuric acid for 1 hr, *G*, which melted at 157 to 159.4°, was obtained. When either *F* or *G* was refluxed with 20% sulfuric acid, *H* which melted at 174 to 176.5° was obtained. Identify *F*, *G*, and *H*.

TABLE 35-3 ACID ANHYDRIDES

Anhydride	B.p. (°C)	M.p. (°C)	Derivatives m.p. (°C) acid	amide	anilide
Acetic	138	-73	16.7	82	115
Propionic	168	-4.5	18	80	105
n-Butyric	191		-6	115	94
*Citraconic	213	7.5	92.6	187d	175
Isovaleric	215			136	
n-Valeric	215	106	-34.5	106	62
Crotonic	248		72	160	118
n-Heptanoic	258		-7.5	96	71
Benzoic	360	42	122	123	160
Lauric		42	43	98	77
Chloroacetic		46	62	120.5	134
*Maleic	198	56	130	181	187(di)
*Itaconic		68	165d	192(di)	190
4-Nitrophthalic		119	165	200d	192
Succinic	261	120	188	242(di)	230(di)
Cinnamic		136	133	147.5	151
*Phthalic	295	131	206	149	170
α-Naphthloic		146	162	205	161
3-Nitrobenzoic		161	140	143	154
*1,2-Naphthalic		169	175d	265d	
d-Camphoric		221	187	192	210

*Will not be assigned as a possible unknown until Chapter 61 is assigned.

TABLE 35-4 ACYL HALIDES
(LIQUIDS)

Compound	B.p. (°C)	M.p. (°C)	Derivatives, m.p. (°C) acid	amide	anilide
Acetyl chloride	55		16.7	82	115
*Oxalyl chloride	64	-12	101	419d	248
Propionyl chloride	80		18	79	103
Acetyl bromide	81		16.7	82	115
Isobutyryl chloride	92		-46.1	129	105
n-Butyryl chloride	101		-6	115	90
Chloroacetyl chloride	105		63	119	134
Dichloroacetyl chloride	107		11	98	118
Acetyl iodide	108		16.7	82	115
Isovaleryl chloride	115		-30	136	109
Trichloroacetyl chloride	115		57.5	141	96
n-Valeryl chloride	127		-34.5	106	63
Chloroacetyl bromide	128		63	119	134
Bromoacetyl chloride	134		50	91	131
Isovaleryl bromide	139		-30	136	109
Bromoacetyl bromide	149		50	91	131
α-Bromopropionyl bromide	153		24	123	99
n-Caproyl chloride	153		-3.5	98.8	95
n-Heptoyl chloride	176		-7.5	96	71
Benzoyl chloride	197	-1	122	129	163
Phenylacetyl chloride	210		76	156	117.7
Benzoyl bromide	218		122	129	163
p-Chlorobenzoyl chloride	222	16	240	179	194
m-Chlorobenzoyl chloride	225		158	134	122
o-Chlorobenzoyl chloride	235	-4	142	142	114
o-Methoxybenzoyl chloride	254			129	131
p-Methoxybenzoyl chloride	262	24	184	162.5	169
*Phthaloyl chloride	276	15	204	220	254

*These compounds will not be assigned as unknowns until Chapter 61 is studied.

TABLE 35-5 ACID HALIDES
(SOLIDS)

Compound	M.p. (°C)	B.p. (°C)	Derivatives, m.p. (°C) acid	amide	anilide
Stearoyl chloride	23		70.6	109	94
p-Methoxybenzoyl chloride	22	262	184	162.5	169
m-Nitrobenzoyl chloride	35	278	140	143	153
o-Iodobenzoyl chloride	36		162	184	
trans-Cinnamoyl chloride	36	258	133	147.5	151
p-Bromobenzoyl chloride	42	246	252	190	197
β-Naphthoyl chloride	43	305	184.6	192	
*Isophthaloyl chloride	43.5	276		280	
2,4-Dinitrobenzoyl chloride	46		183	203	
trans-Cinnamoyl bromide	48		133	147.5	151
β-Naphthoxyacetyl chloride	54		156	147	145
p-Iodobenzoyl bromide	55		269	217	210
Diphenylacetyl chloride	57		148	168	180

*Will not be assigned as a possible unknown until Chapter 61 is assigned.

TABLE 35-6 AMIDES
(LIQUIDS)

Compound	B.p. (°C)	Compound	B.p. (°C)
N,N-Dimethylformamide	153	N-Methylacetamide	206
N,N-Diethylformamide	176	Formylpiperidine	222
Formamide	195d	Acetylpiperidine	226

TABLE 35-7 AMIDES (SOLIDS)

Compound	M.p. (°C)	Compound	M.p. (°C)
Thioformamide	28.4	Cinnamamide	144
Thiopropioamide	42	m-Nitrobenzamide	144
Benzoylpiperidine	48	α-Bromoisovalerylurea	145
Propionamide	79	*Phenylurea	147
*Acetoacetanilide	85	Diphenylguanidine	147
n-Heptamide	94	Cinnamanilide	153
*Semicarbazide	96	α-Phenylacetamide	154
Dichloroacetamide	98	o-Bromobenzamide	155
*Methylurea	101	m-Bromobenzamide	155
n-Caproamide	101	p-Toluamide	160
Palmitamide	106	*Malonamide	170
*Ethyl oxamate	114	o-Nitrobenzamide	176
n-Butyramide	115		
Thioacetamide	116	*Thiourea	182
α-Chloroacetamide	119	3,5-Dinitrobenzamide	183
		o-Iodobenzamide	183
Benzamide	128	m-Iodobenzamide	186
Isobutyramide	128	p-Bromobenzamide	189
*Urea (Carbamide)	132	*o-Tolylurea	192
*Phenacetin	135	*Biuret	192d
*Salicylamide	139	p-Nitrobenzamide	201
α,α,α-Trichloroacetamide	140	p-Iodobenzamide	217
		*Hydantoin	218
o-Toluamide	140	*Phthalamide	219d
*m-Tolylurea	142	*Saccharin	220
		*Phthalimide	233
		*Caffeine	234

*Will not be assigned as a possible unknown until Chapter 61 is assigned.

TABLE 35-8 ESTERS (LIQUIDS)

Compound	B.p. (°C)	D_4^{20}	n_D^{20}	Acid m.p. (°C)	Acid b.p. (°C)	Alcohol b.p. (°C)
Methyl formate	32	0.974	1.3597	8.4	101	65
Ethyl formate	54	0.923	1.3597	8.4	101	78
Methyl acetate	57	0.927	1.3617	16.7	118	65
Isopropyl formate	70	0.873		8.4	101	83
Methyl chloroformate	72	1.223	1.3867			65
Vinyl acetate	72		1.3941	16.7	118	acetaldehyde
Ethyl acetate	77	0.901	1.3721	16.7	118	78
Methyl propionate	79	0.915	1.3778	18	144	65
Methyl acrylate	80	0.960	1.3984	13	140	65
n-Propyl formate	81	0.907	1.3779	8.4	101	97
t-Butyl formate	83			8.4	101	83
Isopropyl acetate	89	0.901	1.3760	16.7	118	83
*Methyl carbonate	90	1.069	1.3686			65
Methyl isobutyrate	93	0.891	1.3840	-46.1	155	65
Ethyl chloroformate	93	1.135	1.3974			78
sec-Butyl formate	97	0.884	1.3832	8.4	101	99
t-Butyl acetate	98	0.869	1.3860	16.7	118	83
Isobutyl formate	98	0.876	1.3856	8.4	101	108
Ethyl propionate	98	0.889	1.3853	18	144	78
Ethyl acrylate	100	0.924	1.4056	13	114	78
Methyl methacrylate	100	0.936	1.4131	16	163	65
Methyl 2,2-dimethylpropanoate	102	0.876	1.4228	35	164	65
n-Propyl acetate	102	0.883	1.3847	16.7	118	97
Methyl n-butyrate	102	0.898	1.3879	-6	163	65
Allyl acetate	103	0.928	1.4048	16.7	118	97
Chloromethyl chloroformate	107	1.464	1.4294			formaldehyde
n-Butyl formate	107	0.886	1.3894	8.4	101	116
Ethyl isobutyrate	110	0.869	1.3903	-46.1	155	78

*These compounds will not be assigned as unknowns until Chapter 61 is studied.

+Crotonic acid is trans-2-butenoic acid.

TABLE 35-8 (cont.)

Compound	B.p. (C°)	D_4^{20}	n_D^{20}	Acid m.p. (C°)	Acid b.p. (C°)	Alcohol b.p. (C°)
Isopropyl propionate	111	0.872	1.3844	18	144	83
sec-Butyl acetate	112	0.866	1.3870	16.7	118	99
n-Propyl chloroformate	115	1.090	1.4035			97
Methyl isovalerate	117	0.881	1.3900	-30	176	65
Isobutyl acetate	117	0.875	1.3901	16.7	118	108
Ethyl 2,2-dimethylpropanoate	118	0.856	1.3912	35	164	78
Ethyl 3-butanoate	119		1.4230	-3.2	168	78
Methyl crotonate +	120	0.944	1.4242	72	186	65
Isopropyl isobutyrate	121	0.872	1.3999	-46.1	155	83
Ethyl n-butyrate	120	0.879	1.3918	-6	163	78
n-Propyl propionate	124	0.881	1.3921	16.7	118	97
t-Amyl acetate	124	0.874	1.3920	16.7	118	102
Allyl propionate	124			18	144	99
Isoamyl formate	124	0.882	1.3975	8.4	101	130
n-Butyl acetate	125	0.880	1.3941	16.7	118	116
Ethyl carbonate	127	0.975	1.3852			78
Methyl n-valerate	128	0.885	1.3973	-34.5	186	65
Isopropyl butyrate	128	0.865		-6	163	83
Isobutyl chloroformate	129	1.042				108
Methyl methoxyacetate	131	1.051	1.3964		203	78
Methyl chloroacetate	132	1.238	1.4221	63	188	65
n-Amyl formate	132	0.885	1.3992	8.4	101	138
2-Pentyl acetate	133	0.868	1.3960	16.7	118	119
3-Pentyl acetate	133	0.871	1.3966	16.7	118	116
Ethyl isovalerate	135	0.866	1.4009	-30	176	78
n-Propyl isobutyrate	135	0.884	1.3959	-46.1	155	83
*Methyl pyruvate	136	1.152		13.6	165d	65
Isobutyl propionate	138	0.865	1.3975	18	144	108

TABLE 35-8 (cont.)

Compound	B.p. (C°)	D_4^{20}	n_D^{20}	m.p. (C°)	b.p. (C°)	b.p. (C°)
Isoamyl acetate	142	0.867	1.4003	16.7	118	130
Isopropyl isovalerate	142	0.852	1.3938	-30	176	83
n-Butyl chloroformate	142	1.079	1.4130			116
n-Propyl n-butyrate	144	0.872	1.4005	-6	163	97
Ethyl crotonate	144	0.918	1.4235	72	186	78
Ethyl methoxyacetate	144	1.011			203	78
Ethyl chloroacetate	145	1.251	1.4227	63	188	78
β-Chloroethyl acetate	145	1.170	1.4234	16.7	118	131
Ethyl n-valerate	146	0.874	1.4009	-34.5	186	78
n-Amyl acetate	147	0.876	1.4031	16.7	118	138
Isopropyl carbonate	147	0.916	1.3932			83
Isobutyl isobutyrate	149	0.875	1.3999	-46.1	155	108
Methyl caproate	151	0.885	1.405	-3.5	205	65
Isopropyl n-valerate	154	0.858	1.4009	-34.5	186	83
Ethyl lactate	155	1.030	1.4141	52.8		78
Ethyl pyruvate	155	1.058	1.4016	13.6	165d	78
n-Propyl isovalerate	156	0.863	1.4038	-30	176	97
Isobutyl n-butyrate	157	0.863	1.4025	-6	163	108
Ethyl ethoxyacetate	158	0.970	1.4029		206.5	78
*Ethyl glycolate	160	1.082		79.3		78
Isoamyl propionate	160	0.858	1.4065	18	144	130
Cyclohexyl formate	163	1.010	1.443	8.4	101	160
n-Butyl n-butyrate	166	0.870	1.4067	-6	163	116
n-Propyl n-valerate	166	0.870	1.4065	-34.5	186	97
Ethyl trichloroacetate	168	1.383	1.4507	57.5	197.5	78
*Methyl acetoacetate	170	1.079	1.4196	-73	140	65
Methyl heptanoate	173	0.898	1.4122	7.5	223	65
Cyclohexyl acetate	175	0.970	1.4421	16.7	118	160

TABLE 35-8 (cont.)

Compound	B.p. (C°)	D_4^{20}	n_D^{20}	M.p. (C°)	b.p. (C°)	b.p. (C°)
n-Butyl chloroacetate	175	1.080		63	188	116
Ethyl β-bromopropionate	179	1.425		62		116
Ethyl acetoacetate	181	1.025	1.4198	-73	140	116
Methyl malonate	181	1.154	1.4140	133d		65
*β-Hydroxyethyl acetate	182			16.7	118	198
1-Pentyl butanoate	185	0.866	1.4119	-6	163	138
*Ethyl oxalate	185	1.078	1.4104	100		78
Propyl caproate	187	0.867	1.4171	-3.7	205	97
Ethyl heptanoate	189	0.869	1.4132	7.5	223	78
Isoamyl isovalerate	190	0.870	1.4125	-30	176	130
*Isopropyl oxalate	190	1.010	1.4100	100		83
*Ethylene glycol diacetate	190	1.114	1.4150	16.7	118	198
n-Heptyl acetate	192	0.869	1.4165	16.7	118	176.5
Methyl caprylate	194	0.890	1.4172	16.3	238	65
*Phenyl acetate	196	1.078	1.5029	16.7	118	182
*Ethyl malonate	198	1.077	1.4162	133d		78
Methyl benzoate	198	1.089	1.5164	122	249	65
Benzyl formate	203	1.080	1.5161	8.4	101	205
n-Amyl valerate	204	0.889	1.4161	-34.5	186	138
*Butyrolactone	204	1.121	1.4371	-17	d to lactone	
*Ethyl levulinate	205	1.014	1.4229	34	246	78
Ethyl caprylate	207	0.887	1.4177	16.3	138	78
Butyl caproate	207	0.865	1.4189	-3.7	205	116
*γ-Valerolactone	207	1.047	1.4325		d to lactone	
*Trimethyl glycol diacetate	210	1.069	1.4325	16.7	118	215
1-Heptyl propanoate	210	0.868	1.4198	18	144	198
*Ethylene glycol dipropionate	212	1.049		18	144	198
*p-Cresyl acetate	212	1.051	1.5003	16.7	118	202

TABLE 35-8(cont.)

Compound	B.p. (C°)	D_4^{20}	n_D^{20}	m.p. (C°)	b.p. (C°)	b.p. (C°)
*Propyl oxalate	213	1.018	1.4168	100		97
Methyl nonanoate	213	0.892		12.3	254.5	65
Ethyl benzoate	213	1.048	1.5057	122	249	78
Methyl o-toluate	214	1.068		103	260	65
Benzyl acetate	216	1.055	1.5199	16.7	118	205
Methyl p-toluate	217			179	275	65
Methyl phenylacetate	220	1.067	1.5071	78	256	65
*Methyl salicylate	224	1.184	1.5369	158		65
1-Heptyl butanoate	226	0.864		-6	163	176.5
Methyl caprate	227	0.873	1.426	31.5	270	65
*Isoamyl carbonate	228	0.909				130
*Isobutyl oxalate	229	1.992	1.4180	100		108
Ethyl phenylacetate	229	1.032	1.4982	78	256	78
Propyl benzoate	230	1.023	1.5003	122	249	99
Allyl benzoate	230	1.060		122	249	97
Methyl m-chlorobenzoate	231			158		65
*Ethyl salicylate	234	1.140	1.5254	158		78
*Butyl oxalate	243	0.991	1.424	100		116
Methyl o-bromobenzoate	244			150		65
*Ethyl adipate	245	1.009	1.4277	152		78
Ethyl caprate	245	0.865	1.4258	31.5	270	78
Isobutyl phenylacetate	247	0.998		78	256	108
Butyl benzoate	250	1.000		122	249	116
Ethyl acetonedicarboxylate	250	1.113		135d		78
Butyl phenylacetate	254	0.994	1.4891	78	256	116
Ethyl pimelate	255	0.993	1.4299	106		78
*Glyceral triacetate	258	1.161		16.7	118	290d
*Isoamyl oxalate	263	0.966	1.4270	100		130

TABLE 35-8 (cont.)

Compound	B.p. (C°)	D_4^{20}	n_4^{20}	m.p. (C°)	b.p. (C°)	b.p. (C°)
Isoamyl benzoate	263	1.004	1.4950	122	249	130
Ethyl laurate	269	0.867	1.4321	43	299	78
Ethyl trans-cinnamate	271	1.049	1.5598	133	300	78
*Methyl phthalate	282	1.196	1.5138	206d		65
*Ethyl suberate	282	0.982	1.4324	140		78
*Glyceral tripropionate	289	1.082		18	144	290d
*Ethyl azelate	291	0.974	1.4351	106		78
*Ethyl citrate	294	1.137	1.4455	100		78
*Ethyl phthalate	295	1.117	1.5019	206d		78
*Ethyl sebacate	307	0.965	1.4366	133		78
o-Cresyl benzoate	307	1.113		122	249	191
*Glyceryl tributyrate	318	1.031		-6	163	290d
Benzyl benzoate	321	1.116	1.5686	122	249	205
Isoamyl phthalate	349	1.021		206d		130

TABLE 35-9 ESTERS (SOLIDS)

Ester	M.p(^{o}C)	B.p.(^{o}C)	Acid m.p.(^{o}C)	Acid b.p.(^{o}C)	Alcohol m.p.(^{o}C)	Alcohol b.p.(^{o}C)
Methyl m-chlorobenzoate	21	231	158		-97	65
Cetyl acetate	22		16.7	118	50	295
Ethyl palmitate	24		62		-117	78
Bornyl acetate	29	221	16.7	118	208	
Methyl palmitate	30		62		-97	65
Methyl m-bromobenzoate	32		155		-97	65
Methyl p-toluate	33	217	179	274	-97	65
*Thymyl benzoate	33		121	249	49.6	233
Methyl cinnamate	36	263	133		-97	65
*Ethyl mandelate	37	254	119d		-117	78
*Methyl sebacate	38	288d	134.5		-97	65
Benzyl cinnamate	39		133		-15.3	205
*Benzyl phthalate	42		206d		-15.3	205
*Phenyl salicylate	42		159		42	182.5
*Benzyl succinate	42		187		-15.3	205
*Cyclohexyl oxalate	42		101		25	161
Methyl p- chlorobenzoate	44		242		-97	65
Cinnamyl cinnamate	44		133		33	257
Methyl anisate	45	255	184		-97	65
Ethyl m-nitrobenzoate	47	296	140		-117	78
*Methyl tartrate	48	280	204		-97	65
*α-Naphthyl acetate	49		16.7	118	94	279
*Methyl oxalate	51	163	101		-97	65
*m-Cresyl benzoate	54		122	249	12	203
Ethyl p-nitrobenzoate	56		241		-117	78
*Methyl mandelate	58		119d		-97	65
*Cresyl carbonate	60				36	202
Ethyl trichloroacetate	66	162	57		-117	78
*Coumarin	70	301	207 d to lactone		207	
*Phenyl benzoate	70	299	121	249	42	182.5
*Phenyl phthalate	70		206d		42	182.5
*Methyl m-hydroxybenzoate	70		201			
*β-Naphthyl acetate	70		16.7	118	123	286
*p-Cresyl benzoate	71		121	249	36	202
*Phenyl cinnamate	72		133		42	182.5
*Ethyl m-hydroxybenzoate	72	282	241		-117	78
*Ethylene glycol dibenzoate	73		122	249	-13	197
*Phthalide	73	290	206d d to lactone		206d	
Methyl o-nitrocinnamate	73		241		-97	65
Methyl m-nitrobenzoate	78		241		-97	65
*Methyl citrate	79	285d	100		-97	65
*Benzyl oxalate	80		101		-15.3	205
Methyl p-bromobenzoate	81		252		-117	78
Ethyl 3,5-dinitrobenzoate	95		202		-117	78
*β-Naphthyl salicylate	95		159		123	286
Methyl p-nitrobenzoate	96		241		-97	65
*β-Naphthyl benzoate	107		122	249	123	286
Methyl 3,5-dinitrobenzoate	108		202		-97	65
Ethyl p-hydroxybenzoate	116	97	213		-117	78
*Resorcinol dibenzoate	117		121	249	110	276

* Will not be assigned as a possible unknown until Chapter 61 is studied.

TABLE 35-10 NITRILES
(LIQUIDS)

Compound	B.p. (^{o}C)	Acid m.p. (^{o}C)	Acid b.p. (^{o}C)
Acrylonitrile	78	13	114
Acetonitrile	81	16.7	118
Propionitrile	97	144	18
Isobutyronitrile	108	-46	155
n-Butyronitrile	118	-6	163
3-Butenonitrile (allyl cyanide)	118	-3	168
*α-Hydroxyisobutyronitrile	120	79	212
Methoxyacetonitrile	120		203
Isovaleronitrile	130	-30	176
n-Valeronitrile	141	-34.5	186
3-Methyl-1-cyanobutane	155		199
n-Capronitrile	164	-4	205
*Mandelonitrile	170d	119d	
*Lactonitrile	182d	80	
1- Cyanohexane	183	7.5	223
Benzonitrile	191	122	249
1-Cyanoheptane	200	16.3	138
o-Tolunitrile	205	103	260
m-Tolunitrile	212	113	263(sub)

*These compounds will not be assigned as unknowns until Chapter 61 is studied.

TABLE 35-11 NITRILES
(SOLIDS)

Compound	M.p.(°C)	Acid M.p.(°C)	Acid B.p.(°C)
Mandelonitrile	28	119d	
p-Tolunitrile	29.5	179	275
α-Naphthonitrile	34	131	294
β-Naphthonitrile	66	184	306
*Cyanoacetic acid	66	133d	
p-Bromobenzonitrile	112	251	
p-Nitrophenylacetonitrile	116	153	
m-Nitrobenzonitrile	118	140	
p-Nitrobenzonitrile	147	241	

*This compound will not be assigned as an unknown until Chapter 61 is studied.

TABLE 35-12 SULFUR DERIVATIVES OF CARBOXYLIC ACIDS

Compound	Structure	M.p. (°C)	B.p. (°C)	D_4^{20}
Thioacetic acid	CH_3COSH	-20	93	1.06
Ethyl thioacetate	$CH_3COSCH_2CH_3$		117	0.98
Thioformamide	$HCSNH_2$	28.4		
Thiopropionamide	$CH_3CH_2CSNH_2$	42		
Thioacetanilide	$CH_3CS\underset{H}{N}C_6H_5$	75.5		
2-Thienylacetic acid	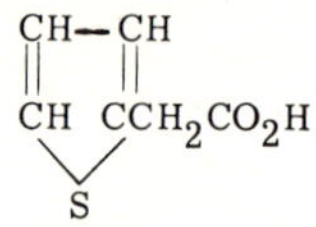	76		
Phenylthioacetamide	$C_6H_5CH_2\underset{H}{C}SNH_2$	98		
Thiobenzanilide	$C_6H_5CSNC_6H_5$	102		
Thioacetamide	CH_3CSNH_2	116		
2-Thiophenecarboxylic acid	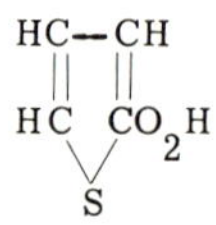	130	260d	
		130	260d	
Thioformanilide	$HCS\underset{H}{N}C_6H_5$	138		
Thiourea	$CS(NH_2)_2$	182		1.405

36

Oxidation of Side Chains of Alkylbenzene Derivatives

Synthesis of *p*-Nitrobenzoic Acid

$$p\text{-}NO_2C_6H_4CH_3 \xrightarrow{[O]} p\text{-}NO_2C_6H_4CO_2H \qquad (36\text{-}1)$$

B.p. 234° M.p. 241°
M.p. 52°

INTRODUCTION

Aliphatic chains of benzene derivatives can be oxidized to the carboxylic group regardless of the nature of the side chain. If the benzene ring is not activated by electron-donating groups, it is not significantly oxidized, and good yields of carboxylic acids are obtained.

EXPERIMENTAL PROCEDURE

In a 250-ml round bottom flask dissolve 15 g of chromic acid in 45 ml of water. Slowly add 20 ml of concentrated sulfuric acid with stirring. Add 6 g of *p*-nitrotoluene and two boiling chips. Attach a Claisen head, thermometer, and condenser to construct the apparatus shown in Figure 36-1.

Heat the mixture and shake the apparatus occasionally. When the mixture becomes hot enough (usually about 75°), an exothermic oxidation begins, and the temperature rapidly increases. Remove the flame at this time. After the temperature stops rising, return the flame and reflux the mixture for 1 hr. Allow the mixture to cool a few minutes and pour with stirring onto 120 g of ice in a 250-ml beaker.

Filter the mixture and wash the precipitate with two 20-ml portions of 5% sulfuric acid. Transfer the solid to a 100-ml beaker, add 35 ml of 5% sulfuric acid, and heat the mixture on a steam or hot water bath for 10 min with occasional stirring. Cool in an ice bath and filter. Wash the precipitate with 10 ml of water. Place the solid in a 100-ml beaker and dissolve it by stirring with 55 ml of 5% sodium hydroxide solution. Clarify the solution by heating to 45° with 1 g of Darco or other activated carbon. Filter the mixture with a gravity funnel. Wash the activated carbon with 10 ml of 5% sodium hydroxide solution. *Add* the basic solution *to*

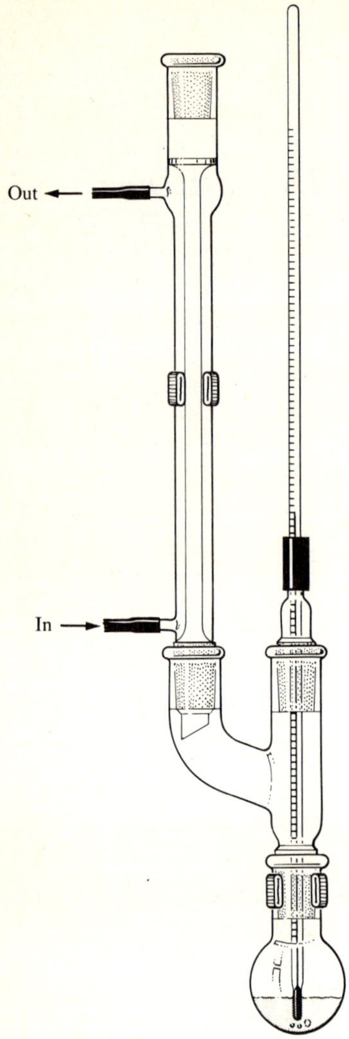

Figure 36-1 Reflux apparatus with thermometer.

25 ml of concentrated hydrochloric acid and 30 g of ice and stir the mixture well. If the resulting mixture is not strongly acidic (use pH paper), add concentrated hydrochloric acid until it is.

$$p\text{-}NO_2C_6H_4CO_2H + HO^- \rightarrow p\text{-}NO_2C_6H_4CO_2^- + H_2O \qquad (36\text{-}2)$$

$$p\text{-}NO_2C_6H_4CO_2^- + H^+ \rightarrow p\text{-}NO_2C_6H_4CO_2H\downarrow \qquad (36\text{-}3)$$

Problem 36-1. What impurities are removed by each purification step outlined in the above paragraph?

Filter, wash the precipitate with water, and recrystallize it from the minimum amount of glacial acetic acid. Dry, determine the melting point, and weigh the crystals in a tared yield bottle. Calculate the % yield, label the bottle, and hand in the product to your instructor.

THIN-LAYER CHROMATOGRAPHY OPTION

Experimental Procedure

Prepare thin-layer chromatography plates with Silica Gel G as the adsorbent following the instructions beginning on p. 202 or use a commercially precoated sheet such as Eastman Chromagram silica gel sheet. Follow the instructions in Chapter 26 for spotting, activating, and developing with the exceptions noted below.

Use one part Silica Gel G and two parts of water to prepare the slurry. Do not add any wheat starch or other binder as the Silica Gel G contains 13% calcium sulfate as a binder. If commercially *precoated* sheets are used, activate at 100° for 45 min. If the precoated sheets are heated much above 100°, they curl and become almost unusable. If a large enough developing tank is available, use glass plates that can accommodate up to six spots. Omit the wick from the developing tank.

If no other developing tank or chamber is available, one can be made from a 800- or 1000-ml beaker and a watch glass or from a mason jar. Place the watch glass on top of the beaker, cover the top and seal it air-tight with a piece of polyethylene secured by large rubber bands. Alternately, six materials could be chromatographed in three of the smaller developing chambers shown in Figure 26-3 (omit the wicks) by placing two slides back to back in each chamber.

The developing chambers should be shielded from drafts and heat radiation; follow the instruction in Chapter 26. The compounds chromatographed in this experiment are not heat, light, or air sensitive. The purges with nitrogen and other precautions described to prevent changes of the sensitive plant pigments can be omitted for this chromatogram.

The developing solvent is made from 25 ml of 95% ethanol, 2 ml of water, and 2 ml of concentrated ammonium hydroxide. Pour the solvent mixture into the developing chamber so that the top of the solvent is from 6 to 10 mm below the bottom of the spots. Place a piece of filter paper into the developing chamber so that it reaches at least half way up one side. When the plate is put in the chamber, make sure this filter paper does not touch the plate above the origin. This paper promotes faster saturation of the atmosphere in the chamber. Seal the chamber. If a mason jar is used, place a rubber gasket between the jar and its lid.

While the vapor-liquid equilibrium is being established in the chamber, spot the activated thin-layer plate or sheet with approximately 1×10^{-5} g each of purified product, crude solid reaction product (just before dissolving in 5% sodium hydroxide solution), and up to four known materials that are or might be in the product. Place each compound in a different spot. Use acetone to dissolve the materials. The spots should be no larger than 5 mm in diameter and the distances between the centers 25 and preferably 30 mm. The centers of the spots should be at least 15 mm from the side of the plate or strip. Take care while spotting so that the spots do not become so large that they merge when the plate is developed.

Place the plate or strip in the developing chamber. Replace the cover as quickly as possible so that the equilibrium between the liquid and vapor in the chamber is not significantly disturbed. Develop until the solvent front comes within 2 to 10 mm of the end of the strip or plate.

Remove the plates and either allow the solvent to evaporate at room temperature for 40 min or place the plates in an oven at 60 to 80° for 10 min or at 80 to 100° for 5 min. Unlike the thin-layer separation of plant pigments, most thin-layer chromatograms separate colorless compounds. Ultraviolet light or various reagents are often used to make the developed spots visible ("visualization"). In a room as dark as possible irradiate the dried strips with short wavelength ($\sim$254 mμ) ultraviolet light. **Danger: Do not look into the light.**

The location of compounds is shown by absorption (dark spots) or fluorescence (colored spots). Mark the spots by outlining them or the plate with a soft graphite pencil. Press hard enough so that the adsorbent is scratched; this outlines the spot even if the graphite fails to take. The rate of elution or advance of each compound relative to the solvent front is the R*f* value,

$$Rf = \frac{\text{Distance of compounds from origin}}{\text{Distance of solvent front from origin}}$$

The origin is the *center* of the originally applied spot. The distances are measured from the origin to the centers of the developed spots unless the spots tail. If there is tailing, measure to the centers the spots would have if there were no tailing.

Record in your notebook the R*f* values for each spot. Indicate whether the spot absorbed the UV light or what was the color of the fluorescence. A rough sketch of a chromatogram with the R*f* numbers is a convenient method of recording the data.

Iodine vapor makes spots from most organic compounds visible. The iodine forms weak complexes with many organic compounds. With aromatic compounds such as are chromatographed in this experiment the iodine forms charge-transfer complexes. After examining the chromatogram with ultraviolet light place the plate or strip in a closed container that has a few crystals of iodine in it. A mason jar is a convenient container; the rubber gasket is unnecessary for this application. Place the jar in a water bath at 50 to 60°. Remove the chromatograms after 8 to 10 min. The areas of the adsorbent that have compounds are a darker brown than the remainder of the adsorbent. Mark where the compounds are evident from iodine that were not evident with ultraviolet light. Record in your notebook the R*f*s of such spots and note if any of the spots made visible by the ultraviolet light fail to show spots with iodine.

Most organic compounds lose iodine (and their brown color) in a fairly short time so the spots should be marked within a few minutes after the plate is removed from the chamber. Sometimes the entire adsorbent is stained a fairly dark brown because of excessive exposure to iodine vapor. If this occurs, wait 3 to 5 min after removing the plates or strips from the iodine container; the adsorbent without the organic compound loses its iodine (and brown color) much more rapidly than the compounds do.

Warning: Always spray thin-layer chromatograms in a hood. The immediate danger is from the spray reagent. Finely divided silica dust is blown off the plate or strip when the siliceous adsorbent is sprayed. Silicosis could be caused by *repeated* inhalation of this dust. Approximately 5 min after the spots have lost the brown color, spray the plates or strips with a neutral solution of 0.04% bromocresol green in isopropyl alcohol. Use one of the spray bottles shown in Figure 59-3 and spray evenly. Yellow spots on the blue-green background show the acids. Note the R*f* values of these spots.

Discussion

One R*f* value of a compound is not sufficient to establish identification. A thin-layer R*f* value is about as useful as a boiling point of a *pure* compound (at a known pressure) for establishing identification and less useful than the melting point of a *pure* compound or the GLC retention time. A chromatographic R*f* value or retention time can be obtained on compounds that are grossly impure or that are in complicated multicomponent mixtures. Melting points are almost never of any use in identifying such materials. Neither are boiling points unless the compounds in the mixture have widely separated boiling points or a 50 or more plate fractionating column is available, the compounds in the mixture do not form azeotropes, *and* 10 to 100 g of the material is available.

R*f* values *and* reaction of the spots to selective visualizing reagents can be more useful for identifying compounds than any of the physical constants mentioned above. Detectors responsive to various groups of compounds can be used on gas chromatographs, but these detectors are less selective than TLC visualizing reagents and cannot be used to indicate specifically the great variety of functional groups that the TLC reagents can. Compounds separated by TLC may, however, be missed completely because the needed visualization reagent was not used or because the general visualization procedures, such as charring with sulfuric acid or spraying with a sulfuric acid-potassium permanganate solution, are not sensitive enough to many compounds. This difficulty does not occur as often with GLC because the two most commonly used detectors are very sensitive to all organic compounds.

Visualizing procedures may be done in series on the same thin-layer chromatograms as illustrated here with ultraviolet light, iodine, and then bromocresol green spray. If, however, more than one of the visualizing reagents reacts with the compounds, several parallel chromatograms can be obtained without much additional effort.

EXERCISES

36-1. Complete the following equations:

(a)

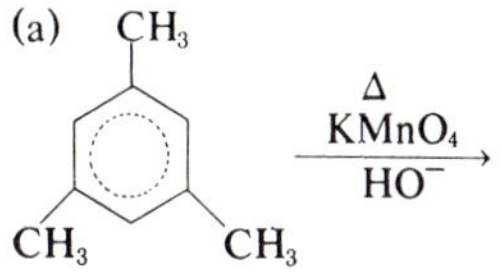

(b)

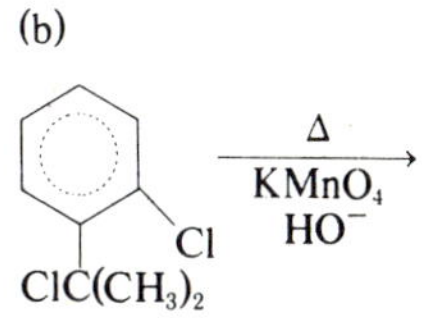

(c)

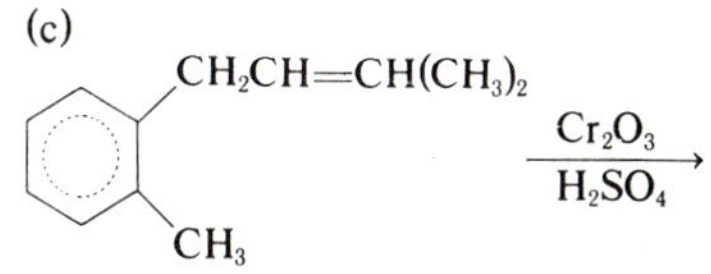

(d)

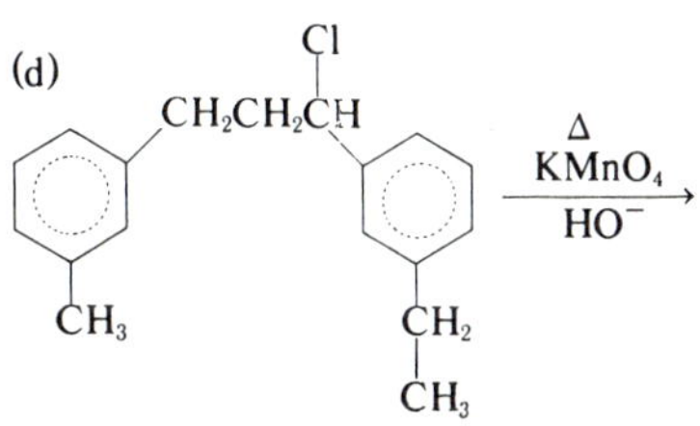

36-2. How would the procedure given in this experiment have to be modified for the synthesis:

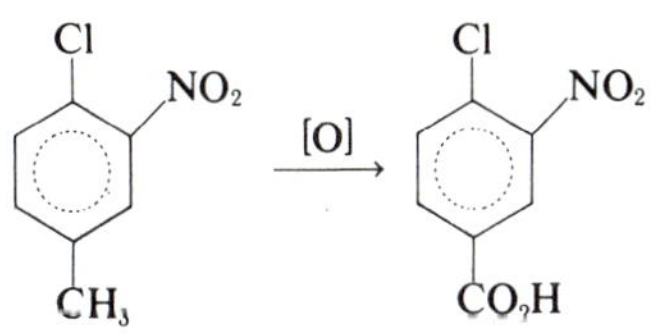

37

Transesterification

Further Studies in Gas Chromatography

$$3CH_3OH + \begin{matrix} RCO_2CH_2 \\ | \\ R'CO_2CH \\ | \\ R''CO_2CH_2 \end{matrix} \overset{CH_3O^-}{\rightleftharpoons} RCO_2CH_3 + R'CO_2CH_3 + R'CO_2CH_3 \qquad (37\text{-}1)$$

INTRODUCTION

Esters undergo base and acid catalyzed alcoholyses or transesterifications which are essentially exchanges of alcohol groups. Among the many applications of this reaction is the preparation of volatile esters from esters of polyols or high molecular weight alcohols with methanol and ethanol. Such a reaction is used in this experiment for the preparation of volatile esters from glycerides.

GAS CHROMATOGRAPHY

Effect of the Liquid Phase on Separation

As was mentioned in Chapter 8, the retention time *RT* of a compound is inversely proportional to its vapor pressure p^o and its activity coefficient a:

$$RT \propto \frac{1}{ap^o}$$

The activity coefficient is dependent upon the relative time the compound spends in the liquid phase. In general, the more soluble a compound is in the liquid phase, the longer it tends to stay there, the lower a, and the longer the retention time.

Water, heptane (b.p. 98°), and 1-propanol (b.p. 97°) have almost identical boiling points; Table 37-1 shows how widely the relative retention times vary on 5-ft columns packed with different liquid phases. Even greater differences in relative retention times could probably be obtained by optimizing the various parameters. The relative retention times and order of elution on one liquid phase can be completely inverted or scrambled by using another liquid phase. Note the relative retention times of heptane and water on paraffin wax and glycerol vary by a factor of 900, and the relative retention times of 1-propanol and *n*-heptane, both organic compounds, vary by a factor of almost 100 on paraffin wax and glycerol.

Problem 37-1. Benzene (b.p. 80°, mol. wt. 78.1) and cyclohexane (b.p. 80°, mol. wt. 84.16) are only partially resolved on a short column that has paraffin wax as the liquid phase, and benzene is eluted before cyclohexane. Benzene's relative retention time is 0.93, and cyclohexane's relative retention time is 1. On a short DEGA (diethylene glycol adipate, structure in Appendix IV) column the two compounds are completely separated, and the retention time of benzene is approximately 1.7 times the retention time of cyclohexane. Suggest an explanation for this including the different order of elution of these two hydrocarbons from these two liquid phases.

Problem 37-2. The retention times of a homologous series of compounds on most liquid phases is in the order of their molecular weights with the lowest molecular weight homolog having the shortest retention time and the higher molecular weight homologs having longer retention times. For example, primary alcohols are eluted from columns of Carbowax, paraffin wax, or DEGA in the order (1) methanol, (2) ethanol, (3) 1-propanol, etc. On a column using glycerol as the liquid phase ethanol has a *shorter* retention time than methanol. Why?

TABLE 37-1 RELATIVE RETENTION TIMES

Compound	Glycerol*	Carbowax+ 400	Diethylene+ glycol adipate	Silicone+ oil	Paraffin* wax
	50°	70°	70°	70°	60°
Water	90	6	1‡	1‡	1‡
1-Propanol	2.3	4.3	4	1.5	2.5
n-Heptane	1‡	1‡	1.2	7	10

*Structure in Appendix IV and Problem 37-3.
+Structure in Appendix IV.
‡Assumed value; other two are relative to this.

Often good separations are favored by a low *a* (p. 55); however, this is not the complete story. If two compounds with similar vapor pressure are to be separated, their *a* values should be as different as possible, not as low as possible. A Carbowax column is excellent for separating 1-propanol, water, and heptane because *a* (Table 37-1) is so different for each one. Usually for homologus series the lower *a* or the higher the solubility of the compounds in the liquid phase, the better the separation; however, there are often marked exceptions to this.

The separation of isomers is more complicated. Phthalate esters (Fig. 37-1) show the following order of efficiency for separation of isomeric heptanes or octanes: dilauryl > didecyl > dinonyl > diisooctyl. The difference in effectiveness is very small. For effectiveness in separations of the xylenes the order of the phthalate esters is reversed, and the difference is larger than was found for the separation of the isomeric alkanes. Apparently for the alkanes the governing factor is the solubility of the alkanes in the liquid phase and the rule of "like dissolves like" controls *a*. The higher the percentage of the nonpolar part of the liquid phase (for alkyl phthalates this is the alkyl moiety), the better the separation.

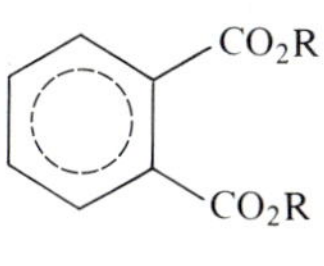

R = —$(CH_2)_{11}CH_3$ Lauryl
—$(CH_2)_9CH_3$ Decyl
—$(CH_2)_8CH_3$ Nonyl
—$CH(CH_2)_4CH_3$ with CH_2CH_3 branch ($CH(CH_2CH_3)(CH_2)_4CH_3$) Isooctyl

Figure 37-1 Phthalate esters.

The separation of the xylenes is dependent upon the differences in complex formation between the relatively electron-rich xylene rings and the electron-poor phthalate rings. Long chain alkyl groups, interacting more strongly with the methyl groups than the smaller alkyl chains interact, reduce the importance of the complex formation between the *pi* systems of the xylenes and the phthalate esters. That is, the longer side chains have a leveling effect on the separation of xylenes by the *pi* system interactions.

From the above it appears that the selection of the "best" liquid phase for a separation is neither easy nor perhaps even likely. The selection of column packing for gas chromatography has been likened to a black art and, indeed, many publications in gas chromatography indicate that the investigator may have been more dependent on occult laws than on scientific principle. In spite of these difficulties the use of principles of structure and attractions usually aids in selection of a good, or even sometimes the optimum, liquid phase and almost always helps avoid a liquid phase that gives very poor resolution.

Problem 37-3. Column 1 lists a series of liquid phases often used in GLC. Column 2 lists groups of compounds or mixtures to be separated. Indicate which of the liquid phases would give the lowest activity coefficients (and, therefore, probably better separation) for the components in each mixture.

1

a. Paraffin wax, C_nH_{2n+2}, where average n is 28
b. Polyphenylether, $C_6H_5\text{-}(OC_6H_4\text{-})_3OC_6H_5$
c. Glycerol, $CH_2OHCHOHCH_2OH$
d. QF-1. 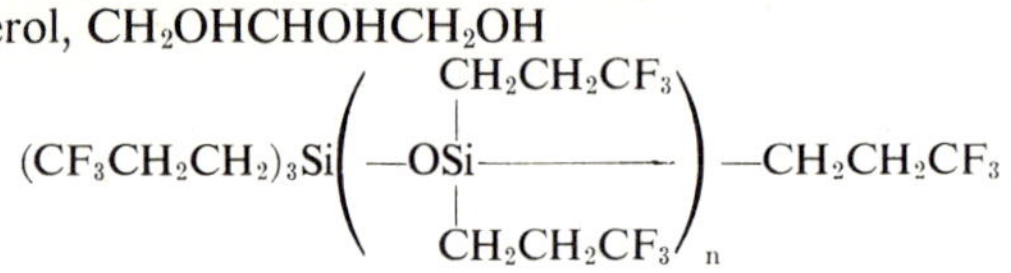
e. Tetraethylenepentamine, $NH_2(-C_2H_4\overset{H}{N})_4H$
f. Porapak,

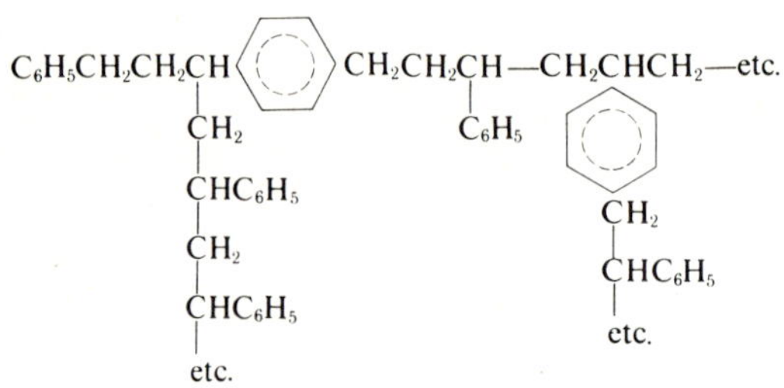

g. Silver nitrate and propylene glycol, $CH_2OHCHOHCH_3$

2

a. Natural gasoline; C_5–C_9 normal, branched, and cycloalkanes
b. A mixture of C_5-C_9 normal alkenes
c. Commercial gasoline (a mixture of a, b, alkylbenzenes, and miscellaneous compounds)
d. Product of: toluene + $CH_3CH_2Cl \xrightarrow[\Delta]{AlCl_3}$
e. A mixture of polyfluoropropanes and -butanes
f. A mixture of C_1-C_5 alkanols
g. A mixture of alkylamines of the general formula $CH_3(CH_2)_nNH_2$ where n = 2 to 8
h. A mixture of polychloroethylenes and -ethanes

Problem 37-4. Which of the liquid phases listed above might be best for the quantitative analysis of benzene and toluene in a mixture of alkanes boiling at 80 to 130° if it is unnecessary to analyze the alkanes?

Problem 37-5. What would the order of retention times of the following compounds be on the liquid phases indicated:

a. Hexane, heptane, octane, pentane on paraffin wax?
b. Ethanol, 1-propanol, hexane, heptane, pentane on paraffin wax? On Carbowax, $H(OCH_2CH_2)_nOH$?
c. 1-Butanol, ethanol, methanol, 1-propanol on paraffin wax? On Carbowax?
d. Ethanol, methanol, 1-pronanol, ethylene glycol (b.p. 198°), 1-octanol (b.p. 195°), 1-decanol (b.p. 231°) on paraffin wax? On Carbowax?

Other Properties of Liquid Phases

In the selection of a liquid phase inertness and volatility must be considered in addition to solubility effects. The liquid phase should be completely inert to the compounds being separated. Ideally the vapor pressure of the liquid phase should be less than 0.1 mm at the *highest* temperature to be used for separation. Liquids of higher vapor pressure bleed excessively from the column into the detector. Excessive bleeding causes the following complications:

a. Fouling of the detector. Even if the detector is operated at a temperature above the column temperature, *excessive* bleeding may deposit considerable amounts of liquid phase and its decomposition products in the detector.
b. High background. If the bleeding is not constant, it causes noise which may hide some of the weak signals.
c. Contamination of any fractions collected.
d. Loss of liquid phase slowly changes column characteristics. Calibrations of peak height and retention times are invalid after a column has been used even a short time if excessive bleeding occurs.
e. Excess bleeding sometimes exposes active sites on the solid support or column walls. This causes adsorption and reaction problems. This becomes more critical when very small percentages of liquid phase are used.

In order to minimize column changes due to reaction and bleeding of the liquid phase the column should be used 20 to 50° below the temperature limits given for each phase in Appendix IV. If you prepare or buy a new column, follow the conditioning instructions given in Appendix IV before using the column for any separation.

Problem 37-6. An ethyl or butyl phthalate column would probably separate the xylenes better than the isooctyl phthalate column used in Chapter 20. Why? Why was not an ethyl or butyl phthalate column used for the separation of xylenes assigned in Chapter 20?

Problem 37-7. Often the retention times of compounds injected in a relatively large quantity are slightly different from the retention times of the compounds injected in much smaller quantities. Alkanes, for example, show this effect on DEGS (for composition of liquid phases see Appendix IV) but not on paraffin wax. Sometimes this effect may be due to overloading of the column, but it also occurs when overloading could not be the cause. Explain. (Hint: see p. 194.)

Solid Support

The support for the liquid phase should be finely divided, of uniform particle size, inert, mechanically stable, and should not adsorb any of the compounds being separated. Often a considerable amount of liquid phase is needed. If this is a requirement, the solid support should be porous so that it can hold a high percentage of liquid phase. Diatomaceous earths are the most commonly used materials that meet most of these requirements. Diatomaceous materials are the siliceous skeletons of microscopic plants.

Adsorption causes tailing (Fig. 37-2). Tailing prevents accurate quantitative analysis by gas chromatography. Tailing also often causes minor peaks to be so spread that their retention times can not be determined and may cause peaks with similar but not identical retention times to merge.

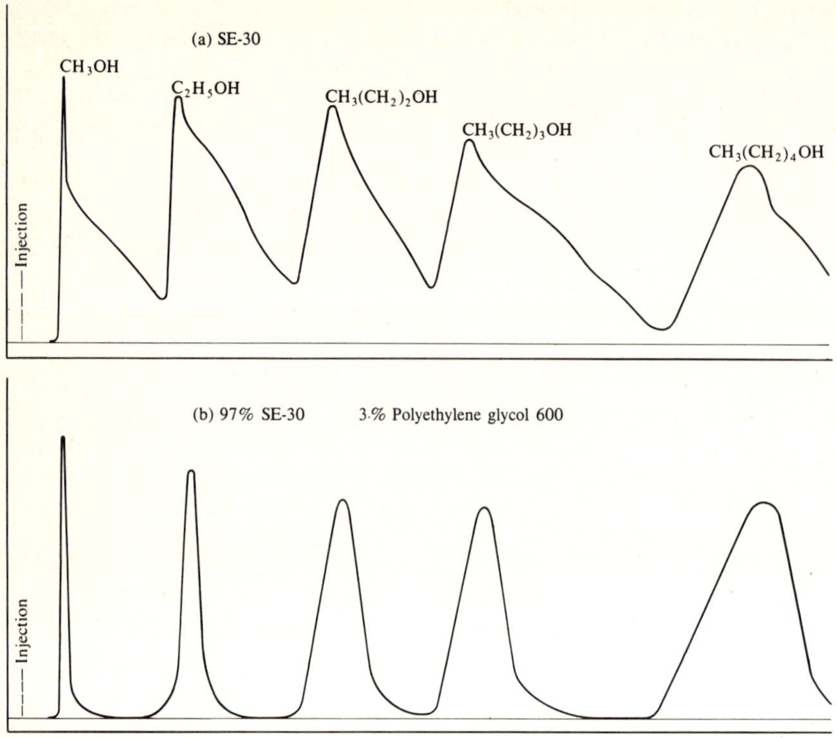

Figure 37-2 Tailing of peaks from adsorption. Chromatography of C_1 to C_5 alkanols on 20% on 50 to 60 mesh firebrick. Note reduction in tailing in (b) from addition of a small amount of polar liquid phase.

Adsorption can be minimized by eliminating polar and reactive sites on the solid support. Acidic sites cause tailing of basic compounds; basic sites cause tailing of acidic compounds. Traces of metals may cause tailing of amines and alcohols because of complex formation. Free hydroxy groups on the siliceous surfaces, $\diagdown\!\!-\!Si\!-\!OH$ (with a third bond $\diagup$), cause tailing by hydrogen bonding or other interactions with polar compounds. Basic sites can be removed or neutralized by acid washing; acid sites can be removed or neutralized by basic washes. Acid washing also removes traces of metal ions. Too strong an acid wash may leave an acid residue which can catalyze dehydrations, rearrangements, and other reactions and cause tailing of bases. A very polar liquid phase can also reduce adsorption and tailing. Compare the chromatograms of some low molecular weight alcohols on paraffin wax, diocytl phthalate, and Carbowax 400 in Figure 37-3. All three use untreated Chromosorb P as the solid support. Often a small amount of nonvolatile, highly polar material mixed with a nonpolar liquid phase reduces tailing. Compare, for example, Figure 37-2a and b.

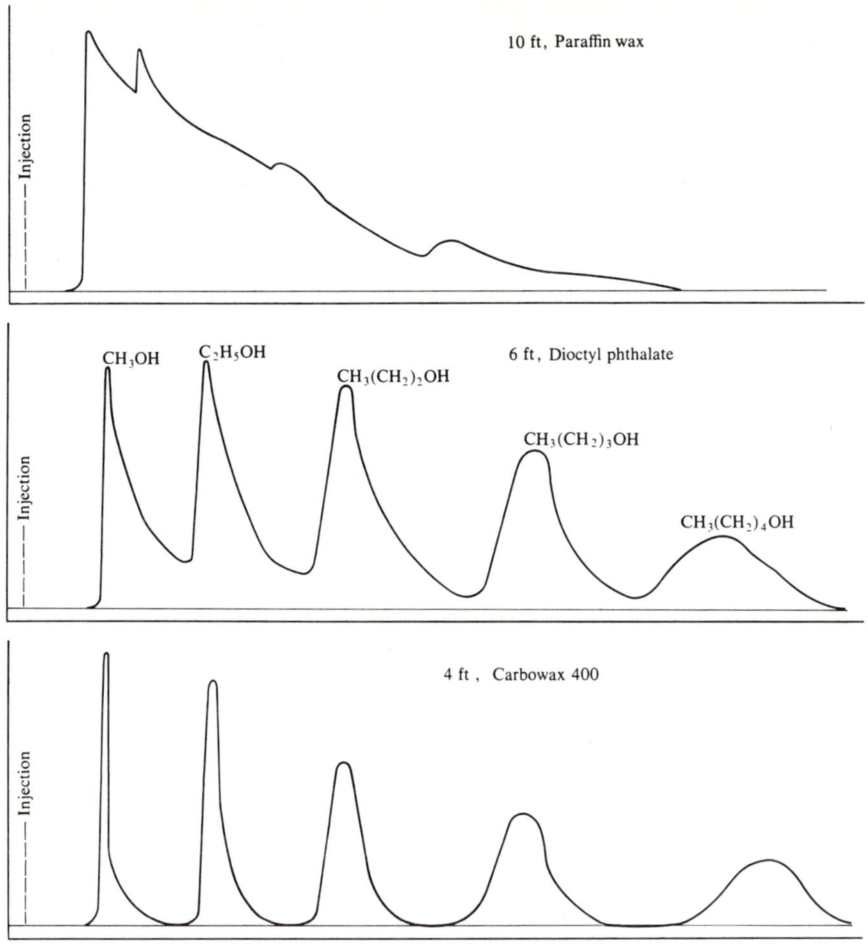

Figure 37-3 Chromatography of homologous alcohols on various liquid phases on untreated Chromosorb P. Liquid phase indicated, 17.5%; solid support, 82.5%. Only shapes are comparable; retention times are not because different temperatures and column lengths were used.

Problem 37-8. Note that the chromatograms in Figure 37-3 were obtained on heavily loaded columns. Would there be any change and if so, which way if all the columns were more lightly loaded (2 to 5%) with liquid phase?

Problem 37-9. Would a mixture of 5- to 10-carbon atom straight-chain alkanes show the same adsorption effects as the mixture of alcohols shown in Figure 37-2?

Solid supports are often silanized to reduce polar adsorption sites. The supports are silanized by converting the polar hydroxy groups to nonpolar organosilyl ether groups.

$$2\ {>}Si{-}OH + \underset{\text{hexamethyldisilazane}}{(CH_3)_3SiNHSi(CH_3)_3} \rightarrow 2\ {>}Si{-}OSi(CH_3)_3 + NH_3 \qquad (37\text{-}2)$$

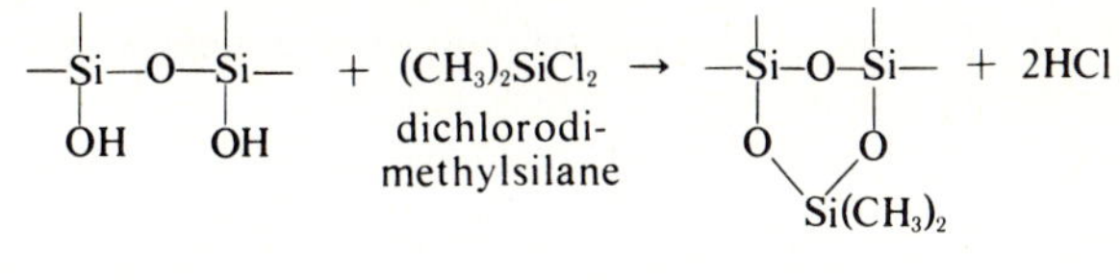

(37-3)

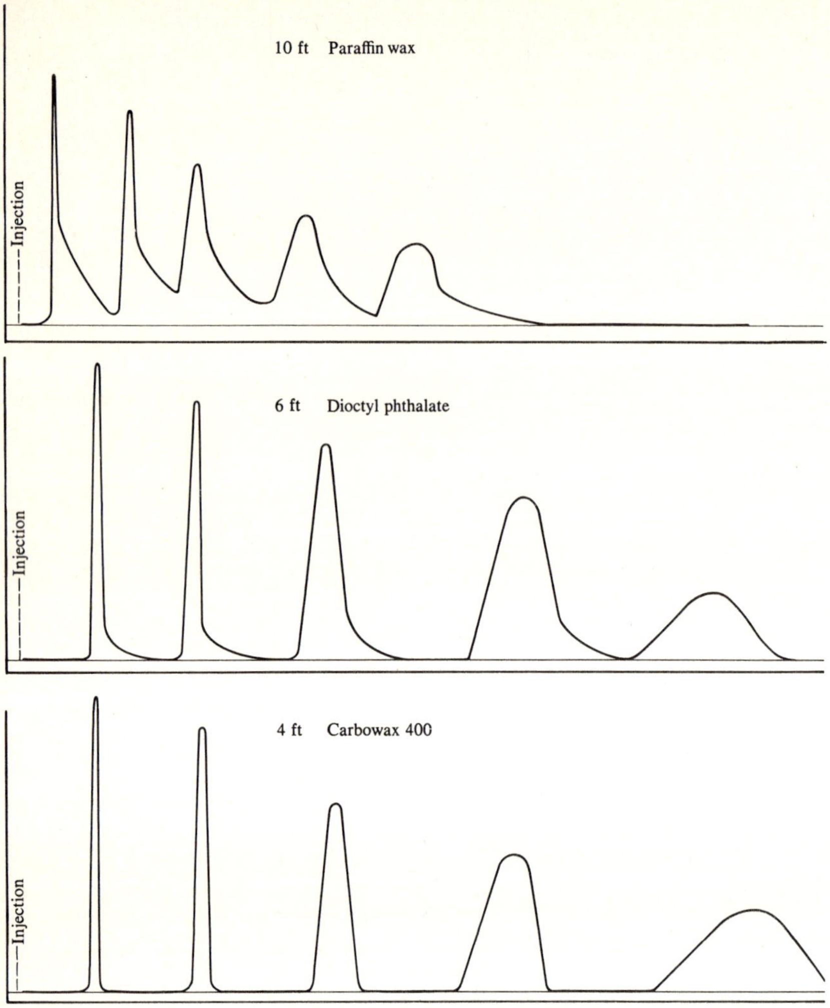

Figure 37-4 Chromatography of homologous alcohols on various liquid phases on silanized Chromosorb P. Liquid phase indicated, 17.5%; solid support, 82.5%. Compare with Figure 37-3.

Figure 37-4 shows chromatograms of a mixture of alcohols on the same liquid phases shown in Figure 37-3, *silanized* Chromosorb P, however, is the solid support.

Amount of Liquid Phase

The higher the percentage of liquid phase, the greater the capacity of the column, that is, the more material it can separate without overloading. Overloading is indicated by broadened peaks (often only the front is distorted) and loss of resolution, see Figures 37-5 and 37-6. Provided a small enough sample is injected, a low percentage of liquid phase usually gives better resolution because equilibrium is established faster in a thin layer of liquid phase. On some solid supports, however, a very thin layer of liquid phase does not cover the active sites sufficiently and significant adsorption occurs.

The quantities that can be separated without overloading on a given amount of liquid phase refer to compounds and not to total amount of material injected. For example, assume a column can separate a mixture of 10 μl of heptane and 10 μl of 2-methylhexane without overloading, but a mixture with 20 μl of

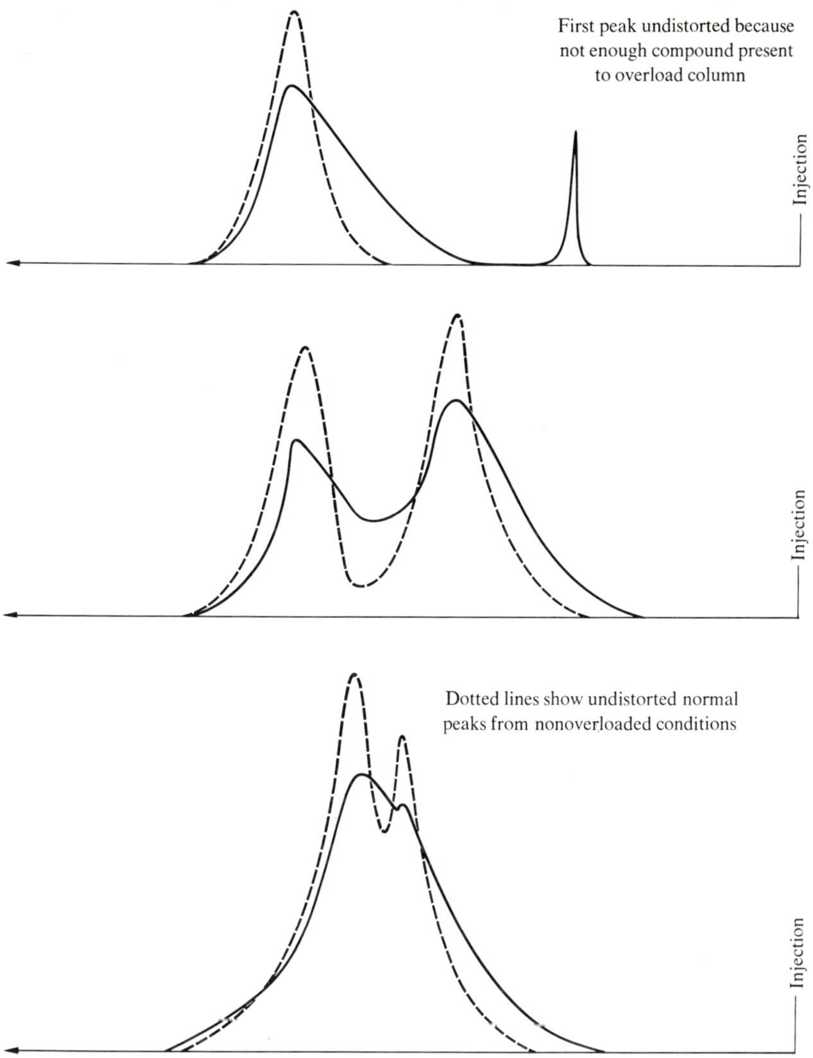

Figure 37-5 Effect of overloading on shape of peaks and resolution. Note that overloading greatly reduces resolution. This is particularly evident when two compounds have very similar retention times.

each of the two hydrocarbons causes overloading. The column could probably separate without any evidence of overloading a mixture of pentane, hexane, 2-methylhexane, heptane, octane, nonane, and decane each in a quantity of 8 μl for a total of 56 μl, considerably more than the 40 μl mixture of the two alkanes that caused overloading.

Problem 37-10. The amount of a compound that can be chromatographed on a stationary phase before overloading occurs is dependent on a or the activity coefficient of the compound in the liquid phase. The lower a or the higher the solubility of the compound in the liquid phase, the smaller the quantity of compound that can be chromatographed before overloading or a significant loss of resolution occurs. For example, when chromatographed on polyethylene glycol, ethanol gives peaks that start to broaden at much lower concentrations than the peaks from acetone or benzene; see Figure 37-7. Suggest an explanation for this.

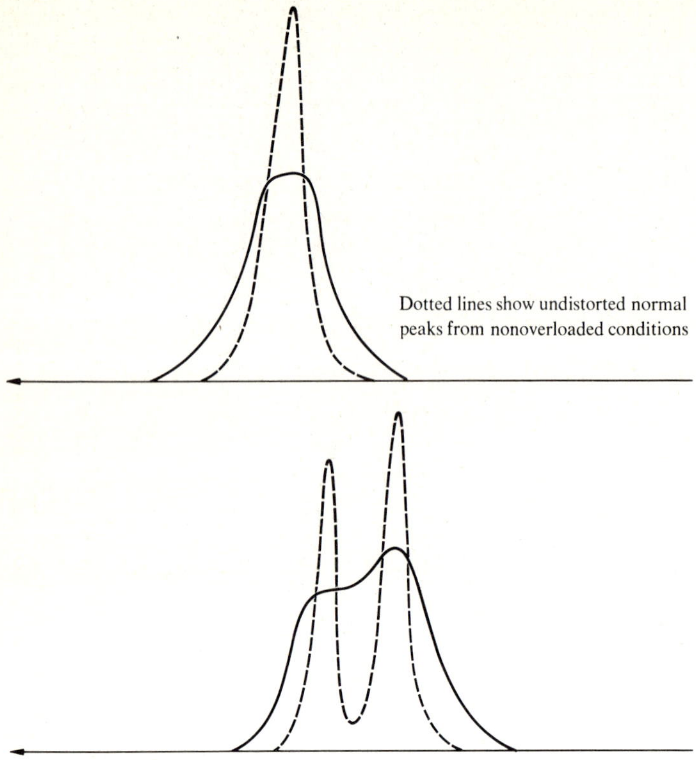

Figure 37-6 Effect of overloading on shape of peaks and resolution. Shapes of peaks shown here are probably more common from overloading on nonpolar phase than on polar phase.

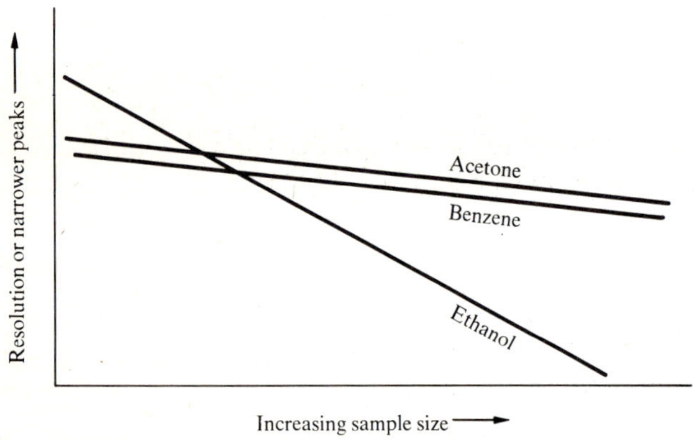

Figure 37-7 Effect of sample size on resolution on 20% polyethylene glycol-80% firebrick.

ANALYSIS OF GLYCERIDES: A CLASS PROJECT

Transesterification

Edible fats are principally glycerides or esters of glycerine and aliphatic carboxylic acids. The number of carboxylic acids in any one fat is usually very large and the number of different glycerides is even greater. Since the glycerides have similar physical properties, they are very difficult, if not impossible, to separate. Their very low vapor pressure at any temperature below their decomposition point precludes the gas chromatography of all but possibly the very lowest molecular weight glycerides.[1] If, however, methanol is exchanged for the glycerine,

[1] The chromatography of intact, high molecular weight glycerides has since been reported. The technique used fast flow rates, short columns, very small samples, and a very sensitive detector. The technique described in this chapter is easier for the student to master and is superior to the chromatography of intact glycerides for the identification or quantitative analysis of the individual carboxylic acids from the glycerides.

esters of sufficient volatility for chromatography are obtained. The average molecular weight of the methyl esters is equal to one third that of the glycerides. The methyl esters have only one polar CO_2 group versus the three of the glycerides. While analysis of the acids via their methyl esters is not the same as analysis of the glycerides, the identity and relative amounts of the fatty acids are usually what is wanted and not their specific arrangement in different glycerides.

Samples

The purpose of this experiment is to identify the aliphatic acids in various fats and to obtain a quantitative estimation of the relative amounts of the acids. The carboxylic acid distributions in the fats are compared with the origin of the fats.

The instructor will assign the transesterification, chromatography, and perhaps the procurement of one fat sample to each student. The samples should include beef and veal fat, lard, at least three different brands of margarine, one of which is a corn oil margarine, one or two brands of the more expensive spread, corn oil, peanut oil, Spry or Crisco, and linseed oil. Other possible samples are lamb fat, cod liver oil, olive oil, safflower seed oil, cottonseed oil, coconut oil, and goat butter. Cream from goat's milk can be used if it is stirred with about twice its volume of calcium chloride and this mixture extracted with petroleum ether. The petroleum ether is evaporated to leave a glyceride residue. Unless there is an exceptionally low student-to-chromatograph ratio (less than 3:1), it will be necessary to spread the calibration and chromatographies of the fats over several laboratory periods.

Analysis

Chromatographic Conditions. Column: 6 ft × $\frac{1}{4}$ in. stainless steel or glass, packed with 8% "stabilized" ethylene glycol adipate on 80/100 mesh Chromosorb W AW-DMCS. Before use, condition the column by heating it to 230° for 10 to 14 hr with approximately 10 ml/min of helium or nitrogen carrier gas (Appendix IV). Condition the column even if it was a commercially prepared column.

Carrier gas: helium at 68 ml/min. If absolutely necessary, nitrogen can be used instead of helium.

Injection port temperature: 290°.

Detector temperature: 300°.

Recorder speed and time: 0.5 in./min, run for 20 to 48 min after injection of fat samples or until the standards are eluted when chromatographing known esters.

Column temperature: three alternatives can be used depending on what chromatography equipment is available. (a) If a programmed chromatograph is available, start at 140°, program at 4°/min to 210°, and operate isothermally at 210° for 30 min. If a 4°/min program is not available, start at 145°, program at 3°/min to 210°, and operate isothermally at 210° for 25 min. See Appendix IV for operation of temperature programmed chromatographs. (b) If a temperature programmed chromatograph is not available, chromatograms at two different temperatures can be obtained, one at 150° for 30 min and one at 200° for 45 min. This can be done by using two chromatographs, one for each temperature, or by using one chromatograph set at the lower temperature for one laboratory period and at the higher temperature for another laboratory period. Make no attempt to operate the chromatograph at both temperatures in the same laboratory period. (c) If only one chromatograph is available and class size precludes its operation at the two temperatures during alternate laboratory periods, set it at 200°. If this option is used, the esters from goat's milk and butter should not be chromatographed.

Standardization and Calibration. Prepare a 5% (by volume) solution in cyclohexane of the ester(s) assigned for calibration. Obtain the peak height and retention times on 5 μl, 10 μl, and 20 μl of solution. Inject a total of 2 to 4 replicate samples; use the reproducibility criteria described on page 159 to determine how many samples should be injected. Average the peak heights for each concentration and plot peak height versus concentration for the methyl ester(s) assigned to you. Give the calibration chart and retention times to the laboratory instructor who will make the data available to the entire class.

If chromatograms at 150° and 200° are to be obtained, standardize all esters from acids of 4 to 14 carbon atoms at 150° and all esters from acids of more than 10 carbon atoms at 200°. If only isothermal chromatograms at 200° are to be obtained, standardize only esters of carboxylic acids with 8 to 20 carbon atoms.

Transesterification. The reaction of sodium with methanol described below is similar to the reaction described for the Williamson synthesis of ethers. Carefully read page 242 before proceeding further. It is important that the sodium be cleaned, cut, weighed, cleaned again if necessary, and quickly added to the flask. The sodium methoxide is a catalyst so the exact amount is not critical. Since a weight range for the sodium is specified, a rapid transfer is possible.

Add 0.04 to 0.10 g of sodium (as two small pieces) to 12 ml of methanol in a 25-ml round bottom flask. Place the flask in a cold water bath, preferably in a hood, and swirl the flask. When all the sodium has reacted, allow the mixture to stand 5 min and then add 2 g of the fat you were assigned to analyze. Reflux the mixture for 45 min. Cool the mixture in an ice bath and transfer it to a separatory funnel containing 2 to 3 g of ice. Rinse the flask with 20 ml of petroleum ether (b.p. 30 to 60°) and use this to quickly extract the material in the separatory funnel. Repeat the petroleum ether wash and extraction and keep the two petroleum ether extracts separate. If Problem 37-11 is assigned, mix the petroleum ether-insoluble material with 5 to 6 g of anhydrous sodium sulfate and press as much liquid out as possible with a glass rod.

Problem 37-11. Infrared option: obtain spectra on the petroleum ether-insoluble material as a capillary film and in a 0.05-mm *demountable* cell. Why is a sealed cell not used? What is the major component? Is there infrared evidence for any minor components?

Wash the two petroleum ether extracts separately and successively with two 15-ml portions of water and one 15-ml portion of saturated calcium chloride solution. Combine the two petroleum ether washes in a 50-ml Erlenmeyer flask. Add 5 g of calcium chloride and stir the mixture. Decant the petroleum solution into a 100-ml round bottom flask, add a boiling chip, and remove the petroleum ether by distillation, preferably under vacuum.

Chromatography of the Esters. Inject a 20-μl sample of the residue into the gas chromatograph. If two chromatographs are used, inject a 20-μl sample into each. When the low temperature chromatograph has been run for 25 min after injection (disregard any peaks with longer retention times), raise the oven temperature to 215°. Keep it at this temperature for 12 to 15 min and cool to approximately 120°. Reset the temperature for 150° and allow the column to equilibrate at this temperature for 10 min before injecting the next sample. This procedure removes any esters that are likely to come off as very broad peaks and cause a drifting baseline during a later chromatography. Although this procedure may not remove some very high molecular weight esters, these come off as such broad, shallow, diffuse peaks that they have little or no effect on the baseline. This high temperature cleaning is unnecessary after the programmed chromatograms, the chromatograms at 200°, and the chromatograms of the standards at 150°.

Set the attenuator as low as it can be set without having the baseline drift and noise level exceed 3%.[2] Be prepared to increase the attenuation quickly when the large peaks come off and decrease it once they are off.

Analysis of the Chromatography Data

From retention times and the peak height-concentration data that you and the others in the class obtained, identify and determine the relative amounts of each of the carboxylic acids in the fat sample you analyzed. Turn in your analysis to your instructor; he will make the results of all the analyses available to the class.

Problem 37-12. Which carboxylic acids are present in higher concentrations, those with an even number

[2] If the temperature is programmed, a somewhat higher baseline of up to 4%/min above 200° is permissible.

of carbon atoms or those with an odd number of carbon atoms? Is there any evidence for branched chain carboxylic acids?

Problem 37-13. What are the principal differences between the plant and animal fats? Between the natural and modified fats (Spry, Crisco, and margarine)? Between mammal and fish fats? Between margarine and butter? Among the various animal fats?

Problem 37-14. Suppose you chromatographed the methyl esters of lauric, tridecanoic, myristic, palmitic, and stearic acids. Could you obtain a close approximation of the retention times of methyl pentadecanoate, methyl heptadecanoate, and methyl nonadecanoate without chromatographing them? Could you obtain reasonable peak height-concentration curves for these esters without chromatographing them?

EXERCISES

37-1. Silanization of solid supports greatly decreases or eliminates adsorption of esters, ketones, ethers, and aromatic hydrocarbons but only moderately decreases adsorption of alcohols and acids. Why?

37-2. Apiezon wax often is used with excellent results as a liquid phase for the GLC separation of methyl esters. For the particular use described in this chapter (a great many separations of complex mixtures) why might the polyester column recommended show smaller changes with use than an Apiezon column?

37-3. One of the methods of reducing tailing caused by adsorption on the solid support is the use of a polar liquid phase (p. 296). Why is this effective?

37-4. Mixtures of silicone oil with small amounts of high molecular weight organic acids such as $HO_2C(CH_2)_{34}$-CO_2H are often used as liquid phases for the gas chromatography of more volatile organic acids. Why?

37-5. Mixtures of Carbowax and small amounts of potassium hydroxide are often used as the liquid phase for the gas chromatography of amines (RNH_2, R_2NH, RN_3) which are organic bases. Why? Would base-washing of the solid support be as effective as adding the strong base to the liquid phase?

37-6. Figure 37-8 shows some chromatograms most or all of which indicate some deficiency in the conditions used to obtain them. Explain what is wrong for each of these chromatograms and how it could be corrected.

37-7. A biochemistry graduate student spent two years isolating a medium molecular weight carcinolytic agent from a rare tropical plant. After demonstrating its potency, he was advised that it would be necessary before he could finish his thesis to attempt another chromatographic separation to verify its homogeneity and to obtain infrared spectra on any component(s). The next morning he prepared a $6' \times \frac{1}{4}''$ dicyclohexyl azelate on Chromosorb G column and chromatographed all his remaining material on it at 165°. He obtained two peaks, trapped them both in micro infrared cells, and obtained the spectra shown in Figures 37-9 and 37-10. What did the chromatogram show about the homogeneity of the sample? What were the two components he isolated?

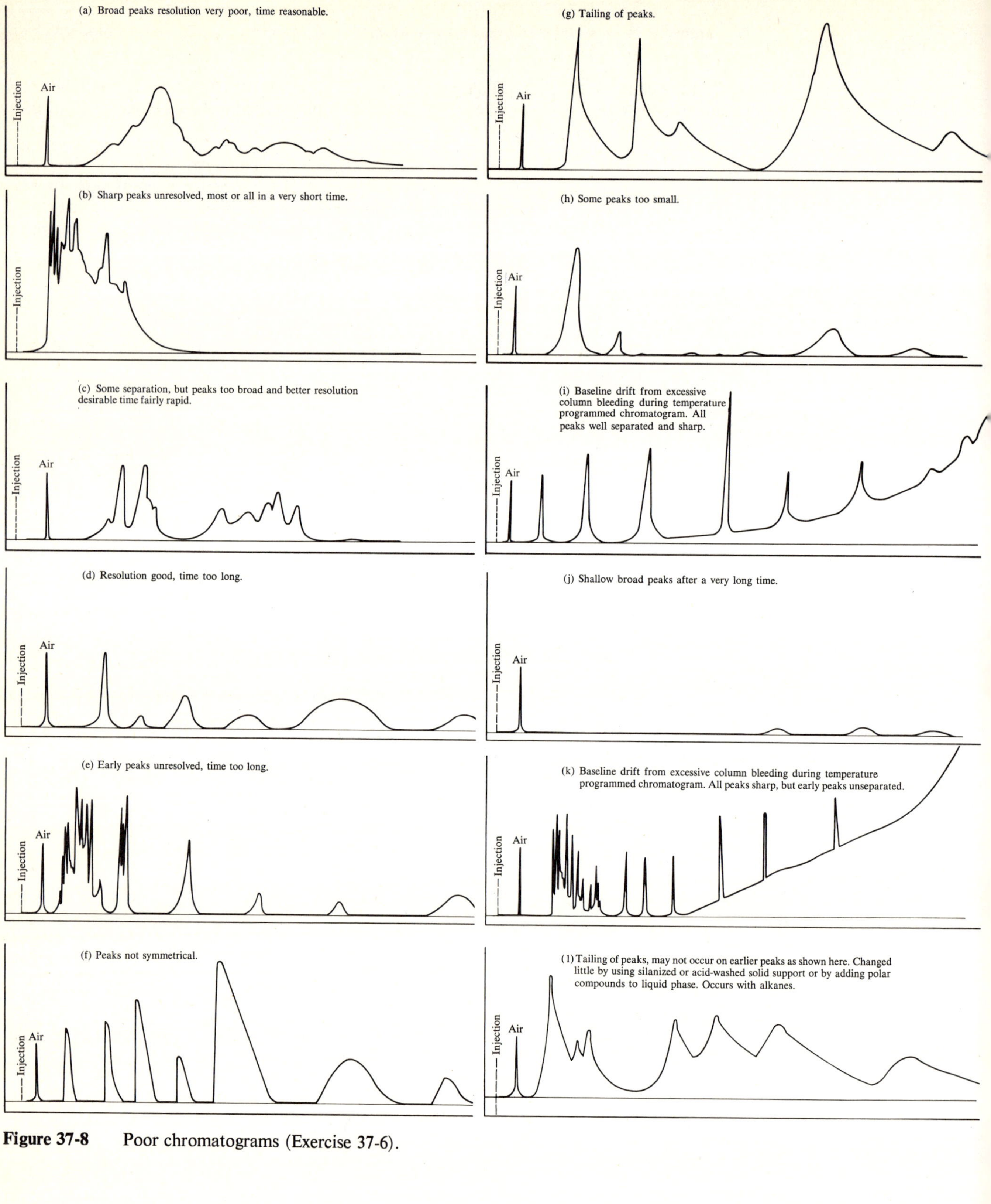

Figure 37-8 Poor chromatograms (Exercise 37-6).

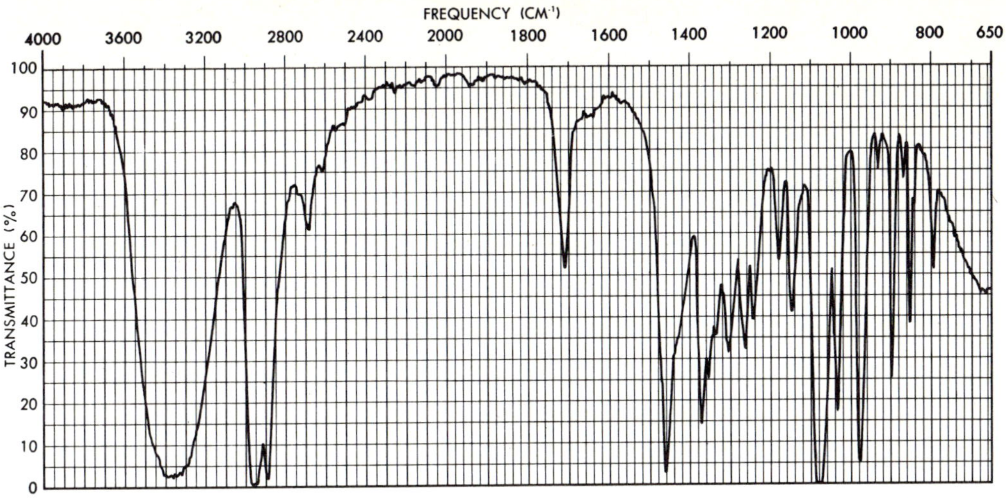

Figure 37-9 Infrared spectrum of first fraction (Exercise 37-7).

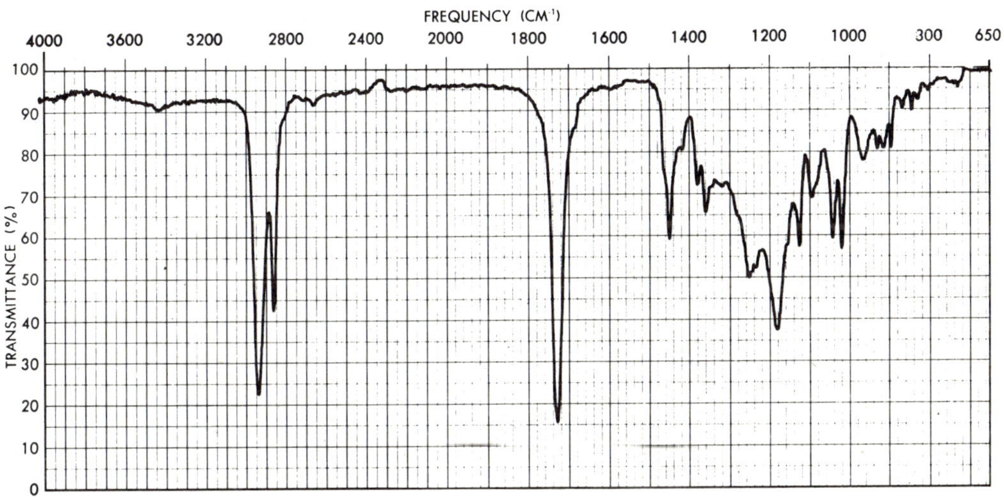

Figure 37-10 Infrared spectrum of second fraction (Exercise 37-7).

38

Synthesis of Aldehydes

Preparation of Propanal

$$3\ n\text{-}C_3H_7OH + Cr_2O_7^{=} + 4H_2SO_4 \rightarrow 3CH_3CH_2CHO + SO_4^{=} + Cr_2(SO_4)_3 + 7H_2O$$

B.p. 97.2°	B.p. 50°
D_4^{20} 0.804	D_4^{20} 0.807

INTRODUCTION

Aldehydes can be synthesized by oxidation of primary alcohols. Since the aldehydes are more easily oxidized than alcohols, the reaction conditions must be carefully controlled and the aldehyde removed from the oxidizing mixture as rapidly as it is formed. This requires more careful control than oxidation of secondary alcohols to ketones (Chapter 34) which does not usually produce any other oxidation products.

Problem 38-1. Aldehydes have much lower boiling points than the corresponding alcohols or acids. An aldehyde (RCHO) synthesized from a primary alcohol can be readily separated by distillation from the starting material and the principal organic by-product (RCO_2H). Why do aldehydes boil so much lower than the corresponding alcohols and acids?

EXPERIMENTAL PROCEDURE

Synthesis

With a 500-ml three-necked flask construct the apparatus shown in Figure 38-1. The thermometer should be about 1 cm from the bottom of the flask. Add 30 ml of 1-propanol to the three-necked flask. Place an ice bath around the receiving flask.

In a 500-ml Erlenmeyer flask dissolve 40 g of sodium dichromate ($Na_2Cr_2O_7 \cdot 2H_2O$) in 215 ml of water. Place the flask in an ice bath and slowly add 28 ml of concentrated sulfuric acid while stirring the mixture with a glass rod. Pour this solution into the dropping funnel. If the dropping funnel capacity is not large enough, the oxidizing solution will have to be added to the funnel in several portions. Each portion should be added just *before* the last one is completely drained.

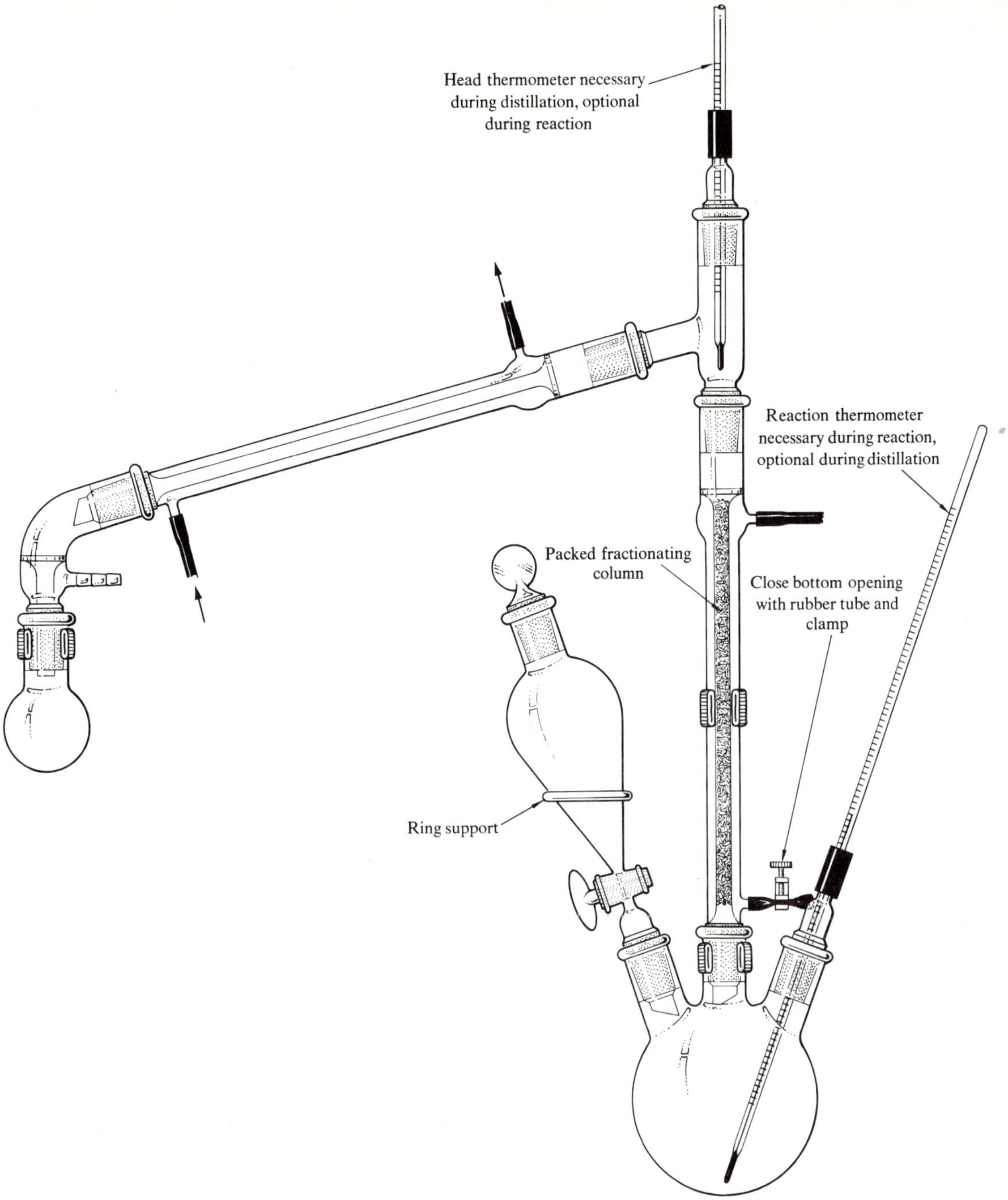

Figure 38-1 Reaction-distillation apparatus.

Heat the mixture to 68°. Have a cold water bath available. Add the oxidizing solution so that the temperature in the flask is maintained at 68° ± 5°. Stir the contents of the flask occasionally by applying a back and forth motion to the two ring stands supporting the apparatus. The addition should take about 15 to 20 min. If the temperature in the flask gets above 75°, use the cold water bath to cool the reaction below 70°. If the temperature falls below 62° (this is more likely to occur near the end of the addition), maintain the correct temperature with a microburner.

Problem 38-2. Why does the color of the reaction mixture change?

If only one thermometer is available, replace the one in the flask with a stopper and position the thermometer in the top of the fractionating column (p. 34). Slowly distil the mixture until the temperature in the top of the column reaches 80°. While allowing the mixture in the distillation flask to cool, add 10 g of potassium sulfate to the distillate and stir the mixture. Return the receiving flask to the ice bath.

Problem 38-3. What is the purpose of the potassium sulfate? Could calcium chloride be used instead? Could potassium carbonate be used?

Gas Chromatographic Analysis of the Reaction Products (Optional)

Problem 38-4. What are the likely volatile by-products of the reaction?

Problem 38-5. Choose a liquid phase and solid support from the following that would allow the best qualitative and quantitative analysis of the likely volatile compounds in the distillate and residue of the mixture in the reaction flask. Review Chapters 8 and 37 if necessary.

Supports[1]	*Liquid Phases*
Chromosorb W	DEGS $(H(O\overset{O}{\overset{\|}{C}}CH_2CH_2\overset{O}{\overset{\|}{C}}OCH_2CH_2OCH_2CH_2)_nOH)$
Firebrick	Paraffin oil
Diatoport P	Silicone oil with dimer acid $(CO_2H(CH_2)_nCO_2H)$ average n = 34
Silanized and acid-washed Chromosorb G	Methylchlorophenyl silicone fluid,
	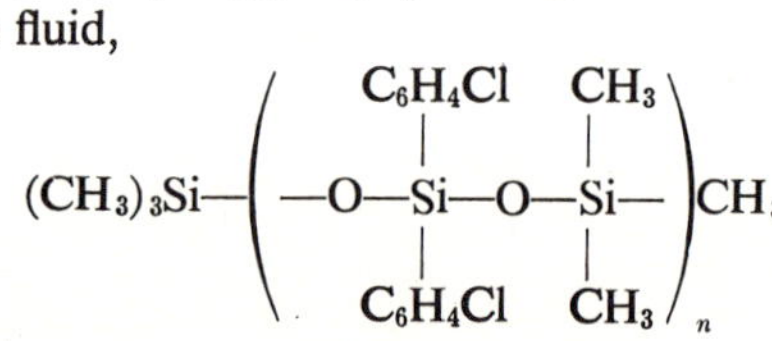
	Propylene glycol-silver nitrate

Pour the distillate into a 100-ml graduate. If any potassium sulfate is undissolved, carefully decant the mixture from it and use a glass rod to press any liquid from it. Assume there is no organic material in the bottom layer. Note the volume of the top layer. Inject 10 μl of the *top layer* into the gas chromatograph. The laboratory instructor will have set the controls; do not touch any of the controls except those used to zero the baseline and attenuate the peaks. The laboratory instructor will supply the retention times for the likely products and a few peak height concentration calibrations. Treat the distillate as described in the section on purification (p.309).

Measure the contents of the three-necked flask with a graduated cylinder. Remove a 30-ml aliquot and dispose of the remainder. Extract the aliquot with three 5-ml portions of methylene chloride. Combine the extracts and stir the mixture. Inject 20 μl of the extract into the gas chromatograph.

Problem 38-6. If the peak due to propanoic acid overlapped with another peak, how could you easily determine without changing columns which part of the peak was due to propanoic acid and which part was caused by the other compound?

Problem 38-7. What factors should be used to convert the gas chromatographic data to percent by weight composition of each phase? Assume distillate D_4^{20} 0.81, residue in three-necked flask D_4^{20} 1.15. What other assumptions are necessary for calculating the composition of the residue?

After verifying your answers to Problem 38-7 do the calculations necessary to complete Tables 38-1 through 38-3.

[1] See Appendix IV for composition of solid supports.

TABLE 38-1 COMPOSITION OF DISTILLATE

Compound	Wt. in g	Percent by wt.
1-Propanol		
Propanal		
Propanoic acid		
Water		

TABLE 38-2 COMPOSITION OF RESIDUE

Compound	Wt. in g	Percent by wt.
1-Propanol		
Propanal		
Propanoic acid		
Water*		

*by calculation.

TABLE 38-3 TOTAL ORGANIC COMPOUNDS AFTER REACTION

Compound	Wt. in g	Moles
1-Propanol		
Propanal		
Propanoic acid		
Total	-	
Starting material	24.1	
Difference	-	

Problem 38-8. Name several possible errors in the analytical procedure that could account for the difference that occurs in Table 38-3.

Purification

If the gas chromatographic analysis showed the distillate to be 95% or more (by weight) propanal, the product may be placed directly in the tared yield bottle after separation from the aqueous layer. If the analysis showed more than 5% water present in a distillate of 95% or more propanal *and* water, the following distillation can be omitted but the product should be dried with calcium chloride.

Separate the two layers of distillate and run the top organic layer into a 50-ml Erlenmeyer flask containing 10 to 15 g of granular calcium chloride. Decant into a 50-ml round bottom flask. Add a few boiling chips and construct the fractional distillation apparatus shown in Figure 5-5. Place an ice bath around the receiving flask. Collect everything that distills up to 55° in a tared yield bottle. Weigh the product, calculate % yield, and hand in the product to your laboratory instructor. Turn in all charts from the gas chromatograph (gas chromatograms) with your product.

Optional. If it was necessary to purify the product by drying or distillation, obtain a gas chromatogram of the purified propanal.

39

Synthesis of Amides

$$C_6H_5NH_2 + (CH_3CO)_2O \rightarrow C_6H_5NHCOCH_3 + CH_3CO_2H \qquad (39\text{-}1)$$

$C_6H_5NH_2$	$(CH_3CO)_2O$	$C_6H_5NHCOCH_3$
B.p. 184°	B.p. 140°	B.p. 305°
D_4^{20} 1.022	D_4^{20} 1.081	M.p. 113.5°
	M.p. −73°	

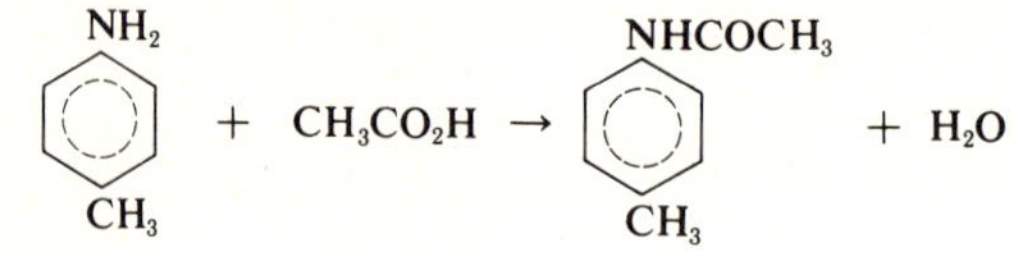

(39-2)

	CH_3CO_2H	
B.p. 200°	B.p. 118°	M.p. 147°
M.p. 45°	D_4^{20} 1.049	
	M.p. 16.6°	

INTRODUCTION

Amides can be synthesized by a great variety of methods. The most common methods involve reaction of acids or acid derivatives with ammonia, amines, or their salts. In the experiments of this chapter amides are synthesized by reaction of amines with a carboxylic acid and with a carboxylic acid anhydride.

DISCUSSION

Amines can be made by the reactions of ammonia or primary and secondary amines with acid derivatives including other amides. Equations for these reactions are listed below in order of their relative rates of reaction. The most rapid reaction is given first (R″ and R′ can be aryl, alkyl, or hydrogen).

$$RCOCl + R_2'NH \rightarrow RCONR_2' + HCl \qquad (39\text{-}3)$$

$$(RCO)_2O + R_2'NH \rightarrow RCONR_2' + RCO_2H \qquad (39\text{-}4)$$

$$RCO_2Et + R_2'NH \rightarrow RCONR_2' + EtOH \qquad (39\text{-}5)$$

$$RCONR_2'' + R_2'NH \rightarrow RCONR_2' + R_2''NH \qquad (39\text{-}6)$$

The amines usually react readily with acid chlorides and acid anhydrides, and the equilibria are far to the right. The reaction of amines with amides and esters occurs less readily, and, if equilibrium is to be reached in a reasonable amount of time, a strongly acidic or basic catalyst is required. The equilibria for the last two reactions usually do not favor the products, and a good yield is dependent upon removal of one of the products from the equilibrium.

The nucleophilic amine adds to the electrophilic carbonyl and the adduct eliminates a halide ion, alcohol, acid, amine, or water molecule in a reaction similar to the hydrolysis or transesterification of esters.

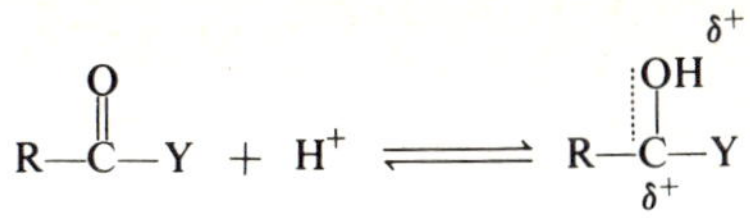

(39-7)

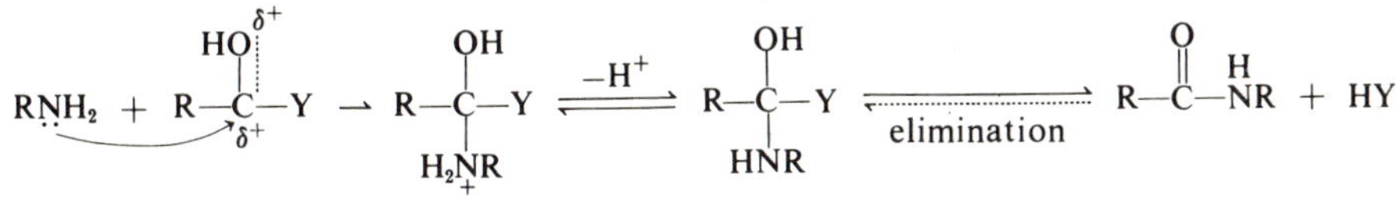

(39-8)

The relative rates of reaction of carboxylic acids and their derivatives with amines are a function of the structure of Y. If the student cannot rationalize the relative rates of the reactions, he should review the appropriate portions of his lecture text.

When an excess of a carboxylic anhydride reacts with an amine, a diacyl derivative may be formed.

$$C_6H_5NH_2 + 2(CH_3CO)_2O \rightarrow C_6H_5N(COCH_3)_2 + 2CH_3CO_2H \quad (39\text{-}9)$$

The diacyl derivatives from monocarboxylic anhydrides are usually rapidly hydrolyzed by water to the monoacyl derivative.

$$H_2O + C_6H_5N(COCH_3)_2 \rightarrow C_6H_5NHCOCH_3 + CH_3CO_2H \quad (39\text{-}10)$$

If water is used either in the purification or the reaction, the product is uncontaminated by diacyl derivative. Many of the impurities in the amine and the reaction products of these impurities with the carboxylic acid and its anhydride are water-soluble. The use of water as a reaction medium or as a solvent for recrystallization keeps these substances from contaminating the product. At room temperature the reaction of acetic anhydride with water is much slower than its reaction with primary amines.

In the procedure for this experiment aniline is dissolved in dilute hydrochloric acid. A slight excess of acetic anhydride is added. Then sodium acetate is added which gives a mixture of the free base and anilium acetate.

$$C_6H_5NH_2 + HCl \rightarrow C_6H_5NH_3^+Cl^- \quad (39\text{-}11)$$

$$C_6H_5NH_3^+Cl^- + CH_3CO_2^-Na^- \rightleftharpoons C_6H_5\overset{+}{N}H_3\ CH_3CO_2^- + Na^+Cl^- \quad (39\text{-}12)$$

$$C_6H_5NH_3^+\ CH_3CO_2^- \rightleftharpoons C_6H_5NH_2 + CH_3CO_2H \quad (39\text{-}13)$$

As the aniline is used, the equilibrium shown in Equation 39-13 shifts to the right to give more aniline. The acetate ion also acts as a catalyst for reaction of the amine with acetic anhydride.

The acylation of amines with carboxylic acids is an equilibrium reaction; good yields are dependent upon removal of one of the products.

$$R'NH_2 + RCO_2H \rightleftharpoons RCONHR' + H_2O \quad (39\text{-}14)$$

In the second experiment of this chapter water is removed. In some procedures precipitation of the amide to the separate solid phase suffices to drive the reaction to completion. This type of reaction is the basis for a commercial synthesis of acetanilide.

Low molecular weight amides can also be prepared from carboxylic acids by first obtaining the ammonium carboxylate and then heating this salt.

$$RNH_2 + CH_3CO_2H \rightarrow CH_3CO_2^-\ H_3\overset{+}{N}R \xrightarrow{\Delta} CH_3CONHR + H_2O \quad (39\text{-}15)$$

EXPERIMENTAL PROCEDURE

Warning! Aromatic amines are very toxic. Avoid spilling *p*-toluidine and aniline on the skin and inhaling their vapors.

Synthesis of Acetanilide

Dissolve 12 ml of aniline by stirring with 290 ml of water and 11.5 ml of concentrated hydrochloric acid in a 500-ml Erlenmeyer flask. If the solution is darker than a light yellow, it should be clarified with activated carbon at 45° and filtered. Put the mixture in an ice bath and immediately add 16 ml of acetic anhydride. Stir for 1 min and then pour in at once a solution of 19 g of sodium acetate in 60 ml of water. Stir the mixture vigorously for 3 to 4 min and allow to stand in the ice bath for 10 min, stirring occasionally.

Filter, wash the precipitate with 25 ml of ice water. Recrystallize the precipitate from the minimum amount (240 to 300 ml) of 5% methanol in water. Collect two crops of crystals by allowing the mixture to cool to room temperature, filtering, chilling the filtrate in an ice bath, and filtering again. Dry the two crops separately and determine their melting points. Place them in *separate*, tared yield bottles. Weigh the products and calculate the % yield. Hand in the labeled yield bottles to your instructor.

Synthesis of *p*-Acetotoluidide

With an air condenser and a 200-ml round bottom flask construct the reflux apparatus shown in Figure 10-1. Place a stopper in the condenser that has tubing leading to a desk hood or to a water trap (Fig. 15-3). Do not run water through the condenser. Reflux 27 g of *p*-toluidine, 105 ml of glacial acetic acid, and 0.3 g of zinc dust for 2 hr. Begin the next experiment during the reflux period.

Slowly pour the *hot* reaction mixture into a 800-ml beaker containing 400 g of crushed ice and 50 ml of 10% sodium hydroxide. Stir several minutes and filter the mixture. Transfer any remaining ice to the Büchner funnel; as it melts, it washes the precipitate. Rinse the beaker with 25 ml of water; use this to wash the precipitate.

Recrystallize the precipitate from the minimum amount of 5% methanol in water. If the hot solution is darker than a light yellow, add an additional 20 ml of 5% methanol in water and clarify with approximately 7 g of activated carbon. Collect two crops of crystals by allowing the mixture to cool to room temperature, filtering, chilling the filtrate in an ice bath, and filtering again. Dry the two crops separately and determine their melting points. Place them in *separate*, tared yield bottles. Weigh the products and calculate the % yield. Label the bottles and hand in the products to your instructor.

INFRARED OPTION

Obtain infrared spectra on: (a) the crude precipitate obtained from the reaction mixture, (b) and (c) on each of the two crops of crystals obtained by recrystallization of the crude precipitate. Obtain the spectra from Nujol mulls or preferably from KBr pellets (Appendix III). In each spectrum assign as many of the absorption bands as possible to the structure of the product.

Problem 39-1. Which sample is the most impure? Which is the purest? What classes of compounds are the impurities? What are the probable structures of the impurities? Note especially the region between 3500 and 2500 cm^{-1}.

EXERCISES

39-1. Simple aromatic amines are colorless. Why are the amines used in these experiments yellow to deep red?

39-2. What is the purpose of the zinc in the synthesis of *p*-acetotoluidide?

40

Synthesis of Sulfanilamide

$$p\text{-}CH_3CONHC_6CH_5 + HOSO_2Cl \longrightarrow p\text{-}CH_3CONHC_6H_4SO_2Cl \xrightarrow{NH_3} p\text{-}CH_3CONHC_6H_4SO_2NH_2$$

p-$CH_3CONHC_6CH_5$	$HOSO_2Cl$		p-$CH_3CONHC_6H_4SO_2Cl$	p-$CH_3CONHC_6H_4SO_2NH_2$
B.p. 305°	B.p.	151°	M.p. 149°	M.p. 220°
M.p. 114°	M.p.	−80°		
	D_4^{20}	1.8		

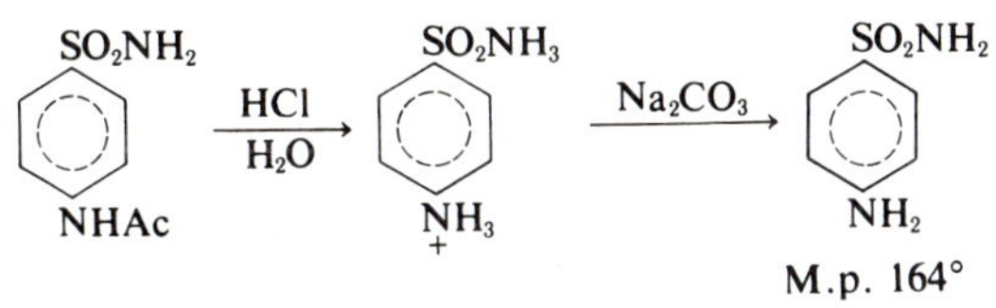

M.p. 164°

INTRODUCTION

Acetanilide is chlorosulfonated and the product treated with ammonia. The hydrolysis of this product to sulfanilamide illustrates that sulfonic acid derivatives are less reactive than carboxylic acid derivatives.

EXPERIMENTAL PROCEDURE

Synthesis of *p*-Acetamidobenzenesulfonyl Chloride

Place a 100-ml flask in a cold water bath (5 to 10°); add 26 ml of chlorosulfonic acid (**Danger, very strong acid**). Add 9 g of acetanilide in **small portions,** stirring the mixture with a glass rod after each addition. Add the acetanilide at such a rate that the heat of reaction maintains the temperature in the flask between 25 and 45°. Stir the mixture for an additional 2 to 3 min after all the acetanilide has been added. To the flask add an air condenser with a drying tube on it[1]. Heat the mixture on a steam bath for 50 min. Pour the mixture onto 150 g of ice in a 250-ml beaker. Stir the mixture with a glass rod. If the acid chloride does not crystallize, grind it with a glass rod until it does. Suction filter, break the clumps, and pull air through the precipitate for 2 to 3 min.

Ammonolysis to *p*-Acetamidobenzenesulfonamide

Transfer the *p*-acetamidobenzenesulfonyl chloride into a 125-ml Erlenmeyer flask, add 35 ml of concentrated ammonium hydroxide solution, and stir the mixture with a glass rod for 10–15 min. Heat the mixture on a steam bath for 15 min and stir the mixture occasionally.

[1] If an air condenser is not available, use a water condenser without any water in it.

Chill the reaction flask in an ice bath while stirring the mixture with a thermometer until the temperature falls below 10° and then add 10 to 15 g of ice. Add 20% aqueous sulfuric acid with stirring until the solution has a pH of 5 to 7. After neutralization allow the flask to stand in the ice bath for 5 to 10 min, filter, and wash the precipitate with three 5-ml fractions of ice water.

If the diamide is to be hydrolyzed to *p*-aminobenzenesulfonamide (sulfanilamide), it can be used directly for the next step without further purification. If the synthesis is to be stopped here and the diamide is to be turned in as the product, dry the precipitate and recrystallize it from 1:5 water-methanol mixture. Determine the melting point and the % yield, and give the product to your instructor.

Partial Hydrolysis to *p*-Aminobenzenesulfonamide

Heat the crude diamide, 15 ml of concentrated hydrochloric acid, and 15 ml of water in a 50-ml round bottom flask with a reflux condenser for 20 min on a steam bath. Pour the mixture into a 125-ml beaker containing approximately 25 g of ice. Place the beaker in an ice bath, add 2 g of solid sodium hydroxide, and stir. Adjust the pH of the mixture to 6 to 8 by adding small amounts of sodium carbonate and stirring slowly to avoid loss of product due to foaming. Stir the mixture in the ice bath for 5 to 7 min after neutralization.

Problem 40-1. Why is it important not to use excess hydrochloric acid nor to heat for a longer time than is called for in the directions?

Problem 40-2. Why is sodium hydroxide not used for the entire neutralization of the excess hydrochloric acid? What would happen if excess base were added? What would happen if the hydrochloric acid were not completely neutralized?

Suction filter the mixture and wash the precipitated sulfanilamide with three 5-ml portions of cold water. Dry the precipitate in the Büchner funnel by pressing it with a small Erlenmeyer flask or beaker with the suction on. Recrystallize the sulfanilamide from 80 to 90 ml of water. If the hot solution is any color deeper than a light yellow, clarify with activated carbon before it cools. Allow the solution to cool to room temperature (overnight if necessary) and suction filter. Wash the precipitate once with 5 ml of cold water and dry the product.

Weigh the precipitate in a tared yield bottle. If the yield is less than 5.0 g, chill the filtrate from the recrystallization in an ice bath. Filter, dry, and add these crystals to the first crop. Determine the % yield and the melting point, label the bottle, and hand the product in to your instructor.

INFRARED OPTION

Make a potassium bromide pellet (Appendix III) of your dried product and obtain an infrared spectrum. Compare your spectrum with Figure III-10. Which spectrum does it more closely resemble? Explain any significant differences between your spectrum and Figure III-10b. Is there any evidence for impurities? If so, what are they?

EXERCISES

40-1. Acetanilide instead of aniline is used as the starting material for this reaction because the amino group reacts with acid chlorides. Suggest at least three *other* ways the presence of the acetyl group may contribute to obtaining a higher yield and a purer product in the synthesis of *p*-acetamidobenzenesulfonyl chloride.

40-2. During ammonolysis of the acid chloride there is more water present than ammonia. Why is the sulfonamide the principal product rather than the sulfonic acid?

41

Sulfonic Acids and Derivatives

Organic Qualitative Analysis VI

INTRODUCTION

The sulfonic acid group is readily characterized because it is the only sulfur-containing functional group that is so strongly acidic and polar. Although sulfonic acids are less reactive than carboxylic acids, amides and esters can be synthesized from them by the proper series of reactions. Sulfonamides are especially useful for the identification of sulfonyl chlorides and sulfonic acids.

PHYSICAL PROPERTIES

Melting and Boiling Points

Acids. The sulfonic acid group is very polar and usually completely or almost completely ionized. The strong ionic and polar intermolecular attractions cause sulfonic acids to be very high boiling and high melting. All sulfonic acids decompose before they boil at atmospheric pressure. Since sulfonic acids are derived from sulfuric acid, it is not surprising that like sulfuric acid they are very deliquescent.

Sulfonic acids usually exist in various states of hydration and even the various hydrates are usually deliquescent. The amount of bound and unbound water in crystals of many compounds, especially sulfonic acids, varies widely with humidity, and methods of preparation, purification, drying, etc. The melting points of most organic compounds vary widely with the amount of bound and unbound water (or other solvent) present. For this reason, before the melting point of a compound is determined, the compound should be dried in such a manner as to drive off all water (or other solvent) or at least give a solvate of known composition. Even when these precautions are taken with sulfonic acids, the amount of water present still may vary greatly, and the melting point is not a constant property. For instance, it is possible to find in the *recent* literature melting points reported that vary by as much as 50° for a common compound like benzenesulfonic acid. Melting points of sulfonic acids are not very useful in identifying the acids. If a sulfonic acid is obtained as an unknown, it should be converted first to a sulfonyl chloride and the melting point of a purified sample determined. The sulfonyl chloride is then converted to one of the amides.

Derivatives. Since the group $-\overset{\displaystyle O}{\overset{\|}{\underset{\underset{\displaystyle O}{\|}}{S}}}-$ is very polar, molecules of sulfonic acid derivatives have very strong dipole-dipole and dipole-induced dipole attractions for each other. Therefore, sulfonic acid derivatives have higher melting and boiling points than have carboxylic acid derivatives of similar molecular weights (see Tables 41-1, 2, and 3) even though these sulfonic acid derivatives do not have the highly ionized $-SO_3H$ group. Because of the absence of the strongly acidic $-SO_3H$ group, sulfonamides and sulfonate esters are not more deliquescent than other similar organic compounds, and they make excellent derivatives with reliable melting points. Sulfonyl chlorides, especially those that are liquids, are hydrolyzed slowly by moisture from the air, and care should be taken to keep them dry.

Solubility

Sulfonic Acids. Because of the highly ionized sulfonic acid group most sulfonic acids are soluble in water and insoluble in ether. Unless basic groups such as NH_2 also are present, sulfonic acids are very strongly acidic to all indicators including pH paper and are usually in Class S_2a. Most Class S_2a compounds that are strongly acidic to pH paper and contain sulfur are sulfonic acids. A few very high molecular weight sulfonic acids and some sulfonic acids having other functional groups are insoluble in water. These water-insoluble sulfonic acids dissolve in aqueous base and are in Class A_1.

Sulfonic Acid Derivatives. Sulfonyl chlorides are less reactive than the corresponding acyl halides. Why? Only a few sulfonyl halides react rapidly enough with water to be in solubility Class S_1a. Sulfonyl chlorides sometimes react so slowly with aqueous base that they often are found in Class M although most sulfonyl chlorides dissolve in dilute base if sufficient time is allowed. Most sulfonamides also dissolve in aqueous base.

AQUEOUS SODIUM BICARBONATE AS A CLASSIFICATION SOLVENT (SOLUBILITY CLASS A_2)

Discussion

Because so many organic compounds dissolve in or react with 5% aqueous sodium hydroxide, it is useful to subdivide acidic compounds into strong acids and weak acids (subclasses A_1 and A_2). This is conveniently done by testing all compounds that are soluble in 5% aqueous sodium hydroxide with 5% aqueous sodium bicarbonate.

Compounds that dissolve in or react with aqueous sodium bicarbonate (and, therefore, are stronger acids than carbonic acid) are in Class A_1. Compounds that dissolve in or react with aqueous sodium hydroxide but not with aqueous sodium bicarbonate (and, therefore, are stronger acids than water but weaker acids than carbonic acid) are in Class A_2 (see Fig. 41-1). Examples of these weaker acids are phenols (Chapter 49), most aliphatic nitro compounds (Chapter 46) and most sulfonamides. Sulfonamides other than N,N-disubstituted sulfonamides dissolve in aqueous sodium hydroxide (but not in water or aqueous sodium bicarbonate) because the hydrogen on the amido nitrogen is more acidic than the hydrogen of water and is removed by the hydroxide ion. Sulfonamides are more acidic than carboxamides because the electron-withdrawing power of the $-SO_2-$ group stabilizes the conjugate base more than the $\rangle C{=}O$ group does. Other than sulfonamides only a relatively few polyfunctional compounds contain both nitrogen and sulfur and are in Class A_2; therefore, the combination of these factors can be taken as strongly indicating an N-monosubstituted or unsubstituted sulfonamide.

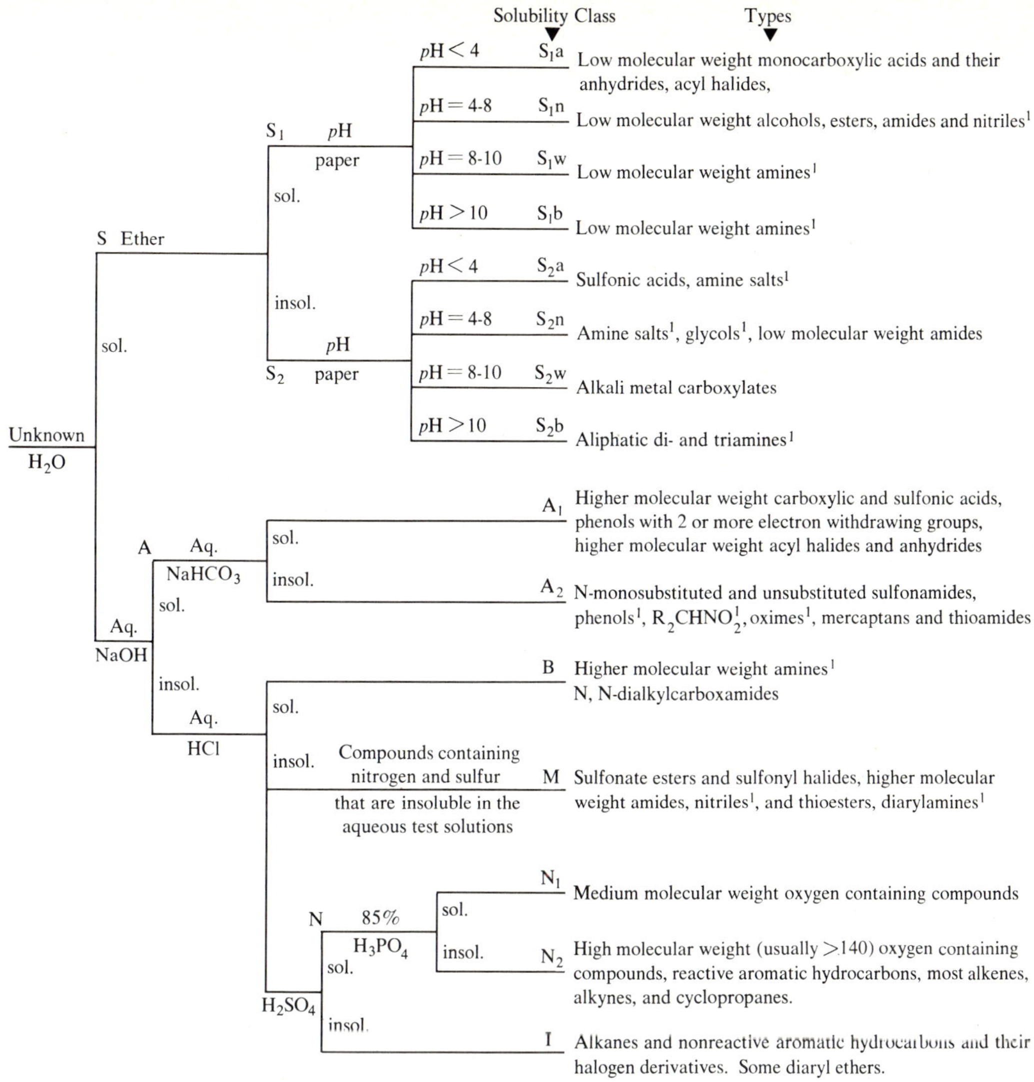

[1] Solubility relationships discussed in later chapters.

Figure 41-1 Solubility classification chart. See also Figures 32-4 and 61-1.

Experimental Procedure

Using the general methods outlined on pages 225 through 226, test: (a) the solubility of the compounds in water, (b) the solubility of the water-soluble compounds in ether, (c) the aqueous solutions of the compounds that dissolve in water with pH paper, (d) the solubility of the water-insoluble compounds in 5% aqueous sodium hydroxide, and (e) the solubility of all aqueous sodium hydroxide-soluble compounds in 5% sodium bicarbonate. If the compound does not dissolve immediately in the bicarbonate solution, note if carbon dioxide is given off. If the compound does not appear to dissolve and carbon dioxide is given off, shake the test tube intermittently for 5 min and then observe to determine if the compound is soluble or insoluble in aqueous sodium bicarbonate. Try this procedure on benzenesulfonamide, *p*-toluenesulfonyl chloride, benzoic acid, cinnamic acid, benzoyl chloride, and benzamide. Write the equations involved.

Problem 41-1. All of these compounds are insoluble in dilute hydrochloric acid. What solubility classes are they in?

To each of the solutions resulting from the dissolving of a compound in aqueous sodium hydroxide or sodium bicarbonate add 20% hydrochloric acid drop by drop until the solution is acidic to pH paper. How do you explain the presence or absence of a precipitate in each case?

CLASSIFICATION TESTS

Sulfonic Acids

The presence of sulfur, the high acidity, and the solubility classification of sulfonic acids allow them to be easily recognized, and no further classification tests are needed. The variations in melting points make it difficult to identify *specific* sulfonic acids. Sulfonic acids may well be the most difficult compounds the beginning organic chemistry student is asked to identify. A good procedure for identification is to convert the acid to the sulfonyl chloride and to determine its melting point if it is solid. The sulfonyl chloride is then converted to the amide and, if need be, to the anilide.

Sulfonyl Chlorides

The halogen on sulfonyl chlorides is sufficiently reactive to always give immediate precipitates with alcoholic silver nitrate. Sulfonyl halides also give immediate precipitates with sodium iodide in acetone. The precipitate is red because of liberation of free iodine (pp. 175–176). Sulfonyl chlorides are easily converted to their amides and anilides for final identification.

Sulfonamides

Most sulfonamides are not affected by alkali metal hydroxide reactions of the type described in Chapter 35. Sulfonamides are usually recognized by their solubility classification, by the presence of both nitrogen and sulfur, and by the fact that other classification tests are negative for these compounds. Sulfonamides can be hydrolyzed with concentrated mineral acids to the sulfonic acid and amine salt. These can then be identified by the usual methods. (See the next section and Chapter 44.)

DERIVATIVES OF SULFONIC ACIDS AND SULFONYL HALIDES

Sulfonyl Chlorides from Sulfonic Acids

$$ArSO_3H + PCl_5 \rightarrow ArSO_2Cl + POCl_3 + HCl \qquad (41\text{-}1)$$

Put a thermometer in a 25-ml distilling flask so that it *almost* but not quite *touches* the *bottom* of the flask. Attach an adapter to the side arm (Fig. 7-3) and place the adapter below the surface of the water in the test tube. The purpose of the water is to trap the hydrogen chloride and other fumes given off. Even with this precaution it is still best to do this reaction and subsequent distillation in a hood.

Put 0.5 g of finely powdered, *dry* sulfonic acid into the flask. Then place 2 g of fine phosphorus pentachloride powder in the bottom of the flask. Phosphorus pentachloride reacts quickly with moisture. This will cause not only the evolution of noxious hydrogen chloride fumes but will also cause the yield of product to be lower and perhaps cause the isolation of pure product to be more difficult. For this reason the phosphorus pentachloride should be quickly pulverized in a dry mortar in a hood and then quickly added to the flask.

Begin to heat the flask immediately with an oil (such as mineral oil from the melting point bath) or sand bath. Very careful use of a micro burner may be substituted for the oil bath. Since a heating bath requires less attention, it is preferred as it frees one to do other operations during the heating. The temperature should be kept at 105 to 110° for ½ hr. If for any reason the heat should be removed or significantly reduced, the test tube must be lowered so that the water surface is below the bottom of the adapter to prevent backing of water into the reaction flask. The phosphorus oxychloride (b.p. 137 to 138°) is removed by distillation. *Do not heat above 155°* as read from the thermometer with the bulb in the *bottom* of the distilling flask.

Allow the flask to cool for 5 min and add 3 ml of *ice* water. Stir the mixture with a glass rod; be sure that all the residue comes in contact with the water. Decant the aqueous layer. Be careful not to pour out the sulfonyl chloride; it usually sticks to the wall of the flask. Repeat the *ice* water wash three more times using the above procedure. If the amide is to be prepared, the product can be used directly in the procedure described in the following section. If the pure sulfonyl chloride is desired or if the melting point of both the sulfonyl chloride *and* one or more amides is needed, follow the above procedure with all quantities doubled or tripled and purify the products.

To purify the crude sulfonyl chloride transfer it to a separatory funnel with the aid of three 6-ml portions of benzene to make the transfer quantitative. Wash the benzene with two 10-ml portions of ice water. Transfer the organic layer into a 50-ml Erlenmeyer flask containing about 4 g of anhydrous calcium chloride and swirl. Filter the dried benzene solution into a distilling flask and remove the benzene by distillation. Crystallize the residue from tetrahydrofuran, chloroform, toluene, or petroleum ether. Sulfonyl chlorides are difficult to crystallize and often must be chilled to ice bath or even to salt-ice bath temperatures and patiently ground before a solid appears (Chapter 12).

Problem 41-2. Why would methanol and ethanol be unsatisfactory solvents for the above recrystallization? Why would benzene be unsatisfactory?

If the material fails to give a solid with the above treatment, distil it under vacuum with the use of a water aspirator as a source of vacuum. Whether or not the initial purification is done by distillation or crystallization, the compound should be purified again by recrystallization as described above unless, of course, the compound is a liquid. If the second recrystallization yields a product with a wide melting point range, the product should be further recrystallized until the melting point range is narrow or, preferably, until the melting point cannot be increased by further crystallization (constant melting point, p. 256).

Sulfonamides from Sulfonyl Chlorides

$$ArSO_2Cl + 2NH_3 \rightarrow ArSO_2NH_2 + NH_4^+ + Cl^- \quad (41\text{-}2)$$

To approximately 0.5 g of sulfonyl chloride (either crude or purified) add 10 ml of concentrated aqueous ammonia and 2 g of finely powdered ammonium carbonate. Stir slowly at first and then vigorously with a glass rod. Allow to stand for about 4 min, stir again, and heat on a water bath for 10 min at 90°. Cool to room temperature; if any sulfonamide precipitates, filter and save the precipitate. Evaporate the filtrate to dryness, dissolve the residue in 10 ml of boiling water (use more if needed to effect complete solution), add activated carbon, stir, and filter. Allow to crystallize, and to the precipitate add any sulfonamide that precipitated when the initial reaction mixture was cooled. Repeat the recrystallization omitting the treatment with carbon black. Dry the crystals and determine their melting point.

Sulfonanilides from Sulfonyl Chlorides

$$ArSO_2Cl + C_6H_5NH_2 + NaOH \rightarrow ArSO_2\bar{N}C_6H_5 + NaCl + H_2O + Na^+ \quad (41\text{-}3)$$

$$ArSO_2\bar{N}C_6H_5 + HCl \rightarrow ArSO_2NHC_6H_5 + Cl^- \quad (41\text{-}4)$$

To 0.5 g of aniline in a test tube add 0.7 g of sulfonyl chloride and 7 ml of 10% aqueous sodium hydroxide. Shake the stoppered test tube gently at first and then vigorously for a few seconds. Test the pH with a drop of the solution from a glass rod. If the solution is not strongly alkaline, add more sodium hydroxide until it is.

Again shake the stoppered test tube vigorously until all the sulfonyl chloride has reacted. Add 20% hydrochloric acid until the solution is slightly acidic. Cool the mixture in an ice bath and filter. If an oil instead of a solid precipitates, grind the oil in contact with the chilled solution until the oil crystallizes. Wash the precipitate with two 5-ml portions of ice water. Recrystallize twice from aqueous ethanol. Dry and determine the melting point of the crystals.

Problem 41-3. What is the purpose of the sodium hydroxide in the preparation of sulfonamides and sulfonanilides?

Problem 41-4. How readily could gas chromatography be used on sulfonic acids? Outline a satisfactory procedure for gas chromatographic analysis of sulfonic acids?

UNKNOWN

The unknown assigned with this chapter can be any compound from the tables at the end of Chapters 14, 23, 28, 32, and 35 and this chapter (Tables 41-1, 41-2, 41-3) except for N-substituted and N,N-disubstituted amides which will not be assigned as unknowns until Chapter 44 is studied.

TABLE 41-1 SULFONAMIDES

Sulfonamide	M.p. (°C)	Acid m.p. (°C)	Sulfonamide	M.p (C°)	Acid m.p. (C°)
2,6-Dimethylbenzene-	96	98	2,5-Dimethyl-3-nitrobenzene	173	200
2-Phenyl-1-ethane-	122	91	p-Nitrobenzene-	180	95
p-Toluene-	137	92	2,4,5-Trimethylbenzene-	181	112
2-Bromo-1-naphthalene-	140		5-Chloro-4-methyl-2-nitrobenzene-	181	128
m-Aminobenzene-	142				
3,4-Dimethylbenzene-	144	64	4-Chloro-2,5-dimethylbenzene-	185	100
p-Chlorobenzene-	144	68	8-Iodo-1-naphthalene-	187	
1-Naphthalene-	150	90	o-Nitrobenzene-	191	70
Benzene-	153	66	1-Bromo-4-naphthalene-	193	
m-Bromobenzene-	154		2,5-Dimethyl-4-nitrobenzene	198	140
p-Bromobenzene-	166	103	1-Nitro-2-naphthalene-	214	105
2,4-Dimethoxybenzene-	167	192	2-Naphthalene-	217	91
			3-Chlorobenzene-1,5-di	224	100d

TABLE 41-2 SULFONIC ACIDS

Sulfonic Acids	Acid M.p. (°C)	Derivatives, m.p. (°C) chloride	amide	anilide
2,5-Dimethylbenzene-	48	25	148	
Diphenylmethane-4,4'-di-	59	124		178
2,4-Dimethylbenzene-	62	34	137	110
3,4-Dimethylbenzene-	64	52	144	
Benzene-	66	14	153	110
p-Chlorobenzene-	68	53	144	104
o-Nitrobenzene-	70	68	191	115
2,4,6-Trimethylbenzene-	77	56	142	109
1-Naphthalene-	90	68	150	112
2-Phenyl-1-ethane-	91	33	122	77
2-Naphthalene-	91	76	217	132
p-Toluene-	92	69	137	103
p-Nitrobenzene-	95	80	180	171
4-Chloro-2,5-dimethylbenzene-	100	50	185	155
1,3,5-Benzenetri-	100d	187	312	237
3-Chlorobenzene-1,5-di-	100d	106	224	
p-Bromobenzene-	103	75	166	119
1,2-Ethanedi-	104	91		
1-Nitro-2-naphthalene-	105	121	214	202
2,4,5-Trimethylbenzene-	112	61	181	
1,6-Naphthalenedi-	125	128	298	
5-Chloro-4-methyl-2-nitrobenzene-	128	99	181	
2,5-Dimethyl-4-nitrobenzene-	140	75	198	131
2,5-Dimethyl-6-nitrobenzene-	145	110	192	182
2,4-Dimethoxybenzene-	192	71	167	
(+)-or (-)-Camphor-10	193	67	132	121
2,5-Dimethyl-3-nitrobenzene-	200	61	173	144
1,6-Anthraquinonedi-	217	198		227d
Anthraquinone-α-	218	214		216

TABLE 41-3 SULFONYL CHLORIDES
(SOLIDS)

Sulfonyl chloride	M. p. (°C)	Derivatives, m.p. (°C) acid	amide	anilide
Benzene-	14	66	153	110
2,5-Dimethylbenzene-	25	48	148	
2-Phenyl-1-ethane-	33	91	122	77
2,4-Dimethylbenzene-	34	62	137	110
2,6-Dimethylbenzene-	39	98	96	
4-Chloro-2,5-dimethylbenzene-	50	100	185	155
3,4-Dimethylbenzene-	52	64	144	
p-Chlorobenzene-	53	68	144	104
2,4,6-Trimethylbenzene-	56	77	142	109
2,4,5-Trimethylbenzene-	61	112	181	
(+)-or (-)-Camphor-10-	67	193	132	121
o-Nitrobenzene-	68	70	191	115
1-Naphthalene-	68	90	150	112
p-Toluene-	69	92	137	103
2,4-Dimethoxybenzene-	71	192	167	
p-Bromobenzene-	75	103	166	119
2,5-Dimethyl-4-nitrobenzene-	75	140	198	131
2-Naphthalene-	76	91	217	132
p-Nitrobenzene	80	95	180	171
1-Bromo-4-naphthalene	83		193	
1,2-Ethanedi-	91	104		
5-Chloro-4-methyl-2-nitrobenzene-	99	128	181	
3-Chlorobenzene-1,5-di-	106	100d	224	
2,5-Dimethyl-6-nitrobenzene-	110	145	192	182
8-Iodo-1-naphthalene	115		187	140
1-Nitro-2-naphthalene-	121	105	214	202
Diphenylmethane-4,4'-di-	124	59		178
1,6-Naphthalenedi-	128	125	298	
1,3,5-Benzenetri-	187	100d	312	237
1,6-Anthraquinonedi-	198	217		227d
Anthraquinone-α-	214	218		216

42

Reduction of Nitro Compounds to Amines

Synthesis of *o*- or *p*-Toluidine

$$NO_2C_6H_4CH_3 \xrightarrow[H^+]{Fe} NH_2C_6H_4CH_3 \qquad (42\text{-}1)$$

	o	*p*			*o*	*p*
B.p.	222°	234°		B.p.	200°	200°
M.p.	−4°	52°		M.p.	−20°	45°
D_4^{20}	1.163			D_4^{20}	1.001	

INTRODUCTION

Amines can be prepared by reduction of many organic nitrogen compounds. A large variety of reagents, catalytic hydrogenation, and electrochemical methods reduce nitro compounds to amines. In this synthesis iron and hydrochloric acid are the reducing agents. Possibly, however, in the first reduction of the nitro group to a nitroso group and in other steps the reducing agent is the ferrous ion.

$$Fe + 2H^+ \rightarrow Fe^{++} + 2[H] \qquad (42\text{-}2)$$

$$2H^+ + ArNO_2 + 2Fe^{++} \rightarrow ArNO + 2Fe^{+++} + H_2O \qquad (42\text{-}3)$$

DISCUSSION

Nitro compounds are usually reduced to amines stepwise in the various reductions. Equations 42-4 and 42-5 show all the intermediates possible although not every one is produced by every reductive

$$ArNO_2 \xrightarrow{[H]} ArNO \xrightarrow{[H]} Ar\overset{\overset{O^-}{\uparrow}}{N^+}{=}NAr \text{ or } ArNHOH \qquad (42\text{-}4)$$

$$ArNHOH \text{ or } Ar\overset{\overset{O^-}{\uparrow}}{N^+}{=}NAr \xrightarrow{[H]} ArN{=}NAr \xrightarrow{[H]} ArNHNHAr \xrightarrow{[H]} ArNH_2 \qquad (42\text{-}5)$$

method, and, since the nitroso group is much more easily reduced than the nitro group, nitroso compounds cannot usually be isolated from these reductions. Electrochemical oxidation of amines to nitro compounds reverses the sequence of reactions given by Equations 42-4 and 42-5.

In addition to these products from incomplete reduction, various by-products may be produced by reaction of the reduction products.

$$ArNH_2 + ArNO \rightarrow ArN{=}NAr \qquad (42\text{-}6)$$

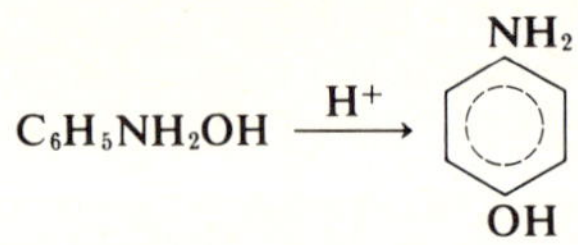

(42-7)

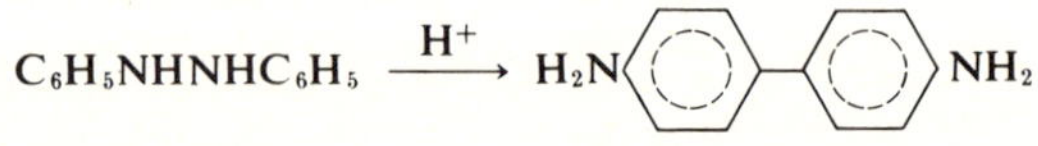

(42-8)

$$ArNHOH + ArNO \rightarrow Ar\overset{\overset{O^-}{\uparrow}}{N^+}{=}NAr \qquad (42\text{-}9)$$

$$ArNHNHAr + ArNO_2 \rightarrow ArN{=}NAr + ArNO + H_2O \qquad (42\text{-}10)$$

Reactions 42-6 and 42-9 do not occur significantly in acidic media, and Reactions 42-7 and 42-8 are not significant except in very strongly acidic media. Further examples of partial or complete reduction of nitro groups are given in Chapters 43 and 46. Chapter 46 also discusses some of the oxidations of these reduction products.

Problem 42-1. Even if a significant portion of the nitro compound reacted according to Equations 42-6 and 42-9, why are these reactions not likely to cause complications if sufficient reducing agent is used?

Problem 42-2. Write reactions corresponding to Equations 42-4 through 42-8 inclusive for *o*- and *p*-nitrotoluene.

Problem 42-3. List all the products from your answer to Problem 42-2 in probable order of relative volatility. List the products from *o*- and *p*-nitrotoluene separately.

EXPERIMENTAL PROCEDURE

Warning. Aromatic amines and nitro compounds are very toxic. The liquids are especially dangerous as they are very easily absorbed through the skin. All precautions should be taken to avoid spilling the compounds and their solutions on the skin and to avoid inhaling their vapors.

In a 200-ml round bottom flask put 17 g of iron powder, 12 ml of either the *ortho* or *para* isomer of nitrotoluene, and 40 ml of water. Add a reflux condenser. Prepare an ice bath because the reaction is often so exothermic that external cooling is needed to keep it under control. Add 1.0 ml of concentrated hydrochloric acid, 7 to 10 drops at a time, through the top of the condenser. Shake the flask vigorously *after each* addition and do not add the next portion of acid until the reaction becomes almost quiescent. If the exothermic reaction does not begin after the first addition of acid, gently heat with a microburner until a vigorous reaction begins. After the first milliliter of hydrochloric acid is added, reflux the mixture 5 min. Then add another milliliter of hydrochloric acid, 7 to 10 drops at a time, through the condenser and reflux the mixture an additional hour. If heating was required to initiate the reaction, the iron powder was probably not active enough. If this occurred, this last reflux period should be lengthened to 1 hr 15 min for the *para* isomer and 1 hr 40 min for the *ortho* isomer. Finish the unknown assigned with Chapter 41 or begin the unknown assigned with Chapter 44 during this reflux time.

Remove the flame and add 4 ml of 6*N* sodium hydroxide. If gas chromatography is to be used, read the optional section at the end of this chapter before proceeding further. Adapt the apparatus for steam distillation and distil the mixture. After 12 min examine a small amount of the distillate in a test tube. If an oily phase accompanies the water, continue the steam distillation until no more oily phase distils.

Problem 42-4. What is likely to contaminate the toluidine obtained from this steam distillation?

Problem 42-5. Outline a method for purifying the toluidine.

Compare your answer to Problem 42-5 with the methods suggested in the answer section before proceeding further. If your method differs considerably from those suggested, you may use it if the laboratory instructor approves, or you may use one of the two methods suggested in the answer section. Note that the first method using a large amount of ether can not be used if steam or heating mantles are unavailable.

When pure the toluidines are colorless. If the product is darker than a light yellow, purify it further by distilling it from 0.5 to 1.0 g of zinc dust. Collect only the first 80% that distils. Use one of the distillation apparatuses shown in Figures 7-3 or 42-1 rather than one with a water condenser. The toluidines have boiling points high enough that a water condenser is not needed if small amounts are distilled, and *p*-toluidine may solidify in a water-cooled condenser and be difficult to remove. If *p*-toluidine solidifies in the adapters, gently heat the area with a microburner until the *p*-toluidine melts and flows into the receiver.

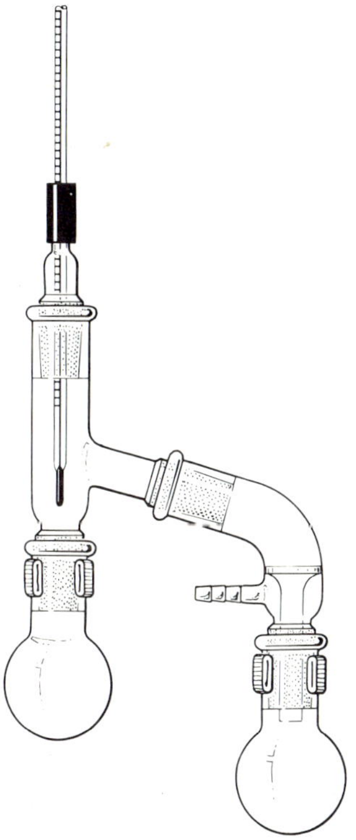

Figure 42-1 Semimicro distillation apparatus with ground glass equipment.

Problem 42-6. What impurities cause aromatic amines to be dark? How does the distillation from zinc dust remove these impurities? What could be done to prevent further contamination of the amine just prepared?

GAS CHROMATOGRAPHY (Optional)

Problem 42-7. Choose from the following liquid phases and solid supports materials for a GLC column that would give good separation of the compounds (answer to Problem 42-2) from the reduction of nitrotoluene. Composition of the liquid phases and the solid supports is given in Appendix IV.

Liquid Phases	*Solid Supports*
(a) QF-1	(a) Teflon
(b) Dimer acid and silicone oil 550	(b) Chromosorb W-AW-DMCS
(c) Bentone 34 and stearic acid	(c) Chromosorb W-AW
(d) Carbowax 600	(d) Diatoport P
(e) Tetraethylenepentamine	(e) Chromosorb G-AW-DMCS
(f) Squalane	(f) Chromosorb G-DMCS

Chromatograph 10 μl fractions of the following: (a) the organic phase obtained after the 6 *N* sodium hydroxide was added (p. 324), (b) the oily phase[1] in the steam distillate (p. 325), and (c) the purified product[1] just before turning it in to your instructor. This last may be omitted if chromatogram b showed 95% or more toluidine and the final purification was omitted.

The laboratory instructor will supply reference chromatograms or retention times and quantitative calibrations of the possible reaction products. Calculate the percent composition of each of the two or three organic mixtures that chromatograms were obtained for. Turn in the chromatograms with your product.

Problem 42-8. Why might there be some problems with the isothermal chromatography of the reaction mixture before steam distillation even though the isothermal chromatograms of the other two organic mixtures are completely satisfactory? How could the difficulties with the chromatography of the first reaction mixture be eliminated?

Problem 42-9. In calculating the percentage composition of the organic mixtures what assumptions were necessary?

[1] If *p*-toluidine was synthesized and the product has solidified, dissolve a few crystals in 2 to 3 drops of ethyl ether and inject this. Since only relative amounts of volatile reaction products are to be determined, the exact amounts used are not critical.

43

Reduction of One Nitro Group on Polynitro Compounds

Synthesis of *m*-Nitroaniline

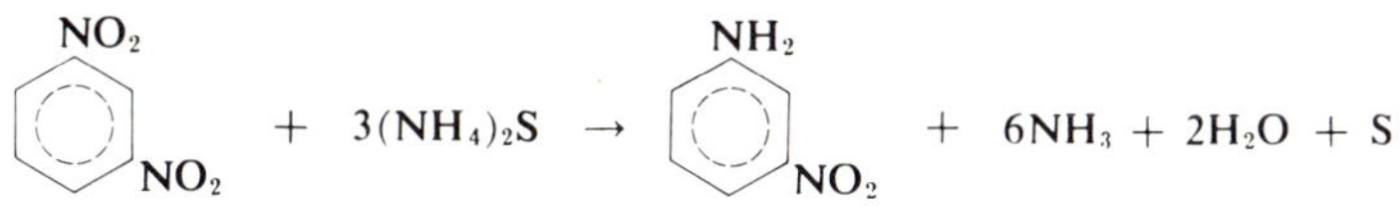

(43-1)

B.p. 301° M.p. 89° B.p. 306° M.p. 114°

INTRODUCTION

Ammonium sulfide is a much weaker reducing agent for the nitro group than the iron powder-hydrochloric acid mixture discussed in the previous chapter. Alkali metal and ammonium sulfides can be used to selectively reduce one nitro group on polynitro aromatic compounds.

DISCUSSION

Partial reduction of polynitro compounds by catalytic reduction or by most chemical reducing systems such as hydrochloric acid and iron powder give a great variety of partially reduced products. If *m*-dinitrobenzene is reduced by exactly the theoretical amount of agent necessary to reduce one group, some of the *m*-dinitrobenzene is not reduced, and the reduction products include *m*-diaminobenzene, *m*-nitroaniline, and many compounds with one or both nitro groups partially reduced.

Problem 43-1. If *m*-dinitrobenzene were reduced by the technique described in Chapter 42 with only enough iron to reduce half of the *m*-dinitrobenzene, what compounds might be isolated in the products?

Problem 43-2. It is easier to reduce an aromatic nitro group if there is another nitro group on the aromatic ring. Suggest an explanation for this.

Weak reducing agents such as ammonium or alkali metal sulfides can be more selective than stronger reducing agents. The electron-withdrawing effect of one nitro group causes the other nitro group to be readily reduced by ammonium sulfide.

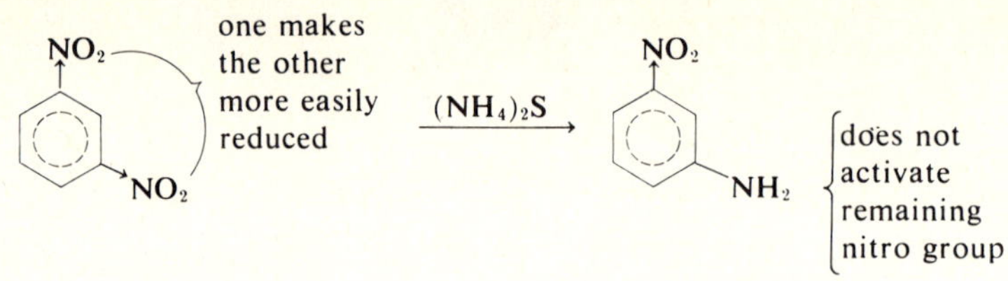

(43-2)

When one nitro group is reduced to an amino group, there is no (or less in polynitro compounds) electron-withdrawing effect to activate the other nitro group to further reduction. The reduction then stops unless there is excess reagent and reaction conditions are sufficiently vigorous.

EXPERIMENTAL PROCEDURE

Dissolve 5 g of *m*-dinitrobenzene in 50 ml of warm 95% ethanol in a 100-ml flask. With a Claisen adapter, dropping funnel, and reflux condenser construct the apparatus shown in Figure 16-1 except that no stopper with hose is needed in the top of the condenser. Place a warm water bath (70 to 75°) around the flask. Transfer 21 ml of ammonium sulfide solution to the dropping funnel. Run the sulfide solution into the reaction flask, 10 to 15 drops at a time, and shake the flask vigorously after each addition. Note that this reaction is not strongly exothermic like the reaction of nitrotoluene with iron and dilute hydrochloric acid. Heat the water bath at 72 to 78° for 5 min after the addition.

Problem 43-3. Why is it important not to reflux the mixture?

Allow the mixture to cool 4 to 5 min and then pour it into 80 g of ice and 60 ml of water in a 250-ml beaker. Stir the mixture until all the ice melts and suction filter immediately. Wash the precipitate with three 10-ml portions of ice water, first using the washes to rinse solids from the reaction flask and beaker to the funnel.

Problem 43-4. What does the precipitate contain? Verify your answer in the answer section before proceeding with Problem 43-5.

Problem 43-5. How could the *m*-nitroaniline be easily separated from the other materials in the precipitate? The answer is below.

Transfer the precipitate to a 125-ml Erlenmeyer flask. Add 12 ml of concentrated hydrochloric acid and 50 ml of water. Heat the mixture on a steam bath and stir with a glass rod. If the solution is any darker than a strong yellow (both *m*-nitroaniline and sulfur are yellow), clarify with ½ g of activated carbon. Gravity filter the hot mixture and cool the mixture in an ice bath. While the mixture is still in the ice bath, add 25 ml of concentrated ammonium hydroxide and stir the mixture.

$$C_6H_4(NO_2)NH_2 + HCl \rightarrow C_6H_4(NO_2)\overset{+}{N}H_3 + Cl^- \qquad (43\text{-}3)$$

$$C_6H_4(NO_2)\overset{+}{N}H_3 + Cl^- + NH_4^+ + HO^- \rightarrow C_6H_4(NO_2)NH_2\downarrow + NH_4^+ + Cl^- + H_2O \qquad (43\text{-}4)$$

Suction filter and wash the precipitate with two 10-ml portions of cold 10% ammonium hydroxide, using the cold solutions to first rinse the flask in which the acid was neutralized.

Air dry the crystals by pressing them on the filter paper while the suction is on. Dry the product further in a vacuum desiccator for ½ hr or by allowing it to stand in the locker until the next laboratory period. Transfer the product to a tared yield bottle. Weigh the product and determine its melting point. Determine the % yield and hand in the product to your laboratory instructor.

THIN-LAYER CHROMATOGRAPHY (Optional)

Follow the general procedures in Chapter 26 and the more specific procedures at the end of Chapter 36 for preparing, activating, spotting, and developing the plates or strips with the exceptions and changes noted below. Some of the compounds separated in this experiment, while considerably less sensitive to air than the plant pigments, are oxidized by oxygen from the air.

The final heating for activation of the Silica Gel G *plates* should be at 105° for 1 hr (100° for the Eastman Chromagram strips). Use 4 to 6 drops of ethanol to dissolve the compounds. Spot two plates or strips with approximately 1×10^{-5} g of (a) the crude reaction product that precipitated in the 250-ml beaker, (b) the final purified dried product, (c) *m*-dinitrobenzene. Also spot each plate with any likely by-products, except do not spot plate (1) with diamines and do not attempt to put sulfur on either plate.

Develop plate (1) in a chamber containing a solvent of 15 parts dibutyl ether, 15 parts ethyl acetate, and 1 part acetic acid. Develop plate (2) in a chamber containing 4 parts 1-butanol, 1 part acetic acid, and 5 parts water. Remove the plates when the solvent front is 2 to 5 mm from the end.

Dry the plates as instructed at the end of Chapter 36. For both chromatograms use the three successive visualizing procedures described in Chapter 36 except use a long wavelength (~365 mμ ultraviolet light if available) and spray with 0.3% ninhydrin instead of bromocresol green for the third visualization. Heat the sprayed chromatograms at 95 to 100° for 15 min. Note the spots as soon as the plates cool to room temperature.

Problem 43-6. How pure is your product? What are the principal by-products (in the crude material)? Which by-products contaminate the purified product?

EXERCISES

43-1. Suppose phenethylamine ($C_6H_5CH_2CH_2NH_2$) were contaminated with sulfur and 2-phenyl-1-nitroethane. Could the purification procedure described above be used to obtain pure phenethylamine? If not, how would the procedure have to be modified?

44

Properties and Reactions of Amines

Organic Qualitative Analysis VII

INTRODUCTION

Amine functional groups are usually detected by their basic properties. The physical properties of amines are predictable from their structures and, in turn, can give information about their structures. Amines undergo many reactions that are useful as classification tests or for the synthesis of derivatives.

PHYSICAL PROPERTIES

Boiling Points

Primary and secondary amines undergo hydrogen bonding with each other (Fig. 44-1). This causes these amines to have higher boiling points than monofunctional organic compounds of similar carbon skeleton and molecular weight whose molecules are incapable of hydrogen bonding with each other. As a result of this association amines boil higher than hydrocarbons, ethers, esters, and acid halides of similar carbon skeleton and molecular weight.

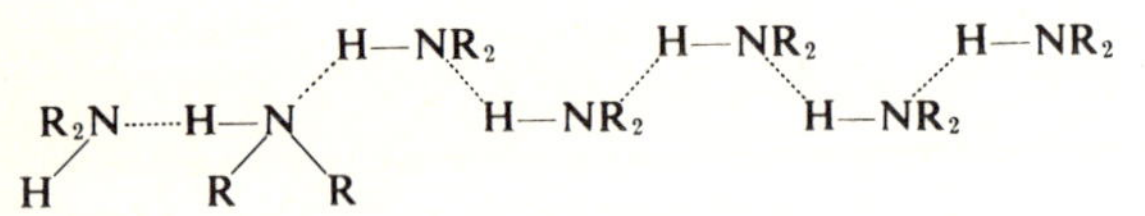

Figure 44-1 Hydrogen bonding in secondary amines.

Since nitrogen is less electronegative than oxygen, the N—H bond is less polar than the O—H bond. This means that the nitrogen atom of an amine is less negative than an oxygen atom of an alcohol and an amino hydrogen is less positive than a hydroxy hydrogen. The hydrogen bonds of amines are between atoms with electrostatic charges weaker than the charges on alcohols and acids. The hydrogen bonds between amines are, therefore, usually weaker than the hydrogen bonds between alcohols (Fig. 44-2). Alcohols have considerably higher boiling points than amines because their stronger hydrogen bonds cause the alcohols to be more strongly associated. The boiling point difference is 50 to 60° for lower molecular weight compounds. This difference decreases as the molecular weights of the amines and alcohols increase; *n*-nonylamine and alcohol have a boiling point difference of only 15°. Table 44-2 compares the boiling points of amines with several groups of compounds. Note that primary and secondary amines

TABLE 44-1 EFFECT OF STRUCTURE ON THE BOILING POINTS OF ISOMERIC AMINES; COMPARISON OF SUCH EFFECTS IN AMINES, ALCOHOLS, AND ESTERS.

Amines and alcohols	B.p. of amines °C	B.p. of alcohol °C	Δ B.p. alcohol−b.p. amine °C	Similar esters	b.p. °C	Δ b.p. amine−b.p. ester, °C
Propyl	49	97	48	Methyl formate	32	17
(Isopropyl	33	83	50			
(N-Methylethyl	36		47			
n-Butyl	78	118	40	Ethyl formate	54	14
Isobutyl	68	108	40			
(sec-Butyl	63	99.5	37	Methyl acetate	57	14
(N-Methylpropyl	64		36			
t-Butyl	46	83	37			
Amyl	103.6	138	34.4	Propyl formate	81	12.6
2-Isoamyl	95	132	37	Isopropyl formate	71	24
2-Methyl-1-butyl	96	129	33			
(2-Pentyl	92	120	28	Ethyl acetate	77	15
(N-Methylbutyl	91		29			
(3-Pentyl	90	116	26	Methyl propionate	79	11
(N-Ethylpropyl	81		35	Methyl propionate	79	2
$(CH_3)_2CH$-CH-CH_3	83	113	30			
N-Methylisobutyl	77		36			
n-Hexyl	130	157	27	Butyl formate	107	23
Isohexyl	125	152.3	27	Isobutyl formate	98	27
2-Ethylbutyl	125	150	25			
Cyclohexyl	134	161	27			
n-Heptyl	155	176	21	Amyl formate	132	23
2-Heptyl	142	158.5	16.5	Butyl acetate	125	17
4-Heptyl	139.5	156	16.5	Ethyl butyrate	120	20
Benzyl	184.5	205	20.5			
2-Phenylethyl	198	221	23			
1-Phenylethyl	186	202	16			
N-Methylbenzyl	181	202	19			
3-Phenylpropyl	222	236	14	Benzyl formate	203.5	18.5
1-Phenylpropyl	208	219	11			

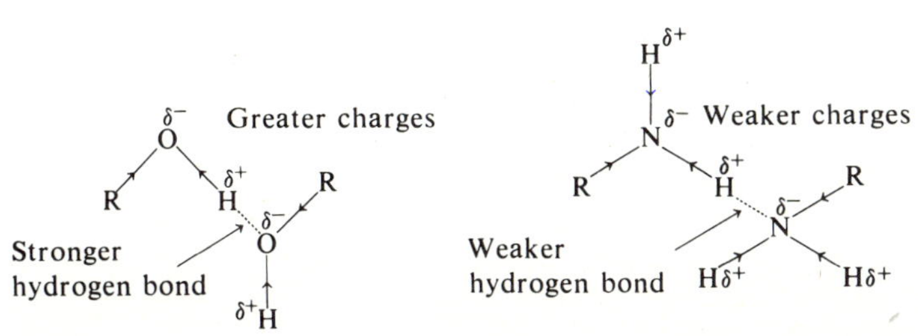

Figure 44-2 Comparison of hydrogen bonding in amines and alcohols.

TABLE 44-2 PHYSICAL PROPERTIES OF AMINES COMPARED WITH HYDROCARBONS AND SOME OF THEIR OXYGEN DERIVATIVES OF SIMILAR MOLECULAR WEIGHT

	B.p. (°C)	M.p. (°C)	D_4^{20}	Solubility (g/100g H_2O, 20°C)	Mol. wt.
Propane	-42	-189.9		sl. sol.	44
Ethylamine	17	-80		very sol.	45
Dimethylamine	7	-96		very sol.	45
Ethanol	78.5	-117.3	0.7893	∞	46
Dimethyl ether	-23.7	-138		very sol.	46
Butane	0	-138	0.622	sl. sol.	58
Isobutane	-12	-159	0.604	sl. sol.	58
Propylamine	49	-83		∞	59
Isopropylamine	33	-101		∞	59
Methylethylamine	36			∞	59
Trimethylamine	3	-117		91	59
1-Propanol	97	-126	0.804	∞	60
2-Propanol	82.5	-86	0.789	∞	60
Methyl ethyl ether	7			very sol.	60
Methyl formate	32	-100	0.974	30	60
Pentane	36	-130	0.626	insol.	72
2-Methylbutane	28	-160	0.620	insol.	72
Butylamine	78	-50	0.740	∞	73
2-Aminobutane	63	-104	0.718	∞	73
Isobutylamine	68	-85	0.728	∞	73
t-Butylamine	46	-67	0.699	∞	73
Diethylamine	56	-50	0.710	∞	73
Methylpropylamine	64			∞	73
Dimethylethylamine	37.5			∞	73
1-Butanol	118	-90	0.896	7.9	74
2-Butanol	100	-114	0.869	12.5	74
t-Butyl alcohol	83	25.5	0.789	∞	74
Diethyl ether	34.6	-116.2	0.708	7.5	74
Methyl propyl ether	39		0.738	sl. sol.	74
Methyl isopropyl ether	32		0.732	sl. sol.	74
Ethyl formate	54	-80	0.923	12	74
Methyl acetate	57	-99	0.927	32	74

boil about 10 to 15° above the corresponding esters and from 80° (methylamine and ethane) to a minimum of 12 to 13° above the hydrocarbons with similar molecular weight and structure. This minimum difference of boiling points between hydrocarbons and amines extends to homologs of considerably higher molecular weight than are shown in Figure 44-3 and Table 44-2. Such a range in difference in boiling points between hydrocarbons and their polar derivatives is to be expected. As the number of methylene groups is increased in a homologous series of polar compounds such as amines, the fraction of hydrocarbon or nonpolar portion of the molecule increases and the fraction of polar portion decreases. The higher homologs, therefore, resemble the hydrocarbon more closely than the lower homologs do. This phenomenon has been encountered before (pp. 218, 268, 269).

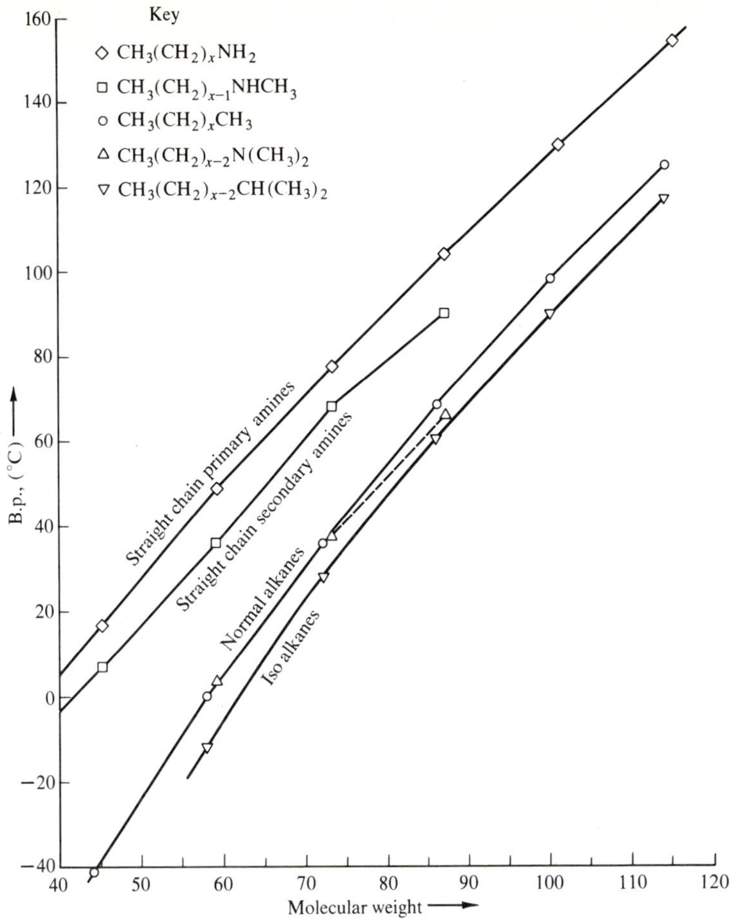

Figure 44-3 Comparison of boiling points of alkylamines and alkanes.

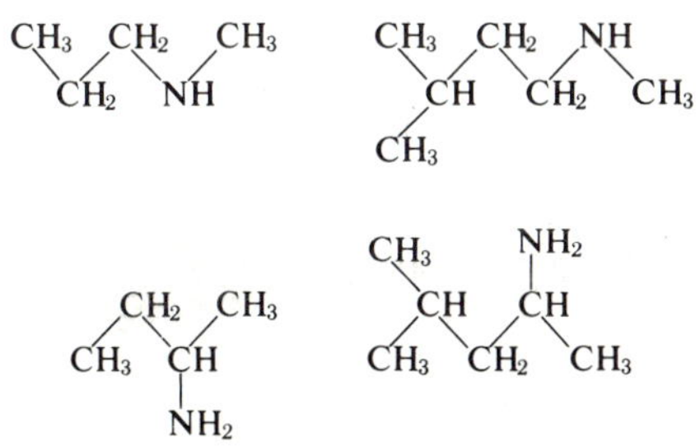

Figure 44-4 Structurally similar primary and secondary amines with similar boiling points.

Table 44-1 shows that boiling points of amines are affected by location of the functional groups on the carbon skeleton, branching of the carbon skeleton, and cyclic and aromatic structures (with the exception noted below) in the same manner as the boiling points of other compounds (pp. 215, 219–223, 248). A secondary amine and its structurally similar primary amine isomer boil at similar temperatures. That is, a secondary amine can be regarded as structurally closest to the primary amine that has the amino group on the same position in the chain that the nitrogen atom occupies in the secondary amine See Figure 44-4 and the bracketed pairs in Table 44-1.

Molecules of tertiary amines cannot form hydrogen bonds with each other. The intermolecular

TABLE 44-3 COMPARISONS OF PHYSICAL PROPERTIES OF PRIMARY AMINES

Amine	B.p. (°C)	M.p. (°C)	D_4^{20}
Cyclohexylamine	134	-17.7	0.867
Aniline	184	-6.2	1.022
Benzylamine	184.5	-1	0.981
o-Toluidine	200	-27	1.001
m-Toluidine	203	-41	0.992
p-Toluidine	200	44	
β-Phenylethylamine	198		0.961
o-Methylbenzylamine	208		0.976
m-Methylbenzylamine	207		0.965
p-Methylbenzylamine	208		0.952
o-Ethylaniline	212	-45	0.985
m-Ethylaniline	211		0.982
p-Ethylaniline	216	-5	0.970
3,5-Dimethylaniline	220	9.8	0.971
2,3-Dimethylaniline	221	3.5	0.993
2-Phenylpropylamine	210		0.978
3-Phenylpropylamine	222		0.953
4-Isopropylaniline	225		0.953
4-Propylaniline	225		
2,4-Dimethylbenzylamine	218		0.978
2,4,6-Trimethylaniline	230		

TABLE 44-4 COMPARISONS OF PHYSICAL PROPERTIES OF SECONDARY AMINES

Amine	B.p.(°C)	M.p.(°C)	D_4^{20}
N-Methylbenzylamine	181		0.941
N-Ethylaniline	205		0.962
N-Methyl-o-toluidine	207		0.972
N-Methyl-m-toluidine	206.5		
N-Methyl-p-toluidine	210		
N-Ethylbenzylamine	199		0.931
N-Propylaniline	222		
N-Ethyl-m-toluidine	219		
N-Cyclohexylaniline	278	16	1.0155
Diphenylamine	302	52.8	

attractions of tertiary amines are weak dipole-dipole, dipole-induced dipole, and induced dipole-induced dipole forces. Primary and secondary amines also have these attractive forces but have in addition the stronger hydrogen bond attractions. Tertiary amines, therefore, have fewer and weaker intermolecular attractive forces and boil and melt lower than primary and secondary amines of similar structure and molecular weight. In their structure and intermolecular attractive forces tertiary amines resemble ethers. Tertiary amines and ethers of similar structure and molecular weight have almost identical boiling points (see Table 44-2). Amines with cyclic structures boil higher than noncyclic amines of similar molecular weight, and amines with aromatic rings boil higher than amines of similar molecular weight without aromatic rings. See Table 44-1 and the two examples in Tables 44-3 and 44-4.

TABLE 44-5 COMPARISONS OF PHYSICAL PROPERTIES OF TERTIARY AMINES

Amine	B.p.(°C)	M.p.(°C)	D_4^{20}
N,N-Dimethylbenzylamine	181		0.900
N,N-Dimethyl-*o*-toluidine	185	-60	0.929
N,N-Dimethyl-*m*-toluidine	212		0.941
N,N-Dimethyl-*p*-toluidine	210		0.929

Problem 44-1. Comparison of Tables 44-3, 44-4, and 44-5 shows that aromatic amines (the amino group is attached directly on the aromatic ring) boil about 15 to 30° higher than aralkylamines (such as $ArCH_2N$—) of identical molecular weight. The aralkylamines have an aromatic ring but do not have the amino group attached directly on it. Suggest an explanation for this boiling point difference.

Problem 44-2. Note that the only exception listed in the tables in this chapter to the observation in Problem 44-1 is N,N-dimethyl-*o*-toluidine. The boiling point of this *ortho* isomer is only 4° above that of N,N-dimethylbenzylamine although N,N-dimethyl-*m*- and *p*-toluidine boil approximately 30° above the tertiary benzylamine, and the *ortho* derivatives of aniline listed in Tables 44-3 and 44-4 have boiling points that do not differ significantly from their *meta* and *para* isomers. Suggest an explanation for this.

Solubilities

Amines show the same *general* solubility characteristics in *nonreactive* solvents that have been encountered with other groups of organic compounds having one polar functional group. These are:

a. High solubility of all monofunctional amines (except possibly those of *very* high molecular weight) in nonpolar solvents such as ether, benzene, and chloroform.
b. Complete miscibility of low molecular weight amines in polar solvents such as water.
c. Decreasing solubility in nonreactive polar solvents such as water with increasing molecular weight in a homologous series.
d. Higher water solubility of branched and cyclic amines than straight chain amines of similar molecular weight.
e. Aromatic rings have about the same effect on water solubility as 3 to 4 methylene groups. However, this rule does not apply if the amino group is attached directly to the aromatic ring. This exception has not been observed in other groups of compounds you have studied thus far.

In addition to increasing with the magnitude of the electrostatic charges (pp. 330 and 331) the strength of a hydrogen bond increases with the basicity of the acceptor atom and the acidity of the hydrogen atom (see Table 44-6). Thus, although amines form weak hydrogen bonds with each other, the hydrogen bond between an amino nitrogen and a hydroxy hydrogen is one of the strongest types of hydrogen bonds. Amine molecules, therefore, form stronger hydrogen bonds with water or alcohol molecules than with each other. Because of formation of these stronger hydrogen bonds, more energy is re-

TABLE 44-6 ENERGIES OF HYDROGEN BONDS

Bond	Examples	Average strength (K cal/mole)
-O....H-O<	Water, alcohols, acids, and mixtures of	6
>N....H-O<	Solutions of amines in H_2O or alcohols	7
>N....H-N<	Primary and secondary amines	2-4
-O....H-N<	Solutions of amines in H_2O, alcohols, or ethers	2-3
-F....H-N<	Crystalline NH_4F	5
-F....H-F	Hydrogen fluoride	6-8
F^-...$\overset{+}{H}$...F^-	KHF_2	27
π....H-O cloud	Solutions of alcohols and arenes	3
H-C≡N....H-CN		4

leased when amines dissolve in water than when most other polar compounds do. This extra energy is available to overcome other forces preventing solution. The limit of water solubility (Class S) of noncyclic amines by the solubility classification scheme is 6 carbon atoms rather than the 4 to 5 carbon atom limit of noncyclic alcohols.

TABLE 44-7 COMPARISONS OF SOLUBILITIES OF CYCLIC ALIPHATIC AND AROMATIC AMINES

ALIPHATIC	Mol. wt.	Solubility H_2O (g/100g 20°C)	AROMATIC	Mol. wt.	Solubility H_2O (g/100g 20°C)
N-Methylpyrrolidine	85	∞	Aniline	93	3.7
Piperidine	85	∞	N-Methylaniline	107	0.1
Cyclohexylamine	99	sl. sol.	N,N-Dimethylaniline	121	1.4
N-Methylpiperidine	99	very sol.	o-Toluidine	107	1.7
Benzylamine	107	∞	p-Toluidine	107	0.7
α-Phenylethylamine	121	4.2	p-Ethylaniline	121	0.4

Problem 44-3. Aromatic amines are somewhat *less* soluble in water than similar oxygen derivatives even though all alkylamines and aralkylamines are more soluble in water than similar alcohols and ethers (Table 44-7). Why?

Problem 44-4. Tertiary amines are significantly more soluble in water than primary and secondary amines of comparable structure and similar or higher molecular weight. Compare N-methylpiperidine with cyclohexylamine, N,N-dimethylaniline with N-methylaniline or *p*-ethylaniline (Table 44-7), triethylamine with hexylamine and diisopropylamine or di-*n*-propylamine, and N,N-dimethylbutylamine 2-ethyl-1-aminobutane and the dipropylamines (Table 44-2). Ethers on the other hand are as or slightly

TABLE 44-8 DIPOLE MOMENTS OF AMINES

PRIMARY	μ in D*	SECONDARY	μ in D*	TERTIARY	μ in D*
n-Propylamine	1.30	Dimethylamine	1.0	Trimethylamine	0.7
n-Butylamine	1.37	Diethylamine	1.0		
Isobutylamine	1.27				
Hexylamine	1.33	Dipropylamine	1.0	Triethylamine	0.8
Aniline	1.56	N-Ethylaniline	1.65	N,N-Dimethyl-aniline	1.57

*At 20°C

TABLE 44-9 DIPOLE MOMENTS OF ALCOHOLS AND ETHERS

ALCOHOLS	μ in D*	ETHERS	μ in D*
1-Propanol	1.72	Methyl ethyl ether	1.22
1-Butanol	1.81	Diethyl ether	1.30
2-Butanol	1.78	Methyl propyl ether	1.24
1-Pentanol	1.66	Methyl butyl ether	1.25
3-Methyl-1-butanol	1.78	Methyl propyl ether	1.16
2-Methyl-2-butanol	1.72	Methyl t-butyl ether	1.22
1-Hexanol	1.65	Ethyl butyl ether	1.22
p-Cresol	1.55	Anisole	1.25

*At 20°C.

less soluble than alcohols of similar structure and molecular weight. Compare ethyl ether with the butyl alcohols (Table 44-2) and the dipropyl ethers with the hexyl alcohols. Explain this divergence in water solubility trends of nitrogen and oxygen derivatives.

BASICITY OF AMINES

Base Strength

Because of the free isolated electron pair on an atom of relatively low electronegativity amines are basic. Aliphatic amines are slightly more basic than ammonia, and aromatic amines are significantly less basic. The amino electron pairs of amines with strong electron-withdrawing groups and of diarylamines do not react readily enough with acids to impart to these amines easily recognized basic properties.

Solubility Class B

All unknowns insoluble in water should be tested for solubility in 5% aqueous hydrochloric acid whether or not they are soluble in dilute base. Compounds that are insoluble in water and dilute sodium hydroxide and soluble in 5% hydrochloric acid are basic and in solubility class B. Compounds that are insoluble in water but soluble in both dilute acid or dilute base are in Solubility Class A_1 (B) or A_2 (B). Almost all organic compounds (and all an undergraduate is likely to encounter) in Class B have an amino group or an N-alkylamido group.

Solubility Classes of Amines

Low molecular weight amines are soluble in water and ether, and their aqueous solutions are strongly basic to Hydrion paper. All other basic water-soluble compounds are insoluble in ether. The borderline between Class S and Class B for aliphatic amines is from 6 to 10 carbon atoms. No aromatic amines are in Class S_1; a very few aromatic amines with other polar groups that are hydrogen bond acceptors are in Class S_2. Diarylamines and di- and trinitroarylamines are too weakly basic to be soluble in dilute hydrochloric acid and are in Class M. Trichloroaniline and *o*-nitroaniline are of borderline solubility in dilute aqueous acid and will be placed in Classes B or M depending on exactly how their solubility is determined.

Solubility Classes of Amides

The borderline of water solubility for noncyclic amides other than N,N-disubstituted amides is at 5 to 6 carbon atoms. The borderline of water solubility for N,N-disubstituted amides is at 7 carbon atoms. Note the similarity to water solubility of primary and secondary amines relative to tertiary amines. Branching, cyclic, and aromatic structures have the same effect on the water solubility of amides as on the water solubility of other groups of compounds that have functional groups capable of hydrogen bonding with water. The water solubility of amides is also discussed on pages 270 and 271.

Low molecular weight water-soluble amides (other than those that are N,N-disubstituted, see Problem 35-1) are usually so polar that they show borderline solubility in ether. Amides such as propionamide and acetamide are borderline between Classes S_1 and S_2. Care must be taken when testing the solubility of many solids including amides in aqueous solutions or ether that a slow rate of solution is not confused with insolubility; i.e., allow enough time so that an unfavorable rate is not confused with an unfavorable equilibrium.

Most N,N-dialkylamides and a very few N-alkylamides that are water-insoluble are basic enough to dissolve in dilute acid. N-Arylamides and amides with several electron-withdrawing groups are never basic enough to be in Class B unless they are polyfunctional compounds with other basic groups. Imides or N,N-diacylamides and sulfonamides, if water-insoluble, are usually soluble in aqueous sodium hydroxide and insoluble in aqueous sodium bicarbonate. Most water-insoluble unsubstituted, N-monosubstituted, and N-arylsubstituted carboxamides are insoluble in either dilute base or dilute acid and are in Solubility Class M.

Problem 44-5. Predict the solubility class of the following compounds: (a) 1-aminoheptane, (b) 1-azaheptane [$\overline{\text{(CH}_2)_6\text{NH}}$], (c) N,N-dimethylbenzylamine, (d) 3,4-dinitroaniline, (e) sulfanilamide, (f) acetanilide, (g) triethylamine, (h) di-*p*-tolylamine, (i) 3,5-dichloro-2-nitroaniline, (j) di-*n*-propylamine, (k) 1-(dimethylamino)pentane, (l) 2,4-dinitroacetanilide, (m) acetamide, (n) *m*-toluidine, (o) N,N-dimethyl-*p*-ethylbenzylamine, (p) N,N-dimethylbenzamide.

Problem 44-6. Verify your answers to Problem 44-5 before doing this problem. Predict the solubility classes of the following compounds: (a) N-methylbenzamide, (b) N-methyl-N-ethylbenzylamine, (c) N,N-diethylbutyramide, (d) N,N-dimethyl-2,4-dinitrobenzamide, (e) N-methylacetanilide, (f) valeramide (g) di-*n*-isopropylamine, (h) 2-amino-2,3-dimethylbutane, (i) triphenylamine.

Problem 44-7. Verify your answer to Problem 44-6 before doing this problem. Predict the solubility classes of the following compounds: (a) N,N-diethylbenzamide, (b) N-methylacetamide, (c) formanilide, (d) N-methylcyclohexylamine, (e) tri-*n*-propylamine, (f) *p*-chloroaniline, (g) 2,4,6-tribromoaniline,

(h) N-(2,4-dinitrophenyl)-*p*-nitrobenzenesulfonamide, (i) N,N-diphenylbenzenesulfonamide, (j) 2-(methylamino)-2-methylbutane.

REACTIONS OF AMINES USED FOR CLASSIFICATION TESTS

All Class B, Class S_1w, Class S_1b, and all other compounds thought to be amines should be tested with both benzenesulfonyl chloride and nitrous acid. The results of one test verify the results of the other, and, although both of these tests serve to distinguish among primary, secondary, and tertiary amines, the two tests give complementary rather than duplicate information about many amines. Many of the classification reactions (those with acid chlorides) give crystalline solids that are useful as derivatives.

Reaction of Amines with Benzenesulfonyl Chloride and Aqueous KOH. The Hinsberg Test

$$RNH_2 + C_6H_5SO_2Cl + 2HO^- \rightarrow C_6H_5SO_2\bar{N}R + Cl^- + 2H_2O \quad (44\text{-}1)$$

$$R_2NH + C_6H_5SO_2Cl + HO^- \rightarrow C_6H_5SO_2NR_2\downarrow + Cl^- + H_2O \quad (44\text{-}2)$$

$$R_3N + C_6H_5SO_2Cl + HO^- \rightarrow \text{no reaction} \quad (44\text{-}3)$$

Discussion. All primary and secondary amines react with benzenesulfonyl chloride to yield N-substituted or N,N-disubstituted benzenesulfonamides. Tertiary amines do not react with sulfonyl chlorides. Since sulfonamides from secondary amines have no amido hydrogen, they cannot react with aqueous alkali. Benzenesulfonamides of most primary amines react with aqueous potassium hydroxide sufficiently to dissolve in this reagent. The principal exceptions are higher molecular weight cyclic and highly branched aliphatic amines such as cyclooctylamine, cyclononylamine, and 2-amino-2,4,4-trimethylpentane. The precipitate from the Hinsberg test is tested for solubility in 5% hydrochloric acid to aid in distinguishing tertiary amines from precipitated sulfonamides. If no solid precipitates in the Hinsberg test, the reaction mixture is acidified to distinguish between N-monosubstituted benzenesulfonamides and water-soluble tertiary amines.

The observation of whether any water-insoluble material from the reaction floats or sinks is often helpful. Tertiary amines are usually less dense than the 10% potassium hydroxide solution used for the test, and all sulfonamides are much denser. Such observations should be interpreted carefully because the presence of halogen atoms and other structural features may cause tertiary amines to be more dense than water (pp. 222, 223), and the benzenesulfonyl chloride may dissolve the tertiary amine and form an oil at the bottom of the test tube.

When a precipitate is obtained from the Hinsberg test, it should be recrystallized, dried, and the melting point determined. This not only aids in determining if a reaction has occurred but also serves as synthesis of a derivative. Often only the melting point of the benzenesulfonamide, the physical properties of the amine, and the results of one or two other classification tests are needed to identify a primary or secondary amine. If one is certain there will be no need to determine the melting point of any precipitate, the quantities given in the experimental section below should be reduced by half.

Experimental Procedure. In a test tube place 0.5 ml of amine if the boiling point is less than 90°, 0.6 ml for amines boiling at 91 to 130°, 0.8 ml for amines boiling at 131 to 170°, 1.0 ml for amines boiling at 171 to 200°, 1.2 ml or 1.2 g for amines boiling above 201° or for solid amines that decompose before boiling. Place the tube in a water bath at 10 to 18°; add 12 ml of 10% aqueous potassium hydroxide and 1.0 ml of benzenesulfonyl chloride. Measure the quantities of amine and sulfonyl chloride carefully to the nearest 0.1 ml; an excess of benzenesulfonyl chloride may make interpretation of the test difficult. Stopper the test tube, shake it vigorously, and test with Hydrion paper. If not strongly basic, add more potassium hydroxide solution, and shake the tube again. Return the test tube to the cold water bath. If no precipitate appears, remove the stopper and place the test tube in a warm water bath (40 to 45°) for 5 min and stir the mixture vigorously with a glass rod.

Cool the mixture in an ice bath and isolate any precipitate by suction filtering or centrifuging. If there is an oil, remove the aqueous phase from it by carefully decanting the first 7 to 8 ml of water solution and using a capillary pipet for the remainder. If a precipitate or water-insoluble oil is obtained, try to dissolve it in 5% hydrochloric acid.

$$R_3N + H^+ \rightarrow R_3NH^+ \qquad (44\text{-}4)$$

If only a single aqueous phase results from the reaction, place the test tube in an ice bath, add a few pieces of ice, and then add chilled hydrochloric acid[1] a few drops at a time until the mixture is strongly acid to Hydrion paper. Keep the test tube in the ice bath and scratch the tube with a glass rod.

$$C_6H_5SO_2\bar{N}R + H^+ \rightarrow C_6H_5SO_2NHR\downarrow \qquad (44\text{-}5)$$

If an oil insoluble in 5% hydrochloric acid is the only organic product observed from the initial reaction, acidify the aqueous layer decanted from the oil with chilled hydrochloric acid as described above.

Reaction of Amines with Nitrous Acid

$$RNH_2 \xrightarrow[0^\circ]{HNO_2} N_2\uparrow + R^+ \qquad (44\text{-}6a)$$

$$R^+ \begin{cases} \xrightarrow{H_2O} ROH \\ \xrightarrow{(-H^+)} \text{alkenes} \\ \xrightarrow{Cl^-} RCl \\ \xrightarrow{NO_2} RNO_2 \\ \xrightarrow{\;\;\sim\;\;} R'^+ \end{cases} \qquad (44\text{-}6b)$$

$$R'^+ \begin{cases} \xrightarrow{H_2O} R'OH \\ \xrightarrow{(-H^+)} \text{alkenes} \\ \xrightarrow{Cl^-} R'Cl \\ \xrightarrow{NO_2^-} R'NO_2 \end{cases} \qquad (44\text{-}6c)$$

$$R_2NH \xrightarrow[0^\circ]{HNO_2} R_2NNO \qquad (44\text{-}7)$$

$$R_3N \xrightarrow[0^\circ]{HNO_2} \text{no reaction} \qquad (44\text{-}8)$$

$$ArNH_2^+ \xrightarrow[0^\circ]{HNO_2} ArN_2^+ \qquad (44\text{-}9)$$

$$ArN_2^+ \xrightarrow[\Delta]{HNO_2} ArOH + N_2\uparrow \qquad (44\text{-}10)$$

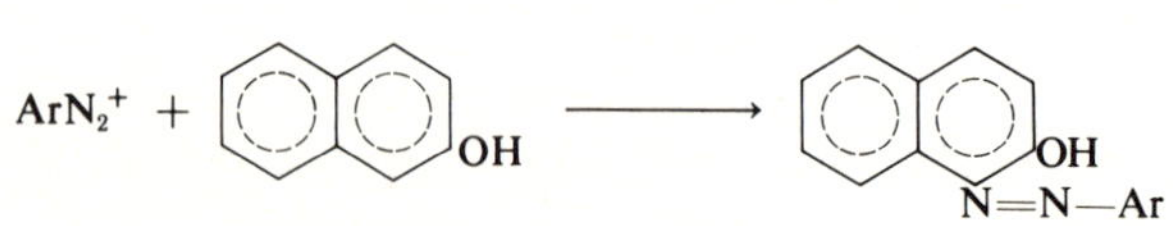

(44-11)

$$ArNHR \xrightarrow[0^\circ]{HNO_2} ArNR \qquad (44\text{-}12)$$
(with NO on N of ArNR)

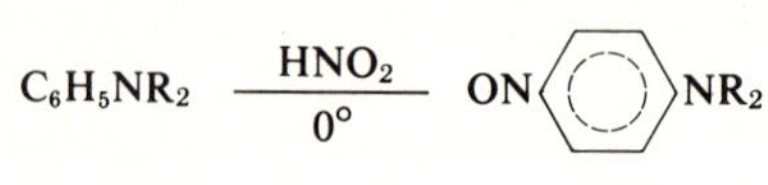

(44-13)

[1] Made by mixing equal volumes of crushed ice and concentrated hydrochloric acid in a test tube in an ice bath.

Discussion. Primary aliphatic amines react immediately with nitrous acid at 0° to give nitrogen, visible as colorless gas bubbles, and typical carbonium ion decomposition products. Secondary aliphatic and aromatic amines react to give yellow N-nitroso derivatives. The N-nitroso derivatives are insoluble in dilute aqueous acid and form yellow solids or oils in the test mixture.

Aliphatic tertiary amines usually do not react with nitrous acid. A few aliphatic amines have chloride or nitrite salts that are insoluble in the reaction mixture. These salts are usually colorless and react with strong base to regenerate the amine.

$$R_3N + H^+ + NO_2^- \rightarrow R_3\overset{+}{N}H\ NO_2^- \tag{44-14}$$

$$R_3\overset{+}{N}H\ NO_2^- + HO^- \rightarrow R_3N + H_2O + NO_2^- \tag{44-15}$$

The electrophilic agent of nitrous acid is believed to be the nitrosonium ion, NO^+. This ion attacks the electron pair of the amino group.

$$R_2\ddot{N}H \xrightarrow{NO^+} R_2\overset{\overset{H}{|}}{N^+}\text{—}NO \xrightarrow{(-H^+)} R_2N\text{—}NO \rightarrow \text{further reaction if one of the R's is H,} \tag{44-16}$$

where R is H, alkyl, or aryl. Since it is an electrophilic reagent, it also attacks aromatic rings. It is a much weaker electrophilic reagent than nitronium (NO_2^+), bromonium (Br^+), or even carbonium ions and as a result attacks only strongly activated aromatic rings such as phenols and tertiary aromatic amines (Eq. 44-13). These ring-substituted nitroso tertiary amines are usually blue or green.

Primary aromatic amines do not evolve nitrogen rapidly when treated with nitrous acid below room temperature. They do, however, rapidly evolve nitrogen when warmed with nitrous acid. Addition of the cold reaction mixture to an alkaline solution of phenol also distinguishes between aromatic and aliphatic primary amines. Aliphatic diazonium salts decompose too rapidly to react with the phenol; aromatic diazonium salts are fairly stable at low temperature and couple with phenols to give brilliant colors, usually orange or red.

$$RNH_2 \xrightarrow{HNO_2} [R\overset{\overset{H}{|}}{N}NO_2] \rightarrow [RN_2^+] \xrightarrow{\text{rapid}} R^+ + N_2 \tag{44-17}$$

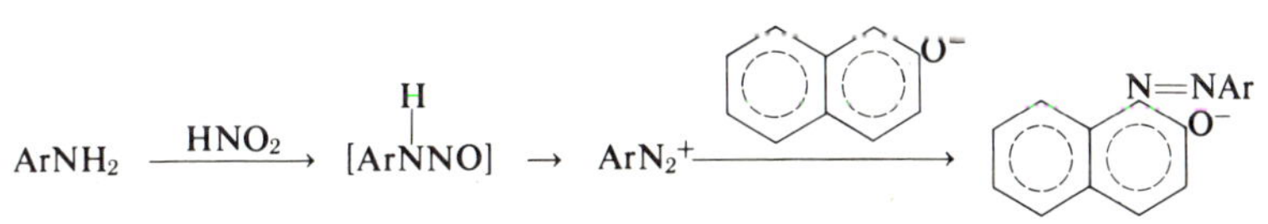

(44-18)

Experimental Procedure. Add 6 ml of ice and water to 4 ml of concentrated hydrochloric acid in a test tube in an ice bath. Dissolve 1 g of amine in the mixture. Add 1 ml of 15% aqueous sodium nitrite to the cold mixture. Add more of the nitrite solution a few drops at a time until the cold mixture gives a blue color with starch iodide paper (a positive test for nitrous acid) after stirring. Do not use excess sodium nitrite solution.

The immediate evolution of a *colorless* gas is a positive test for primary amines. A brown or red gas is due to the presence of oxides of nitrogen and is usually caused by adding excess sodium nitrite. This is to be avoided as such gases mask the evolution of nitrogen. If no nitrogen evolution is observed, add 2 to 3 ml of the cold nitrous acid solution to a mixture of 0.1 to 0.2 g of β-naphthol, 5 ml of 5% sodium hydroxide, and 2 to 3 g of ice. A brilliant yellow, orange, or red voluminous precipitate is a positive test for a primary aromatic amine. If no nitrogen evolved from the cold solution, warm it in a water bath at 45 to 50° for 3 min with stirring. Try this test on butylamine, triethylamine, aniline, N-methylaniline, and N,N-dimethylaniline.

Problem 44-8. Why is it important not to heat the nitrous acid solution too strongly when testing for primary aromatic amines?

Reaction of Amines with Bromine Water

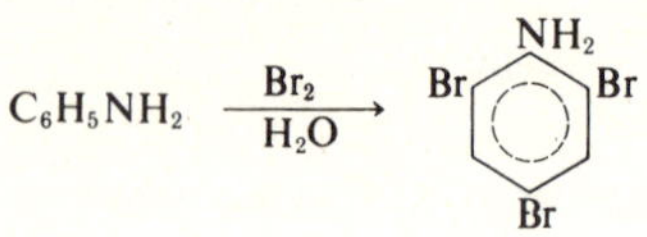

(44-19)

Discussion. Aromatic amines quickly decolorize bromine water and give yellow precipitates that can often be used as derivatives. Many aromatic amines give negative tests because they are too insoluble in water or because they are made inactive to bromine water by several electron-withdrawing groups.

Since most amines and many other compounds decolorize bromine water, the test should be considered positive only if a white or yellow precipitate is obtained and the bromine water is decolorized. Benzylamines, for example, are oxidized by bromine to benzonitriles.

$$ArCH_2NH_2 + 2Br_2 \rightarrow ArCN + 4HBr \qquad (44\text{-}20)$$

Phenols are the only other group of compounds likely to be encountered by the undergraduate student that gives a precipitate with bromine water, and phenols are not likely to be confused with aromatic amines.

If the amine tested is a solid, the melting point of any precipitate should be determined because small amounts of the unreacted amine could precipitate under the conditions described below. If the precipitate is a polybromo derivative, a melting point may aid in identification of the amine. The reaction of amines with bromine in carbon tetrachloride is discussed in the section on interference with other classification tests.

Experimental Procedure. Stir approximately 0.3 g of unknown with 7 to 8 ml of water in a 10-cm test tube in a water bath at 35 to 40°. Transfer most of the aqueous solution with a warm pipet to a warm test tube. If the unknown dissolves *completely*, the compound may be tested in the original test tube without transfer. Immediately add 6 to 7 drops of bromine water and stir. If the bromine water is decolorized, continue adding it a few drops at a time until the bromine color persists after stirring. Cool the tube in a water bath at 15 to 20°. Try this test on butylamine, aniline, and *p*-toluidine. Repeat the test on aniline using cold 5% hydrochloric acid as a solvent instead of warm water. If a precipitate is obtained, wash with water, dry, and determine its melting point.

Problem 44-9. Complete Table 44-10 for all compounds shown. The correct notations are demonstrated with *n*-butylamine as an example.

TABLE 44-10 CLASSIFICATION TESTS OF REPRESENTATIVE AMINES

Type	Water solubility	Example	Reaction with $C_6H_5SO_2Cl$ and aq. KOH	Behavior of product of Hinsberg test with aq. HCl	Reaction with HNO_2: evolution of N_2	Reaction with HNO_2: other	Reaction with bromine water
Aliphatic primary	sol.	n-butyl-amine	dissolves	ppt. from soln.	evoln. when cold	none	no ppt.
High mol. wt. cyclic aliphatic primary	insol.	cyclononyl-amine	ppt.	ppt. doesn't dissolve No ppt. from soln.			
Aliphatic primary	insol.	n-heptyl-amine					
Secondary aliphatic	sol.	diethylamine				yellow ppt.	
Secondary aliphatic	insol.	dibutylamine					
Tertiary	sol.	triethylamine					
Tertiary	insol.	tributylamine					
Aromatic primary	insol.	(1)aniline			evoln. only when warm	red ppt. with β-naphthol	
		(2)2,4-dinitro-aniline					
Aromatic secondary	insol.	N-methyl-aniline					
Aromatic tertiary	insol.	N,N-dimethyl-aniline					

Reaction of Amines with Acyl Chlorides

Discussion. Primary and secondary amines react with acyl chlorides to give amides. These amides usually precipitate as solids and differ in odor and appearance from the amines. The reaction with acetyl chloride is usually very exothermic.

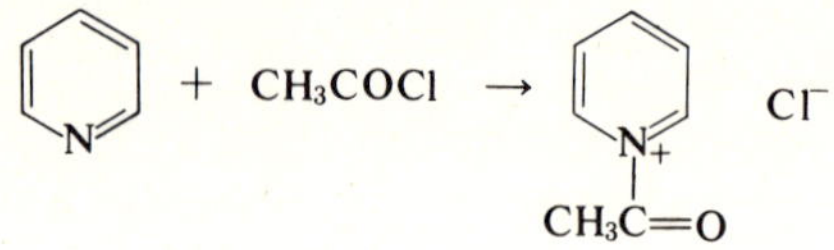
(44-21)

Most tertiary amines, however, also react exothermically with acetyl chloride to give quaternary ammonium derivatives that are hydrolyzed by water to the acid and the amine salt.

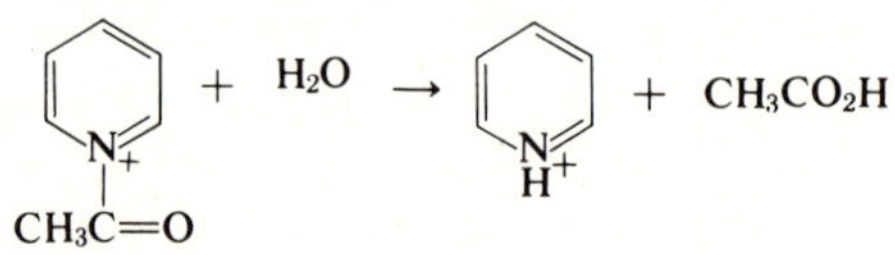
(44-22)

These quaternary ammonium compounds react with other compounds very readily. They are the active acylating agents when pyridine is used as a catalyst for the reaction of acyl halides with alcohols (p. 228).

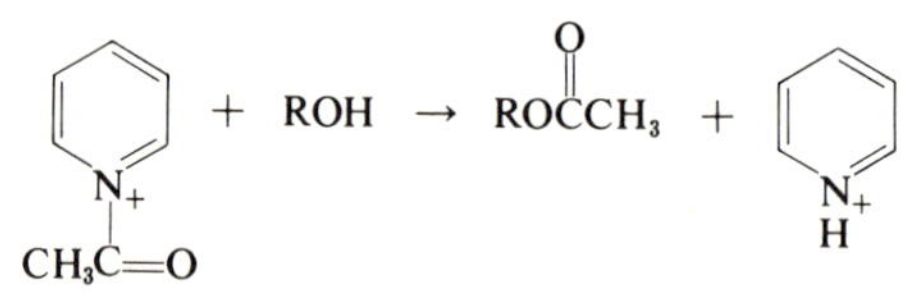
(44-23)

The two methods given below are recommended principally as classification tests although solids are often obtained that serve as derivatives. The methods described later in the chapter are usually superior for the preparation of pure carboxamides for derivatives.

$$CH_3COCl + RNH_2 \rightarrow CH_3CON(H)R + HCl, \quad (44\text{-}24)$$

$$CH_3COCl + R_2NH \rightarrow CH_3CONR_2 + HCl. \quad (44\text{-}25)$$

Experimental Procedure. Acetyl chloride: Put 1.0 ml of the compound to be examined in a 7.5-cm test tube, carefully add 10 drops (about 0.5 ml) of acetyl chloride, and note what happens. Shake the tube, allow it to stand for a few minutes, and pour into 6 ml of 5% sodium bicarbonate solution in a 15-cm test tube. Try this test on aniline, butylamine, diethylamine, and N,N-dimethylaniline. If any precipitate is obtained, wash it with cold water, dry it, and determine its melting point.

Benzoyl chloride (Schotten-Baumann reaction):

$$C_6H_5COCl + R_2NH \rightarrow C_6H_5CONR_2 + HCl \text{ (or } R_2\overset{+}{N}H_2Cl^-) \quad (44\text{-}26)$$

Put 1 ml of the compound to be examined in a 7.5-cm test tube, add 5 ml of 10% aqueous sodium hydroxide and 8 drops (about 0.4 ml) of benzoyl chloride, and shake the stoppered tube vigorously. Place the test tube in an ice water bath and adjust the pH from 8 to 9 by adding concentrated hydrochloric acid. Try this test on aniline and diethylamine. Wash any precipitate obtained with 5% aqueous Na_2CO_3, 5% aqueous HCl, nnd cold water; dry, and determine its melting point.

Problem 44-10. What is the purpose of the aqueous sodium hydroxide in the Schotten-Baumann test? Why do amines and alcohols react with the benzoyl chloride before the sodium hydroxide does?

Problem 44-11. Compare the instructions given above for the Schotten-Baumann reaction with those given on page 228 when amines were not among the possible unknowns. What changes have been made? Why were they made?

Reaction of Amines with Sodium

$$2R_2NH + 2Na \rightarrow 2R_2N^- + 2Na^+ + H_2\uparrow \qquad (44\text{-}27)$$

Like all compounds with a hydrogen atom bonded to an electronegative atom, primary and secondary amines react with sodium to give hydrogen gas. The instructions and comments on pages 227 and 228 should be followed for the test and interpretation of the results. Tertiary amines may give false positives because they often are contaminated with primary or secondary amines or water.

Reaction of Amines with Base and Chloroform

$$RNH_2 + CHCl_3 + 3HO^- \rightarrow RNC + 3Cl^- + 3H_2O \qquad (44\text{-}28)$$

Discussion. Chloroform reacts with primary amines in the presence of base to give isocyanides. These isocyanides are readily detected by their odor unless they are made from amines of very low volatility. Secondary and tertiary amines do not form isocyanides. Because this test is so sensitive, secondary and tertiary amines sometimes appear to give positive tests because they are often contaminated with small amounts of primary amines.

Experimental Procedure. Add 1 drop or 60 to 70 mg of solid amine to 2 to 3 drops of chloroform. Add 1 ml of 5% alcoholic potassium hydroxide and warm for 2 to 3 min on a water bath at 45 to 50° while stirring with a glass rod. Smell cautiously (p. 1). The odor of isocyanides is difficult to describe; adjectives such as *disagreeable* or *nauseating* give some idea. Once smelled, the odor of isocyanides is not readily forgotten. Try this test on aniline and diethylamine.

REACTION OF AMINES THAT INTERFERE WITH OTHER CLASSIFICATION TESTS

Bromine and Carbon Tetrachloride[2]

Almost all amines decolorize bromine in carbon tetrachloride. Equations 44-20, 44-28, 44-29, and 44-30 are examples of such reactions. Although these are mostly substitution type reactions, evolution of hydrogen bromide is often not observed because: (a) excess amine reacts with the hydrogen bromide; (b) perbromides may be formed (Eq. 44-28); and (c) mono and dibromo aromatic amines react with the first mole of HBr formed (Eq. 44-29).

$$R_2NH + Br_2 \rightarrow R_2NBr + HBr \qquad (44\text{-}29)$$

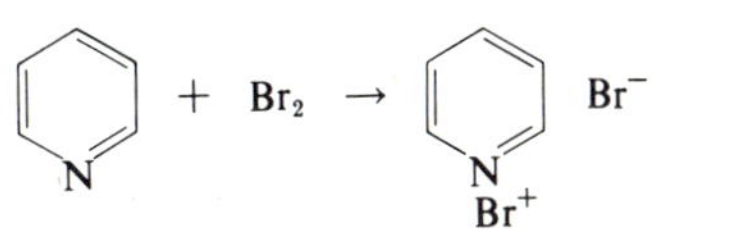
(44-30)

$$C_6H_5NH_2 + Br_2 \rightarrow BrC_6H_4\overset{+}{N}H_3\ Br^- \qquad (44\text{-}31)$$

Dilute Aqueous Potassium Permanganate, Baeyer Test[3]

Aromatic amines that do not have strong electron-withdrawing groups are oxidized by cold, neutral permanganate and give a positive Baeyer test. The first products of oxidation are quinones.

[2] See page 98.
[3] See page 99.

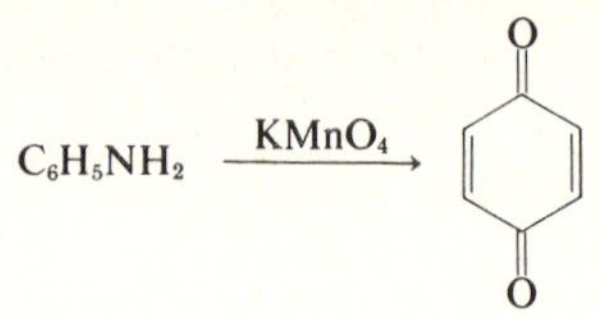

(44-32)

REACTIONS OF AMINES USED FOR THE PREPARATION OF DERIVATIVES

Although three of the classification tests may give satisfactory precipitates that can be used as derivatives, only the Hinsberg test consistently yields products pure enough to be used for derivatives.

Arenesulfonamides

Wash the precipitate obtained from the Hinsberg test (p. 339) with ice water and recrystallize it from ethanol. *p*-Toluenesulfonamides are also prepared by the same procedure. Add 1.5 to 1.7 g of *p*-toluenesulfonyl chloride; note that unlike benzenesulfonyl chloride, it is a solid.

Acetamides

The precipitate from the acetyl chloride test should be washed with ice water and recrystallized from ethanol. If this method does not give a good yield of pure amide, acetamides of water-insoluble amines may be prepared as follows.

Dissolve 1.0 g of amine in 40 ml of 5% hydrochloric acid in a 125-ml Erlenmeyer flask. Add saturated sodium carbonate solution a drop at a time with stirring until the solution becomes turbid, showing that the amine has begun to precipitate from solution. Redissolve the amine, using the minimum amount of 5% hydrochloric acid. Chill the solution in an ice bath and add a few pieces of ice. Add 10 ml of acetic anhydride, stir slightly, immediately add 8 ml of saturated sodium acetate solution, and stir the mixture vigorously.

If no amide precipitates, chill the mixture in an ice bath. If still no amide precipitates, leave the mixture in the refrigerator until the next laboratory period. Wash the precipitate with cold water and recrystallize it from ethanol.

Benzamides

Wash the precipitate from the Schotten-Baumann reaction (p. 344) with cold water and recrystallize it from ethanol.

Polybromo Derivatives of Aromatic Amines

If sufficient precipitate was obtained from the classification test with bromine water (p. 342), wash it with cold water and recrystallize it from ethanol. If this does not give a satisfactory derivative, the following procedure may be used.

Dissolve 3.0 g of potassium bromide in 20 ml of water. Carefully add 0.6 ml of bromine with agitation. *Dissolve* 0.3 g of unknown in 6 ml or less of water, methanol, or dioxane. Add just enough of the brominating solution to the dissolved amine to give a yellow color that persists after agitation (about 5 to 7 ml). Add 15 ml of water and shake the stoppered flask vigorously. If any solid lumps form, break them with a glass rod and agitate the mixture again. Filter the mixture and wash the precipitate with water.

Recrystallize the precipitate from the minimum amount of ethanol. If a good yield is not obtained from the recrystallization, the solution should be heated almost to boiling and water added drop by drop with agitation until a cloudiness just persists. Heat the mixture until the cloudiness disappears and allow the mixture to cool to room temperature. Collect the product by filtration.

Before expending effort on the preparation of the polybromo derivatives of amines, be certain the melting points of polybromo derivatives of the amines you may have are available. Because so many of the polybromo derivatives of the aromatic amines are not recorded and none are in an easily accessible list, these derivatives are not nearly as useful for the identification of amines as they are for the identification of phenols (Chapter 49).

Picrates

Discussion. **Warning: picric acid is an explosive (TNP). Do not heat picric acid or subject it to mechanical shock. Do not increase the quantities recommended in the section that follows.** Amines (aliphatic and aromatic), phenols and their ethers, and aromatic hydrocarbons form picrates that are easily purified. Picrates are useful derivatives for both purification and identification of these compounds.

Experimental Procedure. Dissolve 0.3 to 0.4 g of amine in 8 to 10 ml of ethanol. If the amine does not completely dissolve, warm the mixture in a water bath at approximately 50°. If the amine still does not dissolve completely, be careful *not* to add the ethanol-insoluble material to the picric acid solution. Pour all the solution (decant carefully from any undissolved amine) into 7 to 8 ml of a saturated solution of picric acid in ethanol at 40°. Slowly and carefully heat the solution to boiling and allow it to cool to room temperature. If no crystals appear, remove 3 ml of solution by careful boiling and cool to room temperature. If still no crystals appear, stopper the mixture and allow it to stand until the next laboratory period. Filter, wash the precipitate with cold ethanol, and recrystallize it from ethanol.

Often the picrate is so pure it does not need to be recrystallized from ethanol. The picrates of many aromatic hydrocarbons and ethers dissociate on heating and the picrates of these compounds should not be recrystallized.

Methiodides of Tertiary Amines

$$CH_3I + R_3N \rightarrow R_3\overset{+}{N}CH_3 + I^- \quad (44\text{-}33)$$

Heat 0.5 g of tertiary amine and 0.5 ml of methyl iodide in a steam bath for 3 to 4 min. Remove the test tube from the steam bath for a few minutes and then place the mixture in an ice bath. If no crystals appear, scratch the wall of the test tube with a glass rod. Wash the crystals with isopropyl ether. Recrystallize the product from ethanol, dry it, and determine its melting point.

EXERCISES

44-1. A water-soluble liquid had a boiling point of 48 to 49° and gave a basic reaction to wet Hydrion paper. It reacted with cold nitrous acid solution to give a vigorous evolution of a colorless gas. What compounds might it be? What derivatives should be made?

44-2. A water-insoluble compound (I) dissolved in dilute aqueous base. It had a boiling point of 214.5 to 216.5°. When treated with cold aqueous sodium nitrite and hydrochloric acid, it gave no gas bubbles until warmed; when alkaline β-naphthol solution was added to this cold reaction mixture, a brilliant red precipitate (II) was produced. I reacted with acetyl chloride to give a precipitate (III) that melted at 133.5 to 136°. Identify I, II, and III. Does another derivative need to be prepared to identify I?

44-3. As was mentioned on page 336, tertiary amines are usually more soluble in water than primary and secondary amines of similar molecular weight. Would tertiary amines be more or less soluble in nonpolar solvents like hexane or carbon tetrachloride than primary or secondary amines?

44-4. In the preparation of the picrates of amines why is it important to keep any undissolved amine out of the reaction mixture?

44-5. The infrared hydroxyl stretching absorption is shifted to a lower frequency by hydrogen bonding, page 210. Would the N—H stretching absorption of amines be changed by hydrogen bonding? If hydrogen bonding does change these frequencies, would they be increased or decreased? Would the frequency shift be larger or smaller than was observed with alcohols?

TABLE 44-11 PRIMARY AND SECONDARY AMINES (LIQUIDS)

Amine	B.p. (°C)	D_4^{20}	n_D^{20}	Derivatives, m.p. (°C) acet-amide	benz-amide	benzene sulfon-amide	p-toluene sulfon-amide	picrate
Isopropylamine	33	0.690	1.3751	39		26		
t-Butylamine	46	0.699	1.3786	102	134			198
n-Propylamine	49	0.719	1.3889		84	36	52	135
Cyclopropylamine	50	0.824	1.4211		99	120		149
Diethylamine	56	0.710	1.3865		42	42	60	155
Allylamine	58	0.742	1.4202	69		39	64	140
2-Aminobutane	63	0.718	1.3951		76	70	55	140
Trimethyleneimine (azetidine)	63	0.844	1.4307					167
Isobutylamine	69	0.728	1.3988		57	53	78	150
n-Butylamine	77	0.740	1.4011		42			151
Ethylpropylamine	81	0.772	1.3966					150
2-Aminopentane	92	0.738						
2-Methylpyrrolidine	97	0.838						89
Piperdine	106	0.861	1.4530		48	94	96	152
Di-n-propylamine	109	0.738	1.4046			51		75
*Ethylenediamine	116	0.898	1.4561	172(di)	244(di)	168(di)	360(di)	233(di)
*1,2-Diaminopropane	119	0.876		139(di)	192(di)			135(di)
2,6-Dimethylpiperdine	127	0.819	1.4346		111	50		163
n-Hexylamine	130	0.764			40	96		126
Morpholine	130				75	118	147	146
Cyclohexylamine	134	0.867	1.4372	101	149	89		
*1,3-Diaminopropane	136	0.888		125(di)	147(di)	96	149	251
2,2,6-Trimethylpiperdine	138							195
Bis(2-methylpropyl)amine	139	0.745	1.4093	86		56		121
2-Aminoheptane	142	0.766	1.4195		269			
N-Methylcyclohexylamine	146	0.866			86			170
4-Ethylpiperidine	151	0.872	1.4532		74.5			
*1,4-Diaminobutane	159	0.878		137(di)	177(di)		224(di)	249

*These compounds will not be assigned as unknowns until Chapter 61 is studied, unless instrumental techniques are used for the identification of amines.

TABLE 44-11 (cont.)

Amine	B.p. (°C)	D_4^{20}	n_D^{20}	Derivatives, m.p. (°C) acet-amide	benz-amide	benzene sulfon-amide	p-toluene sulfon-amide	picrate
(Aminomethyl) cyclohexane	163	0.872	1.4638		101			185
2-Fluoroaniline	176			80	113	268		
*1,5-Diaminopentane (cadaverine)	179	0.915		169 (di)	135 (di)	119		237
Aniline	184	1.022	1.5863	114	160	112	103	180
Benzylamine	184.5	0.981	1.5401	65	105	88	118	194
α-Phenylethylamine	186	0.938		57	120			
N-Methylaniline	196	0.989	1.5721	102	63	79	94	145
1-Phenyl-2-aminopropane	196	0.942	1.5202		159			
β-Phenylethylamine	198	0.961	1.5266	51	116	69		172
o-Toluidine	200	1.001	1.5688	111	145	124	186	213
m-Toluidine	203	0.992	1.5686	65	125	95	172	200
N-Ethylaniline	205	0.962	1.5559	54	60		87	133
2-Chloroaniline	207	1.212	1.5895	87	99	129	193	134
N-Methyl-o-toluidine	207	0.972	1.5650	56	66			90
o-Methylbenzylamine	208	0.976	1.5422	69	88			215
p-Methylbenzylamine	208	0.952	1.5364	108	137			204
N-Methyl-p-toluidine	210			83		67		131
o-Ethylaniline	212	0.985	1.5592	111	147			195
2,5-Dimethylaniline	214	0.973	1.5595	139	140	138	232	171
2,6-Dimethylaniline	216	0.984	1.5610	177	168		212	190
p-Ethylaniline	216	0.970	1.5550	94	151		104	
2,4-Dimethylaniline	217	0.978	1.5612	133	192	130	181	209
3,5-Dimethylaniline	220	0.971	1.5581	143	144.5			200
2,3-Dimethylaniline	221	0.993	1.5684	135	189			221
p-Isopropylaniline (p-cumidine)	225	0.953		102	162			
4-t-Butylaniline	229			172	139		180	
o-Ethoxyaniline	229			79	104	102	164	
2,4,6-Trimethylaniline	230			216	204	137	167	190
2-Aminoindane	230			127	155		168	239

TABLE 44-11 (cont.)

Amine	B.p. (°C)	D_4^{20}	n_D^{20}	Derivatives, m.p. (°C) acet-amide	benz-amide	benzene sulfon-amide	*p*-toluene sulfon-amide	picrate
m-Chloroaniline	233	1.220	1.5931	74	120	121	208	177
1,2,3,4-Tetrahydroiso-quinoline	233	1.066	1.5810	46	129	154	169	198
4-Aminoindane	236			126	136		170	
2-Methyl-5-isopropylaniline	241	0.994	1.5387	71	102		168	
N-Butylaniline	241	0.936	1.5381		56		56	
2-Chloro-6-methoxyaniline	246d			123	135			
m-Ethoxyaniline	248			97	103		157	158
p-Ethoxyaniline	250	1.063		137	173	143	106	69
1,2,3,4-Tetrahydroquinoline	250	1.056	1.5949		75	67		
*2-Aminoacetophenone	251			77	98		148	
3-Bromoaniline	251d	1.579	1.6261	87	123			180
3-Bromo-2-methylaniline	254			163	177			
4-Amino-1,2,3,5-tetra-methylbenzene	255			216				200
p-Amino-N,N-diethylaniline	261			104	172			
4-*n*-Butylaniline	261	0.945		105	126			
Dibenzylamine	300	1.027	1.5731		112	68	159	

TABLE 44-12 PRIMARY AND SECONDARY AMINES (SOLIDS)

Amine	M.p. (°C)	B.p. (°C)	Derivatives, m.p. (°C) acetamide	benzamide	benzene-sulfonamide	p-toluene-sulfonamide	picrate
3-Bromo-4-methylaniline	25	255	115	132	117		196
o-Aminothiophenol	26	234	135	154			194
Dodecylamine	27	248		169		73	197
Di-n-heptylamine	30	271		167			119
5-Bromo-2-methylaniline	32	253d	165				
m-Iodoaniline	32		119	157		128	
N-Benzylaniline	37	298	58	107	119	149	48
2-Iodo-4-methylaniline	40		130	161			200
2-Chloro-4,6-dimethylaniline	40		206	148			201
4-Methylaniline (p-toluidine)	44	200	147	158	120	118	182
p-Aminodibenzyl	48			170.5			
o-Aminobiphenyl	49	299	121	102			
2-Bromo-4,6-dimethylaniline	49.5		196.5	185			
4-Chloro-2-methoxyaniline	50	260	150				200
Indole	52	253	157.5	68	254		
o-Aminodiphenylmethane	52		135	116			
Diphenylamine	53.4		101	180	124	141	182
5-Bromo-2-aminobiphenyl	57		130	162			
4-Methoxyaniline (p-anisidine)	58	240	129	155	95	114	
4-Bromo-2-methylaniline	59	240	156.5	115			
o-Iodoaniline	60		109	139			112
2,4-Dichloroaniline	63	246	145	117	128	126	106
*m-Phenylenediamine	63	284	191(di) 88(mono)	241(di) 125(mono)	195	173	184
p-Aminobenzyl-alcohol	65		188(di)	150(N-)			
p-Bromoaniline	66	245	168	204	134		180

* These compounds will not be assigned as possible unknowns until Chapter 61 is studied unless instrumental techniques are used.

TABLE 44-12(cont.)

Amine	M.p. (°C)	B.p. (°C)	Derivatives, m.p. (°C) acetamide	benzamide	benzene-sulfonamide	p-toluene-sulfonamide	picrate
p-Iodoaniline	67.5		184	222			
8-Aminoquinoline	68		103	98		155	
o-Nitroaniline	71	284	93	101	104	142	73
p-Chloroaniline	72	232	178	193	122	97	
*7-Aminoquinoline (hyd)	73		167	189		169	
4-Nitro-2-aminomesitylene	75		191	169	163		
2,4-Dibromoaniline	79		146	134		134	124
4-Chloro-3-methylaniline	83	241	91	119			
2,6-Dibromoaniline	83.4		210				124
5-Chloro-2-methoxyaniline	84		104	77.5			194
3-Iodo-1-naphthylamine	84		207	174			
4-Aminobenzonitrile	86		205	170			150
3-Aminophenanathrene	87.5		201	213			
3,4-Diaminotoluene	89	265	210 (di) 95(3-N) 131 (4-N)	264	179	140(4-N)	
2-Hydroxy-3-bromo-5-methylaniline	93		129(N-) 166(di)	157(N-) 230(di)			
2,6-Dibromo-4-chloroaniline	93		226	194			
7-Aminoquinoline (anh)	94		167	189			
2,4-Diiodoaniline	96		151	181			
5-Bromo-2-methoxyaniline (4-bromo-o-anisidine)	97.5		160	108			
4-Chloro-1-naphthylamine	98		186				
4-Bromo-1-naphthylamine	100		193				
N-p-Tolyl-2-naphthylamine	103		85	139			
8-Nitro-2-naphthylamine	104		196	162			
2-Bromo-4-nitroaniline	105		129	160			
5-Nitro-2-methylaniline	107		151		172		

TABLE 44-12(cont.)

Amine	M.p. (°C)	B.p. (°C)	Derivatives, m.p. (°C) acetamide	benzamide	benzene-sulfonamide	p-toluene-sulfonamide	picrate
*o-Aminobenzamide	110		177	215			
2-Iodo-1,4-diaminobenzene	111		211	254			
5-Aminoquinoline	110	310	178			204	
β-Naphthylamine	112	306	132	162	102	133	195
m-Nitroaniline	114	306d	155	155	136	138	143
6-Aminoquinoline	114		138,75(di)	169		193	
4-Chloro-2-nitroaniline	117		104			110	
2-Nitro-4-methylaniline	117		99	148	102	146	
5-Nitro-1-naphthylamine	119		220		183		
2-Aminotriphenylcarbinol	121		192				123
*3-Hydroxyaniline (m-aminophenol)	122		148,101(di)	174(N-)		157	
2,4,6-Tribromoaniline	122		232	200			
2,4-Dimethyl-5-nitroaniline	123		159	200	149	192	
*Benzidine	127		317(di) 199(mono)	352(di) 204(mono)	232(di)	243(di)	
6-Bromo-2-naphthylamine	128		192	218			
3-Methyl-2-naphthylamine	135		181.5	190			
2-Nitro-1-naphthylamine	144		199	175			
2,5-Dimethyl-4-nitroaniline	144.5		168.5		162	185	
4'-Bromo-4-aminobiphenyl	145		247			174	
4-Chloro-2-hydroxyaniline	154		140(di)				
4'-Nitro 2 aminobiphenyl	159		199			163	
*p-Aminoacetanilide	162		304				
2-Aminofluorenone	163		227				
*4-Hydroxy-3-methylaniline	175		179(N-)	194(di)		110	
4-Chloro-1-aminoanthraquinone	180		203.5				
2,4-Dinitroaniline	183		120	209		219	
2,4,6-Trinitroaniline (picramide)	190		230	196	211		
4-Nitro-1-aminonaphthalene	195		190	224	170	185	

TABLE 44-13 TERTIARY AMINES
(LIQUIDS)

Amine	B.p. (^oC)	D_4^{20}	n_D^{20}	Derivatives, m.p. (C^o) methiodide	picrate
Dimethylethylamine	38	0.721		301	193
N-Methylpyrrolidine	79	0.8188	1.4311		221
Triethylamine	89	0.727	1.4002		173
Pyridine	116	0.978	1.5097	117	167
1-Methylpyrazole	127	0.990		190	148
N-Ethylpiperidine	128	0.824	1.4416		168
2-Methylpyrazine	136	1.029		129.5	133
3-Methylpyridine	143	0.953	1.5057		150
4-Methylpyridine	143	0.956	1.5055	150	167
*4-Chloropyridine	147				146
3-Chloropyridine	149				135
Tri-_n_-propylamine	157	0.753	1.4176	207.5	116
2,4-Dimethylpyridine	158	0.929	1.5002		183
2,5-Dimethylpyridine	160	0.928	1.5001		169
3-Ethylpyridine	163	0.950			129
4-Ethylpyridine	164	0.942			168
2,4,5-Trimethylpyridine	170	0.935	1.5074		133
2-Chloropyridine	170	1.204			
3-Bromopyridine	170	1.641	1.5694	165	154
2,4,6-Trimethylpyridine	172	0.917	1.4979		156
N,N-Dimethylbenzylamine	181	0.900		179	93
N,N-Dimethyl-_o_-toluidine	185	0.929	1.5153		120
N,N-Dimethylaniline	193	0.956	1.5582	226	163
2-Bromopyridine	194	1.656			
N,N-Dimethyl-2,6-dimethylaniline	199	0.912	1.5129		
N-Methyl-N-ethylaniline	201			125	135
N,N-Dimethyl-2,5-dimethylaniline	204				158
N,N-Dimethyl-2,4-dimethylaniline	205	0.916	1.5201		124

TABLE 44-13 (cont.)

Amine	B.p. (°C)	D_4^{20}	n_D^{20}	Derivatives, m.p. (C°) methiodide	picrate
N,N-Dimethyl-2-chloroaniline	207			152	132
N,N-Dimethyl-p-toluidine	210	0.929	1.5361	219	129
Tri-n-butylamine	212	0.776		180	106
N,N-Dimethyl-m-toluidine	212	0.941	1.5492	177	131
N,N-Diethylaniline	217	0.935		102	142
N,N-Diethyl-p-toluidine	229	0.923		184	110
Quinoline	239	1.093	1.6268	133	203
Isoquinoline	243	1.0986	1.6154	159	222
2-Methylquinoline	247	1.058	1.6126	195	193
*1-Methyl-2-(3-pyridyl) pyrrolidine (Nicotine)	248	1.010	1.5281	219	218
3-Chloroquinoline	259			276	182
N,N-Di-n-butylaniline	271				125
3-Bromoquinoline	275				190
8-Chloroquinoline	288			165	
8-Bromoquinoline	303			281	
N-Benzyl-N-methylaniline	306	1.040	1.5969	164	127

TABLE 44-14 TERTIARY AMINES
(SOLIDS)

Amine	M.p. (^oC)	B.p. (^oC)	Derivatives, m.p. (^oC) methiodide	picrate
4,6-Dimethylpyrimidine	25	160	169	141
4-Bromoquinoline	28	270d	268	
4-Chloroquinoline	31	263		212
4-Bromoisoquinoline	40	282	233	
6-Chloroquinoline	41	262	248	
*N,N-Diethyl-p-aminobenzaldehyde	41			
N,N-Dimethyl-β-naphthylamine	47	305		206
5-Bromoquinoline	50	280	205	
2-Iodoquinoline	53		212	
3-Nitro-N,N-dimethylaniline	60		205	119
3,4'-Bipyridyl	62	297		215
N,N-Dibenzylaniline	71		135	
4,4'-Bipyridyl (hyd.)	73			257
*4-Dimethylaminobenzaldehyde	74			
8-Hydroxyquinoline	75		143	204
*p-Nitroso-N,N-dimethylaniline	85			140
1-Phenylisoquinoline	96	300		165
4,4'-Bipyridyl (anh.)	114			257
7-Nitroquinoline	132		232	
6-Hydroxypyrimidine	165			190

*These compounds will not be assigned as unknowns until Chapter 61 is studied, unless instrumental techniques are used.

45

Use of Sulfonic Acid Group to Effect Certain Orientations

Synthesis of 2,6-Dichloroaniline

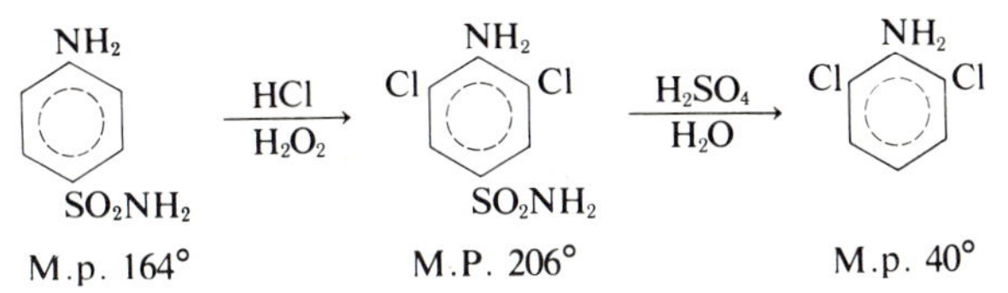

(45-1)

INTRODUCTION

Sometimes groups are introduced temporarily on organic molecules only to effect certain orientations of other entering groups. Other groups are introduced to deactivate portions of the molecule to the reagents, e.g., the acetyl group in the synthesis of sulfanilamide. Easy removal is one of the requirements for a good blocking or orienting group. The sulfonic acid moiety is a choice blocking or orienting group for most aromatic molecules because it is easily introduced and easily removed.

DISCUSSION

The synthesis of 2,6-dichloroaniline from aniline began in Chapter 39 and continued in Chapter 40. The product from Chapter 40 is chlorinated and the sulfonamide group removed.

$$C_6H_5NH_2 + Ac_2O \rightarrow C_6H_5NHAc \xrightarrow{HOSO_2Cl} p\text{-}AcNHC_6H_4SO_2Cl \qquad (45\text{-}2)$$

$$p\text{-}AcNHC_6H_4SO_2Cl + NH_3 \rightarrow p\text{-}AcNHC_6H_4SO_2NH_2 \xrightarrow[H_2O]{H^+} NH_2C_6H_4SO_2NH_2 \qquad (45\text{-}3)$$

Problem 45-1. What are the products of the following reactions:
(a) aniline (1 mole) + chlorine (1 mole) →
(b) aniline (1 mole) + chlorine (2 moles) →
(c) aniline (1 mole) + chlorine (3 moles) →
(d) acetanilide (1 mole) + chlorine (1 mole) →
(e) acetanilide (1 mole) + chlorine (2 moles) →

Problem 45-2. (a) Why is it difficult to isolate and purify *o*-chloroacetanilide from the products of the reaction of 45-1d by recrystallization? (b) By distillation? (c) What technique could be used to separate the *o*- and *p*-chloroacetanilides? (d) How could the yield of the *ortho* isomer be increased without using a blocking agent?

Problem 45-3. By the techniques described in the answers to Problem 45-2c and d, could significant amounts of 2,6-dichloroacetanilide be directly synthesized without the use of blocking groups?

Problem 45-4. Under what conditions might the techniques described in the answers to Problems 45-2c and d be a much more profitable method for the *commercial* production of the *o*-chloroacetanilide than the following series of reactions?

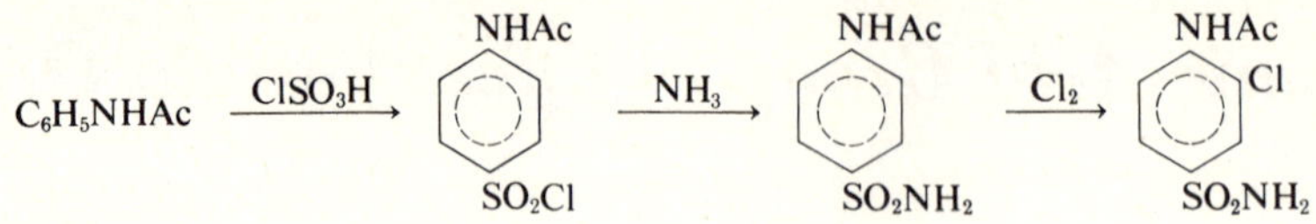

(45-4)

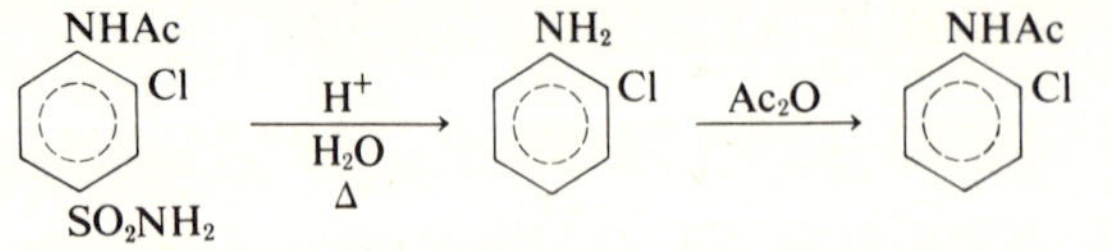

(45-5)

Chlorine gas is usually used for commercial chlorination as bromine was used directly for bromination in Chapter 16. The use of a noxious gas, such as chlorine, in a laboratory causes many problems and, if possible, chlorine is usually generated *in situ* or in the reaction mixture. In this experiment chlorine is generated *in situ* by the reaction of hydrochloric acid with hydrogen peroxide.

$$2HCl + H_2O_2 \rightarrow 2H_2O + Cl_2 \qquad (45\text{-}6)$$

EXPERIMENTAL PROCEDURE

Chlorination of Sulfanilamide

Warning: many materials catalyze the fast, sometimes explosive, release of oxygen from hydrogen peroxide. Make certain the graduated cylinder, reaction flask, etc. are meticulously clean.

In a 200-ml round bottom flask add 5 g of sulfanilamide, 50 ml of water, and 5 ml of concentrated hydrochloric acid. Swirl the mixture until all the sulfanilamide dissolves and then add 45 ml more hydrochloric acid. Prepare an ice bath to cool the mixture if necessary. Add 6.5 ml of 30% hydrogen peroxide, 1/2 to 1 ml at a time, agitating the mixture by swirling the flask after each addition. If after the first addition the temperature is not above 45°, heat it with a burner flame until it is. During the remainder of the addition maintain the temperature between 45° and 60° by heating with a burner flame or cooling with an ice bath as required. After the addition maintain the temperature at 55 to 65° and stir the mixture occasionally for 25 min.

Problem 45-5. In the bromination of benzene (Chapter 16) a Lewis acid catalyst was used. Why isn't a catalyst, such as aluminum chloride, used for this halogenation?

With a thermometer stir the mixture in an ice bath. Suction filter *as soon as* the temperature falls below 28°. The precipitate is 3,5-dichlorosulfanilamide.

Desulfonation of 3,5-Dichlorosulfanilamide

Rinse the flask just used with some water, add 9.0 g of ice, then 16 ml of concentrated sulfuric acid and stir. Transfer all the crude 3,5-dichlorosulfanilamide to the flask, add a reflux condenser, and heat on an oil bath at 180 to 195° for 1 hr 50 min. Remove the flask from the oil bath and allow it to cool a few minutes. Pour the mixture into a 200-ml round bottom flask containing 40 g of ice and 50 ml of water. Attach a Claisen adapter and other glassware to construct the apparatus shown in Figure 16-2; however, use a stopper in place of the steam line running into the adapter. If standard taper glassware is not being used, attach a condenser to the flask with a Kjeldahl adapter as shown in Figure 45-1.

Be sure there is an ice bath around the receiving flask. Steam distil (Chapter 15) the mixture until approximately 50 ml of water collects in the receiver. Suction filter the distillate *after* cooling the Büchner funnel in an ice bath. Press the precipitate dry with some folded filter paper while the suction is on and pull air through it for 5 min to dry it. Weigh the product in a tared yield bottle, determine its melting point, calculate the % yield, and hand in the product to your instructor.

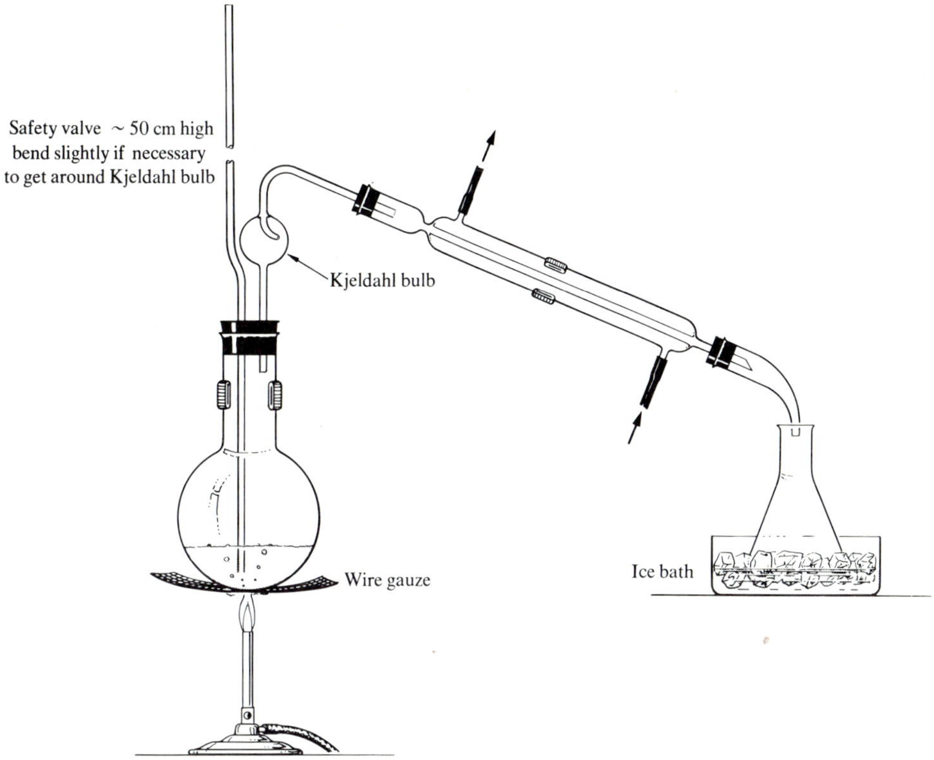

Figure 45-1 Steam distillation apparatus from nonground glass equipment.

THIN-LAYER CHROMATOGRAPHY (Optional)

Follow the directions at the end of Chapter 43 for activation, spotting, elution (or development), and visualization of the chromatogram. Develop the chromatogram with only the 15 parts dibutyl ether, 15 parts ethyl acetate, and 1 part acetic acid mixture. Spot the plate or strip with approximately 1×10^{-5} g of: (a) final purified product, (b) the oily material obtained by pouring the reaction mixture from desulfonation on the 40 g of ice and 50 ml of water, (c) as many of the likely contaminants of (a) and (b) as there is space for on the plate. The spots should be no larger than 5 mm in diameter, and the distances between the centers should be at least 25 mm and preferably 30 or 35 mm. The center of the spots should be at least 15 mm from the side of the plate or strip. Use 5 to 7 drops of ethanol to dissolve the materials that are spotted. Use the three visualization procedures (with the ninhydrin spray last) described at the end of Chapter 43.

Problem 45-6. How pure is your product? What are the principal contaminants?

EXERCISES

45-1. How could this procedure be easily modified to synthesize *o*-chloroaniline uncontaminated by 2,6-dichloroaniline?

45-2. Why was it not necessary to add sodium hydroxide just before the steam distillation to free the amine from its sulfate salt $(C_6H_3Cl_2NH_3)_2SO_4^{=}$? How *might* neutralization of the acid at this point cause a slightly lower yield?

46

Properties and Reactions of Organic Nitro, Nitroso, and Azo Compounds

Organic Qualitative Analysis VIII

INTRODUCTION

Nitro, nitroso, hydrazo, hydroxylamino, and azo functional groups undergo reactions characteristic of each group. These groups usually modify and often radically change the reactions of other functional groups. These groups also markedly affect the physical properties of compounds.

PHYSICAL PROPERTIES

The high dipole moment (3.5 to 4.5 D for nitro hydrocarbons) of these compounds indicate that they have strong polar, intermolecular attractive forces. The strong polar attractive forces cause these compounds to have higher melting and boiling points for their molecular weight than any other group of nonionic compounds incapable of strong hydrogen bonding. Nitro compounds have boiling and melting points very close to those of alcohols and phenols of similar molecular weight and structure (Table 46-1). For a further discussion of the effect of dipole moments on physical properties see Appendix V.

Nitro compounds are considerably less soluble in water than other compounds with oxygen or nitrogen functional groups discussed in this text. Only the two lowest molecular weight monofunctional nitro compounds are significantly soluble in water. Nitro groups cause a large increase in density. Of the functional groups studied thus far, only bromine and iodine cause a larger increase in density.

ACID-BASE REACTIONS

Nitro Compounds as Acids

Nitro groups strongly withdraw electrons by both inductive and resonance effects. This electron withdrawal usually causes nitro compounds to be considerably more acidic than similar compounds without these groups.

TABLE 46-1 COMPARISON OF PHYSICAL PROPERTIES OF VARIOUS NITROGEN COMPOUNDS WITH ALCOHOLS AND PHENOLS

Nitrogen Compound	Mol. wt.	B.p. (^oC)	M.p. (^oC)	D_4^{20}	Solubility (g/100g H_2O, 20^oC)
Nitromethane	61	101.3	-28.6	1.133	9.5
Nitroethane	75	114	-90	1.050	4.5
1-Nitropropane	89	132	-108	1.000	1.4
2-Nitropropane	89	120	-93	1.001	1.7
Nitrobenzene	123	211	6	1.203	0.19
o-Nitrotoluene	137	223	3	1.163	0.05
m-Nitrotoluene	137	232	16	1.157	0.04
p-Nitrotoluene	137	238	52		0.01
Nitrosobenzene	107	184	67.5		insol.
Azobenzene	182	297	68	1.203	insol.
Hydroxy compound					
1-Propanol	60	97	-127	0.804	
1-Butanol	74	118	-90	.810	7.9
1-Pentanol	88	138	-78	0.815	2.3
2-Pentanol	88	116	-9	0.820	5.5
2-Phenyl-1-ethanol	122	221		1.203	1.9
3-Phenyl-1-propanol	136	236		1.008	
4-Propylphenol	136	232	22		
2,4,5-Trimethylphenol	136	232	110		
Benzyl alcohol	108	205		1.045	4.0
p-Benzylphenol	184	315	84		

Aliphatic nitro compounds with hydrogen *alpha* to the nitro group react with and dissolve in dilute aqueous sodium hydroxide.

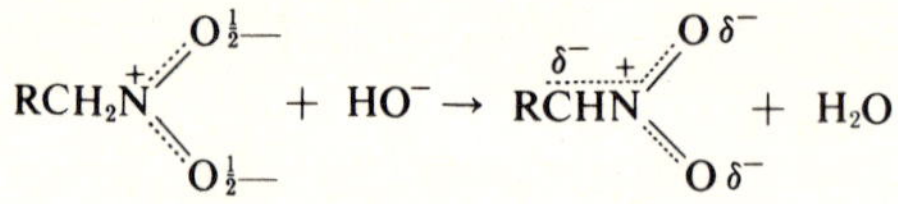
(46-1)

Water-insoluble mononitroalkanes are not, however, sufficiently acidic to dissolve in aqueous bicarbonate. Nitro and other electron-withdrawing groups lower the basicity of amines (p. 337) and increase the acidity of phenols.

Problem 46-1. In what solubility class are the following compounds: CH_3NO_2, $CH_3(CH_2)_3NO_2$, $(CH_3)_3CNO_2$, $C_6H_5NO_2$, 2,4-dinitroaniline, $(NO_2)_3CH$, $CH_3CH{=}CHC(CH_3)_2NO_2$, $CH_3CH{=}C(CH_3)NO_2$?

Reaction of Base with Aromatic Nitro Derivatives

Nitrophenols develop an intense yellow or yellow-orange color as soon as they dissolve in aqueous base. Nitrosophenols develop a greenish yellow color. Nitro- and nitrosoanilines take somewhat longer to develop these colors, probably because the aniline must first react with hydroxide ion to give the phenol.

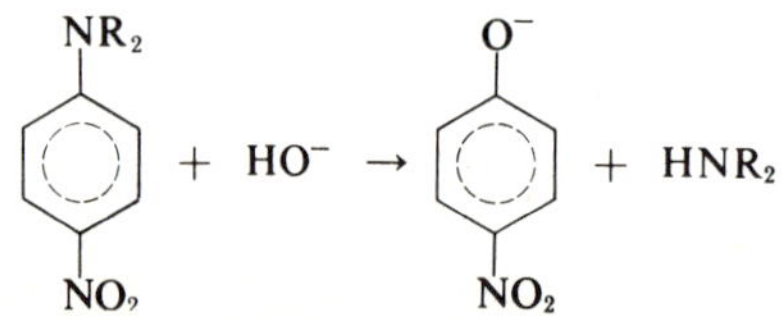
(46-2)

As discussed earlier (pp. 175, 177, 276) electron-withdrawing groups, such as nitro and nitroso, activate aromatic halogen atoms to nucleophilic displacements.

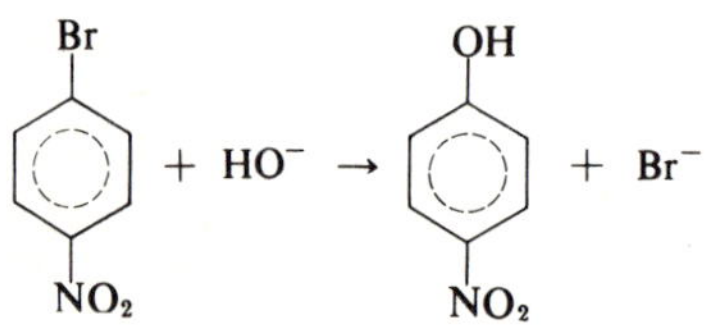
(46-3)

Reaction of Base and Acetone with Polynitro Compounds

When shaken with strong base and acetone, dinitro compounds give a purplish blue color except for *o*-dinitrobenzene which gives very little color change, *p*-dinitrobenzene which turns greenish yellow, and 2,4-dinitroaniline which turns red. Trinitro compounds turn deep red. For this test add 1 to 2 ml of 10% aqueous sodium hydroxide to a solution of 0.05 g of unknown in 5 ml of acetone. Shake the mixture and observe.

Hydrazines

Most water-insoluble hydrazines are sufficiently basic to react with and dissolve in dilute hydrochloric acid.

$$RNHNH_2 + H^+ \rightarrow R\overset{+}{N}H_2NH_2 \qquad (46\text{-}4)$$

$$C_6H_5NHNH_2 + H^+ \rightarrow C_6H_5NH\overset{+}{N}H_3 \qquad (46\text{-}5)$$

Solubility Class M

Most azo compounds, aliphatic tertiary and aromatic nitro and nitroso compounds, and some hydrazines are insoluble in water and neither significantly basic nor acidic. These compounds are in Solubility Class M.

OXIDATION-REDUCTION REACTIONS

Tollens' Reagent

Discussion. Hydrazines and hydroxylamines are among the few types of compounds oxidized by ammoniacal silver nitrate.

$$RNHOH + 2Ag(NH_3)_2^+ \rightarrow RNO + 2Ag\downarrow + 2NH_4^+ + 2NH_3 \quad (46\text{-}6)$$

$$RNHNH_2 + 2Ag(NH_3)_2^+ + H_2O \rightarrow RNO + 2Ag\downarrow + 2NH_4^+ + 3NH_3 \quad (46\text{-}7)$$

This test along with others gives a good indication of whether or not these groups are present.

Experimental Procedure. To prepare the reagent pour 6 ml of 5% silver nitrate into a *clean* 10-cm test tube; then add one drop of 5% sodium hydroxide solution. Add 5% ammonium hydroxide solution drop by drop with shaking until the precipitate just dissolves.

$$2Ag^+ + 2HO^- \rightarrow Ag_2O\downarrow + H_2O \quad (46\text{-}8a)$$

$$Ag_2O + 2NH_4^+ + 2NH_3 \rightarrow 2Ag(NH_3)_2 + H_2O \quad (46\text{-}8b)$$

Add 0.04 g of compound to a clean test tube containing 2 ml of the reagent. Agitate the tube vigorously and allow it to stand for 10 min. If no reaction has occurred, place it in a water bath at 40° for 5 min. A silver mirror on the bottom of the test tube is a positive test. If the test tube is not clean, a black precipitate may be obtained instead of the silver mirror. This black precipitate is always a positive test. **Caution!** On standing, Tollens' reagent often forms very sensitive, **explosive** precipitates. Therefore, Tollens' reagent should be prepared fresh each week it is used. Residues from the reagent or the tests should not be allowed to stand overnight. Test mixtures and all excess reagent should be washed down the drain with a large excess of water. Any test tubes or other containers that held the reagent should be washed with dilute nitric acid at the end of the laboratory period.

Problem 46-2. An excess of ammonium hydroxide should not be used in preparing Tollens' reagent as the excess will decrease the sensitivity of the test; why?

Zinc and Ammonium Chloride Test

Discussion. Nitro and nitroso compounds are reduced by zinc and ammonium chloride to hydroxylamines and hydrazines.

$$RNO_2 \xrightarrow[NH_4Cl]{Zn} RNHOH \quad (46\text{-}9)$$

$$RNO \xrightarrow[NH_4Cl]{Zn} RNHOH \quad (46\text{-}10)$$

The reduction products are easily oxidized by Tollens' reagent to give a visible reaction for these groups.

Experimental Procedure. First test the unknown with Tollens' reagent. If there is a silver mirror or a black precipitate, do not proceed with this test. Dissolve 0.2 to 0.3 g of compound in 5 ml of water. If the compound is water-insoluble, dissolve it in a mixture of 2 ml of water and 3 ml of 95% ethanol. Add 0.7 g of zinc dust and 0.3 g of ammonium chloride. Stir the mixture with a glass rod while heating it to its boiling point. As soon as it reaches the boiling point, remove it from the heat source and allow it to stand 5 to 7 min. Separate the clear liquid by filtering or centrifuging and test it with Tollens' reagent. A silver mirror or black precipitate indicates a nitro or nitroso group (provided, of course, the compound did not reduce Tollens' reagent before it was reduced).

Ferrous Sulfate and Potassium Hydroxide

Discussion. Nitro and nitroso compounds, and hydroxylamines, but not hydrazines or azo compounds, reduce ferrous compounds in basic solutions.

$$RNHOH + 2Fe^{++} + 4HO^- + H_2O \rightarrow RNH_2 + 2Fe(OH)_3\downarrow \quad (46\text{-}11)$$

$$RNO_2 + 6Fe^{++} + 12HO^- + 4H_2O \rightarrow RNH_2 + 6Fe(OH)_3\downarrow \quad (46\text{-}12)$$

A reddish brown (rust) precipitate of ferric hydroxide is a positive test. A green precipitate of ferrous hydroxide even if somewhat brown or darkened by partial oxidation is a negative test.

Experimental Procedure. Two reagents are needed for this test. The first is prepared by dissolving 15 g of ferrous sulfate in 300 ml of distilled water. The water should be boiled about 20 min before preparing the reagent to drive out dissolved oxygen. Add 2 ml of concentrated sulfuric acid and an iron nail to minimize oxidation. The other reagent is a 10% alcoholic potassium hydroxide solution prepared by dissolving 17 g of potassium hydroxide in 18 ml of distilled, freshly boiled water and adding this to 135 ml of 95% ethanol.

Add 10 to 15 mg of compound to 1 to 1.5 ml of ferrous sulfate reagent in a test tube. Place a thin glass tube into the test tube and attach it to a gas jet with a piece of rubber tubing. Bubble gas through the mixture and the test tube for a few seconds to remove as much air as possible. Add 1 ml of alcoholic potassium hydroxide solution, stopper the test tube, and shake it vigorously for 30 sec.

Tin and Hydrochloric Acid

Discussion. Nitro, nitroso, azo, hydrazo, and hydroxylamino compounds are reduced by a variety of reagents to amines (Chapter 42). N-Nitroso compounds are reduced to hydrazines. The reduction of these compounds to amines by tin and hydrochloric acid is the most universal method of making derivatives of these compounds. If the amine is a liquid or a second derivative is needed, any suitable derivative of the amine can be prepared.

$$RNO_2 \xrightarrow[HCl]{Sn} RNH_2 \quad (46\text{-}13)$$

$$RNO \xrightarrow[HCl]{Sn} RNH_2 \quad (46\text{-}14)$$

$$RN{=}NR \xrightarrow[HCl]{Sn} 2RNH_2 \quad (46\text{-}15)$$

$$RNHNH_2 \xrightarrow[HCl]{Sn} RNH_2 + NH_3 \quad (46\text{-}16)$$

$$RNHOH \xrightarrow[HCl]{Sn} RNH_2 \quad (46\text{-}17)$$

$$R\underset{|}{\overset{NO}{N}}CH_3 \xrightarrow[HCl]{Sn} R\overset{NH_2}{N}CH_3 \quad (46\text{-}18)$$

Experimental Procedure. Place 4 g of granulated tin and 1.5 g of unknown in a 100-ml round bottom flask. If the unknown is insoluble in water, add 5 ml of 95% ethanol. Place a reflux condenser on the flask. Add 45 ml of 10% hydrochloric acid, 3 to 4 ml at a time through the top of the condenser. Shake the mixture vigorously after each addition. After all the hydrochloric acid is added, heat the mixture on a steam bath for 14 to 18 min. Immediately after removing the mixture from the steam bath, decant it from any solid residue onto a mixture of

10 to 15 g ice and 3 g of sodium hydroxide. Stir the mixture well. If any solid tin compounds remain, add 40% aqueous sodium hydroxide, a few drops at a time with stirring, until all the solid dissolves. Extract the mixture with three 10-ml portions of ether. Remove the ether by distillation. If the amine is a solid, purify it by crystallization from an ethanol-benzene mixture.

Problem 46-3. The above method is not useful for preparing derivatives of azo and hydrazo compounds of the type RN=NR′ and RNHNHR′ where R and R′ are not the same. Why?

MISCELLANEOUS REACTIONS

Reaction with Nitrous Acid (Red, White, and Blue Test)

Discussion. Primary aliphatic nitro compounds react with nitrous acid to yield nitrolic acids. The nitrolic acids dissolve in aqueous base to give red anions which are intensely colored because of the extensively conjugated system. Secondary aliphatic nitro compounds react with nitrous acid to give blue nitroso derivatives. Since these compounds have no hydrogen *alpha* to the nitroso and nitro groups, they are not acidic and do not dissolve in aqueous base and, therefore, can not form the red nitrolium salts.

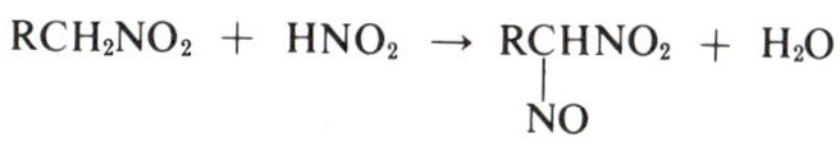

(46-19)

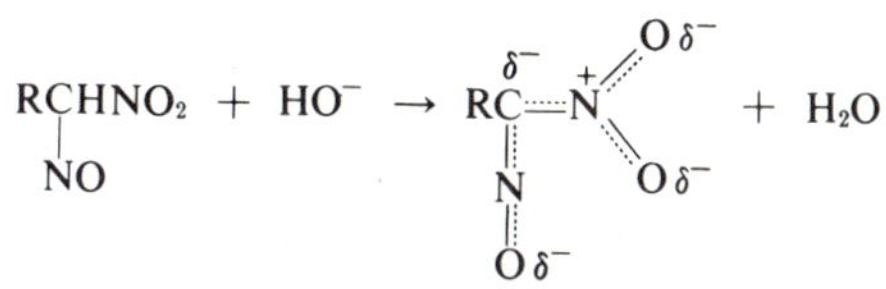

(46-20)

Aromatic and tertiary aliphatic nitro compounds cannot react with nitrous acid. With the procedure described below primary nitro compounds give a red color, secondary nitro compounds give a blue color, and tertiary and aromatic nitro compounds give no significant color change. The test is often called the red, white, and blue test although red, blue, and colorless would be more accurate.

Experimental Procedure. To 3 ml of 10% aqueous sodium hydroxide add 7 drops of unknown and allow the mixture to stand 2 to 3 min. Add 1 to 2 ml of 10% sodium nitrite, follow with 12 drops of 10% sulfuric acid, and then agitate the mixture.

Problem 46-4. This test could be adapted to distinguish among primary, secondary, and tertiary halides. Explain how this could be done. What difficulties might be encountered?

Nitrosation of Phenols

Discussion. Most nitroso compounds nitrosate phenols in the presence of acid. The color of the nitrosophenol gives the visual indication of a positive test.

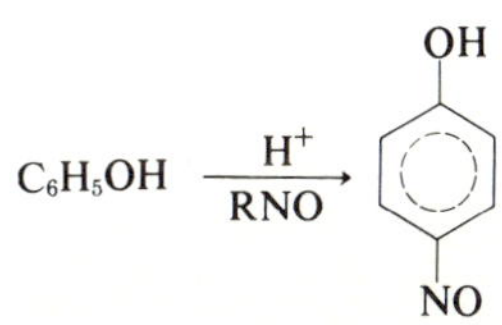

(46-21)

Experimental Procedure. Dissolve approximately 0.05 g of unknown in 3 ml of concentrated sulfuric acid, add 0.07 mg of phenol, shake the tube, and then place it in a water bath at 50°. A green to blue color that becomes reddish on warming indicates that the compound being tested has a nitroso group.

Reaction of Hydrazines with Aldehydes

Discussion. Monosubstituted and 1,1-disubstituted hydrazines react with aldehydes to give hydrazones. Many amines react with aldehydes under the same conditions to give precipitates of aldimines.

$$RCHO + H_2NNHR' \rightarrow \underset{\text{hydrazone}}{RCH{=}NNHR'} \qquad (46\text{-}22)$$

$$RCHO + H_2NC_6H_5 \rightarrow \underset{\text{aldimine}}{RCH{=}NC_6H_5} \qquad (46\text{-}23)$$

Many amines, however, do not reduce Tollens' reagent, and nitro, nitroso, azo, and hydroxylamine derivatives do not give a precipitate with aldehydes.

Experimental Procedure. Dissolve or suspend 0.1 g of unknown in 2 ml of water to which a drop of acetic acid has been added. Add more acetic acid a drop at a time until all the unknown has dissolved. Add 10 drops of a 5% solution of acetaldehyde in water. A precipitate indicates a monosubstituted or 1,1-disubstituted hydrazine or an amine.

UNKNOWN

Your unknown assigned with this experiment can be any compound from Tables 46-2 to 46-6 and any compound from the tables at the end of the other chapters on organic qualitative analysis (Chapters 14, 23, 28, 32, 35, 41, and 44). If known, the melting and boiling point of the corresponding amine and its benzamide are given. Remember that a boiling point is not as useful a constant for identification as the melting point. Therefore, if an amine from the reduction of a nitrogen compound is a liquid and the benzamide is not a satisfactory derivative, another derivative should be prepared. The boiling points (and also the melting points) of the amines given

TABLE 46-2 NITRO DERIVATIVES OF HYDROCARBONS (LIQUIDS)

Compound	B.p. (°C)	D_4^{20}	n_D^{20}	Amine b.p. (°C)	Benzamide m.p. (°C)
Nitromethane	101.3	1.126	1.3816	-6	80
Nitroethane	114	1.050	1.3921	16.5	71
2-Nitropropane	120	1.031	1.3943	33	26
3-Nitropropene	127	1.050	1.410	58	
1-Nitropropane	132	1.007	1.4023	49	84
2-Nitrobutane	140	0.984	1.4013	63	76
2-Methyl-1-nitropropane	140	0.985	1.4022	69	57
2-Methyl-2-nitrobutane	150		1.401	78	
2-Nitropentane	153			92	
1-Nitrobutane	153	0.971	1.4103	77	
1-Nitrohexane	193	0.9396	1.4234	130	40
2-Nitroheptane	196	0.944			
Nitrocyclohexane	205	1.068	1.4604	134	149
Nitrobenzene	211	1.203	1.5531	184	160
o-Nitrotoluene	223	1.159	1.5474	200	145
o-Nitroethylbenzene	224	1.127	1.5401	212	147
2,6-Dimethyl-1-nitrobenzene	226	1.110	1.54	216	168
Phenylnitromethane	226d	1.159	1.5323	184.5	105
m-Nitrotoluene	232	1.157	1.5465	203	125
2,5-Dimethyl-1-nitrobenzene	241	1.132		214	140
2,4-Dimethyl-1-nitrobenzene	246	1.126		217	192
2,3-Dimethyl-1-nitrobenzene	250			221	189
4-Isopropyl-2-nitrotoluene	264	1.074	1.5309	241	102
1-t-Butyl-4-nitrobenzene	267			229	104

TABLE 46-3 NITRO DERIVATIVES OF HYDROCARBONS, HALOHYDROCARBONS, AND ETHERS (SOLIDS)

Nitro compound	M.p. (°C)	B.p. (°C)	Amine B.p. (°C)	Amine m.p. (°C)	Benzamide m.p. (°C)
2-Methyl-2-nitropropane	25.5	127	46		134
4-Fluoro-1-nitrobenzene	25.7	207	185		185
2-Chloro-1-nitrobenzene	32	246	207		99
2,4-Dichloro-1-nitrobenzene	33	258	246	63	117
3-Nitrophenetole	34		248		103
2-Nitrobiphenyl	36		299	49	102
3-Iodo-1-nitrobenzene	37	280		32	157
4-Nitroindane	44		236	-3	136
3-Chloro-1-nitrobenzene	45	236	232		120
2-Iodo-1-nitrobenzene	49	292		60	139
4-Nitrotoluene	52	234	200	45	158
4-Nitroanisole	53		240	58	155
3-Bromo-1-nitrobenzene	56		251d	18	122
4-Nitrophenetole	59		250	4	173
3,4-Dinitrotoluene	61		265	89	264
3,5-Dimethyl-1-nitrobenzene	75	273	220		143
2-Nitronaphthalene	78			112	162
4-Chloro-1-nitrobenzene	84	242		72	192
4-Bromo-1-nitronaphthalene	85			102	
4-Chloro-1-nitronaphthalene	86			98	
8-Chloro-1-nitronaphthalene	94			90	
4-Chloro-2-nitroanisole	98			84	77.5
5-Chloro-1-nitronaphthalene	111			85	
5-Bromo-1-nitronaphthalene	122			68	
2,4,6-Tribromonitrobenzene	122			122	200
4-Bromo-1-nitrobenzene	126		245	66	204
3-Nitrophenanthrene	171			87.5	213
4-Iodo-1-nitrobenzene	173			67.5	222

TABLE 46-4 NITROSO COMPOUNDS

Nitroso compound	M.p. (°C)	Amine B.p. (°C)	Amine M.p. (°C)	Benzamide M.p. (°C)
1-Nitroso-2, 4-dimethylbenzene	41	217		192
p-Nitrosotoluene	48	200	45	158
N-Nitrosoacetanilide	51d			
m-Nitrosotoluene	53	203		125
N-Nitrosodiphenylamine	66	44+	53.4	180
Nitrosobenzene	68	184		160
o-Nitrosotoluene	72	200		145
p-Nitroso-N, N-dimethylaniline	85	262	42	228
1-Nitrosonaphthalene	98		50	160
*p-Nitrosophenol	125d		186	217
p-Nitrosodiphenylamine	144		75	203
*2-Nitroso-1-naphthol	152d			
*4-Nitroso-1-naphthol	194d			

*Will not be assigned as a possible unknown until phenols are studied (Chapter 49).

+Melting point of hydrazine.

TABLE 46-5 HYDRAZINES

Hydrazine	B.p. (°C)	M.p. (°C)	D_4^{20}	Amine b.p. (°C)	Amine m.p. (°C)	Benzamide m.p. (°C)
Isopropyl-	108			33		
1, 2-Diisopropyl-	125		0.789	33		
Allyl-	125			33		
1, 2-Diallyl-	146			58		
1, 2-Diisobutyl-	175		0.800	69		53
Phenyl-	243	20	1.099	184		160
m-Tolyl-	244		1.057	203		125
Benzyl-		26		184.5		105
1, 2-Di-m-tolyl-		38		203		125
1, 1-Diphenyl-		44			53.4	180
1, 1-Dibenzyl-		47		300		112
1, 2 Dibenzyl-		65		184.5		105
p-Tolyl-	244	66		200	45	158
1, 2 Diphenyl-		130		184		160
1, 2 Di-p-tolyl-		134		200	45	158
1, 2 Di-p-naphthyl-		140			112	162
1, 2 Di-p-naphthyl-		153			50	160
1, 2-Di-o-tolyl-		161		200		145
1, 2-Di-p-biphenyl-		170		30	53	230
1, 2-Di-o-biphenyl-		182		299	49	102

in the Tables 46-2 to 46-6 are useful for finding the amines in Table 44-11 so information on the amine derivatives can be obtained.

Tables 46-2 and 46-3 list only nitro derivatives of hydrocarbons, halides, and ethers. Nitro derivatives of other groups of compounds are listed with these compounds, i.e., 2,4-dinitrobenzoic acid is listed with other solid carboxylic acids in Table 32-4. The nitroso, azo, and hydrazo groups are sufficiently unique that all such compounds assigned as possible unknowns are listed in Tables 46-4 to 46-6.

TABLE 46-6 AZO COMPOUNDS

Azo compound	M.p. (°C)	Amine B.p. (°C)	Amine M.p. (°C)	Benzamide M.p. (°C)
o-Azotoluene	55	200		145
m-Azotoluene	55	203		125
Azobenzene	68	184		160
o-Azophenotole	131	229		104
p-Azotoluene	145	200	45	158
o-Azobiphenyl	145	299	49	102
p-Azophenetole	160	250	4	173
1,1-Azonaphthalene	190		50	160
2,2-Azonaphthalene	208		112	162

EXERCISES

46-1. A compound (I) that is water-insoluble and melts at 32 to 33.5° gives a positive ferrous sulfate test and a negative Tollens' test. It reacts with zinc and ammonium chloride to give II which gives a positive Tollens' test. I and II are reduced by tin and hydrochloric acid to III (b.p. 246 to 248°) which reacts with benzoyl chloride to give IV (m.p. 101.5 to 103°). Identify I, II, III, and IV.

46-2. A solid (V) melts at 50 to 52.5°; contains nitrogen, and is insoluble in dilute acid and base. It reduces basic ferrous sulfate but not Tollens' reagent. Treatment of V with zinc and ammonium chloride gives VI. VI reduces Tollens' reagent. Reduction of V with tin and hydrochloric acid gives VII (m.p. 44 to 45°). Identify V, VI, and VII.

46-3. A water-insoluble solid (VIII) melts at 112 to 114.2° and contains nitrogen but no halogen or sulfur. It is soluble in dilute hydrochloric acid and reduces ferrous sulfate in the presence of base. It reacts with benzenesulfonyl chloride to give IX which is insoluble in water and soluble in aqueous potassium hydroxide and melts at 135 to 136°. VIII reacts with tin and hydrochloric acid to give X (m.p. 62 to 64°) which is soluble in water. X reacts with acetyl chloride to give XI (m.p. 85°) and benzenesulfonyl chloride to give XII (m.p. 193 to 194.5°). XII is insoluble in water and aqueous acid and soluble in aqueous potassium hydroxide solution. Identify VIII through XII.

47

Synthesis of o- *and* p-*Nitrophenols*

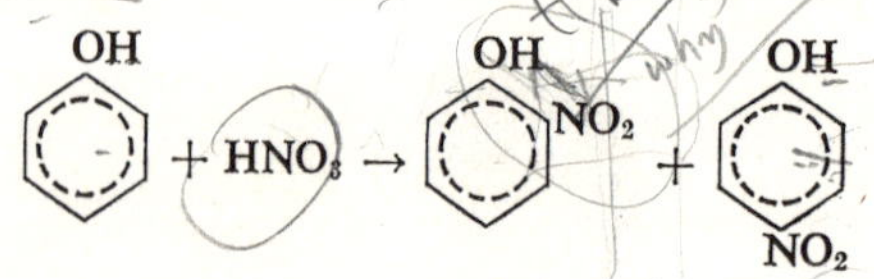

M.p. 41°	M.p. 45°	M.p. 114°
B.p. 182°	B.p. 216°	B.p. 279°d

INTRODUCTION

Because of the strongly activating effect of the hydroxy group, phenol is easily nitrated with mild reaction conditions. The two products of the reaction are readily separated by their different volatilities with steam.

EXPERIMENTAL PROCEDURE

Nitration

Add 13 ml of concentrated nitric acid to 27 g of ice and water in a 125-ml Erlenmeyer flask in an ice bath. Stir the mixture with a thermometer. When the temperature falls below 5°, place a 200-ml round bottom flask containing 12 g of phenol* in the ice bath. Add all the dilute, chilled acid mixture and stir with a thermometer. When the reaction mixture temperature falls below 20°, remove the flask from the ice bath. With a thermometer stir the mixture every 4 to 5 min for 25 min. Cool the mixture with the ice bath when the temperature rises above 30° during the first 15 min of stirring and above 35° during the final 10 min of stirring.

Wash the mixture twice by adding 7 g of ice and 20 ml of water to the flask and stirring. Remove the water washes with a pipet operated with a suction bulb or with vacuum from a water aspirator. If any of the oily reaction mixture is pulled into the pipet, transfer the oily layer (and whatever part of the aqueous acid that is necessary to give complete transfer) to a separatory funnel and return the oily layer to the reaction flask.

Isolation of *o*-Nitrophenol

Place the flask in the steam distillation apparatus shown in Figure 16-2. If steam is obtained from a laboratory steam line rather than from a steam generator with a safety valve, a separate safety valve should be made as shown in Figure 47-1. Note that the glass tubing should be at least 50 cm long and 3 to 4 cm below the surface of the water in the flask. Adjust the water flow through the condenser cooling jacket to approximately one drop a second. Steam distil the mixture. Note the temperature of the distillate after a small amount of *o*-nitrophenol distils. Continue the distillation until no more *o*-nitrophenol comes over. If the *o*-nitrophenol

* First prepare a homogeneous solution by mixing the 12 g of phenol with 3 g of water.

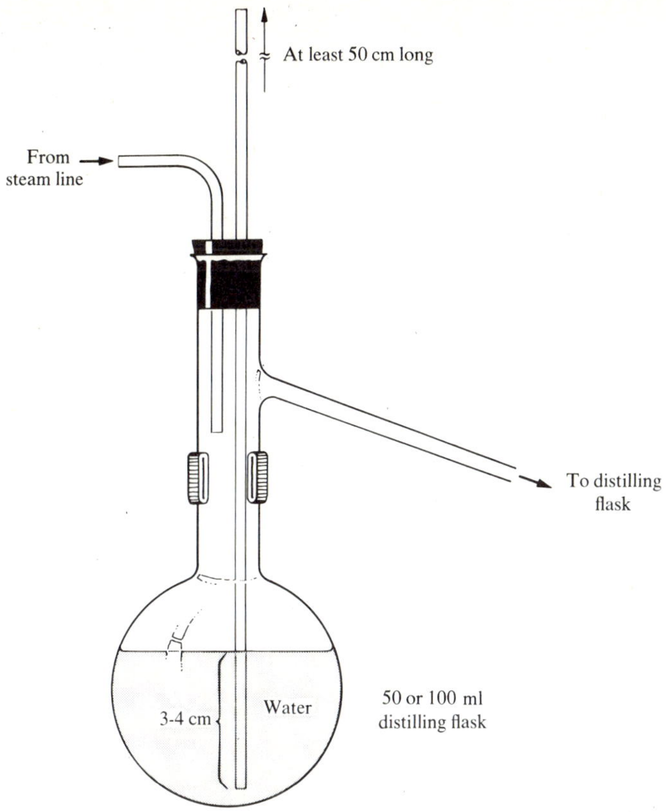

Figure 47-1 Safety valve for steam distillation.

solidifies in the condenser or adapter, shut the condenser water off completely until the solid melts and flows into the receiver. Filter the distillate and recrystallize the product from ethanol.

Problem 47-1. Note the melting point of the product. *Exactly* how should it be recrystallized to obtain maximum yield of pure product? Verify your answer before recrystallizing the product.

Dry the product by pressing the crystals with several pieces of dry filter paper while the suction is on. Weigh the product in a tared yield bottle, determine its melting point, calculate the % yield, and hand the product in to your laboratory instructor. If *p*-nitrophenol is to be isolated, follow the procedure that the instructor assigns.

Isolation of *p*-Nitrophenol by Water Extraction (Optional)

Add a boiling chip, 1 g of activated carbon, and enough water to bring the contents of the flask to 72 to 80 ml (slightly less than half full). Boil the mixture for 5 min. If significant residue has splattered in the Claisen adapter, place the adapter on the flask and heat the flask so water refluxes in the adapter and washes the residue into the flask. Gravity filter, evaporate the filtrate to approximately 15 ml, and allow it to cool almost to room temperature. Chill the concentrated filtrate in an ice bath, filter, and dry the crystals. Weigh the product in a tared yield bottle, determine its melting point, and hand it in to your instructor.

Isolation of *p*-Nitrophenol by Distillation with Superheated Steam (Optional)

Construct the water trap shown in Figure 16-3. This should be installed in the superheated steam apparatus shown in Figure 47-2 just 2 to 3 ins. *ahead* of the superheater and after the steam source. The superheater can

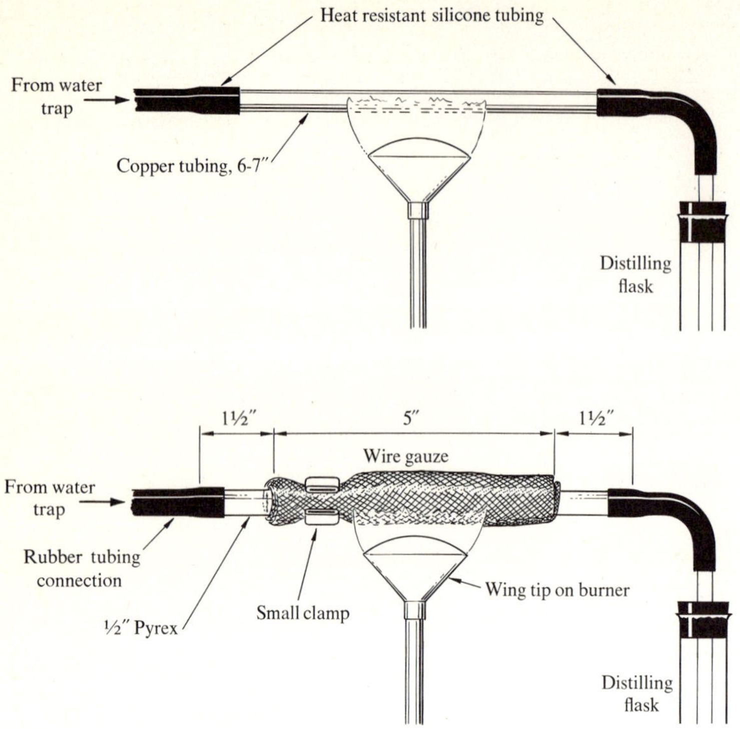

Figure 47-2 Steam super heaters.

be made by tightly wrapping 5 in. of 3/8 or 1/2-in. diameter Pyrex tubing with two winds of wire gauze. The wire gauze is clamped as shown on the steam source side. Carefully observe the dimensions shown in Figure 47-2. Rubber tubing too close to the flame will smolder or burn; too much glass or rubber tubing will cause excessive temperature drop.[1] The superheater should be no farther than 2 in. from the inlet to the distilling flask and preferably closer.

Drain the water from the condenser and remove the tubing. Place an ice bath around the receiving flask. The distillation flask should be heated in an oil bath at 160° or gently heated with a small burner through a wire gauze. If a significant amount of water collected in the distillation flask during the first distillation, heat the flask until most of this water is removed. Steam distil the mixture and adjust the burner flame on the superheater so that the thermometer in the distilling head reads 150 to 160°. Do not heat the superheater unless steam is passing through it and do not heat it so strongly that it melts. The heat on the distilling flask is adjusted so that no water collects there. Continue distillation until no more *p*-nitrophenol collects in the air condenser or receiver.

If *p*-nitrophenol has solidified in the condenser, scrape it with a long spatula or glass rod directly into the receiving flask (after first removing the adapter). Chill the distillate in an ice bath and filter. Dry the product, determine its melting point, and weigh it in a tared yield bottle. Calculate the % yield and hand the product in to your instructor.

Isolation of a By-product (Optional)

Place the distillation flask in a water bath at 35 to 40° and extract the residue with two 20-ml portions of 10% sodium carbonate solution. If significant amounts of residue have splattered on the surface of the Claisen adapter, pour the sodium carbonate solution over these residues into the distillation flask. Be certain the sodium carbonate solution comes into contact with all the residue. Stir the residue with the sodium carbonate solution.

[1] If heat resistant silicone rubber tubing is available, copper tubing can be used as the superheater (Fig. 47-2).

Decant the aqueous solution and use a separatory funnel, if necessary, to separate the aqueous extracts from any organic liquid. Combine the extracts and clarify them with 0.3 to 0.5 g of activated carbon if they are darker than a deep yellow. Chill the extract in an ice bath, add concentrated hydrochloric acid until the mixture is acidic to Hydrion paper, and then add an additional 3 ml. Extract the acidic mixture with three 10-ml portions of chloroform. Evaporate the chloroform and identify the residue after Chapter 49 is studied.

THIN-LAYER CHROMATOGRAPHY (Optional)

Follow the directions at the end of Chapter 36 for activating, spotting, developing, and visualizing the chromatograms with the exceptions and additions noted below. Spot approximately 1×10^{-5} g of as many of the following as your instructor indicates: (a) the nitration mixture after the two ice-water washes and before any further purification, (b) the *residue* from the *o*-nitrophenol distillation, (c) the residue from the *p*-nitrophenol distillation, (d) the recrystallized *o*-nitrophenol, (e) the purified *p*-nitrophenol, (f) the compound obtained in the section on isolation of by-products, and (g) as many of the possible products, by-products, and other contaminants as you are able to get on the plates or strips. Use a few drops of ethanol to dissolve the compounds. Develop with a mixture of 47 parts benzene, 6 parts methanol, and 1 part acetic acid.

As for the chromatogram described in Chapter 43 successively visualize the plates or strips using the spray last. If only one chromatogram is obtained, spray with a 5% ethanolic solution of phosphomolybdic acid and then heat the plates to 80 to 90° for 5 min. The phenols and other reducing compounds are blue on a yellow background. For many compounds the phosphomolybdic acid spray can be made more sensitive by holding the dried plates over an open ammonia bottle. The yellow background fades to white and the blue spots darken or become green or violet. If duplicate chromatograms were obtained, spray the second one with bromocresol green indicator.

Problem 47-2. (a) From only the reactions to the visualization procedures and your knowledge of the reaction how many of the compounds from spots a through f can you identify? (b) How many more can be identified by using in addition to the above the R*f* values of the known compounds that you chromatographed? (c) What can you determine about the purity of your product and what compounds are separated by which steps after the nitration?

EXERCISES

47-1. Using data obtained from this experiment and appropriate information from Appendix I determine the vapor pressure of *o*-nitrophenol at the distillation temperature. Could this method be used for determination of the vapor pressure of *p*-nitrophenol at the temperature of the steam distillation described on page 378? If so, how? If not, why not?

47-2. Why is *o*-nitrophenol so much more volatile than *p*-nitrophenol?

47-3. In the isolation of the by-product why is it important not to use a higher temperature in the sodium carbonate extraction than is called for?

48

The Sandmeyer Reaction

Synthesis of *m*-Nitrophenol

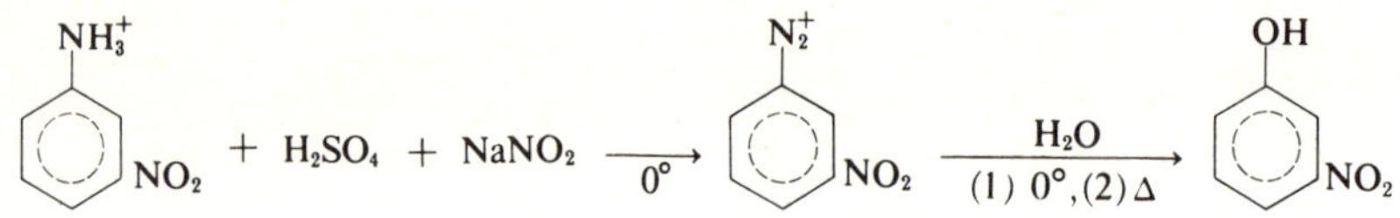

(48-1)

INTRODUCTION

Diazonium groups are readily replaced by hydroxyl groups in warm aqueous solutions. This method is used in this experiment to prepare *m*-nitrophenol. This may be regarded as the last step of a multistep synthesis of *m*-nitrophenol from benzene. Benzene was nitrated in a two-step process in the experiment with Chapter 13. Dinitrobenzene was reduced to *m*-nitroaniline in Chapter 43.

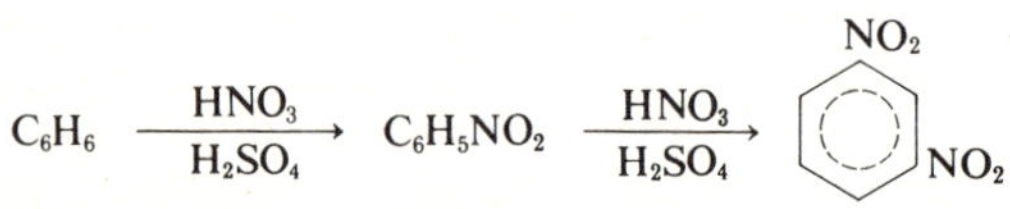

(48-2)

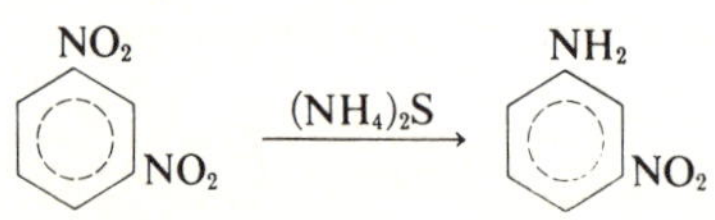

(48-3)

EXPERIMENTAL PROCEDURE

Slowly add, with stirring, 17 ml of concentrated sulfuric acid to 22 ml of water in a 100-ml beaker. While the solution is still hot from the mixing, add 9 g of *m*-nitroaniline and stir. If all the *m*-nitroaniline does not dissolve, warm the mixture on a steam or hot water bath. Pour the mixture into a 400-ml beaker containing 39 g of crushed ice. Stir the mixture vigorously while pouring so that the *m*-nitroaniline sulfate precipitates as a finely divided solid; a colloid would be ideal if it could be obtained. Note that these instructions are opposite to those given in Chapter 12 for obtaining precipitates when purifying a solid by crystallization.

Place the 400-ml beaker in an ice bath and place a thermometer in the beaker. Slowly add 22 ml of 20% sodium nitrite solution in small portions while gently stirring the mixture. Add sufficient ice to the ice bath and stir the solution so the reaction mixture stays below 5° while adding the nitrite solution; if absolutely necessary more ice can be added to the reaction mixture. If all the amine salt has not dissolved, add more sodium nitrite solution a few drops at a time and stir the mixture until the salt dissolves; however, do not add more than 4 ml extra (i.e., a maximum of 26 ml in all).

While adding the amine salt to the sodium nitrite solution, prepare a 70% sulfuric acid solution by *slowly* adding with stirring 41 ml of concentrated sulfuric acid to 32 ml of water in a 500-ml round bottom flask. Equip the flask with a reflux condenser and separatory funnel and stopper the extra neck if it is a 3-neck flask. Heat the solution to boiling. Decant approximately half the solution from the diazonium sulfate crystals into the separatory funnel, and return the 400-ml beaker to the ice bath. Slowly add this solution to the boiling acid mixture. The rate of addition of the supernatant solution and the crystals should be the fastest possible that does not cause excessive foaming; the foam should stay below the bottom of the necks and never rise into the reflux condenser. When the separatory funnel is empty decant the remainder of the supernatant diazonium sulfate solution into the funnel and return the 400-ml beaker with the crystals to the ice bath. Add the remainder of the supernatant liquid to the boiling 70% acid. The addition of all the supernatant liquid should take 10 to 15 min. Then *slowly* and *cautiously* add the *m*-nitrobenzenediazonium sulfate crystals in small portions to the refluxing acid mixture. If necessary up to 5 ml of water can be used to transfer any crystals that stick to the beaker into the refluxing reaction mixture. During the replacement of the diazo group by the hydroxyl group, the mixture turns dark, and usually some *m*-nitrophenol can be observed to form an oil.

Remove the condenser and gently boil the mixture for 15 to 20 min after addition of all the diazonium sulfate. Allow the reaction mixture to cool to 90 to 100°, and then *slowly* pour it into a 600-ml beaker in an ice bath. While pouring the mixture, stir the contents of the beaker thoroughly to cool the reaction mixture rapidly. This promotes formation of a finely divided, homogeneous solid. Chill 50 ml of water in a separate flask in the ice bath while allowing the reaction mixture to stand in the ice bath for 10 to 15 min. Stir the reaction occasionally during this time. Collect the crystals by vacuum filtration. Wash the crystals with three 10-ml portions of the water previously chilled. Use these washes to first rinse any solid remaining in the 600-ml beaker into the Buchner funnel.

Recrystallize *m*-nitrophenol from the minimum amount of 17% hydrochloric acid solution (prepared from 44 ml of concentrated hydrochloric acid and 56 ml of water), approximately 40 to 50 ml will be needed. Clarify the mixture by adding a small amount of decolorizing carbon and then boiling the solution for a minute. Allow the *filtrate* from the clarification to cool to 50 to 55° and then pour the filtrate into a 600-ml beaker in an ice bath, while stirring the contents of the beaker thoroughly for rapid cooling. Allow the mixture to stand in the ice bath for 15 to 20 min; occasionally stir the mixture. Vacuum filter the mixture and wash the precipitate with two 10-ml portions of chilled water. Dry the crystals in an oven set at 70° or in a desiccator. Weigh the product, determine the percent yield and melting point, and hand the product in to your instructor.

THIN-LAYER CHROMATOGRAPHY (Optional)

Follow the directions at the end of Chapter 36 for activating, spotting, developing, and visualizing the chromatograms with the exceptions and additions noted below. Spot as many of the following as your instructor indicates: (a) 2 to 3 drops of the diazotized mixture after heating and before cooling, (b) the product after it precipitates from the concentrated ether, (c) purified product, and (d) as many of the possible products, by-products and other contaminants as you are able to get on the plates or strips. Spot approximately 1×10^{-5} g of (b), (c), and (d) in a few drops of ethanol. Develop with a mixture of 47 parts benzene, 6 parts methanol, and 1 part acetic acid.

As for the chromatogram described in Chapter 36, successively visualize the plates or strips using the spray last. If only one chromatogram of each compound or mixture is obtained, spray with a 5% ethanolic solution of phosphomolybdic acid and then heat the plates to 80 to 90° for 5 min. The phenols and other reducing compounds are blue on a yellow background. For many compounds the phosphomolybdic acid spray can be made more sensitive by holding the dried plates over an open ammonia bottle. The yellow background fades to white and the blue spots darken or become green or violet. If duplicate chromatograms were obtained, spray the second one with ninhydrin indicator.

Problem 48-1. (a) From only the reactions to the visualization procedures and your knowledge of the reaction how many of the compounds from spots a through c can you identify? (b) How many more can be identified by using in addition to the above the R*f* values of the known compounds that you chromatographed? (c) What can you determine about the purity of your product and what compounds are separated by which steps after the diazotization?

49

Properties and Reactions of Phenols

Organic Qualitative Analysis IX

INTRODUCTION

The phenolic functional group, ArOH, is usually readily identified by its acidic properties, reaction with ferric chloride, and activating effects on the aromatic nucleus. The physical properties of phenols are predictable from their structures and, in turn, can give information about their structures. Phenols undergo many reactions that are useful for the synthesis of derivatives and a few reactions that are useful as classification tests.

PHYSICAL PROPERTIES

Boiling and Melting Points

The variation of physical properties of phenols with their structure and molecular weight is discussed in the chapters on alcohols (pp. 214, 233) and nitrogen compounds (pp. 339, 366), and these pages should be reviewed at this time.

Problem 49-1. At 70 mm pressure *o*-nitrophenol has a boiling point of 100°, and *m*- and *p*-nitrophenol have boiling points close to 200°. The *ortho* isomers of the halophenols and dihydroxybenzenes are also considerably more volatile than the *meta* and *para* isomers. The *ortho* isomers of the alkylphenols, however, have about the same boiling points as the *meta* and *para* isomers, e.g., the cresols in Table 49-1. (a) Explain fully. (b) Note that the difference in boiling points between the *ortho* and other isomers is less for the halophenols and dihydroxybenzenes than for the nitrophenols. Why?

Water Solubility

Only phenol and a few simple di- and trihydroxy phenols are more soluble in water than 4 to 5 g per 100 ml of water at 25° (Table 49-1). All other phenols are insoluble in water. Some phenols that show a greater solubility in water than 4 to 5 g/100 ml, which is the solubility limit for Class S, may appear to be insoluble under the conditions of the solubility classification tests because their rate of solution is so slow. If given enough time, however, such phenols do dissolve in water and are in Class S.

TABLE 49-1 PHENOLS

Name	Formula	B.p. (^oC)	M.p. (^oC)	Solubility (g/100g H_2O, 25^oC)	K_a 10^{-9}	Solubility classification
Phenol	C_6H_5OH	182	42	9.3	0.11	S_1a-n
o-Cresol	$CH_3C_6H_4OH$	191	31	2.5	0.06	A_2
m-Cresol	$CH_3C_6H_4OH$	203	12	2.6	0.10	A_2
p-Cresol	$CH_3C_6H_4OH$	191	31	2.3	0.09	A_2
o-Fluorophenol	FC_6H_4OH	152	16		1.5	
m-Fluorophenol	FC_6H_4OH	178	14		0.52	A_2
p-Fluorophenol	FC_6H_4OH	185	48		0.11	A_2
o-Chlorophenol	ClC_6H_4OH	175	7	2.8	7.7	A_2
m-Chlorophenol	ClC_6H_4OH	214	33	2.6	1.6	A_2
p-Chlorophenol	ClC_6H_4OH	217	42	2.7	0.63	A_2
o-Bromophenol	BrC_6H_4OH	195	5		4.1	A_2
m-Bromophenol	BrC_6H_4OH	236	32		1.4	A_2
p-Bromophenol	BrC_6H_4OH	236	64	1.4	0.56	A_2
m-Iodophenol	IC_6H_4OH		40		1.3	A_2
p-Iodophenol	IC_6H_4OH		94		0.63	A_2
o-Aminophenol	$H_2NC_6H_4OH$		174	1.8	0.2	A_2(B)
m-Aminophenol	$H_2HC_6H_4OH$		122	2.6	6.9	A_2(B)
o-Nitrophenol	$NO_2C_6H_4OH$	216	45	0.2	60.0	A_2
m-Nitrophenol	$NO_2C_6H_4OH$		97	1.4	5.0	A_2
p-Nitrophenol	$NO_2C_6H_4OH$		114	1.7	69.0	A_2
2,4-Dinitrophenol	$(NO_2)_2C_6H_3OH$		114	0.6	$6x10^6$	A_1
2,4,6-Trinitrophenol	$(NO_2)_3C_6H_2OH$		122	1.4	$2x4x10^8$	A_1
2,4,6-Trichlorophenol	$Cl_3C_6H_2OH$	245	69		26	A_2
2,4,6-Tribromophenol	$Br_3C_6H_2OH$		95		500	A_1
Catechol	1,2-$C_6H_4(OH)_2$	246	105	45	0.1	S_1n
Resorcinol	1,3-$C_6H_4(OH)_2$	280	110	123	0.3	S_1n
Hydroquinone	1,4-$C_6H_4(OH)_2$	286	173	8	0.2	S_1n
Pyrogallol	1,2,3-$C_6H_3(OH)_3$	309	133	62	100	S_2a
Phloroglucinol	1,3,5-$C_6H_3(OH)_3$		218	1	100	A_2
Hydroxyquinol	1,2,4-$C_6H_3(OH)_3$		140	70	100	S_2a

TABLE 49-1 (cont.)

Name	Formula	B.p. (°C)	M.p. (°C)	Solubility (g/100g H_2O, 25°C)	K_a 10^{-9}	Solubility classification
Chlorohydroquinone	2-Cl-1,4-$(OH)_2C_6H_3$		106			S_1-A_2
2,5-Dichlorohydroquinone	$C_6H_2(OH)_2Cl_2$		168			A_2
m-Hydroxybenzaldehyde	HOC_6H_4CHO	240	106	2.78		A_2
o-Hydroxybenzaldehyde	HOC_6H_4CHO	196	-7	2		A_2
p-Hydroxybenzaldehyde	HOC_6H_4CHO		116	1.38		A_2-N_1
o-Hydrobenzylalcohol	$HOC_6H_4CH_2OH$		87	1.4		A_2
3,4-Dihydroxybenzaldehyde	$(HO)_2C_6H_3CHO$		154	4.5		S_1-A_2
Vanillin	3-(CH_3O)-4-(HO)—C_6H_3CHO	285	81	2.0	$3x10^3$	A_1

Ether Solubility

A molecule with several polar functional groups has stronger dipole attractions for similar molecules than monofunctional molecules do. These polyfunctional compounds are not very soluble in nonpolar solvents such as ethyl ether, benzene, and petroleum ether. This results from the inability of the nonpolar solvent molecules to supply sufficient attractive forces to overcome the strong polar attractive forces that the polyfunctional molecules have for each other. This is merely another manifestation of the rule "like dissolves like."

The number of polar groups necessary to cause insolubility in nonpolar solvents such as ether varies with the type of polar group and with the ratio of the number of polar groups to the number of carbon atoms. The higher the ratio of polar groups to carbon atoms, the less the substance resembles a hydrocarbon and the less soluble it is in nonpolar solvents such as hydrocarbons and ether. Phenols with two hydroxy groups or groups of similar polarity are usually soluble in ether, and those with three such groups are usually insoluble in ether under the conditions of the solubility classification tests.

Problem 49-2. (a) Which of the three dihydroxybenzenes would you expect to be least soluble in water? Why? (b) Which of the three dihydroxybenzenes would you expect to be most soluble in water? Why? (c) Which of the three would you expect to have the highest melting point? Why? Verify your answers first with Table 49-1 and then in the answer section.

Problem 49-3. Phloroglucinol (1,3,5-trihydroxybenzene) is the least soluble in water of all the trihydroxybenzenes. In spite of the high ratio of hydroxy groups to the number of carbon atoms it is even less soluble in water than the dihydroxybenzenes and phenol. Why?

ACIDITY OF PHENOLS

Phenols are stronger acids than alcohols. Substitution of electron-withdrawing groups for hydrogen in phenols causes an increase in acidity. Substitution of electron-donating groups for hydrogen in phenols causes a decrease in acidity.[1]

All phenols are stronger acids than water ($K_a = 10^{-14}$) and dissolve in aqueous sodium hydroxide if the sodium salt is water soluble. That is, the hydroxide ion is a stronger base than all phenoxide ions. An equilibrium involving these two ions, water and a phenol, favors the almost complete reaction of hydroxide ion with the phenol to give the aryloxide ion.

$$Na^+ + OH^- + ArOH \rightleftharpoons ArO^- + H_2O + Na^+ \quad (49\text{-}1)$$

[1] Explanations for these facts are discussed in more detail in any modern organic text. It is assumed the student has studied *thoroughly* this material in his lecture text prior to reading this chapter.

The alkali metal aryloxide is usually soluble in water because it is an ionic compound. As a result phenols (except the few whose sodium salts are water insoluble) dissolve in dilute aqueous sodium hydroxide. The first dissociation constant of carbonic acid is 4.2×10^{-7}.

$$H_2CO_3 \rightleftharpoons H^+ + HCO_3^- \tag{49-2}$$

Acids that have acidity constants of this value or higher displace carbonic acid from its salts such as sodium carbonate. If the sodium salts of the acids are water soluble, the acids dissolve in aqueous sodium bicarbonate.

$$Na^+ + HCO_3^- + QOH \rightleftharpoons QO^- + H_2O + CO_2 + Na^+ \tag{49-3}$$

Carbon dioxide is obtained from the reaction because most of the carbonic acid decomposes at atmospheric pressure.

$$H_2CO_3 \rightleftharpoons H_2O + CO_2 \tag{49-4}$$

Most phenols have K_a's of 1×10^{-9} or less, are too weak to displace carbonic acid from bicarbonate salts, and do not dissolve in aqueous sodium bicarbonate. However, phenols with two or more strongly electron-withdrawing groups, especially if the groups are *ortho* or *para* to the hydroxy group, are acidic enough to cause Equation 49-3 to be shifted far to the right.

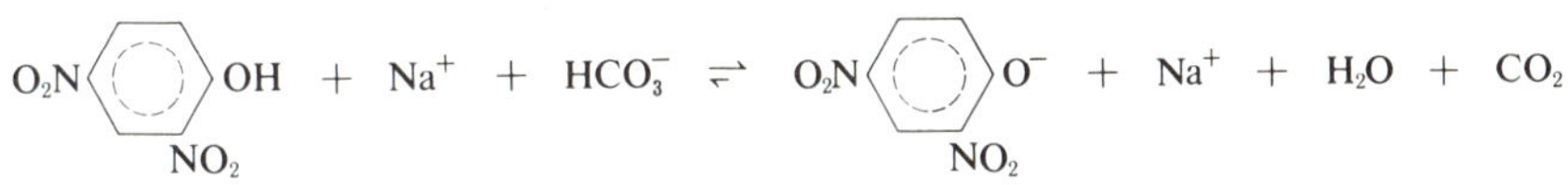
(49-5)

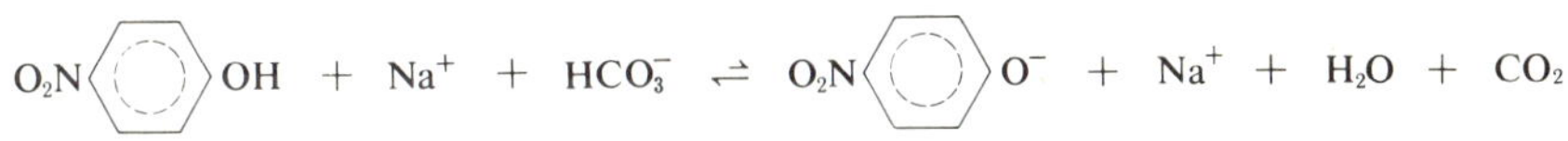
(49-6)

If the sodium salts of these strongly acidic phenols are water soluble, these phenols dissolve in aqueous sodium bicarbonate and are in Class A_1. The solubility classes of some phenols are listed in Table 49-1. Figure 61-1 in the chapter on multiple functional compounds is a solubility classification scheme showing the classes in which the different types of phenols and other compounds studied thus far are found.

REACTIONS OF PHENOLS USED FOR CLASSIFICATION TESTS

Ferric Chloride Test

Most, but not all, phenols form deeply colored complexes with ferric chloride in one solvent or another. Most oximes, hydroxamic acids (Chapter 35), and some polyfunctional compounds such as β-dicarbonyl compounds (Chapter 61) give similar colors with ferric chloride. Most of these compounds are acidic enough to be in Solubility Classes A_1 or A_2 also. If careful use is made of the other classification tests, these compounds are not likely to be confused with phenols.

The method recommended in this manual is to test the unknown by two different procedures: (A) The phenol is dissolved or suspended in water and tested with aqueous 5% ferric chloride solution. (B) The phenol is dissolved in chloroform and tested with a solution of chloroform, pyridine, and ferric chloride. The latter procedure often gives colors with phenols that do not give colors with aqueous ferric chloride. It also does not give colors with most oximes. Procedures a and b usually give different colors for the same phenol.

The colors that this test gives with different phenols are listed in the last two columns of the tables of derivatives at the end of the chapter. Only the abbreviation N or Neg means that no color is produced. A blank means that the results of the test are unrecorded. If any phenols are tested with ferric chloride for which no colors are listed, add the results to the tables in your book.

Procedure A. To prepare the reagent dissolve 5 g of anhydrous ferric chloride in 100 ml of distilled water and filter. Dissolve or suspend 30 mg (1 drop if a liquid) of unknown in 1 ml of water. To dissolve the unknown for procedure 1 or 2 it may be necessary to heat the mixture and then cool it rapidly to room temperature. Add 2 or 3 drops of the ferric chloride reagent and observe any color changes. Sometimes the color appears and disappears quickly.

Procedure B. To prepare the reagent dissolve 1 g of anhydrous ferric chloride in 100 ml of chloroform. Add 7 ml of pyridine, stir the mixture thoroughly, and filter. Dissolve or suspend 30 mg (1 drop if a liquid) of unknown in 1 ml of chloroform. Add 4 drops of the reagent and observe.

Try procedures A and B on phenol, *p*-cresol, β-naphthol, and hydroquinone and compare your results with those listed in the tables of derivatives for these compounds.

Reaction with Bromine Water

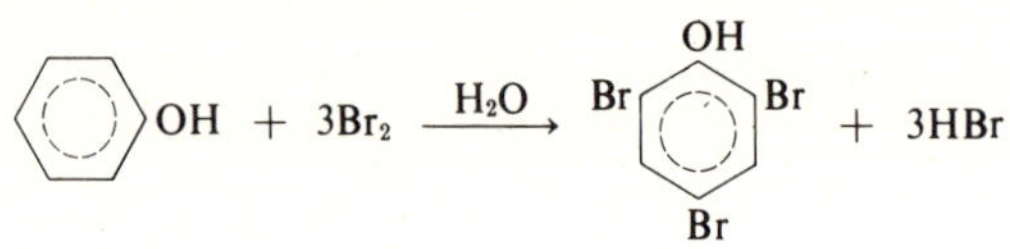

(49-7)

The procedure and interpretation for this test are the same as those given in the chapter on amines (pp. 348, 353). In addition to the tribromo derivative, phenol is converted to a tetrabromo derivative by an excess of bromine water.

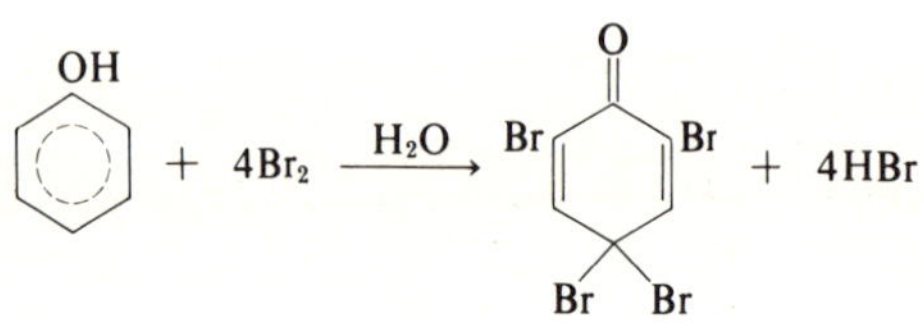

(49-8)

Solid phenols that are *very* insoluble in water give no visible phase change or turbidity with bromine water Phenols with strong electron-withdrawing groups often decolorize bromine water very slowly and may appear not to react. A negative test with a Class A solid should be interpreted cautiously and verified with other tests. The use of this reaction to make derivatives of phenols is described at the end of the section on derivatives.

Coupling Reactions

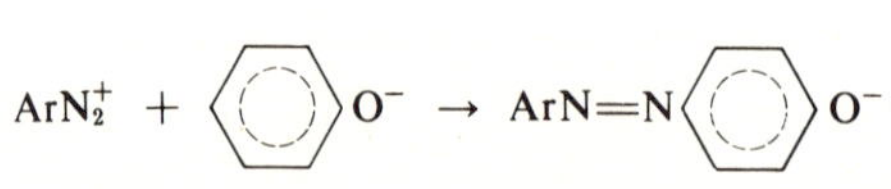

(49-9)

All phenols that do not have strongly deactivating groups react with aryldiazonium salts to give brilliant colors. The experimental procedures for this reaction are given in the section on the nitrous acid classification test for amines (pp. 346 and 347). Aniline should be used as the amine when testing for phenols.

Reaction with Acid Chlorides

$$CH_3COCl + ArOH \longrightarrow ArO\overset{\overset{\displaystyle O}{\|}}{C}CH_3 \qquad (49\text{-}10)$$

$$C_6H_5COCl + ArOH \xrightarrow{NaOH} ArO\overset{\overset{O}{\|}}{C}C_6H_5 \qquad (49\text{-}11)$$

Review pages 228, 350, 351, and 352.

REACTIONS OF PHENOLS THAT INTERFERE WITH OTHER CLASSIFICATION TESTS

Reaction with Bromine in Carbon Tetrachloride

Most phenols decolorize bromine in carbon tetrachloride. The reaction is slower than the reaction of phenols with bromine water because it occurs in a nonpolar solvent.

$$Br_2 + C_6H_5OH \xrightarrow[CCl_4]{} BrC_6H_4OH + HBr \qquad (49\text{-}12)$$

Since hydrogen bromide is evolved, there is little chance of confusing the *substitution* reactions of phenols with the addition reactions of alkenes and alkynes.

Reaction with Cold Dilute Potassium Permanganate

Phenols that do not have strong electron-withdrawing groups give a positive Baeyer test.

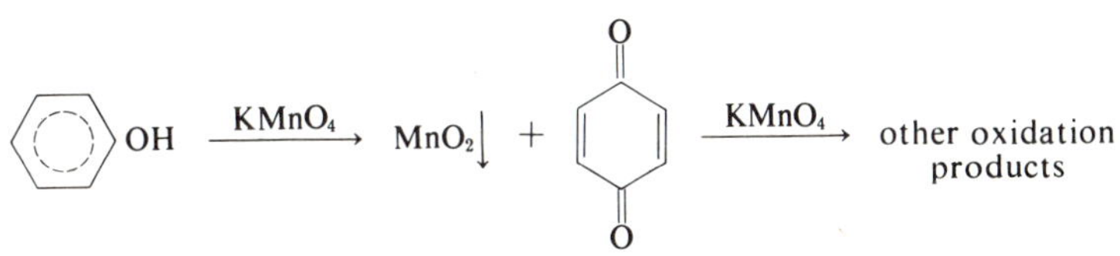

(49-13)

Testing a compound known to be a phenol with Baeyer's reagent or bromine in carbon tetrachloride to see if the phenol has an unsaturated side chain is a waste of time. A few phenols such as α-naphthol are so easily oxidized that they reduce even oxidizing agents as weak as Tollens' reagent.

Problem 49-4. Arrange the following phenols in order of their ease of oxidation, beginning with the most easily oxidized phenol: phenol, hydroquinone, hydroquinone monomethyl ether, pyrogallol, *o*-nitrophenol.

Problem 49-5. Would you expect phenols or their sodium salts to be more easily oxidized? Why?

Problem 49-6. Complete the following equations:

(a) *p*-Nitrophenol $\xrightarrow[H_2O]{Br_2}$

(b) *p*-Nitrophenol + $C_6H_5\overset{+}{N_2}Cl^- \rightarrow$

(c) *m*-Nitrophenol $\xrightarrow[H_2O]{Br_2}$

(d) Resorcinol + $C_6H_5\overset{+}{N_2}Cl^- \rightarrow$

Problem 49-7. Do phenols react more rapidly with bromine in carbon tetrachloride or bromine in water? Why?

REACTION OF PHENOLS AND ALCOHOLS USED FOR THE PREPARATION OF DERIVATIVES

Most of the reactions used for the synthesis of phenol derivatives can also be used for the synthesis of alcohol derivatives. For many derivatives identical procedures are often used.

When alcohols were studied (Chapter 28), no directions for the preparation of derivatives were given although derivatives were listed in Tables 28-6 to 28-8. Because of the limited number of compounds listed as possible unknowns, it was unnecessary to prepare derivatives to identify unknowns at that time. Now that more compounds are assigned as possible unknowns, it may be necessary to prepare derivatives of certain alcohols to confirm identification.

Urethanes

$$ROH + ArNCO \rightarrow RO\overset{\overset{\displaystyle O}{\|}}{C}NHAr \qquad (49\text{-}14)$$

Discussion. With the exception of those prepared from alcohols that contain water, urethanes are the most easily prepared and crystallized derivatives of phenols and primary and secondary alcohols. If possible, other derivatives should be made for alcohols that may contain water (e.g., most water-insoluble alcohols boiling below 170°, all water-soluble liquid alcohols and phenols) or the alcohol should be purified by distillation and then dried over anhydrous potassium sulfate. In fact, the presence of water creates such difficulties that *all* liquid alcohols and phenols should be dried and the test tube flame-dried just before making the urethane.

Water causes so much difficulty because it reacts instantly with isocyanates to give the corresponding disubstituted ureas:

$$H_2O + ArNCO \rightarrow ArNHCO_2H \rightarrow ArNH_2 + CO_2, \qquad (49\text{-}15)$$

$$ArNH_2 + ArNCO \rightarrow ArNH\overset{\overset{\displaystyle O}{\|}}{C}NHAr. \qquad (49\text{-}16)$$

Since the diarylureas are less soluble in most organic solvents than the urethanes, the diarylurea may be what is isolated after crystallization even if it is a minor constituent of the reaction products. In the preparation of urethanes the melting point of the corresponding diarylurea should be kept in mind (N,N′-diphenylurea, m.p. 240°; N,N′-dinaphthylurea, m.p. 297°). Even a small amount of water causes sufficient diarylurea to be produced to contaminate urethanes enough to make their purification and melting point determinations very difficult.

Although the melting points of urethanes of tertiary alcohols are listed in Tables 28-6 to 28-8, the urethanes of tertiary alcohols should be prepared only as a last resort. Urethanes of tertiary alcohols are difficult to prepare and purify because the isocyanate dehydrates a significant portion of the alcohol to give the diarylurea and an alkene.

$$2ArNCO + (CH_3)_3COH \rightarrow ArNH\overset{\overset{\displaystyle O}{\|}}{C}NHAr + (CH_3)_2C{=}CH_2 + CO_2 \qquad (49\text{-}17)$$

Experimental Procedure. Place the *anhydrous* phenol or alcohol (0.4 g if water soluble, 0.6 g if water insoluble) in a *dry* 7.5-cm test tube. Add 0.2 ml of phenyl isocyanate or 0.3 ml of naphthyl isocyanate. **Be careful! Isocyanates are toxic lachrymators!** If the unknown is a phenol, add 2 drops of *anhydrous* pyridine. Mix with a *dry* glass rod; mix especially thoroughly if the unknown is a solid. If no precipitate or only a very small amount of precipitate appears, loosely stopper the test tube with a wad of cotton and heat for 10 min in a hot water bath. Be certain that none of the water vapor gets inside the test tube. Allow the mixture to cool to room temperature. If no precipitate appears, chill the mixture in an ice bath and scratch the inside of the tube.

Remove any liquid from the mixture by vacuum filtration. Wash the urethane with 4 to 5 drops of petroleum ether. Dissolve the urethane in 5 ml of hot carbon tetrachloride or petroleum ether. Some urethanes, especially those of high molecular weight phenols, will not dissolve completely in 5 ml of these solvents even when the solvents are near their boiling points. If this is observed, add benzene or ethyl acetate drop by drop, but not more than 15 drops, to the hot, agitated mixture until the urethane dissolves. Remember that the carbon tetrachloride and benzene used in this recrystallization are very toxic. They should be heated in open containers only when under a hood. A regular fume or desk hood is preferable; however, if no other hood is available, the funnel aspirator hood shown in Figure 12-10 is satisfactory if used correctly.

Often some insoluble material will still remain. What is the probable composition of this impurity? If any such insoluble impurity is present, remove it by gravity filtration or centrifugation of the hot liquid. Allow the mixture to cool to room temperature. If sufficient crystals do not appear, either remove about 2 ml of solvent by evaporation or cool in an ice bath. Use both methods, if necessary, to obtain sufficient crystals.

Esters

Discussion. These esters are prepared by treating the alcohol or phenol with the corresponding acid chloride in the presence of a base.

$$ArCOCl + ROH \rightarrow ArCO_2R + HCl \qquad (49\text{-}18)$$

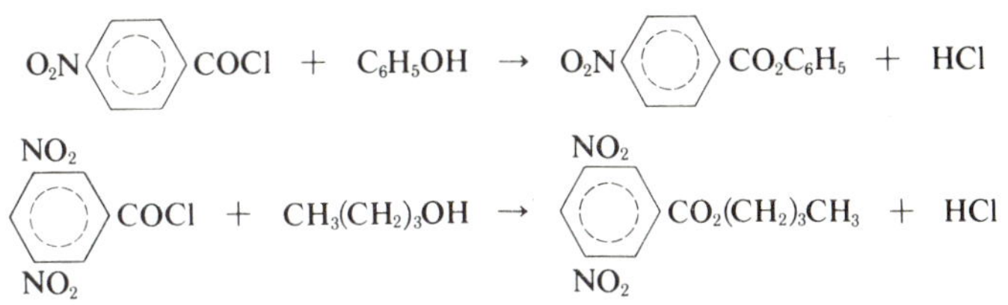

(49-19)

(49-20)

The acid chlorides used in this experiment are hydrolyzed by moist air.

$$H_2O + ArCOCl \rightarrow ArCO_2H + HCl \qquad (49\text{-}21)$$

Therefore, take every possible precaution to keep moisture from the acid chlorides and from the reaction. Do not keep the tops off the acid chloride bottles longer than absolutely necessary and be sure that they are replaced tightly. If the bottles for either the *p*-nitrobenzoyl chloride (m.p. 75°) or the 3,5-dinitrobenzoyl chloride (m.p. 68.5°) have been unsealed for more than a few days, the melting points of the acid chlorides should be checked. If either reagent melts 3.5° or more below the melting point of the pure compound, the reagent should be recrystallized from carbon tetrachloride.

Sometimes the classification reactions with acid chlorides give precipitates suitable for derivatives. Phenols are more likely to do this than alcohols, and the Schotten-Baumann reaction (pp. 228–352) is more likely to give solids than the reaction with acetyl chloride. If precipitates are obtained from these reactions, they should be washed with water and recrystallized from ethanol or aqueous ethanol to a constant melting point.

Experimental Procedure. Dissolve 0.6 g of the alcohol or phenol (use 0.4 g if the unknown is water soluble) in 3 ml of anhydrous pyridine in a 15-cm test tube. If the alcohol is a liquid boiling below 170°, dry it with potassium sulfate before adding it to pyridine; be certain that no potassium sulfate gets into the reaction mixture. Add 0.3 g of benzoyl chloride, 0.4 g of *p*-nitrobenzoyl chloride, or 0.5 g of 3,5-dinitrobenzoyl chloride. Stir the mixture with a glass rod. Stopper the test tube with a slit cork and warm in a water bath at approximately 90° for 2 hr for all tertiary alcohols, 1 hr 30 min for Class S_2 alcohols that are not also tertiary, and 45 min for all other hydroxyl compounds. Heat by placing in the water bath only that portion of the test tube which contains the reactants, allowing the upper tube surface to act as a condenser for evaporating pyridine vapors. If a desk hood is not available, place a funnel aspirator hood over the top of the test tube. After heating the mixture pour it into 8 ml of ice water in a 25-ml Erlenmeyer flask and stir it. Filter and wash the precipitate with two 3-ml portions of 5% sodium carbonate solution, 3 ml of 5% sodium bicarbonate solution, and two 3-ml portions of distilled water.

Recrystallize the precipitate by dissolving it in the minimum amount of hot methanol or ethanol. Agitate the hot solution with a glass rod and add water drop by drop until a cloudiness just persists. Gently heat the solution until the cloudiness disappears, allow to cool to room temperature, and filter the mixture.

The 3,5-dinitrobenzoates and *p*-nitrobenzoates of most naphthols and other polynuclear phenols are only slightly soluble in methanol and ethanol and must be purified as follows: pour the reaction mixture into ice water, filter, and wash the precipitate with two 3-ml portions of *cold* 5% sodium hydroxide solution, 3 ml of 5% sodium carbonate solution, and two 3-ml portions of distilled water. Transfer the precipitate to a 50-ml Erlenmeyer flask and heat to boiling with 20 ml of methanol while agitating the mixture. Allow the mixture to cool 3 to 5 min before filtering. This process is called *digestion* and differs from crystallization in that some of the solid does not dissolve. Wash the precipitate with 3 ml of warm methanol, dry it, and determine its melting point.

The Schotten-Baumann reaction (pp. 228–352) can be used as an alternate procedure for the preparation of benzoates of phenols and primary and secondary alcohols. It is particularly successful with polyhydroxy compounds.

If *p*-nitrobenzoyl chloride and 3,5-dinitrobenzoyl chloride are not available, they can be made from the corresponding acids. Place 0.5 g of 3,5-dinitrobenzoic acid or 0.4 g of *p*-nitrobenzoic acid, and 1 g of phosphorus pentachloride in a dry 5-cm test tube. It is usually necessary to warm the mixture to initiate the reaction. Be careful; the reaction is vigorous. After the reaction has subsided, heat the mixture on a steam bath for at least 5 min and continue heating until no more rapid bubbling occurs. Allow to cool and transfer the mixture to a filter paper on a small Büchner funnel on a suction filter flask. Mash the precipitate on the filter paper with a glass rod while the vacuum is on. Wash the residue with 3 ml of *cold* dioxane. This removes most of the phosphorus oxychloride.

Further purification is unnecessary. The acid chloride should be treated immediately with a phenol or alcohol in the procedure described above. Every delay, no matter how small, will allow some of the acid chloride to be hydrolyzed by moisture from the air.

Bromo Derivatives of Phenols

Because they are formed and purified so easily, the bromo derivatives are often the preferred derivatives for phenols.

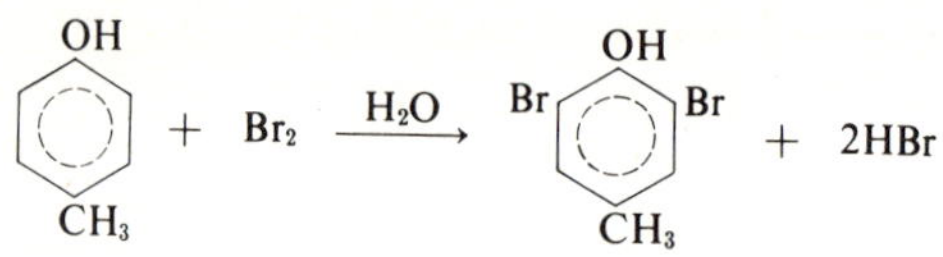

(49-22)

If enough precipitate was obtained from the classification test with bromine water, it should be washed with aqueous sodium bisulfite and water and recrystallized from ethanol. Polybromophenols are best prepared by the same technique used for preparation of polybromoarylamines. The procedure on pages 348 and 353 should be followed except that the polybromophenol should be washed with two 5-ml portions of 10% sodium bisulfite before it is washed with water. The bisulfite wash removes any excess bromine and converts any compounds analogous to 2,4,4,6-tetrabromocyclohexadienone to the "normal" substitution products.

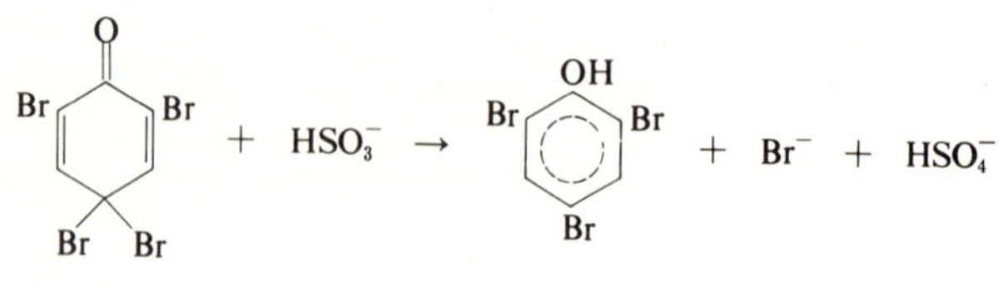

(49-23)

UNKNOWN

Your unknown can be any compound in Tables 49-2 and 49-3 (except those exempted by a footnote) and any previously assigned possible unknown.

TABLE 49-2 PHENOLS (LIQUIDS)

Phenol	B.p. (^oC)	M.p. (^oC)	Derivatives, m.p. (^oC) phenyl-urethane	α-naph-thyl-urethane	benzoate	p-nitro-benzoate	3,5-di-nitro-benzoate	bromo	Color with FeCl3+ aqueous	in $CHCl_3$ with pyridine
2-Chlorophenol	175	7	121	120		115	143	di 78 mono48.5	BP	VP P
Phenol	182	42	126	132.5		127	146	95	BP	P
o-Cresol	191	31	142	142.5	42	94	138	56	B	
2-Bromophenol	195	5	240	129				95	B	
3-Chloro-4-methylphenol	196				71.5					
*2-Hydroxybenzaldehyde (Salicylaldehyde)	197	1.6	133			128			RP	
m-Cresol	203	12	125	127.5	55	90	165.5	84	RP	
o-Methoxyphenol	205	31	136	118		93	141.5	116	R	
o-Ethylphenol	207		142		38.5	56.5	108		B	
2,4-Dimethylphenol	211.5	27.5	112	135	37.5	104	164.6		B	
m-Ethylphenol	217	-4		138	52				V	BV—G
Methyl salicylate (Methyl-o-hydroxybenzoate)	224	-8	117		92	128			RP	V
p-n-Propylphenol	232	22	129		37					
Ethyl salicylate	234	1.3	99		79.5	107.5			RP	
2,4-Dibromophenol	238.5	36			98	183		95	V	
Carvacrol (2-Methyl-5-iso-propylphenol)	240	1	137	116		51	82	46		
3-Methoxyphenol	243.5	-17.5	239	128.5				104		
4-Allyl-2-methoxyphenol (Eugenol)	254.8	3	95.5	122	70	81	131	118	YG	

+ See color code bottom of Table 49-3

* Will not be assigned as an unknown until multiple functional groups are studied or instrumental methods are used to identify unknowns.

TABLE 49-3 PHENOLS (SOLIDS)

Phenol	M.p. (^oC)	b.p. (^oC)	Derivatives, m.p. (^oC): phenyl-urethane	α-naphthyl-urethane	benzoate	p-nitro-benzoate	3,5-di-nitro-benzoate	bromo	Color with $FeCl_3$+: aqueous	in $CHCl_3$ with pyridine
p-n-Propylphenol	22	232	129		36	238				
p-Butylphenol	22	248	115	295	27		67.5			
p-n-Amylphenol	23	250	237		51.2					
2,4-Dimethylphenol	27.5	211.5	112	135	37.5	104	164.6			
o-Ethoxyphenol	28	217	240		31					
*o-Hydroxyacetophenone	28	215			87.5					
o-Cresol	31	191.5	142	141.5		94	138.5	56	B	B
2-Methoxyphenol(Guaiacol)	31	205	136	118	58	93	141.2	116	R	B
3-Bromophenol	32	236		108	86				R P	
3-Chlorophenol	33	214		158	71	99	156		R P	
2-Bromo-4-chlorophenol	33.5				99.5			90		
3-Nitro-o-cresol	34				102					
p-Cresol	36	202	115	146		98	188.6	tri: 95 tetra:108 di: 49	B	B
2,4-Dibromophenol	36	238			98	183			V	
3-Nitro-4-methylphenol	36.5				101.5					
5-Nitro-2-methylphenol (hyd)	38	188			128					
m-Iodophenol	40		138			133	183			
*o-Benzoylphenol	41					124				
2-Nitro-3-methylphenol	41				79					
Phenol	42	182.5	126	133		127	145.8	95	B P	
Phenyl salicylate	42				81	111			N	
p-Chlorophenol	42	217	148.5	166	93	171	186	Mono:33.5 di:90	B V	
2,4-Dichlorophenol	43	209			97			68	B	
o-Iodophenol	43		122		34				B	
o-Nitrophenol	45	216		113		141	155	117	N	R->B
6-Chloro-3-methylphenol	46	196			31					
p-Ethylphenol	47	219	120	128	59	80.5	132.5		V	
2,4,6-Trichloro-m-cresol	47	265			53					
5-Chloro-2-methylphenol	48				71					
2,6-Dimethylphenol	49	205	133	176.5			159	79		
3,5-Dibromo-4-methylphenol	50				94.5	141				
2-Chloro-5-methylphenol	50	194			55.6					
Thymol (3-Methyl-6-isopropyl-phenol)	50.5	233	107	160	33	70	103	55		
o-Benzylphenol	53	312	117.5							
p-Methoxyphenol	55.5	243.5			87				V→ C	
4-Nitro-3-methylphenol	56				77					
o-Phenylphenol	56.6	275			75.6					
*4-n-Caproylresorcinol	56.7	344								
3,5-Dihydroxytoluene (hyd) (Orcinol)	57	288	154	160	88	214	190	104	V	B
3,5-Dibromo-2-methylphenol	57				62	136.7				
2,3,6-Trichlorophenol	58				90				N	

* Will not be assigned as an unknown until multiple functional compounds (Chapter 61) are studied, or instrumental techniques are used to identify unknowns.

+ Color code: bt (bright), dk (dark), B (blue), BP (bluish purple, etc.), Bk (black), Br (brown), Brkr (brick red), Ch (cherry red), G (green), N (negative), O (orange), P (purple), R (red), RP (reddish purple, etc.), V (violet), Y (yellow), C (colorless).

TABLE 49-3(cont.)

Phenol	M.p. (°C)	b.p. (°C)	Derivatives, m.p. (°C) phenyl-urethane	α-naph-thyl-urethane	benzoate	p-nitro-benzoate	3,5-di-nitro-benzoate	bromo	Color with $FeCl_3$ + aqueous	in $CHCl_3$ with pyridine
3,5,6-Trichloro-2-methylphenol	62				110					
3,4-Dimethylphenol	62.5	225	120	141.5			181.6	171	B	
p-Bromophenol	64			169	102	180	191	95	VB	
3,4-Dihydroxytoluene	65	252	166		58					
4-Chloro-3-methylphenol	66	235		153.4	86					
3,5-Dimethylphenol	68	219.5	150		24		195	166		
3,5-Dichlorophenol	68	233			55			189		
2,4,5-Trichlorophenol	68				92.3					
2-Bromo-4,6-dichlorophenol	68.5	268d								
2,4,6-Trichlorophenol	69	245			75.5	105.6	136		N	V
2,4,6-Trimethylphenol	69.6	220	141.5		62			158		
Methyl m-hydroxybenzoate	70	280	115.6						N	
6-Nitro-2-methylphenol	70				42					
2,4,5-Trimethylphenol	71	232	110		63			35		
2,4-Diiodophenol	72				98					
Ethyl m-hydroxybenzoate	74	~290			58				N	
2,5-Dimethylphenol	75	212	163	172.3	61	87	137	178		
N,N-Dimethyl-p-aminophenol	76									
m-Phenylphenol	77	>300			60.5					
4-Chloro-2-iodophenol	78		128		87					
2,3,4,6-Tetramethylphenol	80.5	~240	178.5		71.5					
*4-Hydroxy-3-methoxybenzaldehyde	81	285			78					
4-Chloro-2,6-dinitrophenol	81								N	
*2,4,6-Tribromo-3-methylphenol	82				84.5					
3,4-Diiodophenol	83				123					
2,3,4-Trichlorophenol	83.5				141					
p-Benzylphenol	84	315			87					
2-Nitroresorcinol	85								Br	Br
o-Hydroxybenzyl alcohol	87				51					
2,4,5-Tribromophenol	87				99					
2,4,6-Tribromo-3-nitrophenol	89.5									
3,5,6-Tribromo-2-methylphenol	91				133					
4-Chloro-2,6-dibromophenol	92									
p-t-Amylphenol	92.3	263	108		61					
p-Iodophenol	94		148		119				B	
1-Naphthol	94	279	177.5	152	56	142	217	105	P	G
2,4,6-Tribromophenol	95			153	81	153	174	120	N	BV
*m-Acetylphenol	96	296								
5-Nitro-2-methylphenol	96	188			128					
m-Nitrophenol	97		129	167	95	174	159	91	R	
2,5-Diiodophenol	99.5									
p-t-Butylphenol	100	237	148.5	110	81.5			50		
1,2-Dihydroxybenzene	105	246	169		mono181 di 84	169	152	193	G	
*3-Hydroxybenzaldehyde	106	240			38				G	
3-Chloro-2,4,6-tribromophenol	106									
Chlorohydroquinone	106									
3,5-Dihydroxytoluene	107.5	288	154	160	88	214	190	104	V	B
1,2-Dihydroxynaphthalene	107.5				106					
4-Chloro-2,6-diiodophenol	109									
*p-Hydroxyacetophenone	109								V	

TABLE 49-3(cont.)

Phenol	M.p. (°C)	b.p. (°C)	Derivatives, m.p. (°C) phenyl-urethane	α-naph-thyl-urethane	benzoate	p-nitro-benzoate	3,5-di-nitro-benzoate	bromo	Color with $FeCl_3$+ aqueous	in $CHCl_3$ with pyridine
2,4,6-Trinitro-3-methylphenol	109.5				140					
2,2'-Biphenol	109.7	326	145		101					
Resorcinol	110	280	di 164		117	182	201	112	BV	BV
			mono		135	175				
2,4,6-Tribromoresorcinol	112				mono 120 di 127					
p-Nitrophenol	114			150.5	143	159	187	142	R	
2,3,5-Triiodophenol	114									
2,4-Dinitrophenol	114				132	139			O	R
Ethyl p-hydroxybenzoate	116	297			94				ltV	
p-Hydroxybenzaldehyde	116				90			118	V	V
4-Chloro-3,5-dibromophenol	121				132					
m-Aminophenol	122					143	179		Br	
4-Nitroresorcinol	122				110					
2,4,6-Trinitrophenol	122	often explodes			163				N	N
2-Naphthol	123	286	155.5	156.5	106.5	169	210	84		
4,5,6-Tribromo-2-nitrophenol	123									
2,5-Dihydroxytoluene	124.5				119.5					
6-Nitro-3-methylphenol	129				74					
Methyl p-hydroxybenzoate	131.5		135						N	
p-Cyclohexylphenol	132				118.5	137	168			
Pyrogallol	133	309	tri 173		mono140 di 108 tri 90	tri 230	tri 205	di 158	R	BV
			M.p. below here for acetate							
p-Hydroxybenzophenone	135		81		96				RV	BV
1,2,4-Trihydroxybenzene	140.5				120				R	
3,4-Dihydroxybiphenyl	145		78							
3,5-Stilbenediol	155.6		100.5		150.5					
2,4,6-Triiodophenol	158.5		156		137		181			
Salicylic acid	158.3		135		132	205			V	
4,4'-Dihydroxy-3,3'-dimethylbiphenyl	160		133		185					
p-Phenylphenol	165	307	88		150.5					
2,3,4-Trichloro-1-naphthol	168		123.5							
5-Nitro-1-naphthol	171		114		109					
Hydroquinone	173	286	123		201	258	317	186	N	
3,5-Dinitrosalicylic acid	173.5				163				N	
p-Aminophenol	184d		mono168, di 150				178		V	
Pentachlorophenol	190.2		149.5		163					
m-Hydroxybenzoic acid	200		131						N	
*m-Azophenol	207		144		188					
p-Hydroxybenzoic acid	213		187						N	
1,3,5-Trihydroxybenzene	218		105		173	283	162	151	V	N
5-Nitro-2-hydroxybenzoic acid	299								N	

EXERCISES

49-1. A compound (m.p. 44 to 46°) was insoluble in water and dilute aqueous bicarbonate solution and soluble in aqueous sodium hydroxide. The compound contained chlorine but no other halogen, sulfur, or nitrogen. It gave a precipitate with bromine water, a color with ferric chloride solution, and a positive Schotten-Baumann test. What possible compounds could it be? What would be the quickest step(s) to identification?

49-2. A compound (I) (b.p. 205 to 207°) was soluble in aqueous sodium hydroxide and insoluble in aqueous bicarbonate solution. It gave a color with ferric chloride solution. It reacted with *p*-nitrobenzoyl chloride to give a solid (II) (m.p. 92 to 93°) and with phenyl isocyanate to give a solid (III) (m.p. 235 to 237.5°). It reacted with refluxing hydroiodic acid to give a liquid denser than water (IV) and a water-soluble solid (V) (m.p. 105°). V reacted with phenyl isocyanate to give VI (m.p. 169°) and with benzoyl chloride to give VII (m.p. 100 to 108°). Draw the structures of compounds I through VII.

49-3. A solid (VIII) (m.p. 114 to 116°) was soluble in aqueous sodium hydroxide and insoluble in aqueous bicarbonate solution. It contained no halogen, nitrogen, or sulfur. It gave a light violet color with ferric chloride and reacted under Schotten–Baumann conditions to give IX (m.p. 92 to 93.5°). Identify VIII and IX.

49-4. The *m*- and *p*-cresols have almost identical vapor pressures up to their boiling points. They can not be separated by gas chromatography on silicone oil columns. Their methyl ethers, however, can be easily separated on silicone oil columns. Explain.

50

The Sandmeyer Reaction

Synthesis of *o*-Chlorotoluene

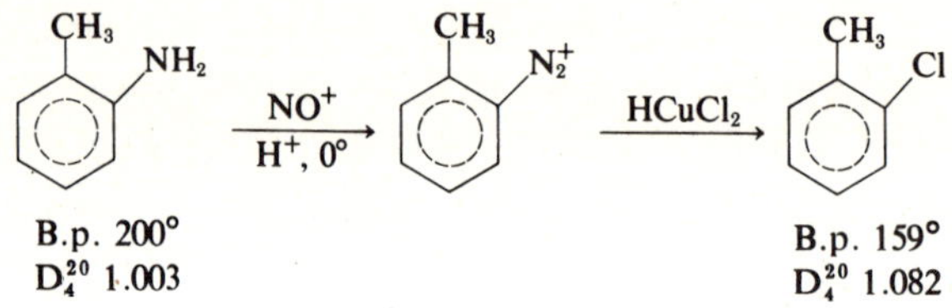

(50-1)

INTRODUCTION

Aromatic halogen compounds are usually synthesized by direct halogenation or from amines *via* the diazonium group by either the Sandmeyer or Gattermann reaction. Although synthesis of aryl halides *via* the diazonium group involves several steps, the yields are high and halogen atoms can be introduced in positions that often are inaccessible to direct halogenation.

EXPERIMENTAL PROCEDURE

The first part of this experiment requires the synthesis of cuprous chloride, a colorless salt which is easily oxidized by air and other oxidizing agents to a blue-green cupric complex. Every precaution should be taken to prevent oxidation of the cuprous salt. The diazonium salt is unstable and decomposes slowly even when cold. This experiment should be started when there is sufficient time (about 1 hr 40 min) to complete the decomposition of the diazonium-cuprous chloride double salt.

Crush 31 g of cupric sulfate ($CuSO_4 \cdot 5H_2O$) with a mortar and pestle. Dissolve the finely divided cupric sulfate and 10 g of sodium chloride in 100 ml of hot water in a 500-ml round bottom flask. In a 50-ml Erlenmeyer flask dissolve 7.2 g of sodium bisulfite and 4.7 g of sodium hydroxide in 60 ml of water. Slowly, over a 5-min period, pour the bisulfite solution into the cupric sulfate solution with stirring. Cool the mixture in a water bath. Do not disturb the mixture while it is cooling so that the cuprous chloride settles out.

Begin the preparation of the diazonium salt while cooling the cuprous chloride. When the cuprous mixture reaches room temperature, carefully decant the aqueous solution from the cuprous chloride. Gently stir the cuprous chloride with two 70-ml washes of water. Stir so that there is a layer of water above the cuprous chloride and so that no air bubbles are made in the solution. After each wash allow the solution to settle for 5 min in an ice bath and carefully decant the clear aqueous solution. Add 15 ml of water and 40 ml of concentrated hydro-

$$\underset{\text{water insoluble}}{Cu_2Cl_2} + 2HCl \rightarrow \underset{\text{water soluble}}{2HCuCl_2} \tag{50-2}$$

chloric acid; gently stir the mixture to dissolve the cuprous chloride. If the diazonium salt is not ready, the flask containing the cuprous chloride should be stoppered and stored in an ice bath.

> **Problem 50-1.** At lower temperatures reactions occur less rapidly. Why is the water used to wash the cuprous chloride not chilled first to slow the rate of oxidation?

Dissolve 11 ml of *o*-toluidine (**Caution! Aromatic amines are very toxic. Do not inhale or spill on the skin.**) in 15 ml of water and 10 ml of concentrated hydrochloric acid in a 250-ml Erlenmeyer flask. Add 25 ml more of concentrated hydrochloric acid after all the amine has dissolved.

$$o\text{-}CH_3C_6H_4NH_2 + HCl \rightarrow o\text{-}CH_3C_6H_4\overset{+}{N}H_3 + Cl^- \tag{50-3}$$

Put a thermometer in the flask and place the flask in an ice bath to cool. The insoluble hydrochloride salt of *o*-toluidine precipitates on this cooling.

While this mixture is cooling, dissolve 7.3 g of sodium nitrite in 21 ml of water in the smaller separatory funnel. When the amine salt mixture has cooled to 0 to 4°, remove the thermometer and place the stem of the separatory funnel as far into the amine salt mixture as possible. The next step is the synthesis of the diazonium salt. The diazonium salt is easily decomposed. To prevent this follow the time limits given in the next paragraph closely. Do not delay the steps even a few minutes. The flask should remain in the ice bath the entire time.

Add a few pieces of ice to the separatory funnel and to the *o*-toluidine hydrochloride mixture. Run the sodium nitrite solution into the flask during a 6 to 9 min period while swirling the flask in the ice bath. Maintain a few small pieces of ice in the flask until the contents are poured into the cuprous chloride mixture. Allow this mixture to stand 4 to 7 min and slowly pour it into the cold cuprous chloride solution while shaking the mixture. A diazonium-cuprous chloride complex precipitates.

$$o\text{-}CH_3C_6H_4N_2^+ + HCuCl_2 \rightarrow o\text{-}CH_3C_6H_4N_2^+\,CuCl_2^- \downarrow + H^+ \tag{50-3}$$

This complex very slowly decomposes at low temperatures to give nitrogen, the product, and cuprous chloride. For the reaction to take place in a reasonable amount of time, it must be heated.

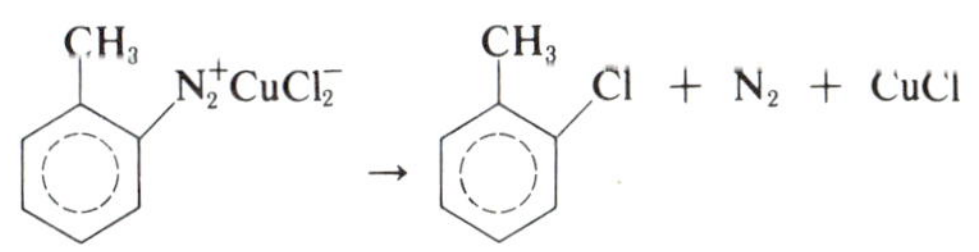

(50-4)

When the addition is completed, remove the ice and swirl the mixture occasionally for 20 min. Then slowly heat the mixture in a hot water bath and agitate the mixture by swirling the flask. The hot water bath should be brought to boiling over about a 15-min period. Heat with the boiling water bath for an additional 5 min. It may be necessary to use a glass rod to break up the solid that forms to prevent loss of product from excess foaming. If necessary, the synthesis may be stopped at this point, the flask stoppered and placed in the locker.

Using the reaction flask, assemble the steam distillation apparatus shown in Figure 16-2.* Collect 45 to 50 ml of distillate. Separate the two layers. Which layer is the product? Be sure not to throw away the product. Add 20 ml of concentrated sulfuric acid to 5 g of ice. Divide this into approximately two equal portions and use them to extract the *o*-chlorotoluene. Wash the *o*-chlorotoluene with three 40-ml portions of water and dry the product over 3 g of calcium chloride. Decant the product from the calcium chloride into a 25-ml distilling or round botton flask. Add a boiling chip and distil. Collect the fraction that boils at 153 to 160° in a tared product bottle. Calculate the % yield, label the bottle, and hand in the product to your instructor.

* It will be necessary also to heat the distillation flask to prevent the accumulation of excess water.

INFRARED OPTION

(a) Obtain infrared spectra in 0.2-mm cells of the organic layer from the steam distillation and of the purified product. Although the 0.2-mm cell is preferable, these spectra can also be obtained in 0.1, or even 0.05 mm cells. (b) Obtain a differential spectrum of these two liquids in 0.2 mm cells (see Infrared Option in Chapter 30). Place the cell with the purified product in the reference beam. (c) Obtain a spectrum of the purified product in a 0.025-mm cell. All of the above spectra should be on the neat liquids, i.e. undiluted with solvents (Chapter 18). Be certain no water droplets are in the sample from the organic layer of the steam distillation.

Problem 50-2. (a) Assign as many of the absorption bands in the spectra as possible to structural features of the product. (b) What types of impurities are evident? (c) From your knowledge of the reaction *and* the spectra what are the probable structures of the impurities? (d) Where did the impurities come from? What steps were taken to minimize their formation? (e) What was done in the purification to remove them? (f) Could anything else be done to minimize their formation? (g) What other steps could be taken to remove them from the product?

EXERCISES

50-1. How is *o*-toluidine synthesized from toluene? What products could be obtained by direct ionic chlorination of toluene? Could this method be used for the synthesis of *o*-chlorotoluene?

51

The Sandmeyer Synthesis of Nitriles

Synthesis of *o*-Tolunitrile

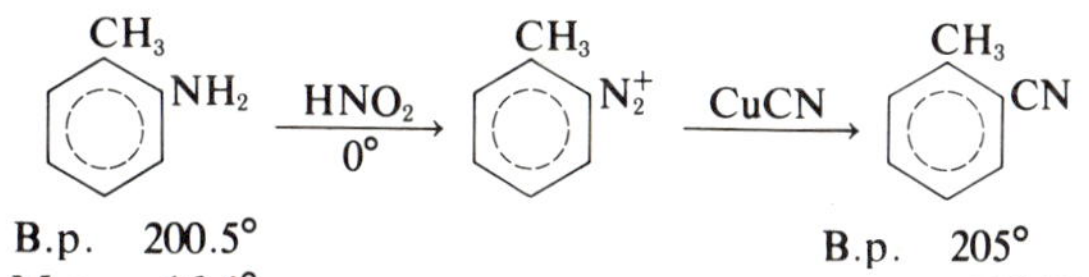

(51-1)

B.p. 200.5°
M.p. −16.4°
D_4^{20} 0.999

B.p. 205°
95°/20 mm.
M.p. −13°
D_4^{20} 0.986

INTRODUCTION

Diazonium salts can be replaced with cyano groups under conditions similar to those used for replacement by halogen atoms. The conditions and materials used in this experiment for the synthesis of *o*-tolunitrile (*o*-cyanotoluene) are very similar to those used for the synthesis of *o*-chlorotoluene. The principal difference is that cuprous chloride reacts to form cuprous cyanide, and the cyanide reacts with the diazonium salt.

DISCUSSION

There are three principal methods for synthesis of nitriles. Nucleophilic displacements of halogen by cyanide ion can be used to synthesize most aliphatic nitriles. As with most nucleophilic displacements the yields from tertiary halides are usually poor. Since aliphatic halides can be synthesized from many other compounds, there are a great many possible starting materials for synthesis of nitriles (and, therefore, acids, amides, amines, etc.) by this method.

$$\mathrm{ROH} \xrightarrow{\mathrm{PBr_3}} \mathrm{RBr} \xrightarrow{\mathrm{CN^-}} \mathrm{RCN} \xrightarrow{\mathrm{H_3O^+}} \mathrm{RCO_2H} \tag{51-2}$$

$$\mathrm{RCH{=}CH_2} \xrightarrow{\mathrm{HX}} \mathrm{RCHXCH_3} \xrightarrow{\mathrm{CN^-}} \underset{\displaystyle\mathrm{CH_3}}{\mathrm{RCHCN}} \xrightarrow{\mathrm{H_3O^+}} \underset{\displaystyle\mathrm{CH_3}}{\mathrm{RCHCO_2H}} \tag{51-3}$$

$$\mathrm{RCH{=}CH_2} \xrightarrow[\text{peroxides}]{\mathrm{HBr}} \mathrm{RCH_2CH_2Br} \xrightarrow{\mathrm{CN^-}} \mathrm{RCH_2CH_2CN} \xrightarrow{\mathrm{H_3O^+}} \mathrm{RCH_2CH_2CO_2H} \tag{51-4}$$

$$\mathrm{C_6H_5CH_3} \xrightarrow[h\nu]{\mathrm{Cl_2}} \mathrm{C_6H_5CH_2Cl} \xrightarrow{\mathrm{CN^-}} \mathrm{C_6H_5CH_2CN} \xrightarrow{\mathrm{H_3O^+}} \mathrm{C_6H_5CH_2CO_2H} \tag{51-5}$$

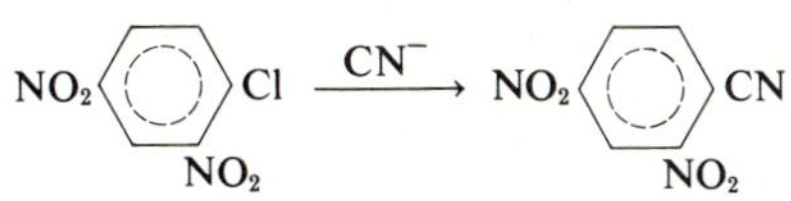

(51-6)

α-Hydroxy nitriles can be synthesized from aldehydes and ketones by the cyanohydrin reaction. These can be hydrolyzed to α-hydroxy acids. The α-hydroxy acids can be dehydrated to α,β-unsaturated carboxylic acids or used to synthesize to give other related compounds.

$$\mathrm{R_2C{=}O} \xrightarrow[\mathrm{CN^-}]{\mathrm{HCN}} \mathrm{R_2C(OH)CN} \qquad (51\text{-}7)$$

$$\mathrm{RR'CHCHO} \xrightarrow[\mathrm{CN^-}]{\mathrm{HCN}} \mathrm{RR'CH}\underset{\mathrm{OH}}{\underset{|}{\mathrm{C}}}\mathrm{HCN} \xrightarrow[\mathrm{H_2O}]{\mathrm{HO^-}} \mathrm{RR'CH}\underset{\mathrm{OH}}{\underset{|}{\mathrm{C}}}\mathrm{HCO_2^-} \qquad (51\text{-}8)$$

$$\mathrm{RR'CH}\underset{\mathrm{OH}}{\underset{|}{\mathrm{C}}}\mathrm{HCO_2^-} \xrightarrow{\mathrm{H_3O^+}} \mathrm{RR'CH}\underset{\mathrm{OH}}{\underset{|}{\mathrm{C}}}\mathrm{HCO_2H} \xrightarrow{\mathrm{H^+}} \mathrm{RR'C{=}CHCO_2H + H_2O} \qquad (51\text{-}9)$$

Aromatic nitriles can be synthesized by the Sandmeyer reaction as illustrated by this experiment.

$$\mathrm{ArH} \xrightarrow{\mathrm{NO_2^+}} \mathrm{ArNO_2} \xrightarrow{[\mathrm{H}]} \mathrm{ArNH_2} \qquad (51\text{-}10)$$

$$\mathrm{ArNH_2} \xrightarrow[0^\circ]{\mathrm{HNO_2}} \mathrm{ArN_2^+} \xrightarrow{\mathrm{CuCN}} \mathrm{ArCN} \xrightarrow{\mathrm{H_3O^+}} \mathrm{ArCO_2H} \qquad (51\text{-}11)$$

Most of the above examples illustrate not only the synthesis of nitriles but also the synthesis of carboxylic acids *via* a nitrile synthesis. An example of synthesis of carboxylic acids by this method is given in the next chapter. In addition to carboxylic acids and their derivatives many other compounds can be synthesized from nitriles; among them are amines, aldehydes, and ketones.

$$\mathrm{RCN} \xrightarrow{[\mathrm{H}]} \mathrm{RCH_2NH_2} \qquad (51\text{-}12)$$

$$\mathrm{RCN + LiAlH_4} \xrightarrow[-80^\circ]{} \mathrm{R}\underset{\mathrm{H}}{\underset{|}{\mathrm{C}}}\mathrm{{=}N^-} \xrightarrow{\mathrm{H_3O^+}} \mathrm{RCHO} \qquad (51\text{-}13)$$

$$\mathrm{RCN + R'MgBr} \longrightarrow \mathrm{R}\underset{\overset{+}{\mathrm{NMgBr}}}{\underset{\|}{\mathrm{C}}}\mathrm{R'} \xrightarrow{\mathrm{H_3O^+}} \mathrm{R}\underset{\mathrm{O}}{\underset{\|}{\mathrm{C}}}\mathrm{R'} \qquad (51\text{-}14)$$

EXPERIMENTAL PROCEDURE

The first part of this experiment requires the synthesis of cuprous cyanide. Cuprous salts are easily oxidized by air. Although cuprous cyanide is less easily oxidized than most cuprous salts, every precaution should be taken to prevent oxidation of the cuprous compounds. The diazonium salt is unstable and decomposes even when cold. This experiment should be started when there is sufficient time (about 2 hr) to complete the decomposition of the diazonium-cuprous cyanide double salt.

Synthesis of Cuprous Cyanide

Crush 31 g of cupric sulfate ($CuSO_4 \cdot 5H_2O$) with a mortar and pestle. Dissolve the finely divided cupric sulfate and 10 g of sodium chloride in 100 ml of hot water in a 500-ml round bottom flask. In a 125-ml Erlenmeyer flask dissolve 7.2 g of sodium bisulfite and 4.7 g of sodium hydroxide in 60 ml of water. Slowly over a 5-min period pour the bisulfite solution into the cupric sulfate solution with stirring. Cool the mixture in a water bath.

When the cuprous mixture reaches room temperature, carefully decant the aqueous solution from the cuprous chloride. Gently stir the cuprous chloride with two 70-ml washes of water. Stir so that there is a layer of water above the cuprous chloride and so that no air bubbles are made in the solution. After each wash allow the solution to settle for 5 min in an ice bath and carefully decant the clear aqueous solution. Suspend the cuprous chloride in 50 ml of water. Dissolve 16 g of sodium cyanide (**Danger! Very poisonous.**) in 25 ml of water and add the solution to the cuprous chloride suspension. Stopper the flask and agitate it. Feel the side of the flask; where does the heat come from? Cool the mixture in an ice bath, and begin the preparation of the diazonium salt while the cuprous cyanide is cooling.

Synthesis of Diazonium Salt

Dissolve 11 ml of *o*-toluidine (**Caution! Aromatic amines are very toxic. Do not inhale or spill on the skin.**) in 30 ml of concentrated hydrochloric acid in a 250-ml Erlenmeyer flask. Put a thermometer in the flask, place the flask in an ice bath, stir the mixture a few minutes, and add approximately 80 g of crushed ice.

While this mixture is cooling, dissolve 7.3 g of sodium nitrite in 21 ml of water in the smaller separatory funnel. Remove the thermometer and place the separatory funnel *above* the flask with the amine salt mixture. The next step is the synthesis of the diazonium salt. The diazonium salt is easily decomposed. To minimize this decomposition closely follow the time limits given in the next paragraph; do not delay the steps even a few minutes. Keep the flask in the ice bath the entire time.

Add a few pieces of ice to the separatory funnel. Run the sodium nitrite solution into the flask during a 6 to 9 min period while swirling the flask in the ice bath. Maintain a few small pieces of ice in the flask until the contents are poured into the cuprous cyanide mixture. While stirring the mixture, slowly add solid, anhydrous sodium carbonate until the mixture gives a basic reaction to *red litmus* paper. Try to avoid an excess of sodium carbonate (about 6 g of the sodium carbonate will be needed to neutralize the acid).

Problem 51-1. *It is vitally important that the acid is neutralized before the next step is attempted.* Why?

Problem 51-2. Give at least two reasons why sodium carbonate instead of sodium hydroxide is used for the neutralization.

Synthesis and Decomposition of Diazonium-Cuprous Cyanide Double Salt

Verify your answers to Problems 51-1 and 51-2 before proceeding further. Add a few pieces of ice and 25 ml of toluene to the cold cuprous cyanide. Then add the neutralized, chilled diazonium solution in 10 to 15 ml portions with vigorous agitation after each addition. Vigorous and thorough agitation is very important here. By adjusting the rate of addition and adding pieces of ice maintain the temperature at 0 to 5°. After all the diazonium salt has been added, keep the mixture in the ice bath 20 min with occasional stirring. Remove the mixture from the ice bath and stir occasionally for 20 min. If necessary, the synthesis can be stopped at this point and the reaction mixture stored in the locker. It is preferable, however, to heat the mixture as described below before interrupting the reaction.

Transfer the mixture to the reflux apparatus shown in Figure 36-1. Heat the mixture to 55° with a hot water bath for 30 min without stirring. Cool by substituting a cold water bath for the hot water bath.

Isolation and Purification of *o*-Tolunitrile

Pour the mixture into a separatory funnel and separate the two layers. Extract the aqueous layer with 15 ml of benzene and add this to the organic layer. With the apparatus shown in Figure 16-2 steam distil the organic layer until no more oily substance distils (approximately 320 ml of distillate). Separate the two layers, extract the aqueous layer with 15 ml of benzene, and add it to the organic layer.

Transfer the mixture to a 100-ml flask. With a Claisen adapter and other glassware construct the distillation apparatus shown in Figure 29-2. Add a boiling chip and vacuum distil the mixture. Immediately discontinue the distillation when the temperature reaches 88°. Change receivers and add a fresh boiling chip. Distil and collect the fraction that boils at 88 to 115°. The boiling point range will be narrower, but the exact range can not be predicted because the vacuum furnished by the water aspirator may vary between 15 and 35 mm pressure. If the product is used to synthesize *o*-toluic acid (Chapter 52), it can be used directly. If it is to be turned in, redistil the product at atmospheric pressure from a 25-ml flask. Collect the fraction that boils 203 to 208° in a tared yield bottle. Weigh the product, calculate the % yield, label the bottle, and hand in the product to your laboratory instructor.

INFRARED OPTION

(a) Obtain infrared spectra in 0.2-mm cells on as many of the following as your instructor tells you: (1) the organic layer from the reaction before dilution with the benzene extract, (2) the organic layer in the steam distillate, (3) the purified product. Although the 0.2-mm cell is preferable, these spectra can be obtained also in 0.1-mm or even 0.05-mm cells if the thicker cell is not available. (b) Obtain differential spectra from (1) and (2); place these in the sample beam. Use (3) as the reference liquid for both samples. See Infrared Option in Chapter 30. Use 0.2-mm cells. (c) Obtain a spectrum of the purified product in a 0.025-mm cell. All of the above spectra should be on the neat liquids, i.e., undiluted with solvents. Be certain no water droplets are in the sample of (1) and (2) that are placed in the infrared cells.

Problem 51-3. (a) Assign as many as possible of the absorption bands in the spectra to structural features of the product (Appendix VII). (b) What types of impurities are evident? (c) From your knowledge of the reaction *and* the spectra what are the probable structures of the impurities? (d) Where did the impurities come from? What steps were taken to minimize their formation? (e) What was done in the purification to remove them? (f) What further steps could be taken to minimize their formation? (g) What further steps could be taken to remove them from the product once the impurities have formed?

Problem 51-4. Compare the by-products and other impurities from this reaction with those from the syntheses in Chapter 50.

52

Synthesis of Carboxylic Acids by the Hydrolysis of Nitriles

Synthesis of *o*-Toluic Acid

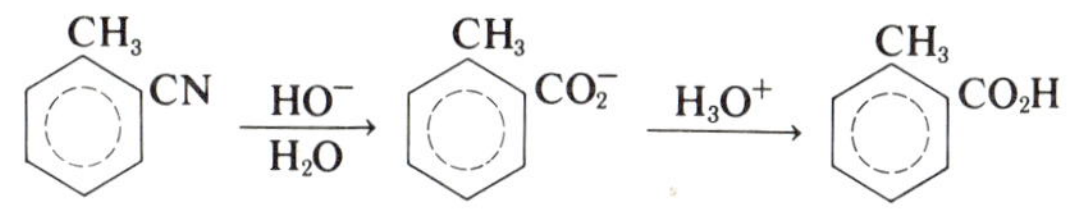

(52-1)

B.p. 205°
M.p. −13°
D_4^{20} 0.986

B.p. 260°
M.p. 104.5°

INTRODUCTION

The hydrolysis of nitriles is a convenient route for the synthesis of carboxylic acids when the nitrile is available or can be easily synthesized. Aromatic nitriles can be synthesized *via* diazonium salts. Aromatic carboxylic acids can thus be synthesized from the corresponding amine, nitro compound, or hydrocarbon.

DISCUSSION

Nitriles can be hydrolyzed by basic or acidic reagents. If groups are present that are sensitive to bases, acidic hydrolysis is indicated.

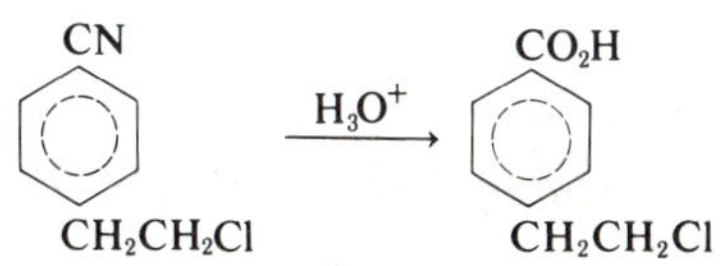

(52-2)

If groups are present that are sensitive to acids, basic hydrolysis is indicated.

$$(CH_3O)_2\overset{H}{C}(CH_2)_7CN \xrightarrow[H_2O]{HO^-} (CH_3O)_2\overset{H}{C}(CH_2)_7CO_2^- \qquad (52\text{-}3)$$

By proper adjustment of the solvolysis conditions amides and esters can be synthesized by solvolysis of nitriles.

$$RCN + H_2O \xrightarrow{H_2SO_4} RCONH_2 \qquad (52\text{-}4)$$

$$RCN + EtOH + H_2O \xrightarrow{H_2SO_4} RCO_2Et \qquad (52\text{-}5)$$

EXPERIMENTAL PROCEDURE

Pour 7.5 ml of *o*-tolunitrile or the entire product from Chapter 51 into a 250-ml flask. Add 140 ml of 10% sodium hydroxide and 20 ml of 95% ethanol. Reflux the mixture for 1 hr 40 min. Remove the reflux condenser and place the flask under a hood or a funnel connected to a water aspirator (Fig. 12-10). Boil the mixture 15 to 20 min, allow to cool a few minutes, and place the reaction flask in an ice bath. Chill 50 ml of concentrated hydrochloric acid in an ice bath. Add a few pieces of ice to the reaction mixture, and then cautiously pour 30 to 35 ml of the chilled acid with stirring into the reaction mixture. With stirring slowly add the remaining chilled acid to the reaction mixture until it gives a strong acid test to pH paper and then add a 3 to 4 ml excess.

Stir the mixture in the ice bath until it is at 0 to 10° and filter. Wash the precipitate with two 15-ml portions of ice water. Recrystallize the precipitate from a methanol-water mixture (p. 85). Dry the precipitate and determine the melting point. Weigh the product in a tared yield bottle and calculate the % yield. Label the bottle and hand in the product to your instructor.

THIN-LAYER CHROMATOGRAPHY (Optional)

Follow the directions at the end of Chapter 36 for activation, spotting, elution, and visualization of the chromatogram. Spot the plate or strips with (a) approximately 3×10^{-5} g of the final purified product, (b) as many of the likely products and contaminants ($\sim 3 \times 10^{-5}$ g each) as there is space for on the plate, and (c) 8 to 10 drops of cooled reaction mixture taken after approximately 20 ml of the chilled hydrochloric acid has been added. Be sure to stir the mixture well before taking this sample. Use a few drops of ethanol to dissolve (a) and (b) for spotting. The spots should be no larger than 5 mm in diameter, and the distances between the centers should be at least 25 mm and preferably 30 or 35 mm. The center of the spots should be at least 15 mm from the side of the plate or strip. Use the three successive visualization procedures (with the bromocresol green spray last).

Problem 52-1. (a) How pure is your product? (b) What are the principal contaminants and by-products? Use your knowledge of the reaction, the R*f* values, and the results from application of the three visualization procedures to *both* separated reaction products a and c.

EXERCISES

52-1. The carboxylic acid group can be attached to a molecule by two different methods: (a) carbonation of a Grignard reagent (Chapter 33); (b) attachment of a cyano group (Chapter 51) followed by hydrolysis to a carboxylic acid group (this chapter). One of these two methods is the last step in the synthesis of the carboxylic acids listed below. Outline how each of the acids below could be synthesized from compounds containing only carbon, hydrogen, and oxygen. Indicate briefly why the nitrile or Grignard synthesis was chosen to introduce the carboxylic acid group for each compound.

(a) $(CH_3)_3CCO_2H$

(b) 

(c) 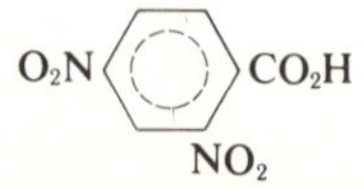

(d) 

(e) $CH_3CH(OH)CH_2CH_2CH_2CO_2H$

53

Use of Acetamido Group to Effect Certain Orientations

Synthesis of *m*-Bromotoluene

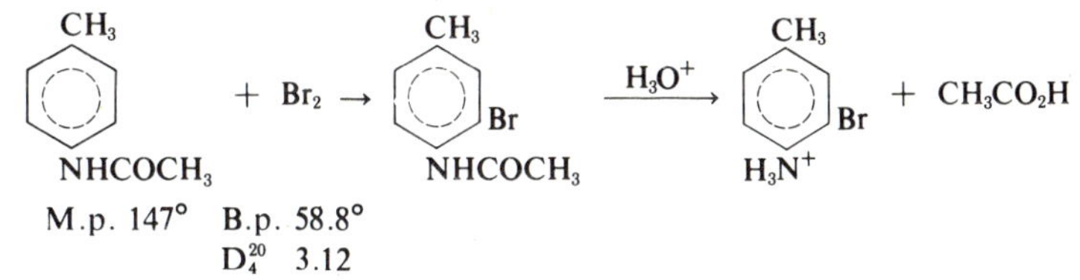

(53-1)

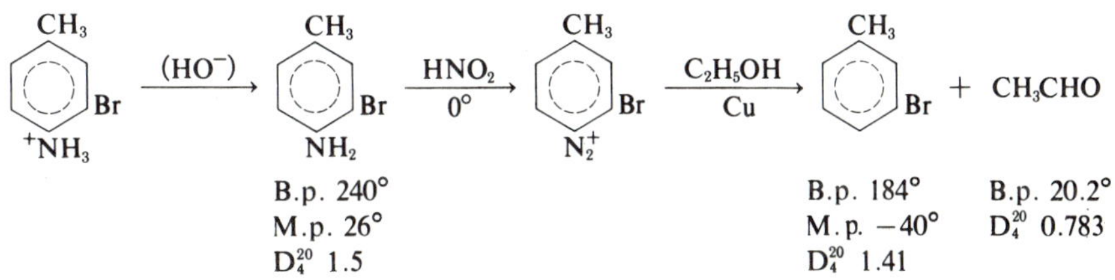

(53-2)

INTRODUCTION

The nitro, amino, and acetamido groups are often used to effect certain orientations of entering groups in aromatic molecules and are removed after the desired orientation is effected. The principle is similar to that used in Chapter 45 where the sulfonamide group blocked the *para* position so that the two chlorine atoms could go only *ortho* to the amino group.

DISCUSSION

Amino groups very strongly activate their *ortho* and *para* positions to electrophilic substitution. After the substitution reaction the amino group can be removed *via* the diazonium salt to give compounds unobtainable by direct substitution.

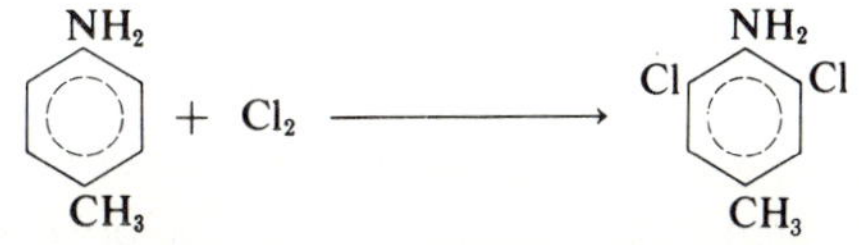

(53-3)

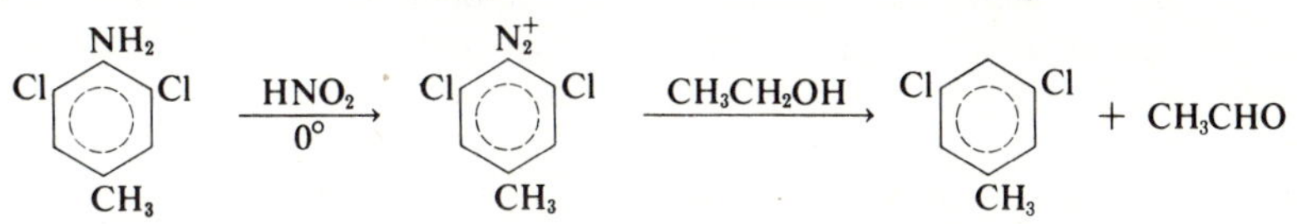

(53-4)

The acetamido group is also *ortho, para*-directing but not as strongly as the amino group. The acetamido group can be easily hydrolyzed to the amino group which can be removed as described above.

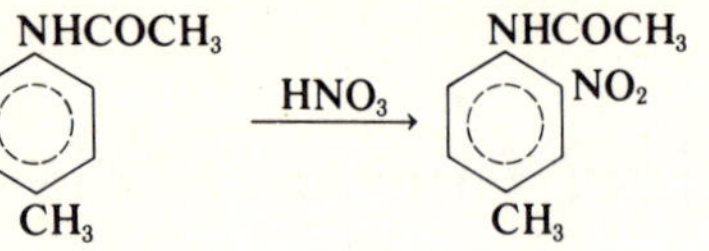

(53-5)

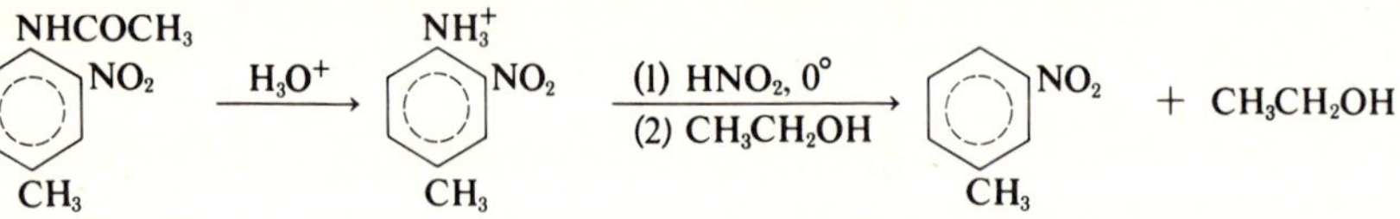

(53-6)

The nitro group can also serve as a blocking or orienting group and be removed after being reduced first to an amino group.

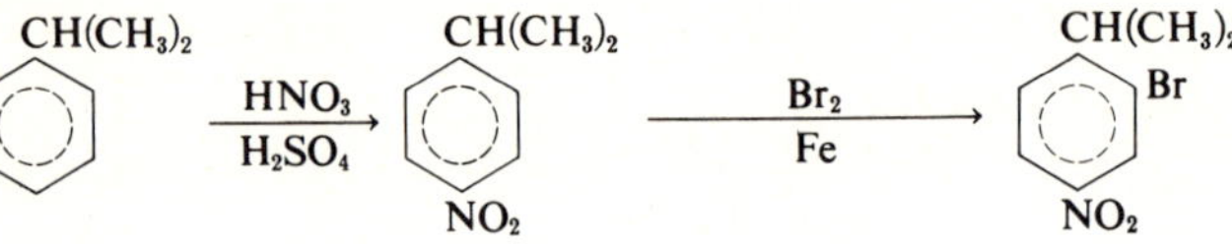

(53-7)

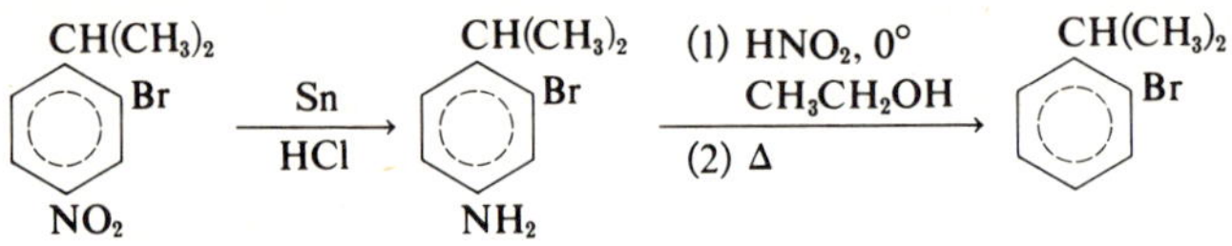

(53-8)

Problem 53-1. Both the sulfonic acid group (Chapter 45) and the nitro group are strongly *meta*-directing and deactivating. When an orienting or blocking group is needed that is *meta*-directing or deactivating, the sulfonic acid or sulfonamide group is usually used rather than the nitro group. Why?

Problem 53-2. Outline a synthesis of 2-chloro-5-cyano-1-ethylbenzene from ethylbenzene that uses as few steps as possible.

The current experiment synthesizes *m*-bromotoluene from *p*-acetotoluidide. This synthesis could be regarded as the final steps in the synthesis of *m*-bromotoluene from toluene or *p*-nitrotoluene. Two of the steps in this synthesis from toluene have been done in earlier experiments.

$$C_6H_5CH_3 \xrightarrow[H_2SO_4]{HNO_3} o\text{- and } p\text{-}NO_2C_6H_4CH_3 \tag{53-9}$$

$$p\text{-}NO_2C_6H_4CH_3 \xrightarrow[HCl]{Sn} p\text{-}NH_2C_6H_4CH_3 \text{ (Chapter 42)} \tag{53-10}$$

$$p\text{-}NH_2C_6H_4CH_3 \xrightarrow[\Delta]{AcOH} p\text{-}(AcNH)C_6H_4CH_3 \text{ (Chapter 39)} \tag{53-11}$$

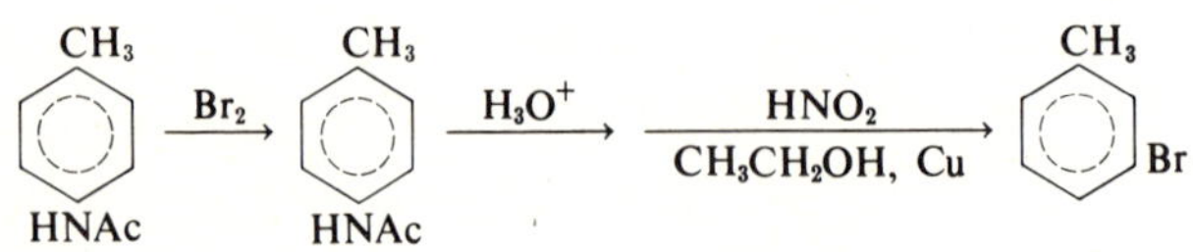

(53-12)

The effect of the acetamido blocking group in this synthesis is similar in principle but different in orientation to the sulfonamide blocking group used in Chapter 45. In both syntheses the groups block the *para* position. The sulfonamide group reinforces the orientation by the amino group, whereas the acetamido activates the position *ortho* to it more strongly than the methyl group activates the positions *ortho* to it. (Fig. 53-1).

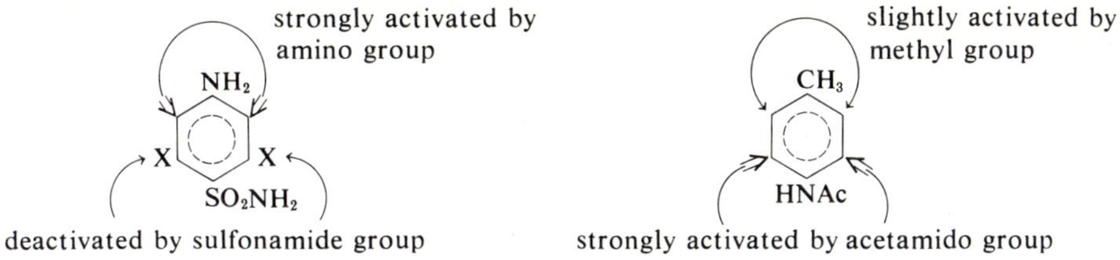

Figure 53-1 Comparison of effects of acetamido and sulfonamide groups.

EXPERIMENTAL PROCEDURE

Bromination

Dissolve 16 g of *p*-acetotoluidide in 45 ml of glacial acetic acid in a 200-ml round bottom flask. Place the flask in the apparatus shown in Figure 16-1. Add 6 ml of bromine to a separatory funnel. **Caution!** See Chapter 16 for techniques for handling bromine. Adjust the stopcock so that the bromine flows into the reaction mixture a drop at a time. Adjust the flow so that all 6 ml of bromine is added in 6 to 9 min. While the bromine is dripping in, swirl the ring stand and the reaction flask so that the bromine mixes well with the acetic acid solution. After all the bromine has been added, rinse the separatory funnel with 5 ml of glacial acetic acid and add this to the reaction mixture.

Continue swirling the mixture for 12 to 15 min. Pour the mixture into a 250-ml beaker containing 80 g of ice, 60 g of water, and 2 g of powdered sodium bisulfite. Stir the mixture; if the amide does not crystallize, scratch the side of the beaker. Stir for a few minutes after the ice melts and allow to stand with occasional stirring at least 5 min after crystalline amide first appears. Filter the mixture and wash the crystals with four 15-ml portions of ice water.

Hydrolysis of Amide

Transfer the precipitate to a 250-ml round bottom flask, add 30 ml of water, 18 ml of concentrated sulfuric acid, and reflux the mixture for 1 hr. Allow to cool approximately 5 min, place in an ice bath, stir, and pour into a mixture of 80 g ice and 35 ml of 50% aqueous sodium hydroxide solution. Stir the mixture and transfer[1] it to a separatory funnel to separate the two layers. Wash the bromotoluidine with two 10-ml portions of water.

Diazotization and Removal of the Amino Group

Chill 55 ml of 95% ethanol in a 500-ml round bottom flask in an ice bath. Slowly add 14 ml of concentrated sulfuric acid with stirring. Stir this mixture a few minutes in the ice bath and add the crude bromotoluidine. While the flask stands in the ice bath, construct the apparatus shown in Figure 36-1 using the 500-ml flask. Cool a solution of 10 g of sodium nitrite in 15 ml of water in an ice bath. Add one-half of this sodium nitrite solution and a few pieces of ice to the dropping funnel; return the other half to the ice bath. The sodium nitrite is added in portions so that it does not become too warm. While swirling the mixture, run the sodium nitrite into the flask a few drops at a time. Set the rate of addition so that the temperature of reaction is maintained below 10°. Repeat this with the remaining sodium nitrite solution, again adding a few pieces of ice to the dropping funnel. Allow the mixture to stand in the ice bath with occasional swirling.

Add 4 g of copper bronze or copper powder that has been washed with ether. Remove the ice bath but

[1] Occasionally in a cold laboratory the bromotoluidine solidifies before purification. If this occurs, it should be filtered and washed with cold water instead of being purified by extraction. If the bromotoluidine solidifies during the *extraction*, wash it with water at 30 to 35° and it will melt.

keep it nearby if moderation of the reaction becomes necessary. Allow the flask to stand 10 min with occasional swirling. Place a water bath under the flask and gradually heat the mixture. If the reaction becomes violent, cool it with the ice bath. After the strongly exothermic reaction subsides, reflux the mixture for 20 to 25 min.

Add 200 ml of water and change the apparatus to that shown in Figure 29-2 and *steam* distil the mixture until no more oil comes over with the water. Separate the oil and wash it successively with: 10 ml of 10% sodium hydroxide solution, cold 60% aqueous sulfuric acid made by adding 9 ml of concentrated sulfuric acid to 10 g of ice, two × 10 ml (i.e. two successive 10-ml portions) of concentrated sulfuric acid, and two × 10 ml of cold 5% sodium carbonate solution (be careful of foaming). The above washes and extractions are facilitated by using two separatory funnels. The product can be drained *directly* into a separatory funnel in all the separations in which it is denser than the wash.

Drain the product into a 25-ml Erlenmeyer flask containing 4 to 6 g of calcium chloride. Swirl a few minutes and decant into a 25-ml round bottom or distilling flask. Distil into a tared yield bottle, collecting the fraction that boils at 180 to 187°. Weigh, calculate the % yield, and hand in the product to your laboratory instructor.

INFRARED OPTION

(a) Obtain infrared spectra in 0.2-mm cells on as many of the following as your instructor tells you: (1) the organic layer after refluxing with the copper and before steam distillation, (2) the organic layer in the steam distillate, (3) the purified product. Although the 0.2-mm cells are preferable, these spectra can be obtained also in 0.1-mm cells or even 0.05-mm cells if the thicker cells are not available. (b) Obtain differential spectra, using 0.2-mm cells from (1) and (2); place (1) and (2) in the sample beam. Use (3) as the reference liquid for both samples. See Infrared Option in Chapter 30. (c) Obtain a spectrum of the purified product in a 0.025-mm cell. All the above spectra should be on the neat liquid, i.e., undiluted with solvents. Be certain no water droplets are in the samples of (1) and (2) that are placed in the infrared cells.

Problem 53-3. (a) Assign as many as possible of the absorption bands in the spectra to structural features in the product (Appendix VII). (b) What types of impurities are evident? (c) From your knowledge of the reaction *and* the spectra what are the probable structures of the impurities? (d) Where did the impurities come from? What steps were taken to minimize their formation? (e) What steps in the purification were taken to remove them? (f) Could any further steps be taken to minimize their formation? (g) Could any further steps be taken to remove them from the product once they have formed?

Problem 53-4. Compare the by-products and other impurities from this reaction with the by-products and impurities from the syntheses in Chapter 50 and 51.

54

Meerwein-Ponndorf-Verley Reduction

Synthesis of Benzhydrol (Diphenylcarbinol)

$$3(CH_3)_2CHOH + Al \xrightarrow{(Hg)} 3(CH_3)_2CHO^- + Al^{3+} + \tfrac{3}{2}H_2 \qquad (54\text{-}1)$$

B.p. 82.5°
D_4^{20} 0.785

$$(C_6H_5)_2CO + (CH_3)_2CHOH \xrightarrow{[(CH_3)_2CHO]_3Al} (C_6H_5)_2CHOH + (CH_3)_2CO \qquad (54\text{-}2)$$

$(C_6H_5)_2CO$: B.p. 305°, M.p. 47°
$(C_6H_5)_2CHOH$: B.p. 288°, M.p. 68°
$(CH_3)_2CO$: B.p. 56°, D_4^{20} 0.791

INTRODUCTION

The Meerwein–Ponndorf–Verley reaction reduces ketones to secondary alcohols. The reaction is specific for ketones; other easily reduced groups such as olefinic double bonds are not usually reduced by this method.

Aluminum isopropoxide is the usual catalyst for this reaction. Synthesis of and efficient catalysis by aluminum isopropoxide require that all apparatus and reagents be very dry.

EXPERIMENTAL PROCEDURE

Scrape the surface of 5 g of aluminum foil with some fine sandpaper or emery paper. Put the scraped foil, 35 ml of isopropyl alcohol, and 0.2 g of mercuric chloride into a 50-ml round bottom flask. Attach a reflux condenser with a drying tube in the top to construct the apparatus shown in Figure 10-lb. Heat the mixture with a water bath until it boils gently. Remove the water bath and add a few drops of carbon tetrachloride. Shake the flask a few times. The reaction should begin and is indicated by a rapid evolution of hydrogen. The reaction may occasionally become violent, and a beaker of cold water should be available to place around the flask to slow the reaction. Sometimes the reaction will not start unless the mixture is heated after the addition of the carbon tetrachloride. When the rate of hydrogen evolution decreases significantly, reflux the mixture until all the aluminum dissolves. This sometimes takes one or more hours. If the experiment is to be continued the next period, replace the hot water bath with a cold water bath; cool the mixture about 10 min before taking apart the apparatus and tightly stopper the flask *as soon as* the condenser is removed.

Problem 54-1. Why is the aluminum foil scraped?

Problem 54-2. What is the purpose of the mercuric chloride catalyst? Of the carbon tetrachloride catalyst?

Cool the aluminum isopropoxide solution to room temperature and decant it from the residue into a 100-ml round bottom flask. Add 4 g of benzophenone and two boiling chips. Attach a fractionating column, condenser, adapters, and a small graduated cylinder to construct the apparatus shown in Figure 5-5. Slowly distil the mixture, collecting *about 1 drop every 2 min.* When the temperature reaches 80°, stop the distillation.

Problem 54-3. What is the maximum volume of acetone that can distil? What may be in the distillate other than acetone? Verify your answers in the back of the text before proceeding further.

If less than the theoretical amount of acetone plus 4 ml has distilled, begin distilling again. Distil *slowly* until this amount is collected.

Problem 54-4. Why is it important to remove all the acetone? Why is the mixture distilled slowly? Why is the acetone distilled through a fractionating column?

Remove the fractionating column and distil 20 ml of the isopropyl alcohol. Add a few pieces of ice and 25 ml of 10% hydrochloric acid. Separate the benzhydrol by filtration. Recrystallize the product from a benzene-petroleum ether mixture (p. 85). Benzhydrol is more soluble in benzene than in petroleum ether.

Dry the product and determine its melting point. Weigh the compound in a tared yield bottle and calculate the % yield. Label the product and hand it in to your instructor.

Synthesis of 1-Phenyl-1-propanol (Ethylphenylcarbinol)

$$C_6H_5COC_2H_5 + (CH_3)_2CHOH \xrightleftharpoons{[(CH_3)_2CHO]_3Al} C_6H_5CHOHC_2H_5 + (CH_3)_2CO \qquad (54\text{-}3)$$

$C_6H_5COC_2H_5$: B.p. 219°; M.p. 19°; D_4^{20} 1.01

$C_6H_5CHOHC_2H_5$: B.p. 219°; D_4^{20} 1.004

From the procedure given above devise a method for reducing 8 ml of propiophenone, the product from Chapter 17, to 1-phenyl-1-propanol. Note that the product is a liquid so that the isolation and purification procedure will differ considerably from that outlined above.

Problem 54-5. What could happen in the isolation and purification procedure to decrease the yield and cause the product to be impure? How could this be prevented? Check your answer.

Briefly outline the synthesis, isolation, and purification of 1-phenyl-1-propanol and show the outline to your instructor before beginning the synthesis.

INFRARED OPTION

Obtain infrared spectra on all products from the reactions before purification. Include the acetone and 2-propanol distillates.

Problem 54-6. How should the samples be placed in the spectrophotometer? Be specific as to cell thicknesses, whether the sample should be examined as a solution, and other details. Verify your answer before obtaining the spectra.

55

The Aldol Condensation

Synthesis of Benzalacetophenone

$$C_6H_5CHO + CH_3\overset{\overset{\displaystyle O}{\|}}{C}C_6H_5 \xrightarrow{HO^-} C_6H_6CH{=}CH\overset{\overset{\displaystyle O}{\|}}{C}C_6H_5 + H_2O \qquad (55\text{-}1)$$

Benzaldehyde	Acetophenone	Benzalacetophenone
B.p. 179°	B.p. 204°	B.p. 347°d
M.p. −26°	M.p. 20°	M.p. 62°
D_4^{20} 1.046	D_4^{20} 1.028	

INTRODUCTION

Benzaldehyde is condensed with acetophenone under the influence of base. Like all aldol condensations that initially produce a compound with a hydroxy group α to an aromatic ring and β to a carbonyl group, this reaction produces an α,β-unsaturated ketone rather than a β-hydroxy ketone.

$$ArCHO + RCH_2\overset{\overset{\displaystyle O}{\|}}{C}R \xrightarrow{HO^-} \left[Ar\overset{\overset{\displaystyle H}{O}}{C}HCHR\overset{\overset{\displaystyle O}{\|}}{C}R\right] \rightarrow ArCH{=}CR\overset{\overset{\displaystyle O}{\|}}{C}R + H_2O \qquad (55\text{-}2)$$

DISCUSSION

If two different aldehydes with α-hydrogen atoms are treated with base, all four possible aldol condensation products are obtained.

$$\begin{aligned} CH_3CHO + CH_3CH_2CHO \xrightarrow{HO^-}\ & CH_3\overset{\overset{\displaystyle H}{O}}{C}HCH_2CHO + CH_3CH_2\overset{\overset{\displaystyle H}{O}}{C}HCH_2CHO + \\ & CH_3CH_2\overset{\overset{\displaystyle H}{O}}{C}H\underset{\underset{\displaystyle CH_3}{|}}{C}HCHO + CH_3\overset{\overset{\displaystyle H}{O}}{C}H\underset{\underset{\displaystyle CH_3}{|}}{C}HCH_2CHO \end{aligned} \qquad (55\text{-}3)$$

Crossed aldol condensations do, however, give good yields of a single compound if (a) one of the carbonyl compounds has no α-hydrogen, (b) the compound that supplies the carbanion does not easily undergo an aldol condensation by itself, as

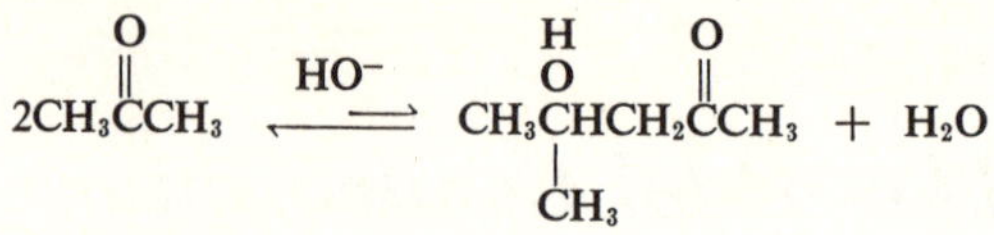
(55-4)

The latter group includes nitro compounds because they have no carbonyl group and most ketones (when the catalyst is a base) because the equilibrium for their condensation is very unfavorable (Eq. 55-4)

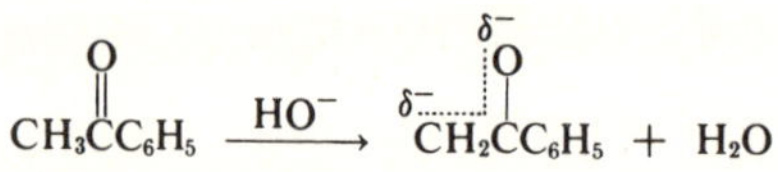
(55-5a)

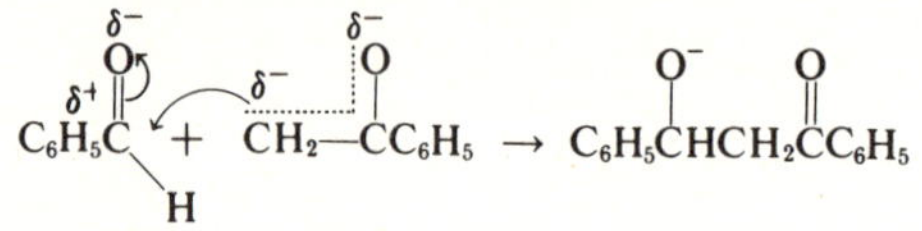
(55-5b)

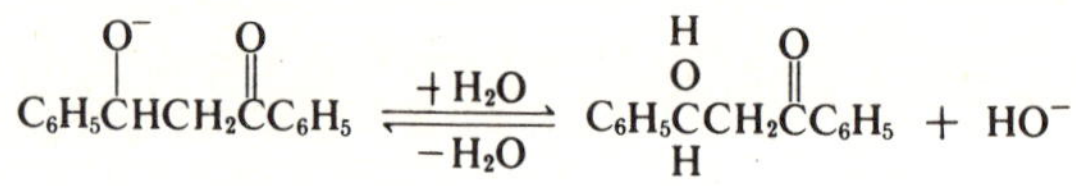
(55-5c)

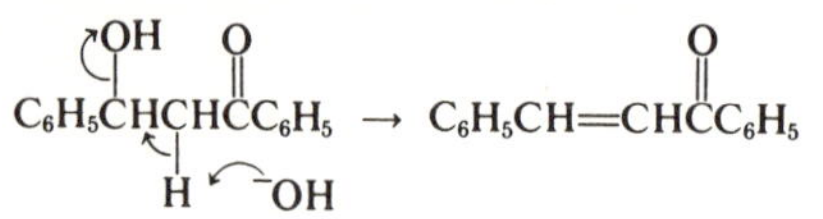
(55-6)

EXPERIMENTAL PROCEDURE

Aldol Condensation

Dissolve 3 g of potassium hydroxide in 22 ml of water in a 125-ml Erlenmeyer flask. Add 20 ml of 95% ethanol and cool the mixture in a cold water bath. Add 8.7 ml of acetophenone and stir the mixture with a thermometer. During the addition of benzaldehyde and the following reaction period the mixture should be maintained at 18 to 29° (ideal temperature 25°). The temperature of the reaction mixture can be increased by removing the flask from the cold water bath and decreased by adding ice to the cold water bath and by stirring the mixture. Pour 7.6 ml of benzaldehyde into the flask and stir it vigorously for 7 to 8 min. Stopper the flask, place it in a water bath maintained at 16 to 23° for 1 hr, and shake it vigorously every 4 to 5 min. After 30 min it is helpful but not necessary to add a few small crystals of benzalacetophenone.

Problem 55-1. Why is the acetophenone added to the basic solution before the benzaldehyde?

Problem 55-2. Why will a temperature of 30° or above cause a decrease in yield? Why will a temperature of 16° or below make purification of the product more difficult and cause a decrease in yield?

Isolation

Warning! The reaction product is a skin irritant. Irritation may persist for hours after contact. If the product is accidentally touched, wash it off the skin with water and then ethanol.

The product now can be isolated by filtering. A slightly better yield, however, is obtained if the reaction mixture is allowed to stand in a refrigerator or a cold room and is filtered during the next laboratory period. Before filtering the mixture, chill it a few minutes in an ice bath. Wash the precipitate with 10 ml of chilled ethanol, 15 ml of chilled 70% aqueous ethanol, and three 15-ml portions of water. Dry the precipitate and determine its melting point. If it begins melting below 56° *or* finishes melting below 59°, recrystallize from 95%

ethanol. Do not heat the ethanol above 50° during the crystallization. If the crude product melted above 50°C chill the mixture in an ice bath before filtering. Weigh and calculate the % yield. Hand in the product to your instructor. If the product was recrystallized, report the melting point of both the crude and purified product.

Infrared Option

Obtain an infrared spectrum on the product. From the spectrum what can you tell about the purity of your product? Note the frequency of the carbonyl stretch absorption. Why is it different from the ketone absorption in acetone or 2-pentanone?

EXERCISES

55-1. Why is the trace of benzalacetophenone added during the reaction?

55-2. Why is the ethanol solution heated to only 50° during the crystallization?

55-3. Propose methods of synthesizing the following compounds from compounds of 8 carbon atoms or less:

(a) 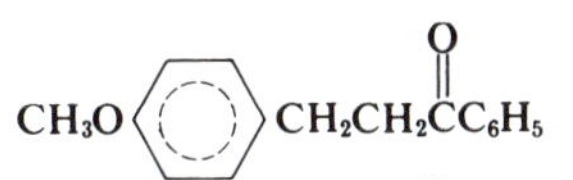

(b) 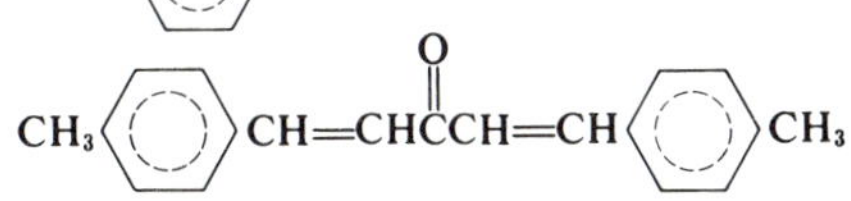

(c) 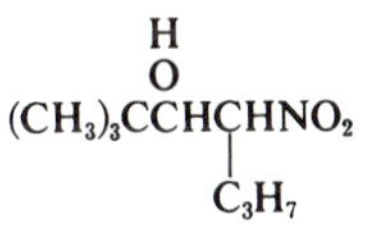

(d) 

56

Aldehydes and Ketones

Organic Qualitative Analysis X

PHYSICAL PROPERTIES

Structure

According to the resonance theory, aldehydes and ketones can be described adequately by electronic structures I and II.

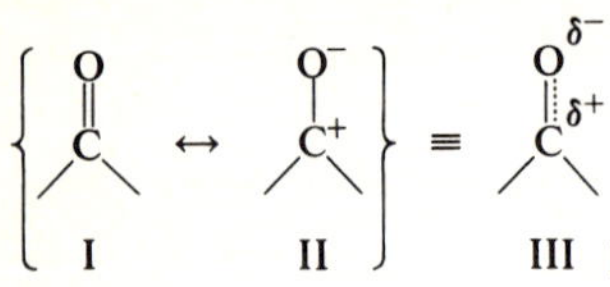

Figure 56-1

The strong contribution of resonance structure II causes the carbonyl group to be polar. Therefore, molecules that have a carbonyl group show strong dipole-dipole and dipole-induced dipole attractions.

Boiling Points

These dipole-dipole and dipole-induced dipole attractions cause aldehydes and ketones (and other compounds with carbonyl groups) to have higher melting points and boiling points and to be better solvents for inorganic compounds than nonpolar organic materials are (Chapter 2). Since molecules possessing only the carbonyl group cannot undergo hydrogen bonding with other identical molecules, such compounds have lower boiling points and melting points than compounds of similar molecular weight that can undergo strong hydrogen bonding, such as alcohols and carboxylic acids. Compounds with high dipole moments, such as aldehydes and ketones, boil at the same or higher temperatures than compounds of similar molecular weight with weak hydrogen bonds, such as amines. In general, where there are no other complicating features such as hydrogen bonding, the higher the dipole moments of a group of compounds the higher their boiling points are. Appendix V has a further discussion on the calculation of dipole moments and the effect of dipole moments on physical properties, especially the boiling points of carbonyl compounds.

Solubilities

Since carbonyl compounds have an oxygen atom which can act as a hydrogen acceptor, these compounds can undergo hydrogen bonding with hydroxylic solvents such as water and alcohols. These compounds, therefore, show about the same solubility in water as alcohols of similar structure (Table 56-1).

TABLE 56-1 SOLUBILITIES OF CARBONYL COMPOUNDS COMPARED WITH ALCOHOLS

Compound	Mol. wt.	Solubility (g/100 ml H_2O, 20°C)	Compound	Mol. wt.	Solubility (g/100 ml H_2O, 20°C)
Propionaldehyde	58	20	Acetone	58	∞
1-Propanol	60	∞	2-Propanol	60	∞
Methyl formate	60	30			
Butyraldehyde	72	4	Methyl ethyl ketone	92	37
1-Butanol	74	7.9	2-Butanol	74	12.5
Ethyl formate	74	12	Methyl acetate	74	32
Isobutyraldehyde	72	11.0			
Isobutanol	74	10.0			
2-Pentanone	86	6.3	3-Pentanone	86	5
2-Pentanol	88	6.0	3-Pentanol	88	5.5
Ethyl acetate	88	7.9	Methyl propionate	88	7.2
3-Methyl-2-butanone	86	7.0			
3-Methyl-2-butanol	88	6.00			
2-Hexanone	100	1.9	3-Hexanone	100	1.7
2-Hexanol	102	1.6	3-Hexanol	100	1.9
Propyl acetate	102	1.92	Methyl butanoate	102	1.65
			Ethyl propanoate	102	1.70
3-Methyl-2-pentanone	100	2.2	2-Methyl-3-pentanone	100	1.7
3-Methyl-2-pentanol	122	2.2	2-Methyl-3-pentanol	102	2.3
4-Methyl-2-pentanone	100	2.6	3,3-Dimethyl-2-butanone	100	2.1
4-Methyl-2-pentanol	102	1.8	3,3-Dimethyl-2-butanol	102	2.7
Isopropyl acetate	102	3.19			
Heptaldehyde	114	0.13	Benzaldehyde	106	0.3
1-Heptanol	116	0.2	Benzyl alcohol	108	4.0

Problem 56-1. The solubility of low molecular weight ethers in water is attributed to the type of hydrogen bonding that causes low molecular weight aldehydes and ketones to be water soluble. Why would ethers be expected to be less soluble than carbonyl compounds?

For most oxygen-containing compounds branching or cyclization of the carbon skeleton and location of the functional group near the center of the molecule have about the same effect on water solubility and other physical properties (pp. 217–222 and 248–250 and Table 56-1). Therefore, like alcohols, branched and cyclic aldehydes and ketones are more soluble in water than their straight-chain isomers.

Problem 56-2. Note in Table 56-1 that esters are usually significantly more soluble in water than alcohols, aldehydes, or ketones of similar molecular weight and structure. Suggest an explanation for this fact.

Solubility Classification

Aldehydes and ketones (except for perfluoro derivatives) like other compounds containing oxygen are soluble in concentrated sulfuric acid because of formation of oxonium ions.

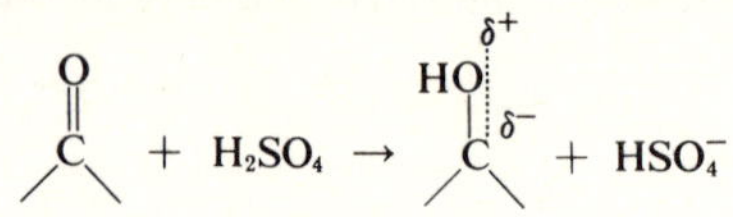

(56-1)

Aldehydes and ketones of 8 to 9 carbon atoms or less are soluble in phosphoric acid. Some compounds at or near borderlines in the solubility classification scheme are given in Table 56-2.

TABLE 56-2 BORDERLINES IN SOLUBILITY CLASSIFICATION OF ALDEHYDES AND KETONES

Isobutyraldehyde	S_1	2-Hexanone	N_1
2-Pentanone	S_1	3-Hexanone	N_1
3-Pentanone	S_1	All methylpentanones	N_1
All methylbutanones	S_1	3,3-Dimethyl-2-pentanone	N_1
Butyraldehyde	S_1-N_1	Benzaldehyde	N_1
Cyclohexanone	S_1-N_1	Acetophenone	N_1
3-Methylbutanal	N_1	5-Nonanone	N_1-N_2
Valeraldehyde	N_1	Benzophenone	N_2

ADDITION REACTIONS OF ALDEHYDES AND KETONES

With Ammonia Derivatives

Aldehydes and ketones react with ammonia and its derivatives to give polymeric products or derivatives of aldimines and ketimines.

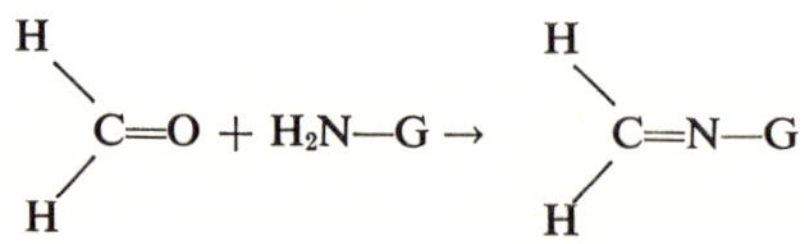

aldimine derivative (56-2)

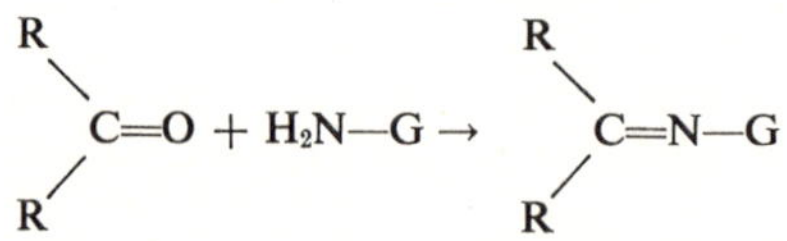

ketimine derivative (56-3)

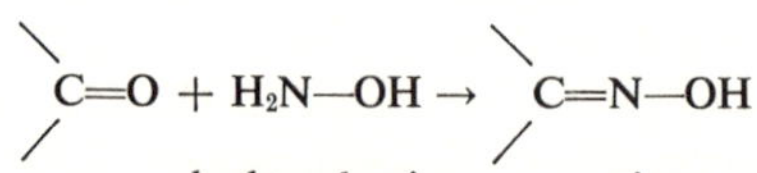

hydroxylamine oxime (56-4)

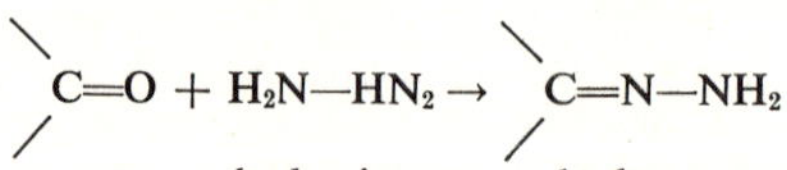

hydrazine hydrazone (56-5)

$$\rangle C{=}O + H_2N{-}NHAr \rightarrow \rangle C{=}N{-}NHAr \qquad (56\text{-}6)$$

arylhydrazine — arylhydrazone

$$\rangle C{=}O + H_2NNH{-}\overset{\displaystyle O}{\overset{\|}{C}}NH_2 \rightarrow \rangle C{=}N{-}NH\overset{\displaystyle O}{\overset{\|}{C}}NH_2 \qquad (56\text{-}7)$$

semicarbazide — semicarbazone

The ammonia derivatives shown in Equations 56-4 to 56-7 are used as classification reagents for aldehydes and ketones and to make derivatives of these compounds.

Arylhydrazones

The arylhydrazines (Eq. 56-6) usually used are phenylhydrazine, *p*-nitrophenylhydrazine, and 2,4-dinitrophenylhydrazine.

Phenylhydrazones and p-Nitrophenylhydrazones. To prepare the reagent, slowly add acetic acid drop by drop to 10 ml of water and 1.0 g of *p*-nitrophenylhydrazine or 1.0 ml of phenylhydrazine until the arylhydrazine dissolves. Agitate the mixture with a glass rod while adding the acid. Note that in some procedures these reagents are used and in others the *pure* arylhydrazine is used.

For water-soluble liquids add 0.4 g of liquid to 4 ml of the reagent and stir it vigorously with a glass rod. If no precipitate appears, warm in a water bath at 75 to 90° for 5 min and then add 2 ml of water. Observe 15 min after removing the test tube from the water bath.

For water-insoluble liquids and all solids dissolve 0.4 g of compound (0.6 g of Class N_2 compounds) in 2 ml of warm ethanol in a 7.5-cm test tube. If the compound does not dissolve, heat the mixture in a water bath at 75 to 80° while agitating the mixture with a glass rod. If the compound still does not dissolve, slowly add more ethanol drop by drop (but add no more than 2.5 ml) with stirring until it does. Add water drop by drop, agitating the mixture after the addition of each drop until a slight cloudiness persists. Warm the mixture in the water bath until the cloudiness disappears. If the cloudiness does not disappear on warming, addition of a drop or two of ethanol will cause it to disappear. Add 0.3 g of *p*-nitrophenylhydrazine[1] or 0.3 ml of phenylhydrazine and agitate the mixture vigorously. The reaction of these reagents with most aldehydes and ketones is exothermic and often the solution will become noticeably warmer if the test is positive. This is sometimes difficult to detect because the solution is already warm, and the decision as to whether the test is positive or not should be based on the presence or absence of a precipitate and the precipitate's melting point. Allow the mixture to cool; if precipitation does not occur within 5 min, add 2 drops of acetic acid, place in the water bath (75 to 80°) for 5 min, remove, and observe the test tube after 25 min.

p-Nitrophenylhydrazine is the better of the two reagents for the classification test because phenylhydrazones are more often oils that sometimes cannot easily be seen or crystallized. When an arylhydrazone is an oil, crystals can often be obtained by grinding the oil against the side of the test tube with a glass rod while the test tube is immersed in an ice bath. Recrystallize the precipitate by dissolving it in the minimum amount of hot ethanol, adding water drop by drop until the cloudiness just persists after agitation, and then warming until the cloudiness disappears. Allow to stand for 1 hr before filtering. Determine the melting point of the solid. If the melting point does not correspond closely (+0.8 or −1.8°) to the melting point of a derivative of a carbonyl compound that could be the unknown,[2] recrystallize the compound until it does or until the melting point increases no more than 0.5° on two successive recrystallizations (constant melting point).

This procedure of recrystallization, determination of melting point, comparison with the tables, and further purification, if necessary, should be done with the other two tests for carbonyl groups listed below before these tests are considered positive. These derivatives, however, are usually obtained fairly pure and seldom are more than two recrystallizations required. Often one recrystallization is sufficient especially for the semicarbazones.

[1] Occasionally when solid *p*-nitrophenylhydrazine is used, only this compound will be found in the precipitate. If this occurs, repeat the test using 3 ml of the *p*-nitrophenylhydrazine reagent.

[2] Compare physical constants, elemental analysis, solubility classification, and the results of other classification tests (pp. 254–256).

$$(NO_2)_2C_6H_3NHNH_2 + R_2CO \rightarrow (NO_2)_2C_6H_3NHN{=}CR_2 + H_2O \qquad (56\text{-}8)$$

2,4-Dinitrophenylhydrazones. To prepare the reagent dissolve 1.2 g of 2,4-dinitrophenylhydrazine in 6 ml of concentrated sulfuric acid in a 50-ml Erlenmeyer flask. Add 10 ml of water with stirring. Add the water slowly enough so that any 2,4-dinitrophenylhydrazine that precipitates redissolves immediately. Do not allow the solution to cool; add 32 ml of 95% ethanol slowly with stirring. Allow the solution to cool to room temperature and remove any precipitate by filtration. Save any unused solution for future tests.

For water-soluble compounds dissolve 0.15 g of unknown in 3 ml of 95% ethanol; for Solubility Class N_1 compounds dissolve 0.20 g in 5 ml of 95% ethanol; and for Class N_2 compounds dissolve 0.3 g in 7 to 10 ml of 95% ethanol. Add the unknown solution to 8 ml of 2,4-dinitrophenylhydrazine solution. Agitate the mixture and observe. If the unknown is a liquid or a low melting solid, be careful not to use an excess of unknown or too little reagent as 2,4-dinitrophenylhydrazones are soluble in aldehydes and ketones.

If only a small amount of solid has precipitated after 20 min, chill to about 15° for 10 min. If an oil forms, grind it for a few minutes and allow it to stand at room temperature, overnight if necessary. Collect the precipitate and recrystallize it from an ethanol-water mixture in a 25-ml Erlenmeyer flask, using the method recommended above for recrystallization of *p*-nitrophenyl- and phenylhydrazones. If the 2,4-dinitrophenylhydrazone does not dissolve in a *maximum* of 12 ml of hot ethanol, add ethyl acetate drop by drop with stirring until it does dissolve. If ethyl acetate is used, the addition of water is unnecessary and the flask should be allowed to stand at room temperature for at least 9 hr before filtration.

Discussion. This reaction is the most dependable of all the carbonyl reactions for giving insoluble solid products. The color of the dinitrophenylhydrazone can often indicate the type of carbonyl compound. Unconjugated carbonyl compounds give yellow dinitrophenylhydrazones. Conjugated carbonyl compounds give red or orange-red dinitrophenylhydrazones. Usually the more extensive the conjugation is the redder the solid is.

There are two possible false positives in use of this reagent that are not encountered when phenylhydrazine or *p*-nitrophenylhydrazine is used. If the student is aware of these possible problems, no difficulty should occur. The first is that 2,4-dinitrophenylhydrazine is more likely to oxidize certain alcohols to carbonyl compounds than either phenylhydrazine or *p*-nitrophenylhydrazine. These alcohols are allyl alcohols of the general structure $G{-}\overset{|}{C}{=}\overset{|}{C}{-}\overset{|}{C}HOH$, where G is aryl or vinyl, and secondary diarylcarbinols. The yield of arylhydrazone in these reactions is always less than 30% and is usually much smaller than this. The second difficulty is that 2,4-dinitrophenylhydrazine, like many polynitro compounds, often forms slightly soluble complexes with many aromatic hydrocarbons, aryl halides, aryl ethers, and phenols (p. 353). If the unknown is suspected of being one of these compounds, a positive test with 2,4-dinitrophenylhydrazine should always be verified by at least one and preferably two other tests.

Problem 56-3. When arylhydrazines oxidize organic molecules (see p. 493), the $-NH-NH_2$ is usually the moiety of the arylhydrazine that is reduced. Why would 2,4-dinitrophenylhydrazine be a better oxidizing agent than phenylhydrazine and *p*-nitrophenylhydrazine?

Semicarbazones

$$R_2C{=}O + H_2NNH{-}\overset{\overset{\displaystyle O}{\|}}{C}{-}NH_2 \rightarrow R_2C{=}N{-}NH{-}\overset{\overset{\displaystyle O}{\|}}{C}{-}NH_2 + H_2O \qquad (56\text{-}9)$$

To prepare the reagent dissolve 2.2 g of semicarbazide hydrochloride in 10 ml of water. For water-soluble liquids mix 0.2 ml of liquid and 1.5 ml of reagent, add 5 drops of pyridine, and heat on a steam bath for 5 min. Allow to cool to room temperature and filter if crystals appear. If this reaction is being used only to prepare derivatives and not as a classification test, the mixture may be chilled to induce crystallization if no crystals appear at room temperature. Recrystallize the semicarbazone from the minimum amount of 33% ethanol in water.

For water-insoluble liquids and all solids dissolve 0.3 g of the compound (use 0.4 g if in Class N_2) in 4 ml of warm methanol; add 6 drops of pyridine. Add this warm solution to 1.5 ml of warm semicarbazide reagent. Heat on a steam bath until crystals appear or for 10 min and continue as instructed in the previous paragraph.

Oximes

$$R_2C{=}O + H_3\overset{+}{N}OH\ Cl^- \rightarrow R_2C{=}NOH + HCl \quad (56\text{-}10)$$

Reflux the aldehyde or ketone (use 0.4 g if water-soluble, 0.5 g if in Class N_1, and 0.6 g if in Class N_2 or M) with 0.5 g of hydroxylamine hydrochloride, 3 ml of methanol, and 3 ml of pyridine for 2.5 hr on a steam bath. Evaporate the methanol and pyridine in a hood; both of these solvents are toxic, especially the pyridine. Grind the residue with 4 ml of cold water while the flask is in an ice bath.

Filter and dissolve the precipitate in the minimum amount of refluxing methanol. Remove the solution from the steam bath and add water drop by drop with agitation until a cloudiness just persists. Then heat on the steam bath until the cloudiness disappears. Allow the mixture to cool slowly to room temperature. If crystallization does not occur, chill in an ice bath and scratch the inside of the test tube with a glass rod. If an oil forms, grind the oil in contact with the supernatant liquid while the mixture is chilled.

Discussion of Use of Arylhydrazines, Semicarbazides, and Hydroxylamines

General Procedure. Of the several ammonia derivatives phenylhydrazine, *p*-nitrophenylhydrazine, and 2,4-dinitrophenylhydrazine give the most reliable tests for aldehydes and ketones, and any compound suspected of being an aldehyde or ketone should be tested with one of these reagents. If the test is negative, the unknown should be tested with at least one other of these reagents. The test with semicarbazide can, if needed, be used as a confirmatory test. For water-soluble aldehydes and ketones, semicarbazones are the preferred derivatives. If any of these tests are to be used only as classification tests with no thought of determining the melting point of any precipitate, the quantities of unknowns, reagents, and solvents can be reduced to one-fourth the quantities stated when the amount of unknown available is small.

Difficulties. The reaction of aldehydes and ketones with hydroxylamine often gives oils or fails to give insoluble products in a reasonable length of time. For this reason the precipitation of the oxime is undependable as a classification test and should be used to make derivatives only when preparation of the semicarbazone or one of the arylhydrazones does not identify the product.

Besides the difficulties described with each reagent at least two possibilities for false negatives exist for these tests. The first, more serious of the two is due to ammonolysis of acid derivatives.

$$R{-}\overset{O}{\overset{\|}{C}}{-}Y + H_2N{-}Z \rightarrow R{-}\overset{O}{\overset{\|}{C}}{-}\overset{H}{\overset{|}{N}}{-}Z + HY \quad (56\text{-}11)$$

$$R{-}\overset{O}{\overset{\|}{C}}{-}OEt + H_2NNHAr \rightarrow R{-}\overset{O}{\overset{\|}{C}}{-}\overset{H}{\overset{|}{N}}\overset{H}{\overset{|}{N}}Ar + HOEt \quad (56\text{-}12)$$

These amide derivatives are insoluble solids usually indistinguishable in appearance from the oximes, arylhydrazones, and semicarbazones. Since these ammonolysis reactions usually require more forcing conditions than the reactions of aldehydes or ketones, one may often avoid formation of the amide derivatives by *not* using longer reaction times, higher temperatures, more reagent, and more catalysts than recommended in the experimental sections above. Since esters and some acid anhydrides are found in the same solubility classes as aldehydes and ketones (Classes S and N), compounds in these solubility classes should be tested with sodium hydroxide, and hydroxylamine and ferric chloride (p. 273) as well as with the arylhydrazines and semicarbazide.

Another source of error with these reagents (and many other reagents) is due to the precipitation of either the reagent or the unknown, if a solid. Therefore, the melting point of the unknown and the reagent (Table 56-3) should be kept in mind. To be certain that any solids formed are not amides, reagents, or other artifacts, the melting point of the derivative should be checked with those of the possible unknowns in Tables 56-6 to 56-9 as outlined in the experimental section on the preparation of phenylhydrazones and *p*-nitrophenylhydrazones. This procedure also has the advantage in that a derivative is prepared during the classification tests with little additional work. The identification of the unknown is then usually possible by running a few additional tests, such as the iodoform test, the bisulfite test, and Tollens' test. A second derivative is necessary for the identification of only relatively few aldehydes and ketones.

TABLE 56-3 PHYSICAL CONSTANTS OF ALDEHYDE AND KETONE REAGENTS

Compound	B.p. (°C)	M.p. (°C)
2,4-Dinitrophenylhydrazine		197.5d
Hydroxylamine		34
Hydroxylamine hydrochloride		151
p-Nitrophenylhydrazine		157d
Phenylhydrazine	243	19
Semicarbazide		96
Semicarbazide hydrochloride		173d

Syn and anti Isomers of Derivatives. Note in Tables 56-6 to 56-9 that two melting points are listed for many of the derivatives. This is usually a result of the possibility of two geometric isomers for the derivatives of all the aldehydes and unsymmetrical ketones.

$$CH_3-CH_2-\overset{\overset{O}{\|}}{C}H + H_2NOH \rightarrow \underset{\substack{anti \\ \text{“}trans\text{-like”}}}{CH_3CH_2-\overset{\overset{HO-N}{\|}}{C}-H} + \underset{\substack{syn \\ \text{“}cis\text{-like”}}}{CH_3CH_2-\overset{\overset{N-OH}{\|}}{C}-H} \qquad (56\text{-}13)$$

The hydrazine derivatives can also show this geometric isomerism.

$$CH_3CH_2\underset{\underset{O}{\|}}{C}CH_2CH_2CH_3 + H_2NNHAr \rightarrow CH_3CH_2\underset{\underset{NAr}{\|}}{C}CH_2CH_2CH_3 + CH_3CH_2\underset{\underset{ArN}{\|}}{C}CH_2CH_2CH_3 \qquad (56\text{-}14)$$

This type of isomerism cannot occur with symmetrical ketones:

$$CH_3-\underset{\underset{O}{\|}}{C}-CH_3 + H_2N-OH \rightarrow CH_3-\underset{\underset{N-OH}{\|}}{C}-CH_3 \text{ the same as } CH_3-\underset{\underset{HO-N}{\|}}{C}-CH_3 \qquad (56\text{-}15)$$

The observation of the two isomers occurs considerably less often than is possible for two reasons. The first is that the high melting isomer is less soluble and, therefore, the one usually isolated by crystallization. For aldehyde derivatives the *syn* isomer is usually the higher melting, and for ketone derivatives the isomer with the hydroxy group *syn* to the smaller organic moiety is usually the higher melting. This results from the stronger crystal structure that can be obtained when the hydroxy group is not shielded by the larger organic moiety.

The other reason is that with many derivatives only one isomer forms because steric factors hinder formation of the other isomer. For example, in the reaction of 2,2-dimethylpropanal with hydroxylamine only the *syn* oxime is formed in significant amounts because the formation of the less stable, sterically hindered *anti* oxime is more difficult.

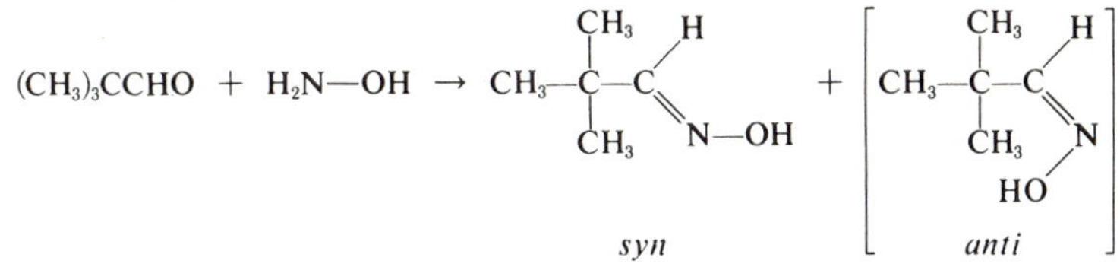

(56-16)

steric repulsion between hydroxy and *t*-butyl group

This situation can be generalized to that shown in Figure 56-2. Where R′ is much larger than R, only the less hindered isomer will be isolated. If the nitrogen substituent, Z, is large, the difference in size between R′ and R need not be large to cause only the less hindered isomer to be produced in significant amounts. If Z is $-NHC(=O)-NH_2$ or —NHAr, it is large enough to cause only the less hindered isomer to be produced in significant quantities with aldehydes and most ketones. Note that the production of the less hindered isomer in greater amount and the lower solubility of the less hindered isomer both lead to isolation of this isomer.

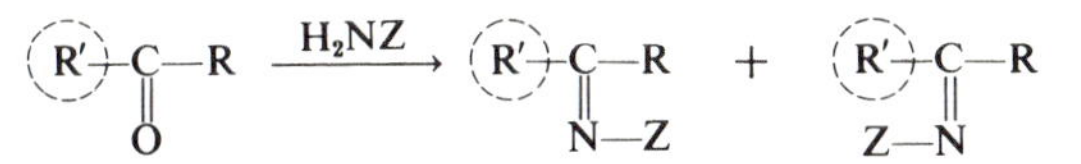

where R′ is larger than R — less hindered isomer, produced in greater amount — more hindered isomer

Figure 56-2 Formation of derivatives of aldehydes and unsymmetrical ketones showing steric hindrance in geometric isomers.

Problem 56-4. Which of the following aldehydes and ketones give oximes that are capable of existing as geometric isomers: (a) 2,6-dimethylbenzaldehyde, (b) furfural, (c) hexanal, (d) 2-hexanone, (e) 3-hexanone, (f) benzophenone, (g) *p*-methylbenzophenone, (h) pinacolone (3,3-dimethyl-2-butanone)? Which are *likely* to give two stereoisomeric oximes? Which are *likely* to give two stereoisomeric 2,4-dinitrophenylhydrazones?

If two isomers are likely to be produced in the preparation of a derivative, a mixture of the stereoisomers may be what is isolated. Such products, like most mixtures, would have a melting point between those of the two isomers or lower than the melting point of either isomer. When such a mixture is possible (indicated in the tables by two melting points listed for one derivative), it is often advisable to prepare another derivative that will give only one isomer. The occurrence of two melting points in the tables of derivatives should not be an absolute prohibition against preparing that derivative but merely an indication that another derivative might be more useful and a warning on the interpretation of the melting point.

o-Dianisidine

Discussion. The *o*-dianisidine spot test is another test that uses the reaction of aldehydes with ammonia derivatives. Two molecules of the aldehyde react with one molecule of *o*-dianisidine to produce a dialdimine with extensive conjugation. This conjugation causes the dialdimine to absorb light and be strongly colored; the color varies with the structure of the aldehyde.

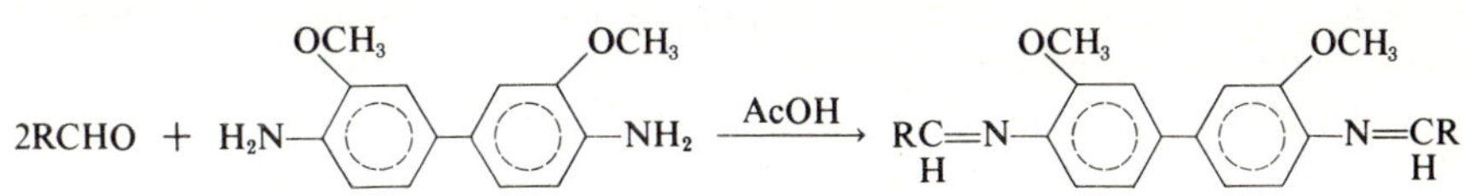

(56-17)

Colors for individual aldehydes are listed in Tables 56-6 and 56-7. The colors are often red, red-orange, or red-brown when the aldehyde group is conjugated with an aromatic ring or a double bond. Other aldehydes produce green, yellow, or orange hues. These colors almost always appear at room temperature. When the test mixture is heated, its color usually shifts to a longer wave length (becomes more red) and becomes more intense.

As with many carbonyl addition reactions the equilibrium is favorable for obtaining products from aldehydes and unfavorable for obtaining products from ketones. Large amounts of some ketones, however, produce yellow, green, or brown colors especially when heated. This probably occurs because large amounts of ketones force the equilibrium to the right (Law of Mass Action), and a small amount of the highly colored product is formed. The test is very sensitive for aldehydes, as little as 0.1 μg (0.0000001 g) of some conjugated aldehydes can be detected. This sensitivity actually causes some difficulties because compounds containing aldehyde impurities may give positive tests. The false positives caused by aldehyde impurities and large amounts of ketone can be minimized by carefully following the instructions given in the experimental section below and not using an excess of unknown. Some cycloalkenes such as pinene and camphene may produce brown colors which could be erroneously interpreted as positive tests. Because of these difficulties this test should be used only to verify the Tollens' test and to help identify an aldehyde by a specific color reaction.

Experimental Procedure. The reagent is prepared by saturating 20 ml of glacial acetic acid with *o*-dianisidine. If the solution is colored, add an additional 3 ml of glacial acetic acid and 1 g of carbon black. Heat on a steam bath for 10 min and filter.

Mix *1 drop* of unknown with 3 drops of reagent on a crucible cover. (A watchglass may be used but the white background of the crucible cover aids in observing the color changes.) If the unknown is a solid, dissolve a small amount (0.05 g) in the minimum amount of methanol. Mix 2 or 3 drops of this solution with 3 drops of reagent. Observe the color. Heat the crucible cover in the tip of a microburner flame for a minute. Observe any color change.

Reaction with Sodium Bisulfite

$$\underset{/\quad\backslash}{\overset{\overset{\displaystyle O}{\|}}{C}} + Na^+HSO_3^- \rightleftharpoons -\overset{\overset{\displaystyle OH}{|}}{\underset{\underset{\displaystyle SO_3^-Na^+}{|}}{C}}- \qquad (56\text{-}18)$$

Discussion. Sodium bisulfite reacts with all aldehydes and ketones to give addition products. With most ketones, however, the equilibrium is so unfavorable that no visible sign of reaction is observed. Ketones are less reactive than aldehydes because of steric and electronic effects. (Review your lecture text to familiarize yourself with these effects.) Only cyclic ketones up to cyclooctanone and some methyl ketones give a positive test with sodium bisulfite. Neither aryl ketones, $Ar-\overset{\overset{\displaystyle O}{\|}}{C}-CH_3$, nor sterically hindered aliphatic ketones, such as pinacolone, give positive tests. All other methyl ketones give positive tests.

Quantitive data for the reaction of aldehydes and ketones with sodium bisulfite are shown in Table 56-4. Note that the percentages of products are for 1 mole of carbonyl compound reacting with 1 mole of sodium bisulfite. Under the test conditions described in this text (a large excess of sodium bisulfite), the yield of addition product would be higher. The amount of time allowed for the test is somewhat shorter so that a few borderline compounds such as isopropyl methyl ketone may not give positive tests unless a longer time is allowed as described in Table 56-4.

Problem 56-5. In the following list of compounds five show positive tests with sodium bisulfite, one compound is borderline, and three show negative tests. Indicate which compounds show these reactions. (a) cyclobutanone, (b) 2,2-dimethylpropanal, (c) 3-hexanone, (d) 2-hexanone, (e) 4-methyl-3-buten-2-one, (f) phenylacetone, (g) *o*-tolualdehyde, (h) 2-methylcyclohexanone, (i) 2,6-dimethylcyclohexanone.

The bisulfite test should be done on all compounds giving positive arylhydrazine or semicarbazide tests. It is used to verify the Tollens' test and is used with the iodoform test to divide ketones into convenient subgroups.

TABLE 56-4 REACTION OF EQUIMOLAR QUANTITIES OF CARBONYL COMPOUNDS AND SODIUM BISULFITE AT 20°C

Compound	Percent of addition compound		Result of test
	in 1/2 hr	in 1 hr	
RCHO	50-90	67-90	+
$(CH_3)_2CO$	47	57	+
$EtCOCH_3$	25	37	+
Cyclohexanone	25	35	+
$CH_3(CH_2)_2COCH_3$	15	24	+
$(CH_3)_2CHCOCH_3$	8	13	+
$(CH_3)_3CCOCH_3$		5.6	-
Et_2CO		2	-
$CH_3(CH_2)_2COEt$		1.8	-
$ArCOCH_3$		0.1-2.2	-
$C_6H_5COC_6H_5$		0.1	-

Experimental Procedure. To prepare the reagent dissolve 10 g of sodium bisulfite in 16 ml of water, add 7 ml of 95% ethanol, mix thoroughly, and allow to stand for 10 min. If any precipitate forms, decant the supernatant liquid into a clean container before using the solution.

Put 0.2 g of unknown (0.3 g if a Class N_2 compound) in a test tube. If the unknown is water-insoluble, mix it with 3 drops of methanol. Add 1 ml of sodium bisulfite solution and vigorously shake the stoppered test tube. A voluminous white precipitate is a positive test. If no precipitate appears immediately, shake the test tube intermittently for 10 min and observe.

Schiff Test

Discussion. The reagent is prepared by the reaction of *p*-rosaniline hydrochloride (fuchsin) with sulfurous acid to a colorless leucosulfonic acid. The colorless compound reacts with aldehydes to give purple to violet aldehyde addition products. Only aldehydes give a purple, violet, or *bluish* wine color.

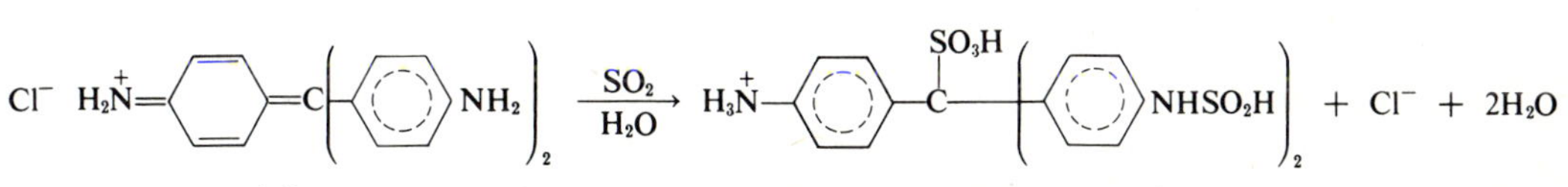

(56-19a)

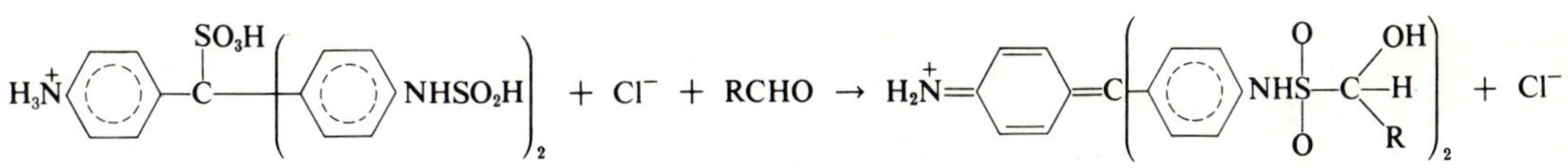

(56-19b)

Some compounds such as α,β-unsaturated ketones reverse the reaction shown by Equation 56-19a to generate a pink color.

Experimental Procedure. The reagent is made by mixing 200 ml of distilled water that is saturated with sulfur dioxide with a solution of 0.2 g of *p*-rosaniline hydrochloride in 200 ml of distilled water. The *p*-rosaniline hydrochloride solution should be filtered just before mixing with the sulfurous acid.

Add 2 drops (or 0.1 g of solid) of aldehyde to 2 ml of the reagent in a test tube. Observe in 3 min; if there is no color change, observe again in 10 min. Try this test on butyraldehyde, acetophenone, and mesityl oxide.

OXIDATION REACTIONS

Aldehydes are easily oxidized by a variety of oxidizing agents, even very weak ones.

$$\mathrm{RCHO} \xrightarrow{[\mathrm{O}]} \mathrm{RCO_2H} \tag{56-20}$$

The most useful of these for the identification of *all* aldehydes is Tollens' reagent. If the corresponding carboxylic acid is a solid and if no other easily oxidizable group is present, the acid is often an excellent derivative.

With the exception of most aromatic ketones which may be oxidized to aromatic carboxylic acids and methyl ketones which are oxidized by alkaline hypohalites, ketones are difficult to oxidize.[3]

$$\mathrm{Ar{-}\overset{\overset{\large O}{\|}}{C}{-}R} \xrightarrow[\mathrm{HO^-},\ \Delta]{\mathrm{KMnO_4}} \mathrm{ArCO_2^-} \tag{56-21a}$$

$$\mathrm{R{-}\overset{\overset{\large O}{\|}}{C}{-}CH_3} + \mathrm{NaOX} \xrightarrow{\mathrm{HO^-}} \mathrm{RCO_2^-} + \mathrm{HCX_3} \tag{56-21b}$$

If very vigorous oxidizing conditions are used, alkanones can be oxidized, but the products are usually a complex mixture, none being produced in good yield. This is illustrated for the oxidation of 5-methyl-3-hexanone with hot nitric acid.

$$\mathrm{(CH_3)_2CHCH_2{-}\overset{\overset{\large O}{\|}}{C}{-}CH_2CH_3} \xrightarrow[\Delta]{\mathrm{HNO_3}} \begin{cases} \mathrm{(CH_3)_2CHCO_2H + CH_3CO_2H} \\ \mathrm{(CH_3)_2CHCH_2CO_2H + CO_2} \\ \mathrm{CH_3CH_2CO_2H + H_2O} \end{cases} \tag{56-22}$$

Tollens' Reagent

The silver ion and ammonia complex in a basic aqueous solution (Tollens' reagent) is a very weak and specific oxidizing reagent. Aldehydes, hydrazines, and hydroxylamines are the only common monofunctional compounds that are universally oxidized by this reagent.

$$\mathrm{RCHO + HO^- + Ag(NH_3)_2^+ \rightarrow RCO_2^- + NH_4^+ + NH_3 + Ag} \tag{56-23}$$

There are a few other organic compounds that are oxidized by this reagent such as aromatic amines, some phenols, and α-hydroxy and α-amino ketones (Chapter 61). If the solubility classification tests are used, there is little chance that these compounds will be confused with aldehydes.

The Tollens' test should be done routinely on *all* compounds giving a precipitate with the arylhydrazines or semicarbazide, with compounds suspected of being oximes, hydrazines, and hydroxylamines, and with all compounds that are to be tested for nitro or nitroso groups by zinc and ammonium chloride (Chapter 46).

[3] There are, however, specialized oxidizing agents such as selenium dioxide and permonosulfuric acid (H_2SO_5) which will oxidize ketones. These are not often encountered in a first year organic chemistry course, and their study is usually postponed to a more advanced course.

Tests on other compounds with Tollens' reagent are a waste of time and may give misleading information. Follow the experimental procedure given on page 369.

Reaction with Aqueous $KMnO_4$

$$2MnO_4^- + 3RCHO \rightarrow 2RCO_2^- + RCO_2H + 2MnO_2 + H_2O \qquad (56\text{-}24)$$

Baeyer's Test (p. 98). Saturated and aromatic aldehydes rapidly decolorize dilute, neutral potassium permanganate solution. Alkenes and alkynes, however, do not give positive Tollens', sodium bisulfite, or arylhydrazone tests so there is no possibility of confusing aldehydes with alkenes and alkynes. Since all aldehydes rapidly decolorize dilute neutral permanganate solution and since the interpretation of the reaction of carbonyl compounds with bromine in carbon tetrachloride is difficult (see following section), the differentiation of alkenals from alkanals by *simple* classification tests is impossible.

Preparation of Acids as Derivatives of Aldehydes.

Discussion: acids are sometimes the preferred derivatives of aldehydes if the acid is a solid and the arylhydrazones, semicarbazone, and oxime can not be prepared easily or are unsatisfactory as derivatives because of their melting points. No attempt should be made to prepare the acid as a derivative if the aldehyde has a double or triple bond, amino group, phenolic group, or other easily oxidized moiety.

Experimental procedure: to a mixture of 0.7 g of aldehyde in 10 ml of water add 3 drops of 5% sodium hydroxide and shake the tube well. Immediately add about 6 drops of saturated potassium permanganate solution and shake the mixture. If the purple color disappears, add more of the saturated permanganate solution a drop at a time until the color just persists. Shake the tube after the addition of each drop. Add 10% sulfuric acid drop by drop until the mixture is strongly acidic as measured by Hydrion paper, then add 2 excess drops. Add 20% sodium bisulfite solution drop by drop with agitation until all purple color (from MnO_4^-) and brown color (from MnO_2) have disappeared.

$$6H^+ + 5HSO_3^- + 2MnO_4^- \rightarrow 5HSO_4^- + 2Mn^{++} + 3H_2O \qquad (56\text{-}25a)$$

$$2H^+ + HSO_3^- + MnO_2 \rightarrow HSO_4^- + Mn^{++} + H_2O \qquad (56\text{-}25b)$$

Filter the mixture and wash the precipitate with two 3-ml portions of cold water. Recrystallize the acid by dissolving it in the minimum amount of hot acetone and adding water a drop at a time until a cloudiness just persists on agitation. Then warm the solution until the cloudiness disappears and allow the mixture to cool to room temperature. If only a little or no precipitate is evident after the addition of the sodium bisulfite solution, extract the mixture with five 6-ml portions of chloroform. Combine the extracts and evaporate the chloroform. Recrystallize the residue from an acetone-water mixture as described above.

The Iodoform Test

$$R{-}\overset{\overset{\displaystyle O}{\|}}{C}{-}CH_3 + NaOH + KI_3 \rightarrow R{-}\overset{\overset{\displaystyle O}{\|}}{C}{-}CI_3 \xrightarrow{HO^-} RCO_2^- + HCI_3 \qquad (56\text{-}26)$$

All compounds that may have a secondary alcohol group, a keto group, or that may be acetaldehyde or ethanol or one of their halogen derivatives should be tested with KI_3 in alkaline solution to establish the presence of the $-\overset{\overset{\displaystyle OH}{|}}{C}HCH_3$ or $-\overset{\overset{\displaystyle O}{\|}}{C}CH_3$ group or their halogen substitution products. The experimental procedure for this test is given on page 232.

TABLE 56-5 RESULTS OF IODOFORM AND BISULFITE TEST

Ketone	Sodium bisulfite	Iodoform
RCH_2COCH_3		
R_3CCOCH_3		
$CH_3(CH_2)_nCO(CH_2)_nCH_3$		
$ArCOCH_3$		
$\overline{\underline{(CH_2)_m C{=}O}}$		
$(CH_2)_m$ $\begin{matrix} CH\text{-}R \\ C{=}O \\ CH\text{-}R \end{matrix}$		

Problem 56-6. Fill in Table 56-5 by putting + where the test is positive and − where the test is negative.

REACTION WITH Br_2 IN CCl_4

Aldehydes and ketones that have an *alpha* hydrogen react with bromine in carbon tetrachloride (see p. 98).

$$-\overset{\displaystyle H}{\underset{|}{\overset{|}{C}}}-\overset{\displaystyle O}{\overset{\|}{C}}- + Br_2 \rightarrow -\overset{\displaystyle Br}{\underset{|}{\overset{|}{C}}}-\overset{\displaystyle O}{\overset{\|}{C}}- + HBr \qquad (56\text{-}27)$$

This reaction is slower than the reaction of most alkenes and alkynes with bromine in carbon tetrachloride, and, since it is a substitution reaction, hydrogen bromide is evolved. The reaction of aldehydes with bromine to yield acid bromides is usually considerably slower than α-substitution under the conditions of the test.

$$RCHO + Br_2 \rightarrow R-\overset{\displaystyle O}{\overset{\|}{C}}-Br + HBr \qquad (56\text{-}28)$$

Because of the relatively slow rate of these reactions and the evolution of hydrogen bromide, there is little chance of confusing the addition of bromine to alkenes and alkynes with the substitution of bromine on carbonyl compounds. Because of the substitution reactions, however, the bromine in carbon tetrachloride test is unreliable for testing for the presence of double or triple bonds in most carbonyl compounds. The Baeyer test should be used to distinguish between alken- and alkynones and alkanones. Because all aldehydes rapidly decolorize Baeyer's reagent, there is no simple classification test that will distinguish between most saturated and unsaturated aldehydes.

Problem 56-7. Could $CH_2{=}CH(CH_3)_2CCHO$ be distinguished from $CH_3CH_2(CH_3)_2CCHO$ by a simple classification test?

REACTION WITH ALKALI METAL HYDROXIDE

Aldehydes, but not monofunctional ketones, react with sodium or potassium hydroxide under the same conditions that esters and amides do. Aldehydes with α-hydrogen undergo the aldol condensation, and those without

α-hydrogen undergo a Cannizzaro reaction. The product from the aldol condensation (Eq. 56-29) is not likely to be confused with any of the products obtained from reaction of acid derivatives with alkali metal hydroxides.

$$2RCH_2CHO \xrightarrow{HO^-} RCH_2\underset{\displaystyle\overset{|}{}}{\overset{\displaystyle OH}{\overset{|}{C}}}HCHRCHO \qquad (56\text{-}29)$$

The products from the Cannizzaro reaction (Eqs. 56-30 and 56-31), however, are the same as those obtained from the saponification of esters.

$$2ArCHO + HO^- \rightarrow ArCO_2^- + ArCH_2OH \qquad (56\text{-}30)$$

$$2(CH_3)_3CCHO + HO^- \rightarrow (CH_3)_3CCO_2^- + (CH_3)_3CCH_2OH \qquad (56\text{-}31)$$

Problem 56-8. What classification tests should be done before unknowns are tested with alkali metal hydroxides for the ester group?

UNKNOWN

Give a clean, stoppered, labeled test tube to your laboratory instructor. Your unknown will be one of the compounds listed in Tables 56-6 to 56-9 or any compound that could have been a previous unknown. Attempt to solve the following exercises before starting your unknown.

EXERCISES

56-1. It has been suggested that a combination of the sodium bisulfite test and the iodoform test be used to differentiate between aldehydes and ketones without attempting the Tollens' or any other test. That is, one would obtain:

	$NaHSO_3$ test	NaOI test
Aldehydes	+	−
Most aliphatic methyl ketones	+	+
All other methyl ketones	−	+

(a) There are at least 15 ketones in Tables 56-8 and 56-9 that would give the same results with these tests as do aldehydes. List 15 such compounds from the tables.

(b) There is 1 aldehyde in Tables 56-6 or 56-7 that would give the same results with these tests as most aliphatic methyl ketones do. What is the aldehyde? Are there other aldehydes not in Tables 56-6 or 56-7 that would give the same results with these tests as most aliphatic methyl ketones do? What are their structures?

56-2. A Class S_1 compound I that contains chlorine boils at 97 to 99°. It gives a yellow precipitate with *p*-nitrophenylhydrazine (m.p. 129 to 130.5°). I reacts violently with water to give a water-soluble solid II that contains chlorine. I reacts with aqueous sodium hydroxide to give a liquid III that is insoluble in concentrated sulfuric acid, contains chlorine, and is heavier than water. Identify I, II, and III.

56-3. A Class N_1 compound boils at 133 to 135°. It gives a precipitate with 2,4-dinitrophenylhydrazine (m.p. 181 to 182°). It gives a positive Tollens' and a negative hydroxamic acid test. What is the compound?

56-4. A Class S_1 compound V has a boiling point of 84 to 85.7° and contains chlorine. It gives positive phenylhydrazine, Tollens', and sodium bisulfite tests. When subjected to the iodoform test, it gives a colorless liquid VI that is denser than concentrated sulfuric acid and insoluble in it. VI can not be made to crystallize even in an ice bath. VI boils at approximately 199° with some decomposition. Identify V and VI.

TABLE 56-6 ALDEHYDES (LIQUIDS)

	B.p. (^{o}C)	M.p. (^{o}C)	Derivatives, m.p. (^{o}C) semi-carbazone	2,4-dinitrophenylhydrazone	p-nitrophenylhydrazone	phenylhydrazone	oxime	acid	o-Dianisidine spot test* cold	hot
Acetaldehyde	20	-123	162	168;157	128.5	57;99	47		O	dk Br
Propanal	48.5	-81	89;154	148;155	125		40		Ol	R
Glyoxal	50	15	270	328	311	180	178			
Acrolein	52.4	-88	171	165	151	50.5	35		RBr	BrV
Isobutyraldehyde	64	-66	125.6	187;182	130.4				OCh	
2-Methyl-2-propenal (methacrolein)	73.5		198	206	155d	74				
n-Butyraldehyde	74.7	-97	95.5;106	123	87,94	94			OCh	
Trimethylacetaldehyde (pivaldehyde)	75	4.5	190	210	119		41	35.5		
Chloroacetaldehyde	85.6		134.5 148					56.0		
2-Chloropropanal	86									
3-Methylbutanal (Isovaleraldehyde)	92.5	-51	107	123	111	oil	48.5		O	Br
2-Methylbutanal	92.4	20,tris-	104	120						
Trichloroethanal (chloral)	98	-57.5	90d	131			56	57.5		
Valeraldehyde	103.4	-91.5		98,107	74		52	−34.5	O	Br
3,3-Dimethylbutanal	103			147				7	O	Br
trans-2-Butenal (crotonaldehyde)	104	-69	199	190	184.5	56	119	72	dk R	dk R
Ethoxyacetaldehyde	106			116.5	113.4			−49		
4-Methyl-2-pentenal	107.5			165						
2-Butynal	108			136						
2,3-Dimethylbutanal	114			124						
n-Propoxyacetaldehyde	120			86						
4-Methylpentanal	122		127	99					OCh	
Paraldehyde (acetaldehyde trimer)	125	12.6	162.3	168;157	128.5	57,99	47		Ol	dk RBr
γ-Chloropropionaldehyde	130							40.5		
Hexanal	131		108	104	90		51		OCh	
3-Methyl-2-ethylbutanal	133			121						
3-Methyl-2-butenal	135		223	182						

* See color code at bottom of Table 56-7.

TABLE 56-6(cont.)

	B.p. (°C)	M.p. (°C)	Derivatives, m.p. (°C) semi-carbazone	2,4-di-nitro-phenyl-hydra-zone	p-nitro-phenyl-hydra-zone	phenyl-hydra-zone	oxime	acid	o-Dianisidine spot test* cold	hot
2-Methyl-2-pentenal	137		207	159		59	48.5			
5-Methylhexanal	144		117	117					R	
Tetrahydrofurfural	145.5		166	134		67		21.0		
2-Chloro-2-butenal	148.6									
3-Hexenal	150		147							
Heptanal	155	-45	109	108	73		57		RBr	R
4-Methyl-2-ethylpentanal	155		98							
Furfural	162	-36.5	202	213,230	54	97	76,92	133.5	dk RV	PV
Cyclohexanecarboxaldehyde	162		174.5	172			90.5	38		
2-Ethylhexanal	163		254	114.5 120.5						
2,2,3-Trichlorobutanal	165						65			
Octanal (caprylaldehyde)	171		99.5	106,96	80		60			
2-Ethyl-2-hexenal	173		152	124						
2,2,2-Tribromoethanal (anh)	174						115	132	N	dkG
Benzaldehyde	179	-26	222,234	237	190,262	158,154	35,130	121	O	RO
Nonanal	185		100,84	100			64		O	
5-Methylfurfural	187		211	212	130	147.7	syn:112 anti:51.5		RP	
Phenylacetaldehyde	194	33	154	121,110		58,62.5	99	76.5	dkBr	dkBr
o-Hydroxybenzaldehyde (Salicylaldehyde)	197	-7	231	249	227	142	57,63	159	O	O
2-Thiophenealdehyde	198			242		119,139		129		
m-Methylbenzaldehyde m-tolualdehyde	199		204,224	212,194	157	91,84	60	11	dkO	R
o-Methylbenzaldehyde	200		209,218	193.4	222	101,111	49	104.5	dkO	R
p-tolualdehyde	204		234,215	234	201	112,121	80,110	180	dkO	R
Citronellal	207		83,92	78					dkG	R
Decanal	208		102	104			69	31.5	paO	dkBr
o-Chlorobenzaldehyde	214	11	146, 229.2	213	237.5, 249	86	75,101	141		
2-Phenoxyethanal	215		145			86	95	98.5		
3,5-Dimethylbenzaldehyde	221	9	201.5					170		
3-Phenylpropanal	224		127	149	122.5		96	48.5		
Citral a (Gerinal)	228d		164	109,116			144		dkR	dkR

TABLE 56-6(cont.)

	B.p. (°C)	M.p. (°C)	Derivatives, m.p. (°C) semi-carbazone	2,4-dinitrophenylhydrazone	p-nitrophenylhydrazone	phenylhydrazone	oxime	acid	o-Dianisidine spot test* cold	hot
2,6-Dimethylbenzaldehyde	229	11	158					116		
m-Methoxybenzaldehyde	230	3	233d		171	76	39	108	dkO	RBr
3-Bromobenzaldehyde	235		205		220	141	72			
p-Isopropylbenzaldehyde	236		211	243	190	129	52 111		dkR	paOP
p-Methoxybenzaldehyde	248	3	210 203	253	160	121	64.5 133	184	dkO	RBr
3-Phenylpropenal (Cinnamaldehyde)	252	-7	216		195	168	64.5 139		dkR	dkR
3,4-Diethoxybenzaldehyde	279						98	165		
2,2-Diphenylethanal	316d		162				120 106			

TABLE 56-7 ALDEHYDES (SOLIDS)

Aldehyde	M.p. (°C)	B.p. (°C)	Derivatives, m.p. (°C): semicarbazone	2,4-dinitrophenylhydrazone	p-nitrophenylhydrazone	phenylhydrazone	oxime	acid	o-Dianisidine* cold	o-Dianisidine* hot
o-Ethoxybenzaldehyde	21	248	219				58	24		
Tetradecanal	23		106.5	108	95		83	54	BrR	R
1-Naphthaldehyde	34	292	221		234	80	90	161		
Phenylacetaldehyde	34	195	156,163	121	151	63,58	99,103	76.5		
4-Methoxy-1-naphthaldehyde	34					113				
5-Hydroxymethyl-2-furaldehyde	35.6		195d,167	184	185	140.5	78,108			
3,4-Methylenedioxybenzaldehyde	37	263	230	266	199	99,106	syn:146 anti :112		btR	dkR
o-Iodobenzaldehyde	37		206			79	108	162		
Octadecanal	38		108.6	101,110	101		89	70.5	BrR	R
o-Methoxybenzaldehyde	38.5	245	215d	253.5	204.5		92	99		
4-Diethylaminobenzaldehyde	41		241d			103	93			
3,4-Dichlorobenzaldehyde	43.4	247.6			276.7		syn:120 anti :114.5	203.5		
2,4,5-Trimethylbenzaldehyde	44		243			127		215.5		
o-Nitrobenzaldehyde	44		256	25d,265		156	syn:154 anti :102	147.5	BrG	BrR
p-Chlorobenzaldehyde	48	216	231.5	245	237	127	110,146	242.5		
2-Pyrrolaldehyde	50	218	183.5		182	139	164			
4-Chloro-2-hydroxybenzaldehyde	52.5		212		257		155			
2,3-Dimethoxybenzaldehyde	54		231d			138	99	180.5		
2,3-Diphenylpropanal	54		125							
3-Chloro-2-hydroxybenzaldehyde	55		242				167.6			
2,3,5-Trichlorobenzaldehyde	56							162.5		
2-Hydroxy-5-methylbenzaldehyde	56	217				149	105			
Phthalaldehyde	56.5					191		203	btY	btY

* Will not be a possible unknown until multiple functional compounds are studied, or spectroscopic techniques are used.

Color code:

pa(pale), bt(bright), dk(dark), Y (yellow), G (green), O (orange), V (violet), B (blue), Br(brown), Bk(black), Ol(Olive green), Ch(Cherry red), P (purple), N (negative), RO(reddish orange), Br R(brownish red), etc.

TABLE 56-7(cont.)

Aldehyde	M.p. (°C)	B.p. (°C)	Derivatives, m.p. (°C) semi-carbazone	2,4-dinitrophenylhydrazone	p-nitrophenylhydrazone	phenylhydrazone	oxime	acid	o-Dianisidine* cold	hot
m-Iodobenzaldehyde	57		226		212	155	62	187.5		
p-Bromobenzaldehyde	57		228.5	128,257	208	113	syn:157 anti:111	252		
2,5-Dichlorobenzaldehyde	58	232				104.5	128	154		
n-Nitrobenzaldehyde	58		246		247	122	121	140.5		
1-Hydroxy-2-naphthaldehyde	60						145			
p-Phenylbenzaldehyde	60		243d	239d		189d	149.5	220		
3-Methoxy-1-naphthaldehyde	60		200		197		102			
3,5-Dichlorobenzaldehyde	65	239				106.5	112	182.5		
*Dibenzofuran-2-aldehyde	68					162				
*Quinoline-2-aldehyde	71				250	204	188			
2,4-Dichlorobenzaldehyde	73.5						136.5	164		
p-Dimethylaminobenzaldehyde	74		222		182	148	185			
*4-Methylthiazole-5-aldehyde	74					160				
*Quinoline-6-aldehyde	76		239			185	191			
3,4,5-Trimethoxybenzaldehyde	76.5		219.5		201.5		83.4			
p-Iodobenzaldehyde	78		224	257	201	121		270		
4-Hydroxy-3-methoxybenzaldehyde	81	285d	240, 230	270	222, 227	105	123, 117			
2-Hydroxy-1-naphthaldehyde	82		240				157			
Isophthaldehyde	89					242	180	348		
Quinoline-8-aldehyde	94.5		238.7			176	115			
3,5-Dichloro-2-hydroxybenzaldehyde	96		227d			153	195.6			
*Hydroxyacetaldehyde	97					162				
2,3,4,6-Tetrachlorobenzaldehyde	97.5									
Benzaldehyde-2-carboxylic acid	98.5		202				120	202		
3-Hydroxybenzaldehyde	105	240	199	259	221.5	131,148	90,88		dkRBr	dkR
9-Anthraldehyde	105		219			207	187			
p-Nitrobenzaldehyde	106		211,221		249	153,159	132,183		OBr	RBr
2,3-Dihydroxybenzaldehyde	108		226d			167				
2,4,5-Trichlorobenzaldehyde	113							163.5		
2,4,6-Trichloro-3-hydroxybenzaldehyde	114				272.5		171			

TABLE 56-7(cont.)

Aldehyde	M.p. (°C)	B.p. (°C)	Derivatives, m.p. (°C) semicarbazone	2,4-dinitrophenylhydrazone	p-nitrophenylhydrazone	phenylhydrazone	oxime	acid	o-Dianisidine* cold	hot
Terephthaldehyde	116	245			281	154,278	200	300		
p-Hydroxybenzaldehyde	116		224 280d	280d 260	266	177.5 184	72 112		dkOR	dkR
1-Bromo-2-naphthaldehyde	118							186		
4-Chloro-3-hydroxybenzaldehyde	121		239		227		108d, 126			
*1,2,3,4-Tetrahydrophenanthrene-9-aldehyde	129									
2,4-Dihydroxybenzaldehyde	135.6		260d	286d		157.3	191			
3-Chloro-4-hydroxybenzaldehyde	139		210d				144.5			
2-Chloro-3-hydroxybenzaldehyde	139		236.5		244.5		149			
2,6-Dichloro-3-hydroxybenzaldehyde	141						174.5			
3,5-Dihydroxybenzaldehyde	157		223				285d			
3,5-Dichloro-4-hydroxybenzaldehyde	158		236.5				185			
*3-Hydroxy-2-ketopropanal	160						135			
m-Carboxylbenzaldehyde	175		265			164	188d	348		
2-Hydroxybenzaldehyde-3-carboxylic acid	179					188	193			
*Indole-3-aldehyde	196				154	198				
Pentachlorobenzaldehyde	202.5					152.5	201			
3,4-Benzpyrene-5-aldehyde	203			197						
3,4,5-Trihydroxybenzaldehyde	212d				226,235		197			

TABLE 56-8 KETONES (LIQUIDS)

Ketone	B.p. (°C)	M.p. (°C)	n_D^{20}	D_4^{20}	Derivatives, m.p. (°C) semi-carba-zone	2,4-di-nitro-phenyl-hydra-zone	p-nitro-phenyl-hydra-zone	phenyl-hydra-zone	oxime
Acetone	56	-95	1.3592	0.792	190	127	148.5	42	59
3-Buten-2-one	80		1.4103		141				
2-Butanone	81	-86	1.3791	0.804	135.6	116	128.5	oil	oil
3-Butyn-2-one	86					181	143		
*Butane-2,3-dione	88	-2	1.3927		mono,235 di,279	314.5	230	134	mono,76 di,240
2-Methyl-3-butanone	94		1.3979	0.805	113	119	108.5	oil	oil
**2-Methyl-1-buten-3-one	98.5		1.4232		173	181			
Cyclobutanone	100		1.4209		204	146			
3-Pentanone	102	-40	1.3922		138.5	156	144		bp. 165
2-Pentanone	102		1.3902	0.806	112,106	143	117		bp. 167
1-Penten-3-one	103		1.4192			129			
3,3-Dimethyl-2-butanone	106	-50	1.3958	0.811	157.5	128	139		77
1-Methoxy-2-propanone	115		1.3981			161	110		
4-Methyl-2-pentanone	117		1.3956	0.801	132	95	79		bp. 176
3-Methyl-2-pentanone	118		1.3990	0.808	94.5	71			
1-Chloro-2-propanone	119				150,165	125			bp. 173
2-Methyl-1-penten-3-one	119		1.4286	0.810	161				
3-Hexanone	125		1.4007	0.815	113	130			oil
2,2-Dimethyl-3-pentanone	127		1.4052			144			
2-Hexanone	129		1.4007	0.811	123	109	88	oil	49
4-Methyl-3-penten-2-one	130		1.4440	0.865	134,164	201	133	142	49
3,3-Dimethyl-2-pentenone	131		1.4100			112			
Cyclopentanone	131	-51	1.4368	0.949	208	145	154	53	56
5-Hexen-2-one	132		1.4194		102	108			
2,2,4-Trimethyl-3-pentanone	135		1.4065		132				144
4-Methyl-3-hexanone	136		1.4071		152				
1-Bromo-2-propanone	136				135d				36
3-Methyl-2-hexanone	136				70				
4-Methyl-5-hexen-2-one	138		1.4217		112				

* Will not be assigned as a possible unknown until multiple functional groups (Chapter 61) are studied.

**Not assigned unless spectroscopic techniques are used.

TABLE 56-8 (cont.)

Ketone	B.p. (°C)	M.p. (°C)	n_D^{20}	D_4^{20}	Derivatives, m.p. (°C) semi-carbazone	2,4-dinitro-phenyl-hydrazone	p-nitro-phenyl-hydrazone	phenyl-hydrazone	oxime
4-Methyl-4-penten-2-one	139d				192 201				76
**3-Methyl-4-penten-2-one	139								
*2,4-Pentanedione	139	-30	1.4487	0.976	mono:122 di:209	209			149
3-Ethyl-2-pentanone	140		1.4073		99				
*3-Hydroxy-3-methyl-2-butanone	140				165	301d			87
4-Methyl-2-hexanone	141		1.4077		126				
3-Methylcyclopentanone	144	-51	1.4336	0.913	185				91.5,68
5-Methyl-2-hexanone	144			0.888					
4-Heptanone	144	-33	1.4073	0.818	132	75			bp. 193
*3-Hydroxy-2-butanone	146		1.4178	0.987	185,202	318		243	
*1-Hydroxy-2-propanone	146	-17	1.4295		196	128	173	103	71
4-Chloro-3-methyl-2-butanone	146		1.4390			116			
**2,5-Dimethyl-3,4-hexanedione	148		1.4206						mono 125 di 172
3-Methyl-2-hexen-4-one	148		1.4490		163				
2-Methyl-2-hexen-5-one	149		1.4289		137				
4,4-Dimethyl-3-hexanone	151			0.830	98				
2-Heptanone	151	-35	1.4005	0.811	126	89,74		207	
2-Methyl-4-heptanone	153			0.813	124				
5-Hepten-2-one	153		1.4292	0.845	105				
3,3-Dimethyl-cyclopentanone	155				178				
1-Bromo-2-butanone	155		1.4670						
Cyclohexanone	156	-16.4	1.4509	0.998	167	161	146.6	81.5	9.
2-Hepten-4-one	156.5				147				
*2,3-Hexanedione	158								175
3,4-Dimethyl-2-hexanone	158				124				
2-Ethyl-1-hexen-3-one	159.5		1.4406		119				
4,5-Dimethyl-5-hexen-3-one	162.2				110				
6-Methyl-3-heptanone	163		1.4209	0.830	132				
2-Methylcyclohexanone	165.5	-14.0	1.4486	0.925	193	136	132	oil	43
*4-Hydroxy-4-methyl-2-pentanone	165.5	(3,5-dinitrobenzoate 55°)			203	209			58
4,5-Dimethyl-4-hexen-3-one	165.5				209				
3-Methyl-2-heptanone	166		1.417	0.822	82				

TABLE 56-8(cont.)

Ketone	B.p. (°C)	M.p. (°C)	n_D^{20}	D_4^{20}	Derivatives, m.p. (°C) semi-carba-zone	2,4-di-nitro-phenyl-hydra-zone	p-nitro-phenyl-hydra-zone	phenyl-hydra-zone	oxime
2,6-Dimethyl-4-heptanone	168		1.4177	0.805	123	66,92			
3-Methylcyclohexanone	170	-73.5	1.4493	0.914	192,180	155	119	94	43
4-Methylcyclohexanone	170	-40	1.4458	0.914	200		128.5	109.5	38
4-Octanone	170				96				
*Methyl acetoacetate	170	-40.6	1.4196	1.077	152				
+2,2-Dimethylcyclohexanone	170.5		1.4482		198	141			
6-Methyl-2-heptanone	171		1.4146		154	77			
trans-2,4-Dimethyl-cyclohexanone	171		1.4413		137				
2-Octanone	173	-21.5	1.4153	0.819	123	58	92.5		
2,5-Dimethylcyclohexanone	173				177				97.5
*Acetoxyacetone	174.5		1.4150	1.075	145		144	60d	
3-Methyl-3-hepten-2-one	175				164				
cis-2,4-Dimethylcyclo-hexanone	176		1.4450		196				98.5
*4-Ethyl-4-hydroxy-3-hexanone	178				177				
2,2,6-Trimethylcyclohexanone	179		1.4480		209	141			
3,3-Dimethylcyclohexanone	179.5		1.4470		219				
5-Ethyl-4-hepten-3-one	180				105				
Cyclohexyl methyl ketone	180		1.4514		177		154		60
trans-3,5-Dimethyl-cyclohexanone	180.5		1.4479		193.5				oil
*Ethyl acetoacetate	181				132	93			
Cycloheptanone	181.5				163	148	137		23
cis-3,5-Dimethyl-cyclohexanone	182.5		1.4407	0.890	202.5				74
5-Nonanone	186.6	-5.9	1.4190	0.822	90				
3-Nonanone	187.5				112				
2,5-Dimethyl-2-cyclohexe-none	190		1.4761	0.940	165				169 92.5
2-Pyridyl methyl ketone	190								121
3-Propylcyclopentanone	190.5		1.4488		178.6				oil

TABLE 56-8(cont.)

Ketone	B.p. (°C)	M.p. (°C)	n_D^{20}	D_4^{20}	Derivatives, m.p. (°C) semi-carba-zone	2,4-di-nitro-phenyl-hydra-zone	p-nitro-phenyl-hydra-zone	phenyl-hydra-zone	oxime
*2,5-Hexanedione	194	-9		0.974	mono 185 di 224	257	211	120	137
**d-Fenchone	195.5		1.4627	0.947	184,172	140			165,123
2-Nonanone	195.5	-8	1.4207		118.5				
*4-Methyl-1-acetyl-cyclohexane	196		1.4501		159,176				58
*p-Fluoroacetophenone	196		1.5101		219				
**2-Ethyl-5-methyl-cyclohexanone	197		1.4485		180				80
2-n-Propylcyclohexanone	200		1.4530		133d				67.5
1-1-Cyclohexenyl methyl ketone	200		1.4892		220				59
m-Trifluoromethyl-acetophenone	202								
β-Thujone	202				174	114			
Acetophenone	204	20		1.028	200	239,250	184.5	105	60
4-Decanone	206.5				51.5				
3,5-Dimethyl-2-cyclohexenone	208.5		1.4827		180			76.5	
1-Menthone	208.5	-6.6	1.4505		187	146		53	59
2-Decanone	212		1.4252	0.824	125				
4-Pyridyl methyl ketone	212								142
**4-n-Propylcyclohexanone	213		1.4539		180				
*2-Acetylthiophene	213	10.5	1.5666		190		181	96	81
o-Methylacetophenone	215		1.5320		208	160			61
1,5,5-Trimethylcyclo-hexen-3-one	215		1.4795	0.926	192d			68	79
6-Bromo-2-hexanone	217		1.4713			81			
Propiophenone	219	19	1.5270	1.011	180	190			
3-Pyridyl methyl ketone	219							137	113
m-Methylacetophenone	220		1.5306	1.007	201	207			56
*1-Phenyl-1,2,3-butanetrione	222		1.5410		231		256	143	166,114
m-Chloroacetophenone	228				232		176		88
n-Butyrophenone	228	12			189	190			50
o-Chloroacetophenone	229		1.687		160	206	215		113
2,4-Dimethylacetophenone	230				186				64
p-Chloroacetophenone	234				203,159	235	239	114	95

TABLE 56-9 KETONES (SOLID)

Ketone	M.p. (°C)	B.p. (°C)	Derivatives, m.p. (°C) semi-carba-zone	2,4-di-nitro-phenyl-hydra-zone	p-nitro-phenyl-hydra-zone	phenyl-hydra-zone	oxime
2-Dodecanone	20.5	246.5	122	81			
1-Phenyl-1-hexanone	25	264	132	168			
3,5-Dichloroacetophenone	26						138
Benzyl methyl ketone	27	216	199	157	145	85	69
p-Methylacetophenone	28	226	205,196	260.4	198	97	87.6
o-Hydroxyacetophenone	28	218	210	213		110	118
Diisopropylideneacetone	28	198	221, 186	117		110	48
2-Tridecanone	28	263	124	69	101.5		56.5
*4-Cyclohexylcyclohexanone	31		216	137			104
*2-Acetylfuran	32.5	171	149	220	185.5	86	104
3-Acetylpropanoic acid	33	245d	187d	206	174.5	108	45.6
2,4-Dichloroacetophenone	33.4						148
2-Tetradecanone	33.5		115.6				
1-Naphthyl methyl ketone	34	302	234			146	139
1,3-Diphenyl-2-propanone	34	330	145	100		127	125
*2,2,6,6-Tetramethyl-4-piperidone	35	206	220				153
p-Chloropropiophenone	35		176				62.5
2-Phenylcyclopentanone	37		214				
p-Methoxyacetophenone	37.6	258	197.5	220,232	195	142	86.5
**1-Phenyl-1-hepten-3-one	39					98	
3,4-Methylenedioxypropiophenone	39		187.5			97	104
m-Bromopropiophenone	40		183				
Benzalacetone	41	212	187	226	166	157	115
**β-Phenylethyl chloromethyl ketone	41			146.5			
1-Indanone	41	243	235	258	234.5	134,124	145
m-Chlorobenzyl phenyl ketone	43						102
**Phenyl undecyl ketone	46		95				63
p-Bromopropiophenone	46		171				91
Benzophenone	48	305	165	235	154	137.5	142.5

* Will not be assigned as possible unknown until multiple functional groups (Chapter 61) are studied.

** Not assigned unless spectroscopic techniques are used.

TABLE 56-9(cont.)

Ketone	M.p. (°C)	B.p. (°C)	Derivatives, m.p. (°C) semi-carba-zone	2,4-di-nitro-phenyl-hydra-	p-nitro-phenyl-hydra-	phenyl-hydra-	oxime
**5-Phenoxy-2-pentanone	50			110			
Phenacyl bromide	50		146				91
o-Bromoacetophenone	51	225	208	232		126	128.5
3-Bromophenacyl bromide	51		163d				
3,4-Dimethoxyacetophenone	51	286.5	218		227	131	140
*Diphenylacetoin	52		169				
*Phenyl-2-thienyl ketone	56						93
**2-Nonadecanone	56						77
*4-Caproylresorcinol	56.5	244					190.5
2-Indanone	57						153
Benzalacetophenone	58	345	170	244d		120	140,68
Phenacyl chloride	59	244	154	212			89
**3-Phenyl-2,4-pentanedione	60		271d				
Desoxybenzoin	60	321	148	204	162	116	98
p-Methylbenzophenone	60	326	121.5	200		109	154,115
**1-Phenyl-1,3-butanedione	61	261.5	301d	151	100.5	150.2	
1,1-Diphenylacetone	61		170			131	165
2,4'-Dibromobenzophenone	62	383					141.5
3-Chlorophenyl benzyl ketone	62						120
2-Phenylcyclohexanone	62		190	139			169
**1-Aceto-2-naphthol	64						
*Benzoylformic acid	66			197d			127,145d
Cinnamalacetone	68		186	222		180	152.5
*Dihydroxyacetone	69			278		132	84
o-Chlorobenzyl phenyl ketone	71						86
**Phenoxymethyl phenyl ketone	74		187				
*9-Aceylphenanthrene	74	170	201				
*2-Benzofuryl methyl ketone	74.5		207			154	
2-Naphthoxyacetone	77.5		203			154	123
p-Chlorobenzophenone	78		271d	185		106	163
1,4-Cyclohexanedione	79		276d				188
Di-n-tridecyl ketone	79						57
m-Nitroacetophenone	81		257	228		128,135	132

TABLE 56-9(cont.)

Ketone	M.p. (°C)	B.p. (°C)	Derivatives, m.p. (°C) semi-carbazone	2,4-di-nitro-phenyl-hydra-zone	p-nitro-phenyl-hydra-zone	phenyl-hydra-zone	oxime
p-Bromobenzophenone	82		348	230		126	169,116
Fluorenone	83	341.5				151.5	195
p-Iodoacetophenone	85						
*α-Hydroxyacetophenone	85		146			112	70
**Stearone (18-pentatriacontanone)	89						66
p,p'-Dimethylbenzophenone	95	335	140	229		100	163
*Benzil	95	347	mono 174 di 244d	188	193 290	134 235	139 237
m-Hydroxyacetophenone	95.5	296	195	257			
p-Chlorophenacyl bromide	97						
*2-Benzoylacrylic acid (anh)	99		190			197	168d
**2-Aceto-1-naphthol	100	325	247			137	168
**Cinnamacetophenone	102			221d	135		
*1,3-Cyclohexanedienone	104						156
Benzoylnitromethane	106					105	96
4-Bromophenacyl bromide	108.5			218			115
p-Hydroxyacetophenone	109		199	261		151	144
1,5-Diphenyl-1,4-pentadien-3-one	112		189	180	173	150	143
**3-Methoxy-4-hydroxystyryl methyl ketone	129			229		127	
*p,p'-Dimethoxybenzil (anisil)	133		254				mono 133 syn. di 217 anti. di 195
**Benzoin	134	344	206	243,238		158.106	151,99
3,4,5-Tribromoacetophenone	134		265d			130	
p-Hydroxybenzophenone	134.5		194	242		144	81,152
2,4-Dihydroxyacetophenone	146		217	218		159	199d
**3,5-Dihydroxyacetophenone	147		205		236		
p,p'-Dichlorobenzophenone	147.5		301d	241			135
**p-Hydroxypropiophenone	148		289d	229,242			171d
**Methone	149		296d				mono;115 di; 176

56-5. A Class N_2 compound contains no halogen, nitrogen, or sulfur and boils at 185.5 to 187°. It gives a positive phenylhydrazine test and negative Tollens', sodium bisulfite, and iodoform tests. What compounds can it be? How could you distinguish among the various possibilities?

56-6. A Class N_2 compound boils at 160.7 to 162.4°. It gives positive 2,4-dinitrophenylhydrazine and Baeyer's tests and negative Tollens', sodium bisulfite, and iodoform tests. What is the compound? How could this be verified?

56-7. Pinacolone like other methyl ketones reacts with alkaline bromine solutions to give bromoform. With alkaline KI_3 solution, however, the reaction stops at the diiodo compound.

$$(CH_3)_3C\overset{\overset{\displaystyle O}{\|}}{C}{-}CH_3 + Br_2 \xrightarrow{HO^-} (CH_3)_3CCO_2^- + HCBr_3 \qquad (56\text{-}32)$$

$$(CH_3)_3C\overset{\overset{\displaystyle O}{\|}}{C}{-}CH_3 + I_2 \xrightarrow{HO^-} \underset{\text{no triiodo compound}}{(CH_3)_3C\overset{\overset{\displaystyle O}{\|}}{C}CHI_2} \qquad (56\text{-}33)$$

The third iodine atom cannot be introduced. Why?

56-8. (a) In the base-catalyzed halogenation of ketones the methyl group is iodinated before the methylene group; that is,

$$RCH_2COCH_3 + X_2 \xrightarrow{HO^-} RCH_2COCH_2X + HX, \qquad (56\text{-}34)$$

but not

$$RCH_2COCH_3 + X_2 \xrightarrow{HO^-} RCHXCOCH_3 + HX. \qquad (56\text{-}35)[4]$$

Why? (Hint: look carefully at the mechanism of the reaction in your lecture text.)

(b) When acetone undergoes the iodoform reaction, why does the second iodine atom substitute on the iodomethyl group rather than on the unsubstituted methyl group. That is, why

$$CH_3COCH_2I + I_2 \xrightarrow{HO^-} CH_3COCHI_2 + HI \qquad (56\text{-}36)$$

and

$$CH_3COCHI_2 + I_2 \xrightarrow{HO^-} CH_3COCI_3 + HI, \qquad (56\text{-}37)$$

and not

$$CH_3COCH_2I + I_2 \xrightarrow{HO^-} CH_2ICOCH_2I + HI \qquad (56\text{-}38)$$

and not

$$CH_3COCHI_2 + I_2 \xrightarrow{HO^-} CHICOCHI_2 + HI? \qquad (56\text{-}39)$$

Why is bromoacetone halogenated on the carbon that already has a halogen atom rather than on the non-halogenated carbon atom?

[4] Reactions 56-35, 56-38, and 56-39 may occur to a limited extent but the other reactions (Eq. 56-34, 56-36, and 56-37) are *much* more rapid.

$$BrCH_2{-}\overset{\overset{\large O}{\|}}{C}{-}CH_3 + X_2 \xrightarrow{HO^-} BrXCH{-}\overset{\overset{\large O}{\|}}{C}{-}CH_3 \qquad (56\text{-}40)$$

(c) Both 1-chloro- and 1,1-dichloroacetone have the methyl ketone ($-COCH_3$) group, but, when tested with alkaline KI_3, they give little or no iodoform. Why? How do these compounds react with aqueous alkaline KI_3?

56-9. When isopropyl methyl ketone reacts with alkaline KI_3 solution, small amounts of isopropyl iodide and sodium iodoacetate are produced in addition to the usual products.

$$(CH_3)_2CHCOCH_3 \xrightarrow[HO^-]{KI_3} (CH_3)_2CHI + ICH_2CO_2^- + HCI_3 + (CH_3)_2CHCO_2^- \qquad (56\text{-}41)$$

Suggest a series of reactions that accounts for these abnormal products.

56-10. *Pure* 3-pentanone, 3-hexanone, and 3-heptanone often give small amounts of iodoform with aqueous alkaline KI_3 even though they do not have the CH_3CO group. Propose a series of reactions that would account for the formation of iodoform from these ethyl ketones. (Hint: how could a methyl ketone be easily formed from these ketones with the iodoform reagent?)

56-11. A Class N_1 compound has a boiling point of 152.2 to 154°. It gives positive *p*-nitrophenylhydrazine and sodium bisulfite tests and negative iodoform, Baeyer's, Tollens', and hydroxamic acid tests. What is the compound? How could this be verified?

56-12. A Class N_1 compound has a boiling point of 147 to 149° and gives positive *p*-nitrophenylhydrazine and Baeyer's tests and negative iodoform, Tollens', and sodium bisulfite tests. What is the compound? How could this be verified?

56-13. A Class N_1 compound X boils at 199 to 202° and gives positive 2,4-dinitrophenylhydrazine, Tollens', and sodium bisulfite tests. When treated with concentrated sodium hydroxide solution, it gave XI, a Class N_1 compound (b.p. 218 to 219.5°, m.p. 33 to 35°), which gave a negative phenylhydrazine test. When the aqueous solution from the reaction of X was acidified, a Class A_1 solid (XII) precipitated. XII has a neutralization equivalent of 136 ± 3. Identify X, XI, and XII.

56-14. A neutral liquid that boils at 103 to 105° contains no nitrogen, halogen, or sulfur and gives negative ferric chloride and hydroxamic acid tests. It gives positive Tollens' and sodium bisulfite tests. What compounds can it be? If no 2,4-dinitrophenylhydrazine were available, what derivative could be made to identify the compound? Could the method described in this chapter be used for making this derivative? If not, what method could be used?

57

The Beckmann Rearrangement

Fractional Recrystallization

$$\underset{\substack{\text{M.p. } 58^\circ \\ \text{B.p. } 326^\circ}}{C_6H_5COC_6H_4CH_3} \xrightarrow{H_2NOH} \underset{\text{M.p. } 154^\circ,\ 115^\circ}{C_6H_5C(NOH)C_6H_4CH_3} \xrightarrow{PCl_5} \underset{\text{M.p. } 144^\circ}{CH_3C_6H_4\overset{O}{\overset{\|}{C}}\overset{H}{N}C_6H_5} + \underset{\text{M.p. } 158^\circ}{C_6H_5\overset{O}{\overset{\|}{C}}\overset{H}{N}C_6H_4CH_3} \qquad (57\text{-}1)$$

INTRODUCTION

Strong acids catalyze the rearrangement of oximes to N-substituted amides. This reaction is called the Beckmann rearrangement. If a mixture of *syn* and *anti* oximes are used as starting materials, the product is a mixture of amides.

Fractional crystallization is a method of separating compounds of different solubilities by repeated recrystallizations.

DISCUSSION

Aldehydes and unsymmetrical ketones react with hydroxylamine to give *syn* and *anti* oximes (review pp. 420–421). The two geometric isomeric oximes are produced in approximately equal amounts if both the electronic and steric effects of the two groups attached to the carbonyl carbon atom are approximately equal. In this experiment the methyl group, which is the only difference between the two groups attached to the carbonyl carbon atom, is remote from the reaction site and cannot affect significantly the relative yields of the two isomeric oximes.

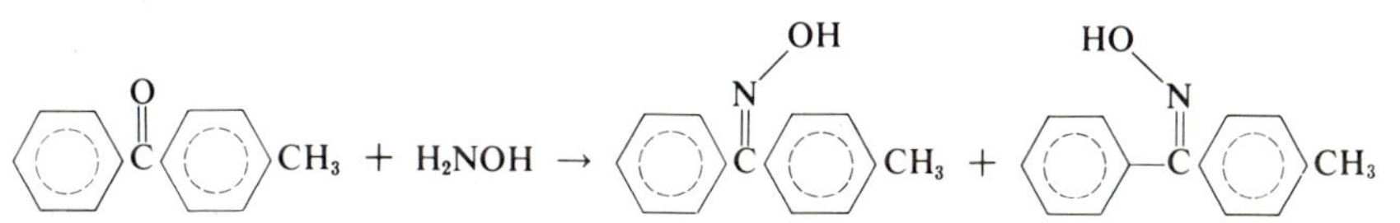

(57-2)

In the presence of strong acids oximes rearrange to amides. The first step in the reaction is production of an electron-deficient nitrogen atom:

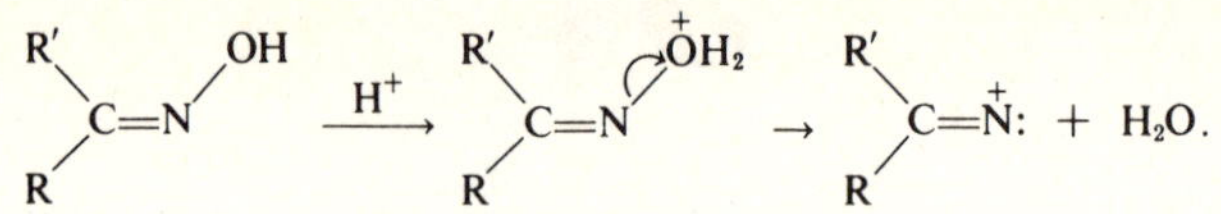

(57-3)

As in similar reactions involving intermediates with electron-deficient atoms these intermediates rearrange.

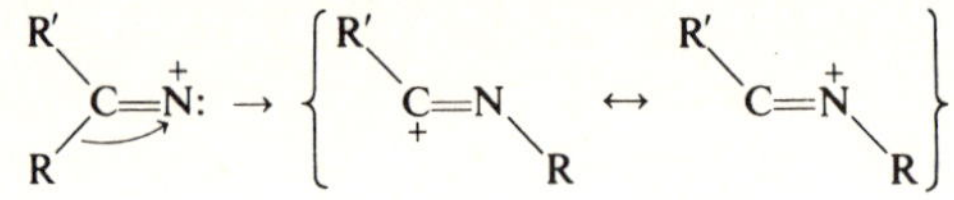

(57-4)

Compare, for example, with the Hofmann rearrangement and carbonium ion rearrangements. Water or its equivalent reacts with the positive ion from the rearrangement.

$$H_2O' + R'-\overset{\delta+}{C}\equiv\overset{\delta+}{N}R \rightarrow R'-\overset{\overset{+}{O}H_2}{\overset{|}{C}}=NR \xrightarrow{(-H^+)} R'-\overset{OH}{\overset{|}{C}}=NR \qquad (57\text{-}5)$$

The product of this reaction then tautomerizes to the amide.

$$R\overset{\overset{OH}{|}}{C}=NR \rightleftharpoons R\overset{\overset{O}{\|}}{C}NHR \qquad (57\text{-}6)$$

The rearrangement is stereospecific. The group that migrates is *trans* (opposite or across) from the leaving hydroxy group. When a mixture of geometric isomeric oximes undergoes the Beckmann rearrangement, the product is a mixture of amides.

$$C_6H_5\underset{\underset{NOH}{\|}}{C}C_6H_4CH_3 \xrightarrow{PCl_5} CH_3C_6H_4CONHC_6H_5 \qquad (57\text{-}7a)$$

$$C_6H_5\underset{\underset{HON}{\|}}{C}C_6H_4CH_3 \xrightarrow{PCl_5} C_6H_5CONHC_6H_4CH_3 \qquad (57\text{-}7b)$$

EXPERIMENTAL PROCEDURE

To save time and reduce the number of Büchner funnels and filter flasks needed the synthesis and isolation of the products can be done in pairs. The oxime can be synthesized by student *A* during spare time in a previous experiment. The rearrangement can be completed by student *B* during spare time of a succeeding laboratory period. The next laboratory period the students can divide the fractional crystallization as shown in Figure 57-1. If enough time is available, five theoretical plates can be used which usually give small amounts of pure amides.

Synthesis of Oxime

(Student *A*, if done in pairs.) In a 100-ml round bottom flask reflux for 2½ hr on a steam bath 14 g of phenyl *p*-tolyl ketone (*p*-methylbenzophenone), 5.0 g of hydroxylamine hydrochloride, 30 ml of pyridine (dried over NA_2SO_4), and 30 ml of *absolute* ethanol. Carefully dry the apparatus before use and place a calcium chloride drying tube in the top of the condenser.

Remove the solvents by vacuum distillation (pp. 239–241) using a steam bath for a heat source. When the solvent is removed, cool the flask in an ice bath and wash the solid in the flask with three 10-ml washes of ice water. Filter the washes so that any solid removed can be recovered and returned to the flask. Recrystallize the oximes from the minimum amount of aqueous ethanol. Chill the mixture in an ice bath before filtering. Wash the precipitate with two 10-ml portions of cold 25% ethanol in water and with 10 ml of cold water. Dry and weigh the precipitate.

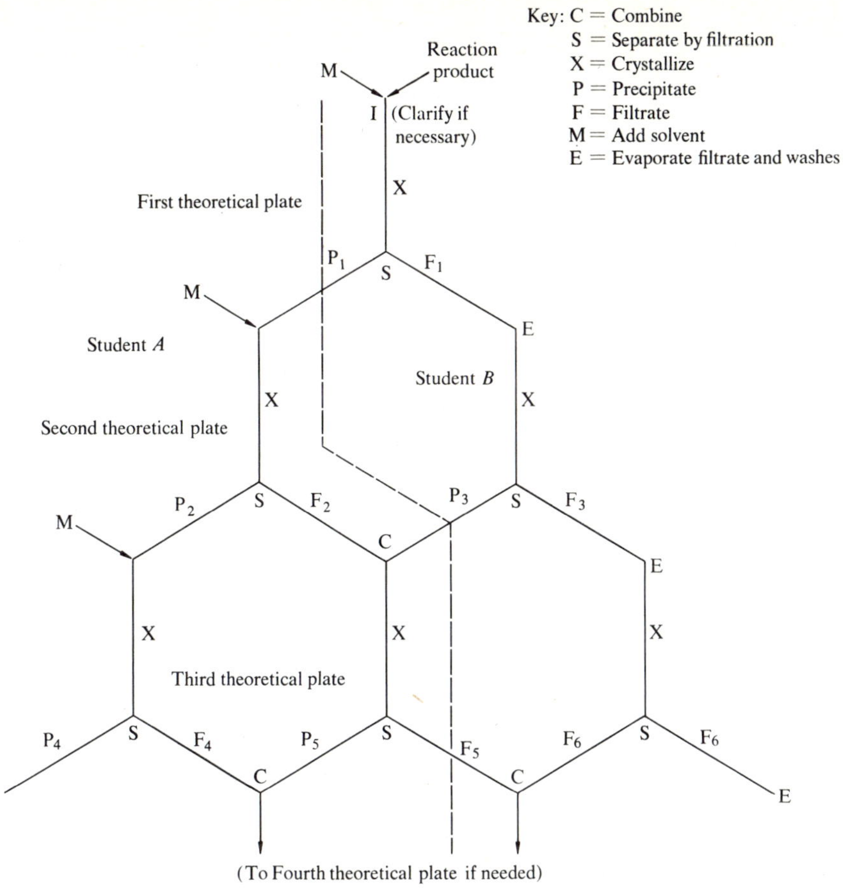

Figure 57-1 Fractional crystallization.

Beckmann Rearrangement

(Student *B*, if done in pairs.) Heat the oximes with 350 g of polyphosphoric acid in a 500-ml Erlenmeyer flask to 130° while stirring the mixture with a thermometer. Maintain the temperature at 130° for 10 min while stirring with the thermometer. Allow the mixture to cool a few minutes and then pour it into a 2000-ml Erlenmeyer flask containing approximately 800 g of crushed ice and 800 ml of water. Stir the contents vigorously with a glass rod.

When almost all the ice has melted, extract the mixture by stirring it vigorously with a mixture of 90 ml of ethyl acetate and 50 ml of toluene. Transfer the toluene-ethyl acetate mixture to a separatory funnel by placing the stem tip just above the organic-aqueous interface and sucking up the top layer with a vacuum from an aspirator or vacuum line (not from your mouth) to the funnel (Fig. 57-2). It is necessary to remove 50 to 100 ml of aqueous solution to be certain all the organic layer has been removed. The reaction product-water mixture can be more easily extracted from a 2000-ml separatory funnel. If a funnel this large is available, extraction in it instead of the extraction procedure described above will save time.

If the stem of the separatory funnel is not long enough, transfer the top layer to the separatory funnel with a 50-ml pipet using a vacuum source or a rubber bulb to suck the organic layer into the pipet. If possible, the direct transfer should be used rather than transfer with the pipet as the latter takes more time. Use the separatory funnel to obtain a very sharp cut between the two phases and return the aqueous layer to the 2000-ml flask.

Wash the organic layer twice with 100 ml of water and then once with 100 ml of saturated sodium chloride solution. Save these three aqueous solutions to wash successively and in the same order the next three organic

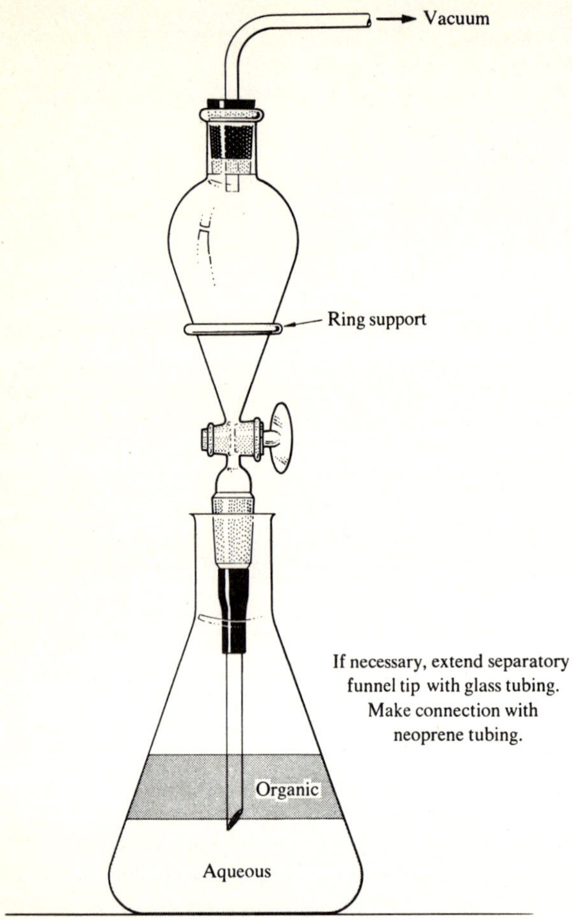

Figure 57-2 Transfer of upper-layer mixture from a large flask into a relatively small separatory funnel.

extracts. Repeat three times the ethyl acetate-toluene extraction of the diluted reaction mixture in the 2000-ml Erlenmeyer flask using the same quantities and transfer procedure outlined above. Agitate the mixture *vigorously* with each extraction. Successively wash each extract with each of the same three 100-ml aqueous washes used to wash the first extract.

While extracting the diluted reaction mixture for the second time, begin distilling the first washed extract from a tared 500-ml round bottom flask. Weigh four boiling chips and add one to the flask before beginning the distillation. If possible, distil under vacuum. Place a beaker of ice around the receiving flask. As each extract is washed for the final time, transfer it to the 500-ml round bottom flask, add a weighed boiling chip each time, and continue the distillation. Do not fill the distillation flask more than half full.

When all the liquid has been removed, heat the flask on a steam bath for 15 min under vacuum from the water aspirator. If any solid has splattered on the Claisen or distillation adapters, scrape it into the flask. Weigh the flask and the dried contents, calculate the weight and the % yield. Determine the melting point of a few crystals that have been pressed on a porous plate. If infrared spectra are to be obtained, read Exercises 57-5 and 57-6 if you have not already done so.

Beckmann Rearrangement, Alternate

If sufficient hood space is available this alternate rearrangement will save some time.

In a hood mix the p-methylbenzophenone oxime thoroughly with 140 ml of anhydrous ethyl ether in a 250-ml round bottom or distilling flask. Add 24 ml of thionyl chloride, 5-10 drops at a time, stirring well after each addition. Stir a few minutes after all the thionyl chloride has been added. Distill the mixture, using a steam bath until all the thionyl chloride and ether is removed.

Add 150 ml of water and boil for 2-3 min. Chill the mixture in an ice bath; decant (and if necessary filter to collect any suspended amide) the supernatent aqueous layer. Repeat the water extraction. Weigh the product, determine the melting point of a few crystals that have been pressed on a porous plate, and obtain infrared an spectrum if that is assigned.

Fractional Crystallization

Crystallize the product from the minimum amount of 95% ethanol. Wash the precipitate once with 5 ml of cold ethanol. Dry the precipitate and determine the melting point of a few crystals that have been pressed on a porous plate. Recrystallize the precipitate from the minimum amount of ethanol. Concentrate the combined filtrate and wash to approximately one-half its original volume and heat the mixture to boiling while stirring with a glass rod. If all the solid does not dissolve, add more ethanol a few drops at a time to the stirred hot mixture until it does.

Repeat this process. Follow Figure 57-1 and make the combinations of filtrates and precipitates shown in the diagram. Always add the wash to the filtrate. The general techniques given in Chapter 12 should be followed. New solvent should be added only at the points indicated or when all the solid does not dissolve when crystallization is being attempted.

Note that only filtrates F_1, F_3, and so on, should be evaporated to half volume or dryness and filtrates that are combined with precipitates should not be concentrated. Filtrates F_1, F_3, and F_6 may be recrystallized in the filter flasks that were used to collect them. It is convenient to use succeedingly smaller filter flasks and Büchner funnels for each successive theoretical plate if these are available.

Problem 57-1. How could this fractional crystallization be extended to four or five plates?

Problem 57-2. Modify your answer to Problem 57-1 so that: (a) more material could be recovered, (b) only as many crystallizations as necessary be done. For example, in the three theoretical plate separation shown in Figure 57–1 crystallization of the combined P_3 and F_2 is unnecessary if only the material isolated from P_4 and F_6 is needed.

TABLE 57-1 DATA FROM FRACTIONAL CRYSTALLIZATION

Fraction	Wt.	M.p.	Fraction	Wt.	M.p.
Reaction Mixture			-	-	-
P_1			F_1	*	-
P_2			F_3		
P_4			F_6		
F_2	*	-	P_3		
F_4	*	-	P_5		
F_5	*	-	P_6		
Ppt. from digestion					

* Filtrate not weighed directly, should be estimated by difference and by assuming no losses.

Verify your answer to Problem 57-2 and extend the crystallization to four or five theoretical plates if you are instructed to do so.

Weigh and dry all precipitates and determine their melting points. Concentrate filtrates F_3 and F_6 to dryness and determine the melting points of the solid residues after first pressing a few crystals on a porous plate. The solid from the last filtrate (F_6 if only three theoretical plates are to be used) should be digested by boiling with 10 ml of petroleum ether for a few minutes. Allow the digested mixture to cool to room temperature, chill in an ice bath, and filter. If more theoretical plates are used, only the residue of the *final* filtrate on the right hand side is digested. Dry, weigh, and determine the melting point of the solid. Prepare a table similar to Table 57-1 and fill in the appropriate data. If four or five theoretical plates are used, extend the table to include data from these fractions. What do the data indicate about the efficiency of the separation?

Calculate the % yield of the two final products based on the original ketone and the % recovery of all the reaction product based on the weight of all the final fractions. Include in your report the % yield of the oximes and the crude mixture of amides.

EXERCISES

57-1. In the Beckmann rearrangement, the group that migrates does so with *retention* of configuration. What does this indicate about the mechanism that was not shown on page 444?

57-2. Fractional crystallization is not used as much as it once was because adsorption liquid-solid chromatography (Chapter 25) and other new techniques can separate most mixtures more efficiently. Why might the mixture of amides separated in this chapter not be easily separated by adsorption chromatography? What other chromatographic methods would probably separate the mixture better than liquid-solid absorption chromatography? Explain your answer.

57-3. Compare Figure 57-1 with Figures 5-1 and 6-7. What are the similarities and the differences?

57-4. What is the purpose of the petroleum ether digestion of the last filtrate? Why is petroleum ether used instead of the ethanol?

57-5. Would you expect the infrared spectra of the two amides to be very greatly different, significantly different, similar with very slight differences in the fingerprint region, or identical?

57-6. Obtain infrared spectra on the reaction product from the Beckmann rearrangement and on the two separated (or partially separated) products.

58

The Pinacol Reaction

Synthesis of Pinacol and Pinacolone

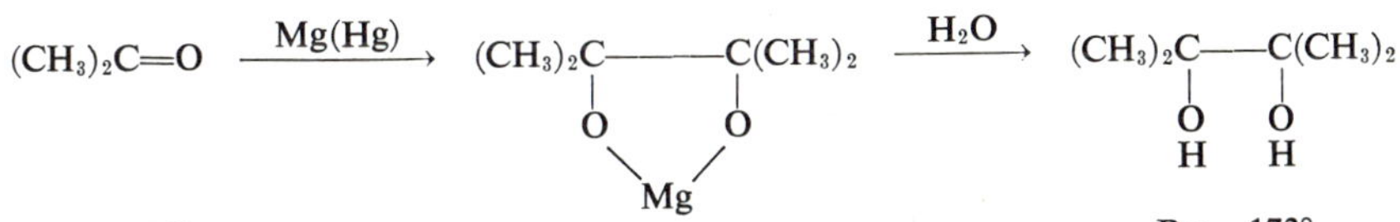
(58-1)

B.p. 56°
D_4^{20} 0.791

B.p. 172°
D_4^{20} 0.965

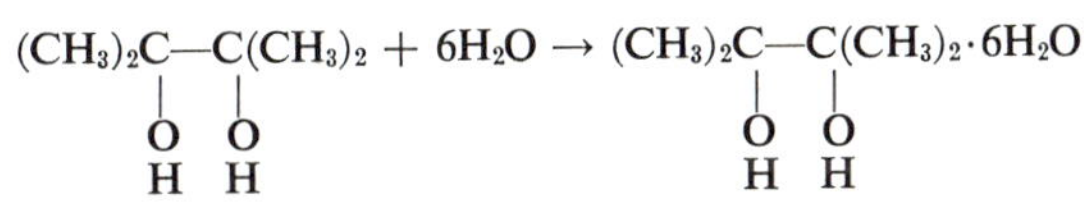
(58-2)

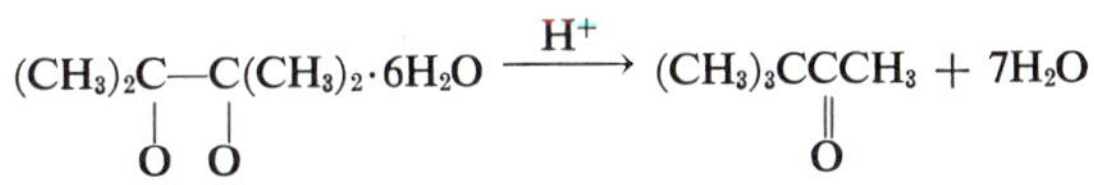
(58-3)

M.p. 45°
B.p. 172°

B.p. 106°
M.p. −50°
D_4^{20} 0.8114

INTRODUCTION

Pinacols (symmetrical *vicinal* glycols) can be made by the bimolecular reduction of carbonyl compounds. This is usually the preferred method of making these compounds. Pinacols are rearranged by strong acids to unsymmetrical carbonyl compounds called pinacolones. Many organic compounds form hydrates. Pinacol hydrate is used in this synthesis to purify pinacol by crystallization.

EXPERIMENTAL PROCEDURE

Pinacol

With a 250-ml round bottom flask construct the apparatus shown in Figure 16-1. Place a calcium chloride drying tube in the top of the condenser instead of the tube to the hood. Place 5 g of magnesium turnings in the flask. The success of this reaction is dependent upon the apparatus and all reagents being completely free from water until the magnesium pinacolate is decomposed. The precautions described on page 189 for drying the apparatus should be followed.

Warning! The benzene and acetone used in this experiment are very flammable and the same precautions described on pages 2 and 188 for handling flammable solvents should be followed. Add 55 ml of dry benzene to the flask. Dissolve 5 g of mercuric chloride in 45 ml of acetone that has been dried over calcium chloride for at least 24 hr. Transfer the acetone solution to the separatory funnel and add 10 to 15 ml of the solution to the flask. Swirl the reaction mixture. Have a cold water bath nearby to moderate the reaction if necessary. If vigorous refluxing does not begin within 5 min, warm the mixture on a hot water bath. Remove the hot water bath as soon as the vigorous reaction begins. When the refluxing becomes less vigorous, adjust the stopcock so that the acetone flows into the reaction mixture in a rapid stream of individual drops. The flow rate should be adjusted to maintain a fairly vigorous refluxing because it is often difficult to start the reaction once it stops and slow refluxing gives a lower yield of product. The refluxing should not be so vigorous, however, that acetone is lost through the top of the condenser. About 5 min after all the acetone has been added, begin heating the mixture with a steam or hot water bath. Gently reflux the mixture for 30 min. Stir it occasionally by applying a rotary motion to the ring stand. Cool the mixture to 40 to 45° and break up the solid mass of magnesium pinacolate with a glass rod. **Careful! Avoid poking the rod through the glass flask.** Reflux the mixture an additional 30 min with occasional agitation.

Decompose the magnesium pinacolate by adding 13 ml of water through the separatory funnel and refluxing the mixture for 30 min with occasional stirring. Cool the mixture to 45 to 50° and suction filter. Return the precipitated magnesium hydroxide to the reaction flask and reflux for 5 to 7 min with 30 ml of benzene. Cool the benzene mixture to 60 to 65° and suction filter it.

Precipitation of Pinacol Hydrate

Rinse all solids from the 250-ml round bottom flask with water and combine the two filtrates in it. Distil until 40 to 45 ml remain and then add 17 ml of water. Place the flask in an ice bath, stir the contents well, and scratch the sides of the flask. Allow the mixture to stand in the ice bath for 25 to 30 min after addition of the water. Stir occasionally. Suction filter the mixture. Wash the precipitate with three 5-ml portions of cold benzene. Use these washes to transfer the small amount of pinacol hydrate remaining in the flask to the Büchner funnel.

Crystallize the product from 15 ml of water. If the solution is yellow, clarify with activated charcoal. Allow to cool to room temperature, chill in an ice bath, stir well, and filter. Dry the precipitate by pulling air through it and determine the melting point of a small crystal that has been pressed on a porous plate. Optional: prepare a potassium bromide pellet (Appendix III) of the pinacol hydrate and obtain its infrared spectrum.

Rearrangement of Pinacol

Add 12 ml of concentrated sulfuric acid to 48 ml of distilled water in a 100-ml round bottom flask. Dissolve the pinacol in the solution, add a boiling chip, and reflux for 15 min. Remove the heat source, add a distilling head, and reattach the condenser to assemble a distillation apparatus. Use an ice bath around the receiver. Distil until no more pinacolone collects. Separate the two layers of chilled distillate and dry the pinacolone over calcium chloride. Transfer the product to a 25-ml round bottom flask and fractionally distil (Fig. 5-5). Collect the very volatile material in a chilled test tube. Collect the compound that distils at 103 to 108° in a tared yield bottle. Calculate the % yield and hand in the product to your laboratory instructor.

Infrared Option

Obtain infrared spectra of the product as a capillary film and of the volatile compound (in a 0.025-mm cell) collected from the fractional distillation prior to collecting the product. Compare the similarities and differences of the two spectra and the spectrum of pinacol hydrate if it was obtained. Assign as many of the absorption bands as you can to functional groups and structural features. Identify the volatile by-product.

59

Paper Chromatography

INTRODUCTION

Paper chromatography is a partition method usually used for analysis and diagnosis. The partition occurs between two liquids, usually a polar and a nonpolar solvent. Except that it is entirely a partition technique and that development of paper chromatograms take much longer than development of other chromatograms, paper chromatography is similar in technique and applications to thin-layer chromatography. The materials are spotted on the paper by virtually the same procedure used for thin-layer chromatography. The chromatograms are developed in an hermetically sealed chamber with an atmosphere saturated by solvent vapor. Both techniques can be used with nonvolatile compounds of any molecular weight. Paper chromatography is usually capable of better resolution than thin-layer chromatography[1]. Very many chromatograms can be simultaneously developed on a large piece of paper. A comparable accomplishment with thin-layer chromatography is awkward because of the size of the plates required. In spite of these advantages thin-layer chromatography because of its speed would have replaced many more applications of paper chromatography than it actually has except for the fact that paper chromatography has been used a lot longer than thin-layer chromatography. In the twenty plus years of application of paper chromatography a considerable number of reproducible standard chromatographic systems have been recorded so that one can usually find in the literature one or more systems for separating almost any mixture and R*f* values for almost all known compounds that can be separated by paper chromatography. Exact R*f* values are sometimes more difficult to reproduce with paper chromatography than with TLC, probably because paper chromatography is more sensitive to temperature. Since development of paper chromatograms takes much longer than development of thin-layer chromatograms, there is a greater possibility of having significant temperature variations occur during the elution.

GENERAL TECHNIQUE

Every possible precaution should be taken to prevent dirt, sweat, skin oils, and foreign material from contaminating the paper. Do not touch the paper any more than absolutely necessary, hold it by the edges whenever possible, and wear rubber gloves if they are available. If it is necessary to leave the paper on the desk top for any length of time, cover it with a *clean* piece of paper, cardboard, or glass.

If several materials or one mixture and several standards are to be chromatographed, draw a line across the top of the paper 4 cm from an edge perpendicular to the direction the paper has been rolled and parallel to this edge. The direction the paper was rolled can be determined by wetting the paper with a drop of water; the long axis of the wet ellipitical spot is in the same direction as that in which the paper was rolled (Fig. 59-1).

[1] The long sheets of paper usually used are responsible for the better resolution often obtained with paper chromatography. For the same distance between origin and solvent front, TLC usually shows slightly better resolution where meaningful comparison can be made.

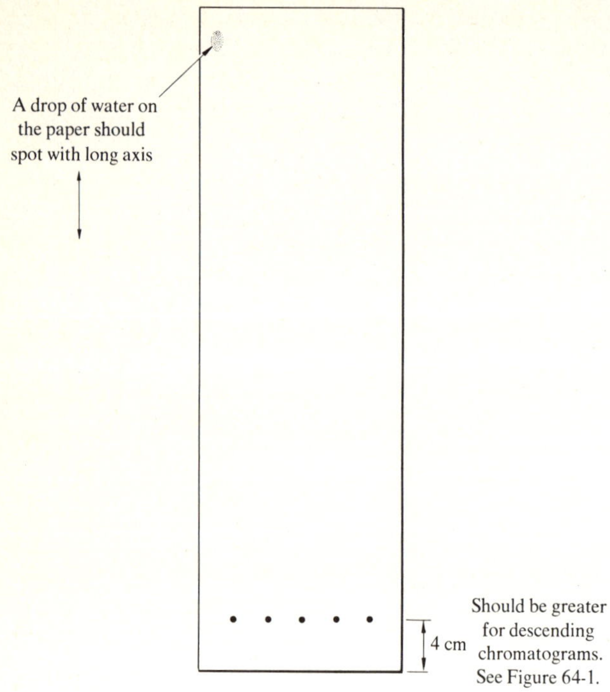

Figure 59-1 Spots on paper ready for chromatography.

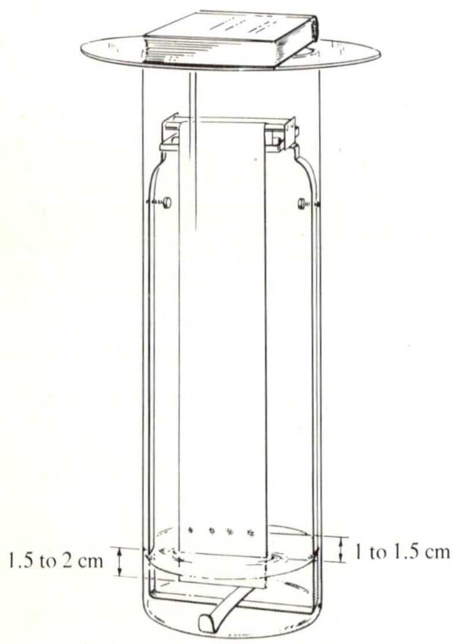

Figure 59-2 Ascending paper chromatography.

Starting 2 to 3 cm from one edge of the paper, spot on the line the materials to be chromatographed 3 cm apart using the same technique described for spotting thin-layer plates (p. 202). Be certain the spots are no larger than 10 mm diameter; the ideal diameter is 4 to 8 mm.

For ascending chromatography used in the identification of dicarboxylic acids described below, the following technique is used.[2] Place the solvent mixture in the bottom of the chromatography tank. Arrange the paper above the solvent layer so that it does not touch the solvent. Place the top on the chamber and allow the solvent to equilibrate with the paper for 20 to 25 min. Lower and fasten the paper so that approximately 1.5 cm of it is immersed in the solvent mixture (Fig. 59-2). Be certain that the top of the solvent mixture is at least 1 cm from the bottom of the spots. Place the cover on the top of the chamber. If there is no gasket, coat the top edge of the

[2] Descending paper chromatography is described in Chapter 64.

chamber with petroleum jelly. Place a weight on the top. The chamber should be in an area that does not have rapid temperature changes and is not exposed to drafts or intense light or heat radiation.

Remove the paper from the chamber after elution and hang it vertically with the edge nearest the original spots at the bottom. Mark the solvent front and allow the paper to dry. The spots are detected on the dried paper by their fluorescence with ultraviolet light or their color reaction with a sprayed reagent. The spots are circled with pencil even when made visible by a sprayed reagent as the color may fade. If a color reagent is used, it usually is sprayed from a device similar to a perfume atomizer (Fig. 59-3). The R*f* values are determined for paper chromatography exactly as they were determined for thin-layer chromatography. Sometimes if the spots are fluorescent or naturally colored, they are cut out, extracted with solvent, and the extracts examined by ultraviolet-visible spectroscopy. This is very similar to the identification of plant pigments separated by thin-layer chromatography (Chapter 26).

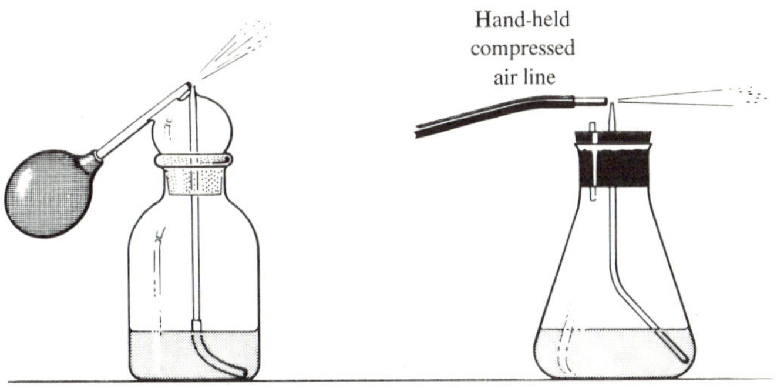

Figure 59-3 Spray bottles.

UNKNOWN DICARBOXYLIC ACIDS

Your unknown will be a mixture of 1 to 6 of the following carboxylic acids: oxalic, malonic, succinic, glycolic, fumaric, malic, tartaric, glutaric, adipic, and sebacic. It is necessary to spot each paper strip with the unknown and with whatever standards are to be placed on the strip. With strips 20 cm wide it is necessary to use 2 strips with 5 standards and the unknown on each strip. Spot the unknown in the center of strip with the standards on both sides. Apply 5 to 10 $\times 10^{-6}$ g of standard and 2 to 3 $\times 10^{-5}$ g of unknown.

Use a solvent mixture of 1 volume of distilled water, 3 volumes of glacial acetic acid, and 13 volumes of ethyl ether to elute the dicarboxylic acids. Elute the acids for 5½ to 7 hrs, but the paper strips should be removed from the chromatography chambers *before* the solvent front gets to within 2 cm of the top.

Hang the paper strips for 16 hr or longer and then spray with bromocresol green. The yellow spots on the dark blue background indicate the positions of the carboxylic acids. Outline the spots in pencil. Determine the identity of the acids in your unknown entirely from the R*f* values.

EXERCISES

59-1. Complete Table 59-1; refer to Chapters 8, 25, 26, 36, and 37 if necessary.

59-2. Is the paper chromatography described in this chapter really a partition technique? If so, what are the *two* liquid phases?

TABLE 59-1 COMPARISON OF CHROMATOGRAPHIC TECHNIQUES

Techniques	Moving phase	Stationary phases	Partition	Absorption	Requirements of compounds being separated	Diagnostic technique	Preparative technique	Time requirements	Equipment requirements	Quant. anal.	Resolution	Other
GLC	Gas, usually He or N_2	A liquid on solid support	100%	None	At least slightly volatile below their decomposition points	Primary	Considerable application. Will be used even more as techniques are developed	Fast	Elaborate and expensive	Precise and accurate	Very high	
GSC (gas solid)			None	100%								
Column			Can be either			Valuable						
Thin layer							Very limited			Approximate unless coupled with spectroscopy or other technique	Good but usually lowest of all chromatographic techniques	
Paper												

60

The Acetoacetic Ester Synthesis

Methyl Isoamyl Ketone and 4-Methylpentanoic Acid

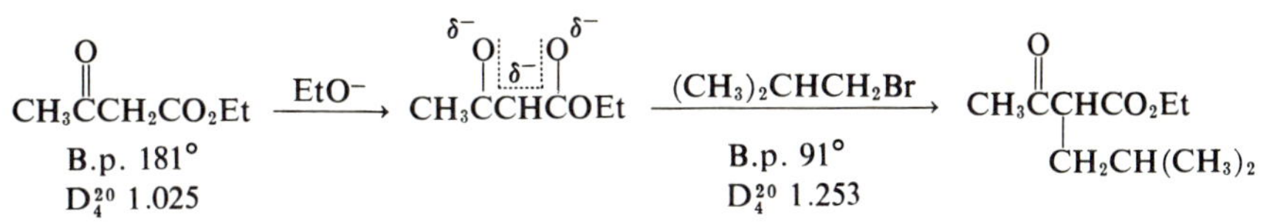

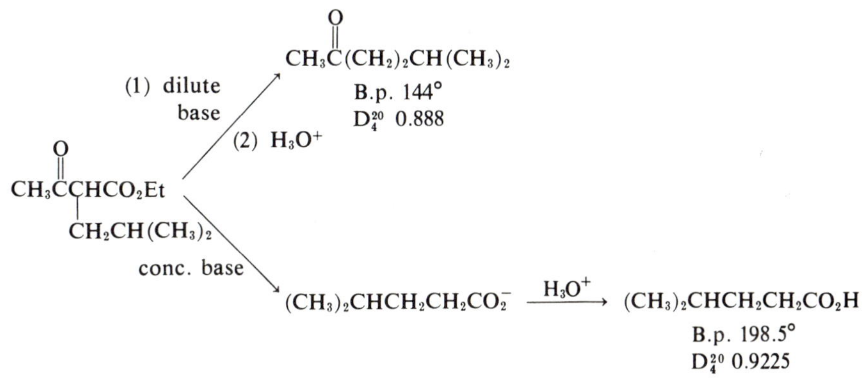

INTRODUCTION

Acetoacetic ester can be used to synthesize a large variety of monofunctional and polyfunctional carboxylic acid and ketone derivatives. This experiment illustrates alkylation of acetoacetic ester and cleavage of the monoalkylated product to an acid and a ketone.

EXPERIMENTAL PROCEDURE

Synthesis of Ethyl 2-(2′-Methylpropyl)-3-ketobutanonate

Synthesis of Ethyl Sodioacetoacetate. Place 80 ml of *absolute* ethanol from a recently opened container into a clean *dry* 200- or 250-ml round bottom flask, and place a reflux condenser on the flask. Add 3.5 $\pm$ 0.05 g of sodium, a few pieces at a time, through the reflux condenser; then place a drying tube on top of the condenser. Follow the instructions on page 242 for handling, cleaning, and weighing the sodium. When all the sodium has reacted, remove the condenser and add 19 ml (19.5 g) of ethyl acetoacetate a few milliters at a time while agitating the flask by gently applying a rotary movement to the ring stand.

Alkylation. After the ester has been added, replace the condenser and add 18 ml of isobutyl bromide through the top of the condenser. The isobutyl bromide should be added immediately after all the ethyl acetoacetate has been added. Stopper flask tightly and store it for two weeks. If the mixture cannot be stored for this period, reflux it for 11 hr, or reflux it for 3 hr and then store in the desk for one week. A better yield is obtained if the mixture is allowed to stand two weeks without refluxing. If the mixture is refluxed, begin the next experiment during this time.

Isolation of Product. Use the appropriate adapters to construct a distillation apparatus (p. 20 and p. 21). Heat the flask on a steam bath until no more liquid distils. Allow to cool a few minutes and then add 100 g of ice, 25 ml of water, and, last, 1.5 ml of concentrated hydrochloric acid. Stir well and transfer the material to a separatory funnel. Use 4 to 5 ml of ether to rinse any material remaining in the flask into the separatory funnel and draw off the sodium halide solution. Wash the organic layer once with 15 ml of cold water and drain it into a 50-ml Erlenmeyer flask containing 5 to 6 g of anhydrous magnesium sulfate.

Stir well, allow the mixture to stand a few minutes, and carefully decant the liquid into a 50-ml Claisen flask or into a 50-ml round bottom flask that can be converted into a Claisen flask with the appropriate adapters. Distil the mixture at 30 mm pressure (pp. 239–241) and collect three fractions: I b.p. to 104°; II b.p. 105 to 135°; III b.p. above 135°.

If there is no precise way of determining the pressure, vacuum distil with a water aspirator until the temperature reaches 100° (95° if the aspirator generates a good vacuum and the water is cold) and collect fraction Ia. Continue distilling to collect fraction IIa until *one* of the following occurs: a large amount of liquid distils and then no more comes over; only 2 to 3 ml remain in the distillation flask; or the distillation temperature reaches 155° (140° if the aspirator usually supplies a good vacuum and most of the material has distilled). Collect fraction II or IIa in a tared yield bottle. Determine the yield and % yield of product.

Infrared Option. Obtain infrared spectra (using the capillary film technique) on the purified product, the material that distilled before the product, any higher boiling fraction, any residue left in the flask after the distillation, and the ethyl acetoacetate used as a starting material. Make as many frequency assignments as possible. If any of the materials are mixtures, attempt to identify the components and their sources by the spectra and any other information available. Identify as many of the by-products as possible.

Problem 60-1. Obtain spectra on 0.5 *M* solutions of ethyl acetoacetate in chloroform and carbon tetrachloride. Compare with each other and with the spectrum obtained from the capillary film of ethyl acetoacetate. Explain the differences.

Synthesis of 4-Methylpentanoic Acid

Reaction. Place 25 ml[1] of 30% aqueous sodium hydroxide solution in a 100-ml* round bottom flask and add 8 ml* of ethyl 2-(2′-methylpropyl)-3-ketobutanoate. Stir the mixture for a short time.

Problem 60-2. What precipitates at this point? Write the equation for the reaction.

Place a reflux condenser on the flask and gently heat the mixture. (If the precipitate does not dissolve within 3 to 4 min, add 3 ml of distilled water through the top of the condenser.) Reflux the mixture for 35 min.

Isolation and Purification of Product. Allow the mixture to cool a few minutes. Add 4 to 5 g* of crushed ice. Stir well and cautiously pour it into a separatory funnel containing 25 g* of crushed ice and 10 ml* of concentrated sulfuric acid. Gently swirl the mixture for a few minutes. Drain the aqueous bottom layer, wash the top layer with 15 ml* of ice water. Use this to wash successively the three chloroform extracts. Transfer the organic layer to a 125-ml* Erlenmeyer flask containing 6 to 8 g of anhydrous magnesium sulfate.

Extract the aqueous layer with three 20-ml* portions of chloroform. After each extraction transfer the chloroform layer into another separatory funnel and, after washing it with the 15 ml* of water, transfer it to the Erlenmeyer flask containing the product and the drying agent. Stir the combined organic layers with the magnesium sulfate and allow the mixture to settle for at least 5 min.

With a 100-ml* round bottom flask construct a distillation apparatus. Carefully decant the product-chloroform solution from the Erlenmeyer flask into the round bottom flask and add a boiling chip. Distil until the temperature reaches 75°. Read the following instrumentation option if that is assigned. Transfer all this distillate

[1] If the following synthesis of methyl isoamyl ketone is omitted, all quantities (denoted by asterisks) should be doubled.

to a waste chloroform container. Distil and collect the fraction that boils from 75 to 139°. If this fraction is larger than 4 ml, dry it with magnesium sulfate. Combine it with the material in the distillation flask, redistil, and collect another fraction of boiling range 75 to 139°.

Continue distilling and collect the fraction that boils at 140 to 147° in a tared yield bottle. Determine the yield and % yield. If more than 4 to 5 ml of liquid remains in the flask, collect a third, higher boiling fraction.

Instrumentation Option. Gas Chromatography:

Problem 60-3. Select a liquid phase and solid support from the following that will best separate the reaction product from the likely contaminates.

Supports[2]	*Liquid Phases*
Chromosorb W	DEGS $[H(O\overset{\overset{O}{\|}}{C}HC_2CH_2\overset{\overset{O}{\|}}{C}OCH_2CH_2OCH_2CH_2)_nOH]$
Firebrick	Paraffin oil
Diatoport P	Silicone oil with dimer acid $[CO_2H(CH_2)_nCO_2H]$ average $n = 34$
Silanized and acid-washed Chromosorb G	Propylene glycol-silver nitrate
	Tetraethylenepentamine

Chromatographic Conditions

Liquid phase, 2.5% (identity given in answer to Problem 60-3)	
Column size	6 ft × $\frac{1}{8}$ in.
Flow rate	25 ml/min
Injection port temperature	250°
Column temperature	170°
Detector temperature	250°

Chromatograph 10-μl samples of each of the fractions and the residue (if any) from the final distillation. Determine the retention times of any compounds available that might be contaminants. Identify as many compounds as possible from the retention times, relative peak heights, the fraction in which they occurred (gives an approximate idea of boiling point and molecular weight), and your knowledge of the reaction.

Infrared: Obtain spectra on each of the fractions and any residue as a capillary film in a demountable cell. Obtain a spectrum on the last 5 to 6 ml of the chloroform distillate in a 0.2-mm cell. Make a frequency-structure correlation for the product. Identify as many compounds as possible from their spectra, the fraction in which they occurred, and your knowledge of the reaction.

Synthesis of Methyl Isoamyl Ketone

Place 70 ml[3] of 5% sodium hydroxide solution and 8 g* of ethyl 2-(2′-methylpropyl)-3-ketobutanoate in a separatory funnel. Shake the mixture every 5 to 6 min for the next 2 hr while proceeding with the next experiment, allow the mixture to stand for an additional 1.5 hr with occasional shaking. The mixture may be stored overnight in the locker if necessary. It should, however, be placed so that a beaker can catch the *entire* contents of the separatory funnel if it should leak (Fig. 60-1) and under no circumstance should it stand more than 48 hr.

Transfer the basic solution to a 200- or 250-ml* round bottom flask. With a Claisen adapter fit a dropping funnel and condenser for distillation (Fig. 60-2). Prepare a cold acid solution from 8 ml* of concentrated sulfuric acid and 16 g* of crushed ice. Slowly add this solution to the basic mixture in the round bottom flask.

Problem 60-4. What is the gas that is vigorously given off? Write the equations for all chemical reactions from the β-keto ester.

[2] See Appendix IV for composition of solid supports.
[3] If the 2-methylpentanoic acid was not synthesized, all quantities (denoted by asterisks) should be doubled.

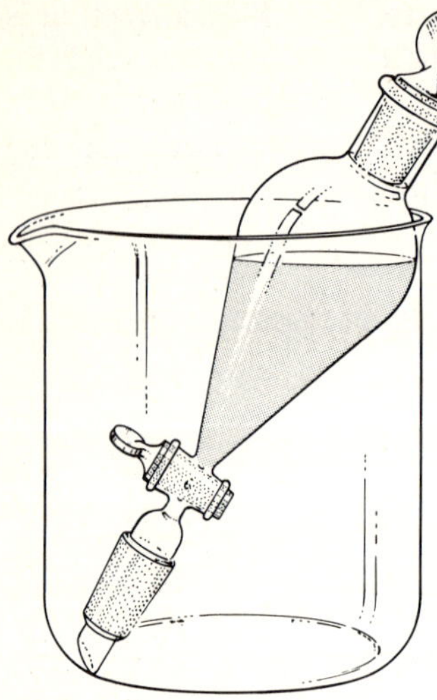

Figure 60-1 Storing a filled separatory funnel.

After all the acid solution is added, the gaseous evolution will decrease significantly; when this occurs, gradually heat the mixture to its boiling point and slowly distil until 35 to 40* ml of distillate collects. Read the instrumentation section below before proceeding further. Discard the residue (residue 1) and transfer the distillate to a 100-ml* round bottom or distilling flask. Add pellets of sodium hydroxide while stirring the mixture until the solution becomes strongly alkaline to Hydrion paper. Then add 1 pellet more. Distil the alkaline mixture collecting 28 to 32* ml of distillate and discard the residue (residue 2). Separate the two layers and redistil the aqueous layer as above; keep repeating this until no more organic material separates from the *cool* distillate. If part b (p. 460) of the infrared option has been assigned, combine all the aqueous residues from the *later* distillations (residue 3). Save and keep residues 1, 2, and 3 separate.

Combine all the organic layers and wash with two 15-ml* portions of 35% aqueous calcium chloride. Transfer to a small Erlenmeyer flask containing 2 to 3 g of magnesium sulfate. Stir the mixture, allow it to settle for at least 5 min, and transfer it to a 10* ml distilling flask. Attach an adapter to give a semimicro distillation apparatus (Fig. 7-3) and distil. Collect the fraction that boils at 140 to 147° in a tared yield bottle.

Problem 60-5. Often the liquid that distills below 140° contains significant amounts of product. What other contaminants are likely? What could be done to recover product from this lower boiling fraction? If the lower boiling fraction from the distillation is greater than 3 ml, attempt to recover product from it. Combine this with the product collected in the tared yield bottle.

Instrumentation Option

Gas Chromatography.

Problem 60-6. Select a liquid phase and solid support from the following that will best separate the reaction product from the likely contaminants.

Supports[4]	*Liquid Phases*
Chromosorb W	DEGS
Firebrick	Paraffin oil
Diatoport P	Silicone oil with dimer acid
Silanized and acid-washed Chromosorb G	Methylchlorophenyl silicone fluid
Glass Beads	Propylene glycol-silver nitrate

[4] See Appendix IV for compositions of solid supports and liquid phases.

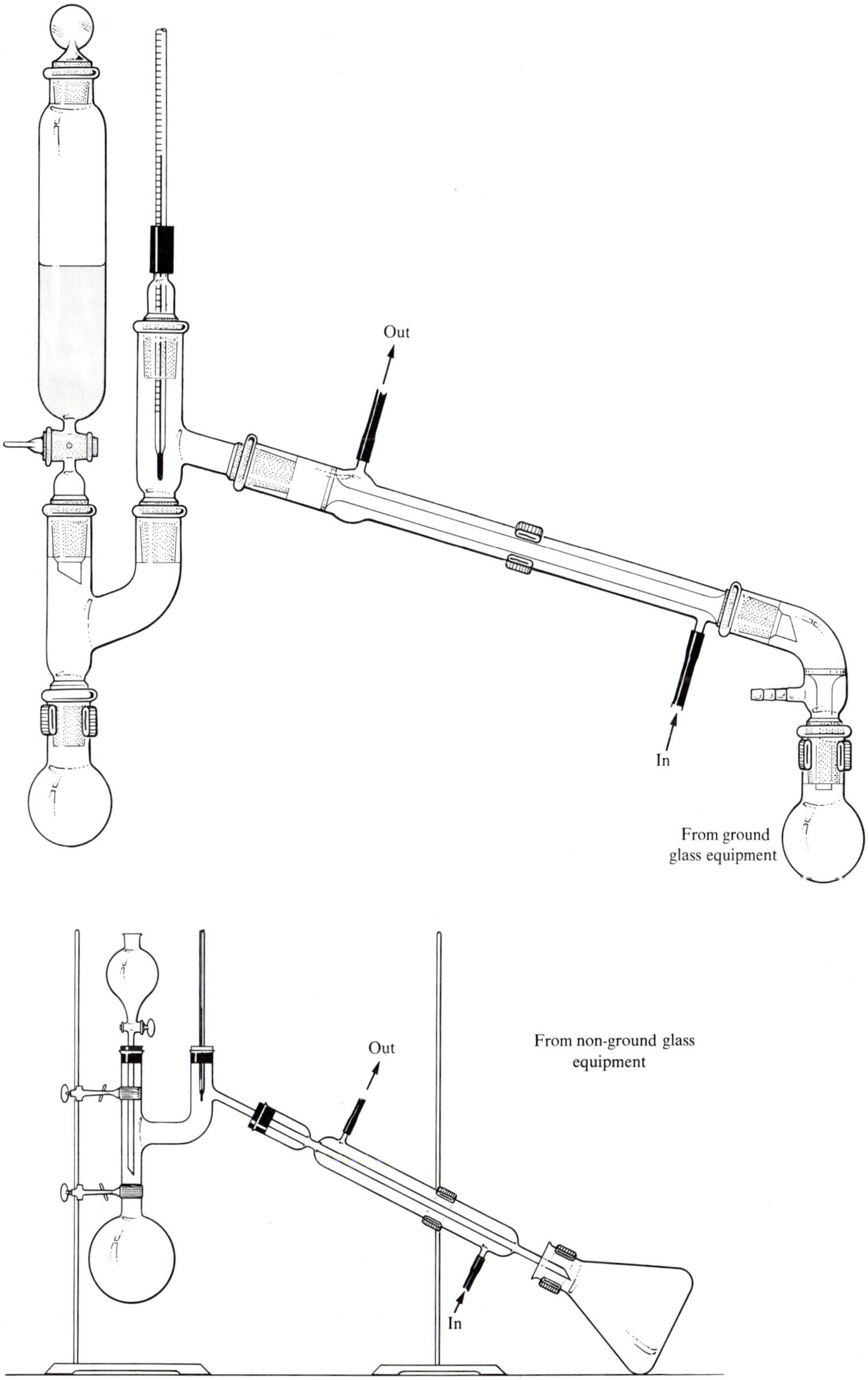

Figure 60-2 Addition-distillation apparatus.

Chromatographic Conditions

Liquid phase, 2.5% (identity given in answer to Problem 60-6)	
Column size	6 ft × ⅛ in.
Flow rate	25 ml/min
Injection port temperature	220°
Column temperature	130°
Detector temperature	190°

Chromatograph 10-μl samples of each of the fractions and the residue (if any) from the final distillation. Determine the retention times of any compounds available that might be contaminants. Identify as many compounds as possible from the retention times, relative peak heights, the fraction in which they occurred, and your knowledge of the reaction.

Infrared. (a) Obtain spectra on each of the fractions and any residue from the final distillation as a capillary film in a demountable cell. (b) **Warning:** carbon tetrachloride is very toxic; avoid spilling the liquid or inhaling the vapor. Separately extract residue 1, residue 2, and residue 3 with two 7-ml portions of spectral grade carbon tetrachloride. Keep the extracts from each residue separate. Dry each extract with 2 to 3 g of magnesium sulfate in a 25-ml Erlenmeyer flask. Distil each extract from a 25-ml round bottom or distilling flask until 2 to 3 ml is left. Using a 0.1-mm cell obtain infrared spectra on each concentrated extract. If possible, use a *slightly* thinner cell with CCl_4 in the reference beam. If any of the absorption bands bottom (except, of course, the 770 cm^{-1} absorption of carbon tetrachloride in uncompensated spectra), obtain spectra of the same solutions in 0.025-mm cells. Occasionally it may be necessary to use 0.015-mm cells or dilute the concentrated extract with carbon tetrachloride. (c) Make a frequency-structure correlation for the product. Identify as many compounds as possible from their spectra, the fraction in which they occurred, and your knowledge of the reaction.

EXERCISES

60-1. In the alkylation of the acetoacetic ester a gas was obtained that had the spectrum shown in Figure 60-3. (a) What is the compound? (b) How was it formed and what could be done to minimize its formation?

60-2. Figures 60-4 to 60-7 show by-products obtained from the ketone or acid synthesis described in this chapter. Most of them resulted from student errors. (a) Identify each of the compounds. (b) Indicate whether it would be expected from the ketone or acid synthesis or from both. (c) Indicate what could be done to minimize the formation of each of these by-products. (d) Where the by-products were caused by a definite student error indicate what the errors were.

60-3. 2-Heptanone (b.p. 151°, D_4^{20} 0.815) and hexanoic acid (b.p. 205°, D_4^{20} 0.9357) were synthesized by the methods outlined in this chapter. (a) Write the equations for formation of the products. (b) Figures 60-5 to 60-13 are spectra of the products and by-products from these reactions. Identify each of the compounds from their infrared spectra and the other information given with the spectra. (c) Indicate how each of the by-products was formed, how its formation could be minimized, and whether it would be likely to be found with 2-heptanone, hexanoic acid, or both.

60-4. How could the following compounds be synthesized from starting materials used in this chapter and any compound with 1 or 2 carbon atoms:

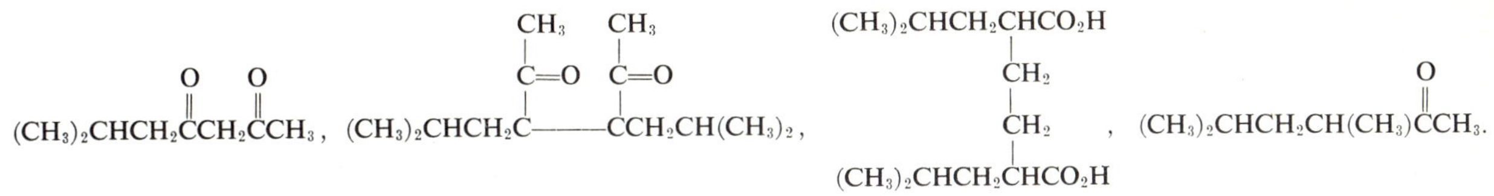

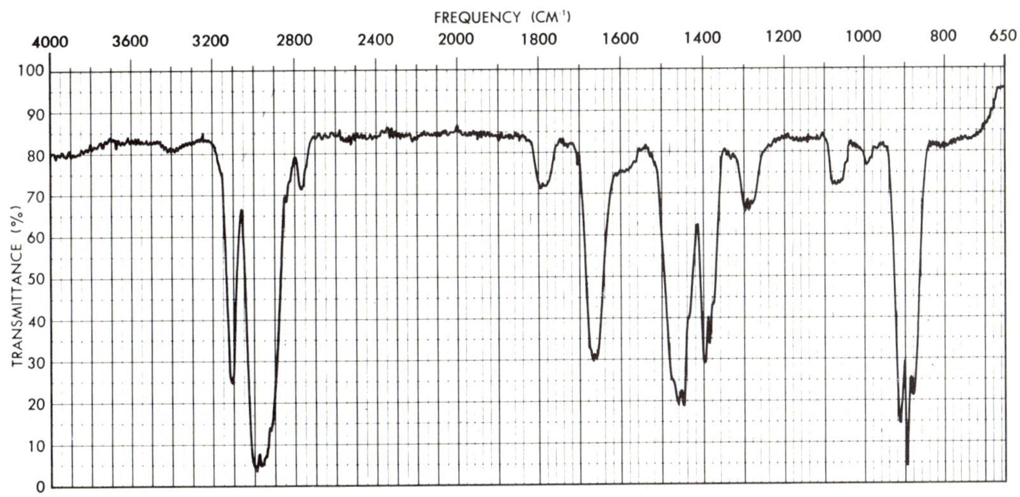

Figure 60-3 Gas by-product in a 10-cm cell. Evacuated 10-cm cell in reference beam.

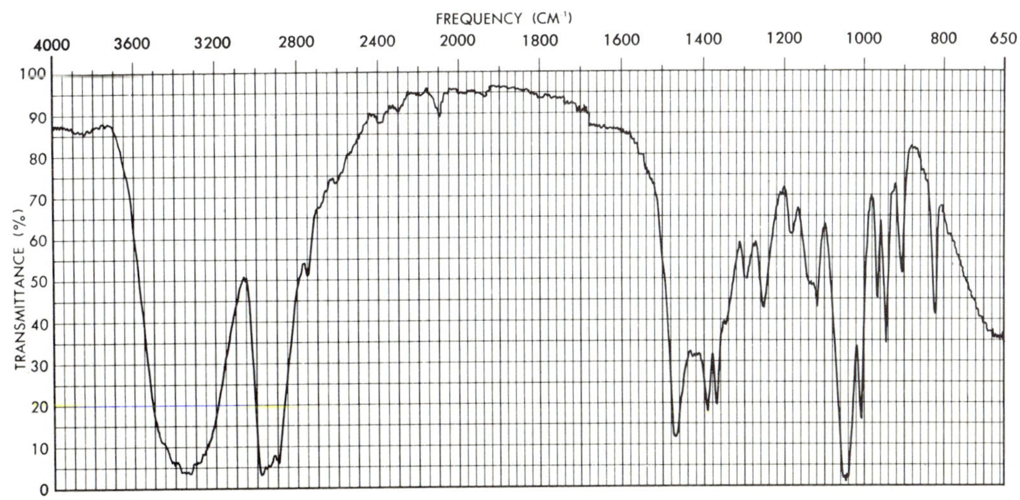

Figure 60-4 Spectrum of by-product described in Exercise 60-2, neat in a 0.015-mm cell, boils near 106°.

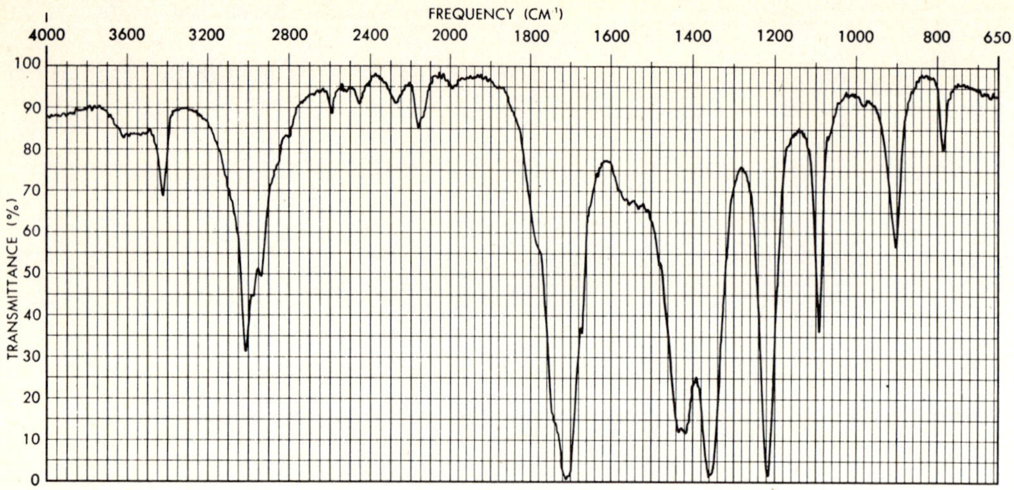

Figure 60-5 Spectrum of a volatile liquid by-product of reactions described in Exercises 60-2 and 60-3. Pure liquid in a 0.015-mm cell.

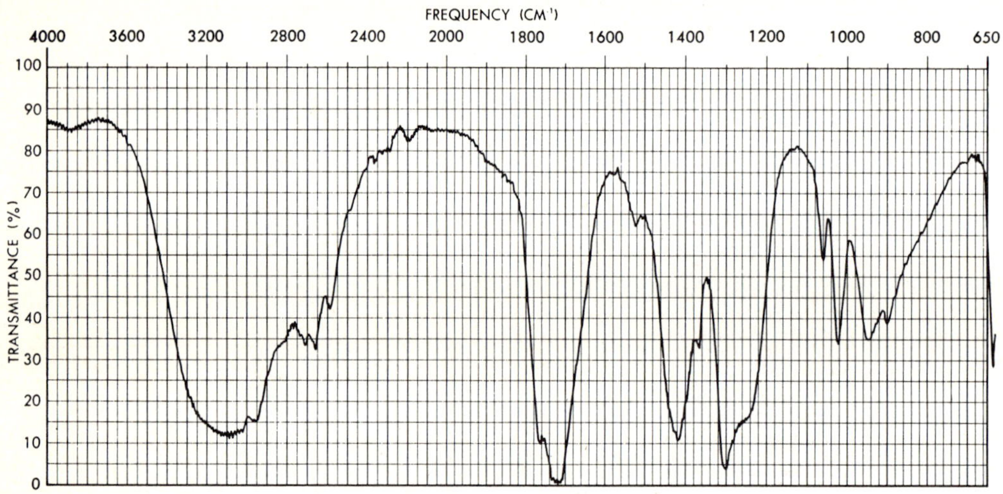

Figure 60-6 Spectrum of liquid isolated during purification of products from reactions described in Exercises 60-2 and 60-3. Pure liquid in a 0.015-mm cell.

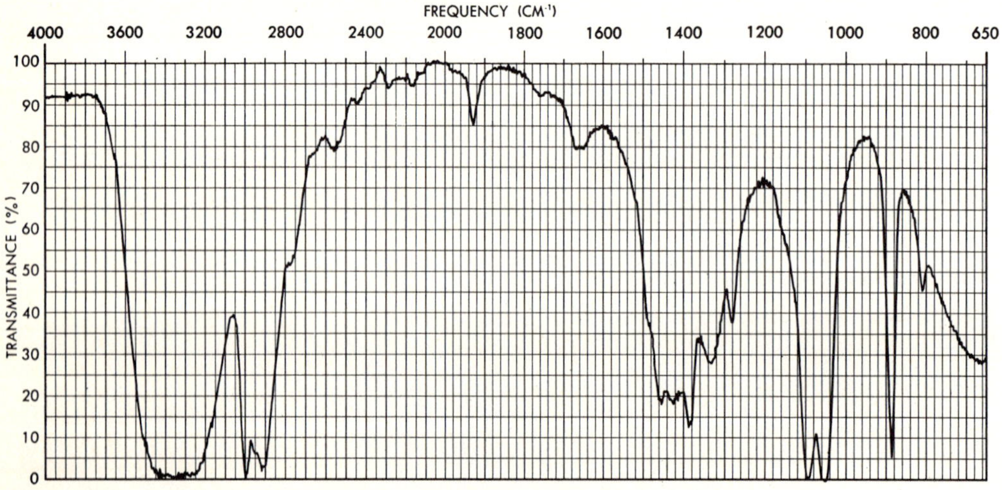

Figure 60-7 Spectrum of liquid isolated during purification of products from reactions described in Exercises 60-2 and 60-3. Pure liquid in a 0.025-mm cell.

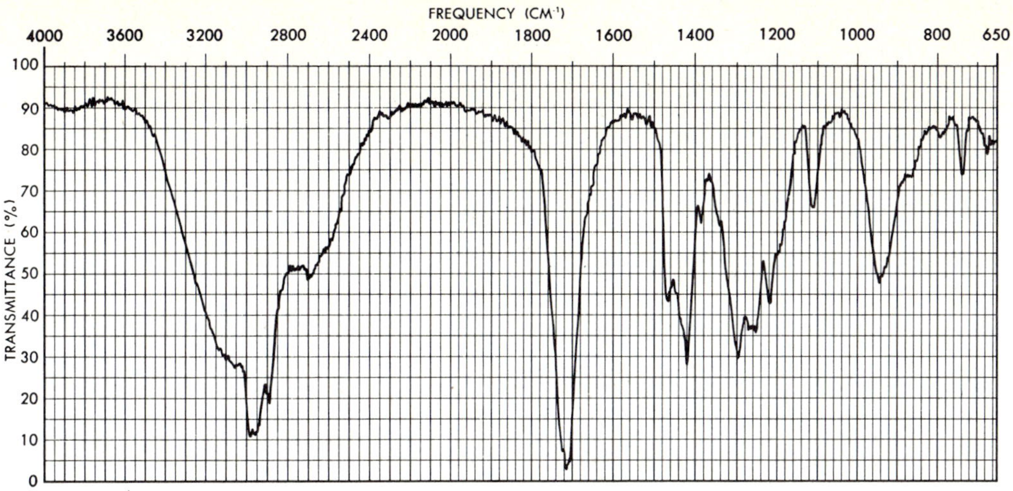

Figure 60-8 Spectrum of liquid isolated from synthesis described in Exercise 60-3, boils near 200°. Pure liquid in a 0.015-mm cell.

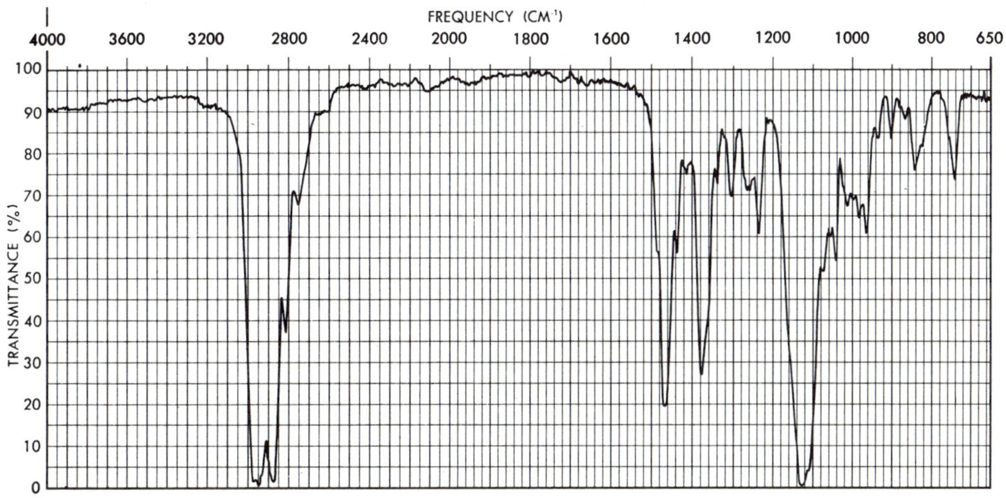

Figure 60-9 Compound isolated from reaction described in Exercise 60-3, boils near 140°. Pure liquid in a 0.025-mm cell.

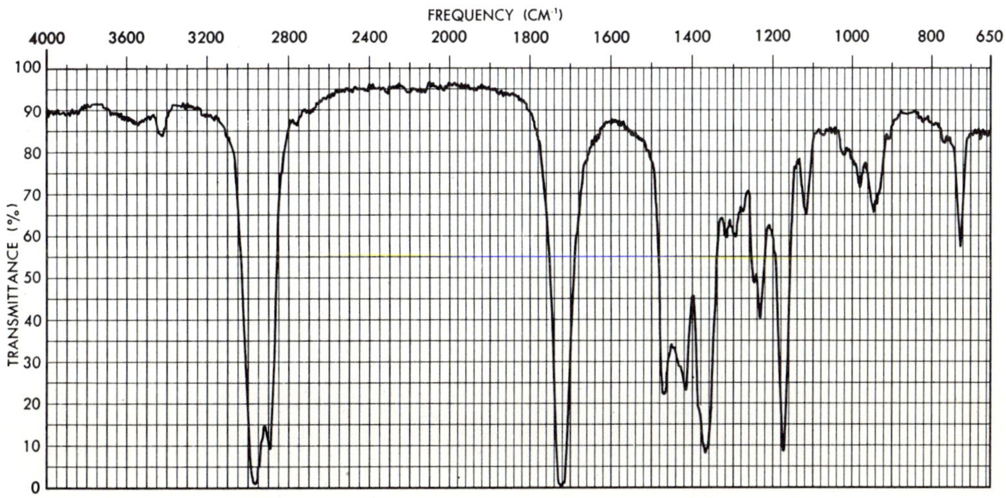

Figure 60-10 Compound isolated from reaction described in Exercise 60-3. Pure liquid in a 0.025-mm cell.

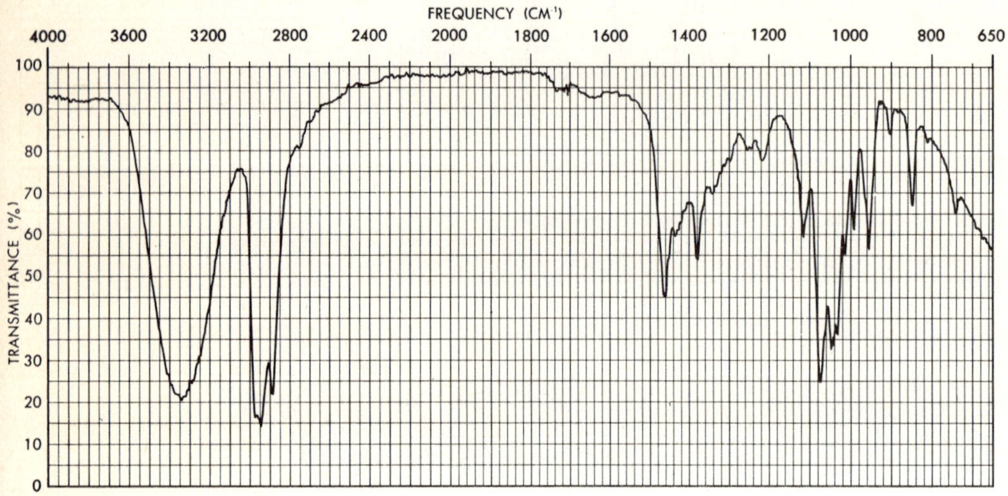

Figure 60-11 Compound isolated from reaction described in Exercise 60-3. Capillary film.

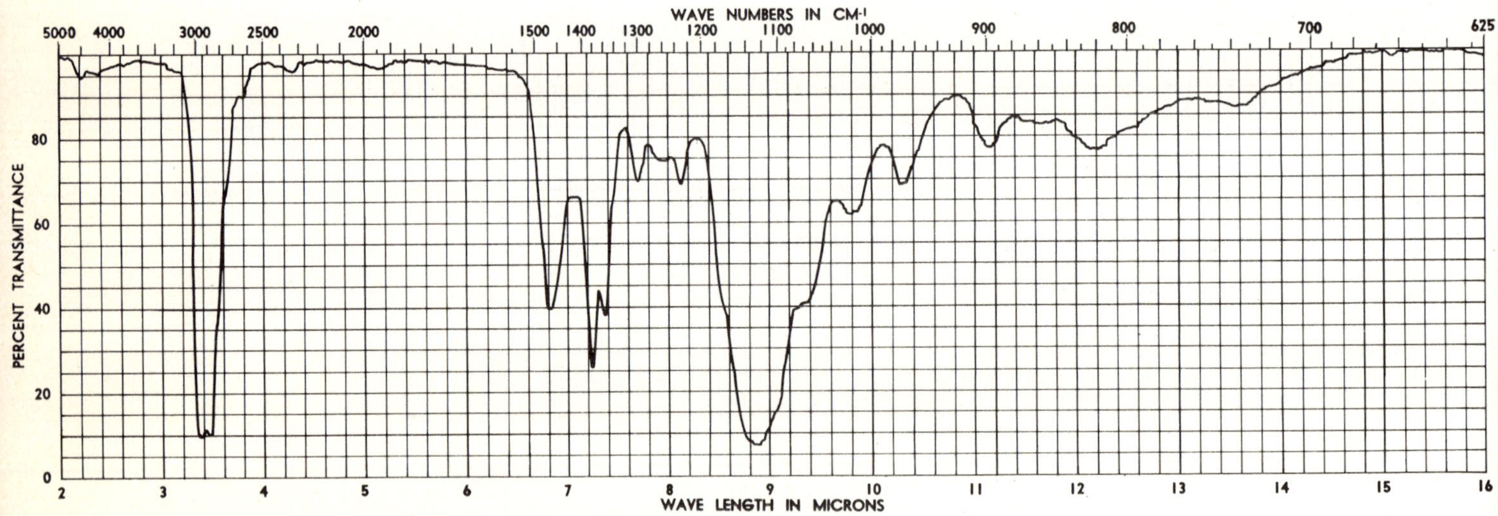

Figure 60-12 Spectrum of liquid (b.p. 91°) isolated from the reactions described in Problem 60-3. Pure liquid in a 0.01-mm cell. Sadtler Research Laboratories, Inc., Philadelphia, Penna. 19104.

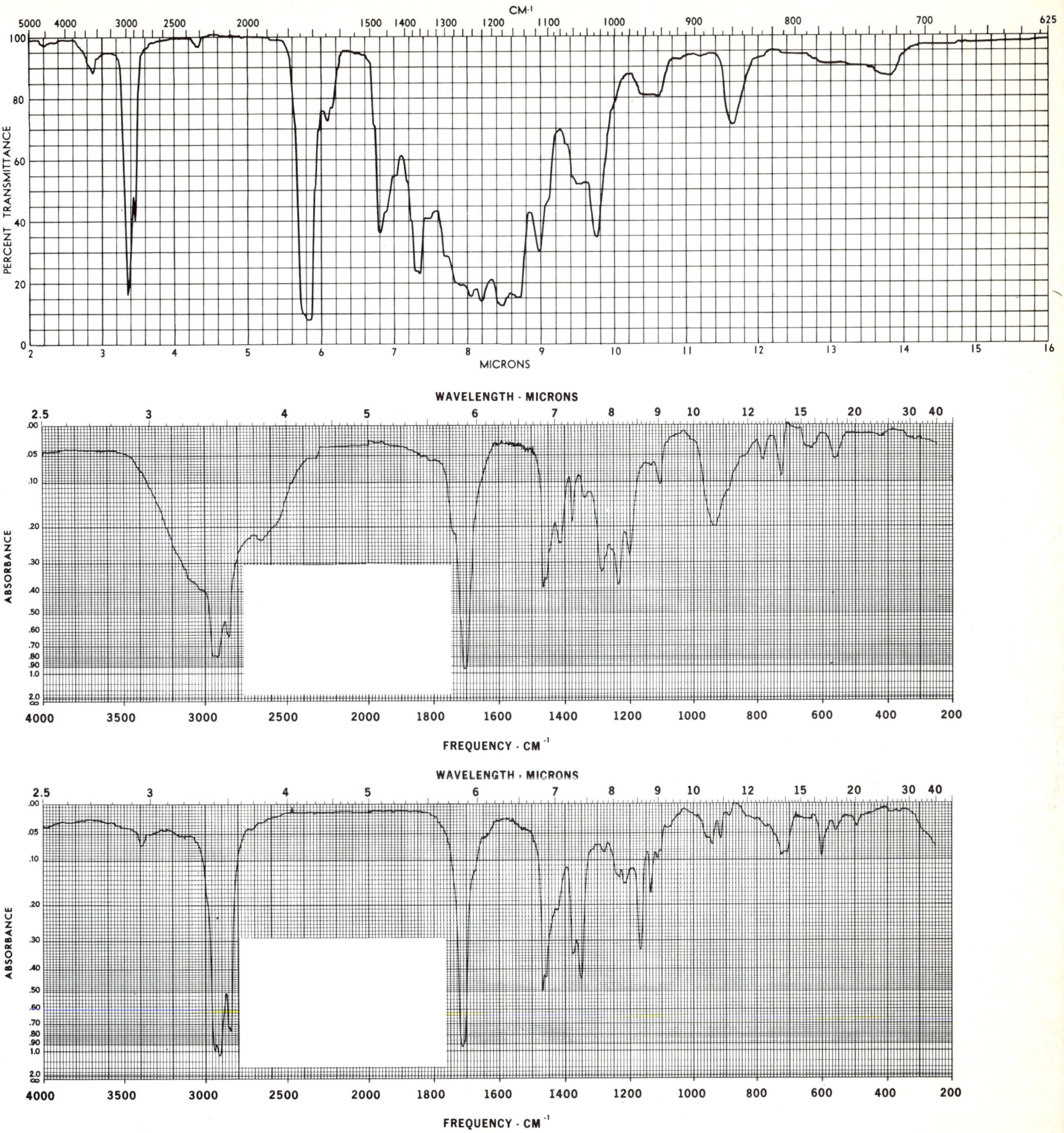

Figure 60-13 High boiling compounds from the reactions described in Problem 60-3. Molecular weights of all three compounds are between 170 and 250. Sadtler Research Laboratories, Inc., Philadelphia, Penna. 19104.

61

Reactions of Polyfunctional Organic Compounds

Organic Qualitative Analysis XI

INTRODUCTION

Many polyfunctional organic compounds react simply as the sum of all the functional groups without the groups' modifying each other's chemical properties to any significant extent. 3-Methyl-7-hydroxy-2-heptanone, $HO(CH_2)_4CH(CH_3)COCH_3$, for example, reacts as a primary alcohol and a methyl ketone. The keto and hydroxy groups have no significant effect on each other's chemical properties, and this compound does not react with any common reagent with which monofunctional alcohols and ketones do not react.

There are, however, many polyfunctional compounds in which two or more groups modify each other's reactivities, and these compounds react with reagents with which the monofunctional equivalents do not react. 3-Methyl-1-hydroxy-2-heptanone, an isomer of the compound discussed, reacts differently to many reagents than monofunctional ketones and alcohols and 3-methyl-7-hydroxy-2-heptanone. This chapter discusses some of the reactions of many of the polyfunctional compounds that react greater or at least differently than the sum of their parts.

PHYSICAL PROPERTIES

Polyfunctional compounds have higher melting and boiling points, higher water solubility, and lower solubility in nonpolar solvents than similar monofunctional compounds. This is illustrated for alcohols in Table 61-1. Note that addition of a primary hydroxyl causes an increase in boiling point of 90 to 120°; addition of secondary hydroxyl group causes an increase in boiling point of 70 to 90°. Many polyfunctional compounds of even moderate molecular weight decompose at temperatures lower than their boiling points at atmospheric pressure. To avoid such decomposition they must be distilled under reduced pressure as was done with the β-keto ester in the previous chapter.

Compounds with two or more polyfunctional groups are less likely to be soluble in ether than monofunctional compounds. The important factor in determining solubility in nonpolar solvents is the ratio of polar to nonpolar moieties. Many compounds with a large number of carbon atoms and three polar functional groups are soluble in ether and other nonpolar solvents. In general, if a compound has more than one carbon atom and has at least one polar group for every 1.3 carbon atoms, it is insoluble in diethyl ether.

TABLE 61-1 COMPARISON OF PHYSICAL PROPERTIES OF ALCOHOLS AND POLYOLS

Alcohol	M.p. (°C)	B.p. (°C)	Sol.* (g/100ml H_2O)	Polyol	M.p. (°C)	B.p. (°C)	Sol. H_2O	in 100g of ether	D_4^{20}	Addition of primary OH	Addition of sec or tert OH
Ethanol	-115	78.3	∞	Ethylene glycol	-16	198	∞	1.0	1.112	120	
2-Propanol	-90	83	∞	1,2-Propanediol		188	∞	8.0	1.040	105	91
1-Propanol	-127	97	∞	1,3-Propanediol		214	∞	insol	1.060	117	
				Glycerol	18	290	∞	insol	1.261	102	76
1-Butanol	-90	118	7.9	1,2-Butanediol		192	sl sol	sol	1.004	90	74
				1,4-Butanediol	18	230	∞	sl sol	1.020	112	
2-Butanol	-114	99.5	12.5	*meso*-2,3-Butanediol	34	183	∞	∞	1.02		83
t-Butyl alcohol	25.5	83	∞	2-Methyl-1,2-propanediol		178.6	∞	sl sol	0.944	94	
Isobutyl alcohol	-108	108	10.0								71
Diethyl ether	-120	34.6	7.5	Ethylene glycol monoethyl ether	-70	135	∞	∞	0.931	100	
Ethyl propyl ether	-79	62.5	sl sol	1-Ethoxy-2-propanol		131	sol	∞	0.903		69
Ethyl isopropyl ether		54	sl sol	2-Ethoxy-1-propanol		140.5	sol	∞	0.904	87	

* All monofunctional liquid alcohols are completely miscible with diethyl ether in all proportions.

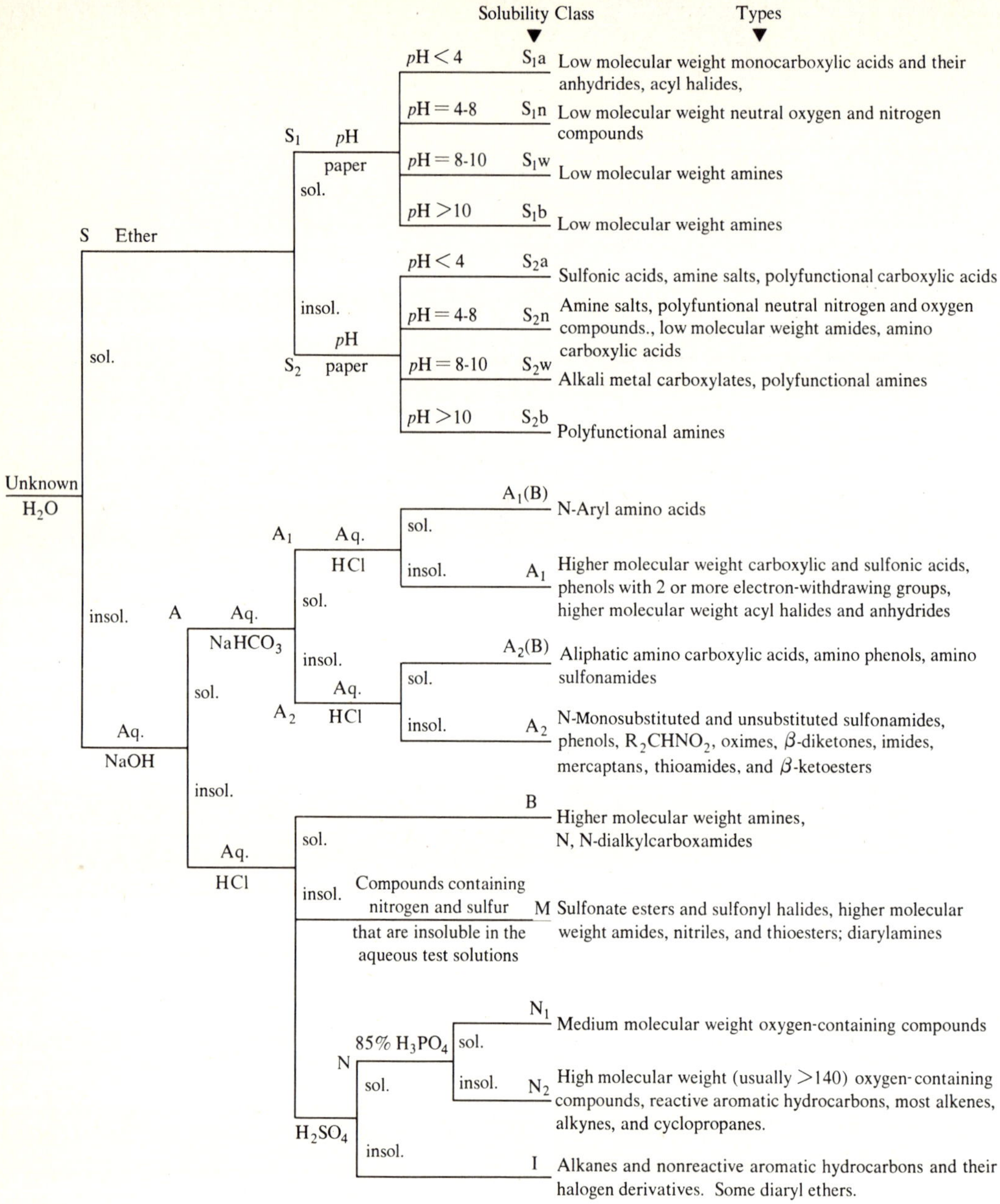

Figure 61-1 Solubility classification chart. See also Figures 32-4 and 41-1.

SPECIFIC CLASSIFICATION TESTS FOR POLYFUNCTIONAL COMPOUNDS

Periodic Acid

Discussion. Periodic acid is a strong oxidizing agent which under forcing conditions will oxidize most organic compounds with oxygen and nitrogen functions. If, however, the conditions are carefully controlled such as the room temperature oxidations specified below, periodic acid is a very selective oxidizing agent. It oxidizes only compounds with certain oxygen or nitrogen functions on adjacent carbon atoms. For example, in the reaction with the two isomeric hydroxyketones:

$$HO(CH_2)_4CH(CH_3)\overset{\overset{\displaystyle O}{\|}}{C}CH_3 + HIO_4 \rightarrow \text{no reaction,} \qquad (61\text{-}1)$$

$$CH_3(CH_2)_3CH(CH_3)\overset{\overset{\displaystyle O}{\|}}{C}CH_2OH + HIO_4 \rightarrow CH_3(CH_2)_3CH(CH_3)CO_2H + HIO_3 + CH_2O. \qquad (61\text{-}2)$$

For qualitative analysis the reaction is made visible by adding silver nitrate and controlling the acidity of the reaction. With the conditions described in the experimental procedure silver periodate is soluble in the reaction mixture and silver iodate is insoluble.

$$R\overset{\overset{\displaystyle O}{\|}}{C}\text{—}CHO + HIO_4 + Ag^+ + H_2O \rightarrow RCO_2H + HCO_2H + H^+ + \underset{\text{white}}{AgIO_3} \qquad (61\text{-}3)$$

Thus, a white precipitate of $AgIO_3$ is a positive test and no precipitate is a negative test.

The reaction apparently proceeds by formation of a cyclic periodate ester. An intramolecular electron transfer causes a combination of ester cleavage, reduction of the periodic acid moiety, and oxidative cleavage of the organic moiety.

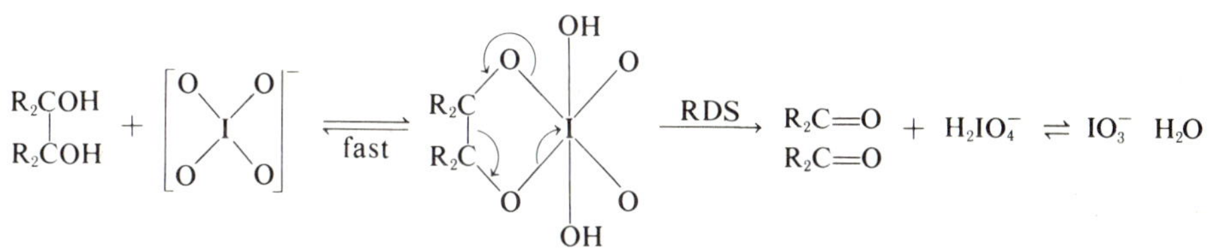

(61-4)

If the oxygen function is an ether, acetal, ketal, or ester, there is no reaction. Similarly, if the nitrogen is a tertiary amine or an amide, there is no visible reaction in the short time allowed.

$$\underset{\displaystyle RC{=}O}{R_2\overset{}{C}OEt} + HIO_4 \rightarrow \text{no reaction} \qquad (61\text{-}5)$$

$$\underset{\displaystyle CHO}{R_2C}\text{—}O\overset{\overset{\displaystyle O}{\|}}{C}CH_3 + HIO_4 \rightarrow \text{no reaction} \qquad (61\text{-}6)$$

$$\overset{\displaystyle RC{=}O}{HC(OEt)_2} + HIO_4 \rightarrow \text{no reaction} \qquad (61\text{-}7)$$

$$\underset{\displaystyle HC{=}O}{\overset{\displaystyle R_2C\text{—}OEt}{R'C\text{—}OH}} + HIO_4 + H_2O \rightarrow R_2\overset{\displaystyle OEt}{C}R'CHO + HCO_2H + HIO_3 \qquad (61\text{-}8)$$

$$\underset{\displaystyle HC{=}O}{\overset{\displaystyle R_2COH}{R'COEt}} + HIO_4 + H_2O \rightarrow \text{no reaction} \qquad (61\text{-}9)$$

$$\underset{\displaystyle R_2COH}{R_2CN(CH_3)_2} + HIO_4 \rightarrow \text{no oxidation} \qquad (61\text{-}10)$$

$$\underset{R_2\overset{|}{C}OH}{R_2C}\overset{H}{\overset{|}{N}}\overset{O}{\overset{\|}{C}}CH_3 + HIO_4 \rightarrow \text{no reaction} \qquad (61\text{-}11)$$

$$\underset{R_2\overset{|}{C}OH}{R_2C}\overset{H}{N}CH_3 + HIO_4 \rightarrow 2R_2C{=}O + CH_3\overset{+}{N}H_3 + IO_3^- \qquad (61\text{-}12)$$

Since the compounds are tested in fairly strong acid solutions, some compounds such as epoxides and acetals and a few esters may give positive tests. Usually, however, enough time is required to hydrolyze these compounds so that very little of periodate-sensitive compound is produced in the short interval allowed for the reaction.

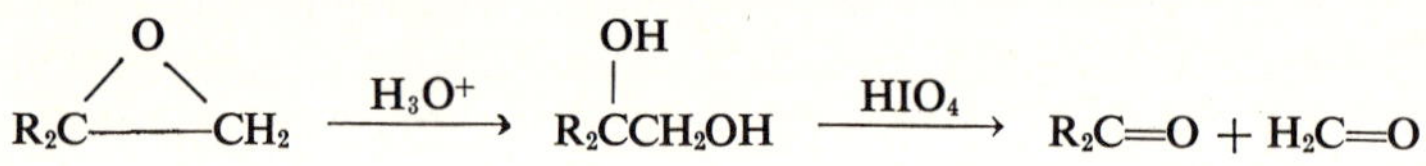

(61-13)

$$RCHOHCH(OEt)_2 \xrightarrow{H_3O^+} RCHOH\overset{H}{\overset{|}{C}}{=}O \xrightarrow{HIO_4} R\overset{H}{\overset{|}{C}}{=}O + HCO_2H \qquad (61\text{-}14)$$

Usually α-hydroxy- and α-ketocarboxylic acids do *not* react with periodic acid with the conditions described below. Sometimes, however, slightly more forcing conditions such as a higher temperature and more periodic acid than is prescribed cause some α-oxycarboxylic acids to give positive tests.

$$R\overset{O}{\overset{\|}{C}}CO_2H + HIO_4 \rightarrow RCO_2H + CO_2 + HIO_3 \quad \text{occasionally} \qquad (61\text{-}15)$$

$$RCHOHCO_2H + HIO_4 \rightarrow RCHO + CO_2 + H_2O + HIO_3 \quad \text{occasionally} \qquad (61\text{-}16)$$

$$HOCH_2CHOHCO_2H \xrightarrow[\text{always}]{HIO_4} CH_2O + OHCCO_2H \xrightarrow[\text{occasionally}]{HIO_4} HCO_2H + CO_2 \qquad (61\text{-}17)$$

Experimental Procedure. The reagent is made by dissolving 0.6 g of periodic acid ($HIO_4 \cdot 2H_2O$) in 60 ml of distilled water. Add 1 drop of concentrated nitric acid and 2 ml of periodic acid reagent to a 7.5-cm test tube and mix well. Add 1 drop (or an equivalent amount of solid) of compound to be tested; shake the tube 15 sec; add 2 drops of aqueous silver nitrate (do not add an excess). If the compound is water-insoluble, dissolve approximately 0.05 g in 1 ml of reagent grade dioxane and test this entire solution. Observe immediately; a white precipitate is a positive test. If a white precipitate forms after standing, an α-oxycarboxylic acid or compound that is hydrolyzed to a compound that reacts with periodic acid at room temperature is indicated. A brown precipitate (especially if it redissolves on shaking) or a black precipitate is a negative test. Try this test on lactic acid, glyoxal, ethylene glycol, and 1,2-propanediol.

Benedict's Test

Discussion. Aldehydes and ketones give three distinct reactions with cupric complexes in aqueous base; these reactions are useful for subclassifying these compounds. For α-hydroxy aldehydes and ketones and α-keto aldehydes the cupric ion is reduced to cuprous oxide and the organic compounds are oxidized to carboxylic acids or more complex products.

$$8HO^- + RCHOHCHO + 4Cu^{++} \rightarrow 2Cu_2O + R\overset{O}{\overset{\|}{C}}CO_2H + 5H_2O \qquad (61\text{-}18)$$

$$8HO^- + 2RCHOH\overset{O}{\overset{\|}{C}}R' + 4Cu^{++} \rightarrow 2Cu_2O + 2R\overset{O}{\overset{\|}{C}}\overset{O}{\overset{\|}{C}}R' + 6H_2O \qquad (61\text{-}19)$$

The visible indications of this reaction are a change of the blue test solution to colorless or very light blue and a yellow to red precipitate of cuprous oxide.

Most aliphatic aldehydes without *alpha* hydroxyl or keto groups such as acetaldehyde, cinnanaldehyde, 2-butenal, and aldol form a red to dark brown tacky precipitate that is *not* cuprous oxide. The supernatant liquid is not colorless as when cuprous oxide is formed but brown, orange, yellow, green, or blue.

Ketones without *alpha* hydroxyl or aldehyde groups, most aromatic aldehydes, and a few aliphatic aldehydes such as isobutyraldehyde cause no change in the reagent, and no color change or precipitation occurs. Hydroxylamines, hydrazines, some aminophenols, and a few other very easily oxidized compounds precipitate cuprous oxide from Benedict's solution.

$$4HO^- + C_6H_5NHNHC_6H_5 + 2Cu^{++} \rightarrow Cu_2O + C_6H_5N{=}NC_6H_5 + 3H_2O \qquad (61\text{-}20)$$

Experimental Procedure. To prepare the reagent dissolve 52 g of sodium citrate[1] and 30 g of sodium carbonate (anhydrous) in 250 ml of hot distilled water. Dissolve 5.2 g cupric sulfate ($CuSO_4 \cdot 5H_2O$) in 35 ml of distilled water. Slowly pour with stirring the cupric sulfate solution into the citrate solution.

Dissolve or suspend 0.1 to 0.2 g (2 to 4 drops if a liquid) of compound to be tested in 4 ml of water. Add 4 ml of Benedict's solution. Shake the test tube a little. If there is no reaction, heat the mixture *just* to boiling. Use propionaldehyde, glyoxal, phenylhydrazine, benzaldehyde, acrolein, and hydroxyacetone for sample tests.

SOME REACTIONS OF MULTIPLE FUNCTIONAL COMPOUNDS WITH OTHER CLASSIFICATION REAGENTS

Iodoform Reaction

Compounds with any combination of secondary alcohol or ketone groups *beta* to each other react with aqueous alkaline halogen solutions (p. 231) to give haloforms even though these compounds may have no methyl groups adjacent to oxygen functions.

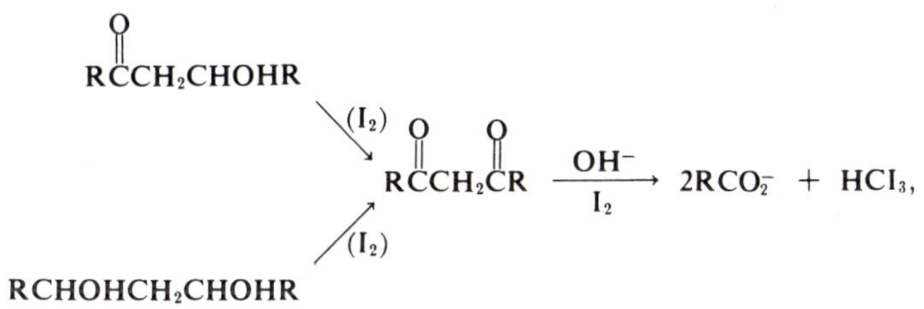

(61-21)

where R is alkyl or most aryl, *o,o*-disubstituted aryl groups being the principal exception.

Problem 61-1. Propose a mechanism for the reaction of β-diketones to haloforms.

Compounds that have methyl groups adjacent to ketone or secondary alcohol groups do not yield iodoform if these groupings are destroyed by the strongly alkaline solution before sufficient iodination occurs. The principal compounds that show this behavior are those that quickly undergo a reversal of the Claisen condensation. For example:

$$CH_3COCH_2CO_2Et \xrightarrow[I_2]{HO^-} 2CH_3CO_2^- + EtOH \quad \text{no significant iodoform formation,} \qquad (61\text{-}22)$$

$$CH_3COCH_2NO_2 \xrightarrow[I_2]{HO^-} CH_3CO_2^- + CH_2NO_2^- \quad \text{no significant iodoform formation,} \qquad (61\text{-}23)$$

$$CH_3COCH_2CN \xrightarrow[I_2]{HO_2^-} 2CH_3CO_2^- + NH_3 \quad \text{no significant iodoform formation.} \qquad (61\text{-}24)$$

[1] Fehling's test is similar except that two solutions are used and there is a cupric-tartrate complex instead of a citrate complex.

Problem 61-2. Complete the following equations:

$$CH_3CH_2\overset{\overset{\displaystyle O}{\|}}{C}CH_2\overset{\overset{\displaystyle O}{\|}}{C}CH(CH_3)_2 \xrightarrow[I_2]{HO^-}$$

$$CH_3CHOHCH_2CO_2Et \xrightarrow[I_2]{HO^-}$$

$$CH_3\overset{\overset{\displaystyle NH}{\|}}{C}CH_2CO_2Et \xrightarrow[I_2]{HO^-}$$

$$CH_3\overset{\overset{\displaystyle O}{\|}}{C}CH_2CH_2CO_2Et \xrightarrow[I_2]{HO^-}$$

Potassium and Sodium Hydroxide Solutions

In addition to the usual hydrolysis and displacement reactions of these reagents with organic compounds aqueous bases (p. 275) also have other reactions with multiple functional organic compounds such as reversal of the Claisen condensation (Equations 61-22, 61-23, and 61-24), decarboxylation, and internal Cannizzaro reactions.

$$CH_3COCH_2CO_2Et \xrightarrow[\text{aq. }OH^-]{\text{dilute}} CH_3COCH_3 + CO_2 + EtOH \qquad (61\text{-}26)$$

$$CH_3\overset{\overset{\displaystyle O}{\|}}{C}CHO \xrightarrow[OH^-]{\text{conc.}} CH_3CHOHCO_2^- \qquad (61\text{-}27)$$

Most β-dicarbonyl compounds are sufficiently acidic (K_a 10^{-9} to 10^{-11}) to dissolve in dilute sodium hydroxide solution even if they are water insoluble. β-Dicarbonyl compounds are not, however, sufficiently acidic to react significantly with aqueous sodium bicarbonate solutions and will not dissolve in this reagent if they are insoluble in water.

Reaction with Sodium

Two carbonyl or other strongly electron-withdrawing groups adjacent to a methylene or methine group make these groups acidic enough to react with sodium (p. 277).

$$CH_3COCH_2CO_2Et + Na \rightarrow CH_3\overset{\overset{\displaystyle \overset{\delta^-}{O}}{\|}}{C}\overset{\delta^-}{C}H\overset{\overset{\displaystyle \overset{\delta^-}{O}}{\|}}{C}OEt + \tfrac{1}{2}H_2 + Na^+ \qquad (61\text{-}28)$$

Reaction with Bromine in Carbon Tetrachloride

β-Diketones and β-ketoesters react with bromine in carbon tetrachloride (p. 98). Since this is a substitution and not an addition reaction, it is accompanied by the evolution of hydrogen bromide. If the reaction is carefully observed, there is little chance of confusing it with the addition of bromine to carbon-to-carbon multiple bonds.

$$CH_3COCH_2COCH_3 + Br_2 \rightarrow CH_3COCHBrCOCH_3 + HBr \qquad (61\text{-}29)$$

Tollens' Reagent

α-Hydroxy ketones, α-alkoxy ketones, and α-dialkylaminoketones reduce Tollens' reagent and give a silver mirror.

Ferric Chloride

Many β-keto esters, β-diketones, and other enolizable compounds give colors with ferric chloride much as phenols do (p. 381).

Effect of Nucleophilic Displacement Reactions

Functional groups affect nucleophilic displacement reactions on other groups in at least two different ways. One is by electron-withdrawing inductive effects; see, for example, the effect of halogen atoms on the Lucas test (p. 474). Another is by actual participation of functional groups in one of the steps of the displacement reactions. For example, a dimethylamino group *beta* to a halogen atom greatly increases the rate of S_N1 reaction of the halogen atom by pushing the halogen atom out (the nitrogen atom is a nucleophilic reagent in this reaction), by stabilizing the resulting cation, or by both of these effects. Hydroxyl, alkoxyl, acyloxy, sulfur functions, carbonyls, hydrogen atoms, and even phenyl and alkyl groups show similar effects, often called neighboring group effects.

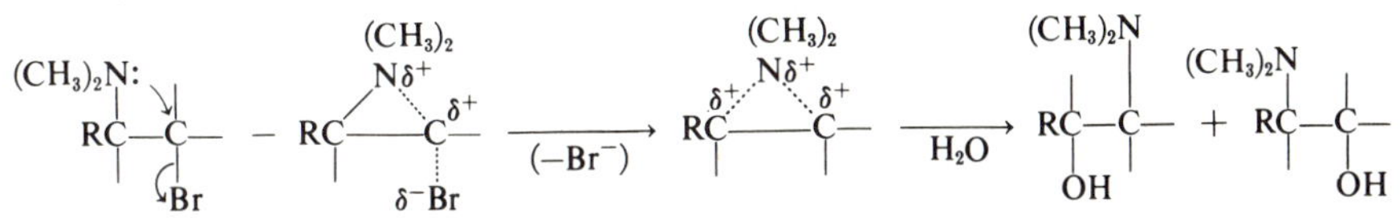

(61-30)

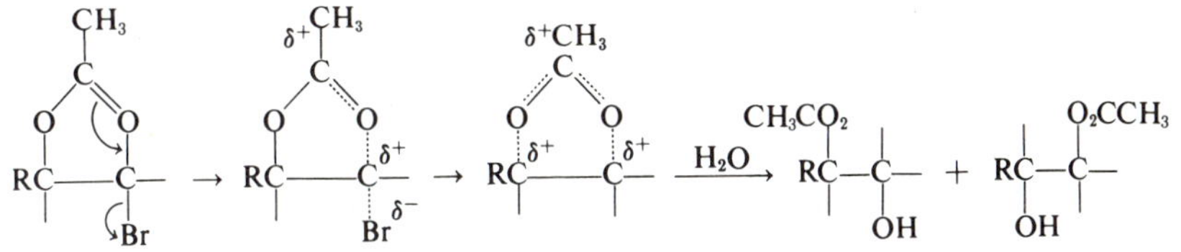

(61-31)

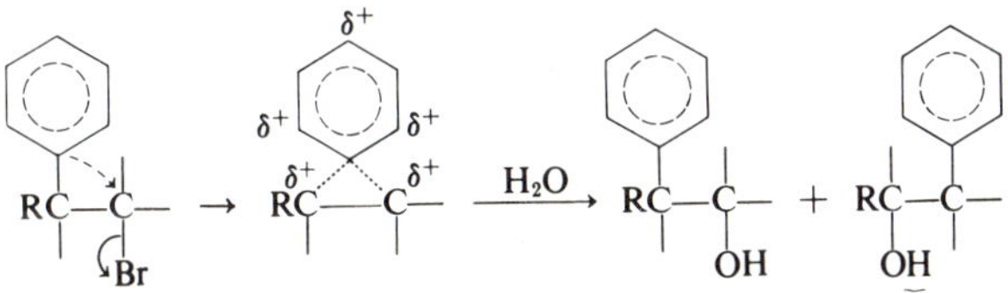

(61-32)

The neighboring group need not be next to the group being displaced as written by conventional structural formulas, but it should be capable of taking part in the displacement reaction without introducing strain in the intermediate or transition state.

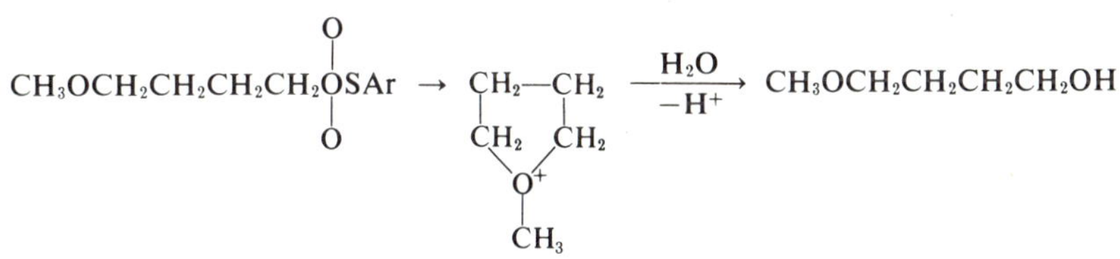

(61-33)

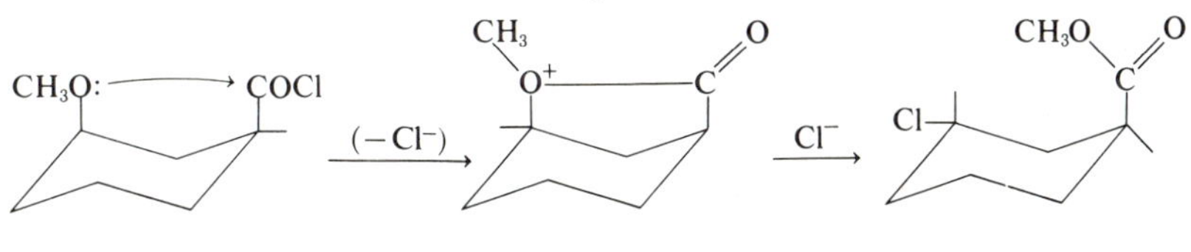

(61-34)

UNKNOWN

There are no tables of possible unknowns in this chapter. Multiple functional compounds are listed in the tables in the other chapters on qualitative analysis. The assignment of most of these polyfunctional compounds as possible unknowns has been delayed until the study of this chapter.

EXERCISES

61-1. Complete the following table:

TABLE 61-2 REACTIONS OF MULTIPLE FUNCTIONAL COMPOUNDS

Compound	Tollens reagent	Benedicts test	2,4-Dinitrophenyl hydrazine	o-Dianisidine	Sodium bisulfite	Iodoform test	HIO_4
CH_3CH_2CHO	+	gummy ppt.	+	+	+	-	-
$ArCHO$	+	-					
$CH_3COCH_2CH_3$							
$CH_3CO_2CH_2CH_3$							
$CH_2OHCHOHCH_3$							
$CH_2OHCOCH_3$							
$CH_3CHOHCHO$							
$ArCOCH_3$							
$ArCOCH_2CH_3$							
$CH_3CH_2CHOHCOCH_2CH_3$							
$CH_3COCH_2CO_2Et$							
$CH_3COCH_2CH_2CO_2Et$							
$CH_3COCH_2COCH_3$							
$CH_3CH_2COCH_2COCH_2CH_3$							
$CH_3\underset{N(CH_3)_2}{CH}COCH_3$							
$CH_3\underset{NH_2}{CH}CHOHCH_3$							

61-2. (1) Indicate whether the following compounds are soluble or insoluble (p. 225) in water and soluble or insoluble in diethyl ether. (2) Indicate in which solubility class each compound is. (a) 1,2-octanediol, (b) hexanehexol, (c) phenoxyacetic acid, (d) succinic acid, (e) sebacic acid, (f) succinimide, (g) pinacol, (h) phthalimide, (i) phthalic acid, (j) *m*-aminoaniline, (k) *p*-aminoaniline, (l) *p*-nitrosodiphenylamine, (m) *m*-dinitrobenzene, (n) ethyl acetoacetate, (o) butanetetraol, (p) ethylenediamine, (q) fumaric acid, (r) glutaric acid, (s) glycine (aminoacetic acid), (t) 2-(diethylamino)ethanol, (u) hydroquinone, (v) benzylmalonic acid, (w) 2,4-diketooctane.

61-3. How would the epimer

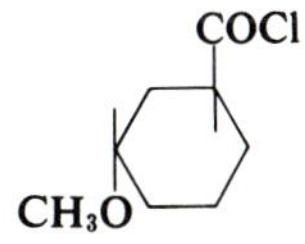

react under the same conditions as in Equation 61-34?

61-4. Make sufficient additions to the solubility classification chart on page 468 to include all the groups of compounds studied so far.

62

The Diels-Alder or Diene Reaction

Addition of Maleic Anhydride to Anthracene

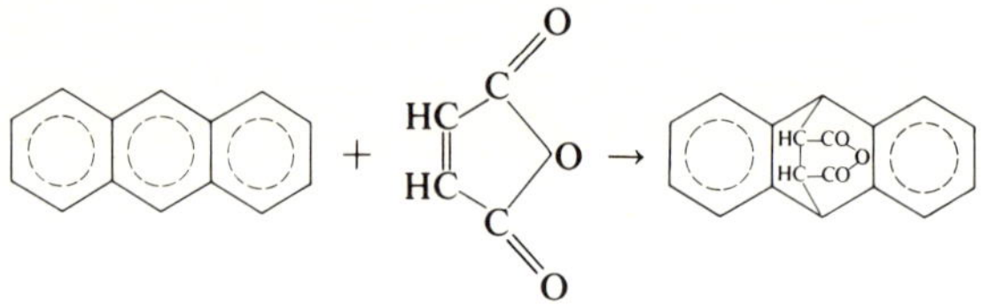

(62-1)

M.p. 216°	M.p. 56°	M.p. 262.2–264°d
B.p. 354°	B.p. 198°	
diene	dieneophile	

INTRODUCTION

Compounds with conjugated double or triple carbon to carbon bonds (dienes) react with other compounds with carbon to carbon multiple bonds (dieneophiles) to generate 6 membered rings. A simple example is the reaction between butadiene and acetylene.

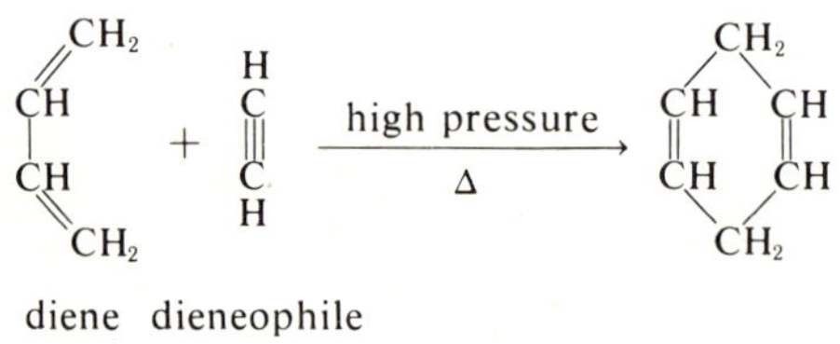

(62-2)

diene dieneophile

If, however, the reaction is to take place at atmospheric pressure and an easily reached temperature, the dieneophile must have one or more carbonyl, cyano, nitro, or aryl groups to activate the olefinic or acetylenic bond.

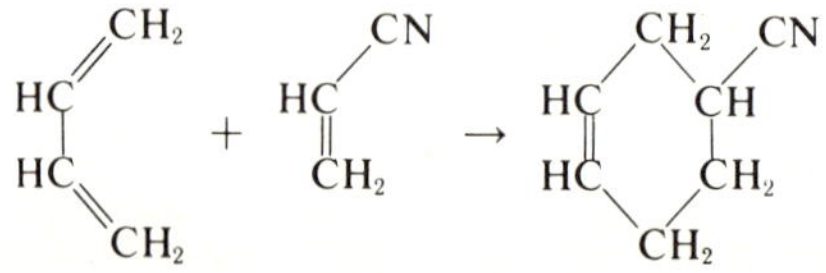

(62-3)

EXPERIMENTAL PROCEDURE

Add 2 g of maleic anhydride, 4 g of purified anthracene, 50 ml of xylene (dried over sodium), and a boiling chip to a 200- or 250-ml round bottom flask. Add a reflux condenser with a drying tube. Reflux for 25 min; agitate the mixture by applying a rotatory motion to the ring stand every 2 to 3 min for the first 15 min. Remove the heat, and *after* the mixture has stopped boiling for at least 3 min, add 1 g of carbon black. Reflux for 5 min, allow to cool to 115 to 125°, and filter through a conical funnel with no stem or a very short one. Heat the funnel to 95 to 130° just prior to filtration. If the funnel is heated on a steam or water bath, thoroughly dry it with a piece of filter paper before filtering. In the round bottom flask heat 20 ml of dried xylene just to boiling. Use this to wash the flask and the filter paper and combine the wash with the filtrate.

In this procedure avoid exposure of the product, 9,10-dihydroanthracene-9:10-*endo*-α,β-succinic anhydride, to water as the acid anhydride is readily hydrolyzed to the dicarboxylic acid. Allow the filtrate to cool to room temperature, vacuum filter, and wash the crystals with a few milliliters of chilled xylene. Chill the filtrate in an ice bath and collect a second crop of crystals. Dry the two crops in separate containers in a vacuum desiccator for 1 to 2 hr. Place a small container with paraffin wax shavings (helps to remove the xylene) and a small container of calcium chloride in the desiccator.

Determine the melting points of the two crops of crystals. If the second crop melts within 1.5° of the first, combine it with the first crop in a tared yield bottle. If it melts 1.5° or more below the first crop, recrystallize it from the minimum amount of dry xylene, dry it in the desiccator, and combine it with the first crop. Determine the yield and % yield.

ABSORPTION SPECTROSCOPY (Optional)

Infrared

Obtain infrared spectra on the starting materials. If facilities are available, obtain the spectra from potassium bromide pellets (1 mg of compound in 280 to 300 mg of potassium bromide). If facilities for preparing KBr pellets are not available, obtain the spectra in carbon tetrachloride solution (0.1 *M* concentration in 0.1-mm cells) using solvent compensation in the reference beam. Correlate the various absorptions bands with structures of the compounds (Appendix VII).

Ultraviolet

Prepare solutions of anthracene (0.001 g/100 ml) and the product (0.01 g/100 ml) in spectral grade cyclohexane. Dilute each 10 to 1 using a 1-ml pipet and 10-ml volumetric flask. Obtain ultraviolet spectra on all 4 of these solutions. For these samples it is necessary to obtain spectra only in the region from 200 to 390 mμ.

Problem 62-1. (a) What are the principal differences between the ultraviolet spectra of the reactant and the product? (b) The ultraviolet spectrum of the product closely resembles the spectrum of what simple hydrocarbon?

Problem 62-2. (a) Does the infrared or ultraviolet spectrum give more structural information about the product? (b) About the anthracene? (c) Which of these two spectroscopic methods gives more information about the structure of the aromatic systems involved?

Problem 62-3. Which of these two spectroscopic methods would probably be better for quantitative analysis of the following: (a) the product in the reaction mixture, (b) maleic anhydride in the reaction mixture, (c) anthracene in the reaction mixture, (d) maleic anhydride in the dried reaction product,

(e) 9,10-dihydroanthracene-9:10-*endo*-α,β-succinic anhydride in the dried reaction product, (f) anthracene in the dried reaction product, (g) succinic acid anhydride impurity in the maleic anhydride, (h) fumaric acid impurity in the dried reaction product, (i) fumaric acid impurity in the maleic anhydride.

Problem 62-4. Indicate which frequency or wavelength would be suitable for each quantitative analysis mentioned in Problem 62-3?

Problem 62-5. (a) Which of the two spectroscopic methods would be better for a kinetic study of this Diels-Alder reaction? (b) Give at least two reasons for your answer to part a. (c) Outline a method for determination of the rate constant for the reaction studied in this chapter. Be specific about sampling technique, wavelength or frequency used, and other details.

THIN-LAYER CHROMATOGRAPHY (Optional)

Chromatography

Follow the general directions on p. 202 for thin-layer chromatography and the specific instructions on p. 295 for activating, spotting, and developing the chromatograms with the exceptions and additions noted below. Use Silica Gel G as the adsorbent. Spot as many of the following as your instructor indicates: (a) A drop or two of the reaction mixture after the first refluxing and before addition of carbon black. Be certain the first drop has evaporated before the second is applied. (b) The second crop of crystals obtained by chilling the filtrate, (c) the purified product, (d) as many standards that may be products, by-products and other contaminants as you are able to get on the plates or strips. Spot approximately 2×10^{-5} g of (b), (c), and (d) in a few drops of acetone. If UV spectra are to be obtained, triple the quantities for (b) and (c) and double them for (a). Develop the chromatograms with a mixture of 9 parts hexane and 1 part acetone.

As for the chromatograms described in Chapter 36 successively visualize the plates or strips, using the bromocresol green spray last. If the section below on ultraviolet analysis is assigned, use only the visualization with ultraviolet light.

Problem 62-6. (a) From the reactions to the visualization procedures, your knowledge of the reaction, and the R*f* values how many of the compounds from spots a, b, and c can you identify? (a) What can you determine about the purity of your product and what compounds are separated by which steps in the isolation and purification?

Ultraviolet Analysis of the Separated Compounds

Follow the directions beginning on p. 209 for obtaining ultraviolet absorption spectra on compounds separated by TLC. Obtain spectra on as many of the separated compounds from spots (a), (b), and (c) as your instructor indicates. It is necessary to obtain spectra on only one of the spots from the 9,10-dihydroanthracene-9:10-*endo*-α,β-succinic anhydride. The compounds separated in this experiment are not nearly so air-sensitive as the plant pigments.

Problem 62-7. From the spectra, R*f* values, and your knowledge of the reaction identify as many of the impurities as possible.

63

Infrared Spectra of Carbonyl Compounds

INTRODUCTION

Variations in structure of molecules cause rationalizable effects on physical properties. This has been illustrated throughout this text by structural effects on boiling points, melting points, solubilities, and densities. Structure has a similar effect on both the intensities and frequencies of spectral absorption bands. The correlations of spectral properties with structure are more predictable and quantitative and have fewer exceptions than the other correlations of physical properties with structure discussed in this text.

EXPERIMENTAL PROCEDURE

Prepare solutions in carbon tetrachloride of as many of the compounds in Table 63-1 as your instructor indicates. The solutions should be 0.5 *M* unless another concentration is specified in parentheses after the compound. If the compounds do not dissolve in carbon tetrachloride, use chloroform. Obtain infrared spectra on each of the solutions you prepared, using a 0.05-mm cell. Be certain that the paper is aligned correctly and that a good trace is obtained, especially from 2000 to 1500 cm^{-1}.

DISCUSSION

The frequency of an infrared absorption band is dependent upon several factors. Among them are the type of vibration (bending, stretching, rocking, etc.) causing the absorption, the mass (actually the reduced mass) of the atoms or groups involved in the vibration, and the force constant of the bond(s) joining the atoms involved in the vibration.

The principal purpose of this experiment is to examine the effect of various groups on the carbonyl stretching absorption, to illustrate the effects of structure on spectra, and to estimate the relative resonance and inductive effects of various groups. Since only one type of vibration is being examined and the masses of carbon and oxygen are the same for all carbonyl groups, any variation in frequency is due to changes in the force constant.[1]

[1] The absorption frequency of a vibrating group may also be dependent on its coupling or interaction with another vibrating group. The amount of coupling is dependent upon how the two vibrating species modify one another's vibrations. Coupling is usually strong if the two groups have similar frequencies and are near each other in the same molecule. Coupling is strongest if the groups have a common atom and identical frequencies. The two bands between 1860 and 1710 cm^{-1} of acid anhydrides are due to coupling of the two carbonyl groups.

TABLE 63-1 CARBONYL COMPOUNDS FOR INFRARED SPECTRA

Butyric acid	Propionic anhydride
α-Bromobutyric acid	Maleic anhydride (0.25M in $CHCl_3$)
Dichloroacetic acid (0.25 M)	Succinic anhydride (in $CHCl_3$)
Trifluoroacetic acid	Phthalic anhydride (0.25 M in $CHCl_3$)
Acrylic acid (0.25 M)	Acetyl chloride
p-Toluic acid (0.25 M)	Benzoyl chloride (0.25 M)
p-Chlorobenzoic acid (0.25 M)	
	N,N-Dimethylforamide (0.25 M)
Butyraldehyde (1.0 M)	N,N-Dimethylacetamide (0.25 M)
Crotonaldehyde	
Benzaldehyde	N-Methylacetamide (0.25 M)
p-Dimethyaminobenzaldehyde	Acetanilide
o-Hydroxybenzaldehyde (0.25 M)	Propionamide (0.25 M)
	Benzamide (0.25 M)
Propyl acetate	Benzanilide (0.25 M)
Isopropenyl acetate	Caprolactam (0.25 M)
Ethyl formate	Acrylamide (0.25 M)
Ethyl chloroacetate	
Butyl chloroformate	Methyl isobutyl ketone
Ethyl trichloroformate	Cyclobutanone
Methyl benzoate (0.25 M)	Cyclopentanone
Ethyl acrylate (0.25 M)	Cyclohexanone
Ethyl acetoacetate	Camphor
Diethyl maleate (0.25 M)	Acetophenone (0.25 M)
Butyrolactone (0.25 M)	p-Chloroacetophenone (0.25 M)
Valerolactone (0.25 M)	Benzophenone
Methyl salicylate (0.25 M)	Chloroacetone
	Benzoquinone (0.10 M)

The force constant is roughly a function of the bond order. That is, triple bonds have higher force constants than double bonds and double bonds have higher force constants than single bonds. If a series of similar bonds has partial double bond character, the force constants are higher for those bonds that have the greater amount of double bond character. The multiple bonds of carbonyl groups have partial double bond character. Two electronic structures, I and II, Figure 63-1, can be considered to

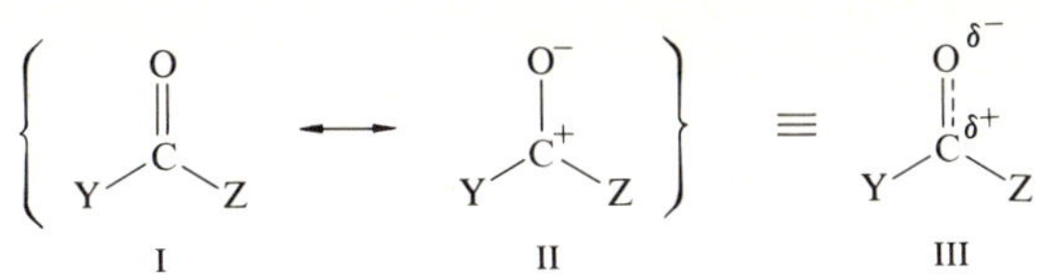

Figure 63-1 Electronic structures of carbonyl groups.

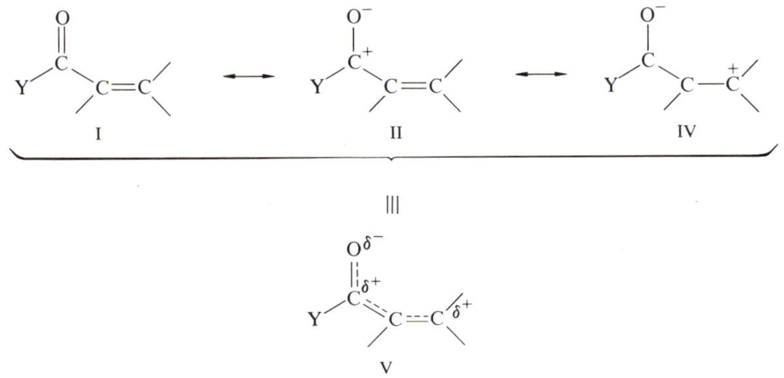

Figure 63-2 Delocalization of π-electrons of a carbon-to-carbon double bond with a carbonyl group.

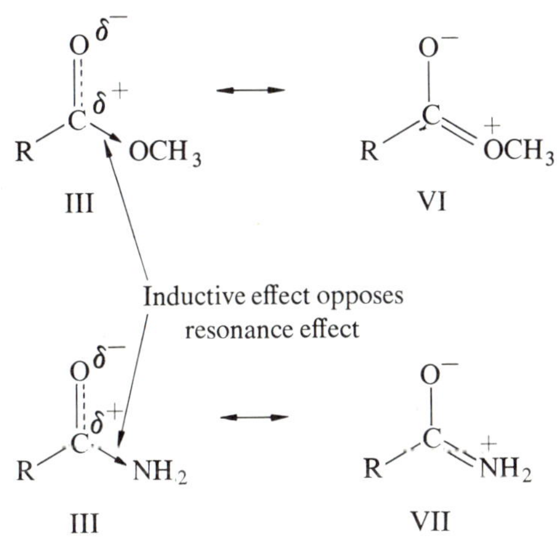

Figure 63-3 Delocalization of nonbonded electrons of methoxy and amino groups with a carbonyl group.

describe the structure of carbonyl groups. (Review pp. 268 and 414.) The greater the contribution from structure I, the greater the double bond character, the larger the force constant, and the higher the frequency of the stretching vibration. By the same reasoning the greater the contribution from the polar structure II, the lower the double bond character and the lower the frequency of the stretching vibration.

Electron-donating groups such as alkyl stabilize electronic structure II and cause it to contribute more so that ketones (two alkyl groups) have carbonyl absorption bands at lower frequencies than aldehydes (one alkyl group). Electron-withdrawing groups such as halogen intensify the charge at the carbonyl carbon atom and cause a decrease in the contribution of electronic structure II. The carbonyl stretching absorptions of acid halides, therefore, are at higher frequencies than the absorptions of aldehydes and ketones.

A group with unbonded or *pi* electrons *alpha* to the carbonyl group increases the contribution of electronic structures like II (Fig. 63-2) because of delocalization of electrons from this adjacent group to the carbonyl group. This causes a decrease in the double-bond character of the carbonyl group, a decrease in the force constant, and causes the carbonyl stretching absorption to be at a lower frequency. Some groups such as alkoxy (RO—), hydroxy (HO—), and amino (NH_2—) are electron-withdrawing by the inductive effect and electron-donating by resonance effects. The force constants and absorption frequencies of carbonyl groups attached to these groups are dependent upon the relative contribution of the two opposing effects (Fig. 63-3).

Problem 63-1. Column 1 lists some structures of carbonyl compounds, column 2 lists in a different order the frequencies of the carbonyl stretching absorptions of these compounds. Determine which frequency belongs with each of the compounds.

	ν cm^{-1}
$CH_3\overset{\overset{O}{\|}}{C}C_2H_5$	1677
$CH_3\overset{\overset{O}{\|}}{C}F$	1706
$F\overset{\overset{O}{\|}}{C}F$	1720
$CH_3\overset{\overset{O}{\|}}{C}Cl$	1808
$CH_3\overset{\overset{O}{\|}}{C}H$	1840
$CH_2{=}CH\overset{\overset{O}{\|}}{C}CH_3$	1928

Problem 63-2. The intensity or extinction coefficient of an infrared absorption band is dependent on the change in dipole during the vibration that causes the absorption. The greater the change in dipole, the higher is the extinction coefficient (i.e., the more intense) of the absorption band. Would a greater contribution by electronic structures I or II (Fig. 63-1) cause a higher intensity for the carbonyl stretching absorption? What broad generalization can be obtained relating frequency differences of various carbonyl groups to their intensities? Predict the relative intensities of the carbonyl stretching absorptions of the compounds listed in Problem 63-1.

From the spectra you obtain determine as accurately as possible the frequency and molar extinction coefficient (pp. 153–155) of each of the carbonyl stretching absorptions. Correlate the various structures with frequencies and extinction coefficients. Using butyraldehyde as a reference, determine the electron-withdrawing or donating character of the various groups. Rationalize the frequency shifts and intensities relative to butyraldehyde with the inductive and resonance effects of the structures causing the shifts and changes of intensity.

EXERCISES

63-1. When a hydrogen atom hydrogen-bonds with a carbonyl oxygen or other electronegative atom, the positive hydrogen pole is close to the electronegative atom. What effect would hydrogen bonding have on the force constant of a carbonyl bond? On the frequency of the carbonyl stretching absorption? On the intensity of the carbonyl stretching absorption? Verify your conclusion from the spectra of ethyl acetoacetate, *o*-hydroxybenzaldehyde, and methyl salicylate.

63-2. Is there always a correlation between frequency shift and relative intensities of carbonyl absorption bands? If not, explain as many of the exceptions as possible.

63-3. (a) Determine from the spectra that you and the others of the class obtained if there is any correlation of the frequencies and intensities of the carbonyl stretching absorptions with dipole moments of the various types of carbonyl compounds (Table V-5, Appendix V). (b) Compare your rationalization of the various

TABLE 63-2 CARBONYL STRETCH ABSORPTIONS FOR ESTERS AND THIOESTERS

Thioesters	ν C=O (cm^{-1})	$\Delta\nu$ (cm^{-1}) from aldehyde	from sat. ester
$R\overset{O}{\overset{\|}{C}}SEt$	1690	-35	
$Ø\overset{O}{\overset{\|}{C}}SEt$	1665	-60	-25
$R\overset{O}{\overset{\|}{C}}SØ$	1710	-15	+20
$Ø\overset{O}{\overset{\|}{C}}SØ$	1685	-40	- 5

Esters	ν C= O (cm^{-1})	$\Delta\nu$ (cm^{-1}) from aldehyde	from sat. ester
$R\overset{O}{\overset{\|}{C}}OR'$	1735	+10	
$Ø\overset{O}{\overset{\|}{C}}OR$	1725	0	-10
$R\overset{O}{\overset{\|}{C}}OØ$	1750	+15	+25
$Ø\overset{O}{\overset{\|}{C}}OØ$	1735	+10	0

electronic effects on the spectra of the carbonyl compounds with the correlation of these same electronic effects on the dipole moments discussed on pages 449–450.

63-4. Predict the relative frequencies of the carbonyl stretching absorptions of the following compounds:

$CH_3CH_2\overset{O}{\overset{\|}{C}}CH_3$, $CH_3CH_2\overset{O}{\overset{\|}{C}}—NEt_2$, $CH_3CH_2\overset{O}{\overset{\|}{C}}CH_2NEt_2$, $CH_3CH_2\overset{O}{\overset{\|}{C}}CH_2F$, $CH_2{=}CH\overset{O}{\overset{\|}{C}}CH_3$, $Et_2NCH{=}CH\overset{O}{\overset{\|}{C}}Et$, $CH_3CH(NEt_2)\overset{O}{\overset{\|}{C}}H$,

63-5. Explain the similarities and differences between the frequency shifts of the esters and thioesters shown in Table 63-2.

64

Carbohydrates

Organic Qualitative Analysis XII

INTRODUCTION

Carbohydrates are a large and important group of polyfunctional compounds. They are polyhydroxy, straight chain aldehydes (aldoses) and ketones (ketoses) and polymers (polysaccharides) which are derived from the monosaccharide units by linkage of their oxygen functions. Many derivatives of these compounds such as methyl ethers and glycosides are also often included in the broad generic classification of carbohydrates even though these compounds do not have the formula $C_x(H_2O)_y$. Since carbohydrates are polyfunctional compounds, most of the discussion of Chapter 61 is applicable to carbohydrates. This chapter discusses the properties and analysis of carbohydrates.

PHYSICAL PROPERTIES

Melting and Boiling Points

Most common carbohydrates are solids although many crystallize with such difficulty that they are often found as oils or syrups. Because of their polyfunctional oxygen groups carbohydrates melt higher than most other organic compounds. Most melt with decomposition and all decompose before they boil at atmospheric pressure.

Solubility

As would be expected, all carbohydrates are insoluble in nonpolar solvents such as hexane, benzene, and ethyl ether. Carbohydrates of low molecular weight are very soluble in water. Some carbohydrates of even very high molecular weight are soluble in water or form aqueous colloids.

Partition Coefficients and Paper Chromatography

Discussion. As might be expected from their solubilities when carbohydrates partition between two liquid phases, the carbohydrates tend to concentrate in the more polar phase. The partition behavior

TABLE 64-1 CARBOHYDRATES AND DERIVATIVES

Carbohydrate	Decom position point (°C)	$[\alpha]^{20}_{D\,H_2O}$	Rf in n-BuOH, HOAc, H_2O	Derivatives, m.p. (°C) osazone	acetate
1,3-Dihydroxyacetone (monomer)	69			132	
Raffinose (hyd)	79	+104.5	0.05		99
1,3-Dihydroxyacetone (dimer)	81			132	
Glucose (hyd)	90	+47.7		205	132(β) 112(α)
L-Rhamnose (hyd)	94	+8.0	0.37	182	
D-Ribose	94	-21.5	0.31	166	
Maltose	100	+129.0	0.11	206	158
D-Fructose	103	-92.0	0.23	205	109(β) 70(α)
L-Rhamnose (anh)	105	+9.4	0.37	182	
D-Lyxose	106	+13.7		163	
Raffinose (anh)	119	+123	0.05		99
D-Mannose	132	+14.1		205	
dl-Glyceraldehyde dimer	142*			132	154
D-Xylose	145	+18.7	0.28	163	141
α-D-Glucose (anh)	146	+112 → +52.8	0.18	205	112
β-D-Glucose	150	+19 → +52.8	0.18	205	132
L-Arabinose	160	+104	0.21	166	
dl-Arabinose	164	0		169	
α-Methylglucoside	165	+157.6			101
Arbutin (hyd)	165	-60.3			136
Maltose	165	+111.7 → +130.4	0.11	206	158
dl-Ascorbic acid	169				
D-Galactose	169	+150.7 → +80.2	0.16	201	142(β) 95(α)
Sucrose	171	+66.5	0.15	205	89
Lactose (hyd)	202	+52.3	0.09	201d	100
Cellobiose	225	+14.2 → +34.6		206	193
Cellulose	d				
Starch	d				

Melts without decomposition.

Figure 64-1 Paper with spots and fringes.

of a compound controls not only its distribution in two immiscible solvents in a separatory funnel but also its chromatographic behavior.

Carbohydrates are not sufficiently volatile to be directly separated by gas chromatography although much has been done recently in the separation of ethers of low to medium molecular weight carbohydrates by gas chromatography.

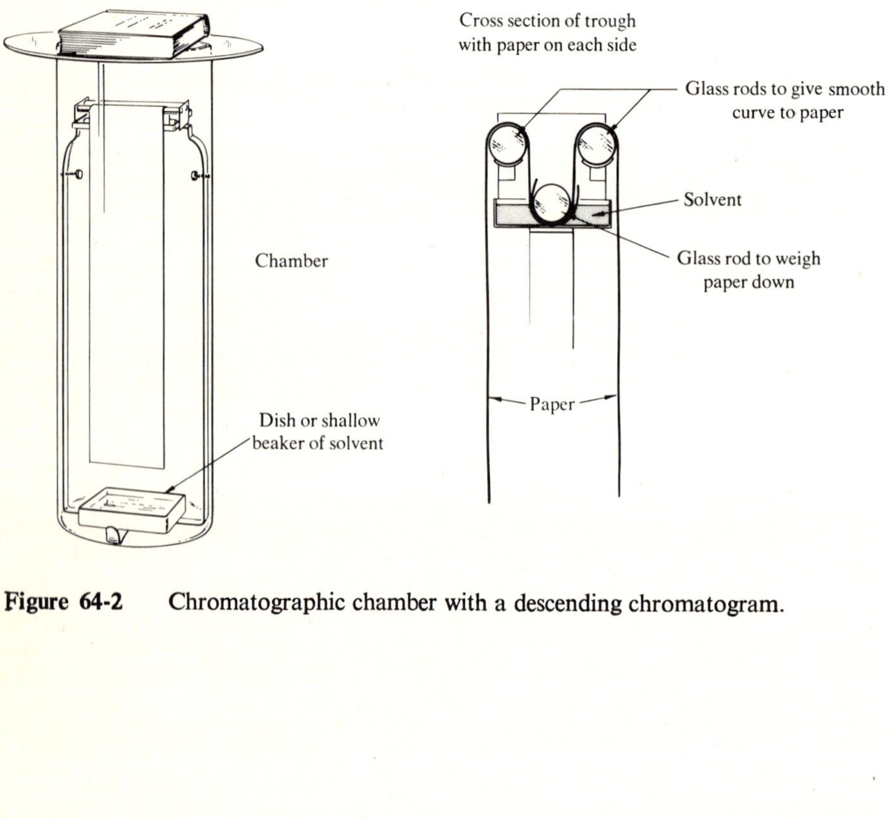

Figure 64-2 Chromatographic chamber with a descending chromatogram.

There has been considerable application of paper chromatography to the separation and analysis of carbohydrates. Many solvent systems have been examined and reported. One of the better systems for differentiating among low and moderate molecular weight carbohydrates is a mixture of water, acetic acid, and 1-butanol. The R*f* values in Table 64-1 are for paper chromatography with this system and a descending technique. (Contrast with the ascending technique in Chapter 59.)

Experimental Procedure. Review the discussion on paper chromatography on pages 451–453 before proceeding with the experiment. In a separatory funnel mix 55 ml of distilled water, 11 ml of glacial acetic acid, and 44 ml of 1-butanol. Shake the funnel well and stopper it. Prepare solutions of 2 g of carbohydrate in 100 ml of distilled water. Using the same technique described for thin-layer chromatography, spot the paper with 0.03 ml of carbohydrate solution and 0.03 ml of glucose solution.[1] Cut fringes 2 to 3 cm deep (Fig. 64-1) at the bottom or end of the paper where the spots were not applied.

Drain the bottom layer from the separatory funnel into an evaporating dish. Carefully place this dish in the bottom of the chromatographic chamber. Place the paper in the trough, using the glass rods as anchors; see Figure 64-2. Place the top on the chamber and allow the solvent mixture to equilibrate with the air in the chamber for at least 20 hr (p. 452). Store the top layer from the solvent mixture in a tightly stoppered bottle or flask.

To develop the chromatogram fill the tray with the top layer from the solvent mixture. The top should not be kept off the chamber more than absolutely necessary; fill the tray as quickly as possible, but be careful not to spill any of this solvent on the paper.

Develop the papers for 20 to 28 hr. The solvent front should be as near the fringe as you can conveniently arrange, but under no condition should it be beyond the fringe. Remove the paper and allow it to dry in a hood for 50 to 90 min.

Spray the paper with anilium phthalate reagent. This reagent is prepared by warming 1.90 g of aniline, 2.38 g of phthalic acid, and 100 ml of distilled water saturated with 1-butanol. This aniline should be colorless or at most a light yellow; it can be conveniently decolored by distilling from zinc. As soon as the paper is sprayed, place it horizontally on glass rods or beakers in an oven at 100 ± 2°. Dry for 10 to 12 min and observe under a fluorescent light. The carbohydrate spots are red, rust, or brown. Circle the carbohydrate spots and determine their R*f* values. If the R*f* of glucose is not 0.17 to 0.19, correct all observed R*f*'s (Eq. 64-1).

$$\frac{\text{Corrected } Rf \text{ of carbohydrate}}{\text{Observed } Rf \text{ of carbohydrate}} = \frac{0.18}{\text{Observed } Rf \text{ of glucose}} \tag{64-1}$$

For establishing identification, an R*f* of a compound determined in this way is a more reliable physical constant than a boiling point and almost as reliable as a sharp melting point. Note that a corrected R*f* is a somewhat more reliable physical constant than an uncorrected R*f* (p. 296).

Specific Rotation

Discussion. Carbohydrates like most natural products are optically active unless the carbohydrate is a racemic mixture. A racemate is unlikely unless the carbohydrate has been synthesized *in vitro* from optically inactive compounds.

The observed rotation, a, of a compound is related to the specific rotation by Equation 64-2:

$$[\alpha]_\lambda^T = \frac{a}{ld} = \frac{100a}{lc} \tag{64-2}$$

where l equals path length in decimeters of the polarized radiation through the material, i.e., the length of polarimeter tube; d equals the density of the material; c equals the concentration in g/100 ml; T is the temperature in °C; and λ is the wavelength of the polarized light (often D for the D line of sodium).

[1] There may also be instructions at this point to spot the paper with other carbohydrate standards.

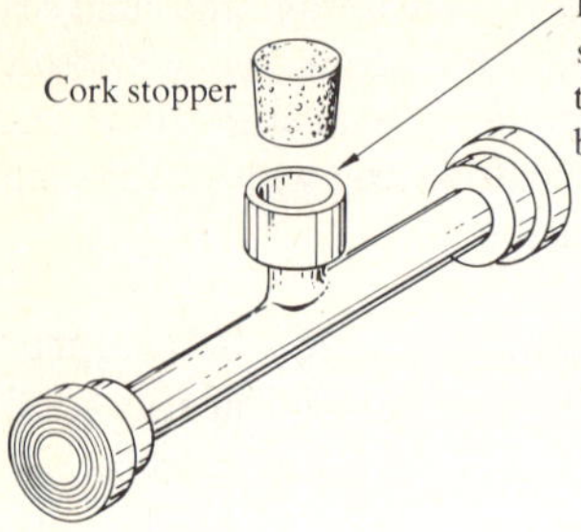

Figure 64-3 Filling a polarimeter tube.

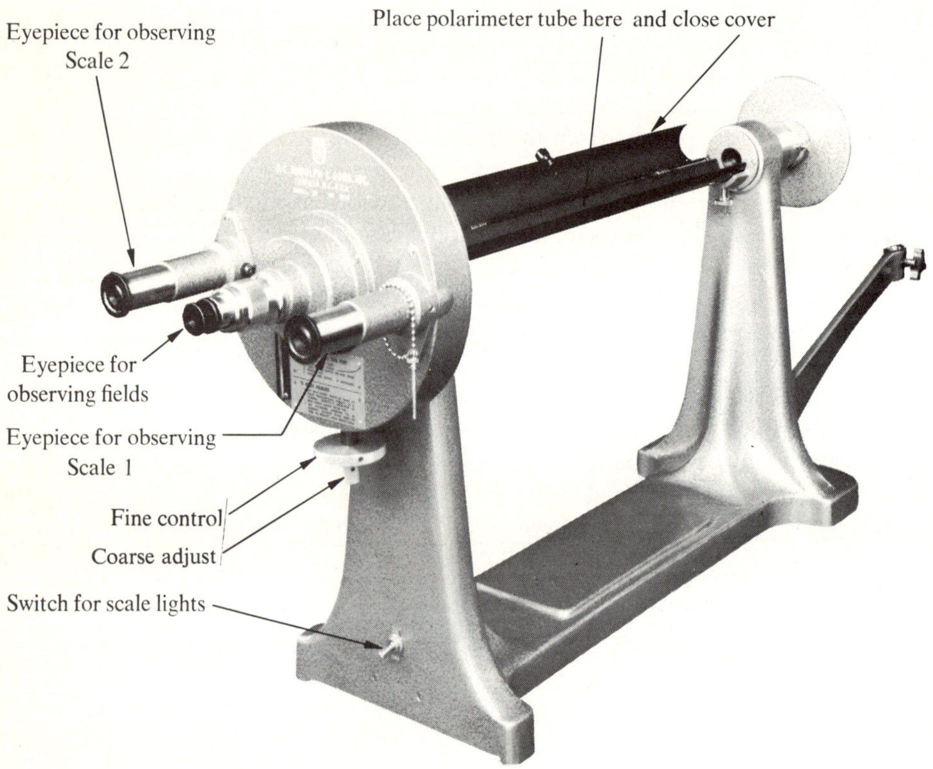

Figure 64-4 End view of polarimeter with adjustment controls. O. C. Rudolph and Sons, Caldwell, N. J.

Experimental Procedure. Review your lecture text for the description of the polarimeter and the measurement of optical rotation. The sodium light should be on at least 10 min before use of the polarimeter. Fill the polarimeter tube with solvent through the top port (Fig. 64-3); be certain there is no air bubble in the tube; and tightly stopper the port. Place the tube in the polarimeter and obtain a zero reading.

Optical rotation should always be observed in a dark room or if this is inconvenient, both the front part of the polarimeter and the observer should be under a black hood. Place the tube in the polarimeter and close the cover. If the polarimeter is similar to that shown in Figure 64-4, move the coarse control knob so that both a very dark and a very light field are observed through the lens in the field eyepiece. Focus the eyepiece to obtain the sharpest possible delineation between the two fields. Turn the coarse control knob until the field intensities are equal or almost equal. Then turn the fine control knob until the two fields are exactly matched. Always match the equally dark fields and not the equally bright fields. Turn on the scale lamp and use the

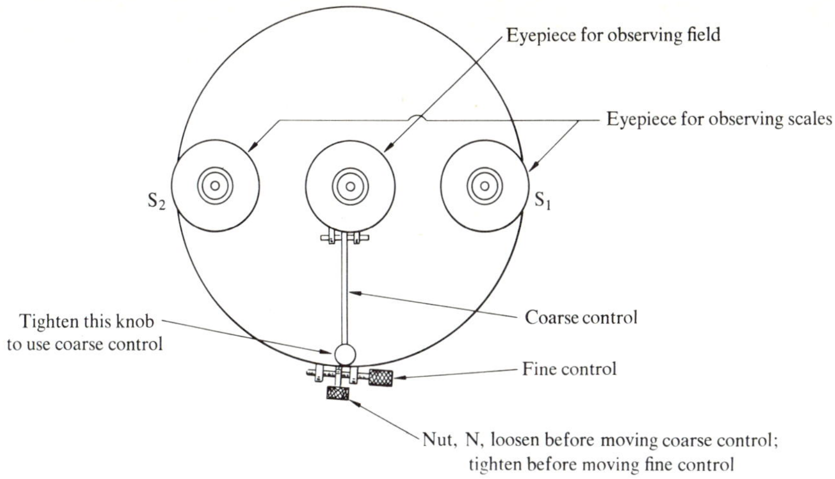

Figure 64-5 Front view of older design polarimeter. Eye pieces for observing scales may be absent.

vernier to read the two scales (S_1 and S_2), focusing the eyepieces if necessary. If S_1 does not equal $S_2 - 180°$, average S_1 and $S_2 - 180°$. The use of the polarimeter shown in Figure 64-5 is only slightly different. The nut, N, must be loosened and the knob on the end of the coarse control lever tightened before the lever can be used to equalize approximately the intensities of the two fields. When the fields are almost matched, tighten the nut, N, and then adjust the fine control.

Take two readings from each direction. If these readings are not within 0.02° of each other, obtain one more reading from each direction. These repeated readings can usually be obtained by changing only the fine control. Average all readings obtained to give the observed rotation, *a*, (with only solvent in the tube this is the zero reading). Drain the solvent from the polarimeter tube and fill the tube with solution. The solution should be prepared in a 5- to 25-ml volumetric flask and the compound weighed to the nearest 0.0001 g on an analytical balance. Obtain four to six readings on the solution as was done for the solvent. Subtract the zero reading from the observed reading and calculate the specific rotation.

Problem 64-1. Determine the observed rotation, *a*, for each of the five scale readings shown in Figure 64-6.

CHEMICAL PROPERTIES

Ignition

Carbohydrates burn easily and leave no residue unless the ignition is not vigorous enough and then a black neutral residue remains. During ignition a caramel odor is apparent.

Taste

Although many carbohydrates are sweet, some are tasteless and some are bitter. Because of this variation, the subjectiveness of the taste test, and the obvious hazards, the use of any test based on taste is not recommended.

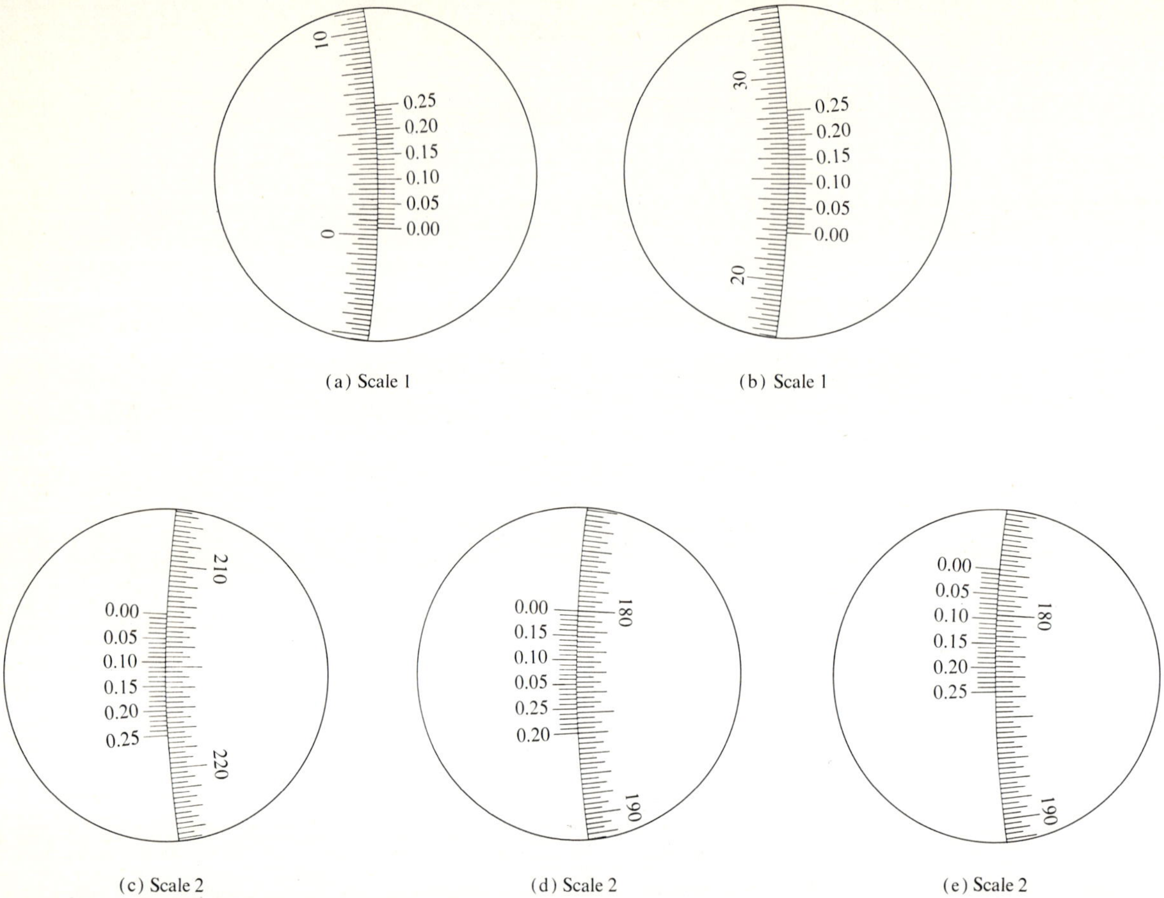

Figure 64-6 Polarimeter scale readings for Problem 64-1.

Reactions as Aldehydes or Ketones

General Reactions. To react as an aldehyde or ketone mono-, di-, oligosaccharides or their derivatives must have a *free* aldehyde or ketone group or a *potentially free* aldehyde or ketone. Almost no carbohydrates or carbohydrate derivative have completely free aldehyde or ketone groups since those with potentially free carbonyl groups exist mainly as cyclic hemiacetals (I) or -ketals.

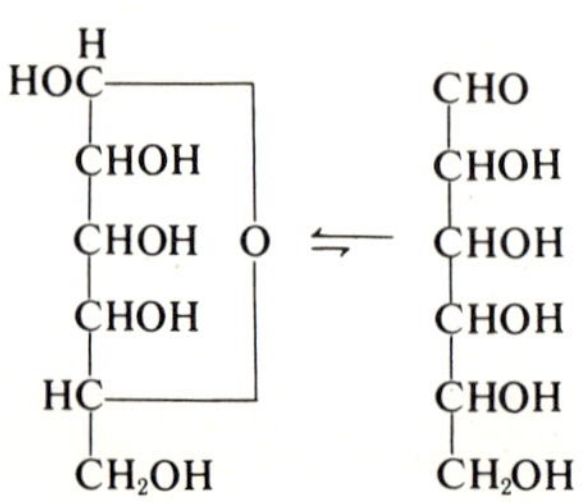

I II

(64-3)

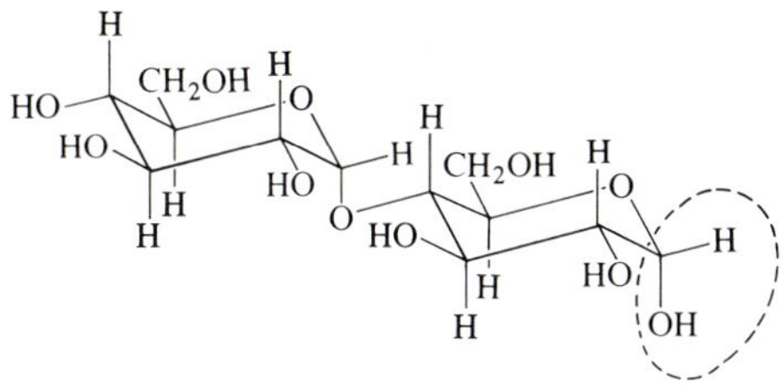

III maltose

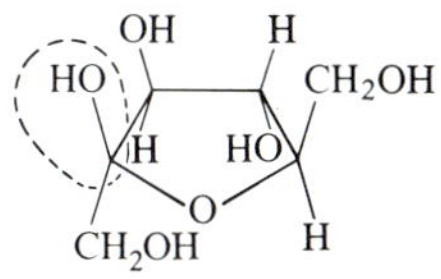

IV fructose

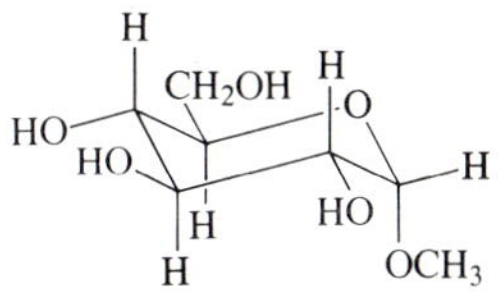

V methyl α-D-glucoside

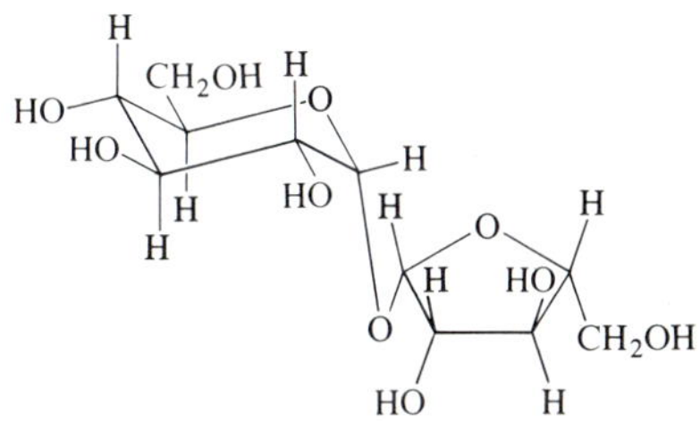

VI sucrose

All these cyclic hemiacetals and hemiketals when in solution are in equilibrium with a small amount of the free aldehyde (II) or ketone. See your lecture text for a full discussion of this equilibrium and mutarotation. Carbohydrates that have these potentially free aldehyde and ketone groups (outlined with dotted lines on III and IV) react with many reagents just like similar hydroxy aldehydes and ketones; they reduce Benedict's and Tollens' reagents, give precipitates with arylhydrazines, and so on. There are many carbohydrates and carbohydrate derivatives that do not react with any of the typical carbonyl reagents. These compounds are full rather than hemiketals or -acetals. Examples are methyl α-D-glucoside (V) and sucrose (VI). Compounds like sucrose are often called nonreducing sugars because they do not reduce many common reagents and give negative Tollens' and Benedict's tests.

Reducing sugars such as glucose do reduce these reagents. Most of the nonreducing sugars are readily hydrolyzed by dilute aqueous mineral acids to carbohydrates that are cyclic hemiacetals and hemiketals and that react with many of the carbonyl reagents.

The equilibria for reactions such as the Schiff test (p. 422) and the sodium bisulfite addition (p. 423) are dependent on a relatively high concentration of the reactant to give a significant amount of product. Since the concentration of free aldehyde or ketone in equilibrium with the cyclic hemiacetal or ketal is always very low, no significant amount of product is formed with carbohydrates. The equilibria in the reaction of carbohydrates with oxidizing agents and arylhydrazines greatly favors the product; the aldehyde or ketone reacts as quickly as it is formed from the hemiacetal or hemiketal.

Reaction with Phenylhydrazine. The reaction of α-hydroxy aldehydes and ketones with phenylhydrazine (p. 419) involves not only the carbonyl group but also the adjacent hydroxy group. No matter how many hydroxyl groups are in the chain, only one next to the carbonyl group is oxidized by the phenylhydrazine.

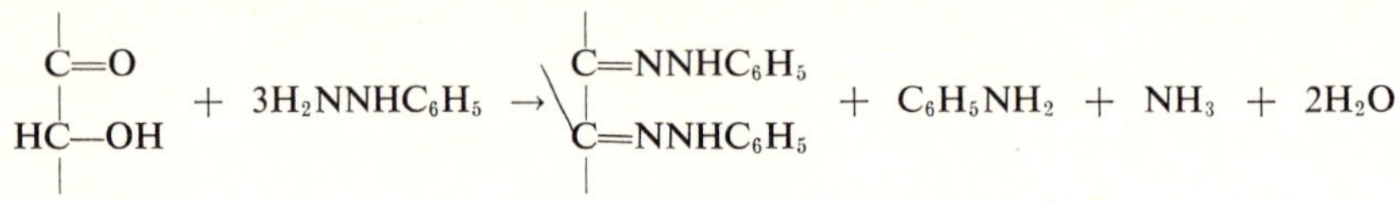

(64-4)

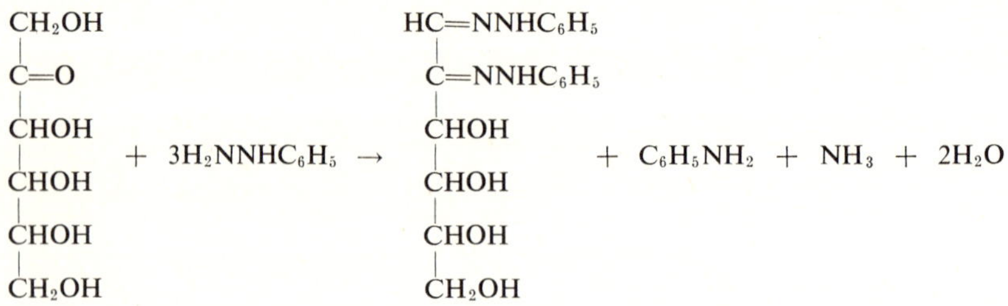

(64-5)

Problem 64-2. Unlike phenylhydrazine, 1-phenyl-1-methylhydrazine reacts with aldoses and ketoses to oxidize all the hydroxyl groups to give alkazones.

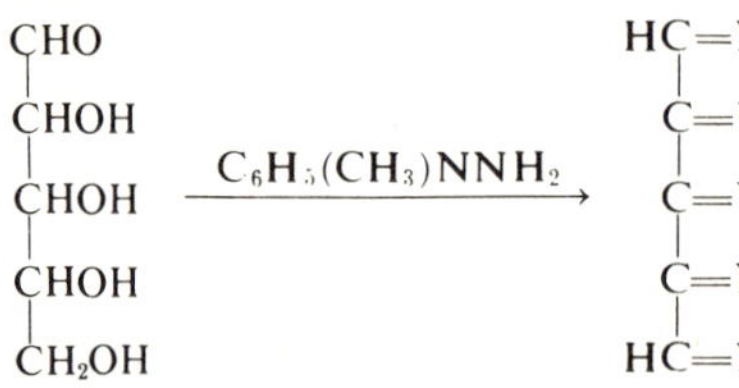

(64-6)

Explain why 1-phenyl-1-methylhydrazine might be expected to oxidize more hydroxyl groups than phenylhydrazine.

Hydrolysis of a Di- and Polysaccharide. Try Benedict's and the phenylhydrazine tests on sucrose and starch olution.

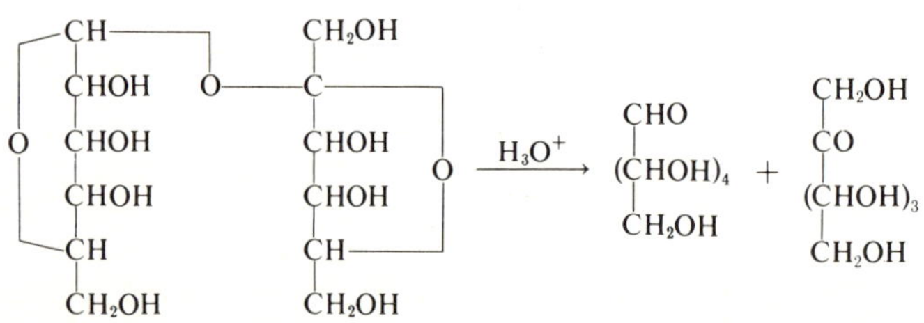

(64-7)

In a test tube place 1 ml of 10% sucrose solution and 3 ml of hydrochloric acid and warm. Shake the test tube and place it in a beaker of hot water. Try Benedict's and the phenylhydrazine tests on a few drops of the solution.

Add 1 ml of concentrated hydrochloric acid to 15 ml of starch solution. Boil a few minutes and test 1 ml of the cooled solution with potassium iodide-iodine solution (see test below), phenylhydrazine, and Benedict's reagent.

Ester Derivatives

The hydroxyl groups of carbohydrates can be esterified by the same techniques used to esterify other hydroxyl groups. Acetates (p. 228) and benzoates (p. 389) are often made. More carbohydrate acetates have been prepared than other carbohydrate esters, and acetates are the only ester derivatives listed in Table 64-1.

Iodine Test

Starches give blue colors with aqueous potassium iodide-iodine solutions. The color becomes purple, then reddish purple, then red, and then colorless as the molecular weight of the starch decreases.

UNKNOWN

The unknown assigned with this chapter can be any compound from any of the tables at the ends of the qualitative analysis chapters in this book including, of course, the carbohydrates in Table 64-1.

EXERCISES

64-1. Complete the following equations:

(a) Methyl α-D-Glucoside (V) + $HIO_4 \rightarrow$

(b) Sucrose (VI) + $HIO_4 \rightarrow$

(c) Sucrose + $Ac_2O \xrightarrow{AcONa}$

(d) Sucrose + Tollens' reagent $\rightarrow$

(e) Sucrose + $C_6H_5NHNH_2 \rightarrow$

(f) Maltose (III) + $C_6H_5NHNH_2 \rightarrow$

(g) Maltose + $Ac_2O \xrightarrow{AcONa}$

65

Rearrangement of α-Phenylallyl Alcohol

A Spectroscopic Kinetic Study

$$C_6H_5CHOHCH{=}CH_2 \xrightarrow{H^+} C_6H_5CH{=}CHCH_2OH$$

INTRODUCTION

Most alcohols of the general formula $Q{-}\overset{|}{C}(OH){-}\overset{|}{C}{=}\overset{|}{C}{-}$ (where Q is unsaturated or aromatic) rearrange to $Q{-}\overset{|}{C}{=}\overset{|}{C}{-}\overset{|}{C}OH$ in the presence of acid. The equilibrium favors the rearrangement product since it is more stable than the reactant because of the loss of delocalization energy by the conjugation of the carbon-to-carbon double bond with the group Q. The additional conjugation of the product causes an increase in the wavelength and extinction coefficient of the electronic absorption bands (Table 65-1). A marked increase in absorption of the reaction mixture at 250 mμ accompanies the rearrangement of phenylvinylcarbinol.

TABLE 65-1 ULTRAVIOLET ABSORPTION MAXIMA OF ARYL-ALCOHOLS

	λ Max. mμ	ε Max.	λ Max. mμ	ε Max.	λ Max. mμ	ε Max.
$C_6H_5CHOHCH{=}CH_2$	207	7,000	258	224		
<u>cis</u> =$C_6H_5CH{=}CHCH_2OH$	241	12,000	282	450	290	120
<u>trans</u> =$C_6H_5CH{=}CHCH_2OH$	250	16,000	282.5	1,000	291	800

EXPERIMENTAL PROCEDURE

Reaction

Mix 60 ml of dried, purified dioxane and 10.2 ml of 1 *M* hydrochloric acid in a 100-ml volumetric flask. Dilute to the calibration mark with distilled water, place the stopper in the flask, and agitate the flask. Pipet 50 ml of the acidic aqueous dioxane into a 125-ml Erlenmeyer flask or 100-ml round bottom flask with a ground glass neck. Stopper the mixture and place in a constant temperature bath at 55 to 60°. The bath should be able to maintain a constant temperature (less variation than ±0.4° for 48 hr). Agitate the flask in the bath every 5 to 6 min for 25 min. This is the reaction flask.

Add 5 mg (weighed on an analytical balance) of phenylvinylcarbinol to a 10-ml volumetric flask. Add purified, dried dioxane to the calibration mark, place the stopper in the flask, and shake it vigorously. With a pipet transfer 1 ml of the phenylvinylcarbinol solution to a 100-ml flask containing exactly 50 ml of solution made by adding distilled water to 30 ml of dried purified dioxane in a 50-ml volumetric flask. Mix the aqueous dioxane and phenylvinylcarbinol solutions well in the 100-ml flask. With a pipet transfer 8 ml of this solution to a 25-ml flask; add 2 ml of 0.5 *M* sodium acetate solution, and mix well. Add 0.8 ml of the 1 *M* HCl and mix the solution well. The resulting sodium acetate solution is used as a blank in the ultraviolet spectrophotometer.

After the acid solution has been in the constant-temperature bath at least 25 min, add 1 ml of the phenylvinylcarbinol solution from the 10-ml volumetric flask, shake the flask vigorously but briefly without removing it from the constant temperature bath. Immediately transfer with a pipet a 4-ml sample from the reaction flask to a 10-ml flask containing 1 ml of 0.5 *M* aqueous sodium acetate and mix well. Determine the precise time the 4-ml aliquot was taken from the reaction flask; this is *time zero*.

Problem 65-1. What is the molarity of the hydrochloric acid in the reaction flask after addition of the phenylvinylcarbinol solution?

Problem 65-2. What is the purpose of mixing the sodium acetate solution with the 4-ml aliquot from the reaction mixture?

Measurement

Place the blank solution into a quartz cell and place this in the reference beam or reference position in an ultraviolet spectrophotometer. As soon as each 4-ml aliquot is taken from the reaction flask and mixed with 1 ml of 0.5 *M* aqueous sodium acetate, immediately transfer it to a quartz cell and determine its absorbance at 250 mμ.

Samples could be collected at 15, 35, 50, and 90 minutes after time 0. Take the samples as close to these times as possible. If, however, the samples are not taken at these times ±30 sec, record the exact time the sample *is* taken. Take one additional sample at the beginning of the next laboratory period. Mix a fresh blank for this last sample.

If a recording ultraviolet spectrophotometer (p. 204) is used, fill a second quartz cell with the blank sodium acetate solution; place this in the sample beam and set the zero control with the wavelength selector at 250 mμ. A quartz cell with the blank solution should be kept in the reference beam for *all* readings and adjustments. Scan from 240 to 260 mμ for each sample and read the absorbance at 250 mμ. If a Model 202 spectrophotometer is used, all the data can be recorded on one sheet of paper by use of the position selector at the bottom of the drum (Fig. 26-7). For this experiment use every second position of the selector. The wavelengths can be read from the indicator at the bottom of the drum.

If a nonrecording ultraviolet spectrophotometer is used, first adjust the dark current if necessary. Use the zero adjust to zero the galvanometer with the blank sample in the beam. If the zero adjust does not have a wide enough dynamic range, use the slit control to coarsely zero the galvanometer for the initial adjustment and the zero adjust for fine tuning. Use the same slit setting for all measurements that you make. With the sample in the beam adjust the transmission (or absorbance) potentiometer to zero the galvanometer, and then read the dial.

Determination of Reaction Order

The class is divided into three, four, or six groups as directed by the laboratory instructor. Group 1 follows the instructions as given above. Groups 2 and 3 follow the instructions given above except in the first paragraph of the Reaction section. Group 2 uses 5.1 ml of 1 *M* hydrochloric acid and Group 3 uses 15.3 ml of 1 *M* hydrochloric acid. Groups 4, 5, and 6 follow the instructions in the Reaction and Measurement sections except *all three* groups dissolve the 5 mg of phenylvinylcarbinol in a 5-ml flask *and* Group 5 uses 5.1 ml of 1 *M* hydrochloric acid and Group 6 uses 15.3 ml of 1 *M* hydrochloric acid.

All those who use 5.1 ml of 1 *M* hydrochloric acid should use 0.4 ml of the acid to prepare the blank. All those (Groups 3 and 6) who use 15.3 ml of the acid should use 1.2 ml to prepare the blank and in addition should use 1 *M* sodium acetate solution any place 0.5 *M* is called for.

Calculation of Rate Constants

The kinetic studies in Chapter 22 used the disappearance of one of the reactants for determining the rate constants. In this reaction the appearance of the product is used to determine the rate constant k:

$$\text{Rate} = \frac{dp}{dt} = k\,(a-x) \tag{65-1}$$

where the symbols are the same as in Equation 22-3. In the same way that Equation 22-6b was derived from Equation 22-3, Equation 65-2 can be derived from Equation 65-1. Calculate the rate constants for each of the four aliquots taken.

$$2.303 \log\left(\frac{A_\infty - A_0}{A_\infty - A_t}\right) = kt \tag{65-2}$$

where A is the observed absorbance $= abc = (\epsilon/mw)\,(bc)$ (p. 153); A_0 is the initial absorbance at time 0; A_∞ equals the absorbance observed when the reaction is completed; and k is the rate constant when absorbance A is observed. Note that if dilutions are necessary to obtain absorptions on scale, the observed absorbance must be multiplied by the dilution factor to keep the concentration factor constant. For A_∞ use the absorbance obtained during the second laboratory period.

Problem 65-3. Derive Equation 65-2 from 65-1. A knowledge of calculus is needed.

Problem 65-4. Derive Equation 65-3 from

Problem 65-5. If rate constants at three temperatures were determined, plot log k *vs* 1/T and determine A and *Ea* graphically. Use the equation

Determination of Energy of Activation

Follow the instructions given above in the Reaction, Measurement, and Calculation of Rate Constant Sections except use temperature baths at 60°, 45°, and 30°. If instructed to do so divide the work into groups, each group using a different temperature. If the laboratory is warmer than 27°, use 35° for the lowest temperature. If the laboratory is warmer than 33°, use only 45° and 60°. For the lower temperatures collect samples at the following times after time 0: 45° – 15 min, 60 min, 2 hr, 3½ hr, *at least* 45 hr; 30 or 35° – 20 min, 2 hr, 4 hr, 7-12 hr (record exact time), *at least* 92 hr. The energy of activation, *Ea*, is determined from the equation

$$2.303 \log\frac{k_2}{k_1} = \frac{Ea}{R}\left(\frac{1}{T_1} - \frac{1}{T_2}\right) \tag{65-3}$$

Where T_1 and T_2 are absolute temperature and R is the gas constant, 1.987 cal/deg mole.
If three temperatures were used, calculate all three possible values for the energy of activation.

$$k = Ae - Ea/RT \tag{65-4}$$

Where A is a constant (*not* absorbance)

$$\log k = \frac{-Ea}{2.303RT} + \log A \tag{65-5}$$

EXERCISES

65-1. Is the reaction first or second order? Write the complete rate equation.

65-2. Propose a mechanism for the rearrangement studied in this chapter. Indicate which is the rate determining step and justify it with evidence from this experiment.

65-3. Outline the principal sources of error in this experiment and indicate what was done to minimize them. Can you think of any other ways to minimize potential error?

Appendix I

Vapor Pressure—Temperature Tables

TABLE I-1 COMMON ALIPHATIC SOLVENTS*
(Vapor Pressure in mm of Hg)

Temp. (°C)	$CHCl_3$	CCl_4	Hexane	Cyclohexane	Formic acid	Acetone	Ethyl acetate
30	246	143	188.6	121	52	283	119
35	301	176.2				346.4	
40	366.4	216	281.5	182	82.6	421.5	186
45	439	262.5				510.5	
50	526	317	408	269	126	600	282
55	625	379.3	490			710	
56.3	650		510			760	
60	739.6	451	576.6	387	190	860.6	415.4
61.2	760.0	465	600	406			
65		530.9	680				
69		600	760	525			576
70	1019	622	796	540	279.6	1190	597
76.8		760		686			749
77.2				696			760
80	1403	843	1062	741.3	398	1611	832.7
90		1122	1407	992	552	2142	1130
100		1463		1304	753.5	2795	1515
100.5					760		

* Similar tables given for methanol and ethanol (pp 27).

TABLE I-2 AROMATIC COMPOUNDS

Vapor Pressure in mm of Hg

Temp. (^{o}C)	Benzene	Toluene	Bromobenzene	p-Dibromo-benzene	t-Butyl-benzene
40	181	60	10		
50	269	93	17		
60	389	138	28		
60.6	400				
70	547	203	43	4	
71.5	585				
72.6	600	240			
78.4	720				
79.5	740				
79.9	750				
80.0	754		66	7	
80.1	760				
80.8	770	297			40
81.2	780	304			
84		335			
87		372.5			
90	1016	402	98	12.0	
92			106	13.0	
94			114	14.0	
95			118	15.1	79.4
96		503	122	15.6	81.7
97			127	16.1	84.0
98		528	131	16.7	86.4
99		542	136	17.2	
100	1344	557	141	18	
103		600			98
110	1748	747	199	28	
110.7		760			
120	2238	965	275	42	
130			372.7		

TABLE I-3 BUTYL HALIDES
(Temperature °C)

VP(mm)	t-Butyl chloride	n-Butyl bromide	2-Chlorobutane	n-Butyl chloride
10		0	-29	-18.5
40		24.8	-5	5
100	-1.0	44.7	14	25
400	32.6	81.7	50	59
600	45	93		
760	51	101.5	68	78

TABLE I-4 WATER

Temp. (°C)	VP (mm)	Temp. (°C)	VP (mm)	Temp. (°C)	VP (mm)
50	92.5	74	277.2	92	567.0
55	118	75	289.1	93	588.6
56.3	125	76	301.4	94	610.9
60	149.5	77	314.1	95	633.9
61	156.4				
61.2	158	78	327.3	96	657.6
62	163.8	78.3	331.0	96.5	672.3
63	171.4	79	341.0	97	682.1
64	179.3	80	355.1	97.5	701
64.7	185	81	369.7	98	707.3
65	187.5	82	384.9	98.5	720
66	196.1	83	400.6	99	733.2
67	205.0	84	416.8	99.3	740
68	214.2	85	433.6	99.5	747
69	223.7	86	450.9	99.6	750
70	233.7	87	468.7	100.4	770
71	243.9	88	487.1	100.7	780
72	254.6	89	506.1	105	901
73	265.7	90	525.8	110	1074.5
		91	546.0		

Appendix II

Dielectric Constants

If a voltage is applied to two conductors separated by an insulator, equal and opposite electrical charges accumulate on the conductors. Such an arrangement of two conductors and an insulator is often called a capacitor or a condenser and the conductors are often called plates even though they may actually have almost any shape. The charge of a capacitor q is proportional to the applied voltage v, the dielectric constant ϵ of the insulator, and the area of plates A and inversely proportional to the distance between them d. For the parallel plate condenser shown in Figure II-1:

$$q = \frac{\epsilon A v}{4\pi d} \tag{II-1}$$

If the voltage, area, and distance between the plates are constant, then the charge is proportional to the dielectric constant:

$$q = k\epsilon \tag{II-2}$$

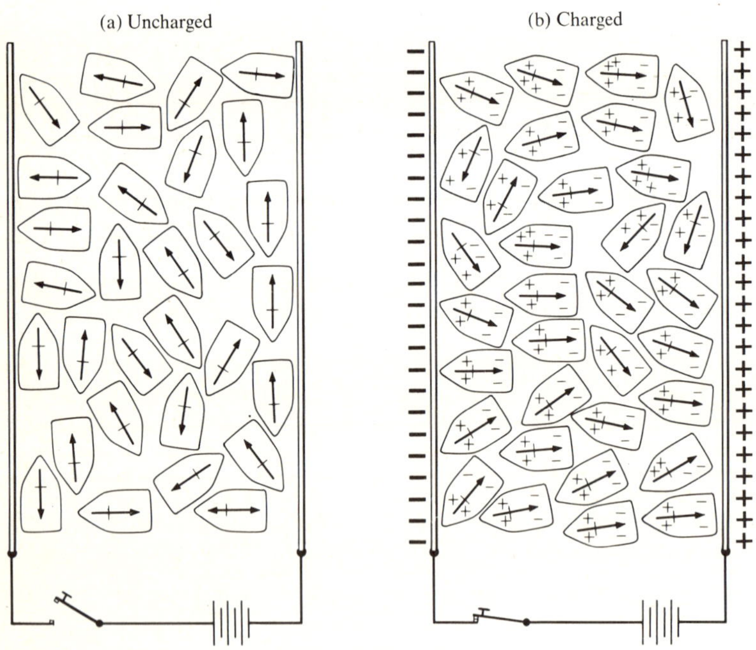

KEY
- —+→ Dipole moment, negative end at arrowhead
- **+ —** Electrical charges on capacitor plates
- + - Charges induced in the applied electrical field

Figure II-1 Alignment of molecules with dipole moments and induction of electrical moments in a charged capacitor.

A perfect vacuum has a dielectric constant of 1 and all matter has a higher dielectric constant. Gases have dielectric constants only slightly greater than 1. Nonpolar liquids and solids have low dielectric constants and polar liquids and solids have high dielectric constants. (See Table II-1.)

TABLE II-1 DIELECTRIC CONSTANTS AND DIPOLE MOMENTS OF SOME SOLVENTS AT 20°

Compound	ϵ	μ in D	Compound	ϵ	μ in D
Hexane	1.89	0	Ethylene glycol	37.7	2.21
Cyclohexane	2.02	0	Ethylene ether	4.34	1.18
Octane	1.94	0	1,4-Dioxane	2.21	0.4
2,2,4-Trimethylpentane	1.94	0	Acetone	21.0	2.8
1-Heptene	2.05	0.34	2-Butanone	18.51	2.7
Cyclohexene	2.22	0.28	3-Pentanone	17.00	2.7
Benzene	2.28	0.0	Cyclohexanone	18.3	3.08
Toluene	2.40	0.4	Formic acid	57.0	1.45
o-Xylene	2.57	0.5	Acetic acid	6.2	1.75
m-Xylene	2.37	0.4	Ethyl acetate	6.3	1.8
p-Xylene	2.27	0	Hydrogen cyanide at 0°	158.1	3.06
Carbon disulfide	2.64	0	Sulfur dioxide at -20°	17.6	1.61
Carbon tetrachloride	2.25	0			
Chloroform	4.81	1.86	Nitroethane	28.7	3.2
Methylene chloride	9.08	1.55	Nitrobenzene	35.0	4.0
			Chlorobenzene	5.71	1.55
Water	80.37	1.82	Pyridine	12.5	2.27
Methanol	33.62	1.71	Piperidine	5.89	1.3
Ethanol	24.5	1.73	Aniline	6.89	1.48
1-Propanol	20.1	1.65			
2-Propanol	18.3	1.71			
1-Butanol	17.8	1.63			

The dielectric constant of a compound is related to the dipole moment (μ) and polarizability (α, the tendency of the electron cloud of the molecule to distort in an electrical field) of the compound by the equation:

$$\frac{\epsilon - 1}{\epsilon + 2} \cdot \frac{M}{\rho} = \frac{4}{3}\pi N\left(\alpha + \frac{\mu^2}{3kT}\right) \qquad \text{(II-3)}$$

where:

M = Molecular weight
ρ = Density
N = Avagadro's number
k = Boltzman constant = 1.38×10^{-16} erg/degree molecule
T = Absolute temperature

The greater the polarizability and the dipole moment, the higher the dielectric constant. Figure II-1 illustrates this. With no charge on the plates of the condenser the molecules and their dipole moments are randomly oriented. The dipole moments are indicated by arrows in the direction of the negative pole. When an electrical field is applied, the molecules tend to align themselves with their negative poles toward the positive plate and their positive poles toward the negative plate. The electrical field tends to distort or polarize the electron cloud of each molecule so that there are electrical moments (shown as small positive and negative signs in Figure II-1) in each molecule caused by the applied electrical field and opposite in direction to it. Note that even in molecules with no dipole moments, such as carbon tetrachloride, methane, and hydrogen, there is an electrical moment from the polarizability of the electron cloud if the molecules are in an applied field. Since positive parts of the molecules tend to line up nearer the plate with the negative charge, and the negative parts of the molecule tend to line up nearer the plate with the positive charge, there is effectively a decrease in the charge separation. If the molecules between the capacitor plates have high dipole moments or their electron clouds are easily distorted, then a higher charge can be accepted by the capacitor plates.

Materials with high dielectric constants not only stabilize charge separation on capacitor plates but also stabilize charge separation in a great variety of chemical and physical changes. A solvent with a high dielectric

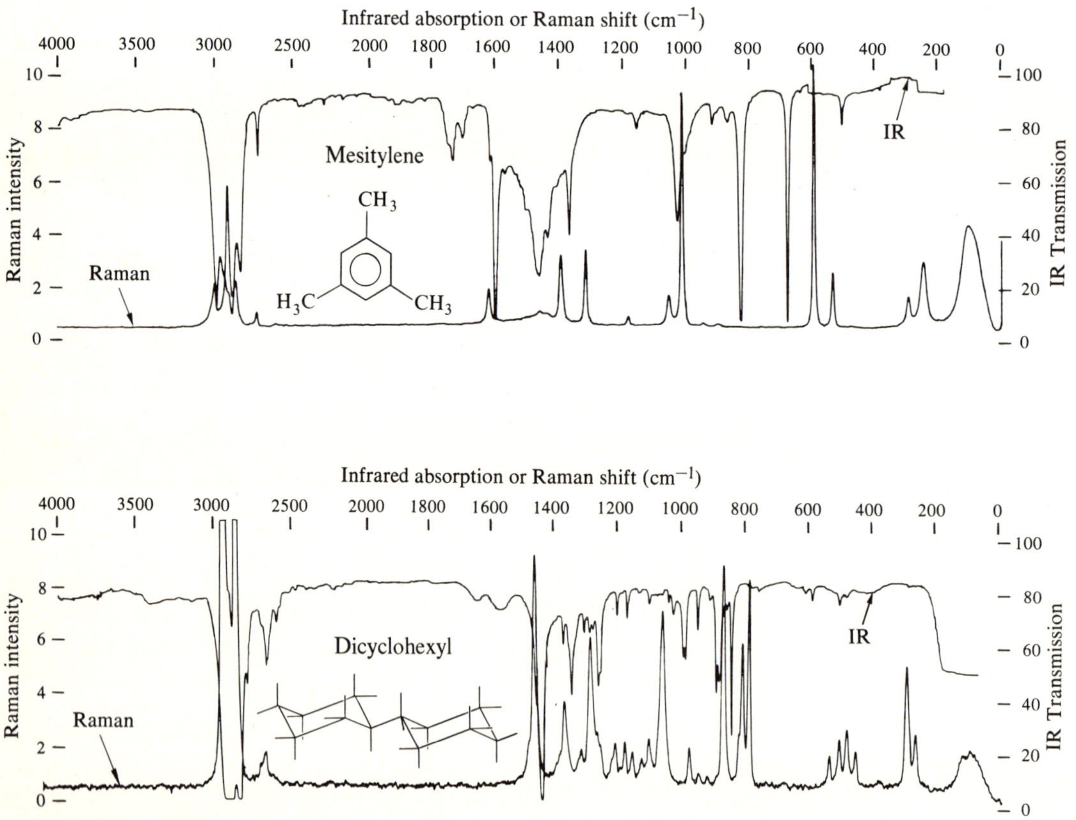

Figure II-2 Comparison of infrared and Raman spectra of two compounds with highly symmetrical molecules.

constant greatly increases the rate of a chemical reaction if the transition state of the rate determining steps has a greater separation of charge than the ground state. Weak electrolytes ionize to a greater extent in solvents with high dielectric constants. Since the dissolving of polar and ionic materials involves separating charges, liquids with high dielectric constants are good solvents for polar and ionic compounds. Some of these phenomena have been explained by polar and ion-pole attractions between solvent and solute (p. 79, 80, 214); the discussion with dielectric constants uses the same concepts to explain the same phenomena in another way.

Dipole moments and polarizabilities are important properties of molecules and chemical bonds. They can, for example, be used to explain the nucleophilicity of reagents and the ease with which various compounds undergo displacement reactions. Molecular vibrations that cause a large change in dipole moments cause strong absorptions in the infrared; the vibrations that cause no change in dipole moment cause no absorption in the infrared. Molecular vibrations that cause a large change in polarizability give strong Raman bands; vibrations that cause no change in polarizability give no Raman bands. Raman scattering is an emission phenomenon. Although both infrared absorption and Raman spectra result from molecular vibrations, they are usually very different for symmetrical molecules (Fig. II-2) but are often very similar for unsymmetrical molecules.

EXERCISES

II-1. It can be seen from Equation II-2 that the dielectric constant of a compound decreases with molecular weight and absolute temperature and increases with density. Explain.

II-2. Gases, even those of polar compounds, have low dielectric constants. Give at least three reasons for this.

II-3. The dielectric constant of compounds with zero dipole moment does not change with temperature. Explain this using Equation II-2 and using the behavior of nonpolar molecules in a charged parallel plate capacitor.

Appendix III

Advanced Topics in Infrared Spectroscopy

INSTRUMENTATION

Radiant Energy

The energy from a source in a spectrometer varies with wavelength and frequency. Infrared sources radiate most of their energy at high frequencies or low wavelengths, and little energy is radiated at low frequencies or long wavelengths. (Fig. III-1.)

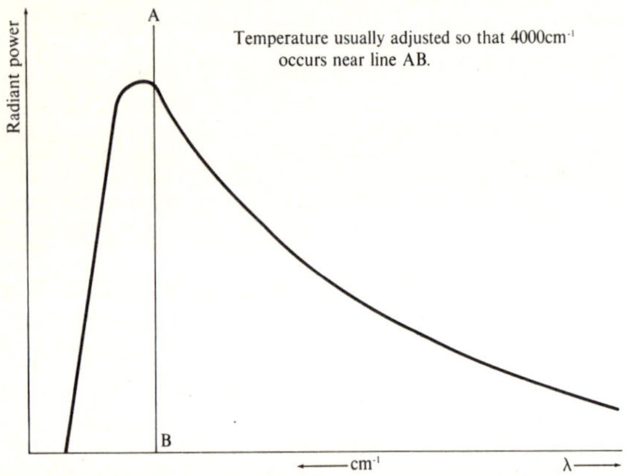

Figure III-1 Variation of radiant power of an infrared source with wavenumber and wavelength.

Problem III-1. What are the limits on the total energy in the servo system of an optical-null, double beam spectrophotometer for optimum operation? (Hint: see pp. 125–127.)

Problem III-2. In a spectrophotometer at a given frequency what is the effect of widening the slit on: (a) radiant energy; (b) total energy in servo system; (c) resolution? (Hint: see pp. 115–117.)

Since the total servo energy in an optical-null, double beam spectrophotometer must be held within certain fairly narrow limits, the beam size as determined by the slit width must be varied over most of the spectrum; at high frequencies the beam must be narrow and at low frequencies it must be wide. The radiant energy from the source at high frequencies must be attenuated to that available at the lowest frequency. Since infrared sources especially at low frequencies have much less energy than sources of other spectrometers and since the detectors are not ultrasensitive, infrared spectrophotometers are severely energy limited. This energy limitation is the primary cause of most problems in infrared spectroscopy from instrument design and cost to obtaining spectra on micro and opaque samples.

Programmed Slits

In infrared spectrophotometers the radiant energy reaching the detector is controlled by changing the width of the slit openings. There is a linkage between the diffracting element (prism or grating) and the slits. As the diffracting element rotates, the slit width changes to compensate for the variation of energy from the source and for other frequency-dependent energy changes. This change of slit width with frequency is called the *slit program.* (Fig. III-2.)

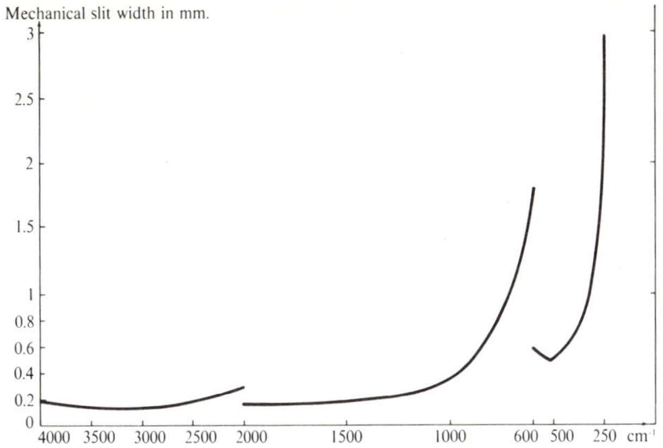

Figure III-2 Normal slit program for the Perkin-Elmer Model 457 multigrating, infrared spectrophotometer.

If there were no other frequency-dependent energy changes, a graphical presentation of slit program would appear similar to that of the energy profile to the right of line AB in Figure III-1 if *decreasing* slit width were used instead of increasing energy for the ordinate, for example, any single curve in Figure III-3. In some prism instruments, the slit program is similar to this. In filter-grating instruments, however, the energy changes with the different gratings and different orders of diffraction so that the actual slit program is more complicated. (Fig. III-2.)

The resolution of an infrared spectrophotometer is dependent upon the width of the slit and the resolving ability of the diffracting element. Fortunately, the resolving ability of a prism or grating increases with decreasing frequency. The decreased resolution caused by increasing the slit width at lower frequencies is approximately compensated for by the higher resolution of the diffracting element at low frequencies.

Many infrared spectrophotometers in organic laboratories do not have a control to change slit programming. Perhaps for the beginner this may be preferable because there is one less control to adjust and one less parameter to worry about, but for greater flexibility in obtaining spectra on a wide variety of samples and for maximum accuracy and precision in quantitative analysis, a selection of several slit programs is needed. The slit width at a given frequency is different in each program, but the change of the slit width is the same for each program (see Fig. III-3). The slit program control should not be confused with a slit override control which is found instead of a slit program control on some infrared spectrophotometers of older design. A slit override control sets the slits for a constant width for a given section of the entire spectrum (dotted line in Fig. III-3).

For most organic applications any slit change will usually be a widening to let more energy through. This may be necessary because the sample or sampling accessory causes excessive losses of radiant energy or because it is necessary to reduce the noise which requires reducing the gain.

Problem III-3. Assume only a variable slit program control and a gain control are available, indicate in columns 2 and 3 of Table III-1 the relative adjustments required on the controls and in column 4 what has to be given up to obtain the change desired (Column 1).

SAMPLE HANDLING TECHNIQUES, SPECTRA OF SOLIDS

General

Solids can be dissolved in various solvents and spectra obtained on the resulting solutions in sealed or demountable cells. Spectra of solutions in carbon tetrachloride or carbon disulfide are more readily compared because the environmental effects of these solvents are very small. Also by always obtaining spectra in the same or very similar solvents any environmental effects are constant. Unfortunately, the spectra of many compounds,

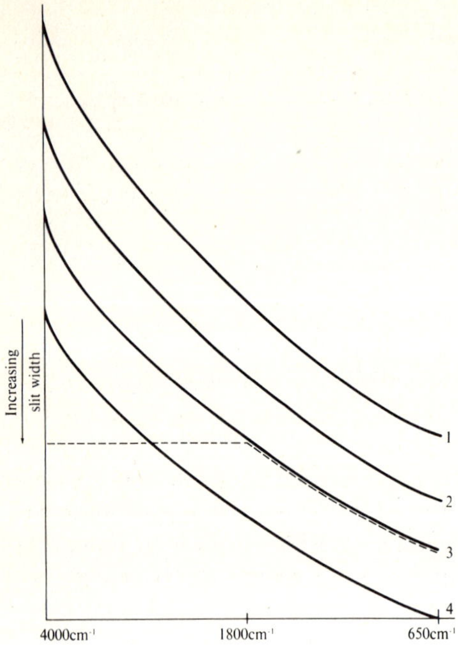

Figure III-3 Four slit programs with slit override (see dotted line) on program 3 from 4000 to 1800 cm^{-1}.

TABLE III-I ADJUSTMENT OF SPECTROPHOTOMETER CONTROLS

Change Desired	Slit program control	Gain control	Spectral quality that is decreased
Decrease Noise			
Improve Resolution			
Increase Energy to Compensate for opaque sample (2 possibilities)			

especially solids, cannot be obtained in these solvents because they do not dissolve many polar compounds. Even when a solid is sufficiently soluble in these solvents, the absorption bands of the solvent may obscure or distort absorption bands of the solute. (See Figs. 18-26 and -27 for the absorption bands of these solvents.) This can be compensated for by obtaining a spectrum in carbon tetrachloride and a spectrum in carbon disulfide and canceling the weak-to-medium absorptions with the same amount of solvent in the reference beam. This usually requires more effort than obtaining spectra by one of the other solid sampling techniques described below.

Potassium Bromide Pellet Technique

Introduction. Under high pressure anhydrous potassium bromide powder can be pressed into thin sheets or wafers that are transparent from the visible region to approximately 300 cm^{-1} (33 μ). Since this includes almost all of the infrared absorption region used by organic chemists and since it exceeds the range of most infrared spectrophotometers, solids are often pressed with potassium bromide (or less often other alkali metal halides) into pellets and spectra obtained on the resulting pellets.

To prevent scattering of the infrared radiation and absorption band distortion by the solid particles the average size of the particles must be much smaller than the shortest wavelength of the infrared radiation used (usually 2.5 μ) *or* the refractive index of the particles should be approximately the same as that of the potassium bromide matrix. Since the latter is not usually true and probably not known when it is true, the average particle size must be less than 1 μ. If the pellet is for *quantitative* analysis using absorptions with wavelengths smaller than 5 μ (greater than 2000 cm^{-1}), the average particle size should be smaller than $\lambda/5$, where λ is the wavelength of the absorption used. Most solids are coarser than this and must be pulverized to smaller particle size. Two general grinding techniques, wet and dry, are used. The wet technique usually gives better results with hard crystals such as those of inorganic or ionic organic compounds, and the dry technique usually gives better results with most nonionic organic compounds. Samples sufficiently soluble in *volatile* solvents such as acetone, methylene chloride, or benzene can often be deposited directly on the potassium bromide; a little mixing but no grinding is necessary for these samples.

Anhydrous Conditions. The sample, the potassium bromide, the die, and everything else involved in making the pellet should be kept as water-free as possible. Even small amounts of water in the pellet cause absorptions in the critical 3600–2900 and 1850–1500 cm^{-1} regions (Fig. 18-19) that may mask or distort absorption bands of the sample. Water may also hinder making a transparent pellet and cause scattering of the radiation. This causes a loss in transmittance especially at high frequencies (Fig. III-10).

Potassium Bromide. Only spectral grade anhydrous potassium bromide of 100 to 200 mesh should be used. The bottle should be open only when it is being dried in an oven or when potassium bromide is actually being withdrawn. Before use the potassium bromide should be dried 12 hours at 105 to 110°, the top replaced, and stored in a desiccator. If possible, after each period of use the bottle should be heated at 105° for 1 to 2 hours before being capped and returned to the desiccator.

Do not grind the potassium bromide either before or after it is mixed with the sample. Mixing with the sample should be gentle but very thorough. If a mechanical agitator has been used to *grind* the sample, use it for no longer than 10 sec to *mix* the sample with the potassium bromide. If the potassium bromide is ground much finer than 100 to 200 mesh, the increased surface area adsorbs water very rapidly from the atmosphere. Very finely divided potassium bromide also adsorbs air and this becomes entrapped in the pellet.

Although more is usually used, pellets can often be obtained with as little as 100 mg of potassium bromide without special microsampling accessories or techniques. The amount of potassium bromide should be such that the concentration of sample is approximately 0.2 to 0.5%. Concentrations as low as 0.05% and as high as 1% sometimes give reasonably good spectra. The concentration is limited on the lower end by the amounts required to give absorption bands of sufficient intensity. High concentrations not only cause strong absorptions to bottom against the 0%*T* line because of too much material in the sample beam but also cause opaqueness and scattering from the introduction of significant amounts of another material into the potassium bromide matrix. These latter effects are often noticed at concentrations lower than those that cause bottoming of the more intense absorptions. A spectrum with more intense absorption bands is preferably obtained by using more of the potassium bromide-sample mixture and pressing a thicker pellet instead of increasing the sample concentration.

Preparing the Sample. Wet Technique: for both the wet and dry grinding techniques it is preferable to grind 5 to 10 mg of the sample in a fairly large mortar (OD 125 mm) and then weigh the needed amount of finely ground material into a clean mortar or mixing vial to be mixed with the potassium bromide (or with Nujol if mulls are to be made). If less than 2 mg of compound is available, avoid transfer losses by weighing directly into a small mortar or mixing vial, grind, and then mix with potassium bromide.

To approximately 5 mg of solid in an agate or mullite mortar add 6 or 7 drops of 95% ethanol. With a firm, rotary motion vigorously grind the sample. Confine the grinding to the bottom half of the mortar bowl. If insufficient sample is available, as little as 0.4 mg of solid can be used. Add approximately 1 drop of ethanol per 1 mg of sample but not less than 3 drops and grind the sample in the bottom 20% of the mortar or preferably use a smaller mortar (OD 25 to 50 mm). Stop grinding when the sample dries. If the sample is sufficiently fine, it will be very glossy and caked to the sides of the bowl. If the sample is not fine enough because of insufficient grinding or excessive sample hardness, again add a few drops of ethanol and repeat the grinding. Weigh 1.5 mg of the ground sample and place in the cleaned mortar. Add 300 mg of potassium bromide and gently but *thoroughly* mix it with the sample. *Do not grind the potassium bromide either before or after mixing it with the sample.*

Dry Grinding Technique: grind the sample in a mullite or agate mortar or in a mechanical agitator such as a Wig-L-Bug. If insufficient sample is available, as little as 0.3 mg *may* give adequate spectra without additional sampling accessories. Vigorously grind with a hard, back and forth motion of the pestle until the sample is glossy and caked on the sides of the mortar. Usually 3 to 4 minutes is needed but up to 12 minutes may be required for some very hard crystals. Once the sample is ground, mix it thoroughly with the potassium bromide. *Do not grind the potassium bromide.*

Evaporation: if the sample is sufficiently soluble, it can be dissolved in 2 ml of chloroform, methylene chloride, or other volatile solvent. Do not use ethanol and, if the solvent is to be evaporated at atmospheric pressure, it should not boil above 85°. Place the solution on 300 mg of potassium bromide and mix well. (*Do not grind.*) Allow the mixture to dry. If less than 1 mg of sample is available, use proportionally smaller amounts of solvent and potassium bromide.

If lyophilization or freeze-drying apparatus is available, a greater variety of solvents can be used. For good spectra the sample should not be dissolved in solvents in which the alkali metal halide is soluble. The freeze-dry method is useful not only because it extends the range of solvents that may be used but also because the solid usually precipitates from the solution more finely divided than with normal evaporation. Sometimes even freeze drying will deposit particles that are too large; when this occurs, one of the two grinding techniques described above must be used. Freeze drying can be used with aqueous solutions, but the resulting potassium bromide matrix tenaciously holds water and the portions of its spectrum where water absorbs are useless. If the compound is both stable and nonvolatile at 110°, the water can be removed by drying the mixture in a vacuum oven (pressure < 2 mm; temperature $\sim 110°$) for 15 min. If any solvent is used in which the potassium bromide is soluble, the resulting pellet may have low transmission at high frequencies from scattering even if all the solvent is removed from the mixture.

Use of an Evacuable Die. These dies are machined to close tolerances; they are durable and can make thousands of pellets if properly used; however, if abused, they are easily ruined. Before the die is used, the parts labeled (b) in Figure III-4 and other parts that have foreign material on them should be cleaned with chloroform. After use the parts labeled (a) or (b) and any other parts with KBr or otherwise dirty should be cleaned with chloroform. Water may be used to clean the die if the wet parts are *immediately* rinsed with methanol. The matching surfaces of the anvil and the plunger have been highly polished so that smooth, transparent pellets can be obtained. Avoid scratching, gouging, or otherwise damaging these surfaces. Many materials including potassium bromide or other alkali metal halides may corrode these surfaces if left in contact with them for more than one hour. Therefore, always wash these surfaces immediately after use. When the die is not in use, store it in a desiccator. In laboratories where the relative humidity is above 50 to 60%, better pellets are often obtained if the die is heated for 20 min at 40 to 45° just before use.

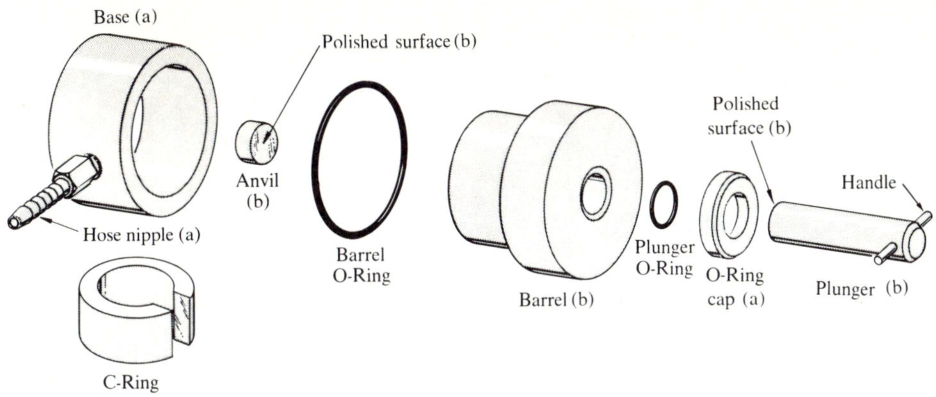

Figure III-4 Explosion diagram of a KBr die.

Place the O-ring cap and the plunger O-ring on the plunger. Force the plunger through the barrel. Note that the plunger goes through the larger end of the barrel so that the O-ring seats in the depression at this end (Figs. III-4, 5, and 6). Place the C-ring on the bench and then place the barrel-plunger assembly on the C-ring with the plunger handle and O-ring *inside* the C-ring (Fig. III-6a). Transfer the sample-potassium bromide down to the shiny surface of the plunger. Unless an especially intense spectrum is desired, do not add more than 300 mg of sample-potassium bromide mixture. If necessary, use a camel's-hair brush to aid in the transfer of the mixture into the die and from the inside of the barrel to the plunger surface. This latter is usually unnecessary if the mixture and barrel are dry. Use the side of the camel's-hair brush whenever possible. Lightly press the anvil into the barrel with its shiny surface toward the KBr mixture (Fig. III-6b). Put the barrel O-ring in its groove, place the base over the anvil end of the assembly, and press firmly. With one finger on the plunger to prevent its falling out remove the assembled die from the C-ring, turn it over, and place it base down on the bench (Fig. III-6c). Use a light, rotary motion of the plunger to level the sample. Press the plunger, O-ring, and O-ring cap firmly against the top of the barrel. Connect a vacuum line to the die and apply a vacuum of 0 to 2 mm. If the sample is an extremely volatile solid, use only 10 mm vacuum. Evacuate the die for 2 min before applying pressure to the die. If the relative humidity is between 60 and 100%, evacuate for up to 5 min before applying pressure.

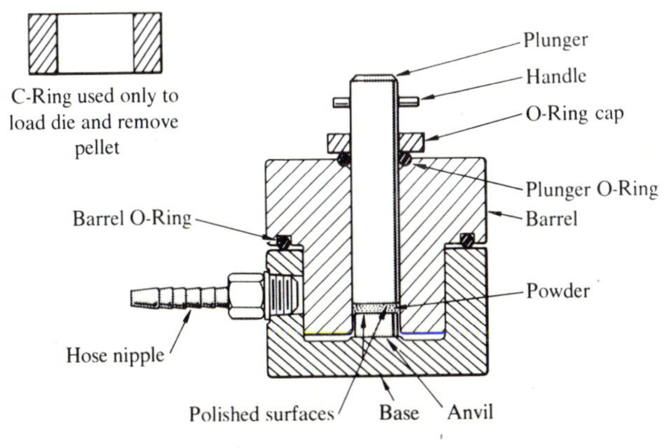

Figure III-5 KBr die cutaway view.

Carefully center the die in the press. If there is a hole in the top press platen that prevents exact centering of the die, cover the hole with a steel block. Be certain the press platens are parallel before applying pressure. An off angle or off center pressure usually gives a poor pellet and may fracture the plunger. Apply 14,000 to 16,000 # gauge pressure for 3 to 6 min. The higher pressures and the 6 min pressing time may be required to obtain transparent pellets with some materials. *Do not exceed 23,000 # total load. The die is easily ruined by*

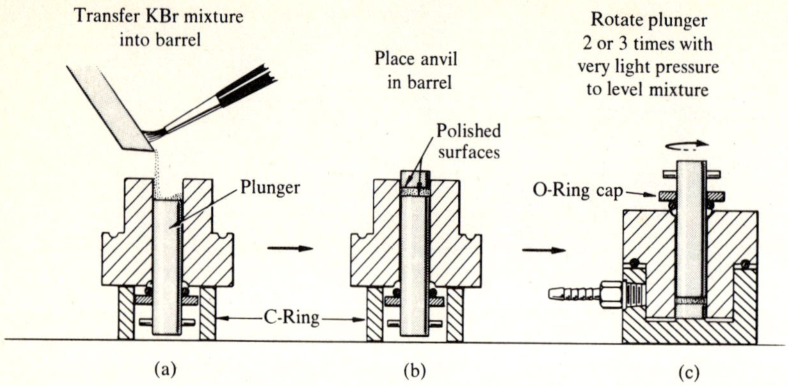

Figure III-6 Transferring the sample KBr mixture to the die and leveling of the mixture.

excessive pressures, and at very high pressures there may be a safety hazard. The limits given here are for the Perkin-Elmer 13-mm pellet die. Other dies may have lower limits. Carefully observe the pressure limitations for the die you are using. Make certain you are determining the total load correctly. On most Carver presses, for example, the scale that reads "load in lbs. on $1\frac{1}{4}$ in ram" gives the total pressure. If the gauge is calibrated only in psi (pounds per square inch), multiply the pressure reading by the *area* of the press ram or piston to determine the total load.

Break the vacuum, then open the press. Remove the barrel from the base, but do not attempt to remove the plunger or the anvil. Place the barrel-plunger assembly upside down in the center of the bottom platen of the press. Place the C-ring on top of this assembly so that the opening faces toward you and you can see the anvil and pellet when they are forced out of the barrel (Fig. III-7). Carefully apply pressure so that the anvil and the pellet are just pushed from the barrel by the plunger. Do not apply so much pressure that the handle of the plunger presses against the O-ring as the handle will be bent or broken by excessive pressure.

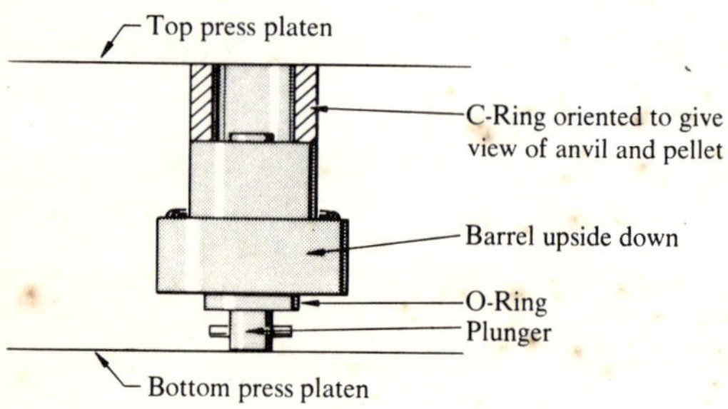

Figure III-7 Using a press to remove a KBr pellet from a die.

Place the inner section of the KBr disc holder on the bench, notched edge down and 13-mm depressed area up (Fig. III-8). Remove the anvil from the top of the pellet. Remove the pellet with a pair of tweezers and place it in the depressed area on the inner section of the disc holder. Hold the inner section so that the disc stays on top in the depressed circle and slide it as far as it goes in the outer part of the disc holder.

Slide the assembled pellet holder with pellet into the opening for microcells in the sample beam cell slide. Align the slots of the pellet holder horizontally so they fit the protrusions in the cell slide opening or the spring just inside it. Gently push the pellet holder back as far as it will go. If it is necessary to keep the pellet any length of time, protect it from moisture and mechanical shock. For example, one to three pellets can be stored in a small, labeled desiccator made from a vial (Fig. III-9).

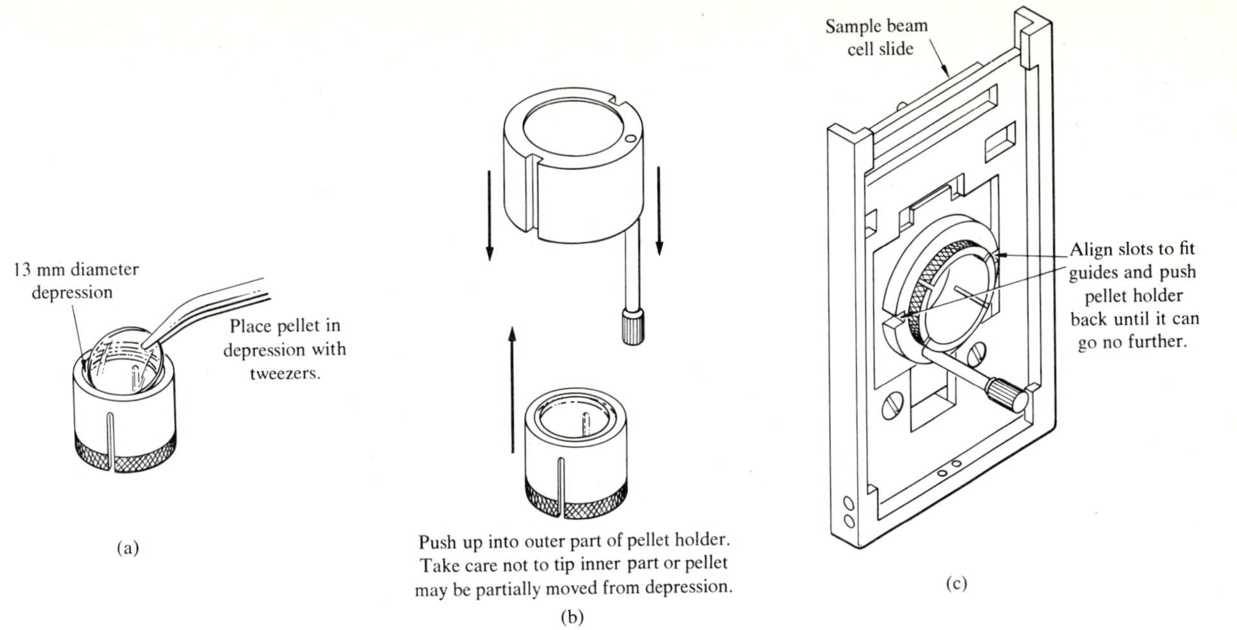

Figure III-8 Placing a KBr pellet in the sample beam of an infrared spectrophotometer.

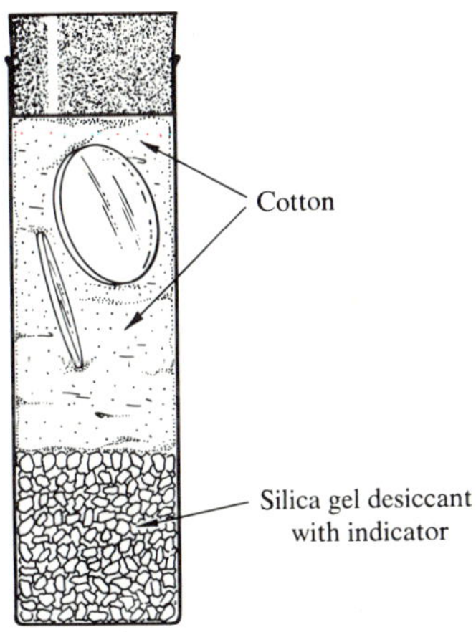

Figure III-9 Desiccator made from a vial, containing two 13-mm KBr pellets.

Table III-2 lists common pellet defects, their cause, and prevention. Note that if the pellet is to be repressed, it should be reground first. Repressing an intact pellet usually gives an unsatisfactory pellet and may damage the polished surfaces of the anvil and the plunger. If the absorption bands are too intense, a more dilute pellet can often be made conveniently by first estimating how much dilution is needed. An appropriate fraction of the pellet can then be crushed and mixed well with the amount of potassium bromide required to give approximately 300 mg and pressed.

The most practical criterion for a good pellet is not the appearance of the pellet but the appearance of the spectrum. If the spectrum has strong, sharp absorptions and absence of scattering, no water absorptions (unless,

TABLE III-2 DEFECTS IN ALKALIMETAL HALIDE PELLETS

DEFECT	CAUSE	REMEDY
Marked scattering of light throughout disc. Poor visibility of distant objects.	Impurities or too much sample. Even mixtures containing as little as 5% of a second alkali metal halide show this effect.	Use pure halides, use less sample, use higher pressure.
Irregular, blotchy, or completely cloudy.	Usually associated with dampness or coagulation of powder or sample.	Dry and break up powder and sample. Apply vacuum longer.
Disc shows number of white spots, otherwise clear.	Few coarse grains in an otherwise fine mixture.	More grinding or perhaps more even grinding of sample.
Translucent or cloudy disc.	Nature of sample.	Spectrum may be satisfactory; if not, grind and repress at higher pressure for a longer time (Do not exceed maximum pressure.)
Disc opaque over part of area.	Insufficient pressure possibly coupled with bad distribution or die not centered properly.	Center die in press more carefully, Grind and repress. Take more care in leveling powder with plunger.
Disc clear when removed from press but develops irregular internal cloudiness after one minute to several hours.	Trapped air.	Check vacuum. Apply vacuum longer before pressing. Do not grind potassium bromide.
Cloudy area in center.	Anvil or plunger faces not flat. Uneven distribution powder.	Replace anvil or plunger. Take more care in leveling powder with plunger.

of course, the solid has water of crystallization), good band shape, good contrast, and a reasonably flat I_0 that is above 80%, then the pellet is satisfactory. If an unsatisfactory spectrum is obtained, then the appearance of the spectrum and the pellet with the aid of Table III-2 and some of the spectra shown in this text can be used to diagnose the problem so that it can be rectified.

Figures III-10b and III-11b are spectra obtained from improperly made pellets. Usually pellets giving spectra with this much distortion or water absorption should be discarded and at least one more attempt made to obtain a better pellet. For much rough survey work a spectrum similar to that in Figure III-11b is satisfactory. There are even a few occasions when a spectrum with as much water absorption as that shown in Figure III-10b might be satisfactory. The spectra in Figure 18-11 have sloping I_0 lines which are probably due to incomplete grinding of the sample. Since there is little band distortion, spectrum b or c of Figure 18-11 is satisfactory for most purposes.

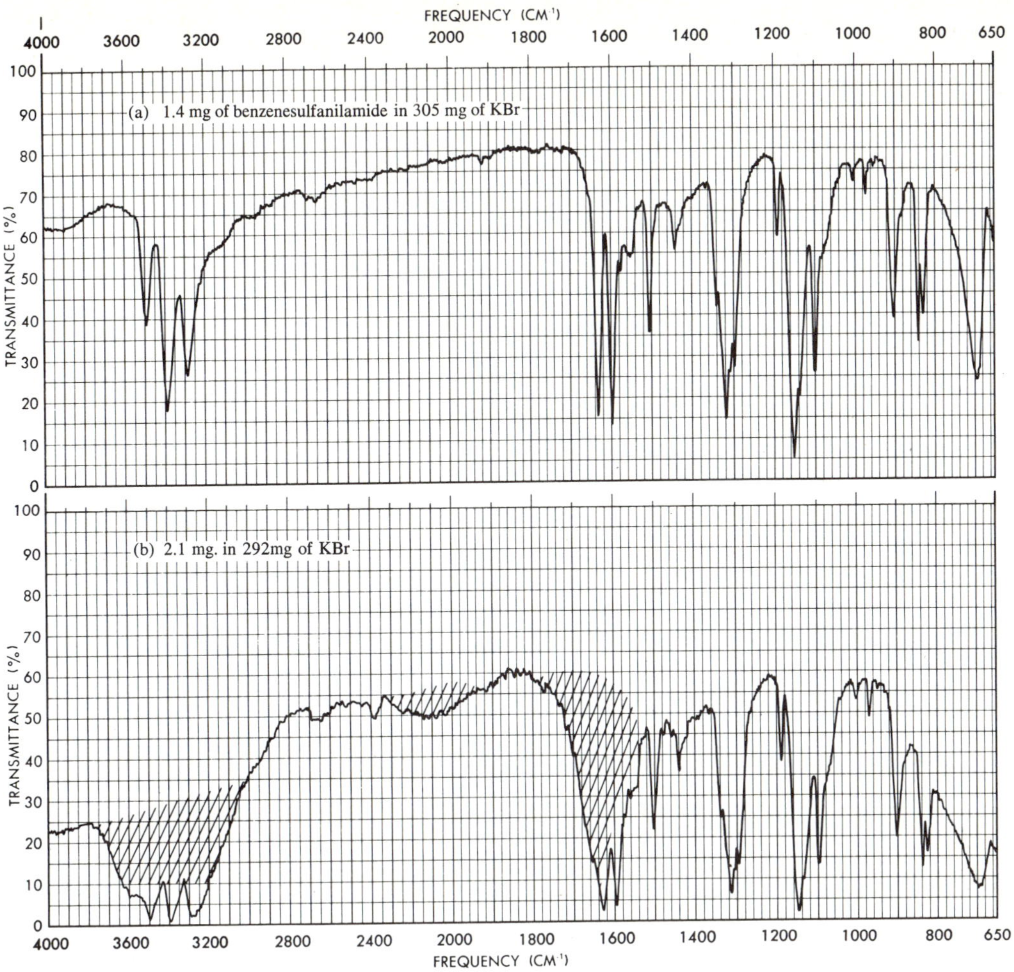

Figure III-10 Spectra of benzenesulfanilamide in KBr pellet (a) Correctly prepared. (b) Correctly prepared except for moisture in the pellet. Water causes problems because of its absorptions (shaded areas in b) in critical regions, and it causes the resulting pellet to scatter considerable radiation, especially at high frequencies. This is evident in the comparison of the lower percent transmittance of the I_0 of b with that of a.

Problem III-4. (a) Why does too large a particle size give a sloping I_0 line as shown in Figure 18-11 rather than more uniform attenuation across the entire spectrum? (b) If the sloping I_0 line were due to coarse sample particles, could it be "straightened" by placing in the reference beam a pure potassium bromide disc prepared the same way? (c) If the sloping I_0 line were due to coarse potassium bromide particles, could it be compensated for by the method outlined in (b)?

Use of a Mini-Press. The preparation of the KBr-sample mixture for a Mini-Press is similar to the preparation for an evacuable die except that smaller amounts of mixture are needed because of the smaller pellets made by the Mini-Press. Mini pellets are made with 60 to 100 mg of KBr and 0.5 to 1 mg sample. As little as 25 mg of KBr and 0.1 mg of sample can be used, although spectra from such pellets are often far from optimum.

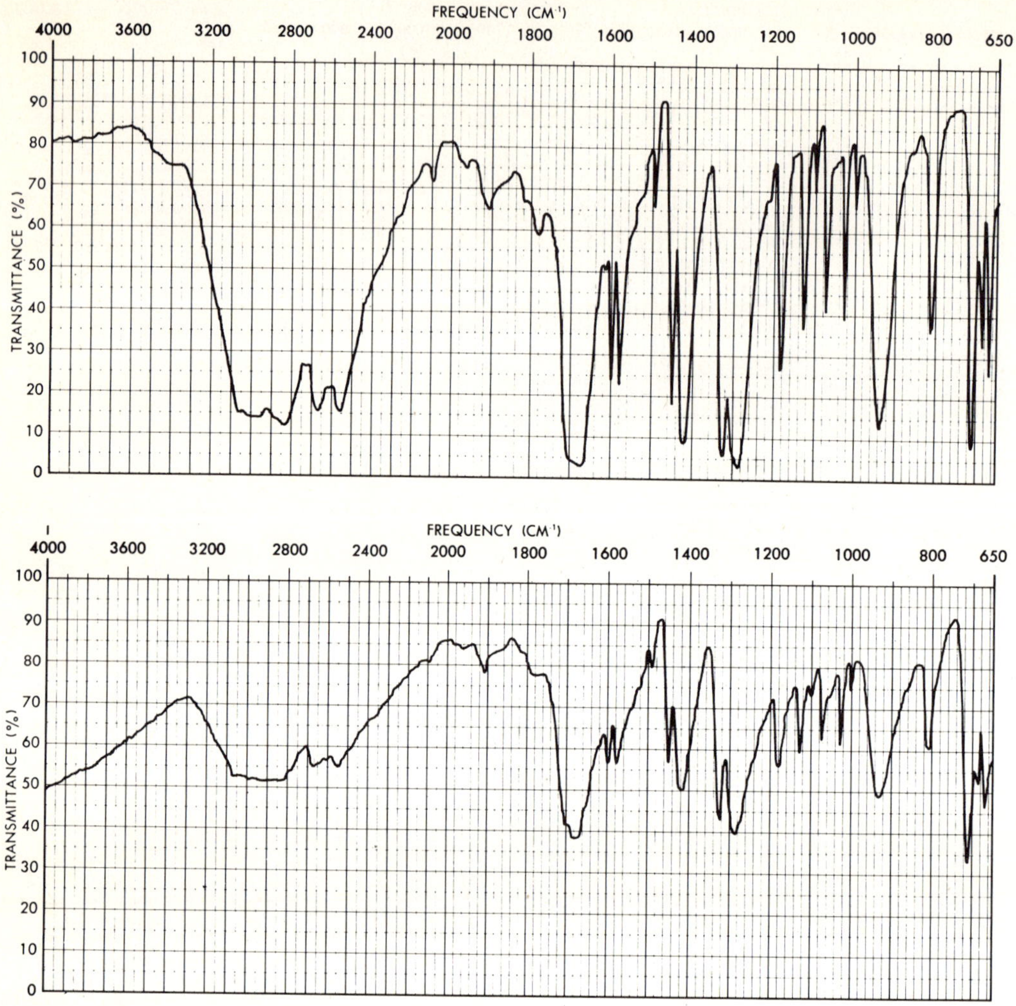

Figure III-11 Spectra of benzoic acid in KBr pellets. (a) Well ground sample. (b) Poorly ground sample.

Warning, be careful not to scratch or otherwise damage the polished surfaces of the two bolts. Since the Mini-Press cannot be evacuated, it should be heated to 40 to 50° just before use in all but the driest environment. Place one bolt in the barrel and rotate it five full turns. Place 60 to 100 mg of sample-KBr mixture on the polished surface of the bolt inside the barrel (Fig. III-12). Keep the open end of the barrel pointing up and the bolt as vertical as possible. The barrel and bolt can be conveniently held in a bench vice or the bench top wrench supplied with some Mini-Presses. Gently tap the press (Fig. III-13) to level the sample-KBr mixture on the bottom bolt. Brush any mixture on the threads to the bottom bolt with the camel's-hair brush.

Insert the other bolt until it is almost finger tight. *Gently* tap the bolt inserted first on the bench top to level the KBr mixture; keep the bolts vertical during the tapping. Tighten the second bolt finger tight and repeat the *gentle* tapping.

With two wrenches or one wrench and a vice or a bench top wrench, gradually apply pressure (Fig. III-14). Tighten as much as possible, then allow the Mini-Press to stand for 1 to 2 min. Remove the bolts; if one bolt cannot be removed, use a larger wrench on the flat sides of the barrel. Look at the pellet; Table III-2 lists some common faults, their causes and remedies.

Figure III-12 Placing the sample KBr mixture in the barrel of a Mini-Press. Wilks Scientific, South Norwalk, Conn.

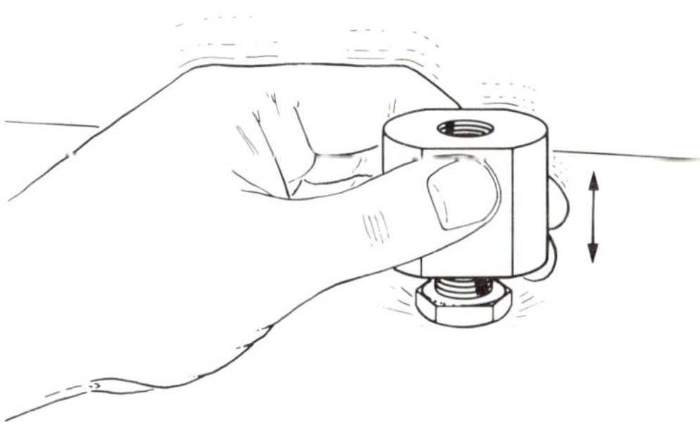

Figure III-13 Gently tapping the Mini-Press to level the powder.

Place the barrel holder in the sample beam cell holder and the barrel with the KBr pellet in the barrel holder (Fig. III-15). Move the 100% adjust all the way out for maximum transmission. Push the barrel with the pellet back towards the cell slide (i.e., away from the open, sample area) as far as possible. Be sure the spectrophotometer is on, and set it at 3950 cm^{-1}. Orient the barrel in the holder for maximum transmission. This usually occurs with one of the flats of the barrel on the "vee" of the holder. The optimum orientation is different for some spectrophotometers so set for maximum transmission. Obtain a spectrum in the usual manner.

Some Mini-Presses have barrels with grooves so that they fit directly in the opening in the sample slides of the Perkin-Elmer 700 and 21 Series spectrophotometers. The black barrel holder is not used. The grooved barrel is pressed back into the opening as far as possible (Fig. III-15b). On the Model 700, I_0 will have a higher percent transmittance if a piece of paper is wadded between the underside of the barrel and the rounded opening.

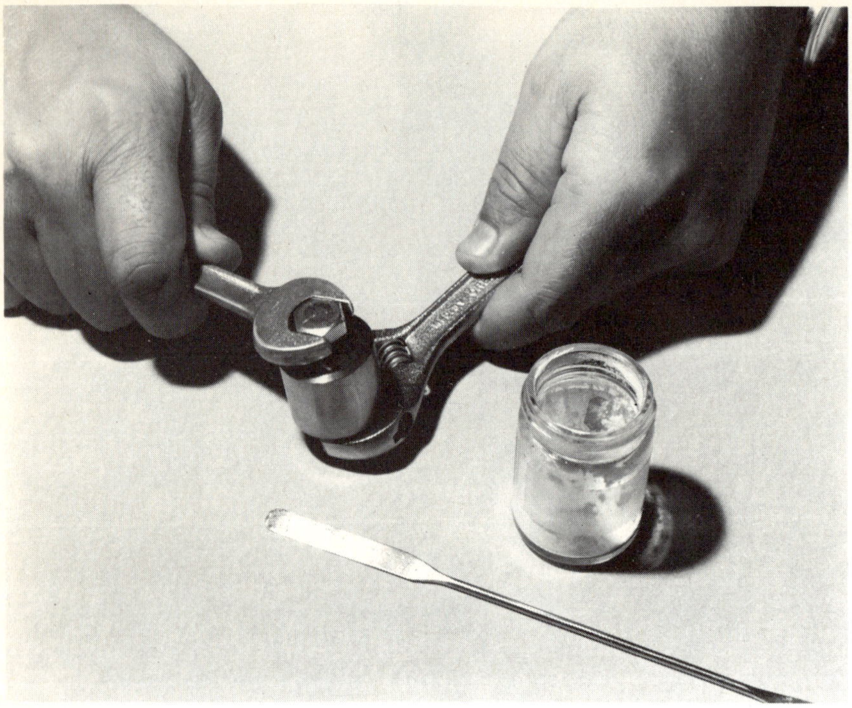

Figure III-14 Tightening the bolts to press a pellet in a Mini-Press. Wilks Scientific, South Norwalk, Conn.

An attenuator can be placed in the reference beam (Fig. III-16) and adjusted to give 85 to 90% transmission at 2000 cm^{-1}. This attenuator removes radiant energy from the reference beam similarly to the removal of radiant energy from the sample beam by vignetting of the barrel and scattering, reflection, and general absorption by the KBr pellet. Use of a reference beam attenuator gives more contrast to the absorption bands. It does this by removing energy from the reference beam and not by adding energy to the sample beam. For the servo loop to operate properly, the gain must be increased to compensate for the loss of the energy. If a variable slit program control is available, the slit width may be increased instead.

> **Problem III-5.** (a) What spectral qualities are not as satisfactory with pellets in a Mini-Press as with pellets from an evacuable KBr die if a reference beam attenuator is used? (b) If no reference beam attenuator is used? Assume for both (a) and (b) that the quality of the KBr pellets is the same and the gain is set for optimum energy in the servo loop.

Clean the Mini-Press immediately after use. Remove the pellet from the barrel by punching it out with a pointed instrument or washing it out with water. Do not attempt to force the pellet out with one of the bolts as this may damage the bolt. Clean the barrel and the bolts well, especially the threads and the delicate polished surfaces. If water is used to clean any parts of the Mini-Press, immediately wash it off with acetone or methanol. Store the cleaned, dried Mini-Press in a desiccator.

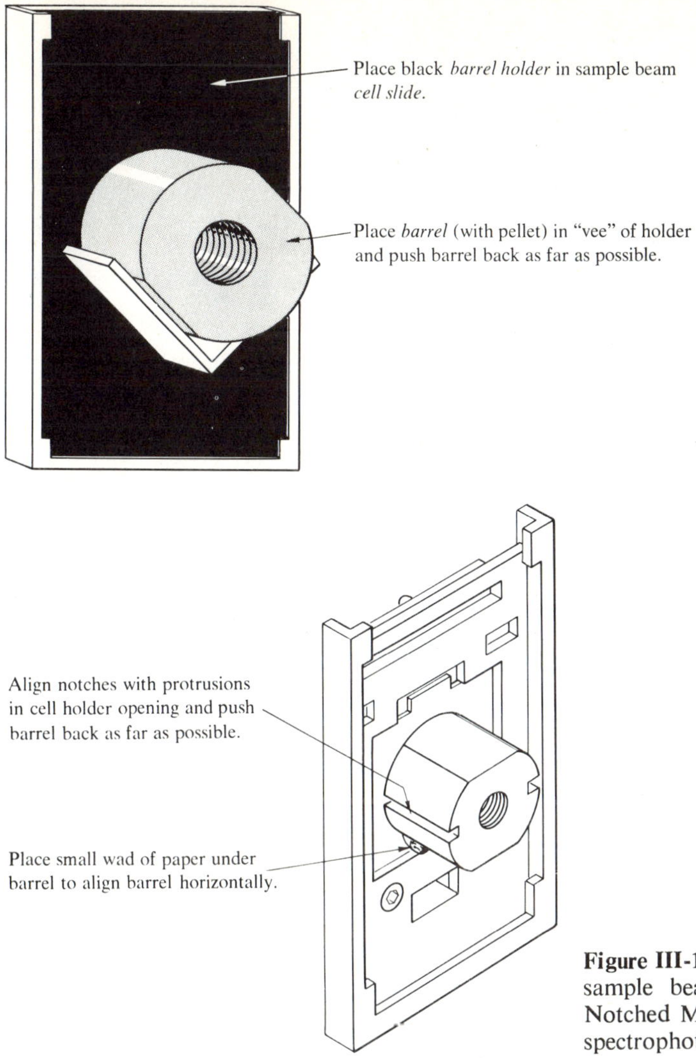

Figure III-15 (a) Barrel and pellet in the sample beam of a spectrophotometer. (b) Notched Mini-Press in a Model 700 infrared spectrophotometer.

MULLS

Spectra are often obtained on solid compounds mulled in high boiling liquids that are transparent to large portions of the infrared spectrum. The two most common mulling agents are mineral oils (such as Nujol) and perfluorocarbons (such as Fluorolube and perfluorokerosene). The spectra of these mulling agents are shown in Figures III-17 and III-18. Mineral oil is transparent over the entire spectrum except where alkanes have

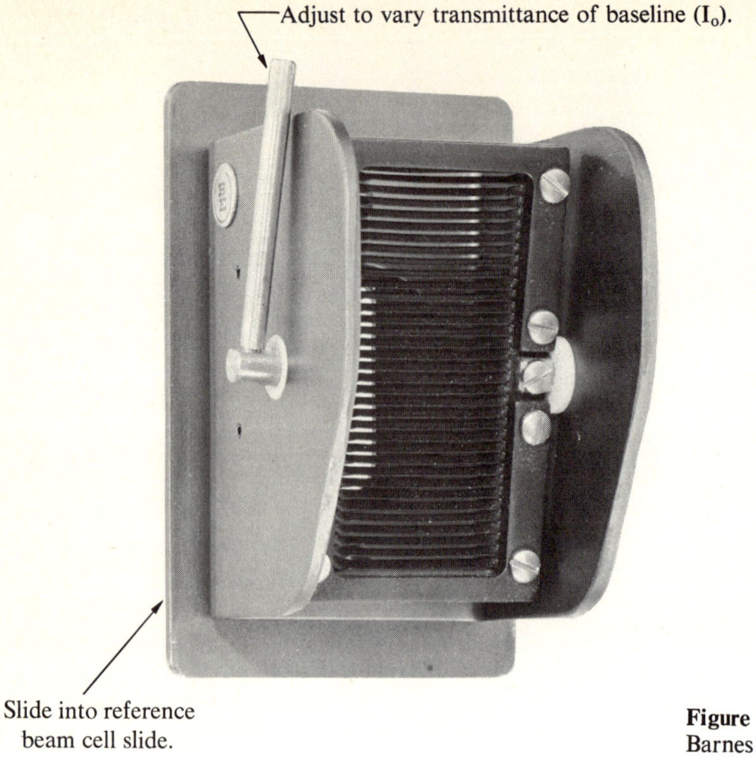

Figure III-16 Reference beam attenuator. Barnes Engineering, Stamford, Conn.

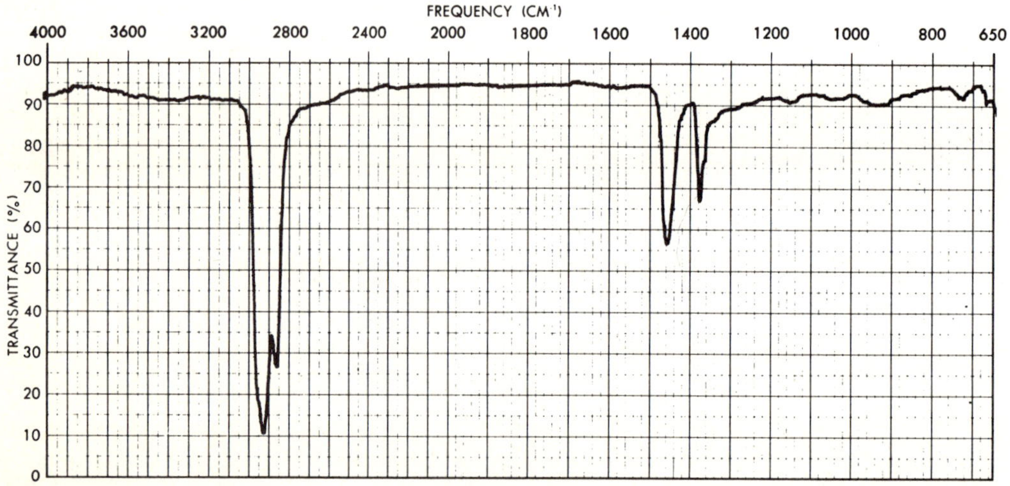

Figure III-17 Spectrum of Nujol, capillary film.

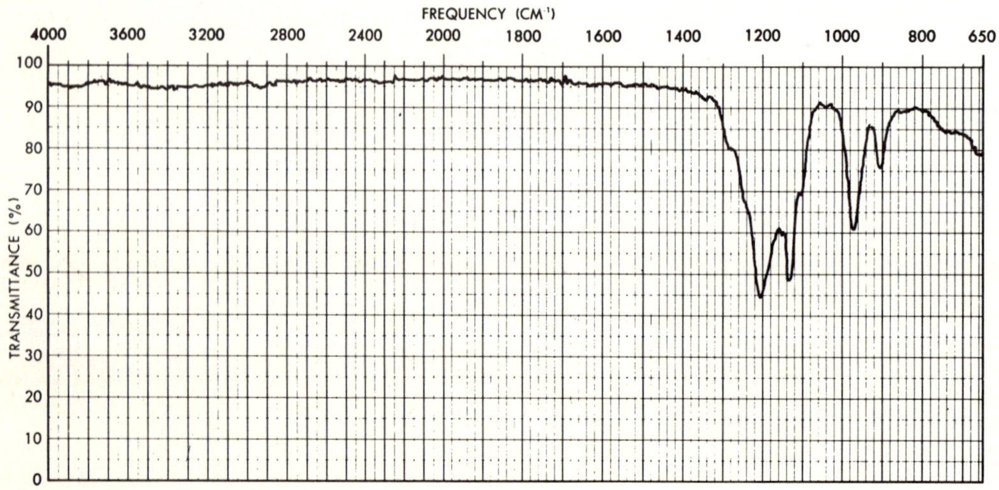

Figure III-18 Spectrum of Fluorolube, capillary film. Hooker Chemical.

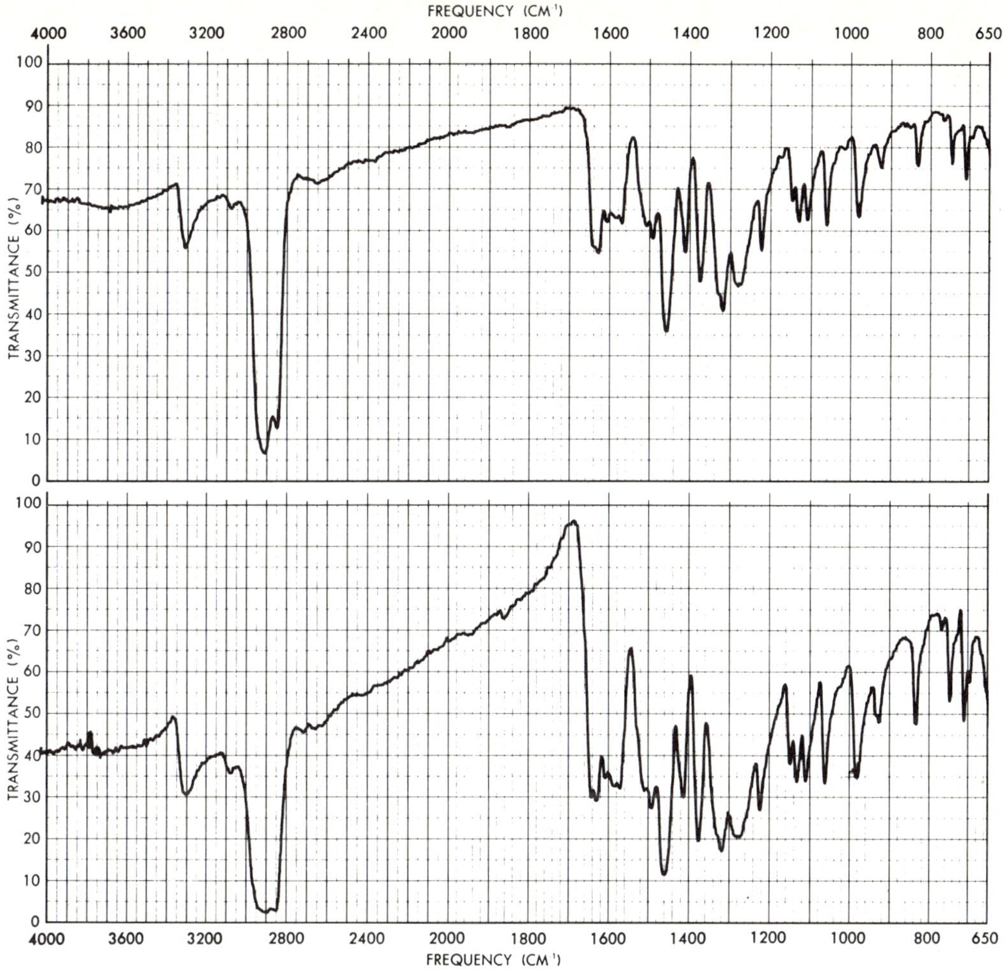

Figure III-19 Spectra of 2,4-dinitrophenylhydrazine in a Nujol mull. (a) A good mull. (b) A poor mull; note the band distortion and low I_0 line especially at high wavenumbers from scattering of radiation. Both these defects are from large particle size of the sample. R. Hannah and J. Swinehart, *Experiments in Techniques of Infrared Spectroscopy*, p. 6–4, Perkin-Elmer, Norwalk, Conn. 1968.

strong absorptions, and perfluorocarbons are transparent at frequencies higher than 1300 cm^{-1}. The two can thus be used to obtain spectral information in their transparent regions, or the spectrum of a solid in each one of these can be used to obtain the equivalent of one complete spectrum unmasked by any mulling agent absorptions. A spectrum obtained from mulls in these two agents is often called a "split-mull spectrum."

The samples can be ground either wet or dry by the same technique used to grind samples for KBr pellets (p. 508) except that approximately 8 to 15 mg of sample is usually used. When the sample has been sufficiently ground, add a drop of mulling agent. Grind with the pestle using a rotary motion until the sample is uniformly suspended. If the mixture is too thick add another drop of mulling agent and grind again. The mull should resemble vaseline in viscosity and consistency. Use a clean rubber policeman to aid in transferring the mull from the mortar to a cell window. Use another window to spread the mull. Spread evenly and take care to remove any air bubbles from the window area through which the sample beam passes. Leave the two windows together,[1] place in a demountable cell, and put the cell in the cell slide.

[1] If silver halide windows are used, attempt to use only one window to reduce reflective losses. Pull the two windows apart, use the window that has retained the most mull or the one that is coated with the most uniform film. With a little luck the mineral oil film will not sag significantly while the spectrum is being obtained. If silver halide windows with depressions are being used, follow the instructions on p. 525.

Proper grinding is as important in the preparation of Nujol mulls as it is in the preparation of alkali metal halide pellets. Figure III-19 shows the spectra of mulls of well and poorly ground samples of 2,4-dinitrophenylhydrazine.

Cast Films

From molten material. Melt a small amount of wax (such as carnauba, paraffin, or beeswax) in a 10 to 50-ml beaker and heat to about 10 to 15° above the melting point. Warm a cell window to 45 to 50° by placing it in the concave surface of a heavy watch glass which has its convex surface on the hot plate that is being used to melt the wax. Do not use a steam bath to warm the cell window.

Be careful not to warm the window too fast or to too high a temperature as it will crack. In general, any surface cool enough to touch for 20 seconds will not crack an alkali metal halide window. The cell windows may be placed directly on a warm glass surface. If, however, it is to be placed on a hot plate, keep a piece of paper towel or filter paper between the metal surface and the cell window. Follow the usual precautions (p. 131) for handling alkali metal halide windows.

(a) Tilt the window by placing one end on the upper edge of the watch glass and the other end toward the bottom. (b) Place a medicine dropper pipet in the beaker of molten wax for a few minutes to warm it. Then use it to transfer a few drops of wax to the warm cell window. (c) If a thin film runs down the window, the window can be removed from the hot plate, allowed to cool, and placed in a demountable cell. (d) If the wax does not form a thin film on the cell window, take another cell window and press it to the first so that the molten wax forms a very thin film. Leave the two windows together, allow them to cool, and place them in a demountable cell.

A spectrum having one or more intense absorptions that bottom at 0% transmittance for more than 10 cm^{-1} shows that the film is too thick, even if it appears to be very thin. If this occurs after instruction c is followed, *gently* heat the cell window until the wax melts and then follow instruction d.

From evaporation of a solution. Use a volatile solvent that does not dissolve alkali metal halides even when warm. Methylene chloride, ether, or chloroform are satisfactory. Methanol, ethanol, water, and dimethyl sulfoxide should not be used unless the film is cast on Irtran or silver halide windows or cast on a mercury pool and then removed. Xylene is borderline because of its high boiling point; if possible, a lower boiling hydrocarbon like toluene or preferably benzene should be used instead.

Dissolve the solid in slightly more solvent than necessary to effect solution. Place a cell window on a paper towel on a hot plate at 45 to 50°. Before the window warms significantly, add 2 to 3 drops of the solution to the cell window so that the solution spreads evenly. Allow the solvent to evaporate for 4 to 5 minutes, remove the cell window from the hot plate and place in a demountable cell. Some solids, especially polymers, will retain significant amounts of solvent after this procedure. More time is necessary for complete evaporation in this case or the cell window may be placed in a vacuum desiccator or oven for a few minutes *after* most of the solvent has evaporated. If the spectrum shows the film is too thin, add a few more drops of solution and evaporate the solvent. If the spectrum shows the film is too thick, remove the film and redeposit a new one by using a more dilute solution or fewer drops of the original solution.

Comparison of Solid Sampling Techniques

General Comparison. All these solid sampling techniques have at least one deficiency when compared with solution spectra: spectra of the same compound may show small, significant differences from variations in sampling techniques and the spectrum of a solid sample may have small, significant differences from its solution spectrum. These differences are usually due to differences in crystal structure although with the compressed disc technique, they may also be due to reaction of the alkali metal halide with the sample. The differences from crystal structure are usually changes in the position or intensity of NH, OH, or SH stretch absorptions, small changes in the position and intensity of carbonyl and imine stretch absorptions, and changes, occasionally large, in the position, intensity, and occurrence of medium to weak absorptions in the fingerprint region (1300 to 650 cm^{-1} and below). In spectra of organic compounds where these changes from different crystal structures occur, there is little chance of misinterpretation since sufficient absorptions show little or no change with sampling techniques. The potential for error does exist so the scientist using solid sampling techniques should be aware that such differences are possible; with this awareness, errors from these differences are not likely.

Cast Films. This is the easiest of the three solid sampling techniques. It is not applicable to a very large number of solids because the crystals obtained are large enough to cause scattering. The spectra of some solids deposited from solution show solvent absorption bands even after what is apparently thorough drying. It is useful for compounds that form microcrystals such as waxes or for compounds that are more or less amorphous as many polymers are.

Mulls. Mulls share the advantage with cast films that no expensive sampling apparatus is required. The preparation of a good mull is, however, the most difficult of the three solid sampling techniques. The mulling agents have significant absorptions and two mulls must be made to obtain a complete spectrum. Quantitative analysis with mulls is much more difficult and less accurate than quantitative analysis with potassium bromide pellets.

Alkali Metal Halide Pellets. Compressed pellets are more likely than mulls to give spectra that differ from solution spectra. The samples for both mulls and pellets are ground which may change the crystalline structure, but the samples in pellets may also change crystalline form when the pellet is formed under high pressure. The potassium bromide or other alkali metal halide may also react with the sample. This occurs very seldom with organic compounds and not too often with inorganic compounds. The compressed disc technique allows the use of smaller amounts of sample than any other technique. Because of the relative ease of preparation and the absence in alkali metal halides of absorptions above 600 cm^{-1}, the compressed disc technique is used more by organic chemists than any other solid sampling technique.

MEASUREMENT OF THICKNESS OF CELLS AND THIN LAYERS

If different cells are to be used for quantitative analysis, the cell thickness must be accurately known. If the same cell is used for both calibration and repeated analyses, the cell thickness must be measured occasionally to determine if there have been any changes from dissolving the surface of the cell window or from distortion of the cell by excessive pressures during filling or cleaning.

There is also a need to measure the thickness of films and other thin layers. The thickness of films that are transparent to a portion of the infrared spectrum and the thickness of cells can be measured from the distance between fringes in their infrared interference patterns.

Whenever a beam of radiation travels from a medium of one refractive index into a medium with a differetn refractive index, part of the energy is transmitted and part is reflected. Figure III-20 depicts a few of the multiple reflections that occur when radiation passes through an empty cell or a thin film. Note that the ray ABCD has traveled a path longer than ray AB by twice the thickness, *d*, of medium 2. These two rays interfere with each other, and, depending on their wavelength, either reinforce one another or interfere destructively with resulting attenuation.

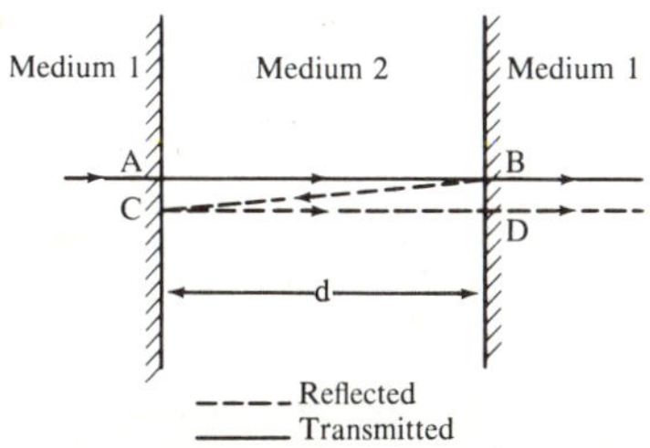

Examples

	Media #1	Media #2
Transparent film	air	film
Empty cell	window	air
Filled cell	window	liquid

Figure III-20 Path of radiation between inner surfaces of a film or cell. The path of reflected radiation is drawn at an angle to separate it from the transmitted radiation. Perkin-Elmer, Norwalk, Conn.

The relation between cell or film thickness, *d*, and frequency (or wavelength) will determine whether there is reinforcement or attenuation, and as the spectrophotometer scans through successive frequencies, a series of fringes is produced as reinforcement and attenuation alternate. To obtain easily observed fringes, the film or the cell thickness should be uniform and without large irregularities. The lower the frequency of radiation, the greater the amplitude of the fringes; the greater the difference in the refractive index of the two media, the greater the amplitude of the fringes. Figure III-21 is a transmission spectrum of a film of polyethylene and exhibits typical fringes superimposed on the characteristic absorption spectrum. Fringe maxima are labeled *r* (for reinforcement) and fringe minima *a* (for attenuation).

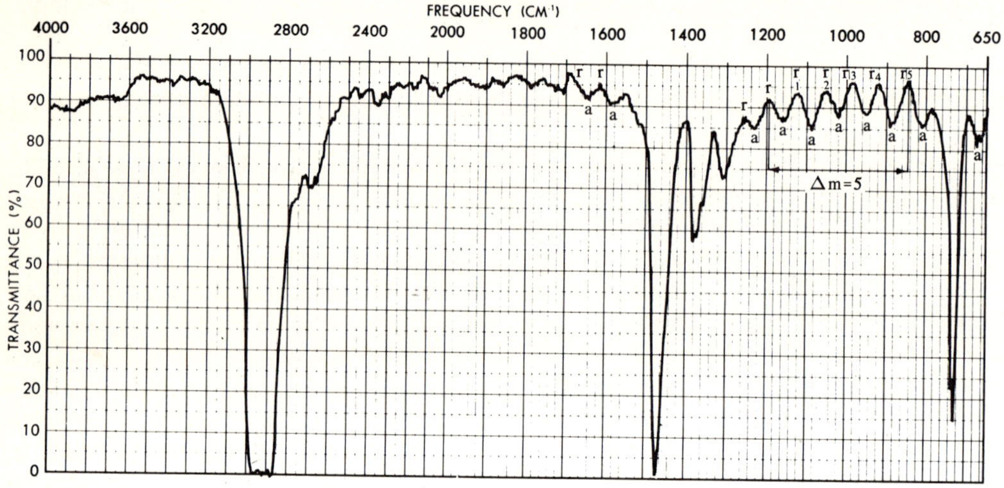

Figure III-21 Spectrum of a thin, uniform film of polyethylene.

Interference fringes can be differentiated from absorption bands because of their uniform appearance and because the maxima and minima are evenly spaced on a linear frequency scale. From these interference fringes the thickness (d) of cells and transparent layers can be calculated. This is done by substituting the number of wavenumber units ($\nu_1 - \nu_2$) spanned by a given number of interference fringes (Δm) in the equation:

$$d \text{ (in cm)} = \frac{\Delta m}{2n(\nu_1 - \nu_2)}$$

where:

n = the refractive index of the layer (1.44 for polyethylene and 1.0 for air).

In the spectrum of Figure III-21 there are five fringes (i.e., five complete cycles) between 1197 and 847 cm^{-1}.

$$d = \frac{5}{2(1.44)(1197 - 847)} = \frac{5}{2.88 \cdot 350} = 4.96 \times 10^{-3} \text{ cm}$$

Note that in making the fringe measurement the regions where the sample has absorption bands is avoided. When counting fringes, if possible choose a region free of absorption which might otherwise mask the exact location or the actual number of the fringes.

Problem III-6. Cells filled with a liquid *usually* do not give interference fringes although they clearly meet the requirements outlined in Figure III-20. Why?

Problem III-7. The interference fringes for two empty cells are shown in Figure III-22. Calculate the thickness of each cell.

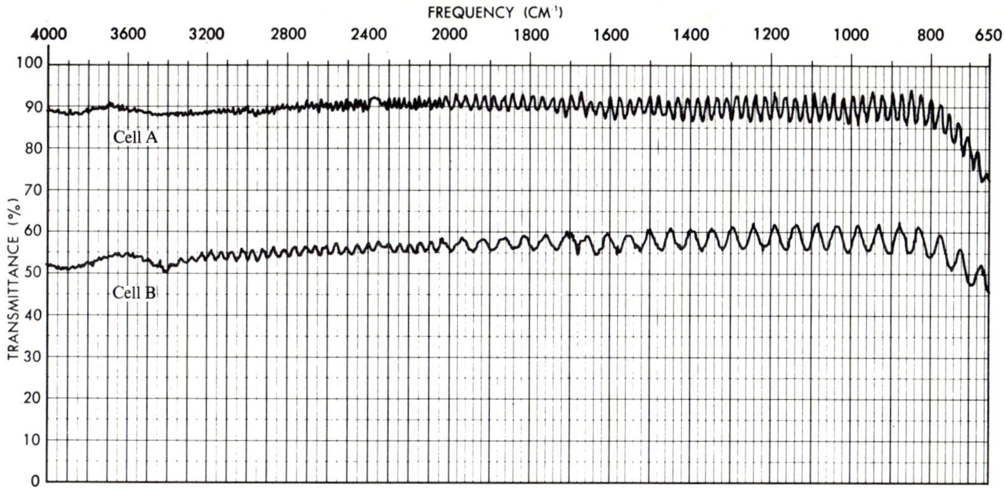

Figure III-22 Interference fringes for two empty cells.

USE OF THE MINI-CELL

General

The Mini-Cell is a small, plastic demountable cell with silver chloride windows (Fig. III-23). The windows have flat circular depressions 0.025 mm deep so that the cell may be used for greases, mulls, capillary films, or as a 0.025 or 0.05-mm demountable cell (Fig. III-24). The silver chloride windows are not damaged by moisture and are not easily broken. They are light sensitive and when not in use should be stored in an opaque container. They are soft and may flow or bend under pressure.

Problem III-8. (a) The silver chloride windows are damaged by ammonia, amines, or their solutions. Why? (b) What other solutions may damage the Mini-Cell windows? (c) The silver halide windows may be etched if placed in contact with some metals. The metals in turn may become corroded by the windows. Why?

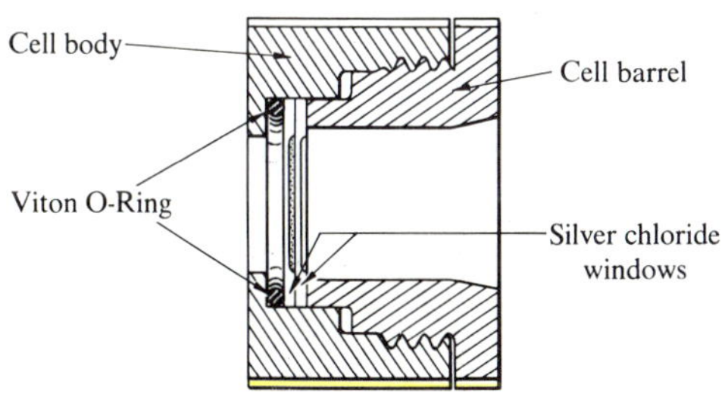

Figure III-23 Cross section of an assembled Mini-Cell with windows set for 0.025 mm thickness.

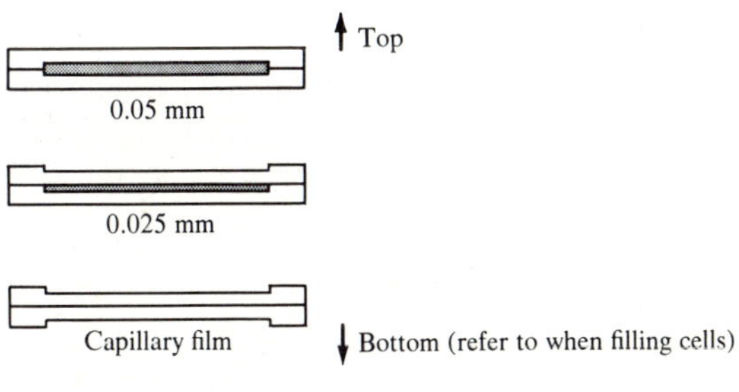

Figure III-24 Orientation of windows for Mini-Cell.

Capillary Films

Place the body of the cell (internally threaded part) on the bench top with the threaded opening up. Be certain the Viton O-ring is in place. Lay the first window in the cell with the depressed area down. With a capillary or micro pipet place a few micro drops of liquid on the window. Lower the top window (depressed area up) into place by touching its edge as close as possible to the rim of the bottom window and then gently lowering the opposite edge (Fig. III-25). This procedure helps to spread the film and avoid bubbles. Do not drop the top window flat on the bottom window. Screw the barrel (externally threaded portion of the cell with the words "Mini-Cell" on the end) into the cell body finger tight. Do not use excessive force in tightening.

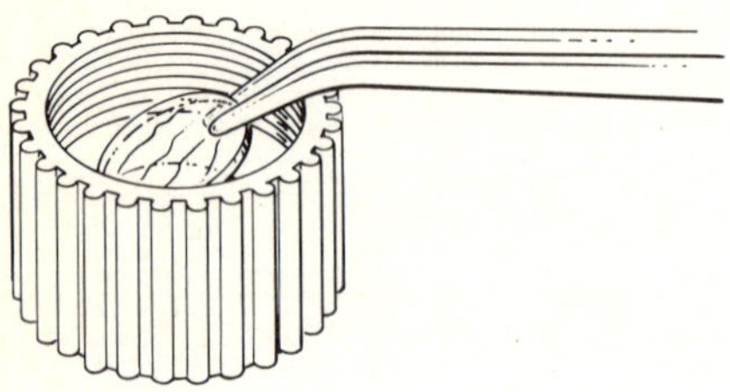

Figure III-25 Touch the bottom window at one edge and lower the top window upon it.

If the liquid is volatile, the top window should be quickly placed and the barrel screwed in as soon as possible. When any demountable cell is used to contain volatile liquids, the spectrum should be obtained as soon as possible before leakage occurs. With capillary films the Mini-Cell may leak less than other demountable cells because the thin silver halide windows can bend around the liquid and form a tighter seal than is possible with alkali metal halide windows.

Liquids of 0.025 or 0.050 mm Thicknesses

Follow the general directions given for capillary films except that windows should be oriented as shown in Figure III-24 for the desired cell thickness. If the cell is to be filled with a volatile liquid, the following procedure may help avoid bubbles and evaporation. Remove both cell windows from the cell but leave the Viton O-ring in place. Place one window on a paper towel with the depressed area up. Add a few drops of solvent, then slide the second window over the top. Immediately place the two windows on the Viton O-ring in the cell body and screw the barrel in finger tight.

Mulls and Greases

To reduce reflective losses from surfaces of the silver halide windows an attempt should be made to use only one window. This can often be accomplished by filling the depressed area of one window with the mull or grease and using the flat side of the other window to smooth it. If this procedure gives too thick a layer of mull or grease, follow the directions in the footnote on page 520.

Placement of the Mini-Cell in the Spectrophotometer

The Mini-Cell does not vignette the sample beam as much as many sampling devices; however, because of the reflective losses from the silver halide windows, every attempt should be made to get as much radiant energy through the sample as possible. This can be achieved by placing the Mini-Cell in the most condensed portion of the sample beam. In the sample compartment of most spectrophotometers this occurs immediately behind the cell slide away from the open sample area. Therefore, push the Mini-Cell as far back in the cell slide as possible. In many spectrophotometers this means *not* using the black plastic cell holder that comes with most Mini-Cells.

For example, in the Perkin-Elmer Model 700 remove the square micro cell adapter and place the Mini-Cell (with the lettering pointing out towards the source) as far back in the opening as possible (Fig. III-26). The Mini-Cell used in this manner with the 700 should have six outer ridges, three each from exactly opposite sides, removed from the cell body. The cell should then be pushed in the opening with these flats oriented at the top and bottom of the opening. On Infracords hold back the spring on the right-hand side of the cell slide with a small Allen wrench or bent paper clip or hair pin while pushing the Mini-Cell (lettering out) as far back as possible (Fig. III-27).

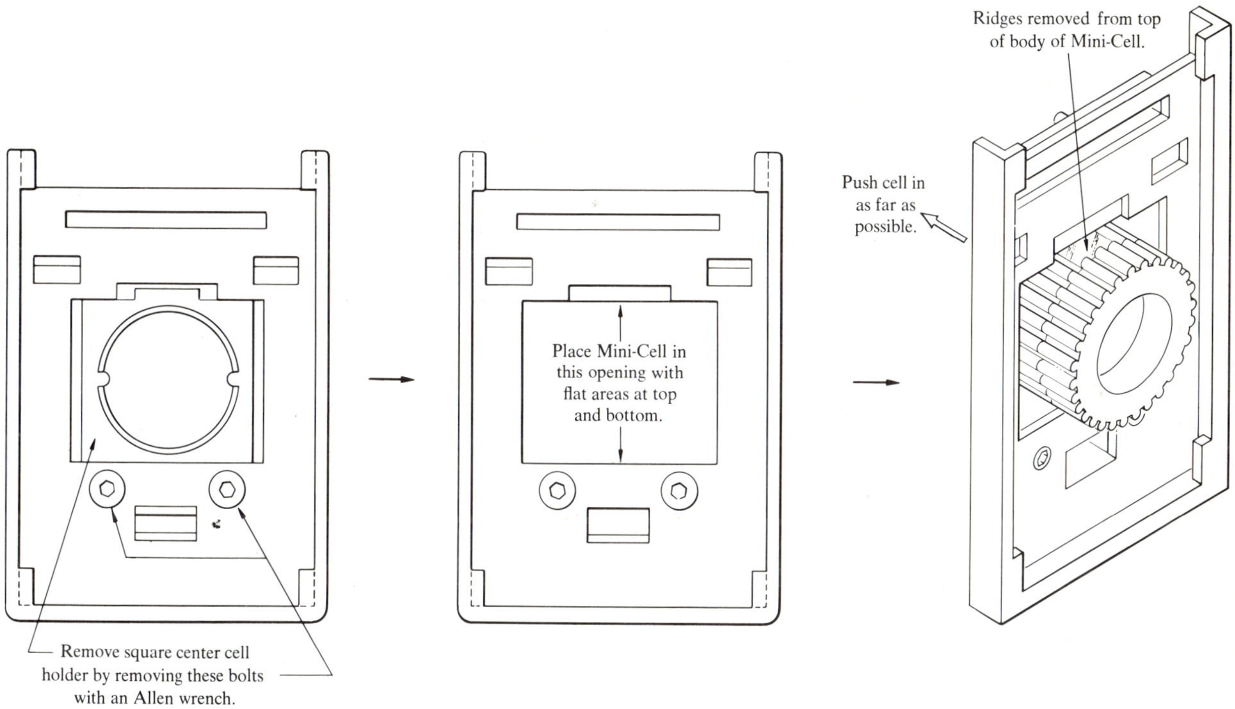

Figure III-26 Placement of a Mini-Cell in a Model 700 infrared spectrophotometer.

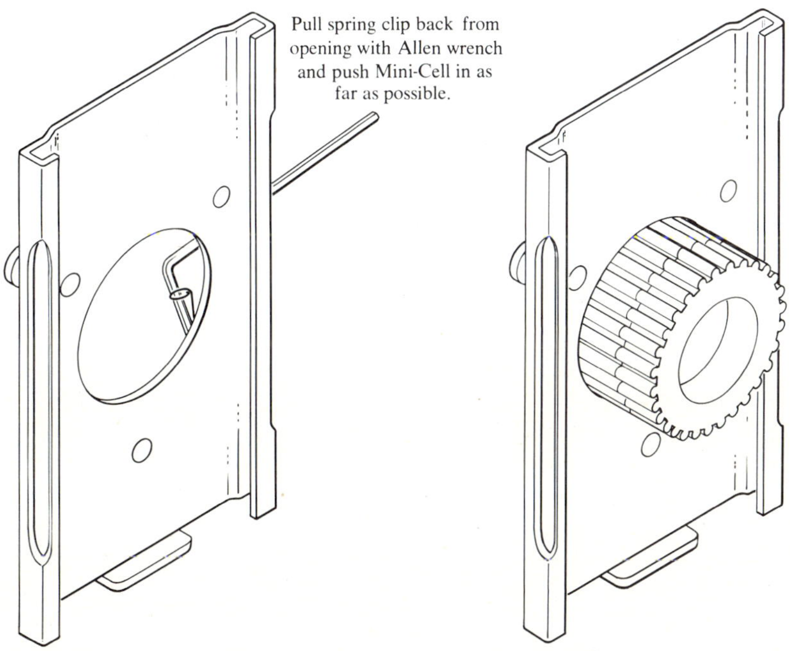

Figure III-27 Placement of a Mini-Cell in an Infracord.

EXERCISES

III-1. (a) Assign all the absorption bands in the spectrum of polyethylene (Fig. III-21) to structural features of this polymer. (b) Repeat the assignment for mineral oil (Fig. III-17). (c) What are the structural similarities of mineral oil and polyethylene? (d) What are their differences? (e) How can the similarities and differences of these two materials be determined from their infrared spectra?

III-2. Compare the relative advantages and disadvantages of silver chloride and sodium chloride cell windows.

III-3. Figure III-28 shows the effect of change of slit width on the intensities of the 905 and 860 cm^{-1} absorptions of cyclohexane. Note that the area of each absorption does not change with slit width. (a) The peak intensities of these (and all other) absorptions as measured by percent transmittance or absorbance decreases with increasing slit width; why? (b) The relative intensities of these two absorptions reverse as the slit width increases; why? (c) There are several "fingerprinting" methods for identifying organic compounds from large libraries of reference spectra without use of any frequency-structure correlations. Most of these methods use relative intensities of prominent absorption bands as searching parameters. What are the limitations of such an approach? (d) In general, for the identification of compounds by comparison of their infrared spectra with spectra of known compounds what rules should be followed for comparison of relative band intensities?

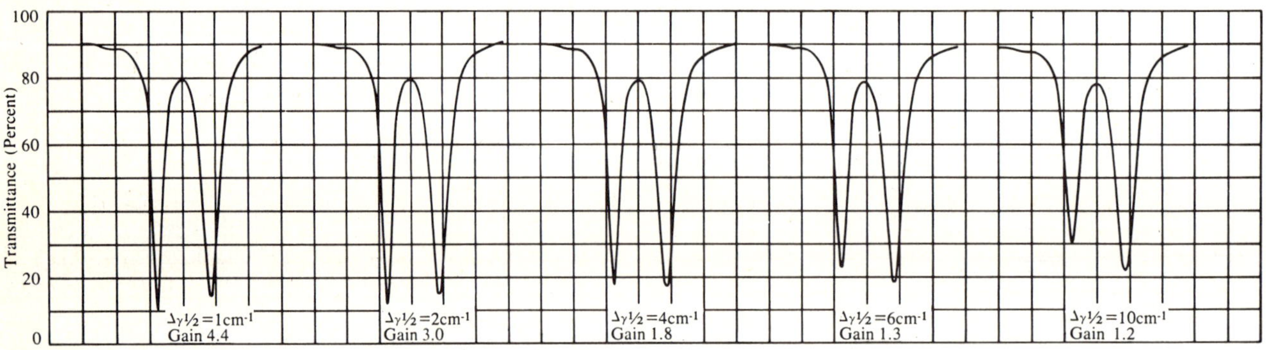

Figure III-28 Effect of slit width on intensities of the 905 and 860 cm^{-1} absorption bands of cyclohexane. Note how gain is decreased as the slits are opened to keep constant energy in the servo loop.

III-4. As indicated in Chapter 20, it is sometimes difficult to select the base line (I_0) for quantitative analysis if there are several overlapping bands. In such cases it usually makes little difference what method is used to choose I_0 so long as the same method is consistently used, i.e., an error in determining I_0 does not cause an error in quantitative analysis if the same error is made each time provided a graphical calibration is used and a sufficient number of points are taken. (a) Why? (b) A given error in I_0 in transmittance units, even if not consistent, causes a *smaller* error in quantitative determinations; why? (c) What is the effect of a similar error in I? For (b) and (c) use as an example I_0 true 90% T, erroneous value 92.5% T; I true 20% T erroneous value 22.5% T.

III-5. Fluorolube is similar in structure to Nujol differing only in a somewhat smaller average number of carbon atoms per molecule and in having all the hydrogen atoms replaced with fluorine atoms. The infrared spectra of these two materials are very different. Explain the spectral differences on the basis of structural differences.

Appendix IV

Special Topics in Gas Chromatography

CONDITIONING OF COLUMNS

Most columns for GLC are conditioned before use to assure that all of the solid support is covered with liquid phase, to make less likely redistribution of the liquid phase during the first chromatograms, and principally to remove relatively volatile material that might deposit in the detector, add to the background, and contaminate any fractions that are collected. These volatile materials are usually residual solvent, adsorbed water, and impurities in the liquid phase. For example, in *n*-octyl phthalate there might be small amounts of *n*-hexyl phthalate, 1-octanal, and unesterified 1-octanol and 1-hexanol. Some column-packing material such as waxes, diethylene glycol succinate, and the Igepals contain many molecular species of various volatilities. The most volatile of these materials will bleed from the column if it is operated at a temperature anywhere near the upper temperature limit (Table IV-1) of the liquid phase.

Problem IV-1. Give at least two reasons why conditioning of columns leads to more reproducible results in GLC.

In general, columns are best conditioned for 8 to 16 hrs with a carrier gas flow rate of approximately 10 ml/min 20° above the highest anticipated temperature at which the column will be used. Carrier gas should be passed through the column for 15 to 20 min before heat is applied to remove all the air. To avoid decomposition or excessive loss of liquid phase the conditioning temperature should be at least 30° below the upper temperature limit of the liquid phase. For noncapillary columns 2 ml of water should be injected sometime after the first 4 hrs of conditioning and before the final hour (or final 2 hr if the conditioning temperature is below 90°). During conditioning the column should be *disconnected* from the detector and the opening to the detector from the oven closed. The column should, of course, be connected to the injection port so that carrier gas will flow through the column. It is preferable that a small piece of tubing lead from the exit of the column to outside the oven.

SELECTION OF SOLID SUPPORTS

The preparation, treatment, and conditioning of solid supports is shrouded in secrecy because most of the processes are apparently unpatentable. The companies preparing and marketing these supports are, therefore, understandably reluctant to release many of the details concerning the preparation of the supports. Much of the information in Table IV-2 was obtained from promotional literature describing the supports and from discussions with analytical chemists who have been able to determine some of the proprietary details. All the information placed in Table IV-2 has been obtained from sources which are believed reliable, and most, if not all, the information is correct. Under the circumstances, however, it would not be too surprising if a few of the details were in error. If any errors do occur, they are probably from omission rather than misstatements.

Many equivalents are indicated among the brand names. For example, Gas-Chrom S, Chromosorb W, and Anakrom-U are similar. For most purposes these three could be used interchangeably or even mixed together and used in the same column. However, if exact reproducibility of a chromatogram is required, the specific support of the original chromatogram should be used.

One should not switch *types* of solid supports unless the property of the new type might give better results. For example, in all but the most exacting repeats of a chromatogram, the substitution of Anakrom U for Chromosorb W (or vice versa) would probably cause no significant difference, but Chromosorb G should not be substituted for either one unless it was probable that the lower pH, greater inertness, higher density, or other properties of the Chromosorb G would improve the analysis and not significantly reduce the separation because of its lower efficiency.

TEMPERATURE PROGRAMMING

Many gas chromatographs are equipped with devices that heat the column oven at definite, reproducible, controlled rates. Figure IV-1 illustrates a control panel of a temperature programmer. On some chromatographs these controls are not on a separate panel but are placed alongside the injection port temperature, detector, and other controls. The appearance and arrangement of the controls may differ in other ways, and some chromatographs may be more complicated offering greater flexibility and more automation. These more complicated types are not likely to be found in student laboratories. If the application and use of the controls shown in Figure IV-1 are understood, the operation of other temperature programmers will be evident. Regardless of the differences in controls most programmers for gas chromatography are basically the same in function and design even though there may be large variations in the engineering details.

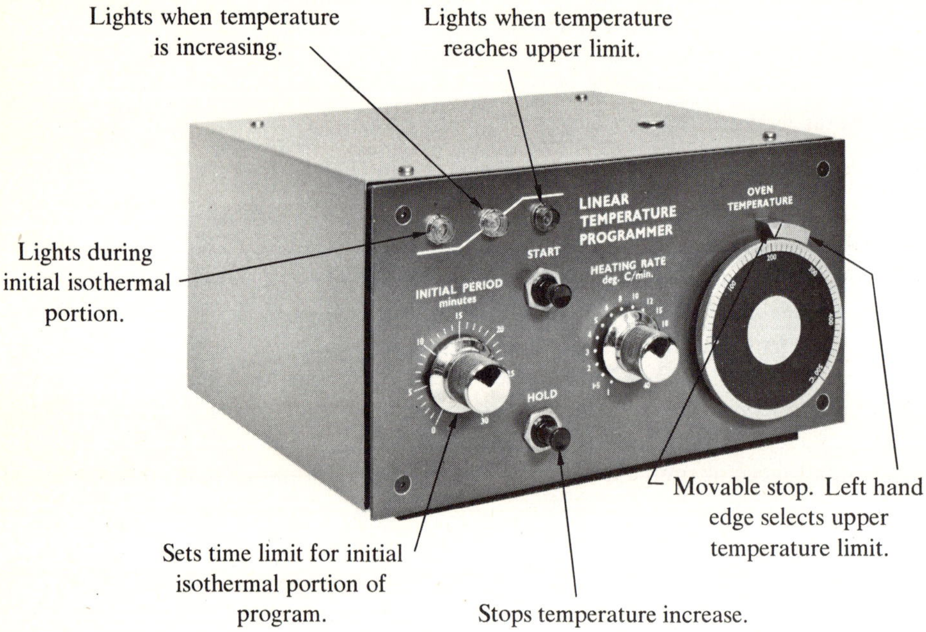

Figure IV-1 An F-11 temperature programmer. Perkin-Elmer, Norwalk, Conn.

The oven temperature control is set by turning the dial so that the pointer indicates the desired starting temperature. To prevent difficulties from backlash turn the dial approximately 40° *higher* than the starting temperature, then immediately return the dial to the starting temperature. Setting the oven temperature in this manner without pressing the start button operates the chromatograph isothermally at the selected temperature.

The left hand edge of the adjustable curved stop selects the higher temperature limit of the program. For example, in Figure IV-1 the starting temperature limit is set at approximately 208° and the upper temperature limit is set only 1 to 2° higher. The heating rate selects the degrees per minute the oven temperature is raised. The higher rates (20°/min and above) often do not give linear increases when large bulky columns or temperatures above 280° are used. This is not a serious restriction as temperatures above 280° are not common for the separation of organic compounds, and heating rates faster than 20°/min are not useful in very many applications.

The final control to adjust is the "initial period" which selects how long the chromatograph operates isothermally at the starting temperature before the temperature increases at the selected heating rate. Some temperature programmers do not have this feature. The same effect can be achieved by first injecting the sample, waiting for the time needed for isothermal operation, and then pressing the start button. For temperature programmers like the one illustrated here set the initial time dial for the needed time, then inject the sample and press the start button simultaneously.

Pressing the hold button stops the temperature program, and the chromatograph operates isothermally at the temperature indicated by the oven temperature control until the start button is pressed again, resuming the program where it had been stopped. Figure IV-2 illustrates the function of the hold button. Some chromatographs do not have a hold button; some have controls for completely automating this function. For those chromatographs without a hold button or its equivalent this function can be accomplished by setting the upper temperature limit for the first temperature at which the program is stopped. When the program stops, reset the upper temperature limit for the next stop in the program and press the start button when the isothermal portion has been completed.

Problem IV-2. (a) Indicate what control settings and start and hold button manipulations on the F-11 temperature programmer would produce the temperature programs illustrated in Figure IV-3. (b) If

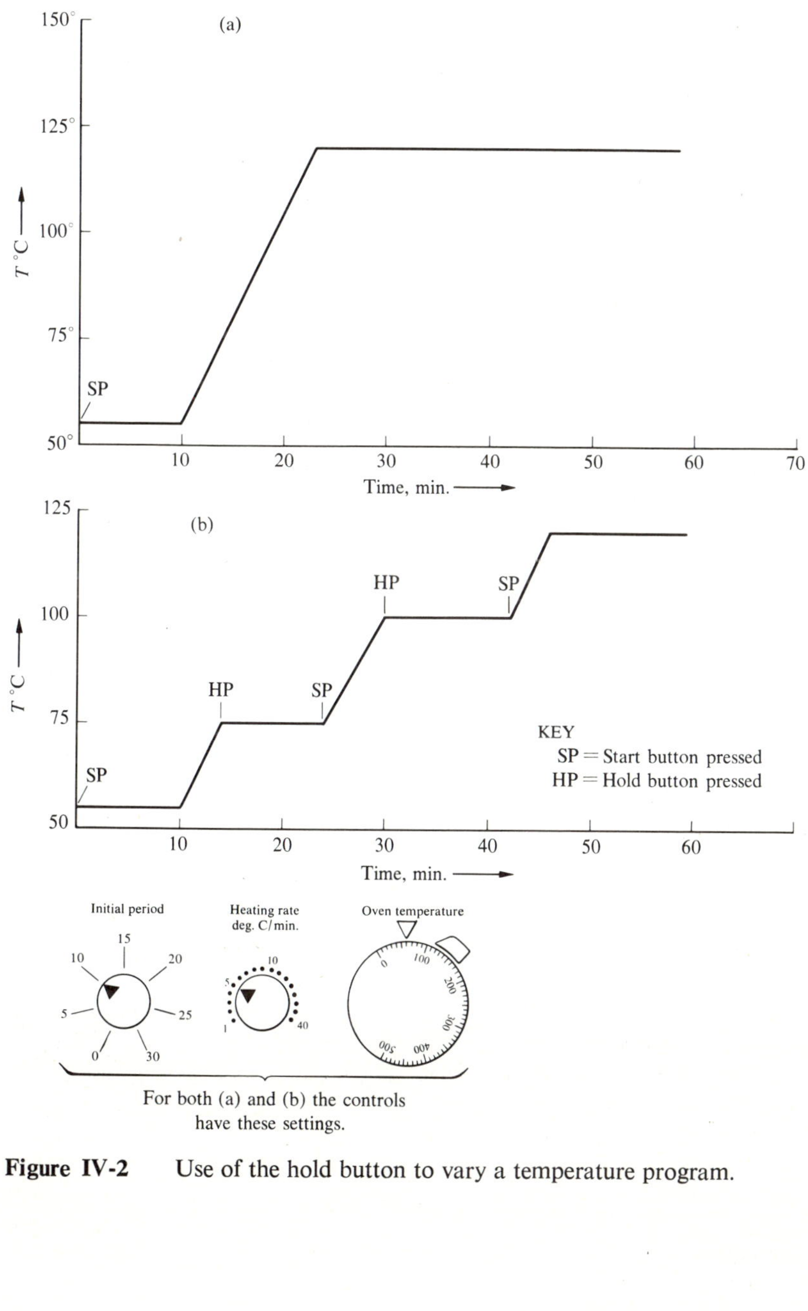

Figure IV-2 Use of the hold button to vary a temperature program.

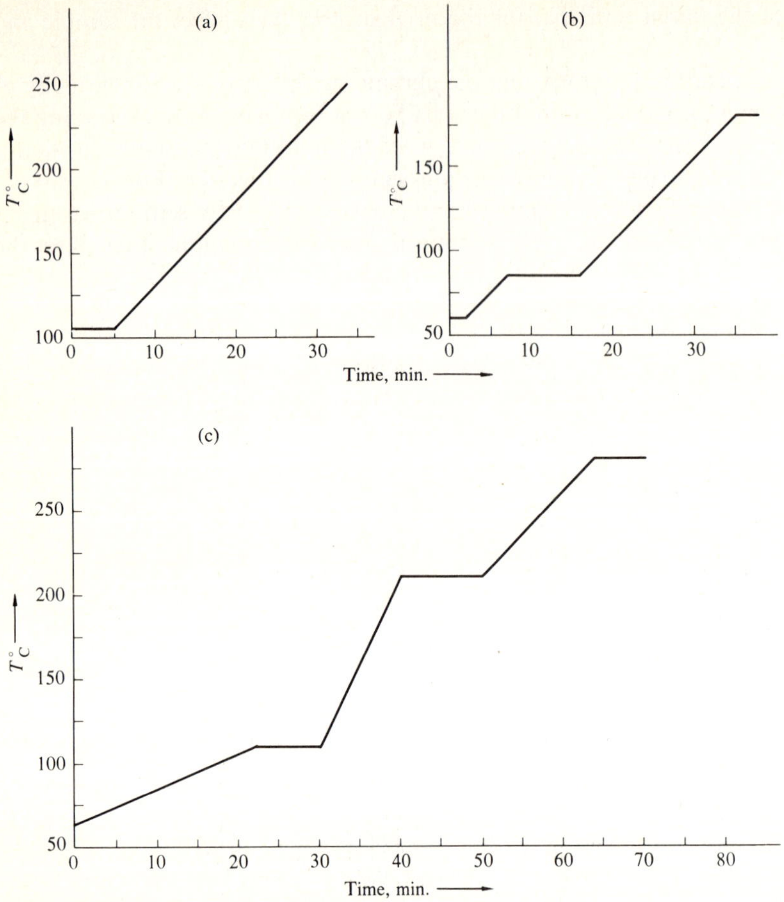

Figure IV-3 Temperature programs for Problem IV-2.

the temperature programmer in your laboratory has controls different from those of the F-11, indicate which settings would produce the programs illustrated in Figure IV-3.

When programmed chromatograms are obtained in rapid succession, it is very important that the starting temperature by reproduced exactly. The column packing and not just the oven air should be at the starting temperature. Cooling the oven approximately 20° below the starting temperature and then heating the oven and column at the starting temperature for 4 to 5 min before injecting the next sample usually ensures that the column packing is at the selected temperature. The manipulations that accomplish this on most gas chromatographs are: (a) Turn the oven temperature control back to 25° but do not turn the power off as the fan must be left on. (b) Open the oven door until the temperature drops to 20° below the starting temperature. If there is no oven pyrometer, leave the door open for 3 to 4 min if the difference between the starting and terminal temperature is less than 50°, 15 min if this temperature difference is more than 220°, and intermediate time intervals for intermediate temperature differences. (c) Close the oven door. (d) Set the oven temperature for the starting temperature taking the usual precautions against backlash on setting the dial. (e) Allow the oven air and column to equilibrate for 4 to 5 min before injecting the sample.

Problem IV-3. With a microscope examine Chromosorbs P, W-NAW, W-AW-DMCS, G, and A, Gas-Chrom S, Anakrom U, and any other solid support listed in Table IV-2 that can be obtained. Examine the solid supports as obtained from the bottle and after *slightly* crushing them with a spatula on a glass plate. Use 100× magnification. How do each of the supports differ? Explain the effect of the method of preparation on the appearance of the supports. Correlate the appearance of each support with its effect on GLC separation.

TABLE IV-1 LIQUID PHASE COMPOSITION, UPPER TEMPERATURE LIMIT, AND RECOMMENDED SOLVENTS FOR APPLICATION TO SOLID SUPPORTS

Liquid phase	Composition (if not obvious from name)	Upper temperature limit (^oC)	Solvent
Acetyl tributyl citrate		180	Acetone
Adiponitrile	$NC(CH_2)_4CN$	50	$CHCl_3$
Amine 220	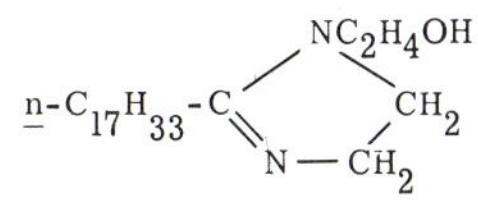	180	$CHCl_3$
Apiezon H and M	Mainly high mol. wt. aliphatic hydrocarbons and a low percentage of aromatic rings.	275	Benzene
Apiezon J, L, and N		300	Benzene
Armeen 2S	$[CH_3(CH_2)_n]_2NH$	125	Toluene
Arneel S. D.	$CH_3(CH_2)_nCN$	100	Methanol
Beeswax	Principally a mixture of esters of long chain aliphatic alcohols and acids as $CH_3(CH_2)_{14}CO_2(CH_2)_{13}CH_3$, long chain acids as $CH_3(CH_2)_{24}CO_2H$, and smaller amounts of long chain paraffins.	200	$CHCl_3$
Bentone 34	$[CH_3(CH_2)_{17}]_2N(CH_3)_2$ (clay)- clay-aluminum silicates with Fe and Mg	200	Toluene
7,8-Benzoquinoline		120	$CHCl_3$
Benzylamine adipate		125	Methanol
Benzyl cyanide		50	$CHCl_3$
Benzyl cyanide-silver nitrate (2:1)		50	Acetone
Benzyldiphenyl	$C_6H_5CH_2C_6H_4C_6H_5$	110	Acetone
Benzyl ether		80	$CHCl_3$
Benzyl cellosolve	2-(Benzyloxy) ethanol	80	Methanol
Bis(2-butoxyethyl) phthalate		150	$CHCl_3$
Bis(2-ethoxyethyl) sebacate		150	$CHCl_3$
Bis(2-ethylhexyl) phthalate		150	Methanol
Bis(2-ethylhexyl) sebacate		125	Acetone
Bis(2-ethylhexyl) tetrachlorophthalate		150	$CHCl_3$
Butanediol adipate	$HO[CH_2CH_2CH_2CH_2O_2C(CH_2)_{2\ or\ 4}CO_2]_nH$	225	$CHCl_3$
Butanediol succinate (Craig ester)		225	$CHCl_3$
Butanediol borate		200	Methanol

TABLE IV-1(cont.)

Liquid phase	Composition (if not obvious from name)	Upper temperature limit (^{o}C)	Solvent
Carbitol	$CH_3CH_2(OCH_2CH_2)_2OH$	55	Methanol
Carbowax 200		100	$CHCl_3$
Carbowax 400		125	$CHCl_3$
Carbowax 600	A series of polyethylene glycols,	125	$CHCl_3$
Carbowax 750	$HO(CH_2CH_2O)_nH$, The number after	125	$CHCl_3$
Carbowax 1000	the name indicates the average mol.	125	$CHCl_3$
Carbowax 1000 monostearate	wt. Carboxylate name after the	150	$CHCl_3$
Carbowax 1500	number indicates one (mono-) or	200	$CHCl_3$
Carbowax 1540	both (di-) of the hydroxy groups	200	$CHCl_3$
Carbowax 1500 monostearate	are esterified with that acid.	200	$CHCl_3$
Carbowax 4000		200	$CHCl_3$
Carbowax 4000 dioleate		220	$CHCl_3$
Carbowax 4000 monostearate		220	$CHCl_3$
Carbowax 6000		200	$CHCl_3$
Carbowax 20000 (20M)		250	$CHCl_3$
Chlorowax 70	Chlorinated paraffin	140	$CHCl_3$
Cyclohexanedimethanol adipate	$HO\left[CH_2C_6H_{10}CH_2O_2C(CH_2)_{2\text{ or }4}CO_2\right]_nH$	210	$CHCl_3$
Cyclohexanedimethanol succinate		210	$CHCl_3$
Di - see also bis			
Dibutyl phthalate		30	$CHCl_3$
Dibutyl tetrachlorophthalate		150	$CHCl_3$
Dicyclohexyl azelate	$C_6H_{11}O_2C(CH_2)_7CO_2C_6H_{11}$	125	Acetone
Diethylene glycol adipate (DEGA)		220	$CHCl_3$
Diethylene glycol sebacate (DEGSB)	$HO\left[CH_2CH_2OCH_2CH_2O_2C(CH_2)_{2\text{-}8}CO_2\right]_nH$	220	$CHCl_3$
Diethylene glycol succinate (DEGS)		220	$CHCl_3$
Diethylene glycol monethyl ether	$CH_3CH_2(OCH_2CH_2)_2OH$	45	Methanol
Diethyl d-tartrate		75	Methanol
Didecyl phthalate		170	Acetone
Diglycerol	$(HOCH_2CHOHCH_2)_2O$	120	Methanol
Diisodecyl adipate		150	Methanol
Diisodecyl phthalate		176	Methanol
Di-n-octyl adipate		125	Acetone
Dimer acid-silicon oil 550 (1:1)	$HO_2C(CH_2)_{34}CO_2H$ and see silicon oil, DC 550	150	Acetone
Dinonyl phthalate		150	$CHCl_3$
Dioctyl phthalate (DOP)		145	$CHCl_3$

TABLE IV-1(cont.)

Liquid phase	Composition (if not obvious from name)	Upper temperature limit (^oC)	Solvent
Ethylene glycol adipate (EGA)	$HO(CH_2CH_2O_2C\text{-}Y\text{-}CO_2)_nH$	220	Acetone
Ethylene glycol phthalate (EGP)		250	$CHCl_3$
Ethylene glycol isophthalate (EGIP)		250	$CHCl_3$
Ethylene glycol sebacate (EGSB)		220	$CHCl_3$
Ethylene glycol succinate (EGS)		220	$CHCl_3$
Fluorolube HG 1200	$(CF_2CFCl)_n$	100	Acetone
Fluorolube S 30		100	Acetone
Fluoro-ester	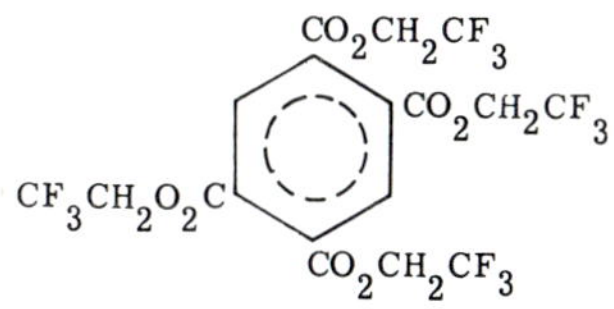	250	Acetone
Fluorosilicone QF-1-0065	See QF-1	250	Acetone
Hallcomid M18	Dimethylstearamide	150	Methanol
Hallcomid M18OL	Dimethyloleylamide	150	Methanol
n-Hexadecane		50	Benzene
1,2,3,4,5,6-Hexakis(2-cyanoethoxy)cyclohexane		200	$CHCl_3$
Hypose SP 80	Octakis(2-hydroxypropyl)sucrose	210	Methanol
Igepal CA630	$C_8H_{17}C_6H_4O(CH_2CH_2O)_nH$	200	Methanol
Igepal CO880		200	Methanol
Igepal C0990	$C_9H_{19}C_6H_4O(CH_2CH_2O)_nH$	200	Methanol
β,β-Iminodipropionitrile	$HN(CH_2CH_2CN)_2$	90	Methanol
Kel-F oil #3	$(CF_2CFCl)_n$	50	Acetone
Kel-F oil #10		100	Acetone
Kel-F wax		200	Acetone
LAC-2-R-446	Crosslinked DEGA	190	$CHCl_3$
Lanolin	Mainly esters of cholesterol and isocholesterol with n-alkanoic acids such as stearic acid.	200	$CHCl_3$
Marlex	Polyethylene	240	Xylene
Methylchlorophenyl silicone fluid	See silicons		Acetone
Microcrystallinewax bottoms	Mainly high mol. wt. alkanes	250	$CHCl_3$
Neopentyl glycol adipate (NPGA)	$H[OCH_2C(CH_3)_2CH_2O_2C(CH_2)_{2,4,\text{ or }10}CO]_nOH$	240	$CHCl_3$
Neopentyl glycol sebacate (NPGSb)		240	$CHCl_3$
Neopentyl glycol succinate (NPGS)		240	$CHCl_3$
Nitrobenzene		60	Methanol

TABLE IV-1 (cont.)

Liquid phase	Composition (if not obvious from name)	Upper temperature limit (°C)	Solvent
Nujol	Alkanes	170	Toluene
Oronite polybutene 32	$(C(CH_3)_2CH_2)_n$	210	Toluene
Oronite polybutene 128		210	Toluene
OV-1, OV-17, OD-1	See silicones		
β,β-Oxydipropionitrile	$O(CH_2CH_2CN)_2$	100	Methanol
Paraffin wax	Alkanes	200	$CHCl_3$
Paraffin oil	Alkanes	140	Toluene
Phthalonitrile		150	Acetone
Polyethylene glycol 600	See Carbowaxes		
Polyethylene		240	Xylene
Polyphenyl ether (OS-124)(5P4E)	$C_6H_5(OC_6H_4)_3OC_6H_5$	250	Acetone
Polyphenyl ether (6P5E)	$C_6H_5(OC_6H_4)_5OC_6H_5$	350	Acetone
Polypropylene glycol	$HO(CH_2CH(CH_3)O)_nH$	150	Methanol
Polyvinyl pyrrolidinone(PVP)	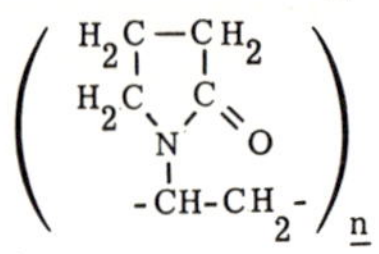	225	Methanol
Propylene glycol silver nitrate(3:1)		75	Methanol
QF-1	$(CF_3CH_2CH_2)_3Si[OSi(CH_2CH_2CF_3)_2]_nCH_2CH_2CF_3$	250	Acetone
Quadrol	N,N,N'-tetrakis(2-hydorxypropyl)ethylenediamine	150	Acetone
Reoplex 400	Polypropylene glycol adipate	190	$CHCl_3$
Renex 678 surfactant	See Igepal	200	Methanol
Silastic LS(420)	$[CF_3CH_2CH_2Si(CH\text{-}CH_2)O]_n$ [1]	275	Acetone
SILICONES[2]	$\begin{matrix} R \\ R'SiO \\ R'' \end{matrix}\left(\begin{matrix} R' \\ SiO- \\ R'' \end{matrix}\right)_n\begin{matrix} R' \\ SiR \\ R'' \end{matrix}$ [1]	high	
Methyl	$(CH_3)_3SiO[Si(CH_3)_2O]_nSi(CH_3)_3$		
DC[3] 200		250	Toluene
DC 410 (Gum)		290	Toluene
DC Vac grease		350	EtOAc
GE[4] SF 96		270	Toluene
GE SE 30 (Gum rubber)		380	Toluene
OD-1		375	$CHCl_3$
OV-1		350	$CHCl_3$
UC L45		300	Toluene

1 Many of silicones have a more random arrangement of the various siloxy groups and organic radicals than some of these general formulas indicate.

2 Fluids unless stated otherwise

3 Dow-Corning

4 General Electric

TABLE IV-1 (cont.)

Liquid phase	Composition (if not obvious from name)	Upper temperature limit (°C)	Solvent
Methyl vinyl	$(CH_3)_2\underset{CH_2=CH}{SiO}[\underset{CH=CH_2}{Si}(CH_3)O]_nSi(CH-CH_2)(CH_3)_2$ [1]		
GE SE 31 (Gum rubber)		300	Toluene
GE SE 33 (Gum rubber)		300	Toluene
UC[5] W 98 (Gum rubber)		310	Toluene
UC W 96 (Gum rubber)		300	Toluene
Methyl phenyl	$(CH_3)_x\phi_ySi-O[Si(CH_3)\phi O]_nSi(CH_3)_x\phi_y$ [1]		
DC 550		275	Acetone
DC 555		260	$CHCl_3$
DC 703		225	$CHCl_3$
DC 710		300	Acetone
GE SE-52 (Gum rubber)		300	Toluene
DC 720	$CH_3\left(\overset{CH_3}{\underset{\phi}{SiO}}\right)_2SiCH_3\phi_2$ [1]	260	Toluene
DC 730	Same as 720 but one more phenyl and one less methyl	270	Toluene
OV 17		350	$CHCl_3$
Methyl chlorophenyl	$\left(-\overset{CH_3}{\underset{C_6H_4Cl}{Si}}O-\right)_n$		
GE F50		275	Toluene
F60 (DC 560)		250	Acetone
Methyl cyanoethyl			
GE XE 60 (Gum rubber)	$\left(-\overset{CH_3}{\underset{CH_2CH_2CN}{Si}}-O-\right)_m-\left(-\overset{CH_3}{\underset{CH_3}{Si}}-O-\right)_n$ [1]	270	Acetone
GE XF 1105	As above, cyanoethyl 5%	200	Acetone
GE XF 1112	As above, cyanoethyl 12%	200	Acetone
GE XF 1125	As above, cyanoethyl 25%	200	Acetone
GE XF 1150	As above, cyanoethyl 50%	200	Acetone
Miscellaneous			
GE SF 54 (Gum rubber)	$\left(\overset{CH_3}{\underset{\phi}{Si}}-O\right)_n\left(\overset{CH_3}{\underset{CH=CH_2}{Si}}-O-\right)_m$ [1]	300	Toluene
FS-1265 or DC QF-1	See QF-1	250	Acetone
DC LSX-3-0295 (Gum rubber)	$\left(-\overset{CH_3}{\underset{CH=CH_2}{Si}}-O-\right)_n-\left(\overset{CH_3}{\underset{CH_2CH_2CF_3}{Si}}-O-\right)_m$	275	Acetone
Sorbitol	$H(CHOH)_6H$	150	Methanol

TABLE IV-1 (cont.)

Liquid phase	Composition (if not obvious from name)	Upper temperature limit (^{o}C)	Solvent
Span 60	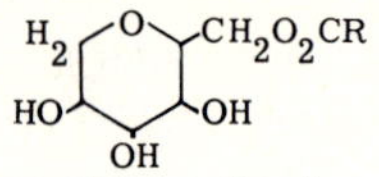 RCO_2-stearate for 60, -oleate for 80; or other cyclic ether of sorbitol	150	Methanol
Span 80	(same as Span 60)	150	Methanol
Squalane	$H(CH_2(CH_3)CHCH_2CH_2)_3(CH_2CH_2CH(CH_3)CH_2)_3H$	140	Toluene
Squalene	$H(CH_2(CH_3)C{=}CHCH_2)_3(CH_2CH{=}C(CH_3)CH_2)_3H$	120	Toluene
Sucrose acetate isobutyrate (SAIB)		225	Toluene
Surfonic N 300	Alkylaryl polyether alcohol similar to Igepal.	200	Toluene
Tergitol NPX	Same as Igepal CO 880 or 990	200	Methanol
Tergitol (NP-35)	Same as Igepal CO 880 or 990	200	$CHCl_3$
Tetracyanoethyl pentaerythritol ether	$C(CH_2OCH_2CH_2CN)_4$	140	$CHCl_3$
Tetraethylene glycol (TEG)	$HO(C_2H_4O)_4H$	90	Toluene
Tetraethylene glycol dimethyl ether (Ansul ether)		75	Methanol
Tetraethylene pentamine	$H_2N(C_2H_4NH)_4H$	100	Methanol
Tetrahydrofuryl phosphate	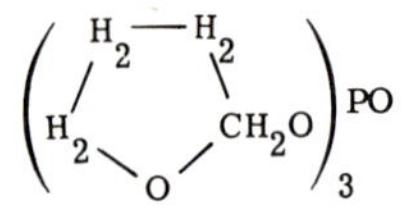	125	Methanol
Tetrahydroxyethylethylenediamine (THEED)	$(HOC_2H_4)_2NCH_2CH_2N(C_2H_4OH)_2$	135	Acetone
1,2,3,4-Tetrakis(2-cyanoethoxy)butane		200	$CHCl_3$
β,β'-Thiodipropionitrile	$S(CH_2CH_2CN)_2$	100	$CHCl_3$
Triacetin	Glycerol triacetate	85	$CHCl_3$
Tributyl citrate		150	Methanol
Tributyrin	Glycerol tributyrate	100	Methanol
Tricresyl phosphate (TCP)	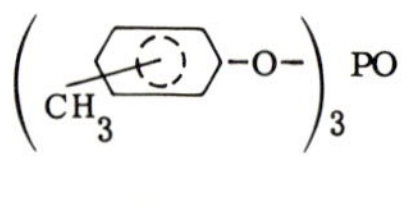	125	Acetone
Triethanolamine	$(HOCH_2CH_2)_3N$	75	$CHCl_3$
Triethylene glycol	$HO(C_2H_4O)_3H$	70	Methanol
Trimer acid	$HO_2C(CH_3)_{16}HC[(CH_2)_{17}CO_2H]_2$	200	Methanol
Triphenylmethane		150	$CHCl_3$
1,2,3,-Tris(2-cyanoethoxy)propane		180	Methanol

TABLE IV-1(cont.)

Liquid Phase	Composition (if not obvious from name)	Upper temperature limit (^oC)	Solvent
Trixylol phosphate	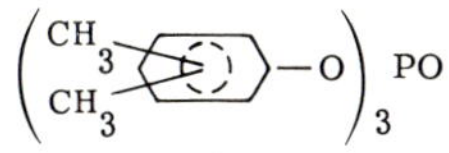	190	Methanol
Triton X 100	Alkylarylpolyether alcohols,	190	Methanol
Triton X 305	similar to Igepals	250	Methanol
Tween 80 (polysorbate)	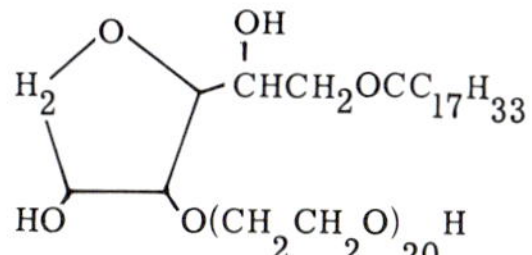 and other cyclic ethers of sorbitol	160	Methanol
Ucon[5] (water soluble) (polar) (HB or H)		230	Methanol
Ucon (water insoluble) (nonpolar) (LB)		230	Methanol
Ucon 50 HB 270X		215	Methanol
Ucon 50 HB 660		215	Methanol
Ucon 50 HB 2000		180	Methanol
Ucon 1800 XMP 1018	Polyalkylene glycols	170	Methanol
Ucon 50 HB 5100	and diesters similar	200	Methanol
Ucon 50 LB 300X	to the carbowaxes	215	Methanol
Ucon 50 LB 550X		215	Methanol
Ucon 50 LB 1200X		200	Methanol
Ucon 75-H-90,000		230	Methanol
Versamid 900 (polyamide resin)	$H(HNYNH\overset{O}{\overset{\|}{C}}(CH_2)_x\overset{O}{\overset{\|}{C}})_nOH$	385	$CHCl_3$ and
			$\underline{n}$-C_4H_9OH
Versamid 940 (polyamide resin)		210	1:1 and hot
Zinc stearate		150	Methanol

5 Union Carbide

TABLE IV-2 PROPERTIES AND COMPOSITIONS OF SOLID SUPPORTS

Support	Inertness[1]	Max. cap. of liq. phase before marked efficiency loss %	Approx. pH	Free fall density	Efficiency[2]	Composition[3] and comments
Aeropak[4]	3.8	12	9.2	0.18	high	Similar to Chromosorb W-DMCS except that "fines" have been removed
Anakrom[5]A	3.5	20	8.5	0.18	high	Acid washed Anakrom U
Anakrom AB	3.8	20	9	0.18	high	Acid and alcoholic base washed Anakrom U
Anakrom ABS	4	12	9.4	0.18	high	Silanized Anakrom AB
Anakrom AS	4	12	9.0	0.18	high	Silanized Anakrom A
Anakrom C22	1	30	6.5	0.38	very high	Similar to Chromosorb P
Anakrom P	1	30	6.5	0.38	very high	Similar to Chromosorb P
Anakrom PA	1.5	30	5.8	0.38	very high	Acid washed Anakrom P
Anakrom Q	4.5	12	9	0.18	high	Acid washed, silanized Anakrom U but process carefully controlled, similar to Gas-Chrom Q.
Anakrom SD	3.8	12	8.5	0.18	high	Acid washed and possibly chemically deactivated Anakrom U
Anakrom U	3	20	9	0.18	high	Similar to chromosorb W
Anakrom 545	3	20	9	0.18	high	See Celite 545
Anaprep[5]A	3.5	20	8.5	0.18	high	Bulk packed, preparative grade support similar to Anakroms of the same code letter. Less expensive (Anaprep A, ABS, P, U)
Anaprep ABS	4	12	9.4	0.18	high	
Anaprep P	1	30	6.5	0.38	very high	
Anaprep U	3	20	9	0.18	high	
Celite[6]545	3	20	9	0.18	high	A filter aid made by flux calcining salt water diatomite to 1600°F with 2 to 3% $NaCO_3$. Similar but not identical to Chromosorb W
Chromosorb[6]A	3	25	7.1	0.40	high	Flux calcined salt water diatomite made for preparative scale GLC
Chromosorb G						Somewhat similar to Chromosorb W except much harder and denser

NAW - untreated	4	5	9.5	0.47	medium high	Diatomite aggregate made especially for GLC. Less adsorptive, more inert than Chromosorb P or W. Very hard and, therefore, less subject to mechanical breakdown than Chromosorb W or P. Carefully controlled proprietary process may include chemical treatment. Fired to as high a temperature as Chromosorb P. Sodium carbonate added before firing but probably less used than in preparation of Chromosorb W. Although maximum liquid phase capacity is 5%, its higher density allows the same amount of liquid phase per given length of column as 12% on Chromosorb W and 6.5% on Chromosorb P.
AW - acid-washed	4.5	5	8	0.47	medium high	
AW-DMCS - acid-washed silanized	5.5	3	8.5	0.47	medium high	
Chromosorb P						
NAW - untreated	1	30	6.5	0.38	very high	Celite which has been crushed, blended, and pressed into bricks and fired at 2200°F or higher. The bricks are crushed and the pieces sieved to give small ranges of particle size. More adsorptive than Chromosorbs A, G, and W. Relatively hard.
AW - acid-washed	1.5	30	6	0.38	very high	
DMCS - silanized	2	18	7	0.38	very high	
AW-DMCS - acid-washed and silanized	2.5	18	6.5	0.88	very high	
Chromosorb T see polytetrafluoroethylene						

1. As measured by adsorption, peak tailing, and other effects on peak symmetry. Numbers are relative, quantitative, and arbitrary. Highest number is most inert and most desirable for GLC; however, 4 is not necessarily four times as inert (or 1/4 as adsorptive) as 1.

2. As measured by peak sharpness, separation of adjacent peaks, and height per equivalent theoretical plate (HETP). Assumes no adsorption or other nonlinear interactions. For separation of all but completely nonpolar compounds both "inertness" and "efficiency" (and "pH" for acidic and basic compounds) columns should be used to predict the effect of the solid support on efficiency, e.g., very polar compounds such as H_2O, HCN, and HCl are most efficiently separated with a Teflon or Kel-F support even though these supports have a low "efficiency" rating in the Table. Usually, the liquid phase has a much greater effect on efficiency than solid support does.

3. Diatomite unless stated otherwise.

4. Trademark of Varian-Aerograph, Inc.

5. Trademark of Analabs, Inc.

6. Trademark of Johns-Manville Company.

TABLE IV-2(cont.)

Support	Inertness[1]	Max. cap. of liq. phase before marked efficiency loss %	Approx. pH	Free fall density	Efficiency[2]	Composition[3] and comments
Chromosorb W						Diatomite calcined at 1600° with 2 to 3% Na_2CO_3. Soft, subject to mechanical breakdown.
NAW - untreated	3	20	9	0.18	high	
AW - acid-washed	3.5	20	8.5	0.18	high	
DMCS - silanized	3.8	12	9.2	0.18	high	
AW-DMCS - acid-washed and silanized	4	12	9	0.18	high	
Diatoport[7] P	1	30	6.5	0.38	very high	Similar to Chromosorb P-NAW
Diatoport W	3	20	9	0.18	high	Similar to Chromosorb W-NAW
Diatoport S	3.8	12	9.2	0.18	high	Similar to Chromosorb W-DMCS
Firebrick see Chromosorb						
Fluoropak see polytetrafluoroethylene						
Gas-Chrom[8] A	3.5	12	8.5	0.18	high	Acid-washed Gas Chrom S
Gas-Chrom CL	3	20	9	0.18	high	Similar to Celite 545
Gas-Chrom CLA	3.5	20	8.5	0.18	high	Acid-washed Gas-Chrom CL
Gas-Chrom CLH	3.8	12	9.2	0.18	high	Silanized Gas-Chrom CL
Gas-Chrom CLP	3.6	20	9	0.18	high	Acid and base-washed Gas-Chrom CL
Gas- Chrom CLZ	4	12	9	0.18	high	Acid-washed silanized Gas-Chrom CL
Gas-Chrom P	3.6	20	9	0.18	high	Acid and base-washed Gas-Chrom S
Gas-Chrom Q	4.5	12	9	0.18	high	Similar to Gas-Chrom Z, but processing more carefully controlled. Similar to Anakrom Q.
Gas-Chrom R	1	30	6.5	0.38	very high	Very similar to Chromosorb P
Gas-Chrom RA	1.5	30	6	0.38	very high	Acid-washed Gas-Chrom R
Gas-Chrom RP	2.0	30	7	0.38	very high	Acid and base-washed Gas-Chrom R
Gas-Chrom RZ	2.5	18	6.5	0.38	very high	Acid-washed and silanized Gas-Chrom R
Gas-Chrom S	3	20	9	0.18	high	Similar to Chromosorb W except made from fresh water diatomite, but the original source of the diatomite apparently causes little, if any, difference in GLC performance of the support.

Gas-Chrom Z	4	12	9	0.18	high	Silanized Gas-Chrom A; silanized acid-washed Gas-Chrom S.
Glass beads[6] also called Glassport[7] M	6	0.2	7.5	1.7	very low	Greater density compensates to a limited extent for low capacity for Liquid phase.
Glass beads silanized	7	0.1	8	1.7	very low	
Glass, porous,	6	0.7	7.5	1.7	low	From Corning type 7930
Haloport[7] see Kel-F and polytetrafluoroethylene						
Kel-F[11] also called Haloport-K	6.5	10	7	0.5	very low	$(CF_2CFCl)_n$ Cannot be used as a solid support above 150° (includes conditioning temperature.) Harder than Teflon; therefore, packing is easier and less critical.
Polytetrafluoroethylene						
Fluoropak[10] 80 and Haloport F	7	4	7	0.42	very low	$(CF_2CF_2)_n$ Cannot be used as a solid support above 250°, nor should column conditioning temperature exceed 250°. Efficiency is apparently greatly increased by careful packing. This includes not using heat when removing the solvent and keeping the packing out of contact with glass by using plastic funnels and dishes. Column and packing should be chilled below 18° when packing. Do not crush, compress, or vibrate vigorously.
Teflon[9] 6 also called Chromosorb T and Fluoroport T	7	10	7	0.42	low	

7. Trademark of Hewlett Packard, F&M Scientific Division.
8. Trademark of Applied Science Laboratories.
9. Trademark of Dupont.
10. Trademark of Fluorocarbon Company
11. Trademark of 3M company

Appendix V

The Dependence of Physical Properties on Polarity

In general, the more polar a compound the higher its melting point and boiling point, the greater its solubility in polar solvents, and the smaller its solubility in nonpolar solvents (Chapters 2, 6, and 12). Boiling points and solubilities of liquids and gases are more dependent on polarity than are other physical properties. This can be explained by the low or nonexistent organizational requirements of liquids and gases and the high organizational requirements of solids. Where other factors such as hydrogen bonding and molecular weight are relatively constant, physical properties are usually directly dependent upon the magnitude of the dipole moment. The data in Table V-1 for three series of compounds that have similar molecular weights show that the boiling points increase as the dipole moments increase. It can be seen that carbonyl compounds have moderately high dipole moments, lower than the dipole moments of the higher boiling cyano and nitro compounds and higher than the dipole moments of the lower boiling ethers and hydrocarbons.

TABLE V-1 CORRELATION OF BOILING POINTS WITH DIPOLE MOMENTS

	Mol. wt.	Dipole Moment	B.p. °C
C_6H_5CN	103	4.1	191
C_6H_5CHO	106	2.8	179
$C_6H_5OCH_3$	108	1.35	154
o-$C_6H_4(CH_3)_2$	106	0.5	143
m-$C_6H_4(CH_3)_2$	106	0.4	139.1
p-$C_6H_4(CH_3)_2$	106	0.0	137.5
p-$CH_3C_6H_4CN$	117	4.43	217.5
m-$CH_3C_6H_4CN$	117	4.3	209.5
$C_6H_5NO_2$	123	4.0	209
p-$CH_3C_6H_4CHO$	120	3.5	204
$C_6H_5COCH_3$	120	3.0	200
$C_6H_5N(CH_3)_2$	121	1.61	193
C_6H_5OEt	122	1.35	177
C_6H_5SEt	124	1.27	160.5
$C_6H_5CH(CH_3)_2$	120	0.65	152.6
p-EtC_6H_4CN	131	4.53	
p-$CH_3C_6H_4NO_2$	137	4.4	238.5
m-$CH_3C_6H_4NO_2$	137	4.17	232.6
o-$CH_3C_6H_4NO_2$	137	3.66	221.7
$C_6H_5CO_2CH_3$	136	1.9	198
$C_6H_5O_2CCH_3$	136	1.5	195.6
$C_6H_5OCH_2CH_3$	136	1.35	189
o-$CH_3C_6H_4N(CH_3)_2$	135	0.96	185.5
$C_6H_5C(CH_3)_3$	134	0.8	169.1

Aldehydes and ketones boil above esters of similar molecular weight and structure (Table V-2). Aldehydes and primary amines of similar molecular weight and structure have almost identical boiling points. Ketones boil somewhat above analogous primary and secondary amines. Primary and secondary amines have lower dipole moments than the carbonyl compounds but are capable of undergoing weak hydrogen bonding. This indicates that the intermolecular association caused by these weak hydrogen bonds is approximately the same as that caused by the strong dipole-dipole attraction of the carbonyl compounds.

For carbonyl compounds whose molecules can not undergo hydrogen bonding with each other the general order of dipole moments is N,N-disubstituted amides > ketones > aldehydes > esters > formate esters.

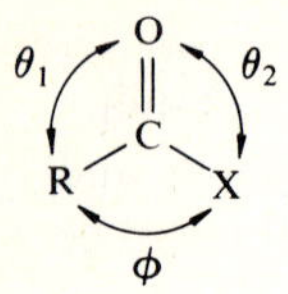

Actual bond angles in degrees for various carbonyl compounds

	H_2CO	R_2CO*	$RC(=O)Cl$*	$Cl_2C=O$
θ_1	121	122.5	127	125
θ_2	121	122.5	123	125
ϕ	118	115	110	111

*±1° varation in noncyclic compounds. Exact figure depends on nature of R group.

Figure V-1 Geometry of carbonyl compounds.

TABLE V-2 COMPARISON OF PHYSICAL PROPERTIES OF CARBONYL COMPOUNDS AND AMINES

Compound	Mol. wt.	Dipole moment D	B.p. (°C)	M.p. (°C)	Sol. in g./100 ml. of H_2O at 20°C	d_4^{20}
Propionaldehyde	58	2.75		-81	20	0.807
Methyl formate	60	1.7	33	-99	30	0.974
1-Aminopropane	59	1.3	49	-83	∞	
Acetone	58	2.8	56	-95	∞	0.791
2-Aminopropane	59		33	-101	∞	
Butyraldehyde	72	2.7	75.7	-99	4	0.817
Ethyl formate	74	1.7	54	-80	12	0.923
1-Aminobutane	73	1.32	78	-50	∞	0.74
Methyl ethyl ketone	72	2.7	79.8	-87	37	0.805
Methyl acetate	74	1.75	57.3	-99	32	0.933
2-Aminobutane	73	1.28	63	-104	∞	0.724
Valeraldehyde	86	2.6	103.4	-39		0.81
Propyl formate	88	1.7	80.9	-93	2.8	0.901
1-Aminopentane	87	1.5	103.6	-55	~40	0.765
2-Pentanone	86	2.7	102.4	-78	6.3	
Ethyl acetate	88	1.8	77.2	-87	7.9	0.910
2-Aminopentane	87		91.5		∞	0.749
3-Pentanone	86	2.7	102	-40	5.40	0.810
Methyl propionate	88	1.86	80		7.2	0.915
3-Aminopentane	87		90.2		∞	0.749
β-Phenylpropionaldehyde	134	2.8	224			
Benzyl formate	136		203.5	3.6		1.07
3-Phenyl-1-aminopropane	135		222			
Propiophenone	134	2.9	219	21		1.014
Methyl benzoate	136	1.9	199.5	-13	0.15	1.09
1-Phenyl-1-aminopropane	135		208			
Benzyl methyl ketone	134	2.7	216			1.003
Phenyl acetate	136	1.52	196		0.318	1.075
1-Phenyl-2-aminopropane	135		196.5			

This is also the general order of boiling points for these compounds if the comparison is made among compounds of similar structure. Note that if an aldehyde has a dipole moment greater than a ketone of identical molecular weight, it may have a higher boiling point (see *p*-isopropylbenzaldehyde and isobutyrophenone, Table V-3).

TABLE V-3 VARIATION OF BOILING POINT WITH DIPOLE MOMENT IN CARBONYL COMPOUNDS

Compound	Molecular wt.	Dipole moment, D	B.p. (°C)
Ethyl formate	74	1.7	54
Methyl acetate	74	1.75	57.3
Isobutyraldehyde	72	2.5	64
Methyl ethyl ketone	72	2.7	80
Nitroethane	75	3.2	115
N, N-Dimethylformamide	73	3.7	153
Isopropyl formate	88	1.7	68
Ethyl acetate	88	1.8	77
3-Methylbutanal	86	2.5	92
Methyl isopropyl ketone	86	2.7	94.5
2-Nitropropane	83	3.3	120.3
N, N-Dimethylacetamide	87	3.8	166
Ethyl benzoate	150	1.9	213
Isobutyrophenone (Isopropyl phenyl ketone)	148	3.0	222
p-Isopropylbenzaldehyde	148	3.5	236
N, N-Dimethylbenzamide	149	4.0	272

The difference in dipole moments of the various carbonyl compounds are due to inductive and resonance interactions of the different groups with the carbonyl groups. Electron-donating groups cause an increase in the dipole moment. Electron-withdrawing groups cause a decrease in the dipole moment (Tables V-4 and 5).

The carbonyl carbon atom is sp^2 hydridized, and the groups attached to it are coplanar with it and the oxygen atom and approximately 120° apart (Fig. V-I). From this fact and the electronic effects summarized in Table V-4, the predicted magnitudes of the dipole moments of various types of carbonyl compounds are those shown in Table V-5.

TABLE V-4 EFFECTS OF VARIOUS GROUPS ON THE DIPOLE MOMENTS OR CARBONYL COMPOUNDS

Structure	Electronic Effects+ Inductive	Resonance	Type of resonance contribution of group	Net effect	Effect on dipole
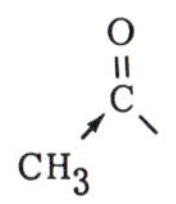	*Weak electron-donating	*Weak electron-donating	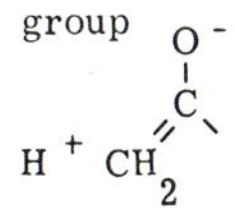	weak electron-donating	small increase
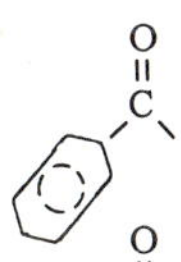	Very weak electron-withdrawing	electron-donating	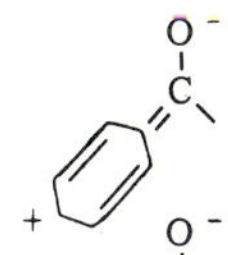	electron-donating	increase
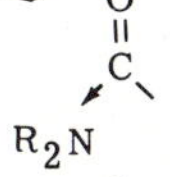	Weak electron-withdrawing	VERY STRONG Electron-donating	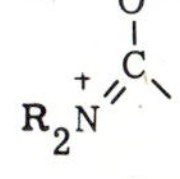	STRONG electron-donating	large increase
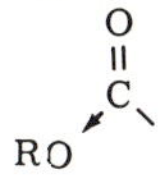	Strong electron-withdrawing	Electron-donating	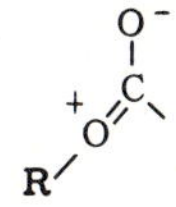	WEAK electron-withdrawing	small decrease

+Compared to hydrogen.

*Same for other alkyl groups.

TABLE V-5 PREDICTED RELATIVE DIPOLE MOMENTS OF CARBONYL COMPOUNDS COMPARED WITH ACTUAL VALUES

Predicted order of dipole moments

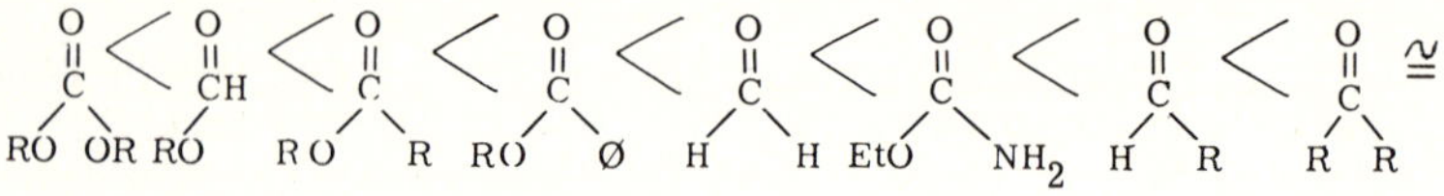

Actual values	0.91 lowest	1.2	1.8	1.9	2.27	2.59	2.7	2.8

Predicted order of dipole moments

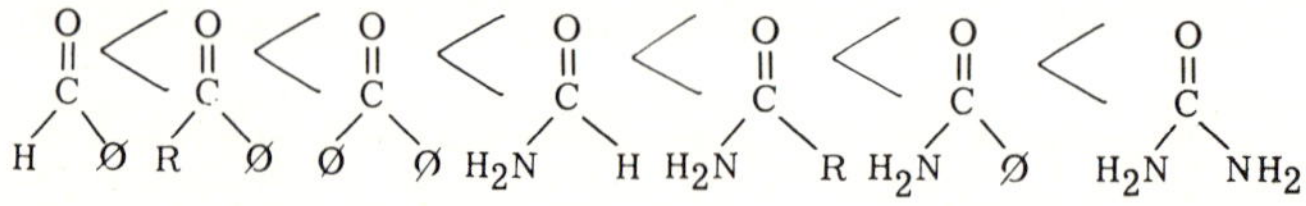

Actual values	2.8	2.9	3.0	3.71	3.75	3.85	4.56 highest

Problem V-1. The dipole moments of most molecules are the vector sum of all the bond dipole moments. In molecules that have only two significant bond moments (μ_1 and μ_2) the dipole moment of the molecule μ_m can be calculated from the formula

$$\mu_m = \sqrt{\mu_1^2 + \mu_2^2 + 2\mu_1\mu_2 \cos\theta}$$

where θ equals the angle between the two bond moments (Fig. V-2). For the student without the requisite mathematical background, the basic idea that must be understood for this question is that bond moments in the same direction ⇸, ⇸ are added arithmetically ($\mu_m = \mu_1 + \mu_2$) and bond moments in exactly opposite directions ($\theta = 180°$) are subtracted arithmetically ($\mu_m = \mu_1 - \mu_2$). These situations occur in *para* disubstituted benzene derivatives.
(a) Calculate from the dipole moments given in Figure V-2 the dipole moments of N,N-dimethyl-*p*-nitroaniline, *p*-nitroaniline, *p*-chloronitrobenzene, and *p*-nitrotoluene.
(b) The observed dipole moments are N,N-dimethyl-*p*-nitroaniline $\mu = 6.87$, *p*-nitroaniline $\mu = 6.10$, *p*-chloronitrobenzene $\mu = 2.65$, *p*-nitrotoluene $\mu = 4.4$. Of all the observed dipole moments quoted in this problem and in Figure V-2 only those for N,N-dimethyl-*p*-nitroaniline, *p*-nitroaniline, and possibly *p*-chloronitrobenzene are significantly higher than the calculated values. Why?

With these concepts in mind the physical properties of organic compounds, especially those studied in Chapters 35, 46, and 56, can be rationalized and often predicted with reasonable accuracy.

Problem V-2. α,β-Unsaturated aldehydes boil significantly higher than their saturated analogs (Table V-6). (a) Why? (b) Explain the differences in boiling points of the isomers: *p*-methylbenzaldehyde (b 205°) and 2-phenylethanal (b 194°); 2-phenoxyethanal ($C_6H_5OCH_2CHO$, b 215°) and *p*-anisaldehyde (*p*-$CH_3OC_6H_4CHO$, b 248°).

Problem V-3. From the information in this Appendix determine the dipole moments of the *para* isomers of: (a) $C_6H_4Br_2$, (b) $C_6H_4(CN)_2$, (c) $C_6H_4(OCH_3)_2$, and (d) $C_6H_4(NH_2)_2$. The answer, of course, is 0 D for all four, yet only (a) and (b) actually have dipole moments of 0. The dipole moment of (c) is ≅1.7 D and (d) is 1.5 D. Explain. Hint: what differences in conformations would you expect between a, b and c, d?

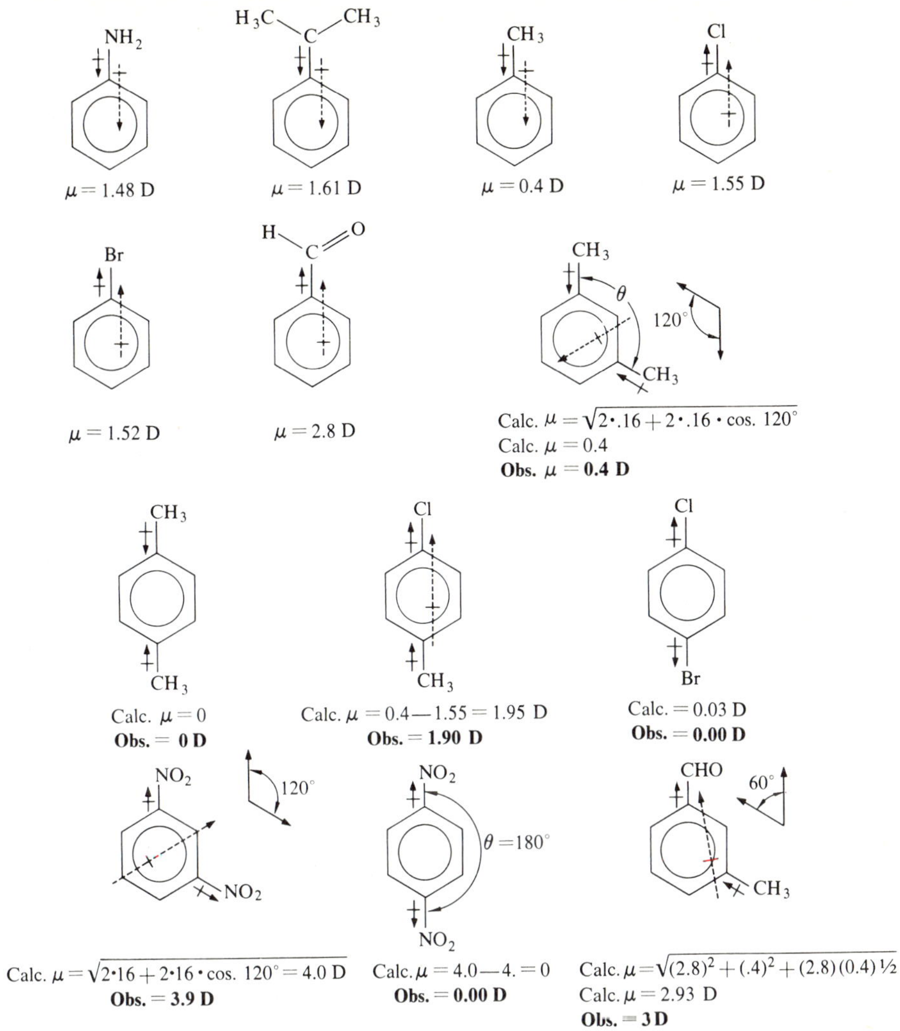

Figure V-2 Observed and calculated dipole moments. Solid arrows ⟷ show moments of individual bonds; dashed arrows -+-----→ show dipole moments of molecules.

Problem V-4. Consider the following compounds and their dipole moments: (a) $CH_2{=}CHCl$ (1.41 D), (b) $CH_2{=}CHCN$ (3.4 D), (c) $CH_2{=}CHCO_2H$ (1.84 D), and (d) $CH_2{=}CHCO_2Et$ (1.6 D). (1) Using the information obtained in this Appendix calculate the dipole moments of the *cis* and *trans* isomers of (e) $ClHC{=}CHCl$, (f) $NCHC{=}CHCN$, (g) $HO_2CHC{=}CHCO_2H$, and (h) $EtO_2CHC{=}CHCO_2Et$. (2) The actual values are (e) *cis* 1.85 D, *trans* 0.0 D, (f) *cis* 5.5 D, *trans* 0.0 D, (g) *cis* 3.17 D, *trans* 2.54 D, (h) *cis* 2.54 D, *trans* 2.38 D. Note that your calculated values are (or should be) close to the actual values except for the *trans* isomers of (g) and (h). Explain why the dipole moments of *trans* g and h are so much higher than expected.

TABLE V-6 COMPARISON OF BOILING POINTS OF SATURATED AND UNSATURATED COMPOUNDS

Compound	B.p. (°C)	Unsaturated analog	B.p. (°C)	Δ B.p. unsat. - B.p. sat.
Propanal	48.5	Acrolein	52.4	4
Butanal	74.7	2-Butenal	104	39.3
2-Methylpropanal	64	2-Methyl-2-propenal	73.5	8.5
Pentanal	103.4	2-Pentenal	125	21.6
2-Methylbutanal	92.5	2-Methyl-2-butenal	116.5	24.5
3-Methylbutanal	92.5	3-Methyl-2-butenal	135	42.5
Hexanal	131	2-Hexenal	150	19
2-Methylpentenal	118	2-Methyl-2-pentenal	139	21
2-Ethylhexanal	163	2-Ethyl-2-hexenal	173	10

Appendix VI

Electronic Absorption Spectra of Some Plant Pigments

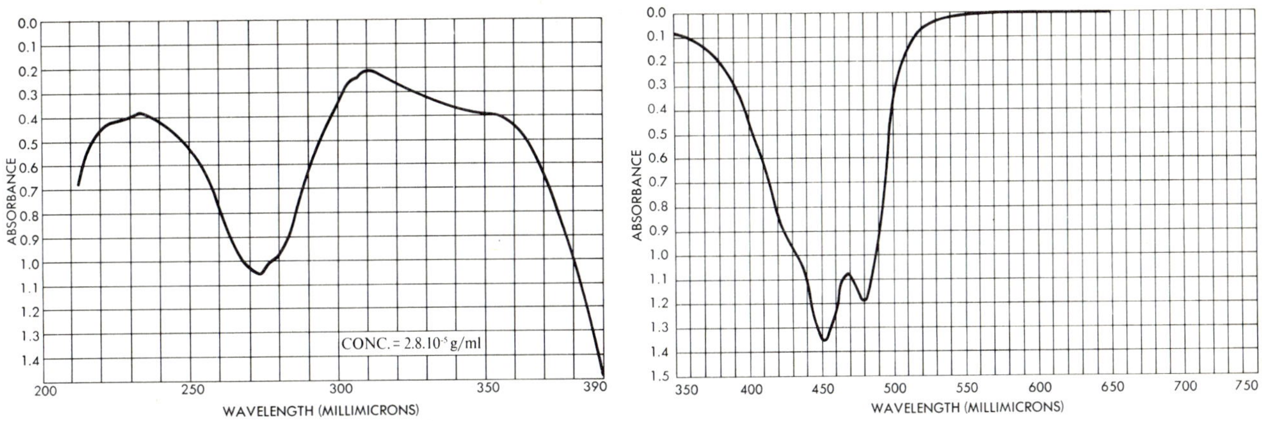

Figure VI-1 Electronic absorption spectra on β-carotene in isooctane in a 1-cm cell. (a) Concentration—2.8×10^{-5} g/ml, 212–390 mμ. (b) Concentration—5.6×10^{-6} g/ml, 350–650 mμ. See Figure 26-9 for α-carotene.

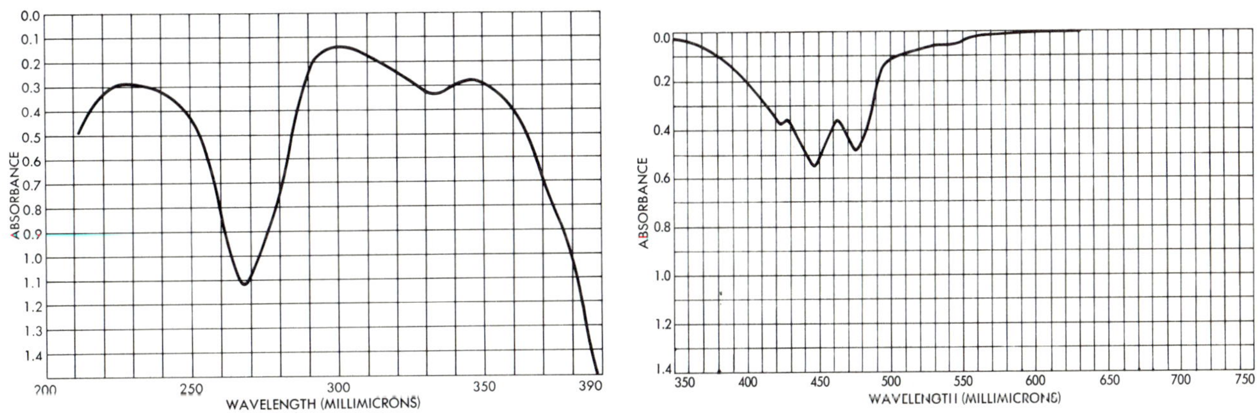

Figure VI-2 Electronic absorption spectra of α-cryptoxanthin in isooctane in a 1-cm cell. (a) Concentration—2.4×10^{-5} g/ml, 212–390 mμ. (b) Concentration—2.4×10^{-6} g/ml, 350–650 mμ.

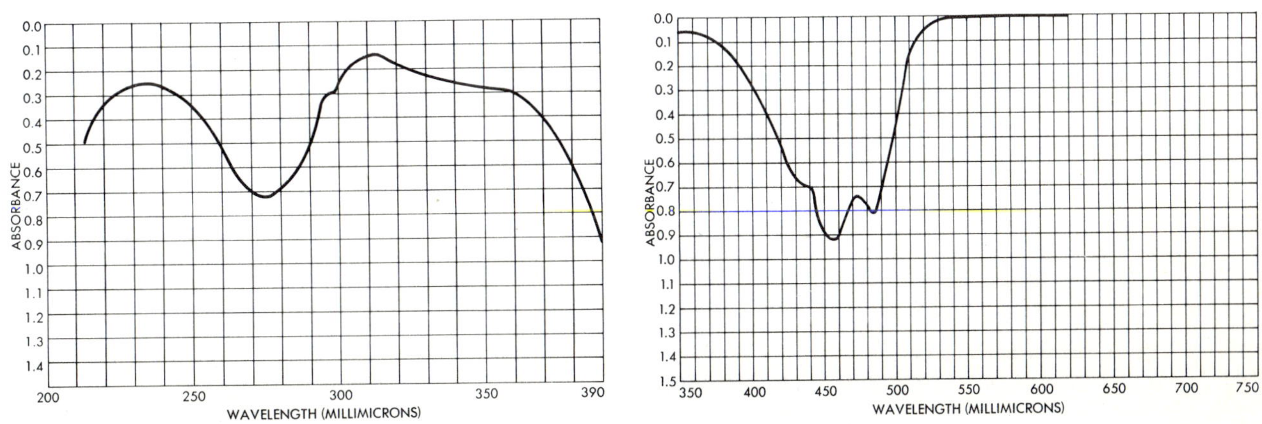

Figure VI-3 Electronic absorption spectra of β-cryptoxanthin in 1:1 ether-isooctane mixture in a 1-cm cell. (a) Concentration—2×10^{-5} g/ml, 212–390 mμ. (b) Concentration—4×10^{-6} g/ml, 390–620 mμ.

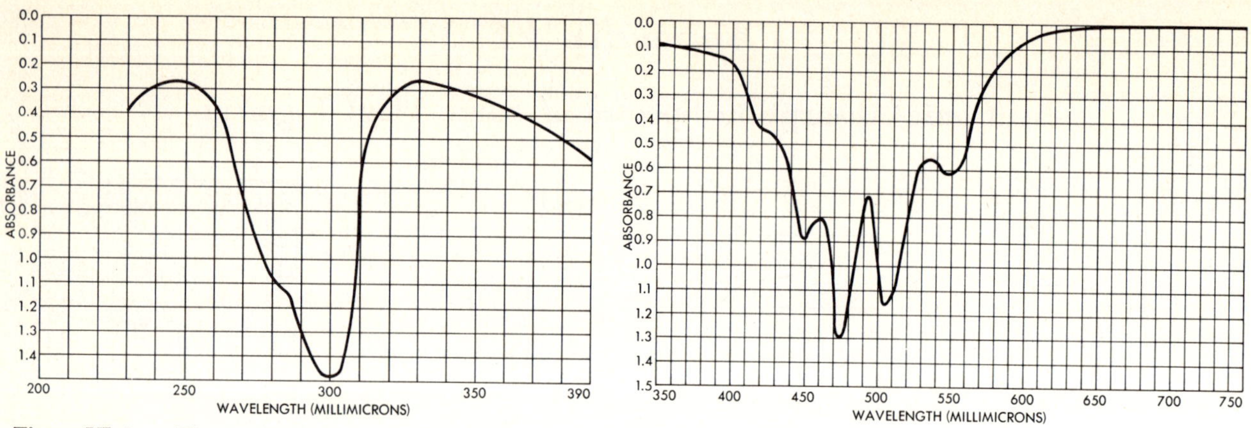

Figure VI-4 Electronic absorption spectra of lycopene in 1:1 isooctane-ethyl ether mixture in a 1-cm cell. (a) Concentration—1.48×10^{-5} g/ml, 230–390 mμ. (b) Concentration—3.7×10^{-6} g/ml, 350–750 mμ.

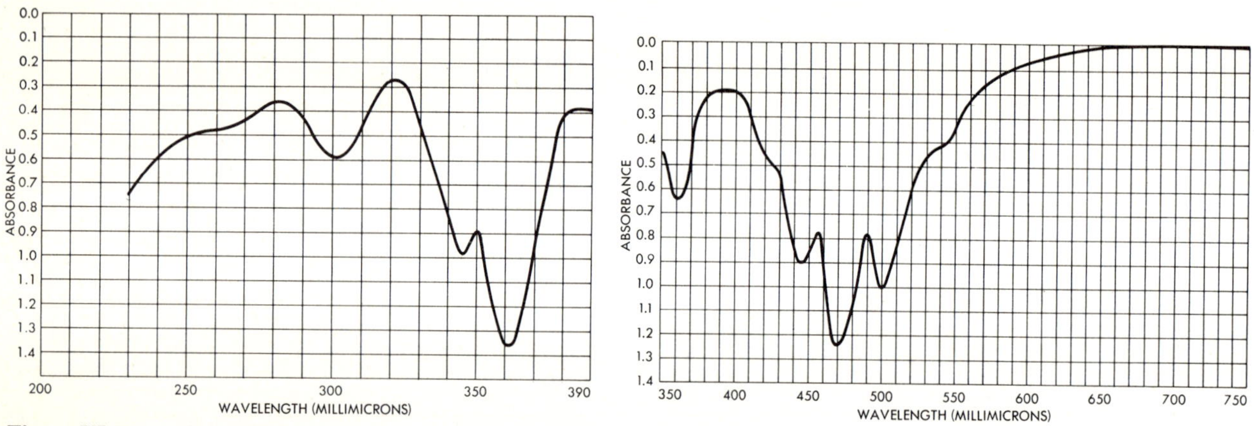

Figure VI-5 Electronic absorption spectra of neolycopene in 1:1 isooctane-ethyl ether mixture in a 1-cm cell. (a) Concentration—1.07×10^{-5} g/ml, 230–390 mμ. (b) Concentration—5.37×10^{-6} g/ml, 350–750 mμ.

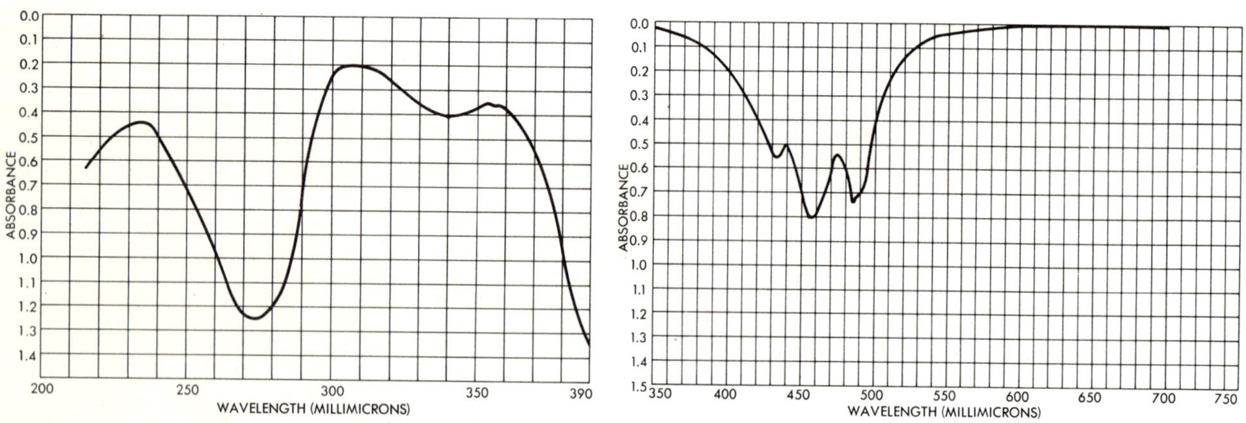

Figure VI-6 Electronic absorption spectra of xanthophyll (lutein) in 1:1 ether-isooctane mixture in a 1-cm cell. (a) Concentration—3.6×10^{-5} g/ml, 215–390 mμ. (b) Concentration—3.6×10^{-6} g/ml, 350–700 mμ.

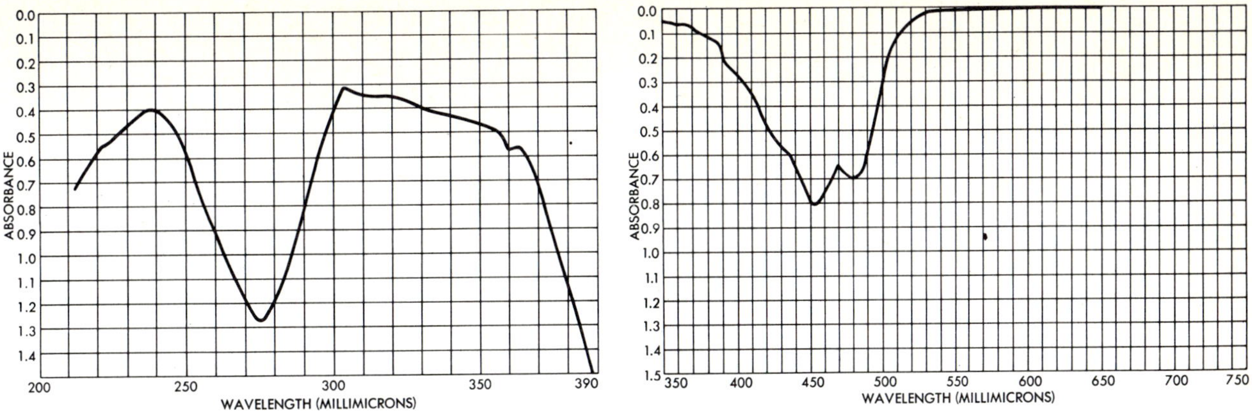

Figure VI-7 Electronic absorption spectra of zeaxanthin in 1:1 isooctane-methanol mixture in a 1-cm cell. (a) Concentration—3.8×10^{-5} g/ml, 212–390 mμ. (b) Concentration—3.8×10^{-6} g/ml, 350–650 mμ.

Structures of carotenoid plant pigments

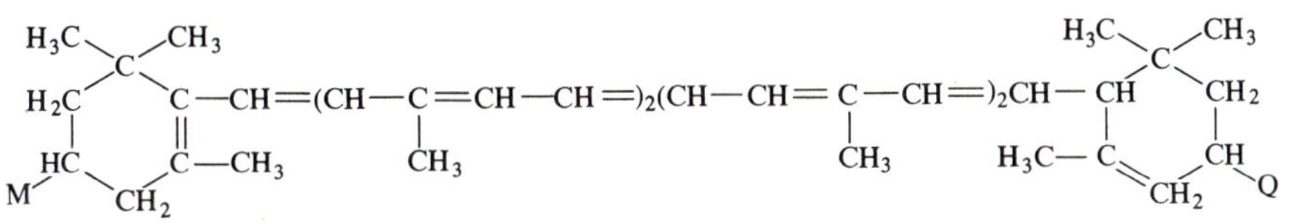

α-Carotene; M, Q = H
α-Cryptoxanthin; M = H, Q = OH
Xanthophyll; M, Q = OH

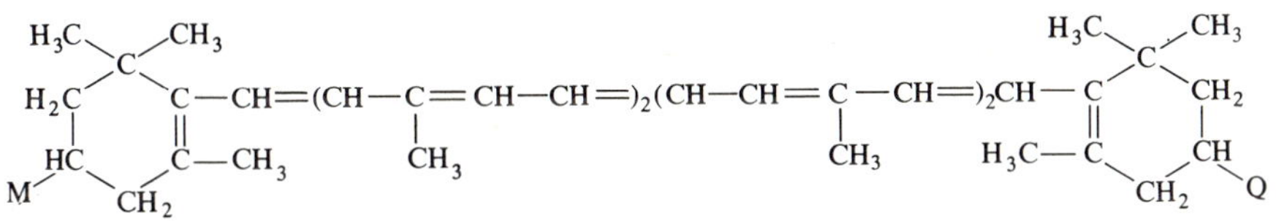

β-Carotene; M, Q = H
β-Cryptoxanthin; M = H, Q = OH
Zeaxanthin; M, Q = OH

For all six compounds shown by the above two generalized formulas, all double bonds (except the ones in the six-membered rings) have *trans* geometry; i.e., the main chain of the molecules is located on the opposite sides of the double bonds as are the hydrogen atoms or the hydrogen atoms and methyl groups.

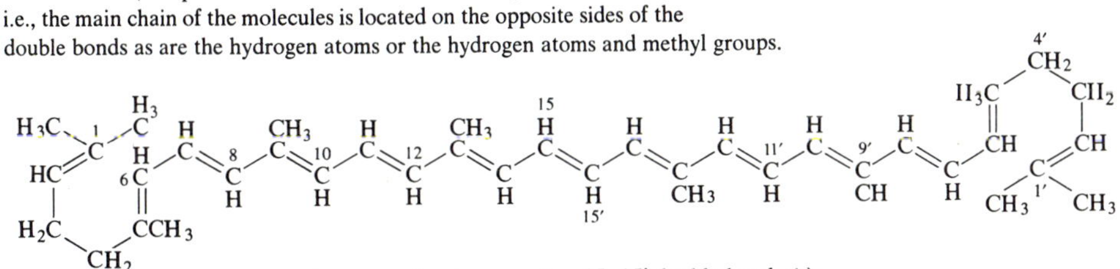

Lycopene (neolycopene has 15, 15′ double bond *cis*)

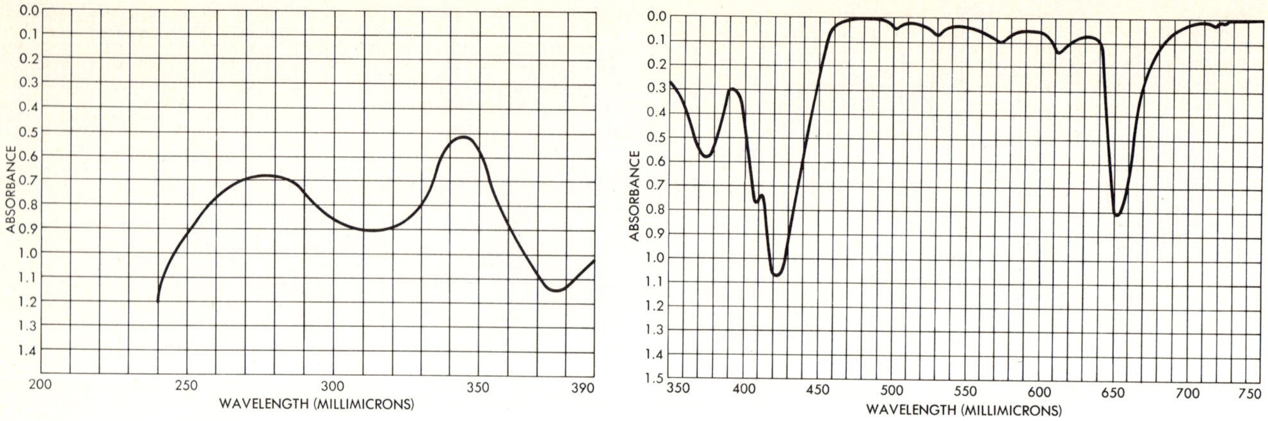

Figure VI-8 Electronic absorption spectra of chlorophyll a in 1:1 isooctane-ethyl ether mixture in a 1-cm cell. (a) Concentration—8.7×10^{-6} g/ml, 240–390 mμ. (b) Concentration—4.35×10^{-6} g/ml, 350–750 mμ.

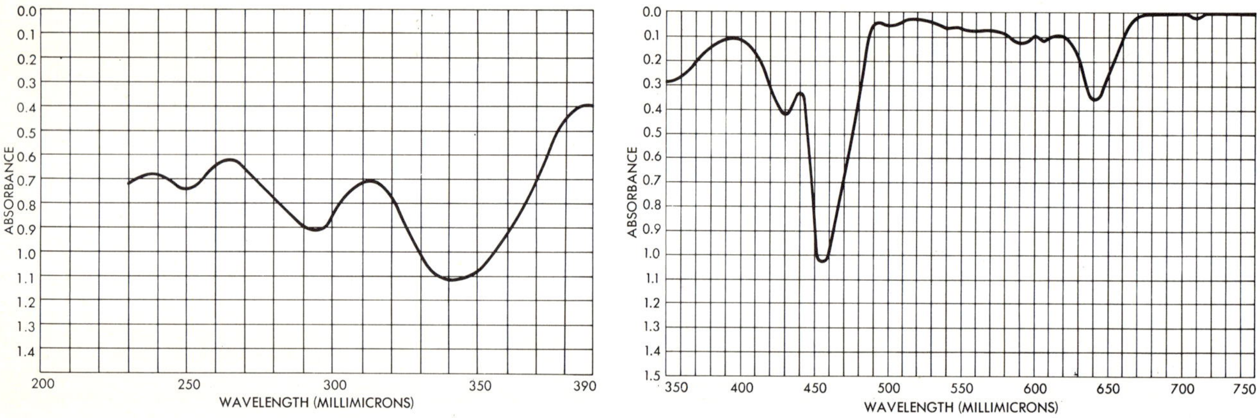

Figure VI-9 Electronic absorption spectra of chlorophyll b in 1:1 isooctane-ethyl ether mixture in a 1-cm cell. (a) Concentration—8.8×10^{-6} g/ml, 230–390 mμ. (b) Concentration—2.2×10^{-6} g/ml, 350–750 mμ.

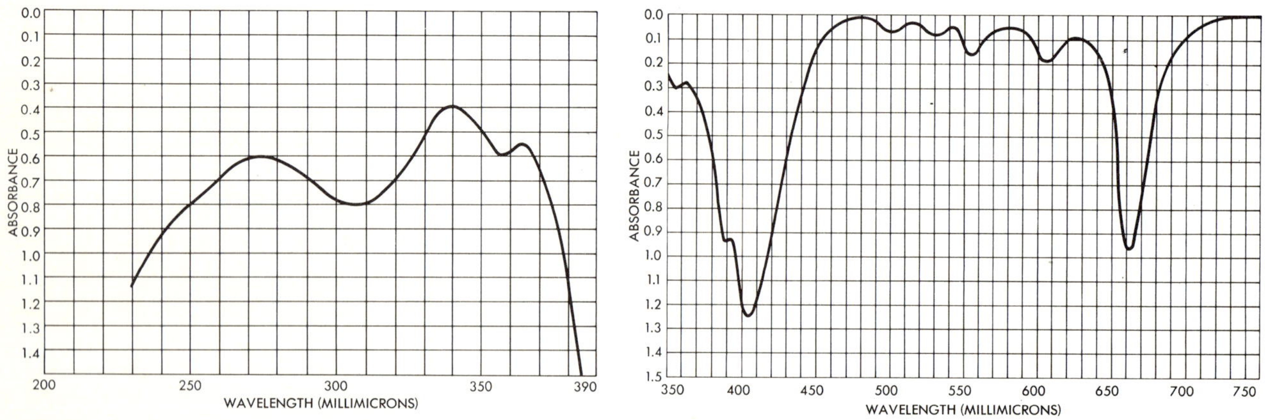

Figure VI-10 Electronic absorption spectra of pheophytin a in 1:1 isooctane-methanol mixture in a 1-cm cell. (a) Concentration—1.2×10^{-5} g/ml, 230–390 mμ. (b) Concentration—6.0×10^{-6} g/ml, 350–750 mμ.

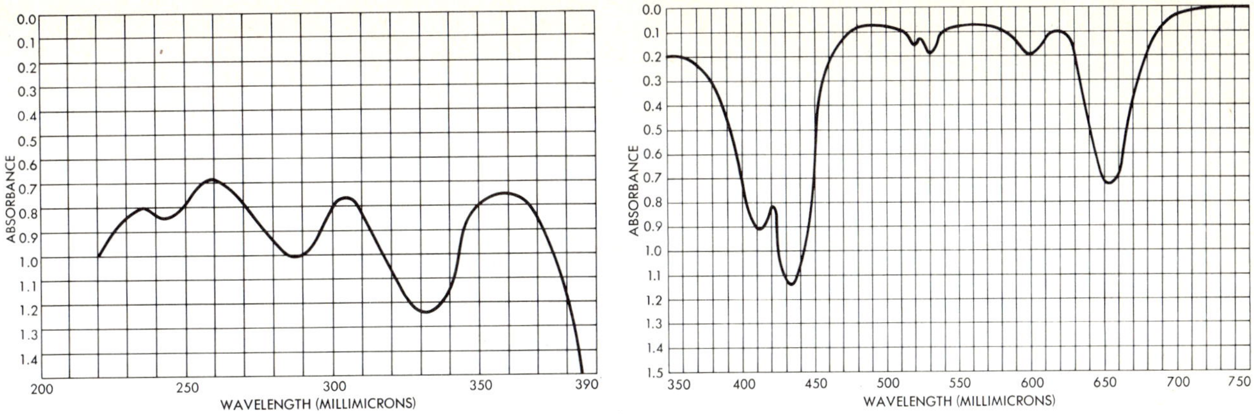

Figure VI-11 Electronic absorption spectra of pheophytin b in 1:1 iso-octane-methanol mixture in a 1-cm cell. (a) Concentration—1.2×10^{-5} g/ml, 220–390 mμ. (b) Concentration—3.0×10^{-6} g/ml, 350–750 mμ.

Structure of phorbine plant pigments

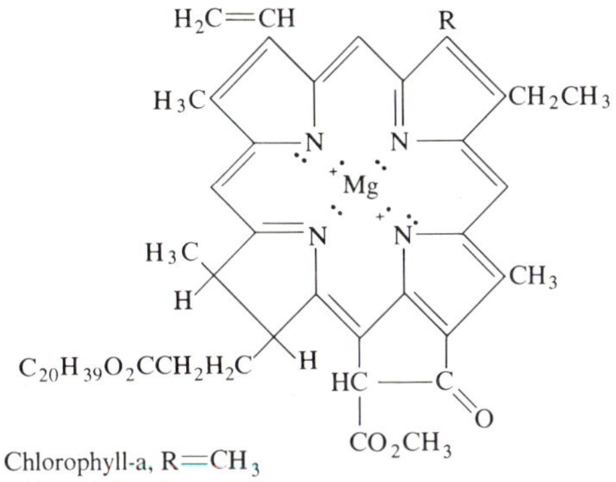

Chlorophyll-a, R=CH_3
Chlorophyll-b, R=CHO

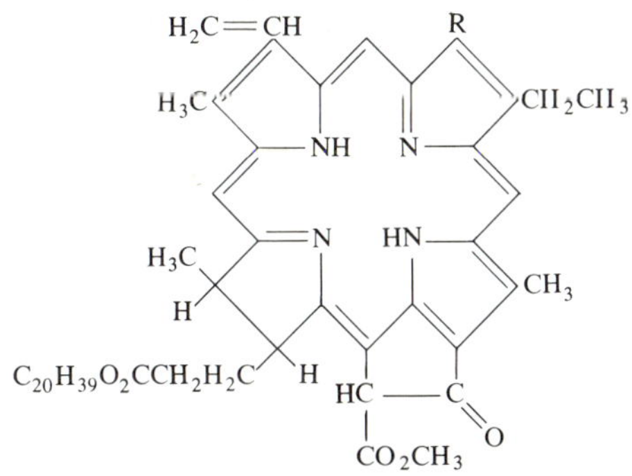

Pheophytin-a, R—CH_3
Pheophytin-b, R—CHO

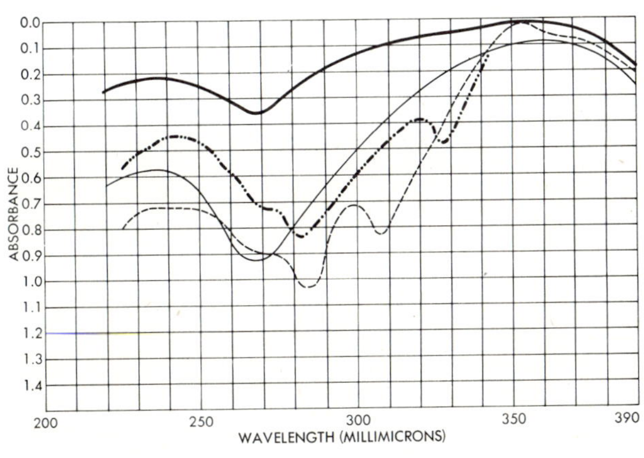

Figure VI-12 Generalized ultraviolet absorption spectra of anthocyanidins. The two solid curves show the limits of intensities for the concentration 2×10^{-4} g/ml in methanol containing 0.1 g/liter of HCl. The broken lines indicate the spectra of anthocyanidins that have hydroxy groups esterified with aromatic acids. The broken lines do not define the limits of the spectra of the aryl acyl derivatives; however, they are close to these limits.

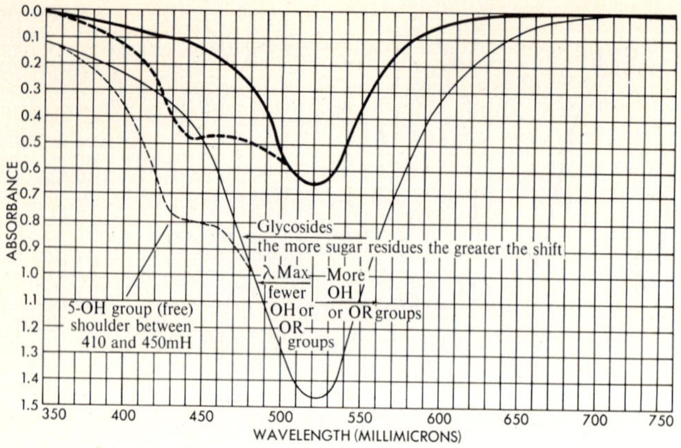

Figure VI-13 Generalized visible absorption spectrum of anthocyanidins. The solid lines indicate limits of intensities for the concentration 2×10^{-4} g/ml in methanol containing 0.1 g/liter of HCl. The broken lines indicate limits of intensities for a free 5-hydroxy group. The actual λ_{max} is between 475 and 550 mμ depending on the number of oxygen functions. See also the generalized formula.

General formula for anthocyanidins

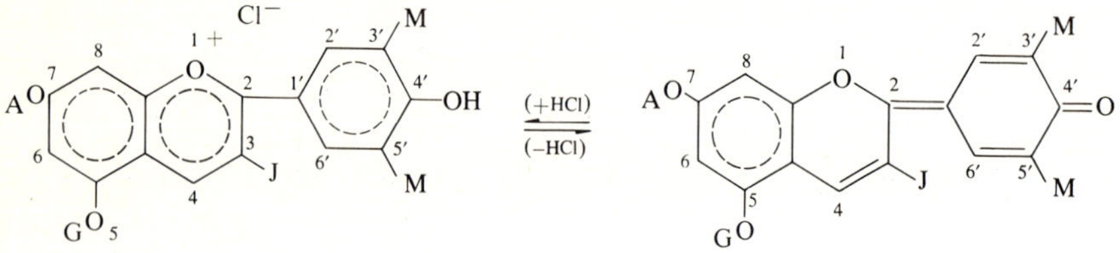

Anthocyanidin base Quinonoid base

A = H— or $ArCO_2$—

G = H—, CH_3—, glucoside—, or $C_6H_5CO_2$—

J = H—, HO—, CH_3O—, or glycoside—O—

M = H—, HO—, or CH_3O—

For example

where	visible λ_{max} in mμ
A, G, M, and J = H	476
A, G, M = H; J = OH	520
A, M = H; J = OH; G = glucoside	513
A, M = H; J = O—glucoside; G = glucoside	503
A, G, 5′—M = H; J, 3′—M = OH	535
A, 5′—M = H; 3′—M = OCH_3; G = glucoside; J = —OH	528
A, 5′—M = H; 3′—M = OH; G = glucoside—O—; J = rhamnoglucoside—O—	523

Appendix VII

Spectra-Structure Correlations

ABSORPTION IN DIFFERENT REGIONS OF THE INFRARED SPECTRUM

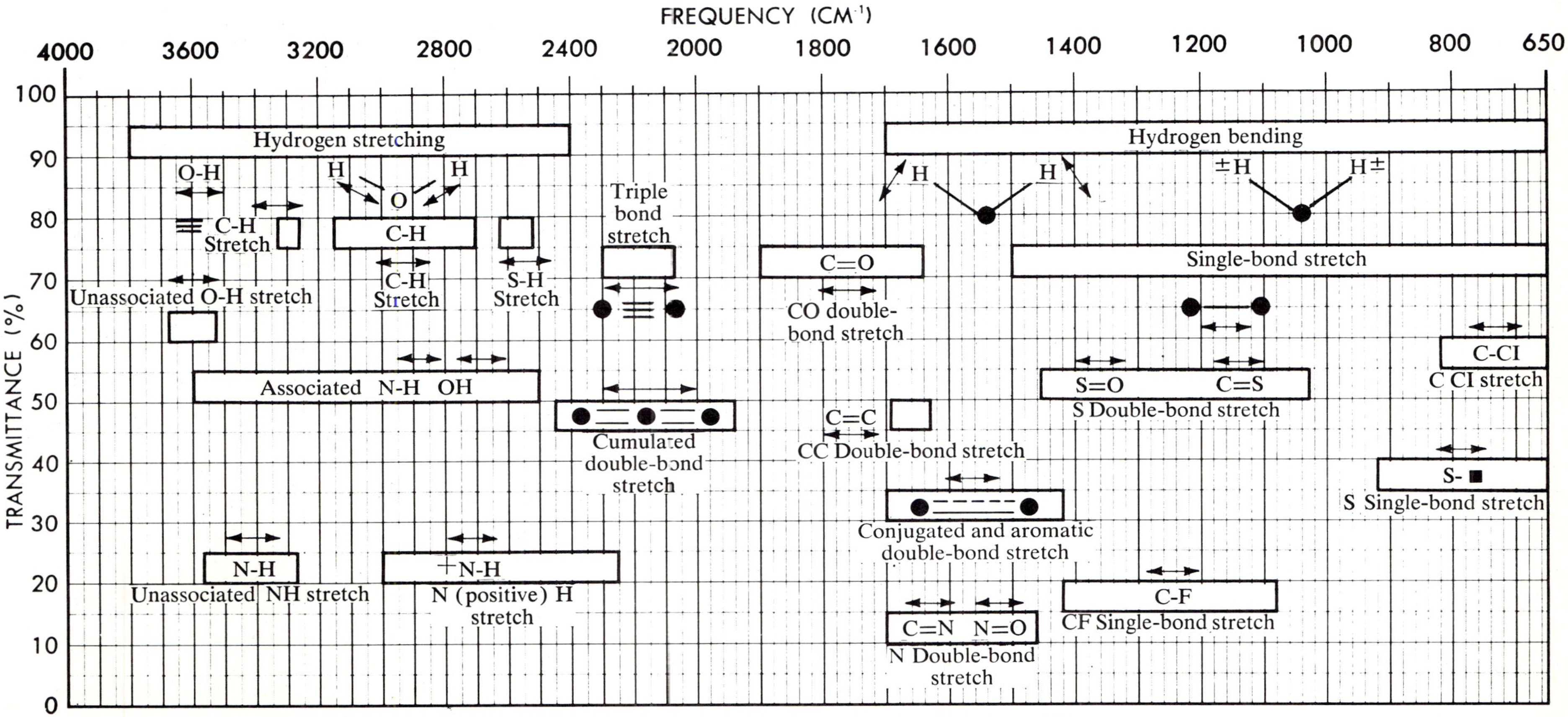

From R. W. Hannah and J. S. Swinehart, *Experiments in Techniques of Infrared Spectroscopy*, Perkin-Elmer Corp., 1967.

SPECTRA-STRUCTURE CORRELATIONS

PROBABLE POSITIONS OF CHARACTERISTIC INFRARED ABSORPTION BANDS

STAMFORD RESEARCH LABS.
AMERICAN CYANAMID CO.

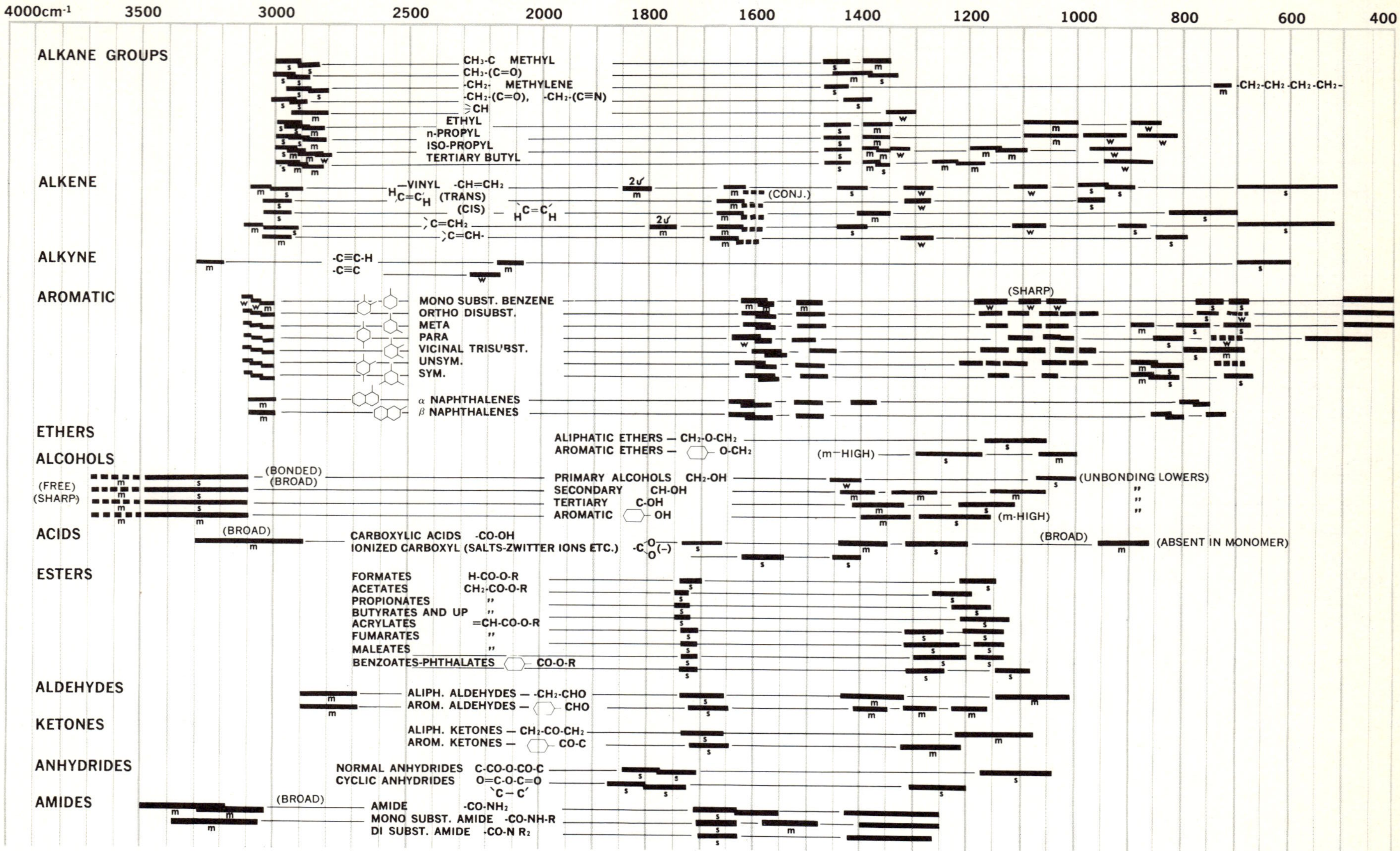

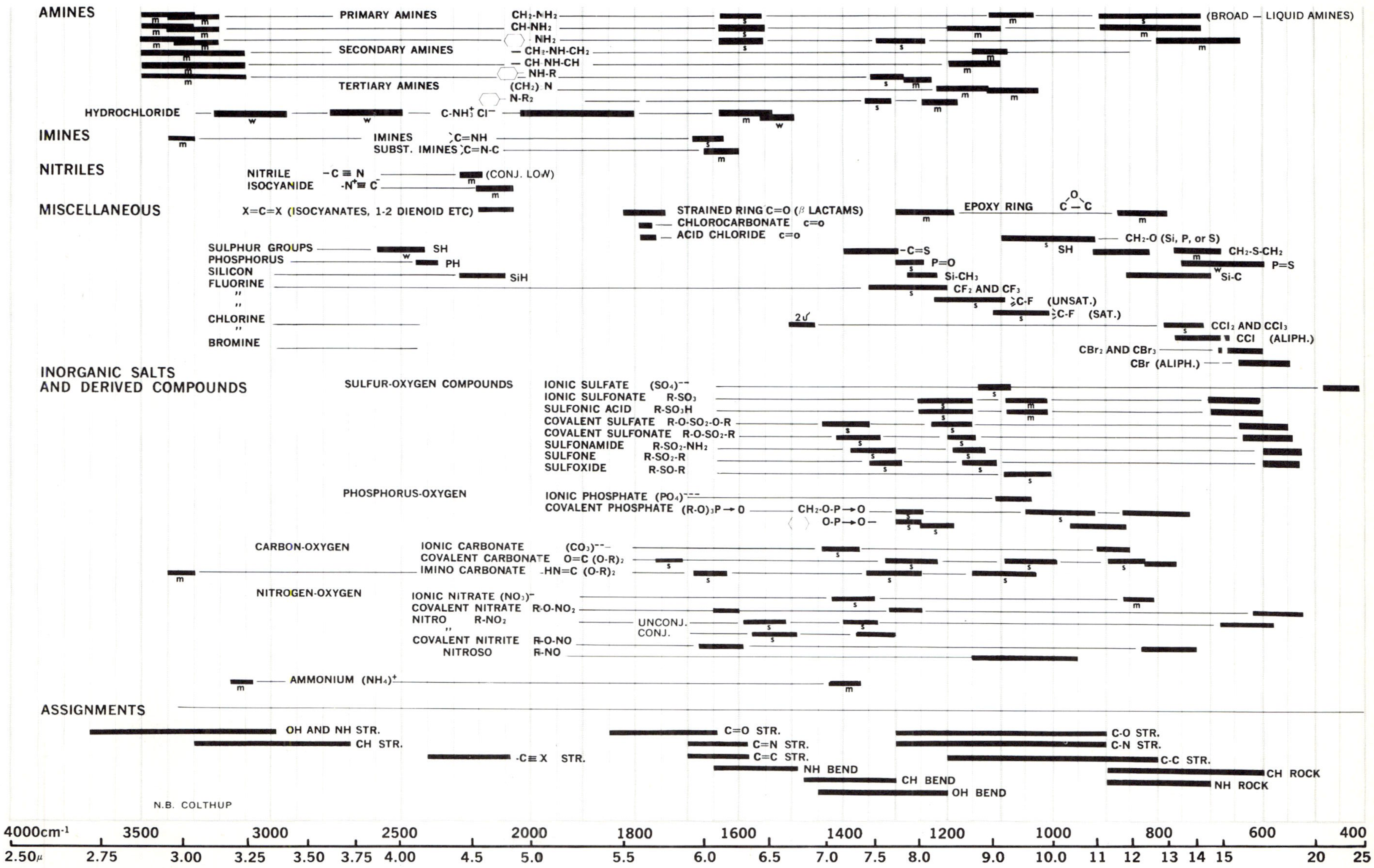

From N. Colthup, *J. Opt. Soc. Amer.* **40,** 397 (1950). Revised 1967 by N. Colthup.

Answers to Selected Problems

CHAPTER 1

1-1. Teflon's coefficient of thermal expansion is considerably larger than that of glass in the 0 to 50° range. Teflon undergoes a crystal structure transition near 19° that is accompanied by a significant reduction in volume.

CHAPTER 3

3-1. Start with a somewhat larger quantity in the outer tube. Obtain a micro boiling point as described in the text. Remove the inner or capillary tube and evaporate 50 to 70% of the sample. Cool the bath at least 15°, insert a capillary tube (the original tube will do if there is no liquid in it), and obtain another micro boiling point. A difference greater than 2° between the two boiling points indicates that the liquid is impure.

EXERCISES

3-1. Heating a liquid above its boiling point without causing it to boil is superheating.

3-2. The surface of a boiling chip is covered with many small indentations or capillary holes. When the boiling chips are heated in a liquid, minute bubbles of air are evolved from the holes. These bubbles serve as nuclei for vapor bubbles and ensure smooth boiling.

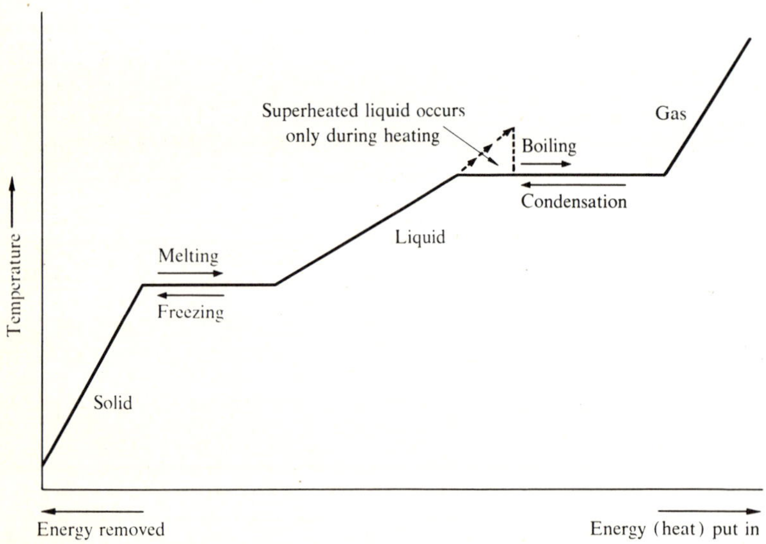

Figure 2-7 Energy and phase changes. Modified to show a superheated liquid.

3-3. When a boiling chip is heated and then cooled in a liquid, liquid partially fills the holes as the chip cools. This liquid will not be driven out as soon as the boiling point is reached and, therefore, the boiling chip is unable to supply the minute air bubbles as soon as they are needed. It may, however, supply the air bubbles after the liquid becomes superheated, often with very violent results. The liquid may be heated above its boiling point. If a boiling chip is dropped into a superheated liquid, the liquid will boil violently, almost explosively, and splatter hot liquid wherever possible. Note that a flammable or toxic, superheated liquid poses hazards in addition to scalding.

CHAPTER 4

4-1. (a)

Liquid N_{CH_3OH}	Composition of atmosphere above mixture (at 760 mm) N_{H_2O}	N_{CH_3OH}	(at 400 mm) N_{H_2O}	N_{CH_3OH}
0.0	0.032	0.000	0.061	0.0
0.1	0.030	0.024	0.057	0.045
0.3	0.026	0.054	0.049	0.103
0.5	0.023	0.087	0.044	0.165
0.7	0.016	0.115	0.030	0.229
0.9	0.013	0.150	0.025	0.285
1.0	0.0	0.18	0.0	0.342

(b) Higher in the vapor phase. Yes, in a binary mixture that approximately follows Raoult's law, the vapor has a higher relative (to the less volatile compounds) concentration of the more volatile compound.

4-2. 80°: liquid—toluene 0.16 *N*, 10 wt%, CCl_4 0.84 *N*, 90 wt%
vapor—toluene 0.06 *N*, 3.7 wt%, CCl_4 0.94 *N*, 96.3 wt%
100°: liquid—toluene 0.79 *N*, CCl_4 0.21 *N*
vapor—toluene 0.59 *N*, CCl_4 0.41 *N*

4-3. Use Raoult's law. If you did not use Raoult's law, attempt the problems again using this law. At 70°: $760 = 234x + 928(1 - x)$ where x = mole fraction of H_2O. The first number after the temperature is the calculated mole fraction of water in the vapor. The second number (in parentheses) is the actual mole fraction of water in the vapor at that temperature. 64.7°, 0; 70°, 0.24 (0.20); 80°, 0.58 (0.48); 85°, 0.70 (0.60); 90°, 0.82 (0.75); 100°, 1.00.

4-4. Use the actual values in parentheses.

4-5. Both liquid and vapor are 77.5 wt% formic acid and 22.5 wt% H_2O. 88 wt% H_2O and 12 wt% formic acid or 4 wt% H_2O and 96 wt% formic acid. 104.5°. Vapor = 30 wt% formic acid, 70 wt% H_2O. As the mixture is distilled, the concentration of formic acid in the vapor and liquid phases increases and the boiling point increases until 107.3° is reached. When the boiling point reaches this temperature, the composition of the vapor and the liquid is identical, and the mixture continues distilling at 107.3° until the entire liquid has distilled.

4-6. 96.7°. Dioxane 88 wt%, H_2O 12 wt%. As the mixture is distilled, the concentration of dioxane increases in both the liquid and the vapor phases and the temperature increases to a maximum of 101.4°. 90°. Dioxane 71 wt%, H_2O 29 wt%. As mixture is distilled, the concentration of water increases in both the liquid and vapor phases and the temperature increases to a maximum of 100°.

EXERCISES

4-2. A liquid with a narrow boiling point range is a relatively pure compound, an azetropic mixture, or a mixture of two or more liquids that boil within 3° of each other. If a liquid has a wide boiling point range, it is impure; if it has a very narrow boiling point range, it probably is a pure compound but the exceptions mentioned above are not uncommon.

CHAPTER 5

5-1. D_4. D_6. 6. 9.

5-2. D_6. The distillate from fraction 7 would have approximately the same composition as fraction 3.

5-3.

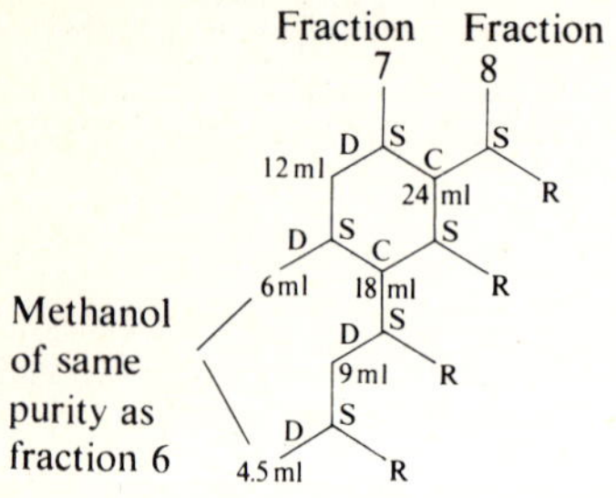

Fraction 9 is not needed.
See key Figure 5-1 for symbols.

5-4. 10. 15.

CHAPTER 6

6-1. More soluble.

6-2. Calculate the distribution coefficient K from your results and the equation on page 40. The equation in Exercise 6-4 at the end of this chapter can be used to determine the number of extractions. The volume V_{H_2O} needed to obtain 95% of the butyric acid from the ether can be calculated by

$$K = \frac{\dfrac{95}{V_{H_2O}}}{5/45}$$

CHAPTER 7

7-1. An azeotropic mixture of water and ethylene bromide. Add to this fraction approximately half its volume of calcium chloride, stir, and decant the liquid into the tared yield bottle.

7-2. 7.46 g.

$$PV = nRT;\ V = \left(\frac{7.46}{28}\right) \cdot \frac{(0.0821)(273)}{1};\ V = 5.97 \text{ liters}$$

7-3. $(10)(2.180) = 21.8$ g; $(21.8/50.4)(100) = 42.5\%$.

7-4. 13.8 ml (43 g).

7-5. Bromine.

7-6. (a) 14.8 g, (b) 17 g, (c) 51 g, (d) 38.4 g.

CHAPTER 8

8-1. In counter-current extraction and fractional distillation both phases are mobile. In chromatography one phase is mobile and one is stationary. In fractional distillation the compounds being separated make up a large percentage (usually 100%) of the two phases in equilibrium. This causes an overlap fraction that is at least slightly greater than the hold-up volume of the column. In chromatography (and counter-current extraction) the compounds being separated make up a small part of the two phases in equilibrium; the selection of phases provides another variable that can be optimized for maximum separation. One of the phases is a gas in both fractional distillation and gas chromatography.

8-2. Helium has a higher thermal conductivity than any of the compounds being chromatographed; thus, every time one of these compounds goes through the detector the thermistor is cooled less rapidly than it would be with helium alone. Nitrogen has a thermal conductivity that may be higher or lower than the thermal conductivity of the compounds being chromatographed. As the compounds are eluted through the detector, they may cool the hot wire either more or less rapidly than the nitrogen. They may give an increase or decrease in voltage from the bridge and a pen deflection in either direction.

8-3. Octane > heptane > hexane > pentane.

8-4. Carrier gas *must* be flowing through the detector.

8-5. So that no liquid phase or residual material from other chromatograms is deposited in the detector.

8-9. When a large gas sample is injected, it does not go in a small sharp slug. Quite a few seconds after the first part of the gas sample has entered the column, there is still material in the injection port. This is somewhat the effect that injecting several small samples in very rapid succession would produce. The separated compounds are eluted from the column over a longer time than they would be if injected in a narrower slug. Since the peak area is proportional to the amount of compound present, this increase in the peak base causes a decrease in the peak height.

CHAPTER 9

9-1. $CH_2{=}C(CH_3)CH_2CH_3$, $C_2H_5(CH_3)_2CCH(CH_3){=}C(CH_3)_2$, and other $(C_5H_{10})_n$

9-2. See answer to 9-1. The compounds should have retention times in the same order as their molecular weight and for alkenes of the same molecular weight in the same order as their dipole moments. Lowest retention time: 2-methyl-1-butene (μ = 0.5 D), next lowest: 2-methyl-2-butene (μ = 0.9 D), higher retention times: $(C_5H_{10})_2$ isomers.

CHAPTER 11

11-1. Because of its low vapor pressure.

CHAPTER 12

12-1. (a) The solvent-solute interactions do not supply enough energy to break the intracrystalline forces. The difference between the lattice energy of the solid and the energy from the solvent-solute interaction comes from the thermal energy of the solution. (b) If a solution becomes hot when a solute dissolves, the solvent-solute interactions have supplied significantly more energy than needed to break the intracrystalline forces. The excess energy raises the temperature of the solution.

12-2. To dissolve in water the nonpolar molecules must slip between the water molecules. To do this they must break some of the strong intermolecular dipole-dipole attractions between the water molecules. The relatively weak dipole-induced dipole attractions between water and nonpolar molecules cannot supply sufficient energy to break these strong polar intermolecular attractions of water.

12-3. (a) 25 g of *A* and 32 g of *B*. (b) 267 ml (c) 21.3 g in solution, 18.3 g in precipitate. 46% recovery.

12-4. *A* must be at least five times as soluble as *B*.

12-5. Because the thermal motion of the molecules is less at lower temperatures, several adjacent molecules are less likely to assume an orientation necessary for crystal formation at the same time.

12-6. The faster the rate of cooling, the more likely it is that oil formation or supersaturation will occur.

12-7. Remove the first crop by filtration. Cool to 0°, and remove the second crop. Obtain melting points on both crops. If the second crop melts more than 1° lower or over a wider range than the first crop, the first crop is significantly purer than the second crop.

12-8. A solution heated at its boiling point may become superheated. Addition of finely divided carbon causes a super-heated solution to boil, often very violently. See answer to Exercise 3-2.

12-9. An unheated funnel would absorb heat from the solution, and the temperature of the solution would drop. If the solution is not kept hot, some of the solute may precipitate before it is filtered.

12-10. Approximately 6%; however, much higher concentrations of acetanilide can be crystallized from water by cooling slowly and stirring vigorously until crystals appear.

12-11. By adding more water or a small amount of a water-miscible solvent (ethanol or methanol would probably be best) or by following the instructions in the answer to 12-10. The directions from Problem 12-10 would probably give the highest yield of purest material.

CHAPTER 13

13-1. Polynitration is reduced by controlling the temperature and not using fuming nitric acid. Sulfonation is reduced by using a relatively low temperature and by using a nitrating reagent vigorous enough to nitrate the benzene before it can be sulfonated.

13-2. Some water has formed from the reaction diluting the acid mixture.

13-3. It contains ether-soluble impurities.

CHAPTER 14

14-1. Compounds that are soluble in concentrated sulfuric acid give only positive tests with $H_2SO_4 \cdot SO_3$; no further information on these compounds could be obtained by testing then with $H_2SO_4 \cdot SO_3$.

14-2. (a) $C_2H_5C(CH_3){=}C(CH_3)C_2H_5 > C_2H_5CH{=}C(CH_3)C_2H_5 > CH_2{=}C(CH_3)C_2H_5 > C_2H_5CH{=}CHC_2H_5 > CH_2{=}CHCH_2CH_3 > CH_2{=}CH_2 > C_6H_5C(CH_3){=}C(CH_3)C_6H_5 > C_6H_5CH{=}CHC_6H_5 > (C_6H_5)_2C{=}C(C_6H_5)_2$.
(b) p-$CH_3OC_6H_4CH{=}CH_2 > p$-$CH_3C_6H_4CH{=}CHCH_3 > p$-$CH_3C_6H_4CH{=}CH_2 > C_6H_5CH{=}CHCH_3 > C_6H_5CH{=}CH_2 > p$-$ClC_6H_4CH{=}CH_2 > m$-$ClC_6H_4CH{=}CH_2 > p$-$NO_2C_6H_4CH{=}CH_2$.

14-3. There is greater resonance stabilization of the carbonium ion of styrene, $C_6H_5CH{=}CH_2$, and the transition state leading to the ion (Reaction 14-9) than there is of styrene; for stilbene, $C_6H_5CH{=}CHC_6H_5$,the reverse is true, see Morrison and Boyd, *Organic Chemistry*, 2nd ed., pp. 401–403. Also, in styrene there is a moderate electron deficiency in the double bond because of the electron-withdrawing effect of the phenyl group. The phenyl group, however, stabilizes the carbonium ion and the transition state leading to the carbonium ion by delocalization of its π electrons [I].

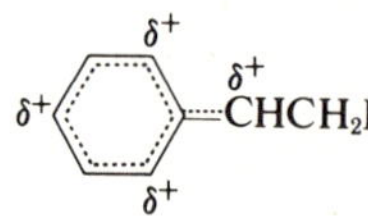

This more than overcomes the rate retardation from the electron-withdrawing inductive effect. Stilbene, $C_6H_5CH{=}CHC_6H_5$ has two phenyl groups withdrawing electrons from the double bond, but the intermediate ion is no more stabilized (in relation to the reactants) from delocalization or resonance than is the ion [I]. There may also be some steric hindrance from the two phenyl groups.

14-4. $(C_6H_5)_2C{=}CHC_6H_5 > C_6H_5CH{=}CHC_6H_5 > (C_6H_5)_2C{=}C(C_6H_5)_2$

14-5. Light promotes free radical substitution by bromine. Compounds without double bonds react with bromine radicals and decolorize bromine solutions.

14-6. Test with the dilute permanganate solution. If there is no decolorization, the ethanol is sufficiently pure.

EXERCISES

14-1. Cyclohexane, benzene. Test with $H_2SO_4 \cdot SO_3$ or $AlCl_3$ and $CHCl_3$.

CHAPTER 15

15-1. (a) 361.5 mm for both mixtures. (b) 185.5 mm for both mixtures. (c) 10 mole % benzene: 110.6 mm, 20 mole % benzene: 131.2 mm.

15-2. (a) 61.25° for both mixtures. (b) 84.3° for both mixtures, (c) 10 mole % hexane: approximately 100°; 50 mole % hexane: approximately 83°. It is difficult to obtain exact figures for (c) because there is insufficient data in Tables I-2 and I-3; however, these answers are probably within 1.5° of the correct value.

15-3. Vp of benzene = 531 mm, vp of H_2O = 229 mm. Moles of benzene per mole of H_2O = (531)/(229) = 2.32.

$$100 = x \cdot 18 + 2.32 \cdot x \cdot 78$$
$$x = \text{moles of } H_2O = .503$$
$$\text{moles of benzene} = 1.166$$
$$\text{Water} = 9.05 \text{ g and ml}$$
$$\text{Benzene} = 90.95 \text{ g or } 103.3 \text{ ml}$$

15-4. Vp of H_2O = 158.5 mm, vp of hexane = 601.5 mm. Moles of hexane per 2 moles of H_2O = 601.5/159.5·2 = 7.54 *M*.

$$7.54\ M = 603 \text{ g or } 912 \text{ ml}$$

15-6. (a) To remove the water by codistilling it with benzene. The mixture of the two that distils is sometimes called a benzene-water azeotrope since it boils lower than either component. (b) Water is codistilling with the benzene. (c) No water is codistilling with the benzene because there is no water in the benzene.

15-7. Acetone removes any traces of water. Benzene removes the residual acetone.

15-8. (a) To prevent the reaction from becoming uncontrolled, to reduce disubstitution on the benzene ring, and to favor *para* instead of *meta* for the disubstitution that does occur. (b) Pure benzene freezes below 6°; once it is contaminated with some *t*-butylbenzene, the freezing point is lower. The reaction may be difficult to start if the temperature is too low.

15-9. The aluminum chloride would catalyze disproportionation of the product to di- and tri-*t*-butylbenzenes and would catalyze isomerization of *p*-di-*t*-butylbenzene.

$$C_6H_5C(CH_3)_3 \xrightarrow{ACl_3} C_6H_6 + C_6H_4[C(CH_3)_3]_2 + C_6H_3[C(CH_3)_3]_3$$

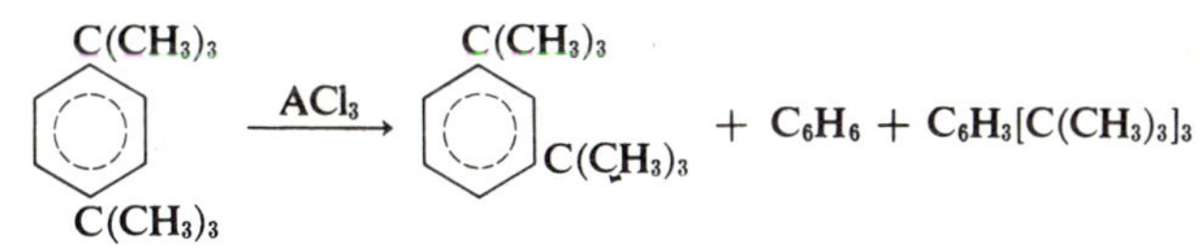

15-10. It destroys the aluminum chloride and thus prevents the reactions outlined in the answer to Problem 15-9.

$$AlCl_3 + 3H_2O \rightarrow Al(OH)_3 + 3HCl$$

The reaction is very exothermic. Energy is absorbed by melting of the ice, and this prevents the reaction from becoming too violent.

15-11. Approximately 34 ml or 0.39 moles of benzene would remain. 0.168 *M* or 3 ml of water is needed to distil all the benzene. Theoretical yield = 0.064 moles. Vp *t*-butylbenzene = 83.4 mm, vp H_2O = 676.6 mm. Moles of H_2O = (0.064) (83.4)/(676.6) = 0.00788 *M* or 0.14 ml of water is needed to steam distil the *t*-butylbenzene.

CHAPTER 17

17-1. The product, propiophenone, is much less reactive than benzene; alkylbenzenes are more reactive than benzene. The propiophenone can not be acylated further nor can it undergo $AlCl_3$ catalyzed rearrangements or disproportionations.

CHAPTER 18

18-1.

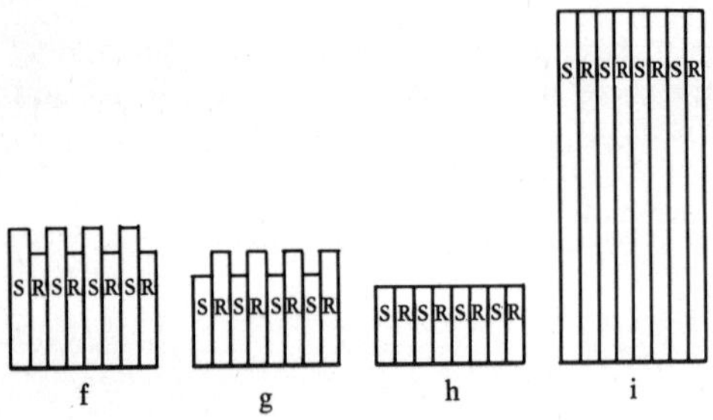

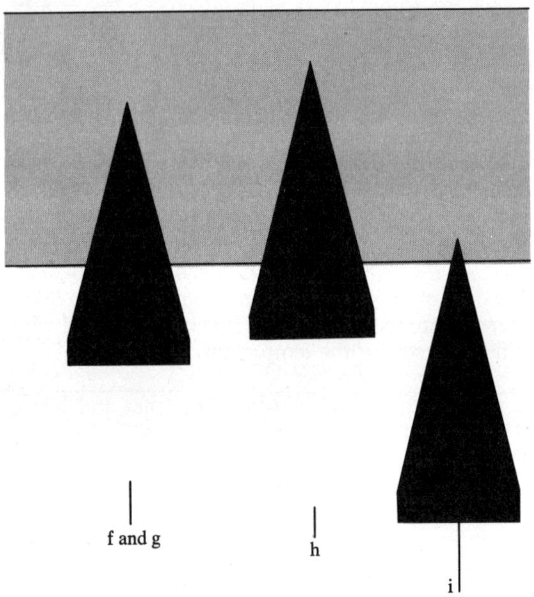

18-2. The pen would travel upscale toward higher transmittance, and the optical attenuator would move out of the reference beam. Except for the very short time that the attenuator is moving, the reference beam energy would be unchanged. During the time the attenuator is moving, the reference beam energy would be slightly lower. The sample beam energy would be unchanged. The DC voltage would be unchanged. There would be no AC voltage except during the very short time the attenuator is moving.

18-4. The absorptions specifically due to the CH_3OAr grouping ($\sim$2850, $\sim$1240, and $\sim$1040 cm^{-1}) are not present in toluene, and most of the absorptions of the anisole spectrum are considerably more intense than the absorptions of the toluene spectrum. The intensity of an absorption band increases as the *change* in dipole moment caused by the vibration (that causes the absorption) increases. Anisole (μ = 1.35D) is more polar than toluene (μ = 0.4D). Since the dipole moment of anisole is greater than that of toluene, it is not surprising that many of the molecular vibrations of anisole cause a greater change in dipole moment than the molecular vibrations of toluene. In general, polar compounds have more intense infrared spectra than nonpolar compounds.

18-5. Increase the cell thickness. If solutions are being used, increase the concentration of the solutions. For less intense bands, decrease the cell thickness and use dilute solutions.

18-6. The anisole-carbon tetrachloride solution has a lower concentration of carbon tetrachloride. The same amount of carbon tetrachloride should be in both beams for exact cancellation. This can be accomplished by using a slightly longer path length for the lower concentration of carbon tetrachloride.

18-7. It is more intense.

18-8. No.

18-9. (a) Energy is removed from both the reference and sample beams at frequencies where carbon tetrachloride absorbs (1530, 1130, 995, and 795 cm^{-1}). The energy of the servo loop is reduced, and it responds more slowly to correct the energy imbalance between the two beams. The more energy that is removed, the slower the response of the servo system. If 100% of the energy is removed, as at 795 cm^{-1} in this experiment, the instrument becomes completely dead.

18-10. The increase in electrical energy compensates for the loss in radiant energy. If all the radiant energy is removed, even a very large increase in the gain would not allow the servo loop to operate. Zero times even a large quantity equals zero.

CHAPTER 20

20-1. (a) I_0 94%, I 23%; (b) I_0 86%, I 25%; (c) I_0 68%, I 0% (could not be used for any quantitative measurement); (e) I_0 89%, I 11%; (f) I_0 88%, I 33%.

20-2. (a) %T 24.5, A 0.61, molar extinction coefficient, ϵ, 61; (b) %T 29, A 0.538, ϵ 53.8; (c) cannot be determined; (e) %T 12.4, A 0.912; (f) %T 37.5, A 0.426.

20-3. (a) The servo system is in equilibrium at absorption maxima. On the slopes of absorptions it is slightly behind the actual transmittance. The amount it is behind depends on the instrument settings and the sample. When a reading is taken at an equilibrium position, the precision and accuracy is likely to be higher. (b) An absorption maximum is an easily located point. A small wavenumber or wavelength error due to misalignment of the paper, paper shrinkage, or instrument error could cause a large error in I, the observed intensity. There can be no error in I from wavenumber inaccuracy if an absorption maximum is used.

20-4. (a) The optical attenuator is less likely to be linear for the extremes of transmission. (b) At very high transmittance small deviations in the baseline, I_0, would cause a greater error than at lower transmittance. (c) As transmittance decreases, the error in reading in *absorbance* units (which is what concentration is proportional to) increases because A = log $1/T$. (d) At *very* low transmittance the servo loop has low energy. Of the four reasons (c) is usually the most important; (d) is very important for I of 0 to 3% T.

20-5. Compound 1 (a) I_0 94%; I 20%, %T 21.3, A 0.672; (b) I_0 85%, I 39%, %T 46, A 0.337; (c) I_0 92, I 84%, %T 87.5, A 0.058. Compound 3 (a) I_0 80.5%, I 47%, %T 57.4, A 0.24; (b) I_0 82%, I 50%, %T 61, A 0.214.

Compound	*Band*	*Mol. Wt.*	*a*	*ϵ (Molar)*	*Concentration (g/liter)*
1	a	100	6.72	672	10
1	b	100	3.37	337	10
1	c	100	0.58	58	10
3	a	200	2.4	480	10
3	b	200	0.3	60	10*

* Absorbance of CCl_4 at this frequency (approximately 0.184, see Fig. 18-21) must be subtracted from 0.214. So that 0.03 = (60)·(10/mol.wt.)(0.01).

20-6. *Ortho* (top), *meta* (middle), *para*.

20-7. The photometric precision (or reproducibility) of the individual infrared spectrophotometers is excellent, but the accuracy *may* not be too good The variations from instrument to instrument may be significant. These differences in optical null instruments are principally from nonlinearities in the optical attenuator.

20-8. The cells may be of different thicknesses. If the thicknesses are accurately known, the data may be corrected to a single thickness. Multiply the calibration data by 48/53 *or* the determinations by 53/48.

20-9. (a) Because the same cell can be used for calibration and unknown scans. (b) Do not take the cell apart until all data is collected. Do not use excessive pressure in filling or cleaning the cell.

20-10. Detector 120°, injection port 160°.

20-11. In general: *Definitely no!* Establish conincident times on three or more liquid phases of different types or collect fractions and obtain spectra.

21-1. Butyl bromide is the top layer in the sulfuric acid extraction and the bottom layer in the other extractions.

21-2. To remove unreacted 1-butanol and by-product 1- and 2-butene.

21-3. A water-butyl bromide azeotrope. Add calcium chloride, stir, and decant into the distilling flask. Distil and collect the fraction that boils at 97 to 103° in the tared yield bottle.

21-4. Dehydration to isobutene. Low temperature and high concentration of chloride ion help to minimize this.

21-5. Hydrolysis to *t*-butyl alcohol and dehydrohalogenation to isobutene. The use of *cold* water, a *weak* base, and *minimum* contact with the washes minimize these reactions.

CHAPTER 22

22-1. (a)

$$CH_3CH_2CH_2I + HO^- \rightarrow CH_3CH{=}CH_2 + I^- + H_2O$$

$$CH_3CH_2CH_2I + EtO^- \rightarrow CH_3CH_2CH_2OEt + I^-$$

$$(CH_3)_3CCl \xrightarrow[(-Cl^-)]{} (CH_3)_3C^+ \begin{cases} \xrightarrow{HO^-} (CH_3)_2C{=}CH_2 + H_2O \\ \xrightarrow{EtOH} (CH_3)_3COEt + H^+ \end{cases}$$

(b) The relative concentrations of the products are proportional to their rates of formation modified by loss of any of the products. Loss may be important only for the two gaseous alkenes. The concentration of ethyl propyl ether is low because of the low concentration of ethoxide ion relative to hydroxide ion. (c) The rates of elimination and substitution vary with the concentration of base. The more concentrated the base, the greater the rate of elimination relative to the rate of substitution. An exception is a reaction mixture where only E_2 and S_N2 mechanisms occur. For this condition the rates of elimination and substitution change by the same amount with changes in the concentration of base.

22-2. (a) The inorganic materials in the reaction mixture would leave residues in the injection port and the column. (b) These other solvents would not have the long retention time needed to avoid overlaps with all possible products.

22-3. The retention times of the alcohols would be too much longer than the retention times of the other compounds. Even if this difficulty were circumvented by temperature programming, there would still be the problem of overlap of the *p*-cymene with the alcohols.

22-4. There should be no effect on the rate of reaction of *t*-butyl chloride. The rate of reaction of 1-iodopropane should be proportional to the concentration of the base.

22-5. For the major errors see the discussion and problems in the text immediately following this question. Also contributing are: errors in measurement and titration, other reactions with 1-iodopropane, the small change in the dielectric constant of the medium as the reaction progresses (see Appendix II), and further reaction of the mixture during titration.

22-6. Equation 22-4 gives:

$$\int_0^x \frac{dx}{(a - x)(b - x)} = k \int_0^t dt$$

Use of partial fractions yields

$$\frac{1}{a - b} \int_0^x \left(\frac{-dx}{a - x} + \frac{dx}{b - x} \right) = k \int_0^t dt$$

or

$$\frac{1}{a - b} [\ln (a - x) - \ln (b - x)]_0^x = kt$$

which rearranges to Equation 22-6a.

22-7. The reaction of *t*-butyl chloride follows first-order kinetics. $\Delta x/\Delta t = k(b - x)$. Or for most accurate results the first-order equivalent of Equation 22-6b should be used: $-2.303 \log (b - x)/b = kt$.

CHAPTER 23

23-1. The dipole moments of these compounds are less than those of other aliphatic halides because of contributions of electronic structures similar to $-\bar{C}-C{=}\overset{+}{Cl}$ (with single bonds downward from each carbon). Weaker dipole-dipole attractions would be expected with the lower dipole moments.

23-2. The inversion of rates compared with aliphatic compounds may be from $ArCHCl_2$ and $ArCCl_3$ reacting by mechanisms other than *S*N1 such as *via* carbenes or carbanions. If the reaction of these halides does proceed by an *S*N1 mechanism, the formation of a carbonium ion with its empty *p* orbital would allow delocalization of one pair of *p* electrons from each chlorine atom present with the *pi* electrons of the benzene ring. This extra delocalization energy apparently more than overcomes the deactivating inductive effect of the chlorine atoms by stabilizing the transition state. Note that because the benzyl carbon atom takes on sp^2 hybridization in the carbonium ion any halogen atom attached to the benzyl carbon atom can be coplanar with it and the aromatic ring. Therefore, maximum overlap of the *p* orbitals of both the halogen(s) and the sp^2 benzyl carbon atom and the *pi* cloud of the aromatic ring can occur.

23-3. (a) The bulky $(CH_3)_3C$ on the primary carbon atom prevents the approach of I^- so that no *S*N2 reaction can occur. (b) No reaction. (c) The same as for any primary chloride and bromide. Neopentyl chloride would react with silver nitrate solution only if boiled, and neopentyl bromide would react with silver nitrate solution within 5 min at room temperature. Adjacent large groups have little steric effect on rates of *S*N1 reactions except in extremely bulky molecules. Neopentyl halides often undergo *S*N1 reactions *slightly* slower than other primary halides such as *n*-alkyl halides.

23-4. The rate would show a moderate decrease because the charge is dispersed in the transition state of the rate determining step.

$$R{-}\overset{+}{O}H_2 \longrightarrow [\overset{\delta+}{R}\cdots\cdots\overset{\delta+}{O}H_2]$$

23-5. The groups attached to the bridgehead are tied back and cannot assume the planar conformation necessary for a carbonium ion. With NaI and acetone: no reaction under the conditions of the test. It would react less readily than most tertiary bromides because the bulky groups are tied back where they would most effectively block attack by a nucleophile from behind.

EXERCISES

23-2.

Possibilities	*Distinguishing tests*	
	Behavior with H_2SO_4	*Alcoholic* $AgNO_3$
2-Bromo-2-methylbutane	floats	ppt. within 3 min at 25°
1,1-Dibromoethane	sinks	ppt. only if boiled

Note that sodium iodide in acetone probably would not distinguish between the two because of the low reactivity of the *gem*-dibromide.

23-3.

Possibilities	*Distinguishing tests*			
	Behavior when shaken with water	$H_2SO_4 \cdot SO_3$	NaI *and acetone*	*Alcoholic* $AgNO_3$
1,2,3-Trichloropropane	sinks	NR*	NR	NR
1-Chloroheptane	floats	NR	ppt. within 5 min at 50°	ppt. if boiled
2-Chlorotoluene	stays dispersed and then slowly sinks	dissolves	NR	NR

*NR = No visible reaction under test conditions.

CHAPTER 24

24-1. Reaction of Grignard reagent with CO_2, O_2, or H_2O. The reagents and atmosphere over the reaction are kept as dry as possible. The use of a sodalime-calcium chloride mixture in the drying tube would prevent more carbon dioxide from entering the reaction mixture. Handling the reagents under and purging the atmosphere in the flask with dry nitrogen would minimize all three reactions.

24-2. It fills the flask with ether vapor which keeps out most of the air with its O_2, H_2O, and CO_2.

CHAPTER 25

25-1. Hexamethylbenzene (first), naphthalene, anthracene, chrysene, 3,4-benzpyrene (last).

25-3. (a) Methylene chloride > chloroform > carbon tetrachloride. In general, the more polar the solvent, the stronger its eluting power. The above are in the order of their dipole moments. The differences between these three are small, and very often chloroform is a stronger eluting agent than methylene chloride because chloroform can act as a very weak hydrogen-bond donor and is more polarizable than methylene chloride.

25-5. Ferrocene, acetylferrocene, resinous by-products, diacetylferrocene, and some sodium phosphates.

25-6. In the order listed above with ferrocene first and some of the resinous by-products eluted after the diacetylferrocene.

25-7. Benzene is a more powerful eluting agent than petroleum ether. Too much benzene might elute the ferrocene and acetylferrocene as one band or as adjacent, overlapping bands.

CHAPTER 26

26-1. Develop the leaf extract spots with 100% isooctane.

26-2. Their color indicates absorption maxima in the visible which usually results from an extensively conjugated electronic system. An idea of polarity can be obtained from their order of elution. The carotenes must be nonpolar since they are eluted so rapidly. Since they are nonvolatile, at least the nonpolar compounds must have molecular weights above 150.

26-3. The monochrometer is before the sample. The ultraviolet and visible radiation is of very much higher energy than that usually encountered in infrared spectrophotometers. If the entire spectrum from the source passed through the sample all the time it was in the spectrophotometer, considerable photodecomposition would be likely.

26-4. Acetone is in the solvent mixture and it absorbs in the ultraviolet.

CHAPTER 27

27-1. The Beer–Lambert law (Equation 20-1).

27-2. $A = \log 100/20 = 0.699$; $0.699 = a \times 100 \times 5 \times 10^{-3}$, $a = 1.4$; for 5% conc. 0.1 mm; for 2% conc. 0.25 mm.

27-3. 1.622 mm, 2.117 mm.

27-4. Hint: how would you expect the *concentration* to affect relative amounts of the vari ous species? Review the chapters on equilibria and association in your general chemistry text.

EXERCISES

27-1. In CCl_4 no interaction with the solvent. In $HCCl_3$ there is some weak hydrogen bonding between $HCCl_3$ and the oxygen of methanol. The absorption near 3600 cm^{-1} is from H—O(CH_3)---H—CCl_3. In dioxane most of the methanol hydroxy protons are hydrogen-bonded to the oxygen of the cyclic ether linkage.

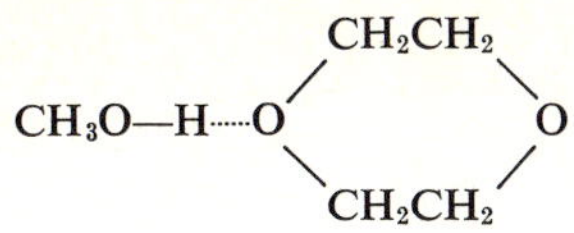

This —O—H---O linkage is similar to that of the methanol dimer and absorbs infrared radiation at the same frequency.

27-3. Hint, see the answers to Problem 27-4 and Exercise 27-1 and continue with the same reasoning.

27-4. Because it can be set to compensate *exactly* for the solvent in the sample beam. Set the spectrophotometer at a frequency (or wavelength) where the solute does not absorb and where the solvent has a moderately intense absorption (between 20 and 65%*T*, preferably between 30 and 50%*T*). If possible, use an absorption maximum. If too strong an absorption intensity is selected, there is not enough energy to easily match the absorption by the solvent. With too weak an absorption the matching requires too fine tuning of the path length of the variable path cell. For Problem 27-4 use the absorption maximum near 1250 cm^{-1}. With nothing in either beam set the 100% control so that the pen reads 95%*T*. Place the 0.05 mm-cell in the sample beam. Fill the variable path cell with carbon tetrachloride, set the cell thickness for slightly less than 0.05 mm, and place the cell in the reference beam. Turn the gain up and adjust the thickness of the variable space cell while it is in the reference beam until the pen reads 95%*T*. Be certain the spectrometer has stayed at the original frequency or wavelength setting during this entire adjustment. Return the gain to its original setting and obtain the spectrum.

27-6. Monomer > terminal > dimer > polymer.

CHAPTER 28

28-1. Polar: water, dimethylformamide, methanol, ethanol, acetic acid, ethylene glycol, 1- and 2-propanol, pyridine, acetone, and methyl ethyl ketone. The last two might be considered to have medium polarity. Nonpolar: petroleum ether, any liquid hydrocarbon as toluene or hexane, carbon disulfide, carbon tetrachloride, methylene chloride, chloroform, and ethyl ether.

28-2. (a) Pentane, isopentane, neopentane. Neopentane would probably be the most soluble in all three solvents at temperatures below 8°. (b) 2-Methylpentane, 3-methylpentane, 2,3-dimethylbutane, and 2,2-dimethylbutane. 2,2-Dimethylbutane would probably be the most soluble in all three solvents. (c) 2,3,4-,3,4,5-, and 2,3,5-trichlorotoluenes (Table 23-8). 2,3,4-Trichlorotoluene would probably be the most soluble in the three solvents. For solids the lowest melting is usually the most soluble. The three trichlorobenzenes (1,2,4-b.p. 213°, mp 17°, others in Table 23-8). The lowest melting, 1,2,4-trichlorobenzene, would probably be the most soluble in the three solvents.

28-3. (a) Formation of simple oxonium ions probably followed by cyclization. Friedel-Crafts reactions similar to Equations 28-7 or 28-8 are unlikely because of deactivating effect of CO_2CH_3.

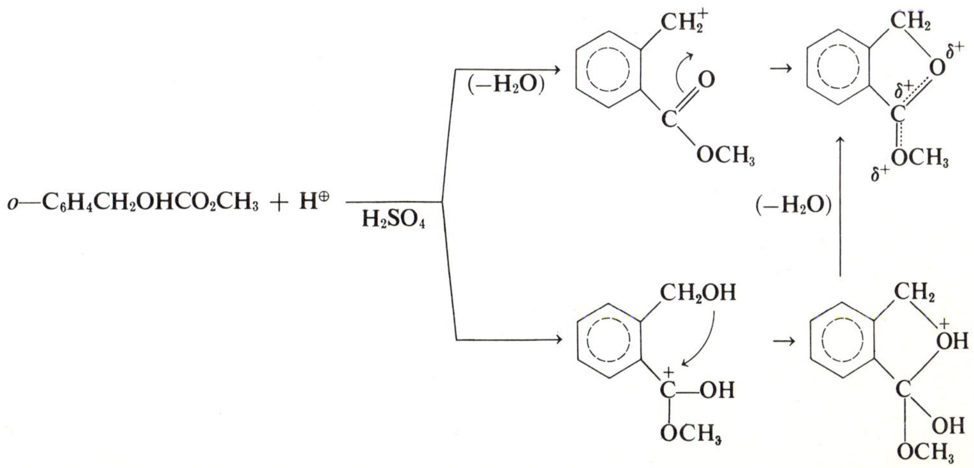

(b) $p\text{—}ClC_6H_4CH_2OH \xrightarrow{H_2SO_4} p\text{—}ClC_6H_4CH_2^{\oplus} \xrightarrow[\text{many steps}]{p\text{—}ClC_6H_4CH_2OH}$

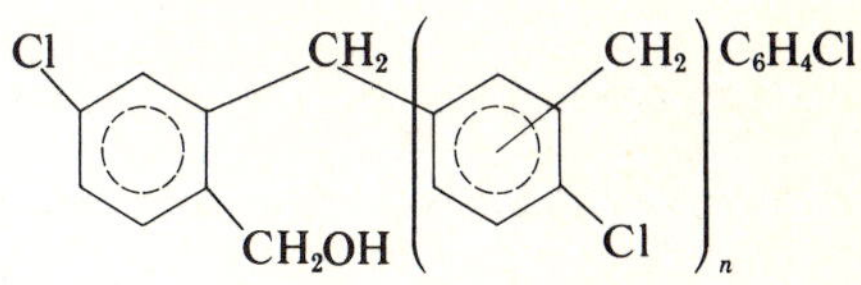

(c) $2,4,6\text{—}Cl_3C_6H_2CH_2OH \xrightarrow{H_2SO_4} Cl_3C_6H_2CH_2\overset{+}{O}H_2$

28-4. Probably because the double bond forms a weak π-complex with protons from the H_3PO_4, $\overset{H^+}{C \rightleftharpoons C}$. This introduces a polar group into the alkene molecule so that it is now soluble in polar solvents. The pentylenes (C_5H_{10}) are the lowest molecular weight alkenes listed in Table 14-2. In phosphoric acid they have a relatively high ratio of polar to nonpolar groups. For higher alkenes this ratio is less. Apparently the crossover point for solubility of alkenes in 85% H_3PO_4 is between 5 and 6 carbon atoms just as the crossover point for solubility of alkanols in water is at 5 carbon atoms.

28-5. The refractive indexes are very close or identical.

28-6. The sodium salt of the organic compound which is insoluble in nonpolar solvents like ether. If this salt precipitates, the organic compound obviously reacted with the sodium.

28-7. Calcium chloride, like water, forms complexes with alcohols.

28-8. To react with the HCl and remove it from the organic phase so that Equation 28-11 is forced to the right-hand side.

28-9. Allyl alcohol forms the carbonium ion, $\overset{\frac{1}{2}+}{C}H_2\text{---}CH\text{---}\overset{\frac{1}{2}+}{C}H_2$, some of which does not react further but remains in ion pairs with anions as $ZnCl_4^{=}$ in the strongly acidic medium. The allyl chloride that does form remains in solution because of a π-complex between it and the $ZnCl_2$ and H^+. Higher molecular weight allylic halides have larger nonpolar groups, and these cause their corresponding π-complexes to be less soluble in the acidic medium. See answer to Problem 28-4. The allyl chloride would precipitate if some distilled water were added or possibly if the test mixture were chilled.

28-10. Allyl chloride is less soluble in cold hydrochloric acid than in warm hydrochloric acid. See answer to Problem 28-9.

28-11. By intensifying the positive charge, electron-withdrawing groups adjacent to the positive carbon atom in a carbonium ion destabilize the carbonium ion and make its formation more difficult. Alcohols with halogen atoms in the *beta* position react *via* S_N1 mechanisms more slowly than similar alcohols without the halogen atoms. 1-Chloro-2-propanol and 1,3-dibromo-2-propanol resemble primary alcohols in their reaction with zinc chloride-hydrochloric acid mixture.

28-12. It evaporates; b.p. 36.5°. Stopper the test tube to prevent it from evaporating. Do not chill; while this would slow the evaporation of isopropyl chloride, this would also slow its rate of formation.

28-13. Ethanol.

CHAPTER 29

29-1. 30 mm.

29-2. $NaBH_4$ reacts with water in the presence of acids, and the $AlCl_3$ would react with the water.

29-3. It is used to rinse any product that may have remained on the magnesium sulfate into the distillin g flask.

29-4. It is heavier than water and allows the successive extraction of the reaction mixture and water wash without the draining of the aqueous layers.

CHAPTER 30

30-1. Both sodium and sodium methoxide react with water to give sodium hydroxide. The sodium that reacts with water cannot form sodium methoxide. If enough sodium reacted with water, there would be insufficient sodium methoxide to react with all the butyl bromide added, and the yield would be lower. Any unreacted butyl bromide would contaminate the product.

30-2. (a) Butanol from; $HO^- + n\text{-}BuBr \rightarrow n\text{-}BuOH + Br^-$, 1-butene from elimination instead of displacement;

$$CH_3O^- + n\text{-}BuBr \rightarrow CH_3CH_2CH{=}CH_2 + CH_3OH + Br^-$$

(b) 1-Bromobutane. (c) The formation of butanol is minimized by preventing formation of sodium hydroxide. See footnote on p. 242 and Problem 30-1. The formation of 1-butene is minimized by keeping the competitive *E*2 reaction to a minimum. This is done by using a solvent of fairly high polarity (methanol instead of a methanol-dioxane mixture, for example), by using butyl bromide instead of butyl iodide, and by heating the mixture only after most of the reaction has occurred. High temperature increases the rate of *E*2 reactions more rapidly than it increases the rate of *S*N2 reactions, apparently because *E*2 reactions have a higher energy of activation. See your lecture text. (d) Instead of refluxing, heat only to 40°. This, however, would require a longer reaction time because the rate of both the *S*N2 and the *E*2 reactions would be decreased. The rates of the *S*N2 reactions decrease more slowly than the rates of the *E*2 reactions as the temperature is lowered, but the rates of *S*N2 reactions *do decrease*. A good compromise would be to heat the reaction to 40° for about one hour and then reflux for another hour, the higher temperature being used only after most of the butyl bromide reacted. Butyl chloride could be substituted for butyl bromide. Here again, however, the amount of time required to complete the reaction may become a limiting factor, i.e., although the rates of *E*2 reactions decrease more rapidly in the series I > Br > Cl than the rates of *S*N2 reactions, the rates of *S*N2 reactions decrease also. If enough time is available, the chloride could be used by heating at 35° for 1 hr, 45° for 1 hr, and refluxing for 1 hr. Another advantage of the chloride is that it is considerably less expensive than the bromide. React sodium butoxide with methyl bromide. Methyl bromide reacts more rapidly in *S*N2 reactions

$$CH_3Br + C_4H_9O^- \, Na^+ \rightarrow C_4H_9OCH_3 + NaBr$$

than butyl bromide, and methyl bromide cannot undergo elimination reactions. Methyl bromide, however, boils at 5°, and a pressure reaction vessel would be needed. (e) No, unless they form azeotropes with the methyl butyl ether. 1-Butene is much more volatile and 1-butanol much less volatile than the product.

30-3. Extract with three 10-ml portions of 25% aqueous calcium chloride. Dry the product (organic layer) over 2 g of calcium chloride. Filter into a 25-ml round bottom flask and proceed as described in the text after Problem 30-3.

30-4. Any low boiling liquid obtained at this point is probably the methanol-methyl butyl ether azeotrope. Wash any such fraction with two 8-ml portions of 25% aqueous calcium chloride, dry over 1 g of calcium chloride, filter into the flask containing the product, and proceed as described after Problem 30-4 in the text.

30-5. (a)

$$(CH_3)_2CHCH_2CH_2ONa + EtBr \searrow$$

$$(CH_3)_2CHCH_2CH_2OEt$$

$$(CH_3)_2CHCH_2CH_2Br + EtONa \nearrow$$

(b)

$$CH_3CH_2C(CH_3)_2{-}ONa + EtBr \rightarrow CH_3CH_2C(CH_3)_2OEt$$

and not by

$$CH_3CH_2C(CH_3)_2{-}Br + NaOEt \rightarrow (CH_3)_2C{=}CHCH_3$$

(c)

$$C_6H_5ONa + (CH_3)_2CHBr \rightarrow C_6H_5OCH(CH_3)_2$$

and not by

$$C_6H_5Br + (CH_3)_2CHONa \rightarrow \text{no reaction}$$

Neither (d) nor (e) can be synthesized by the Williamson synthesis. The following might give reasonable yields of product.

(d)

$$(CH_3)_2C{=}CH_2 + (CH_3)_3COH \rightarrow (CH_3)_3C{-}O{-}C(CH_3)_3$$

(e)

$$(CH_3)_2C{=}CH_2 + C_6H_5OH \rightarrow (CH_3)_3CO{-}C_6H_5$$

BF_3 or a similar catalyst would be needed.

30-6. To intensify the weak absorption bands so that minor components can be identified.

30-7. The instrument is dead *in these regions* where the sample has 0%*T* or almost 0%*T* just as if the source were turned off or a book or other opaque object were placed in both beams.

30-8. *No!* There can be no information obtained in regions of the spectrum where the sample completely or almost completely absorbs no matter what compensation there is in the reference beam. The other compounds in the sample cell could have no absorptions in these regions or have many strong absorptions and the region would appear exactly the same. See Problems 18-9 and 18-10 and the paragraph preceding these problems.

30-9. (a) Because the solvent is more concentrated when it is not diluted with solute. (b) 1. The cell in the reference beam is too thick. 2. The solution in the sample beam is very concentrated with respect to solute and, therefore, dilute with respect to solvent. 3. There are impurities or other compounds in the reference beam that are not present (or are present at lower concentration) in the sample beam cell. 4. Upscale drift of spectrophotometer. Occurs only at frequencies where both sample and reference material have less than $2\%T$ and when the balance control is not correctly set.

CHAPTER 32

32-1. These two carboxylic acids are solids. The higher polarity of the acids that causes liquid carboxylic acids to be more soluble in water than similar alcohols also causes stronger intracrystalline forces. These stronger intracrystalline forces (i.e., high lattice energies) are much less readily broken by energy from the water-carboxylic acid dipole-dipole attractions than are the forces between carboxylic acid molecules in the liquid phase.

32-2. Trichloroacetic acid is a very strong acid because the chlorine atoms withdraw electrons, dispersing the negative charge in the conjugate base and thereby stabilizing it.

32-3. Well over 80 (including geometric but not optical isomers) of all the possible cycloalkenes are included. The structures given below are for acyclic compounds with one double bond in the 1- or 2- position only.

$(CH_3)_2CH(CH_2)_3CH{=}CH_2$, $CH_3CH_2CH(CH_3)CH_2CH_2CH{=}CH_2$, $CH_3CH_2CH_2CH(CH_3)CH_2CH{=}CH_2$, $CH_3(CH_2)_3CH(CH_3)CH{=}CH_2$, $CH_3(CH_2)_3(CH_3)C{=}CH_2$, $(CH_3)_3C(CH_2)_2CH{=}CH_2$, $(CH_3)_3CCH(CH_3)CH{=}CH_2$, $(CH_3)_3CCH_2(CH_3)C{=}CH_2$, $(CH_3)_2CHCH(CH_3)CH_2CH{=}CH_2$, $(CH_3)_2CHCH_2CH(CH_3)CH{=}CH_2$, $(CH_3)_2CHCH_2CH_2(CH_3)C{=}CH_2$, $(CH_3)_2CHCH(CH_3)(CH_3)C{=}CH_2$, $(CH_3)_2CHC(CH_3)_2CH{=}CH_2$, $CH_3CH_2C(CH_3)_2(CH_3)C{=}CH_2$, $CH_3C(CH_3)_2CH(CH_3)CH{=}CH_2$, $CH_3C(CH_3)_2CH_2(CH_3)C{=}CH_2$,

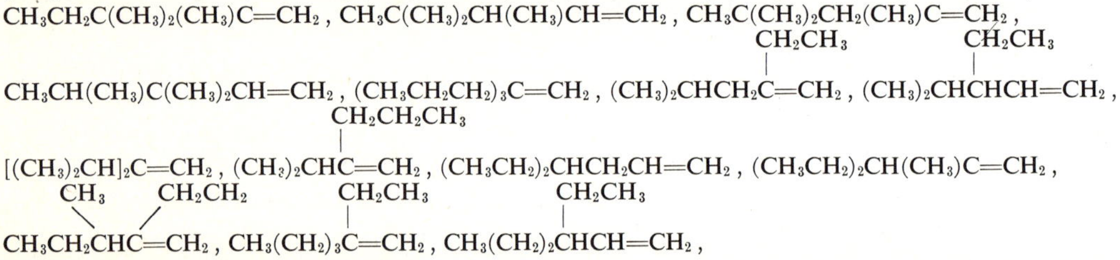

All of the following have *cis* and *trans* isomers.

$CH_3CH_2CH(CH_3)CH_2CH{=}CHCH_3$, $CH_3CH_2CH_2CH(CH_3)CH{=}CHCH_3$, $CH_3(CH_2)_3(CH_3)C{=}CHCH_3$, $(CH_3)_2CH(CH_3)CHCH{=}CHCH_3$, $(CH_3)_2CHCH_2(CH_3)C{=}CHCH_3$, $CH_3CH(CH_3)CH(CH_3)CH{=}CHCH_3$, $CH_3CH(CH_3)_2CH_2(CH_3)C{=}CHCH_3$, $CH_3CH_2CH(CH_3)(CH_3)C{=}CHCH_3$, $(CH_3)_3CCH_2CH{=}CHCH_3$, $(CH_3)_3C(CH_3)C{=}CHCH_3$,

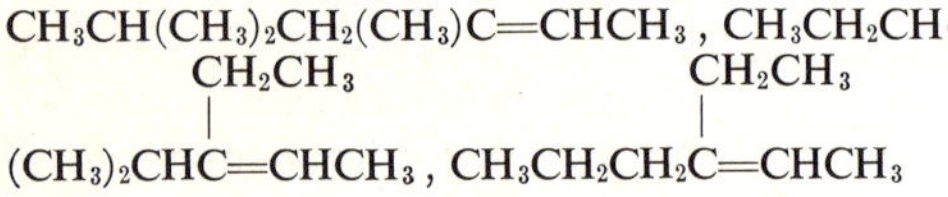

A *few* of these compounds *may* not have boiling points from 112 to 125°, but, if there are such compounds, their boiling points would not be too far outside this range.

32-4. (a)

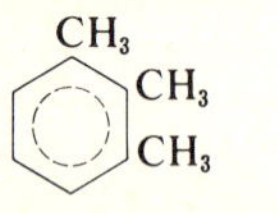

I (In Table 14-4A)
b.p. 176°
Acid m.p. 190°d

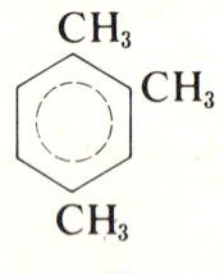

II
b.p. 169°
Acid m.p. 216°

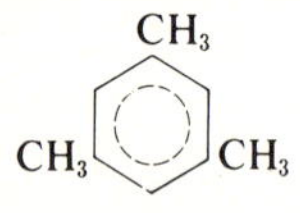

III (in Table 14-4A)
b.p. 164°
Acid m.p. 379°

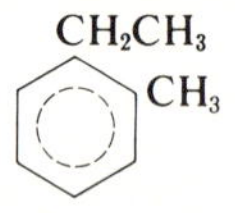

IV
b.p. 165°
Acid m.p. 206°

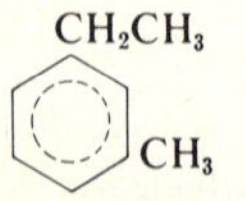

V
b.p. 161°
Acid m.p. 315°

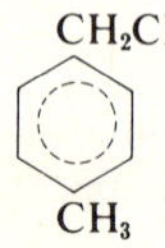

VI
b.p. 162°
Acid m.p. 300°

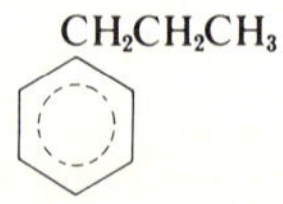

VII
b.p. 160°
Acid m.p. 121°

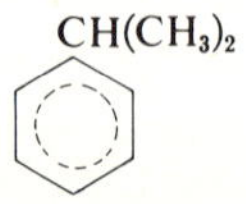

VIII (in Table 14-4A)
b.p. 153°
Acid m.p. 121°

(b) I, II, and III are in Solubility Class N_2 ; IV, V, and VI are borderline between Classes N_2 and I; VII and VIII are in Class I. (c) On the basis of boiling points and solubility one could not distinguish among III, IV, V, and VI; among V, VI, and VII; and between II and IV. The greatest number of these isomers that boil between 160 and 169° that could be identified on the basis of solubility and classification tests and boiling points is three. The ethyltoluenes give borderline solubility tests and could not be distinguished from the trimethylbenzenes or propylbenzenes on the basis of solubility classification. (d) Oxidation to the acid would allow each of these to be identified. Nitro derivatives are apparently unrecorded for I, V, VI, and VII. Note that the melting points of the acids could not be used to distinguish among III, V, and VI with a mineral oil melting point bath. The 4.5° differential of uncorrected boiling points for distinguishing between compounds may be optimistic for single plate distillations with beginning students and atmospheric pressures much different than 760 mm. The syntheses of many of the aromatic polycarboxylic acids from alkylbenzenes and their purification are not too easy. If all alkylbenzenes in this boiling range are possible unknowns, their identification without instrumental methods becomes even more difficult. *t*-Butylbenzene, b.p. 169°; 4-Isopropyltoluene, b.p. 177°; Isobutylbenzene, b.p. 173°; 2-Phenylbutane, b.p. 173°. (e) The out-of-plane Ar-H bending and aromatic overtone-combination infrared absorptions (Figs. 18-34 to 18-44) would differentiate among most of the alkylbenzenes boiling below 180°. If boiling points and the singlet or doublet absorption near 1380 cm^{-1} were also used, all the alkylbenzenes boiling below 180° could be easily and unambiguously identified without reference to any other chemical and physical properties including other regions of the infrared spectrum. If reference spectra of all possible unknowns were available, all aromatic hydrocarbons (not just alkylbenzenes) with boiling points to at least 500° (molecular weights to 300 amu) could be identified by comparison of the spectra of unknowns with those of known compounds (fingerprinting) without use of any other properties including boiling points. The only exception to this might be *n*-alkyl derivatives with a chain length greater than 11 methylene groups. Such compounds that are homologs and differ by only 1 or 2 methylene groups might have what appear to be identical infrared spectra.

32-5. Two, the amide and the anilide.

32-6. The *p*-bromophenacyl ester could be used to distinguish between furoic and sebacic and between furoic and cinnamic acids. The amide should not be used to distinguish between acetonedicarboxylic and any other acid, and, if more suitable derivatives can be found (for example, the anilide), the amide should not be used to distinguish between furoic and acetylsalicylic acids and between furoic and cinnamic acids. The *p*-nitrobenzyl ester should not be used to distinguish among β-campholytic, acetonedicarboxylic, 2,5-dimethylbenzoic, and any other acids, between sebacic and methylmalonic acids, and between malonic and phenylpropiolic acids. The anilide should not be used to distinguish between the cinnamic and *d*- or *l*-mandelic acids (see below), between 2,5-dimethylbenzoic and acetylsalicylic acids, or between furoic and acetylsalicylic acid. Note that the amide would have to be used to identify acetonedicarboxylic acid. Fortunately, acetonedicarboxylic acid does melt 4° above the acids with which it might be confused. This is sufficient differential if great care is used and the derivatives are recrystallized to a constant melting point. Classification tests would also help to distinguish among these three acids.

32-7. The ice water wash removes most of the H_2SO_3 (or SO_2) and HCl as the acids or as the salts of the amine (or alkali metal if an alkali metal carboxylate was used). The HCl wash removes all the amine and its salts that may be left. The NaOH wash removes the last traces of HCl, H_2SO_3, and any carboxylic acid that did not react. The final water washes remove any sodium hydroxide and salts that remain behind.

CHAPTER 33

33-1. Methanol is probably the best. Too large a volume of the hydrocarbon solvents would be needed (the crystallizing solution should not be heated much above 35°). Benzene and water would be unsatisfactory because they freeze at too high a temperature.

33-2. Do not heat the methanol to too high a temperature when dissolving the solid. As high as 35 to 40° may allow crystal formation without oils, but if an oil appears, add more methanol and warm the solution to only 30°. Save a few crystals of the crude product and seed the mixture as it cools. Allow the mixture to cool to room temperature; then chill below 0° by placing the flask in a mixture of calcium chloride and ice. Stir the mixture and scratch the sides of the flask until crystals appear. Allow the flask to sit in the calcium chloride-ice mixture for 8 to 10 min after crystals appear; stir every 2 to 3 min. Filter through a chilled Büchner funnel.

CHAPTER 34

34-1. If the temperature is much above 30°, the oxidation conditions are too vigorous and a significant portion of the cyclohexanone will be oxidized to adipic acid.

CHAPTER 35

35-1. More soluble because the intracrystalline forces of N,N-disubstituted amides are weaker.

35-2.

	Water	*Ether*	H_2SO_3	H_3PO_4	*Sol. Class*
Methyl acetate	+	+			S_1
Methyl decanoate	−	+	+	−	N_2
Methyl pyruvate	+	+			S_1 or S_{1a}*
sec-Butyl pyruvate	−		+	+	N_1
p-Nitrobenzoyl chloride	borderline*	+			S_{1a}* or A*
Butyric anhydride	+*	+			S_{1a}*
N,N-Dimethylformamide	+	+			S_1
N-Methylpentanamide	−				M
Ethyl formate	+	+			S_1 or S_{1a}*

* Shows this solubility characteristic because of hydrolysis in water or aqueous base test solutions.

CHAPTER 36

36-1. The acid and water washes remove the inorganic by-products and water-soluble materials. The activated carbon removes polymeric and colored, highly conjugated by-products; any *p*-nitrotoluene and other water-insoluble nonacidic materials are also left behind on the filter paper with the carbon black.

CHAPTER 37

37-1. There are only weak van der Waals attractions between paraffin oil and hydrocarbons such as benzene and cyclohexane, and the attractions of the liquid phase for these two hydrocarbons are almost identical. With only these weak attractions between the liquid phase and the compounds being separated the relative retention times of the compounds is in order of increasing molecular weight. Note that the attractions of the benzene (or cyclohexane) molecules for one another are not important since the hydrocarbons are in very dilute solution in the liquid phase. The fact that the induced-dipole induced-dipole attractions between benzene molecules (from the high polarizability (pp. 215) of the *pi*-electron clouds) are stronger than the similar attractions between cyclohexane molecules is unimportant here as most of the benzene molecules are not sufficiently close to one another. Note that benzene's stronger intermolecular attractions are important in the pure liquid since it has almost the same boiling point as cyclohexane even though its molecular weight is lower than cyclohexane's by 6 amu. On the diethylene glycol adipate column both cyclohexane and benzene have dipole-induced dipole interactions with the liquid phases. The attractive forces are stronger with benzene because of its easily polarized *pi* electrons.

37-2. The retention time is inversely proportional to the activity coefficient, *a*, and the vapor pressure (Equation 8-1). For chromatography of a homologous series of polar compounds on a polar phase *a* often increases slightly with the addition of each methylene group. This results from the polar group's becoming a smaller part of the molecule as the number of nonpolar methylene groups increases. The vapor pressure usually decreases much more than *a* increases so that the retention time increases with the number of methylene groups. Glycerine is a very polar liquid; most nonpolar and slightly polar organic liquids are insoluble in it. A polar compound such as methanol has a very low activity coefficient in glycerine. Apparently ethanol (33% polar OH) has an activity coefficient in glycerine sufficiently higher than the activity coefficient of methanol (50% polar OH) to more than compensate for the higher vapor pressure of the lower homolog.

37-.3

1	2
a	a
b	d, h (maybe c)
c	f
d	e, h
e	g
f	c, d, h (maybe a, b)
g	b

37-4. f and maybe b

37-5. (a) Pentane < hexane < heptane < octane. (b) On paraffin wax ethanol < 1-propanol < pentane < hexane < heptane; on Carbowax pentane < hexane < heptane < ethanol < 1-propanol. (c) On both liquid phases methanol < ethanol < 1-propanol < 1-butanol. (d) On paraffin wax methanol < ethanol < 1-propanol < ethylene glycol < 1-octanol < 1-decanol; on Carbowax methanol < ethanol < 1-propanol < 1-octanol < 1-decanol < ethylene glycol.

37-6. The shorter the alkyl ester chain, the less interaction between these ester groups and the methyl groups on the xylene. The less the nonselective interactions of the side chain (p. 300) dilute the selective interactions between the electron rich xylene rings and the electron poor phthalate rings, the better the separation of the three xylenes. The ethyl or butyl phthalate columns would bleed excessively at the temperature needed to chromatograph the xylenes in a reasonably short time.

37-7. In extreme cases the compounds may be injected in such large quantities and may be so different from the liquid phase (as alkanes and DEGS) that they slightly change the partition (or solvent) properties of the liquid phase.

37-8. More tailing and probably slightly longer retention times.

37-9. The alkanes would have little or no tailing.

37-10. The lower a, the more the compound concentrates in the liquid phase (and the less it is in the vapor phase). Therefore, when identical amounts of two compounds with different a values are injected, the one with the lower a value will tend to concentrate more in the liquid phase and overload it more than the compound with the higher a value.

37-14. Plot log of retention time versus number of carbon atoms and determine the retention times graphically. The plot should be a straight line. Plot peak height for a given concentration versus number of carbon atoms. This cannot be accurately extrapolated as far as the retention time plots can, but it should give good results for the esters mentioned in this problem.

CHAPTER 38

38-1. Aldehyde molecules do not form hydrogen bonds with each other.

38-2. The chromate is reduced to Cr^{+++} as it oxidizes the alcohol. This ion has a different color than the chromate ion.

38-3. To dry the mixture. Calcium chloride could be used if the only purpose of the experiment is to produce propanal. Calcium chloride, in fact, would probably help give a purer product. If, however, the distillate is to be analyzed by gas chromatography, an inaccurate analysis would be obtained because of removal of 1-propanol by calcium chloride. Potassium carbonate is unsatisfactory as a drying agent because it is a moderately strong base and might cause condensation of the aldehyde.

38-4.

$$CH_3CH_2CH_2OH + H^+ \longrightarrow \begin{cases} CH_3CH{=}CH_2 \\ CH_3CH(OH)CH_3 \end{cases}$$

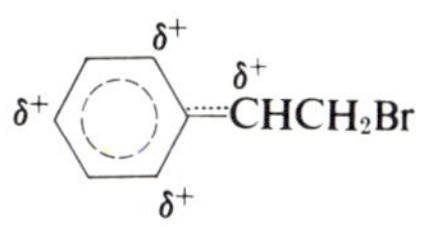

$$CH_3CH_2CH_2OH \xrightarrow[H^+]{[O]} CH_3CH_2CO_2H$$

38-5. For the liquid phase DEGS is probably the most satisfactory. Silicone oil with dimer acid might be preferred if accurate determination of the propanoic acid were the primary consideration. Methylchlorophenyl silicone fluid would probably be satisfactory. The other liquid phases should not be used. Why? For the solid phase only the acid washed Chromosorb G would be satisfactory as the basic solid supports would cause adsorption of the acids and probably of the other compounds. The basic supports might also generate small amounts of artifacts from the aldehydes. The glass beads do not hold enough liquid phase.

38-6. Dissolve a small aliquot of the distillate in ten times its volume of methylene chloride, extract with two small portions of aqueous sodium bicarbonate, and then chromatograph the methylene chloride solution. The propanoic acid is extracted by the bicarbonate, and the chromatogram of this methylene chloride solution has no carboxylic acid peaks.

38-7. $$W_x = \text{Weight per fraction} = \frac{\text{Peak ht.}}{\text{Peak ht. to volume ratio}} \cdot D_4^{20} \text{ of } \textit{compound}$$

Note that weight is $\times 10^{-3}$ g if volume is in μl.

$$\%\text{ of X in distillate} = \frac{W_x \times (100) \times (\text{wt of organic layer})}{1 \times 10^{-2} \cdot 0.81\ (\text{total wt of distillate})}$$

$$\%\text{ of X in residue} = \frac{W_x(100)}{2 \times 10^{-2}\ 1.15} \cdot \frac{1}{2}$$

if W_x is in grams

It is necessary to assume that all the organic material is extracted into the three methylene chloride washes.

38-8. (a) Different errors in calibration and determination with the gas chromatograph. Note that identical errors in each would cancel each other. (b) Evaporation of some propanal before analysis. (c) Densities other than that assumed. (d) Change of volume in the aqueous layer of the distillate because it dissolves some potassium sulfate. This error can be largely eliminated, how? Verify your answer with the instructor and then adjust the calculations for this correction. (e) Some of the organic compounds may remain in the aqueous potassium sulfate solution from the distillate or may remain in the aqueous solution after the methylene chloride extracts. (f) Errors in measuring the volume of the aliquots and extracts. (g) A small amount of propylene may be formed (answer to 38-4) and lost. If any 2-propanol is formed, it would be detected in the chromatogram.

CHAPTER 40

40-1. Some of the —SO_2NH_2 might be hydrolyzed by higher concentrations of HCl or longer reaction times.

40-2. An excess of sodium hydroxide might cause displacement of the amino group because it is *para* to a strong electron-withdrawing group.

$$HO^- + NH_2C_6H_4SO_2NH_2 \rightarrow NH_3 + {}^-OC_6H_4SO_2NH_2$$

Excess base would also react with the sulfonamide group to give a water-soluble salt.

$$HO^- + NH_2C_6H_4SO_2NH_2 \rightarrow NH_2C_6H_4SO_2\bar{N}H + H_2O$$

At low pH the amine forms a water-soluble salt.

$$H^+ + NH_2C_6H_4SO_2NH_2 \rightarrow \overset{+}{N}H_3C_6H_4SO_2NH_2$$

CHAPTER 41

41-2. Alcohols react with sulfonyl chlorides. Benzene would freeze.

41-3. The sodium hydroxide reacts with the HCl, removing it from the equilibrium:

$$ArSO_2Cl + C_6H_5NH_2 \rightleftharpoons ArSO_2NHC_6H_5 + HCl$$

and shifting the equilibrium to the right to give more product. It also reacts with the sulfonamide to give a water-soluble salt and thereby removes product from the organic phase and shifts the equilibrium to the right; Sodium hydroxide, of course, cannot form salts with N,N-disubstituted sulfonamides.

41-4. Sulfonic acids are very difficult to chromatograph as the free acids. They are not volatile and are strongly adsorbed by most solid supports regardless of the liquid phase used. The acids could be chromatographed by first converting them to the ester *via* the sulfonyl chloride. The esters could be chromatographed on a nonpolar liquid phase in a 6 to 10 ft column or on a weakly or moderately polar phase (such as DEGS) in a 2 to 4 ft column. Silanized, acid-washed Chromosorb G or a similar solid support should be used.

CHAPTER 42

42-1. They are reduced to the amine.

42-2 and 42-3. Shown for the generalized nitrotoluene formula; reactions of the *ortho* and *para* isomers that differ are shown separately. Numbers are the *predicted* relative volatilities with 1 indicating the most volatile. The relative volatilities of compounds with the same number are probably similar. The *relative* volatilities of the *ortho* and *para* series should be the same.

$$\underset{3}{CH_3C_6H_4NO_2} \xrightarrow{[H]} \underset{1}{CH_3C_6H_4NO} \xrightarrow{[H]} \underset{6}{CH_3C_6H_4\overset{+}{N}(O^-){=}NC_6H_4CH_3} \text{ or } \underset{4}{CH_3C_6H_4NHOH}$$

$$CH_3C_6H_4NHOH \text{ or } CH_3C_6H_4N^+(O^-){=}NC_6H_4CH_3 \xrightarrow{[H]} \underset{5}{CH_3C_6H_4N{=}NC_6H_4CH_3} \xrightarrow{[H]} \underset{6}{CH_3C_6H_4NHNHC_6H_4CH_3} \xrightarrow{[H]} \underset{2}{CH_3C_6H_4NH_2}$$

$$CH_3C_6H_4NH_2 + CH_3C_6H_4NO \longrightarrow CH_3C_6H_4N{=}NC_6H_4CH_3$$

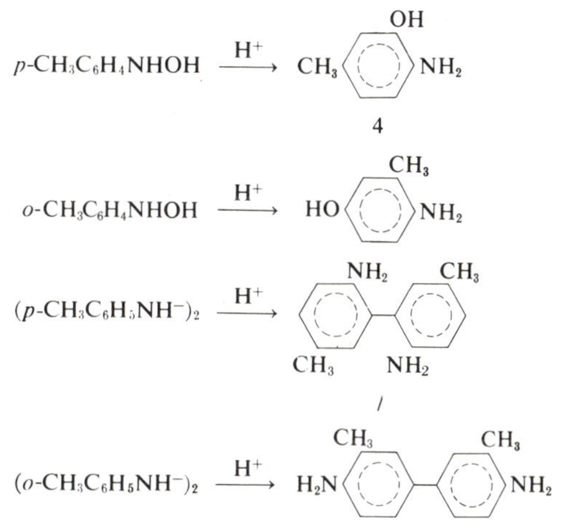

42-4. Nitrotoluene is the most likely contaminant. Large amounts of other reduction products are less likely as they are all more easily reduced than nitrotoluene. The rearrangement products indicated in Equations 42-7 and 42-8 hopefully have been minimized by keeping the concentration of hydrochloric acid low. Any aminophenol present reacts with the sodium hydroxide to form the nonvolatile sodium salt which does not steam distil.

42-5. Mix the distillate with 8 ml of concentrated hydrochloric acid; then follow (a) or (b): (a) Place the mixture in a small distilling flask or round bottom flask with a distillation head adapter. Attach a small condenser (Fig. 3-1, do not use an external source of steam) and steam distil the mixture until no more oily liquid (mainly nitrobenzene) collects in the distillate. (b) Or extract the acidified distillate with two 20-ml portions of ethyl ether (**Fire hazard**). Your laboratory instructor will tell you how to discard the extracts.

Chill the acidified toluidine solution that remains from the extraction or steam distillation. To this slowly and carefully add 20% chilled sodium hydroxide solution until the mixture has a pH greater than 11; then add a few drops more. To reduce the solubility of the amine in water dissolve as much sodium chloride in the solution as possible. Extract the aqueous solution with two 15-ml portions of benzene (**Fire hazard**). After the water has been drained from the benzene layer, drop 2 pellets of NaOH into the separatory funnel and shake the funnel. This dries the benzene layer. Take care that the pellets stay in the funnel when draining the benzene layers into a 50-ml round bottom or distilling flask. Distil the benzene solution and collect the fraction that boils between 193 and 202°.

42-6. Most aromatic amines are easily oxidized. Some of the products are those obtained by the *reverse* of reactions 42-4 and 42-5. The dark impurities are principally quinoid type compounds such as

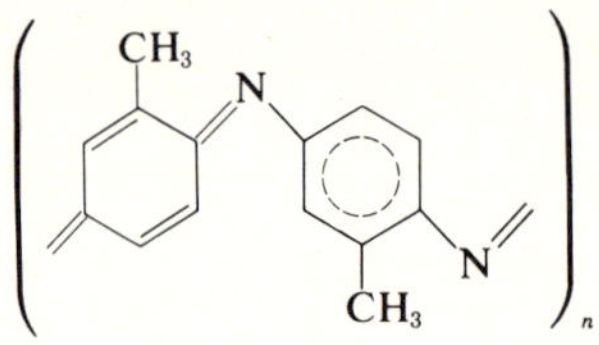

the two nitrogen atoms on each ring may also be *ortho*. The zinc dust reduces the oxidation products, usually to the amine. Contamination of the amine by oxidation products can be greatly reduced or eliminated by filling the bottle with nitrogen. Light-catalyzed reactions can be greatly reduced by storing the amine in a dark bottle away from light.

42-7. Tetraethylenepentamine on Chromosorb G DMCS (*not* AW-DMCS) is the best choice. QF-1 and Carbowax would be acceptable, in fact, these would be preferred if only enough NaOH were added to exactly neutralize the HCl, why? Liquid phases b and c would be very poor choices.

42-8. Compounds with two aryl rings, such as $CH_3C_6H_4NHNHC_6H_4CH_3$, are less volatile than the other materials and would have much longer retention times than the other compounds. Detectable quantities of these less volatile compounds do not steam distil if the directions in this chapter are followed. The long retention times can be avoided by temperature programming or by obtaining two chromatograms at temperatures approximately 40° apart.

42-9. There was no significant polymeric material which would not be eluted from the chromatograph. No significant amounts of phenolic by-products were produced; these products would remain in the first aqueous sodium hydroxide solution. The assumed densities are correct if they are not determined experimentally. Peak height to weight (or volume) ratios are correct. No significant amount of organic compounds are lost by dissolving in the aqueous phases.

CHAPTER 43

43-1. m-$C_6H_4(NH_2)_2$, m-$C_6H_4NO_2NH_2$, m-$C_6H_4(NO_2)_2$, m-$C_6H_4NH_2NHOH$, m-$C_6H_4NO_2NHOH$ plus many other possibilities; see Equations 42-4 to 42-10.

43-2. Reduction is often defined as the gain of electrons. If another electron-withdrawing group (such as $—NO_2$) is on the ring, the nitro group is more electron-poor and can more easily gain electrons.

43-3. This temperature would make the reduction too vigorous, and considerable amounts of *m*-phenylenediamine would be produced.

43-4. *m*-Nitroaniline, sulfur, a small amount of *m*-dinitrobenzene, and possibly very small amounts of some of the other reduction products listed in the answer to 43-1.

CHAPTER 44

44-1 and **44-2.** Hint: read Appendix V. Note in Table V-1 the dipole moments and boiling points of o-$CH_3C_6H_4N(CH_3)_2$ and $C_6H_5N(CH_3)_2$; also *p*-toluidine, μ 1.44D; *m*-toluidine, μ 1.55D; benzylamine, μ 1.28D; aniline, μ 1.58D; p-$CH_3C_6H_4N\cdot(CH_3)_2$, μ 1.30D; *m*-nitroaniline, μ 4.95D.

44-3. Amino nitrogens of aromatic amines are considerably less negative than amino nitrogens of aliphatic amines because the lone pair of electrons delocalize from the nitrogen to the aromatic ring. Thus aromatic amino groups are much weaker hydrogen bond acceptors than aliphatic amino groups. The hydrogen bonds are considerably weaker than the 6 to 8 kcal/mole shown in Table 44-6 for hydrogen bonds between hydroxylic protons (as in water) and aliphatic amines. Since these hydrogen bonds are weaker, aromatic amines are less soluble in water.

44-4. Tertiary amines are more soluble in water than primary and secondary amines because the intermolecular attractions of tertiary amines are significantly weaker than the intermolecular attractions of other amines. This results from lack of hydrogen bonding *between* amine molecules and the lower dipole moment and weaker polar attractions of most tertiary amines. Tertiary amines, however, can still have the relatively strong hydrogen bond with hydroxylic protons, $\diagdown\!—N\cdots H—O\!\diagup$. For aliphatic amines this hydrogen bond is slightly stronger than the hydrogen bond between adjacent water molecules. Ether molecules cannot form hydrogen bonds with one another, and while their dipole moments are lower than the moments of similar alcohols, they are still moderately high. Ethers are not more soluble than similar alcohols in water because the hydrogen bonds between ether and water molecules are no stronger than the hydrogen bonds between adjacent water molecules.

44-5. (a) B, (b) S_1b, (c) S_1b or B, (d) M, (e) A_2, (f) M, (g) S_1b, (h) M, (i) M, (j) S_1b or B, (k) S_1b, (l) A_2, (m) S_1 or S_2, (n) B, (o) B, (p) M.

44-6. (a) M, (b) S_1b or B, (c) M, (d) M, (e) M, (f) B or M, (g) S_1b, (h) B or possibly S_1b, (i) M.

44-8. When heated, the nitrous acid solution evolves gases which might be mistaken for nitrogen.

44-10. To remove the HCl and shift Equation 44-26 to the right and to prevent the HCl from reacting with the amine to give the unreactive salt. The organic compounds are miscible with the benzoyl chloride and are, therefore, in much better contact with it.

44-11. The reaction product is partially neutralized with hydrochloric acid because some amides are soluble in very strong bases. This test is also used to make derivatives of amines. The remaining steps ensure a good yield of the pure benzamide derivative. The chilling in ice water increases the recovery of amide. Without the chilling the reaction mixture remains very warm from heat of reaction between the acids and bases. The Na_2CO_3, dilute acid, and water washes are to remove any benzoyl chloride, benzoic acid, unreacted amine, sodium chloride, and any other water-soluble salts.

CHAPTER 45

45-1. (a)

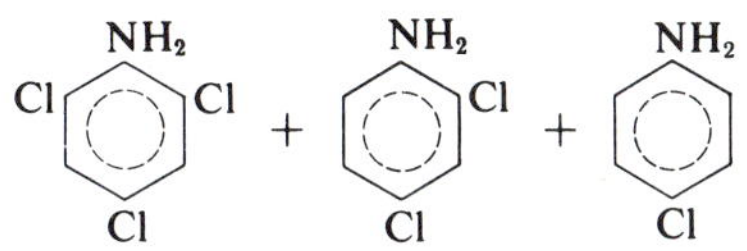

+ unreacted aniline

(b)

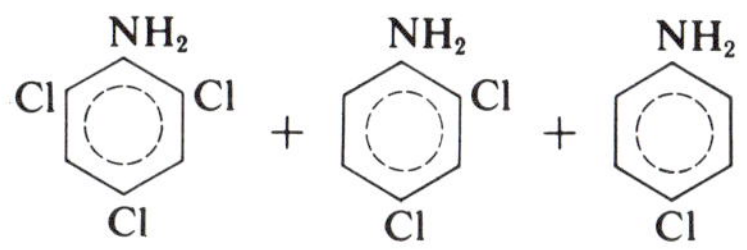

+ unreacted aniline

(c)

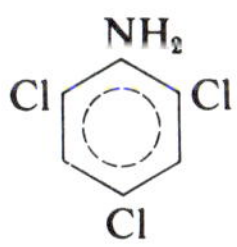

(d)

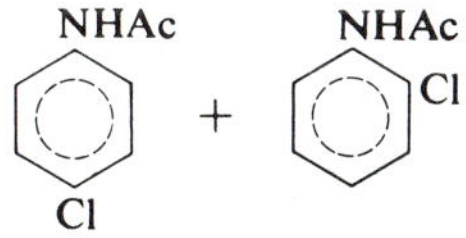

(e)

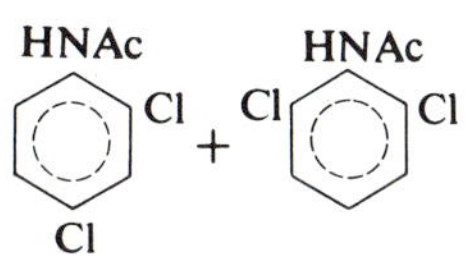

(almost none)

45-2. (a) The *ortho* isomer is more soluble than the *para* isomer in most, if not all, solvents. (b) The boiling points are close together and too high. There probably would be significant decomposition from fractional distillation at atmospheric pressure. Fractional distillation at reduced pressure would require a more efficient column and much more time. (c) Column adsorption chromatography. (d) Use a higher reaction temperature.

45-3. No.

45-4. The *para* by-product may have some commercial value or the extra steps involved in Equations 45-4 and 45-5 cost more than the loss caused by separation and discard of the *para* isomer.

45-5. Sulfanilamide is more reactive to electrophilic halogenation than benzene.

CHAPTER 46

46-1. CH_3NO_2, S_1n-A_2; $CH_3(CH_2)_3NO_2$, A_2; $(CH_3)_3CNO_2$, M; $C_6H_5NO_2$, M; 2,4-dinitroaniline, M; $(NO_2)_3CH$, S_1a; $CH_3CH{=}CHC(CH_3)_2NO_2$, M; $CH_3CH{=}C(CH_3)NO_2$, A_2.

46-3. A mixture of amines would be obtained which would probably be difficult to separate. It would be satisfactory if R or R′ were H or if they were *very* different.

46-4. Treat the organic halide with alcoholic silver nitrite:

$$RX + AgNO_2 \rightarrow RNO_2 + AgX\downarrow$$

The primary halides might rearrange to secondary and tertiary nitro compounds, and secondary halides might rearrange to tertiary compounds. Sufficient halide may react to give the nitrite ester to make the test too insensitive:

$$RX + AgNO_2 \rightarrow RONO + AgX\downarrow$$

CHAPTER 47

47-1. Heat the ethanol mixture to only 42 to 43° to dissolve the *o*-nitrophenol. Allow to cool to room temperature, then chill in an ice bath for 10 to 15 min, and then filter the mixture.

CHAPTER 49

49-1. Phenols substituted in *ortho* positions with hydrogen bond acceptors have lower boiling points than their *meta* and *para* isomers because the hydroxylic proton is hydrogen-bonded to the *ortho* group and not to other molecules. The *o*-hydroxyphenols and *o*-halophenols do not have the requisite geometry (5 or 6 atoms in the cyclic structure *not* including the protons) for the formation of a strong cyclic-type hydrogen bond. The halogen atoms (except fluorine) are poor hydrogen

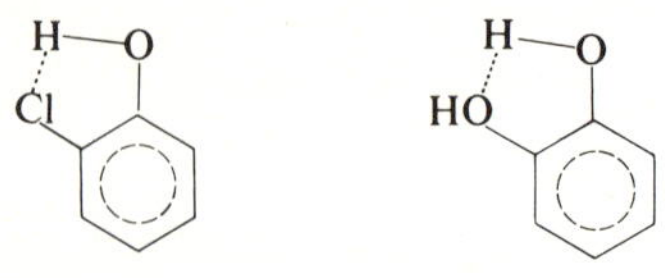

bond acceptors. Since the internal cyclic hydrogen bonds are weaker in these compounds, they tend to undergo intermolecular hydrogen bonding more than compounds such as *o*-nitrophenol but less than *m*- or *p*-XC_6H_4OH (where X = halogen or OH) and, therefore, are somewhat more volatile than their *meta* or *para* isomers.

49-2. (a) and (c) *para*, because of its symmetrical structure, it forms stronger crystal lattices which are less readily broken by dissolving in solvents or by melting. (b) The *meta* isomer is more soluble than the *ortho*. Both hydroxy groups of the *meta* isomer are free for intermolecular hydrogen bonding, but in the *ortho* isomer only one hydroxy proton is available for intermolecular hydrogen bonding.

49-3. The 1,3,5-trihydroxybenzene is very symmetrical with three hydroxy groups that are oriented 120° to each other. This gives three readily accessible, evenly spaced points at which the phenolic molecules can be hydrogen-bonded to one another. This results in a compact crystal structure with a relatively high lattice energy.

49-4. Pyrogallol > hydroquinone > hydroquinone monomethyl ether > phenol > *o*-nitrophenol.

49-5. Oxidation is the loss of electrons. The negative oxide (—O^-) is a much stronger electron-donating group than the neutral hydroxy. Since phenoxides have a higher electron density in the aromatic ring than phenols do, they are more easily oxidized.

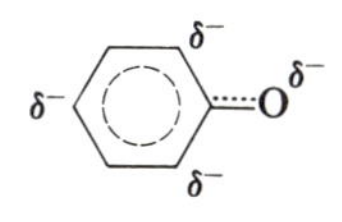

49-6. (a)

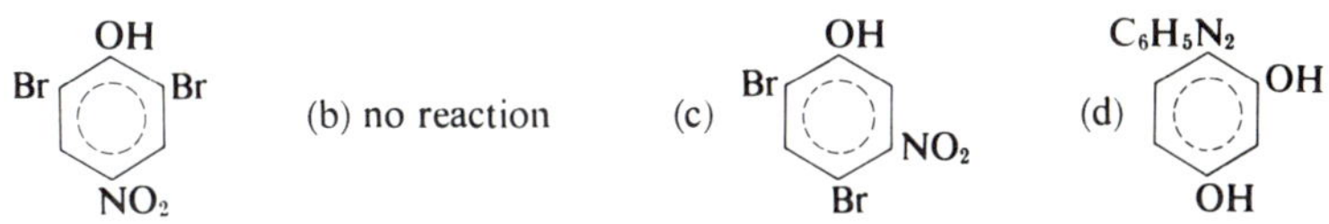

49-7. More rapidly with Br_2 in H_2O. Bromination of phenols is an electrophilic substitution and involves a development of charge in the transition state. The rate of reaction is increased if the development of this charge is stabilized. The charge development is stabilized by solvents with a high dielectric constant such as that of water.

CHAPTER 50

50-1. The colder the water, the more oxygen it can dissolve and the more it can transfer oxygen to the cuprous chloride while they are stirred together.

CHAPTER 51

51-1. Acid reacts with cyanide salts to give hydrogen cyanide, a very poisonous gas.

$$CuCN + H^+ \rightleftharpoons Cu^+ + HCN$$

If the solution becomes warm, HCN bubbles out of the solution forcing the equation to the right-hand side until all the CuCN has reacted.

51-2. A small excess of Na_2CO_3 will not make the solution extremely alkaline as a small excess of NaOH would. While a small excess of Na_2CO_3 will cause a small drop in yield and should be avoided, a small excess of NaOH would cause a large decrease in yield. The reaction of sodium hydroxide with the acid solution is much more exothermic than the reaction of sodium carbonate with the acid solution. It is, therefore, easier to keep the temperature lower when sodium carbonate is used to neutralize the solution.

CHAPTER 53

53-1. The —SO_3H or —SO_2NH_2 group can be removed in one step. The nitro group requires two steps for removal. The —SO_3H group is much larger than the nitro group and blocks the *ortho* positions more effectively. This latter is a steric effect and is in addition to the strong electronic deactivation of the *ortho* and *para* positions by both the nitro and sulfonic acid groups.

53-2.

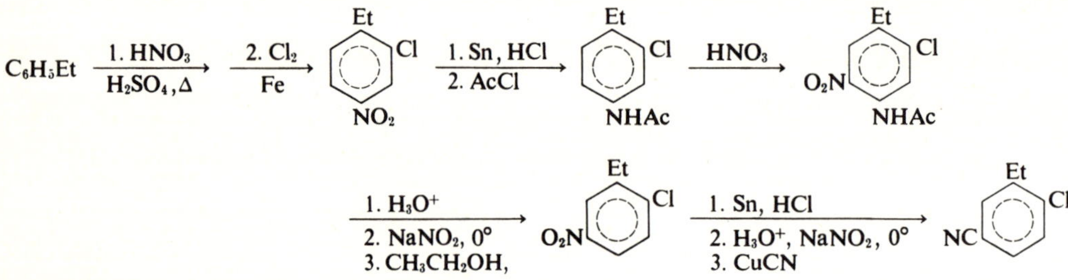

CHAPTER 54

54-1. To provide a portion of the aluminum surface with a much thinner coating of oxide than is usually found.

54-2. Some of the aluminum displaces mercury from the mercuric chloride. The mercury forms an amalgam with some of the aluminum. The isopropyl alcohol apparently reacts more readily with the aluminum in the amalgam. A mercury-aluminum couple is formed. The carbon tetrachloride reacts with the aluminum to give aluminum chloride. This exposes fresh aluminum without any oxide coating which reacts much more rapidly with the isopropyl alcohol.

54-3. 1.6 ml. Isopropyl alcohol.

54-4. To complete the reduction and increase the yield of benzhydrol. The mixture is distilled slowly to allow sufficient time for the reaction to occur. The fractionating column is necessary to remove *all* of the acetone without removing most of the isopropyl alcohol.

54-5. Dehydration of the product by hydrochloric acid.

$$C_6H_5CHOHCH_2CH_3 \xrightarrow{H^+} C_6H_5CH{=}CHCH_3 + H_2O$$

Before the 10% hydrochloric acid is added, chill the reaction mixture and the acid. Add the acid slowly with vigorous stirring but only enough to give a pH of 8 ± 1.

54-6. Use 0.025-mm cell and obtain spectra on the liquids (as the pure liquids or neat). If it is necessary to locate exactly the intense absorptions of the isopropyl alcohol and the acetone, obtain spectra on 10% solutions of these compounds in carbon tetrachloride in a 0.05-mm cell. Even if these solution spectra are obtained, the spectra on the pure liquids should also be obtained so that minor components may be detected. The diphenylcarbinol spectrum can be obtained from a split mull, from *two* 5% solutions in CS_2 and CCl_4 in a 0.2-mm cell, or preferably from a KBr pellet.

CHAPTER 55

55-1. The base would catalyze a Cannizzaro reaction, but if the acetophenone is added first, some

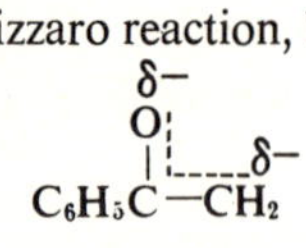

is formed. This reacts with the benzaldehyde (Equation 55-5b) before any significant Cannizarro reaction can occur.

CHAPTER 56

56-1. There is a more intense negative charge on the oxygen in the carbonyl group than on the oxygen in ether linkages. The hydrogen bond to the ether linkage should, therefore, be stronger. There should also be a stronger polar attraction between the carbonyl carbon

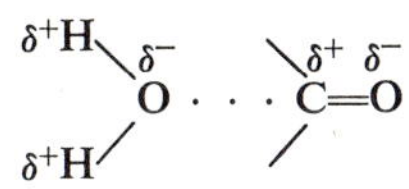

and water than between the *alpha* carbon of ethers and water. Some of the increase in solubility may be from hydration of a portion of the carbonyl compound

$$R_2CO + H_2O \rightarrow R_2C(OH)_2$$

56-2. The esters have twice as many oxygen atoms as the other compounds.

56-3. The nitro groups on the ring withdraws electrons from the hydrazine group. In oxidizing compounds the hydrazine group is reduced. Since reduction involves gaining electrons, the more electron deficient a group is, the more readily it can gain electrons and be reduced, i.e., act as an oxidizing agent.

56-4. (b), (c), (d), (e), (g), (h); oximes (d), (e), (g), maybe (b), and (c), 2,4-dinitrophenylhydrazones; (e), (g), and probably (d).

56-5. Positive (a), (b), (d), (f), (g), borderline (h).

56-6.

Results of Iodoform and Bisulfite Test

Ketone	*Sodium bisulfite*	*Iodoform*
RCH_2COCH_3	+	+
R_3CCOCH_3	−	+
$CH_3(CH_2)_nCO(CH_2)_nCH_3$	−	−
$ArCOCH_3$	−	+
$(CH_2)_mC{=}O$ (cyclic)	+*	−
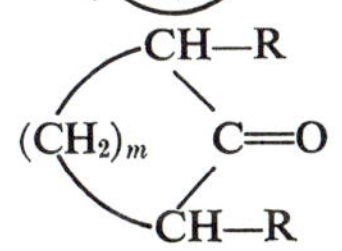	−	−

* If $m <$ than 7. If $m > 7$ usually negative.

56-7. Probably, because the alkanal would decolorize the Br_2 solution more slowly than the alkenal would. Note that both would slowly evolve HBr. The answers to this question depends on how "simple" is defined. The two aldehydes could easily be distinguished by their infrared spectra. The ultraviolet spectra of these two compounds would be identical above 220 mμ.

56-8. Solubility classification in aqueous NaOH and $NaHCO_3$ and classification tests with 2,4-dinitrophenylhydrazone and Tollens' reagent.

CHAPTER 57

57-1.

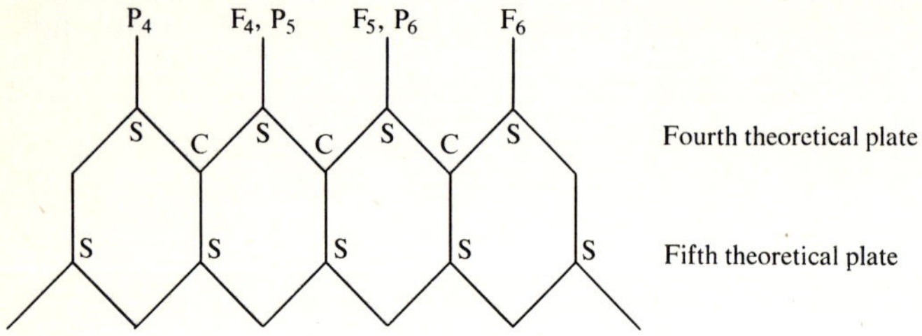

57-2.

(a)

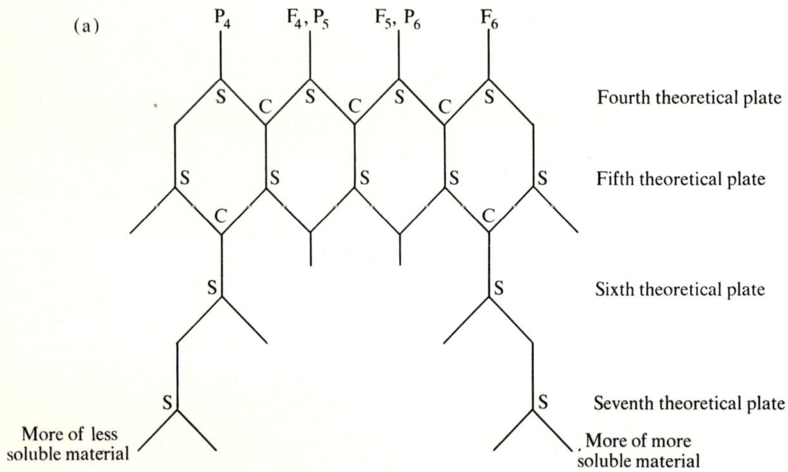

(b)

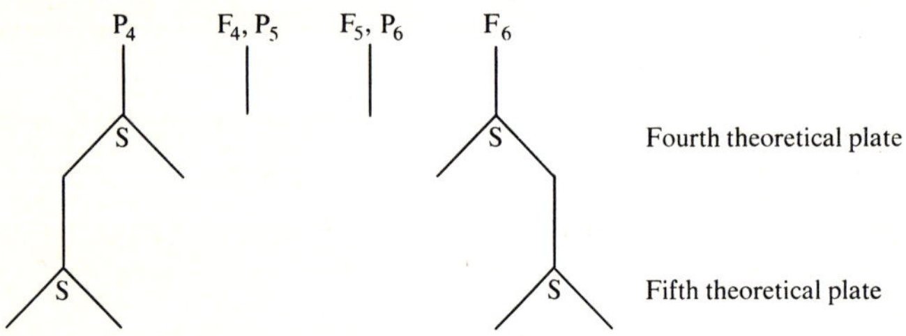

CHAPTER 59

59-1.

EXERCISE 59-1 COMPARISON OF CHROMATOGRAPHIC TECHNIQUES

Techniques	Moving Phase	Station-Phrases	Partition	Absorption	Requirements of Compounds being separated	Diagnostic Technique	Preparative Technique	Time Requirements	Equipment Requirements	Quant. Anal.	Resolution
GLC	Gas usually He or N_2	A liquid on solid support	100%	None	At least slightly volatile below their decomposition point	Primary	Considerable application. Will as techniques are developed	Fast	Elaborate and expensive	Precise and accurate	very high
GSC (gas)	"	Adsorbent	None	100%	Fairly volatile	"	"	"	"	"	"
Column*	Solvent	Solid or liquid on solid support	Can be either		Not extremely volatile	Valuable	Primary	Slow	Inexpensive	"	high
Thin	"	Adsorbent solid	Little	Most	"	Primary	Very limited	Fast	Inexpensive	Approximate unless coupled with spectroscopy or other technique	Good but usually lowest of all chromatograph technique
Paper	"	Liquid on paper	100%	Almost none	"	"	"	Slow	Inexpensive	"	Good to high

* New instrumental techniques are fast and primarily used as diagnostic methods. The equipment, however, is elaborate and expensive.

CHAPTER 60

60-2.

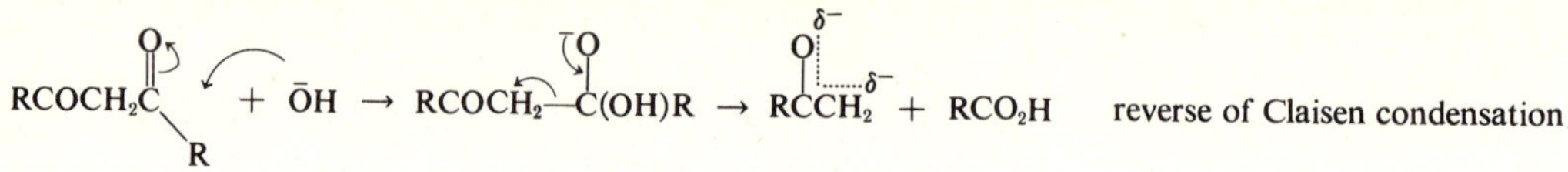

$$RCOCH_2^- + I_2 \rightarrow RCOCH_2I + I^- \xrightarrow[I_2]{HO^-} \text{etc., like usual haloform reaction}$$

60-3. Silicone oil with dimer acid on Chromosorb G AW-DMCS.

60-4. $CH_3COCH(CH_2CH(CH_3)_2)CO_2Et + HO^- \rightarrow CH_3COCH(CH_2CH(CH_3)_2)CO_2^- + EtOH$

$$CH_3COCH(CH_2CH(CH_3)_2)CO_2^- + H^+ \rightarrow CH_3COCH(CH_2CH(CH_3)_2)CO_2H \xrightarrow{H^+} CH_3COCH_2CH_2CH(CH_3)_2 + CO_2\uparrow$$

(gas given off)

60-5. Water, ethanol, and possibly isobutyl bromide. Add calcium chloride, stir, decant into a small distilling flask, and distil.

60-6. DEGS on Chromosorb G AW-DMCS. Silicone oil with dimer acid or methylchlorophenyl silicone fluid on Chromosorb G AW-DMCS would also give satisfactory separation.

CHAPTER 61

61-1.

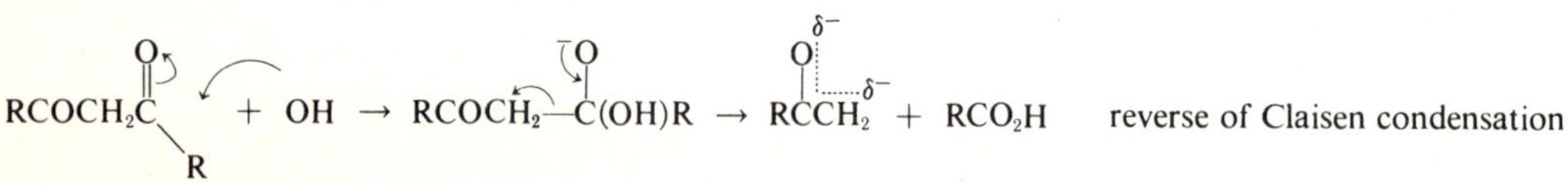

$$RCOCH_2^- + I_2 \rightarrow RCOCH_2I + I^- \xrightarrow[I_2]{HO^-} \text{etc., like usual haloform reaction}$$

61-2. $CH_3CH_2COCH_2COCH(CH_3)_2 \xrightarrow[I_2]{HO^-} CH_3CH_2CO_2^- + (CH_3)_2CHCO_2^- + HCI_3$

$$CH_3CHOHCH_2CO_2Et \xrightarrow[I_2]{HO^-} 2CH_3CO_2^- + EtOH$$

$$CH_3C(NH)CH_2CO_2Et \xrightarrow[I_2]{HO^-} 2CH_3CO_2^- + EtOH + NH_3$$

$$CH_3COCH_2CH_2CO_2Et \xrightarrow[I_2]{HO^-} HCI_3 + {}^-O_2CCH_2CH_2CO_2^- + EtOH$$

For the last three reactions $EtOH + I_2 \xrightarrow{HO^-} HCI_3 + HCO_2^-$

CHAPTER 62

62-1. (a) Anthracene has more maxima (ignore the vibrational structure), maxima at longer wavelengths, and much higher extinction coefficients for its maxima. (b) *o*-Xylene.

62-2. (a) IR. (b) Almost equal but probably UV. (c) UV.

62-3 and **62-4.** (a) IR carbonyl absorption at 1871 cm^{-1} will be more sensitive if xylene is removed. *If* it was certain that all anthracene had reacted and the xylene could be removed, the UV absorption maximum at 263 mμ would give greater accuracy, but even very small amounts of anthracene would cause very large errors. Xylene also would cause errors but not if present in trace amounts. (b) IR, 1850 or 900 cm^{-1} more sensitive if xylene is removed or if *o*-xylene is used instead of xylene mixture. (c) UV, λ_{max} at 355 or 375 mμ. (d) IR, 1850, 900, or 700 cm^{-1}. (e) IR, 1871 cm^{-1}. (f) same as (c). (g) IR, 1871 cm^{-1}; 2850 cm^{-1} would probably be best if a thick cell were used. (h) 985 cm^{-1}; 1690 cm^{-1} (maleic acid also has an intense absorption here but it is not a likely contaminant). (i) 985 cm^{-1}; 1690 cm^{-1} would give total maleic and fumaric acids.

62-5. (a) UV. (b) It is more sensitive and more accurate, and with use of the absorption maxima at 355 or 375 mμ there is no danger of even very large amounts of the other materials having any absorption in this area. (c) Begin with double quantities. Remove 0.5 ml samples at accurately measured time intervals. Use the same pipet for withdrawing each sample and take care to obtain exactly the same amount each time. Immediately after withdrawing the sample transfer it to 9 ml of spectral grade cyclohexane in a 10-ml volumetric flask. Bring the volume to 10 ml with spectral grade cyclohexane. Accurately dilute this as required so that the absorption maximum at 375 mμ has an absorbance between 0.4 and 1.2. Be certain to correct your reading for all dilutions. Only the first dilution to 10 ml need be done immediately. The other dilutions and spectrophotometer readings should be accomplished within 20 min after withdrawing the sample. Slightly better results might be obtained if the reaction were conducted in a constant temperature bath 3° below the boiling point of xylene.

CHAPTER 63

63-1. CF_2O 1928 cm^{-1}, CH_3COF 1840 cm^{-1}, CH_3COCl 1808 cm^{-1}, CH_3CHO 1720 cm^{-1}, $CH_3COC_2H_5$ 1710 cm^{-1}, $CH_2{=}CHCOCH_3$ 1677 cm^{-1}.

63-2. Highest first: $CH_2{=}CHCOCH_3 > CH_3COC_2H_5 > CH_3CHO > CH_3COCl > CH_3COF > CF_2O$.

CHAPTER 64

64-1. (a) +0.33°. (b) +22.37°. (c) +32.37°. (d) 0°. (e) −1.26°.

CHAPTER 65

65-1. 0.10 *M*

65-2. To neutralize the acid without using strong base

$$H_3O^+ + Cl^- + Na^+ + CH_3CO_2^- \rightarrow Na^+ + Cl^- + CH_3CO_2H + H_2O$$

Very little of the acetic acid ionizes. With the acid neutralized there can be no further acid-catalyzed rearrangement, and an accurate determination of the concentration of the product at the time of neutralization is obtained even if the absorbance is determined 5 to 40 min later.

APPENDIX III

III-1. The radiant energy reaching the detector and energy supplied by the electronic amplifier.

III-2. (a) Increases. (b) Increases, but of course the gain would have to be decreased so that the servo system would operate correctly. (c) Decreases or becomes poorer.

III-3.

Adjustment of Spectrophotometer Controls

Change desired	*Slit program control*	*Gain control*	*Spectral quality that is decreased*
Decrease noise	open	decrease	resolution
Improve resolution	close	increase	signal to noise ratio
Increase energy to compensate for opaque sample (2 possibilities)	open unchanged	unchanged increase	resolution signal to noise ratio

III-4. (a) The scattering is noticeable only when the particle size is greater than $\lambda/5$ and significant only when greater than $\lambda/3$ and increases as the particle size becomes larger than $\lambda/3$. If the particle size is constant as with a given sample, then the scattering decreases with decreasing wavenumber (or increasing wavelength). (b) No. (c) Yes, but a better solution would be to grind the KBr to smaller average particle size.

III-5. For both (a) and (b) trapped air may be a problem as the press cannot be evacuated. (See Table III-2.) (a) Increased noise. (b) Loss of contrast.

III-6. There is insufficient difference between the refractive indexes of the windows and the liquid.

III-7. (a) $d = 50/2(1996\text{-}790) = 2.07 \times 10^{-2}$ cm
(b) $d = 23/2(1996\text{-}801) = 9.6 \times 10^{-3}$ cm or 0.096 mm

III-8. (a) Because ammonia or amines form complexes with Ag^+. The reaction is similar to the dissolving of silver oxide by ammonia in the preparation of Tollens' reagent. (b) Solutions of CN^- and $S_2O_3^=$ and *concentrated* solutions of HCl or soluble ionic chlorides. (c) If the metal is below silver in the electromotive series, it will displace silver from its salts.

$$Al + 3AgCl \rightarrow AlCl_3 + 3Ag$$

APPENDIX IV

IV-1. In a well-conditioned column the distribution of the liquid phase does not change significantly, and there is little bleeding of the liquid phase if the column is used well below its conditioning temperature. Bleeding may cause a change in composition of the liquid phase as well as a decrease in the amount of liquid phase.

IV-2. (a)

Figure letter	*Initial period* (min)	*Initial temp.* (°C)	*Heating rate* (deg/min)	*Upper limit* (°C)	*Time at highest temp* (min)	*Other*
(a)	5	105	5	250	0	
(b)	2	60	5	180	3	Press hold button at 7 min from start. Press start button 9 min after hold button was pressed.
(c)	0	66	2	280	6	Press hold button at 22 min from start. Reset heating rate to 10 deg/min. Press start button 8 min after hold button was pressed. Then press hold button 10 min after start button was pressed for the second time. Reset heating rate for 5 deg/min. Press start button 10 min after hold button was pressed.

IV-3. The supports differ in density and in the amount of the diatome fine structure that remains after the calcining. Acid washing and silanization have little or no effect on appearance at 100× magnification.

INDEX